Springer Protocols Handbooks

For further volumes:
http://www.springer.com/series/8623

Huangui Xiong • Howard E. Gendelman
Editors

Current Laboratory Methods in Neuroscience Research

Springer

Editors
Huangui Xiong
Department of Pharmacology
and Experimental Neuroscience
University of Nebraska Medical Center
Omaha, NE, USA

Howard E. Gendelman
Department of Pharmacology
and Experimental Neuroscience
University of Nebraska Medical Center
Omaha, NE, USA

ISSN 1949-2448 ISSN 1949-2456 (electronic)
ISBN 978-1-4614-8793-7 ISBN 978-1-4614-8794-4 (eBook)
DOI 10.1007/978-1-4614-8794-4
Springer New York Heidelberg Dordrecht London

Library of Congress Control Number: 2013950899

Printed on acid-free paper

Springer is part of Springer Science+Business Media (www.springer.com)

Foreword

Neuroscience is arguably the most exciting of all research fields at the current time. Beyond the fascination about "how we make sense of our world" there is a very real interest in the "hardware and software" of the human brain. Important new insights into human brain function under normal conditions and in the setting of neurological and psychiatric diseases are emerging with increasing frequency. Accordingly, scientists, physicians, and the people and patients they serve, all anticipate a rich new harvest of discoveries that will transform our view of cognition, movement, affect, and its disorders.

While excitement, and true optimism, is justified, neuroscience is also among the most challenging of disciplines. Consider, if you will, the need for neuroscientists to examine and define events that extend from individual genes to the regulation of individual to societal behavior. Imagine the challenge that attends understanding how the activity at a particular synapse acts in concert with millions of others. Such individual neural connections are both similar and quite different in structure and function enabling the activation of but a single neuronal circuit. At every level of nervous system organization, one is confronted by complexity as well as the very real challenge of linking the levels—from genes to molecules, molecular assemblies to neurons, individual neurons to circuits and networks, and from each of these levels to a thinking-behaving organism. The scales defining the nervous system are indeed daunting. The length scale extends from events that occur on the scale of nanometers to those that are carried out over meters (i.e., across a dimensional difference of 10^9), temporal scales that range from ~10 μs to years (at least 10^9), and a scale of complexity whose difference in scale is almost certainly much greater.

Understanding the brain requires that we "mind the scales." This is accomplished by exploring and defining events in great detail at each level followed by studies to link one scale to the next. The former domain is populated by researchers with specific expertise—for example, the geneticists, molecular biologists, and cell biologists, among others. The latter is the province of the computational neuroscientists whose work, together with scale specialists, is to build and validate models that allow us to decipher how activity at one scale is informed and regulated by the scales immediately above and below. Thus, for example, molecular complexes exist in relation to the genes that encode their proteins and the structures of these proteins as well as cellular tasks that they perform. The picture that emerges is not just one of great beauty but also great complexity.

In *Current Laboratory Methods in Neuroscience Research*, Drs. Huangui Xiong and Howard Gendelman and their colleagues have assembled a particularly helpful neuroscience resource structure and function as well as providing very helpful insights into spanning research scales. It is at once practical and comprehensive, clearly detailing research methods while also providing insights that come from skillfully practicing such methods and interpreting findings.

The book features 36 chapters covering topics ranging from methods that define gene expression to those ideal for examining behavior in disease models. Detailed methods will make it possible for even the novice to carry out studies of gene expression. The novice and more experienced researcher will benefit from chapters on proteomic studies and metabolic measures. Tools for in vitro studies are nicely explicated, thus enabling the isolation of individual cell types as well as organelles. I found especially useful chapters that addressed immune cells that attend neural inflammation, a nice and perhaps unexpected part of a book on neuroscience methods but one that follows naturally from the editors' interest in such disorders. Students will be rewarded by chapters that address routine and advanced cell and tissue imaging. This includes both confocal microscopy and multiphoton imaging. Routine electrophysiological methods are outlined as well as procedures for various patch-clamp recording. Importantly, chapters devoted to performing and analyzing behavioral and motor tests will be appreciated, especially one that defines how it is that rodent testing can be tailored to address cognitive functions impacted by human disorders.

Readers will appreciate an excellent review of the various models of neurological disorders, citing the strengths and weaknesses. I found these most helpful in putting existing models in context. Finally, it was rewarding to learn about the methods currently in use to "make meaning" of the large databases that address structure and function of the nervous system. Still in its infancy, bioinformatics methods will provide a critical resource for linking the scales of brain. Their addition to this volume is important and useful.

In summary, readers of *Current Laboratory Methods in Neuroscience Research* will be treated to a thoughtful, comprehensive review of the tools in use for exploring the nervous system and its disorders. I recommend that one read not only the chapters needed for today's studies, but those that not immediately relevant. Doing so may well reward you not only by expanding your toolbox but also by providing new insights into deciphering the ever challenging and complex nervous system.

La Jolla, CA William Mobley, M.D., Ph.D.

Preface

We are pleased to offer you *Current Laboratory Methods in Neuroscience Research*, a research manual designed for both students and seasoned researchers. Our goals are to provide a "blue print" for performance of bench to bedside research. The works, in composite, were designed to facilitate broad multidisciplinary neuroscience experiments hampered by vision not techniques. The chapters herein describe methods for research performed in invertebrates and vertebrate species with a singular goal of translational science. Major methodological improvements of established methods are articulated. The chapters, as a whole, include those from the molecular and biochemical to the cellular, computational, and animals. The collective includes most of what is considered as contemporary neuroscience, including anatomical, behavioral, biochemical, cellular, computational, molecular, imaging, physiology, and behavior. The goal is to discover better means for the diagnosis, tracking and treatments of neuroinfectious, neuroinflammatory, and neurodegenerative diseases. All are thoughtfully linked to future clinical research investigations. Inclusions relevant to neuropharmacology emphasize neuronanomedicine and as such serve to bring the composite ever closer to the patient's bedside.

For easy access and retrievals, the sections are defined in logical order. This enables ready access and organization in technical divisions. The first assists the researcher in preparing nervous system tissue for specimen recovery, stereology, light and electron microscopy. The second focuses on the isolation, cultivation, and identification of a broad range of normal and diseased neural tissues and cells, vertebrate and invertebrate, for molecular, physiologic, and cellular studies. An interface between the immune and nervous system is seen by investigations into peripheral immunocyte isolation and identification through the techniques of centrifugal elutriation and flow cytometry. The text proceeds from descriptive mechanistic analysis using RNA and protein assays that utilize polymerase chain reaction and Western blot methods. A balance between mechanistic studies and futuristic therapies is also seen. Animal investigations serve as a bridge between laboratory experiments to bedside evaluations and further highlighted through a section on imaging including descriptions of cells, tissues, and whole organisms. A series of chapters on electrophysiology provides functional insights into anatomical and metabolite evaluations that are extended into immunohistochemical and autoradiographic tests. Opportunities and directives for research into "omics" technologies is also offered for bringing science from the molecule through how it may interact, one with another, as well as provide avenues for therapeutic studies that could be designed to halt or attenuate disease processes. This inevitably leads to the concluding section on neurodegenerative models that serve to bring together the prior methods making them disease relevant.

Through the 20 years of professional experiences at the University of Nebraska Medical Center, we have seen the development of interdisciplinary research programs. These are defined as investigators working in a collective effort in open collaboration. The environment balances individual research pursuits with cross-disciplinary collaborations. This ensures that the science would be cross validating while permitting development into new areas asking questions for collective benefit. The excitement and opportunity of such a collective environment provided the theme for this methods book.

We hope you will not only enjoy the reading but also benefit from the collective methods to provide new opportunities for your research and whatever directive you are taking or take in the future. To this end, we would be remiss not to thank all of our collaborators, students, fellows, and administrators, past, present, and future for their insights and encouragement for without them the contents of this article series would not have been realized.

Omaha, NE

Huangui Xiong, M.D., Ph.D.
Howard E. Gendelman, M.D.

Acknowledgements

We gratefully acknowledge the contributions from all our chapter authors and reviewers whose expertise and dedication proved invaluable for the book's completion. A special appreciation is extended to Robin Taylor whose meticulous attention to detail and continuous strive towards excellence has made a significant imprint for the book's organizational outcome.

Our heartfelt thanks go to Ann H. Avouris, Senior Editor and Michael Koy, Project Coordinator for Springer, USA. These individuals have inspired us to complete this content as part of the Springer Protocols and medical communications. Their leadership directed our works and served as a "blue print" for the multidisciplinary neuroscience experimental contents included.

Special thanks go to our spouses and families, for their abundant support, for their patience and understanding, and for their unconditional love and trust.

Contents

Part I

Neurohistological Techniques

Brain Tissue Preparation, Sectioning, and Staining

1

Jingdong Zhang and Huangui Xiong

Abstract

This chapter will summarize commonly used methods of brain tissue preparation, sectioning, and staining. History, development, and novel application of those methods in neuroscience will be presented herein so students can fully grasp the topic of brain tissue preparation, sectioning, and staining. The selectively introduced methods in this chapter are as follows: (1) Transcardial perfusion, especially on rats and mice, with a supplemental video referenced online to show perfusion procedure for adult rats. (2) Cryoprotection and frozen section cutting, with emphasis on the importance of cryoprotection for obtaining superior histological staining. Examples will be based on the author's own experiences. (3) Paraffin embedding and sectioning, in which a general procedure is introduced. (4) Vibratome sectioning and application in histochemical staining for electron microscopy (EM) studies. In this chapter, procedures introduced will be based on cutting 50 μm sections for EM histochemical stains and plate embedding. We described detailed protocols for tissue treatment using freeze and thaw and ABC kit to reveal immunostain, two key steps in all immuno-EM. (5) Following histochemical staining, the stained structures need to be further observed under EM. These ultrathin sections are generally 1.5 mm long and 0.5 mm wide. To overcome the obstacle of extracting such a small area while ensuring it contains the desired labeling, a plate embedding method is used, which is described in this chapter. We will also illustrate a protocol for production of self-made siliconized slides used for plate embedding. (6) A protocol for a well-known traditional neural stain, Golgi's stain, is presented here, with representative microim-

J. Zhang (✉) • H. Xiong, M.D., Ph.D.
Department of Pharmacology and Experimental Neuroscience, University of Nebraska Medical Center, Durham Research Center, Durham Research Plaza, DRC I 8034, 45th Street at Emile Street, Omaha, NE 68198-5880, USA
e-mail: Jingdong.zhang@unmc.edu; hxiong@unmc.edu

H. Xiong and H.E. Gendelman (eds.), *Current Laboratory Methods in Neuroscience Research*, Springer Protocols Handbooks, DOI 10.1007/978-1-4614-8794-4_1,

ages. This section also highlights examples of new applications of this long-established method with Golgi's stain kit in modern neuroscience research with representative images. (7) Protocols for Nissl staining, another commonly used neuron stain, and relevant photos. Two Nissl stains detailed here are cresyl violet and neutral red stains. (8) The methodology of hematoxylin and eosin (HE) stain, a broadly used histology stain method in both research and clinical labs, will be discussed here with detailed protocol. (9) Finally, we introduce a myelin sheath stain, Luxol fast blue, with practicable protocol. Due to demyelination being a highly researched area, this method is very applicable in current day studies of demyelinating diseases and animal models thereof.

Keywords
Brain tissue preparation • Brain tissue sectioning • Brain tissue staining • Classical neuron staining • Electron microscopy preparation

1.1 Introduction

Preparation of brain (or spinal cord) tissue is the first and most important step in all relevant studies on brain structure and function. This chapter focuses on morphological studies. In this case, procedure of the brain tissue preparation generally includes fixation, pre-embedding, and embedding process. In the case of in situ hybridization or in situ preservation of some highly dynamic protein antigen, the brain is decapitated, and fresh tissue is fixed without fixative by direct freezing in dry ice or liquid nitrogen or by immediately transferring to a −80 °C freezer (Pinaud et al. 2008; Pinaud and Jeong 2010). The aim of your study and the desired level of detail you intend to see from the tissue will determine the method of fixation and embedding necessary. Examples of studies that may require different fixation and/or embedding processes include light microscopy (LM) level or electron microscopy (EM; only transmission EM is mentioned here) level. Formaldehyde is the most common fixative for LM study, and formaldehyde plus glutaraldehyde is the basic fixative for EM preparation. The method of fixation can be in situ or ex situ. The most widely used in situ fixation is transcardial perfusion of fixative; while the ex situ method is simply placing the tissue in fixative solution, however, the size of the tissue and the time of incubation will affect the quality of fixation. Quality fixation of tissue is a prerequisite for decent tissue embedding, sectioning, and staining. Tissue for LM observation can be embedded in paraffin or celloidin and plastic or be cryoprotected and then frozen. Celloidin embedding strengthens the tissue and is excellent for the tissues that will be subjected to repeat or prolonged chemical treatment such as Golgi's stained tissue. But celloidin embedding had extinguished because of complication and time-consuming of the celloidin method. In current neuroscience studies, paraffin embedding and freezing formaldehyde-fixed tissue are the most widely used preparations for LM studies. Gelatin or agarose embedding of Golgi's stained tissue, embryos, and newborn tissues for vibratome cutting are still widely in use today. For EM observation, the tissue is mostly embedded with plastic and occasionally frozen. Equipment for acquiring ultrathin sections from frozen tissue is expensive and requires a complicated procedure, making this method only transiently popular. The majority of the current EM studies aim to detect ultrastructural characteristics or changes revealed by immunostaining; thus, aqueous resin such as Durcupan or LR white (polyhydroxy substituted bisphenol A dimethacrylate resin with C12 methacrylate ester and dimethyl para toluidine) embedding is widely employed in EM studies (Hasegawa et al. 2008; Soontornniyomkij et al. 2010). The immunostaining in this case could be either pre- or post-embedding. The most common

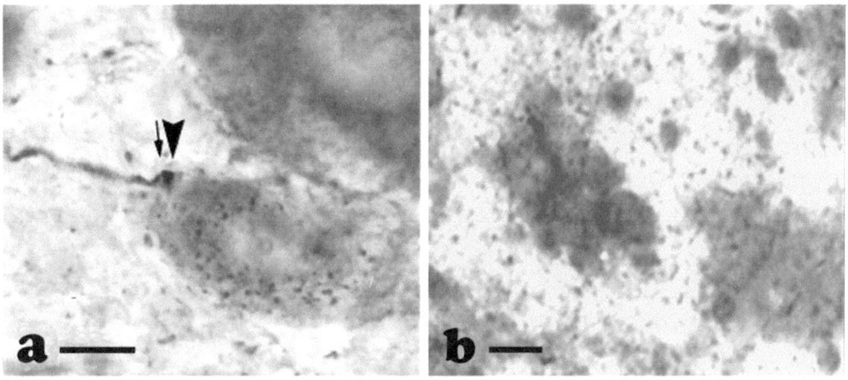

Fig. 1.1 Panel (**a**) shows better cryoprotection (but not the best) than (**b**). Scale bars = 10 μm

techniques of immunohistochemistry for EM studies include the ABC (avidin–biotin complex) kit histochemistry, immunogold, and gold intensified with silver.

In parallel, the most commonly used sectioning done today is paraffin sectioning, cryostat, and vibratome cutting. Vibratome sections can be used in both LM and EM studies, but vibratome sections are mostly used for pre-embedding histochemical or immunohistochemical stains in EM studies. The quality of the final tissue section is determined by the quality of the embedding and the skill of the researcher performing the cutting. Currently, program-controlled paraffin embedding systems are ubiquitously applied in all research and clinical institutes, instead of manual processing of paraffin embedding; hence, as long as the tissue blocks are in the same, or of similar, size and texture, there should be no problems encountered while embedding for paraffin sectioning. In those cases, the skill of the individual cutting and mounting is the key point for sections of excellent quality. However, in the situation of frozen sectioning, your embedding, i.e., cryoprotection procedure, is not without flaws. In this case, when viewing your tissues under higher magnification, you may see many tiny holes even though you feel the cutting is perfect (Fig. 1.1). The problem is most people transfer a whole brain directly into 20 or 30 % sucrose after fixation or postfixation. This is sufficient for loose or small tissues, such as lung or young mouse brain, but not for larger tissues. The optimal cryoprotection for large brain tissue, such as adult rat's or rabbit's brain, is to separate it with a "Brain Blocker" and perform increasing gradient sucrose and finally optimal cutting temperature compound (OCT) infiltration. *Pia mater* should have been removed prior to postfixation but is ignored in most of cases. Completely eliminating of *pia mater* will markedly increase infiltration. However, perfect cryoprotection cannot replace proficient cutting skills. For vibratome sectioning in EM studies, the tissue does not need any pretreatment except excellent fixation, generally by formaldehyde plus glutaraldehyde. Vibratome sections are mostly used as pre-embedding immunohistochemical stain for EM study. They are also used for reconstruction of intracellular enzyme (e.g., horseradish peroxidase, HRP) labeled single-neuron profiles with dendrite, axon, and terminals under LM; it was believed sucrose would reduce histochemical reactivity of enzymes, resulting in missed morphological details if frozen sectioning is performed (Yabuta et al. 1996; Yoshida et al. 1999). The other type of cutting skill is to cut ultrathin sections with an ultratome for transmission EM study. Operation of an ultratome to cut ultrathin sections will not be described here in any detail.

In recent decades, immunofluorescent, immunocytochemistry, or immunohistochemistry staining is ubiquitous in published research articles. Its key principle is binding of a specific antibody to a relevant antigen and then visualization by using a fluorescent dye-conjugated antibody

Fig. 1.2 *Left*: Santiago Ramon Cajal; *right*: Camillo Golgi

or by compounds that can be visualized after certain histochemical reaction, referred to as histochemical stain. Principle and methodology of immunostains will be introduced in Chap. 27. However, EM pre-embedding immunostaining will be presented in this chapter because preparation of tissue for EM study is discussed here, which is closely associated with histochemical revealing of enzyme labeling and immunostain. Nonetheless, this chapter will introduce some well-known classical nervous tissue stains, followed by a more in-depth discussion of several which will be accompanied by a detailed protocol that is widely used to date.

Until a century ago, people could not see a neuron in significant detail, although silver stain could unveil soma with short primary dendrites. The first efficacious stain that enabled an entire contour of neuron to be visualized (see Figs. 1.3, 1.4, and 1.6 in Jones 2010) was developed by Camillo Golgi, an Italian physician and medical researcher. Named after the person who developed it, Golgi's stain is still used in neuroscience researches to this day. Golgi's stain consists of silver intensified with potassium, allowing for selective staining of all soma, dendrite, and axons of a small portion of neurons in each stain. Thus, it generates a result in which a single neuron, with its delicate dendrite trees and axon, can be visualized without too much overlapping or conflict of other cells. The mechanism of this magic stain is still not clear. By applying Golgi's staining method, morphological classification of different neurons was originally delineated (see Figs. 1.5 and 1.7 in Jones 2010) by Santiago Ramon Cajal, a Spanish anatomist, who was honored as the father of the neuroanatomy. By observing neurons with the Golgi stain technique, Golgi postulated that the central nerve system was a great net, which was termed as "Reticular Theory." Contrary, Cajal hypothesized that the nervous system was composed of numerous individual neurons, eventually being referred to as the "Neuron Doctrine" (Jones 2010; Triarhou and Del Cerro 2012). Then, Cajal's idea was verified by the following neuroscience studies but network idea became fashionable later. The two confrontational scientists eventually won the Nobel Prize in Physiology and Medicine together in 1906 (Fig. 1.2). Following the Golgi method, Nissl stain was widely used in staining of neuronal soma that is enriched with Nissl body—granules of rough endoplasmic reticulum. The method was created by Franz Nissl (Rocchietta 1968), a German neurologist from the same era as Golgi and Cajal. Nissl, also a good friend of Alois

Alzheimer, was not as lucky as Golgi and Cajal, in that he continuously moved from lab to lab leaving some promising research projects behind. The most commonly used Nissl stains up to date are cresyl violet and neutral red stains. Before modern immunofluorescent stains as we know them today, fluorescence histochemical stains of biogenic monoamines were the earliest fluorescent tissue stain (Bjorklund 1983). Great advances in immunofluorescent stains have rendered these histochemical stains obsolete. Hematoxylin and eosin (HE) stain is another very classical stain that was not designed for specifically staining of neurons, but widely used in all of research and clinical labs including in neuroscience studies. Thus, HE stain will be discussed later on in detail. Enzyme histochemistry methods are still commonly used today, especially for neuronal pathway tract tracing. However, enzymes histochemical stains such as ABC conjugated HRP labeling can be used in pre-embedding immunostaining for EM study of neurodegeneration and demyelinating diseases today (Steencken et al. 2009; Tang et al. 2009). If the tissue slice is embedded with aqueous resin like Durcupan or LR white, a combination of pre-embedding with post-embedding immunostaining is a superior tool for any kind of EM study. Accordingly, this method will be introduced here with a practicable protocol. In addition, some myelin stains have recently become more commonly used, due in part to demyelination and its mechanisms being a big topic for today. The most well-known, and original, myelin stains are Weigert stain (Wohlrab and Henoch 1988) and Luxol fast blue (LFB) method (Kluver and Barrera 1953). The predominant staining substance in the former consists of iron (or equivalent metal) plus hematoxylin, in the latter is the Luxol dye. Countless reformulations have been generated since the two dyes were first described, but all formulas are fundamentally similar, assuming a metal plus hematoxylin or Luxol dye. Currently, original or reformed LFB method or LFB combined with other stains is frequently applied in studies of demyelinating disease (Pistorio et al. 2006; Carriel et al. 2011).

Therefore, in this chapter the practicable protocol will be introduced as follows: Transcardial In situ Fixation (Sect. 1.2); Cryoprotection and Cryostat Sectioning (Sect. 1.3); Paraffin Section Cutting and Mounting (Sect. 1.4); Vibratome Section and HRP Conjugated ABC Kit Stain for EM Study (Sect. 1.5); Epon 812 or LR white plate embedding for EM study (Sect. 1.6); Golgi's Stain (Sect. 1.7); Nissl Stain (Sect. 1.8); Hematoxylin and Eosin Stain (Sect. 1.9); and Kluver-Barrera LFB Stain (Sect. 1.10).

The objective of this chapter is to enable the user to start the morphological part of a research project based on what has been learned here. When you apply a protocol described here or find out another ready-made protocol, you may have to modify it to better fit your own experiment. Hopefully, the principles here will help you judge whether the morphological part of the experimental design is correct, effective, and feasible.

1.2 Transcardial Perfusion of Fixation

1.2.1 Preparation of Facility and Instruments

- Set up a perfusion pump in a fume hood, a pipe system with bi- or tri-pass valve and needle (blunt needle such as feeding needle is recommended), a tray to collect biohazard fixative (Fig. 1.3a).
- Instruments: See the instruments shown in Fig. 1.3b.

1.2.2 Rinse Solution and Fixative

- Saline (0.9 % NaCl) either fresh made or commercial ready-made.
- Phosphate buffer (PB; pH 7.4), either 0.2 M PB or commercial 10× PBS.
- Paraformaldehyde (PFA; powder) or formalin (37–40 % FA; liquid), 4 % PFA = 10 % formalin. To make 4 % PFA: Weigh and add PFA powder into distilled water (DW), heat the solution, and stir continuously. The solution is cloudy at the beginning because the formalde-

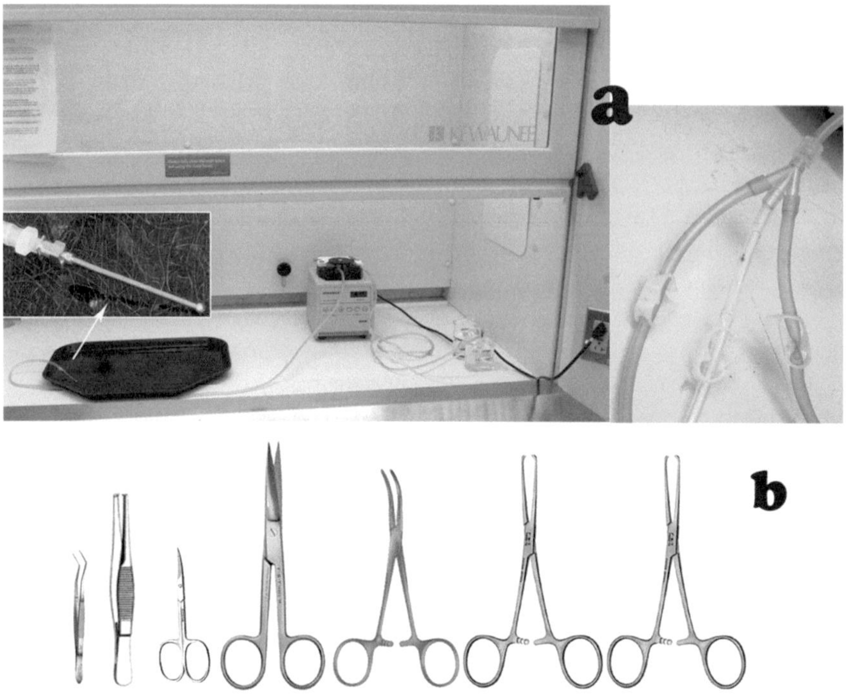

Fig. 1.3 Facility and instruments for transcardial perfusion

hyde is a polymer at this time. It will depolymerize during the temperature is increasing, resulting in the solution becoming clearer and clearer. Adding a small NaOH pellet when DW start to steam but not boiled yet, the solution will become completely clear without boiling. Boiling is undesirable due to the fact that formaldehyde molecules will evaporate quickly during boiling. After the PFA has cooled down, filter the solution with filter paper and funnel. After filtering, add half 0.2 M PB (final concentration is 0.1 M) or 1/9 of commercial 10× PBS (final concentration is 1×).

Note: Do not use PBS directly to make 4 % PFA. Small amounts of formic acid will be produced in solution when paraformaldehyde is depolymerizing, causing salt to be formed with sodium and formic acid that will be precipitated out and hard to dissolve.

- Fixative for EM tissue preparation:
  2 % PFA and 1 % glutaraldehyde; however, if the tissue will be immunostained for EM study, mostly 0.075–0.2 % glutaraldehyde is applied with 4 % PFA.

1.2.3 Transcardial Perfusion

1. Anesthesia of the animal, using any anesthetizer available in your animal protocol. For transcardial perfusion, deep anesthesia by double dose is required. The signs for deep anesthesia include deep irregular breaths; cyanosis (blue color) on footpads, ears, and tail; failure of pupils to respond to light; and slow irregular heartbeat. In this status the blood vessels are completely relaxed.

   Note: Excellent anesthesia and perfusion does not require an anticoagulant and vasodilatant. But some researchers administer heparin or sodium nitrite (SN) or both to avoid blood coagulation and reflexive vessel contraction. Drugs are generally administered in one of these ways:
   (a) Add heparin (10 U/mL) in saline (4–8 °C) prior to perfusion.
   (b) Injection (*i.p.*) of SN (35 mg/kg) prior to anesthesia.
   (c) Apply both heparin and SN.
2. Rapid surgery to open chest and expose heart and ascending aorta (Gage et al. 2012).

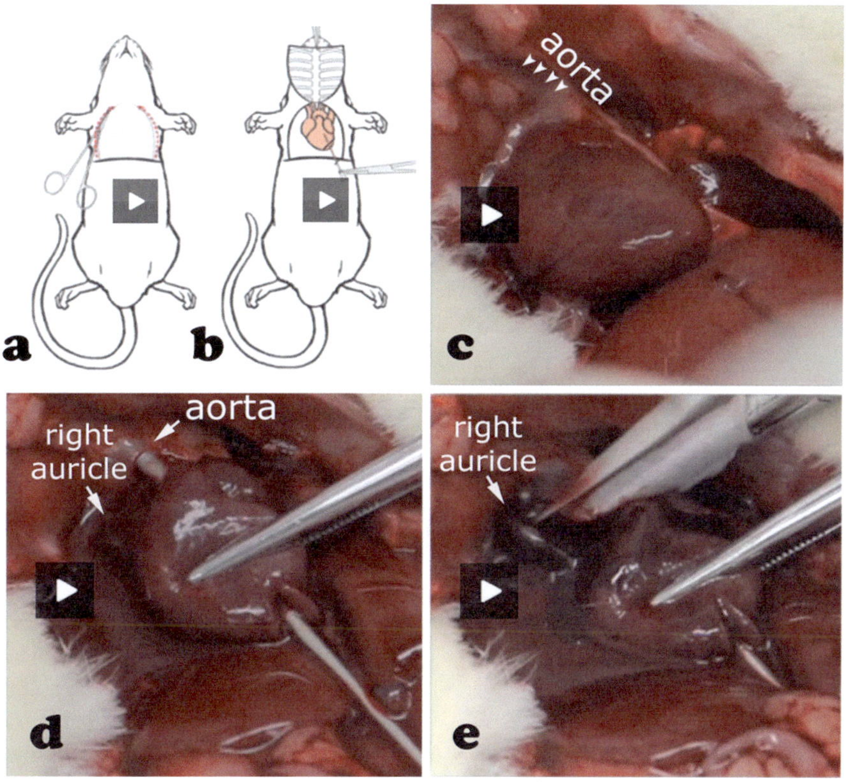

Fig. 1.4 Pictures cited from video published online by Gage et al. (2012)

When animal is deeply anesthetized and laid supine, clamp skin of upper abdomen with a serrated-tip forceps and horizontally cut the skin widely to expose diaphragm and scissor diaphragm quickly and widely, and then cut ribs as shown in Fig. 1.4a, b, to expose the heart and aorta (Fig. 1.4b, c. Video of perfusion procedure is available from Gage et al. 2012).

Note: All descriptions for surgery as shown by pictures and video cited, the perfusion speed, and time course are based on adult rats. For smaller animals like mice and large animals such as rabbits or cats, there are a number of differences in respect of instrument preparation, surgery procedure, perfusion needle (e.g., in cat it is metal or glass pipe), perfusion speed, and time of perfusion.

3. Start to pump saline at very slow rate (1–2 mL/min) and insert feeding needle into left ventricle until the tip is in the aorta (Fig. 1.4d). Slow pumping will make right auricle plump and easy to see. Cut the auricle (Fig. 1.4e) and increase the speed on the pump to the fastest rate (e.g., 40 mL/min). Close the saline pipe as soon as the blood flowing out of the auricle looks clear and open fixative pipe in the meantime.

   Note: Make sure no air bubbles are at the tip of feeding needle prior to pushing it into the heart. Also examine all pipes to make sure no bubble is inside; if any, try to get rid of them before perfusion. Air bubbles can enter into small blood vessel to block flow of perfusion; if this happens in the brain, the brain tissue will not be evenly fixed.

4. Perfuse animal with fixative at a fast rate (about 40 mL/min). After approximately 5 min of fast perfusion, turn pump back to a rate of about 5 mL/min for 15–20 min, and then reduce flow rate again to about 2 mL/min for an additional 30–35 min or until fixative runs up.

Note: If the needle has been placed in the right ventricle, saline will be spilt out from the nose, a sign of bad perfusion for the brain but good perfusion for the lung. A sign for good perfusion is that animal raises up the front limbs and stretch out the rear limbs shortly after fast perfusion of fixative.

5. The amount of saline and fixative to be prepared: Saline should be 5–10 times blood volume (averagely 7.5 times). For example, blood volume of a rat is about 7% of body weight, the blood volume of a 300 g rat is about 21 g (~21 mL). In this rat, the saline should be 21 mL × 7.5 = 157.5 mL.

   Note: Some lab use large amounts of fixative, such as 1 L for a 300 g rat. By our experience, three times the amount of saline is sufficient for fixation. In the above example, approximately 470 mL of fixative would be adequate.

1.3 Cryoprotection and Cryostat Sectioning

1.3.1 Cryoprotection Procedure

(a) Make 10, 20, and 30 % sucrose solution with 1× PBS (diluted from commercial 10× PBS) and mixture of half 30 % sucrose and half OCT.
(b) Cut fixed brain (e.g., rat) either coronal or sagittal with a "Brain Blocker" into about 2-mm-thick slices. Or carefully remove all *pia mater* for small brain (e.g., young mouse).
(c) Put tissue slices or block in 10 % sucrose overnight (in fridge), in 20 % sucrose a whole day with a little shaking (in room temperature, RT), and in 30 % sucrose overnight (in fridge) and then in 1:1 of 30 % sucrose/OCT solution 4–5 h followed with pure OCT 1–2 h in RT.
(d) Mount tissue onto cryostat chucks and ready to be cut.
(e) If a whole brain is required by experiment design, you can do transcardially perfuse 5 % sucrose in sequence of fixative perfusion and place the brain in increasing gradient sucrose after it is extracted.

Note: The sucrose solution had better be at least ten times the tissue size or more; in this case a sign of good infiltration is that the tissue block completely sank onto the bottom.

1.3.2 Cryostat Sectioning and Mounting

(a) Mount a tissue block onto a cutting chuck with OCT and put in cryostat to be frozen until completely solid.
(b) Adjust temperature of cryostat to one that is appropriate for cutting. Generally, the thinner the section is cut, the lower the temperature should be. For example, −21 to −25 °C is appropriate for 5–10 μm section, and −18 to −20 °C is suitable for 15–30 μm section. When using the cryostat, keep the lid closed as much as possible to maintain a constant temperature inside (Fig. 1.5a).
(c) When the temperature has adjusted to the set point, insert and stabilize the chuck with the tissue block onto the headstage (Fig. 1.5b). Then install a degreased (by 70 % alcohol) blade into the knife holder and adjust the surface of the tissue block to be paralleled to the blade edge. Trim any extra OCT away from tissue block until tissue is just exposed. Finally, adjust the chuck once again to ensure the tissue surface is parallel to the blade (Fig. 1.5a, b).
(d) During cutting, a flat section is desired; however, the sections tend to roll up. One way to achieve the desired flattened section is to use an anti-rolling glass plate (Fig. 1.5c). Alternatively, one can use a small oil-painting brush to gently press down the section (Fig. 1.5d) and avoid curving or rolling.
(e) After a section has been cut, there are two ways to collect them for future experiments. The first way is to directly mount sections onto glass slides. When sections are thinner than 15 μm, it is best to directly mount the sections onto glass slides. The second way is

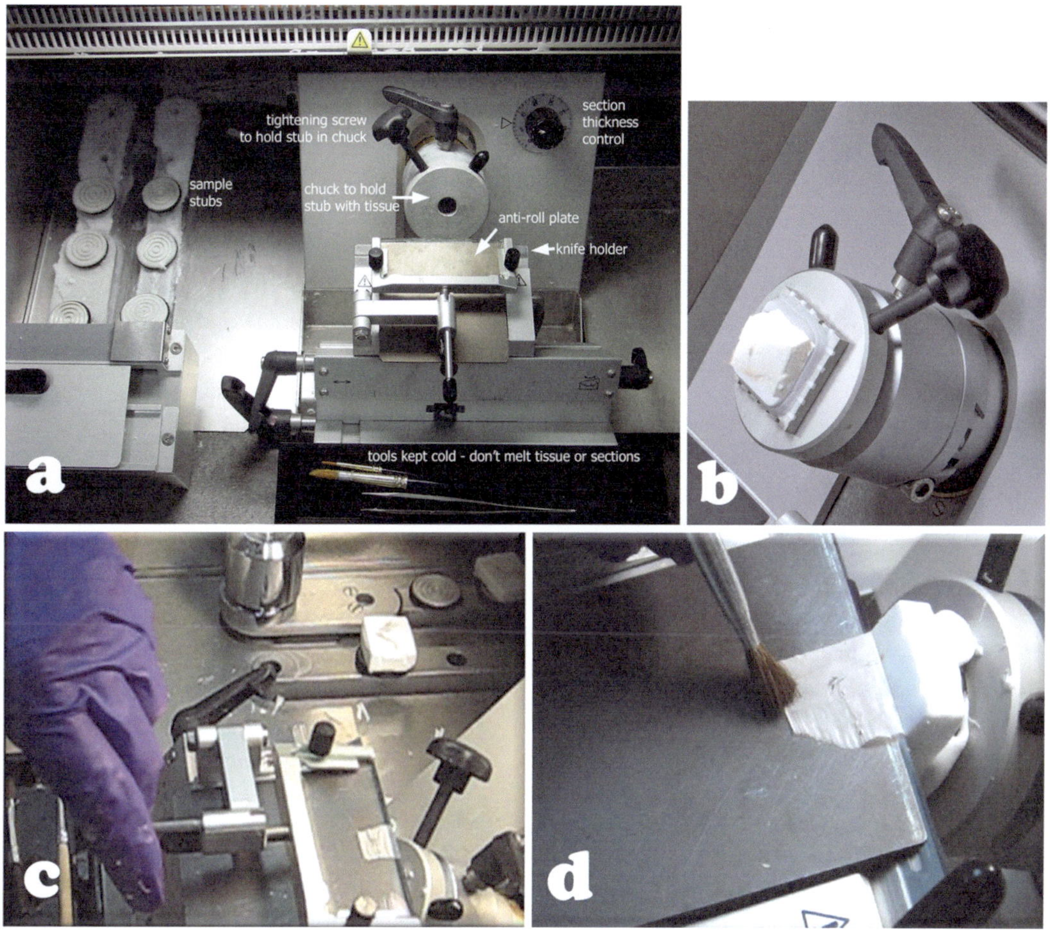

Fig. 1.5 A cryostat chamber (**a**), a tissue chuck on the headstage (**b**), anti-rolling plate (**c**), and unrolling the section by oil-painting brush (**d**). Image reproduced by courtesy of Donna J. Emge, HT (ASCP) of the Mouse Histology and Phenotyping Laboratory of the Robert H. Lurie Comprehensive Cancer Center, Feinberg School of Medicine, Northwestern University, Chicago, Illinois

to transfer the sections to a grilled tray filled with 1× PBS.

Note: If the section is proving difficult to be flattened, the temperature may be too warm; if the section edge is easily crashed, the temperature is too low. Expensive cryostat has temperature control for both chamber and headstage; in this case keep the chamber thermostat and adjust headstage temperature only. If the section has visible scratches on it, either edge of blade is imperfect and dirtied with tiny particles or the anti-rolling glass plate is dirty. It is important to periodically clean the blade edge and the glass plate for cutting sections of superior quality.

1.4 Paraffin Section Cutting and Mounting

1.4.1 Preparation of Tissue and Materials

1. Mark cassettes appropriately and place sliced tissues (<3 mm thick) in paraffin embedding cassettes (Fig. 1.6a).

   Note: It is best to consult with the individual(s) or lab(s) that are performing the embedding prior to beginning an experiment, especially if your tissues are unique or if you have used peculiar chemicals during any previous steps.

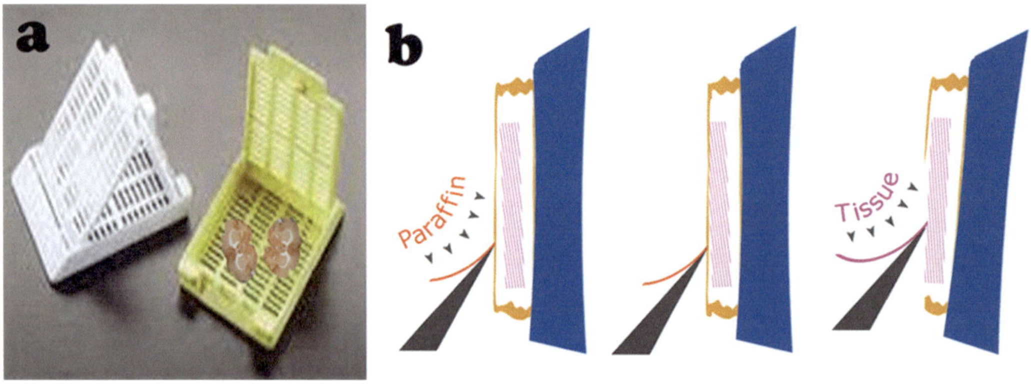

Fig. 1.6 Most commonly used paraffin embedding cassette (**a**) and procedure of trimming block (**b**)

They may require you to prepare tissues in a specific size to fit their embedding program. Avoid wasting your tissue and time without consideration of embedding parameters. The following is an example of the program in an embedding machine:

(a) 70 % ethanol for 1 h
(b) 95 % ethanol (95 % ethanol/5 % methanol) for 1 h
(c) First absolute ethanol for 1 h
(d) Second absolute ethanol 1½ h
(e) Third absolute ethanol 1½ h
(f) Fourth absolute ethanol 2 h
(g) First clearing agent (xylene or substitute) 1 h
(h) Second clearing agent (xylene or substitute) 1 h
(i) First wax (paraplast X-tra) at 60 °C for 1 h
(j) Second wax (paraplast X-tra) at 60 °C 1 h

2. Materials for cutting and mounting:
   (a) Microtome with blades
   (b) Several small oil-painting brushes
   (c) Forceps
   (d) Thin bamboo stick
   (e) Flat paper tray
   (f) Ice wrapped in gauze or 10 % ammonia solution
   (g) Paraffin section water bath
   (h) Glass slides

      Note: Glass slides must be pre-coated with adhesive. Negatively charged slides such as Fisher Superfrost®*/Plus are recommended.

1.4.2 Procedure of Cutting and Mounting

1. Cutting of Paraffin Section
   (a) Mount paraffin-embedded block on a microtome (Fig. 1.7a), adjust clearance angle to 1–5° (Fig. 1.7b, c), and adjust the face of the paraffin block so that it is parallel to the edge of the blade (Fig. 1.6b). Set up thickness for section.
   (b) Trim the paraffin block until the tissue appears (Fig. 1.6b), and then adjust face of the paraffin block again to optimize the parallel relationship between the edge of the blade and plane of the tissue (Fig. 1.6b) instead of the plane of the paraffin block. Trim away extra paraffin on the tissue. The real tissue section is now ready to be cut.
   (c) Cool the paraffin-embedded tissue by attaching coolant against the surface of the block; this is a key step to cutting a good section. Either applying ice wrapped in gauze or applying 10 % ammonia solution on tissue surface can cool the paraffin. Generally, tissue is cooled for 10–15 min.
   (d) Cut the paraffin block. The sections should form a ribbon (Fig. 1.8a), and either forceps or thin bamboo sticks can be used to lift the ribbon (Fig. 1.8b). Gently transfer the ribbon to a flat paper tray (Fig. 1.8c).

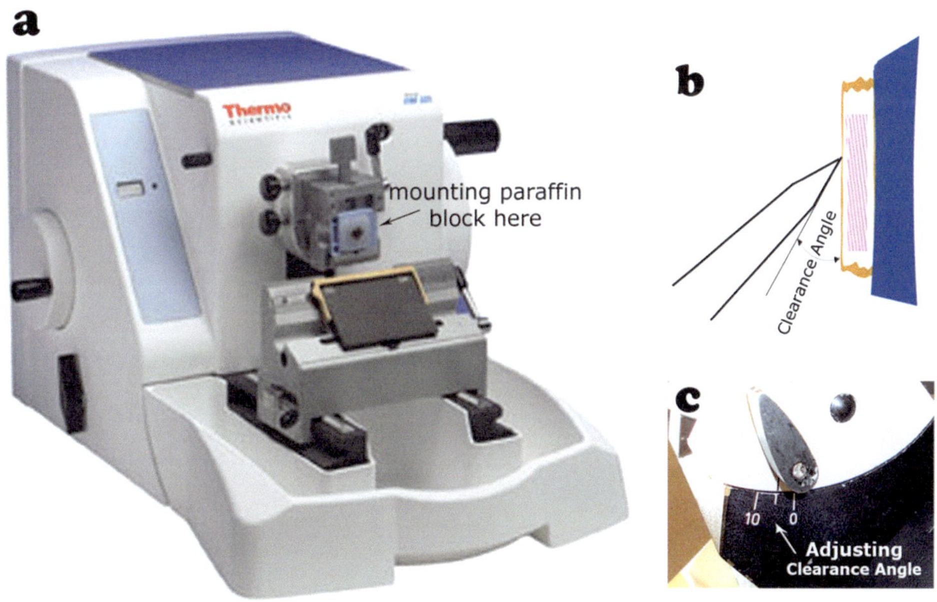

Fig. 1.7 A paraffin section microtome (**a**), clearance angle (**b**), and set up a clearance angle (**c**)

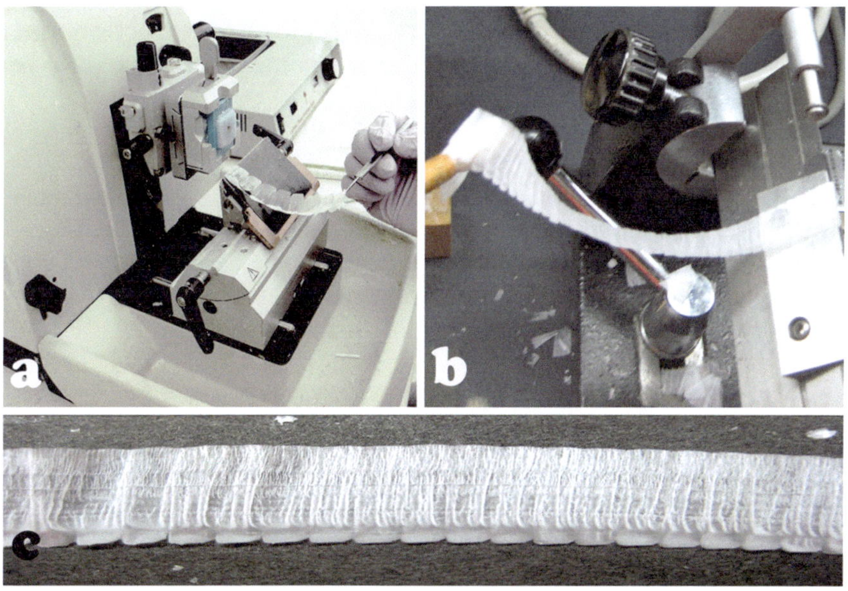

Fig. 1.8 A correct cutting results in good ribbon of sections (**a**), method of handling of ribbon (**b**), and placing a ribbon in a tray (**c**). Image provided friendly by Jenna Judge, from Department of Integrative Biology, University of California, Berkeley

Note: When sections become more and more fragile, the surface needs to cool again. If sections cannot form a good ribbon, the edge of the blade and surface of the block may not be clean or the edge of the paraffin block may not be evenly aligned. It is important to periodically clean the blade edge and the tissue block for cutting sections of superior quality.

2. Mounting of Paraffin Section
   (a) Turn on and set up the water bath to around 45 °C if sections are 7–10 μm; set to

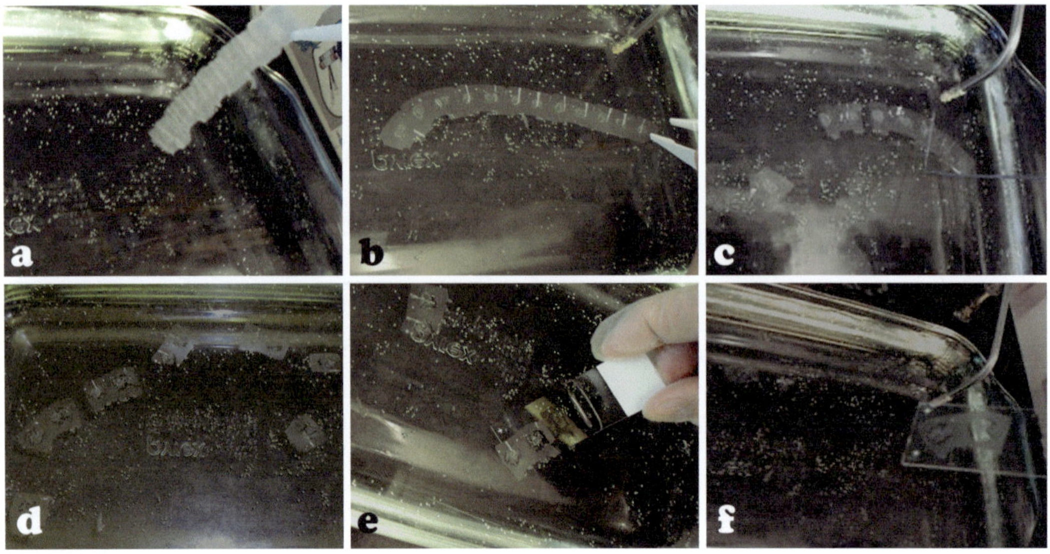

Fig. 1.9 Procedure of bathing and mounting the paraffin sections

37–40 °C if sections are 3–5 μm. Ribbons of paraffin sections can be separated either to individual section or to shorter ribbons.

(b) Transfer paraffin sections into warm water and gently place the section onto surface of the water (Fig. 1.9a, b). Avoid ribbon twisting and formation of bubbles underneath. The previously rumpled section should stretch gradually until it is completely flat (Fig. 1.9b, c) in a short time. If the section or ribbon stretches unevenly or too fast, the temperature is too high.

(c) Disconnecting the ribbon when sections become entirely extended (Fig. 1.9c, d), protrude glass slide carefully underneath the section and lift up (Fig. 1.9e), then the section is mounted on the slide (Fig. 1.9f). The ribbon can be separated with tipped bamboo sticks or by xylene wetted thin wire (Fig. 1.9c).

(d) Air-dry slides mounted with paraffin sections for several hours to overnight, and then place them in a thermostat oven at 60 °C, for 1–1.5 h. After slides have cooled down, the paraffin sections are ready to be stained.

1.5 Vibratome Section and ABC Histochemistry Stain for EM

1.5.1 Vibratome Sectioning

1. Preparing the materials for vibratome cutting:
   (a) Vibratome (Fig. 1.10a)
   (b) Tissue block chuck (Fig. 1.10a)
   (c) Razor blade degreased with 70 % alcohol
   (d) Instant glue
   (e) 1× PBS (enough to fill the tissue holding chamber)
   (f) Small oil-painting brush for collecting sections (Fig. 1.10b)
   (g) Grilled tray with 1× PBS to hold sections (Fig. 1.10b)
2. Procedure of sectioning:
   (a) Rinse fixative off of the tissue block with 1× PBS, absorb PBS liquid under the base of block, and glue the tissue onto the chuck and fix the chuck in the cutting chamber (Fig. 1.10a). Fill 1× PBS into the chamber until the tissue is submerged.
   (b) Install the degreased razor blade on the knife holder (Fig. 1.10a), align and tighten the blade upon the holder, and then adjust

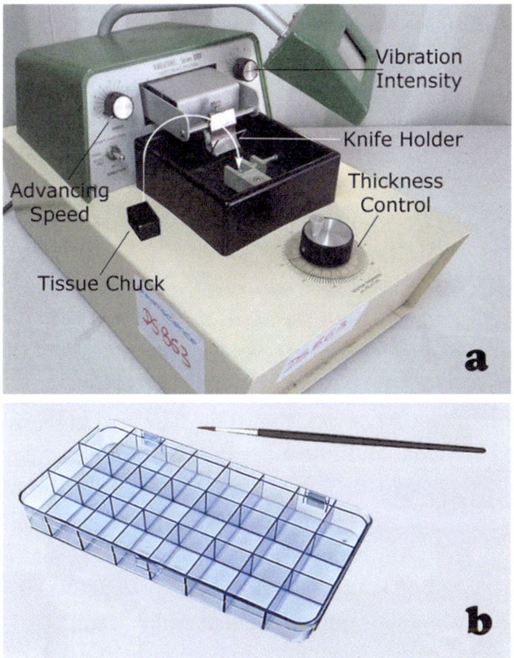

Fig. 1.10 A vibratome (**a**) and section collecting oil-painting brush and grilled tray (**b**)

clearance angle to about 2–8° (Fig. 1.7). For Gillette razor blades used in many labs, a 4° clearance angle is sufficient.

(c) Adjust height of the blade to top of the tissue block and then start to cut or trim slowly; adjust vibrating strength and advancing speed until even and intact sections come out.

(d) Usually, the thinner the sections, the more difficult they are to cut, so do not waste time to try to cut good sections that are thinner than 30 μm. For EM plate embedding, sections of 50 μm thick are perfect (Zhang et al. 2003). For reconstruction of either intracellular-labeled or Golgi-stained cells, 80–120 μm sections are mostly assumed (Yoshida et al. 1999; Li et al. 2007).

Note: Generally no tissue pretreatment is needed for cutting vibratome sections, but in the case of cutting Golgi's (or another metal) stained tissue by vibratome, the tissue block needs to be embedded with gelatin or agarose to strengthen the tissue (see Golgi's stain in the following section).

(e) Collect sections serially into a grilled tray with 1× PBS (Fig. 1.10b). These sections can be first immunostained with a primary antibody and visualized with ABC kit and then embedded with LR white or Epon 812 for EM study. If sections are embedded with LR white, secondary immunogold staining can be performed post-embedding.

1.5.2 Tissue Processing for Immunohistochemical Stain of EM Study

1. Preparing the materials for tissue processing:
   (a) Liquid nitrogen (LN).
   (b) Freezing medium: isopentane is recommended.
   (c) Aluminum cans with bars capable of holding on the can in the LN.
   (d) Metal mash with handles.
   (e) Small oil-painting brush.
   (f) Filter papers.
2. Preparing the reagents and solutions for tissue processing:
   (a) 0.05 M PB (pH 7.4).
   (b) Sucrose.
   (c) Glycerol.
   (d) Cryoprotection medium is made by combining 25 % sucrose and 10 % glycerol in 0.05 M PB (pH 7.4).
3. Procedure of tissue processing:
   (a) Transfer vibratome sections into cryoprotection medium. They will sink to the bottom in 1–3 h for a 50 μm thickness section.
   (b) Put freezing medium in aluminum can and insert the can into the LN; the medium should acclimate and become the same temperature as LN or close it.
   (c) Wash the sections by 0.05 M PB (pH 7.4) after all sections have sunk to the bottom. Rinse sections shortly and gently move

them onto the metal mesh with a small oil painting brush; dry the mesh by placing it on top of a filter paper.

(d) Quickly insert and lift the mesh with sections into the freezing medium in the aluminum can that is still inside of LN. The sections undergo quick freeze and thaw. This procedure should be repeated for three times. Finally, rinse the sections with 0.05 M PB multiple times. The sections are now ready to be immunostained.

Note: There are several freezing mediums: absolute alcohol, isopentane, and toluene to name a few. One of these may or may not have negative effects on your immunostain. Try different mediums if isopentane cannot give rise to excellent result. Some investigators insert sections directly to LN, the section is easy to split in this way, so you had better trim your section to small pieces. The risk of doing so is that you may lose something due to difficulty of localizing the targeted anatomical region and labeling in small tissue; otherwise, the method of direct freeze-thaw through LN is a good method.

1.5.3 Immunohistochemical Stain for EM

1. Preparing the materials for immunohistochemical stain:
   (a) Orbital shaker
   (b) Grilled tray with plastic mesh underneath
   (c) Oil-painting brush
   (d) Erlenmeyer flasks and beakers
   (e) Funnel and filter paper
2. Preparing the medium and chemicals for immunohistochemical stain:
   (a) 0.05 M PB (pH 7.4).
   (b) 0.05 M Tris–HCl buffer (TB; pH 7.6).
   (c) Normal goat serum (NGS) or bovine serum albumin (BSA).
   (d) Vector Labs biotinylated secondary antibodies (also available from Invitrogen, Sigma, Abcam, Millipore, and Santa Cruz).
   (e) Vector Labs *Elite* ABC kit.
   (f) Diaminobenzidine tetrahydrochloride (DAB).
   (g) Nickel ammonium sulfate (NAS).
   (h) 30 % H_2O_2.
   (i) Follow instruction of Vector Labs ABC kit to mix and dilute A and B stock medium.
   (j) Make DAB–NAS reactive solution immediately before reaction: Weigh DAB for 0.05 % (e.g., 0.05 g in 100 ml) and NAS for 0.02–0.025 % in 0.05 M THB.

   Note: DAB must be dissolved in DDW first and filtered with semi-compact filter paper, then mixed with NAS in TB. Based on our experience, DAB from Sigma and NAS from Baker would be the best selection (Luo et al. 2001; Zhang et al. 2003).
3. Protocol of immunohistochemical stain:
   (a) Wash away cryoprotection medium with 0.05 M PB several times.
   (b) Block sections with either 1 % NGS or 1 % BSA in 0.05 M PB (pH 7.4) at room temperature (RT) for 1 h.
   (c) Rinse the sections in the same 0.05 M PB three times, 5 min each.
   (d) Incubate sections in the primary antibody at 4 °C at least for 24 h.
   (e) Rinse the sections in 0.05 M PB three times for 5 min each.
   (f) Incubate sections in biotinylated secondary antibodies at 4 °C at least overnight.
   (g) Rinse the sections in 0.05 M PB three times for 5 min each.
   (h) Bathe sections with ABC solution at RT for at least 5–6 h.
   (i) Rinse the sections once with 0.05 M PB (pH 7.4) and twice in 0.05 M TB (pH 7.6).
   (j) Transfer sections into a mixture of 0.05 % DAB and 0.02–0.025 % NAS in 0.05 M TB (pH 7.6).
   (k) Add 0.007–0.015 % 30 % H_2O_2 gradually in above solution when incubation has started (about 7–15 μL added to each 100 mL reactive solution). Histochemistry reaction starts when H_2O_2 has been added in solution.

(l) Stop reaction after 20–30 min by removing reactive solution and exchanging with 0.05 M TB (pH 7.6) to wash.

(m) Rinse in 0.05 M TB (pH 7.6) once and twice in 0.05 M PB (pH 7.4). Sections are now ready to be processed for EM plate embedding.

Note: During histochemical reaction, examine a section under LM after total 10 μl 30 % H_2O_2 has been dropped into the DAB–NAS solution, generally about 10 min after reaction has started. If the labeling is dark enough as well as precise and clear (if everything is correct it is about 15 min after reaction started), the reaction can be stopped; otherwise drop a little bit more H_2O_2 in solution and check the section again until excellent labeling can be observed.

Figure 1.11 displays EM microimages from ABC histochemistry stain combined with silver-intensified immunogold labeling cited with permission by authors from their published articles (Li et al. 2000).

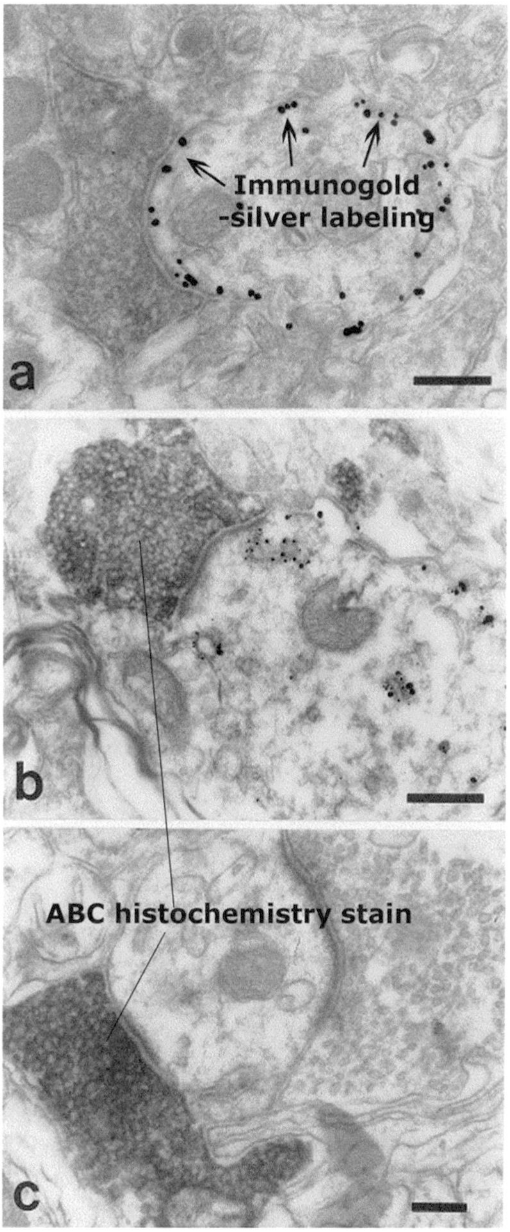

Fig. 1.11 An immuno-EM microimages of double labeling by ABC histochemistry stain and *gold–silver* method. ABC histochemistry stain showed SP-positive terminals, and *silver*-intensified immunogold displayed neurokinin-1 receptor. Scale bars are 1 μm in **a**–**c**

1.6 Plate Embedding for EM Study

1.6.1 Preparation and Siliconization of Glass Slides

1. The prerequisite for plate embedding is placing sections in between siliconized slides in order to visualize the labeled structure under LM and select adequate labeling for further EM observations.
2. Preparing the materials and silicon for siliconization of glass slides:
   (a) Original glass slides (none coated) and coverslips
   (b) Metal or plastic frames for slides and coverslips
   (c) Powerful detergent
   (d) Large trays
   (e) Distilled water
   (f) Ready-made liquid silicon
3. Procedure for siliconizing slides and cover slides:
   (a) Pour some detergent dissolved in tap water in a big tray.

(b) Emerge frames with slides and cover slides in detergent solution and shake the tray vigorously.
(c) Leave them in detergent water for a couple of hours.
(d) Shake the tray vigorously again.
(e) Dump the detergent and pour hot water (just boiled) into the tray and shake again for several minutes.
(f) Move slides into another tray with DW; shake and rinse until the DW becomes clear.
(g) Air-dry the slides at least overnight.
(h) Bathe the slides with liquid silicon for about 30 min.
(i) Air-dry the slides until all slides are dried; collect them into suitable box for future use.

1.6.2 Epon 812 Plate Embedding

1. Preparing the materials for Epon 812 plate embedding:
   (a) Desiccator jar with indicator DRIERITE inside
   (b) Petroleum jelly to seal desiccator
   (c) Digital stirring hot plate
   (d) Heating oven
   (e) Small glass beaker and stir bar
   (f) Large and small glass cylinders
   (g) Disposable syringes
   (h) Cover glass
   (i) Forceps (curved)
   (j) Siliconized slides and coverslips
2. Preparing the chemicals for Epon 812 plate embedding:
   (a) M PB (pH 7.4).
   (b) Sucrose.
   (c) Osmium tetroxide (OsO_4, crystal).
   (d) Uranyl acetate (UA, powder).
   (e) Absolute ethanol.
   (f) Propylene oxide (PO).
   (g) Epon 812 kit (include Epon 812, DDSA, NMA, and DMP-30).
   (h) Follow instruction of Epon 812 kit to mix Epon 812, DDSA, and NMA stock medium and to add DMP-30 (accelerator)—embedding medium.
   (i) Make 1 % OsO_4 in 0.1 M PB (pH 7.4) with 7 % sucrose.
   (j) Make 5 % UA in 70 % ethanol.
   (k) Make 50 %, 80 %, 90 %, and 95 % ethanol; 1:1 of pure ethanol and PO just before use; 1:1 of PO and embedding medium; and 1:3 of PO and embedding medium just before use.
3. Protocol of tissue processing and embedding:
   (a) Mix Epon 812, DDSA, and NMA (transferred by disposable syringes) in a relevant-size beaker and stir to an entire mixing; add DMP-30 until all other compounds have evenly mixed. After all elements, including accelerator, have evenly mixed, move the beaker in a desiccator jar with indicator DRIERITE inside, and seal the jar with petroleum jelly between lid and jar. Tiny bubbles in the embedding medium will be eliminated gradually during a couple of hours.
   (b) Transfer sections mentioned in the last chapter (histochemistry-stained sections) into the 1 % OsO_4–7 % sucrose solution.
   (c) Incubate them in 37 °C oven for about 40–45 min. Monitor the section's color until it turns dark brown but not totally black. Move the container out of oven.
   (d) Dump the solution and rinse the sections gently with 0.1 M PB three times.
   (e) Dehydrate the sections with 50 % ethanol followed with 5 % UA in 70 % ethanol for 8 h at 4 °C.

   Note: UA treatment can be skipped here and performed to stain grids. A new generation of digital EM, not requiring UA staining, can be programmed to change the original contrast.
   (f) Rinse away UA by 0.1 M PB for three times and continue the dehydration with 80, 90, 95, and twice 100 % alcohol, about 3–5 min in each alcohol.
   (g) Follow last alcohol dehydration (second 100 % ethanol), treat sections with 1:1 of PO and pure ethanol shortly.
   (h) Pure PO treatment of sections twice for 1–3 min each.
   (i) Then, bathe sections in 1:1 and 1:3 of PO: embedding medium, each bathing for 1 h in the desiccator jar. Then sections are

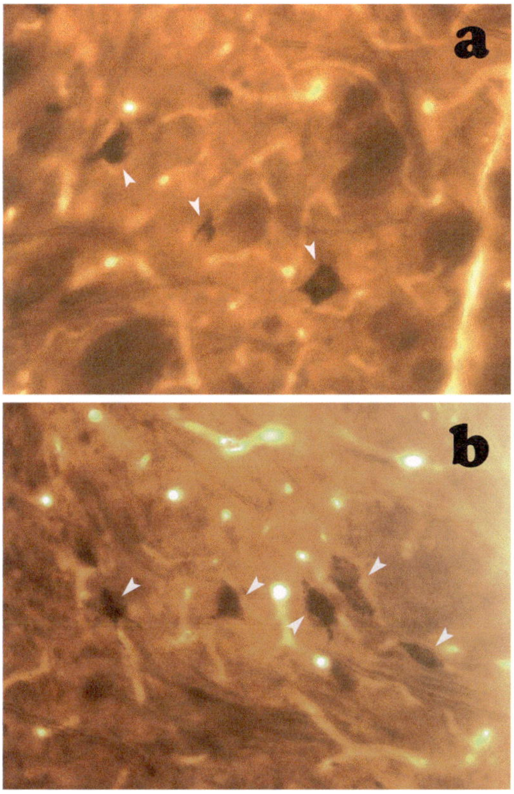

Fig. 1.12 Epon 812 plate embedded osmificated sections, labeled neurons can be observed after embedding (*arrowheads* in **a** and **b**)

transferred into pure Epon 812 embedding medium and bathing 2 h in the jar.

(j) Prepare a hot plate at 45–60 °C, place siliconized slide on the plate, mount Epon 812-incubated section on it, and cover sections with siliconized coverslip; get rid of all air bubbles.

(k) Move all slides into an oven and top coverslip with 30–50 g weight, to conduct resin polymerization that takes generally 24 h at 45 °C and 48 h at 60 °C.

4. Observe and find labeled structures on embedded slides. Figure 1.12 shows LM observation of labeled neurons (arrowheads in a and b) after osmification and plate embedding.

1.6.3 LR White Plate Embedding

1. Preparing materials for LR white plate embedding:
   (a) Slide duplicating molds (Fig. 1.13a)
   (b) Transparent plastic paper
   (c) Forceps (curved)
   (d) Tube rotator
   (e) Heating oven
   (f) Glass beakers and stir bar
   (g) Glass cylinders
   (h) Siliconized slides and coverslips
2. Preparing the chemicals for *LR* white plate embedding:
   (a) 0.1 M PB (pH 7.4).
   (b) Sucrose.
   (c) OsO_4, crystal.
   (d) UA, powder.
   (e) Absolute ethanol.
   (f) LR white kit (include liquid form of white resin in a large bottle and powder accelerator in a small bottle). Chemicals in LR white: 80 % polyhydroxy substituted bisphenol A dimethacrylate resin, 19.6 % C12 methacrylate ester, and 0.9 % dimethyl para toluidine. Accelerator: 9.9 g benzoyl peroxide.
   (g) Mix liquid form of white resin and powder accelerator; when powder is completely dissolved, store it in 4 °C and until ready to be used as embedding medium.
   (h) Make 1 % OsO_4 in 0.1 M PB (pH 7.4) with 7 % sucrose.
   (i) Make 5 % UA in 70 % ethanol.
   (j) Make 50, 80, 90, and 95 % ethanol; make 1:1 of 100 % ethanol and embedding medium just before use.
3. Protocol of tissue process and embedding:
   (a) Take certain amount of cold stoked embedding medium and warm it to RT.
   (b) Transfer histochemistry-stained sections into the 1 % OsO_4 + 7 % sucrose solution.
   (c) Incubate them in 37 °C oven for about 40–45 min. Monitor the section's color until it turns dark brown but not totally black. Move the container out of oven.
   (d) Dump the solution and rinse the sections gently with 0.1 M PB three times.
   (e) Dehydrate the sections with 50 % ethanol followed with 5 % UA in 70 % ethanol for 8 h at 4 °C.

   Note: UA treatment can be omitted here and performed to stain grids.

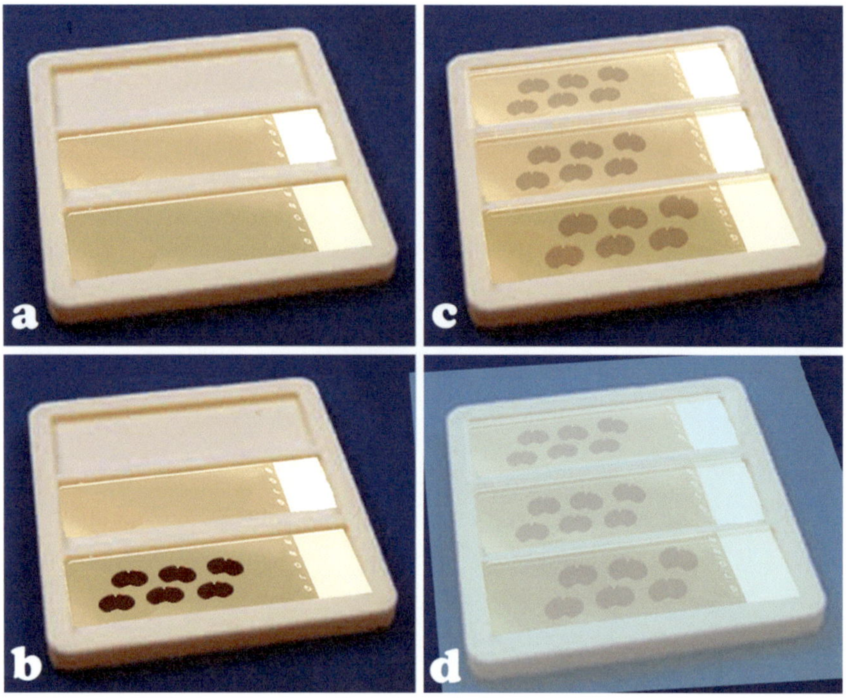

Fig. 1.13 Procedure of plate embedding with LR white by using *slide duplicating molds*

(f) Rinse away UA by 0.1 M PB for three times and continue the dehydration with 80, 90, 95, and twice 100 % alcohol, about 3–5 min in each alcohol.

(g) Incubate section with 1:1 of 100 % alcohol and LR white mixture in a small bottle on a tube rotator overnight at RT.

(h) Bathe and rotate sections at RT in 100 % embedding medium for 1 h in a tube rotator. Replace embedding medium and bathe and rotate sections for an additional hour.

(i) Lay a siliconized slide on a slide duplicating molds (Fig. 1.13a) and fill the molds with embedding medium. Lay tissue sections on the siliconized slide submerged by embedding medium (Fig. 1.13b), top sections in the medium on the slide with another siliconized slide (Fig. 1.13c), and seal the mold with hard plastic paper (Fig. 1.13d).

(j) Carefully position sealed mold into the oven and top a 300 g weight on it. Polymerization will take 24 h at 60 °C.

Note: Mold must be kept sealed and no air should contact LR white medium or sections inside of the mold.

1.7 Golgi's Stain (Traditional Way and Modern Kit)

1.7.1 Traditional Golgi's Stain

1. Preparing materials for traditional Golgi's stain:
   (a) Opaque glass jars with tight covers
   (b) Plastic strainers and forceps (avoid using metal jar and tools)
   (c) Thermostat oven
2. Preparing the chemicals for traditional Golgi's stain:
   (a) Potassium dichromate ($K_2Cr_2O_7$).
   (b) Silver nitrate ($AgNO_3$).
   (c) 1× PBS (from commercial 10× PBS).
   (d) Gelatin (eros powders).
   (e) BSA (eros powder).
   (f) Make 3 % $K_2Cr_2O_7$ in DW.

(g) Make 2 % $AgNO_3$ in DW.
(h) Make 0.5, 1, 2, and 4 % gelatin with 25 % BSA in 1× PBS.

3. Protocol of traditional Golgi's stain:
(a) Place tissue blocks in 3 % $K_2Cr_2O_7$ solution in an opaque jar in the dark, tighten the cover firmly, and position it in the 37 °C oven.
(b) Replace with fresh $K_2Cr_2O_7$ solution every day for 7 days. Keep tissues at 37 °C in the dark for the duration.
(c) Take out the tissue and dry it with filter paper in the dark. Transfer tissue into 2 % $AgNO_3$ solution in a new opaque glass jar, tighten the cover firmly, and place the jar in the 37 °C oven in the dark.
(d) At the beginning, a brown-red precipitation will occur. Change the solution several times in the dark until almost no precipitation is visualized. This procedure will take about 24 h.
(e) Once precipitation is no longer seen, place tissue jar in 37 °C oven in the dark, for 3 additional days.
(f) Embed tissue in gelatin: Bathe the tissue block in 0.5, 1, 2, and 4 % gelatin plus 25 % BSA in 1× PBS in the 37 °C oven. For each of the gelatin concentrations, it will take 30 min, 1 h, 2 h, and 5 h, respectively. This can vary by the size of the tissue.

Note: If the frozen section will be cut, the gradient sucrose infiltration will be conducted as described in Sect. 1.3, and tissue will be frozen and ready to cut sooner after OCT incubation. Gelatin embedding stated here is for vibratome sectioning.
(g) Move the tissue in the gelatin/BSA medium out of the 37 °C oven, and transfer the tissue into a square embedding mold; pour a small amount of medium into the mold, holding on the tissue in an orientation you want with a pair of chopstick-like bamboo tool until the medium becomes sticky and tissue can stand itself. Fill the mold and place it in a 4 °C refrigerator.
(h) The tissue and medium will be completely peptonized in short time in refrigerator. Once this occurs, peel away the mold, trim tissue block in an appropriate size for vibratome cutting, and place it in 4 % PFA overnight.
(i) Dry the tissue block with a piece of filter paper and glue it on a vibratome chuck, and then fix the chuck in tissue chamber, and fill the chamber with 1× PBS. It is ready to be cut. Cut 100–120 μm sections.
(j) Mount sections on slides coated with strong adhesive and conduct routine dehydration and defat.

4. Figure 1.14 shows pictures, from olfactory bulb (a, c–e) and front cortex (b). "b" is stained with traditional Golgi method, and "a, c–e" are stained by using commercial kit. Golgi's stain reveals both neurons (a–d) and glial cells (e), as well as clearly viewed dendrite spines (c, d). Figure 1.14 a, c–e were gradient sucrose infiltrated and frozen sectioned, while "b" was celloidin embedded and sectioned.

1.7.2 FD Rapid GolgiStain Kit

Detailed protocol and notes for tissue processing are available from the kit application instructions and from the website: www.fdneurotech.com. Figure 1.15 cited with permission by authors (Li et al. 2007) demonstrates the use of FD Rapid GolgiStain Kit to perform a Golgi's stain. This article also shows an example of an application of a traditional method to modern neuroscience research.

1.8 Nissl Stain

1.8.1 Cresyl Violet Stain

1. Preparation of stain medium:
(a) 1 % cresyl violet stock solution in DW. Once solution is sufficiently mix and

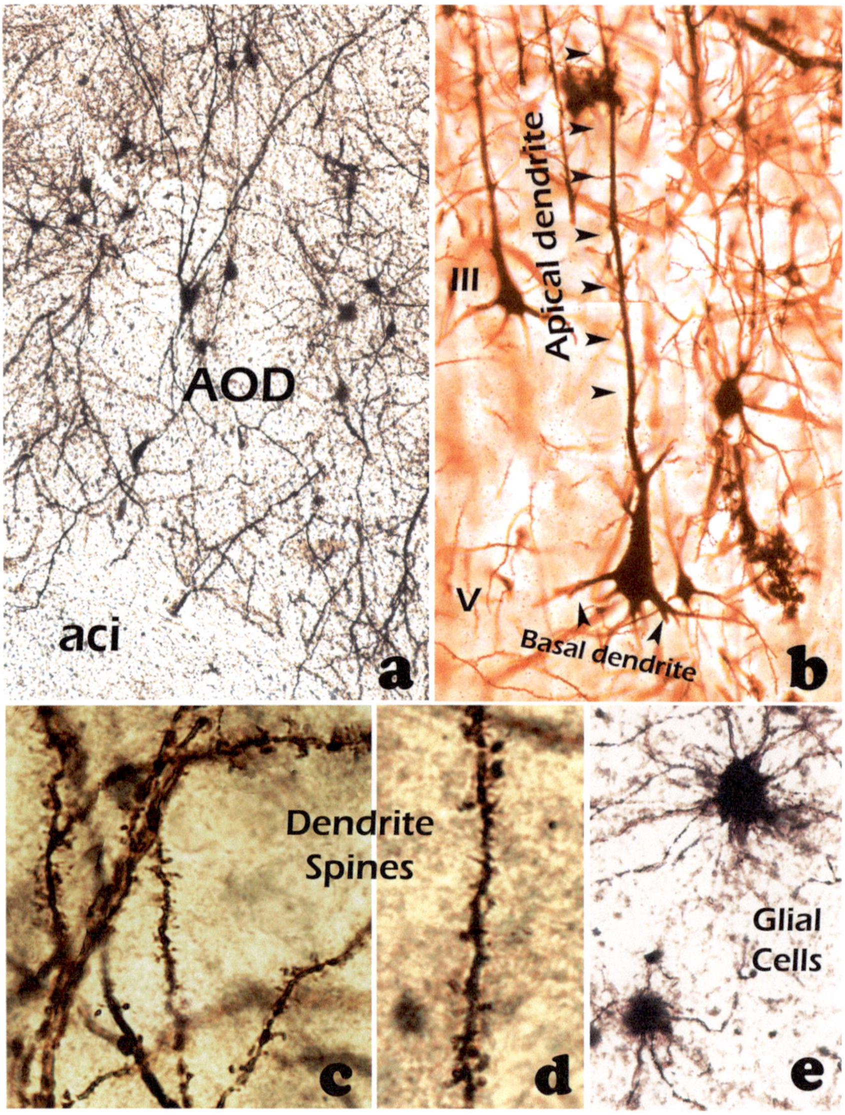

Fig. 1.14 Microimages showing staining by Golgi's stain kit (**a**, **c–e**) and traditional Golgi's stain method (**b**). (**b**) and (**c**) were montaged respectively, from different focuses photos taken from the same scope-fields on the same sections

completely dissolved, filter it with a medium-porosity filter paper.

(b) 0.1 M glacial acetate buffer at pH 3.8–4.2 (AB). Make working solution by adding 5–20 mL stock solution in 95–80 mL AB before using.

(c) Differentiation medium: 1:1:1 of ether, chloroform, and absolute alcohol.

2. Staining protocol:

   (a) Stain frozen sections for 5 or 20 min for paraffin sections in working solution.

   (b) Rinse the sections in DW for a few seconds then dehydrate them in 70 % alcohol for 1–3 min, judging by thickness of the sections.

   (c) Transfer the slide into color differential medium and observe sections under a LM.

   (d) Repeat these steps until the violet color in background has gone but Nissl bodies are visualized.

   (e) Stop color differentiation and conduct routine dehydration (80, 90, or 95 % ethanol).

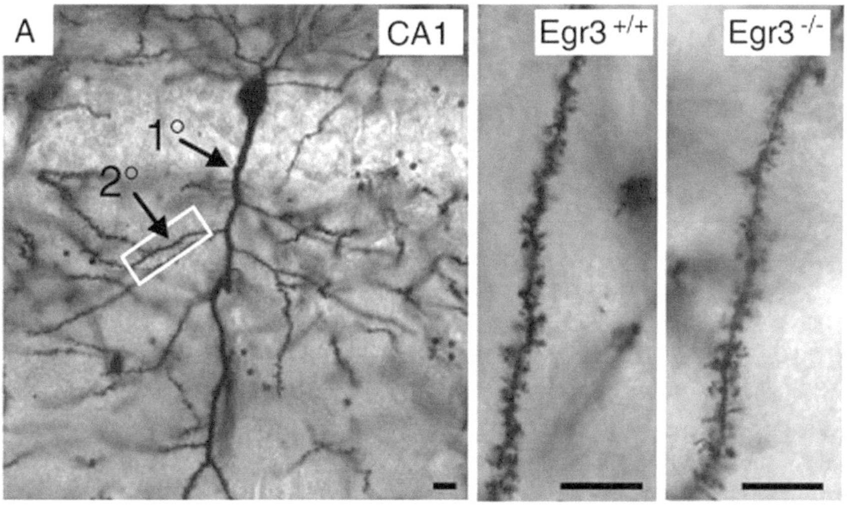

Fig. 1.15 A sample of application classical stain (Golgi's stain) in modern neuroscience, cited from Li et al. (2007). See Fig. 1.2 of the original publication for legend

(f) Bathe the slides in 100 % ethanol twice each for 1–3 min. Adding a couple of drops of chloroform into 100 % alcohol is recommended to stable the section and stain.

(g) Clarify through xylene I, II, and III routinely, and seal the sections on the slide with either DPX (distrene 80 + dibutyl phthalate + xylene) or Permount.

3. In most of cases of current neuroscience research, cresyl violet stain is used as a counterstain for LFB stain or tract tracing. Figure 1.16 shows cresyl violet-stained motoneurons in human abducens nucleus (a) and neurons in inferior olive nucleus (b).

1.8.2 Neutral Red Stain

1. Preparation of stain medium:
   (a) 2 % neutral red stock solution in DW: Stir until it is completely dissolved and sufficiently mixed and then filter it with a medium-porosity filter paper.
   (b) 0.2 M AB at pH 4.8 or 0.2 M phosphate buffer (PB) at pH 4.8.
   (c) Make working solution by making a 1:1 solution of neutral red stock solution and either AB or PB at pH 4.8 before using.

   Note: By our experiences, working solutions made with PB are stable at RT longer, with repeated uses in a couple of months; you may have to add a small amount of acetate acid when the stains become less vibrant.

2. Staining protocol:
   (a) Bathe frozen sections for 3–5 min and paraffin sections for 10–15 min in working solution.
   (b) Rinse the section for second in DW; dehydrate them in 50, 70, 80, and 90 % alcohol for 1–3 min in each solution, judging by thickness of the sections.

   Note: Neutral red stain does not need color differentiation and the background is not stained noticeably, but the dye diminishes in each concentration of alcohol, so a slight overstain is desired as it will fade during dehydration.

   (c) Bathe the slides in 100 % alcohol twice each for 1–3 min, judging by thickness of the sections.
   (d) Clarify tissues through xylene I, II, and III routinely, and seal the sections on the slide with either DPX (distrene 80 + dibutyl phthalate + xylene) or Permount.

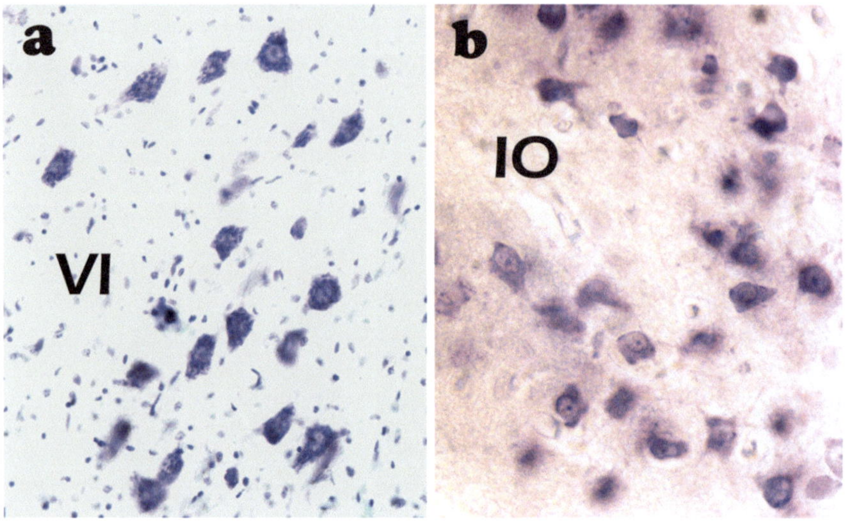

Fig. 1.16 Cresyl *violet* Nissl-stained human motoneurons in abducens nucleus (**a**) and relay neurons in inferior olive nucleus (**b**)

(e) In most of cases of current neuroscience researches, neutral red stain is used to be a counterstain after immunohistochemistry stain or tract tracing. Figure 1.17 shows neutral red counterstained motoneurons in trigeminal motor, facial, hypoglossal, and ambiguous nuclei (Vmo, VII, XII, and Amb) (Zhang et al. 2012).

(f) There are several other kinds of Nissl stains, some of them are still in use today. These include thionine (Lauth's violet) stain that is usually a substitute of cresyl violet, toluidine blue stain, and Einarson's (gallocyanin) stain. Those stains require much lower pH ranging from 1.64 to 2.5, so these may not be good for immunohistochemistry or tract tracing counterstain.

1.9 Hematoxylin and Eosin Stain

1.9.1 Preparation of Stain Medium

- Sigma Mayer's hematoxylin solution (certified by hematoxylin 1.0 g/L, sodium iodate 0.2 g/L, aluminum ammonium sulfate • 12 H_2O 50 g/L, chloral hydrate 50 g/L, and citric acid 1 g/L)
- Sigma Harris hematoxylin solution (certified hematoxylin 7.0 g/L, sodium iodate 0.2 g/L, aluminum ammonium sulfate • 12 H_2O 50 g/L)
- Differential acidic alcohol: 0.3 % (v/v) concentrated HCl in mixture of 70 % ethanol and 30 % DW (pH < 3.0)
- Scott's tap water: sodium bicarbonate ($NaHCO_3$) 2 g/L and anhydrous magnesium sulfate ($MgSO_4$) 10 g/L or $MgSO_4 \cdot 7H_2O$ 20 g/L in DW (pH 8.3)
- Sigma Eosin Y Alcoholic (certified Eosin Y, 0.5 % (w/v) in acidified 90 % ethanol)
- All rehydration and dehydration alcohols (50–100 %) and xylene

1.9.2 Methodology Classification

1. HE stain is the most common histological stain in all clinical and research labs. When preparing an HE stain, the first step is to select an HE stain protocol suitable for your lab or your own study. There are two types of hematoxylin stain mediums, progressive and

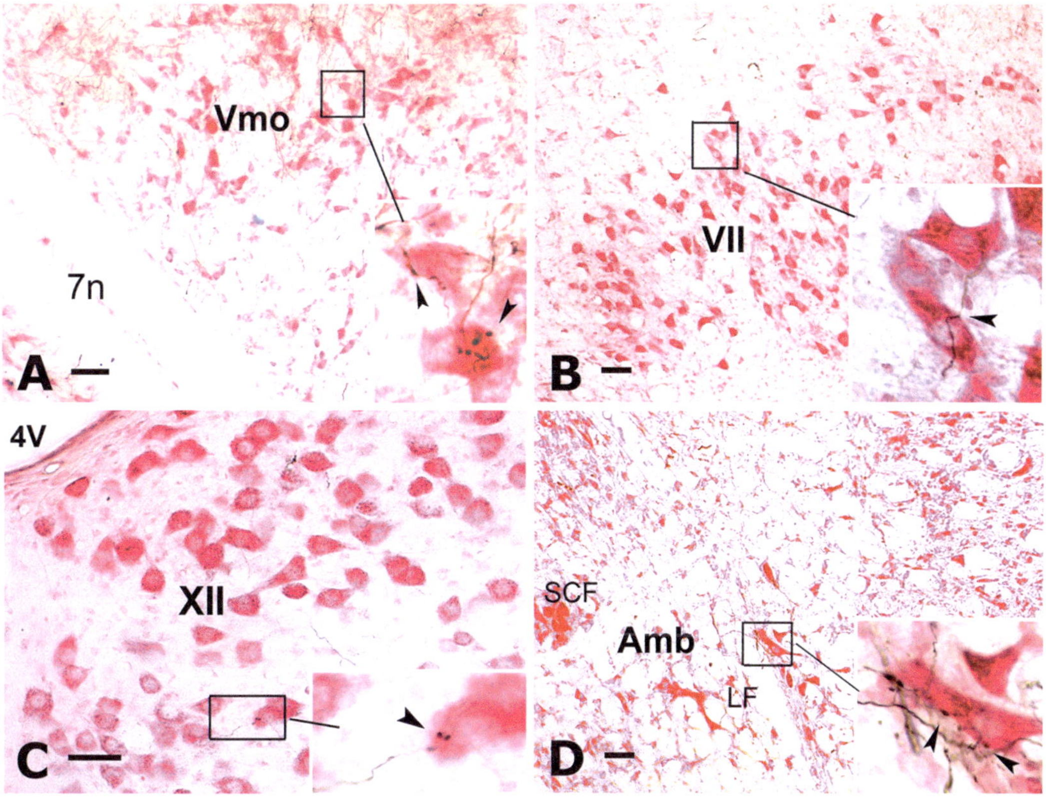

Fig. 1.17 Neutral red Nissl-stained motoneurons in rat orofacial motor nuclei. *7n* facial nerve root, *Amb* ambiguous nucleus, *LF* loose formation of the Amb, *SCF* semi-compact formation of the Amb, *VII* facial nucleus, *Vmo* trigeminal motor nucleus, *XII* hypoglossal nucleus. Scale bars = 50 μm in **a**–**d**. Cited from Zhang et al. (2012)

regressive H stain, and both are still widely used. There are also two kinds of eosin stain mediums, water-based and alcohol-based solutions; most people prefer the latter.

2. Progressive H stain medium contains less hematoxylin with other substances to slow down oxidation of hematoxylin into hematin in situ and generates weaker initial H stain. Regressive H stains contain higher concentrations of hematoxylin without a defending substance and give rise to a stronger H stain that requires color differentiation before E stain.
3. Those stain protocols originally stem from classical Gill's H stain, Mayer's H stain, and Harris's H stain; the former two are in the category of progressive H stain and the latter is regressive H stain. Ready-made Gill's, Mayer's, and Harris' H stain mediums are commercially available from companies such as Sigma-Aldrich or Thermo Fisher Scientific to name some.
4. We will apply Sigma-Aldrich HE stain solution as the example here to elucidate the usage of these kinds of medium. Accordingly, in the following introduction of the protocols, branded Sigma-Aldrich stain solution will be used as H and E stain medium. Regardless of the company from which the stain solution is purchased, the protocol is the same.
5. Sigma-Aldrich Gill's 1 is mainly used to stain cultured cells and is the weakest H medium. Gill's II and III stains and Mayer's H medium are stronger and can be applied to tissue staining, i.e., progressive stain. Sigma-Aldrich Harris's hematoxylin is used for regressive H stain.
6. Sigma-Aldrich Eosin Y and Eosin Y with phloxine are aqueous solutions, the former is

weak and the latter is the strongest dye because phloxine can overwhelmingly color the tissue red. Sigma-Aldrich Eosin Y Alcoholic is a stronger dye but in between, which is the most commonly used E stain medium, i.e., for standard E stain. In this chapter we introduce primarily the progressive or regressive H stain together with Eosin Y Alcoholic E stain.

1.9.3 Staining Protocols

1. Progressive H stain with Alcoholic E stain
   (a) De-wax and rehydrate paraffin sections, then bathe them in DW for 30 s; or simply bathe frozen section in DW for 30 s.
   (b) Stain sections in Mayer's H medium, 5 min for frozen sections and about 10 min for paraffin sections. Adjust time based on results.
   (c) Wash slides in DW for 1–2 min.
   (d) Bluing (intensifying blue stain), move slides into Scott's tap water and stay for 3–5 min; watch the sections turned to be blue and stable.
   (e) Note: Saturated lithium carbonate solution or 1 % ammonia water could be substitute of Scott's water, but by our experience Scott's tap water is much more stable.
   (f) Wash slides in DW for 1–2 min.
   (g) Stain sections in Alcoholic E medium for 20–30 s.
   (h) Note: The older the solution is, the longer time is needed. For new solution, 20 s is sufficient.
   (i) Dip slides in 50 % ethanol five times and then 70 % alcohol five times.
   (j) Equilibrate in 95 % ethanol for 30 s.
   (k) Dehydrate in 100 % ethanol for 1 min two times.
   (l) Clarified tissue shortly in two sets of xylene 2–3 min total, then cover and seal with either DPX or Permount.
2. Regressive H stain with Alcoholic E stain
   (a) De-wax and rehydrate paraffin sections, then bathe them in DW for 30 s; or just bathe frozen section in DW for 30 s.
   (b) Stain sections in Harris H medium, 1 or 2 min for frozen sections, 5 min for paraffin sections. Adjust the time based on the results.
   (c) Wash the slides in DW for 1–2 min.
   (d) Differentiation: Set up two jars of acidic alcohol—dip the slides a couple of seconds in first jar and another couple of seconds in second jar one by one.
   (e) Wash slides in DW for about 1–2 min.
   (f) "Bluing," move slides into Scott's tap water and stay for 3–5 min; watch the sections turned to be blue and stable (color no more change).
   (g) Wash slides in DW for about 1–2 min.
   (h) Stain sections in Alcoholic E medium for 20–30 s. For new Eosin solution, 20 s is good enough.
   (i) Dip slides in 50 % ethanol and 70 % alcohol for a total of 20–30 s.
   (j) Equilibrate in 95 % ethanol for about 30 s.
   (k) Dehydrate in 100 % ethanol for 1 min two times.
   (l) Clarified tissue shortly in two sets of xylene 2–3 min total, then cover and seal with either DPX or Permount.

1.9.4 Technique Note

1. Hematoxylin (oxidized into hematin) is actually a pH indicator, in circumstance of pH value <5; it is soluble red and not attached to the tissue firmly; when pH value is between 5 and 11, it turns to blue and becomes insoluble. Above this point (pH 5) hematin will more firmly attach to alkaline proteins or cellular elements in the tissue.
2. "Bluing" by Scott's tap water with pH 8.0–8.2 is to fix hematin to the tissue sections. Therefore, "bluing" step is important; otherwise soluble hematin will be lost as soon as the sections have moved into Eosin solution with pH 4–4.5. However, without bluing step may result in what looks like only E stain (Fig. 1.18a).

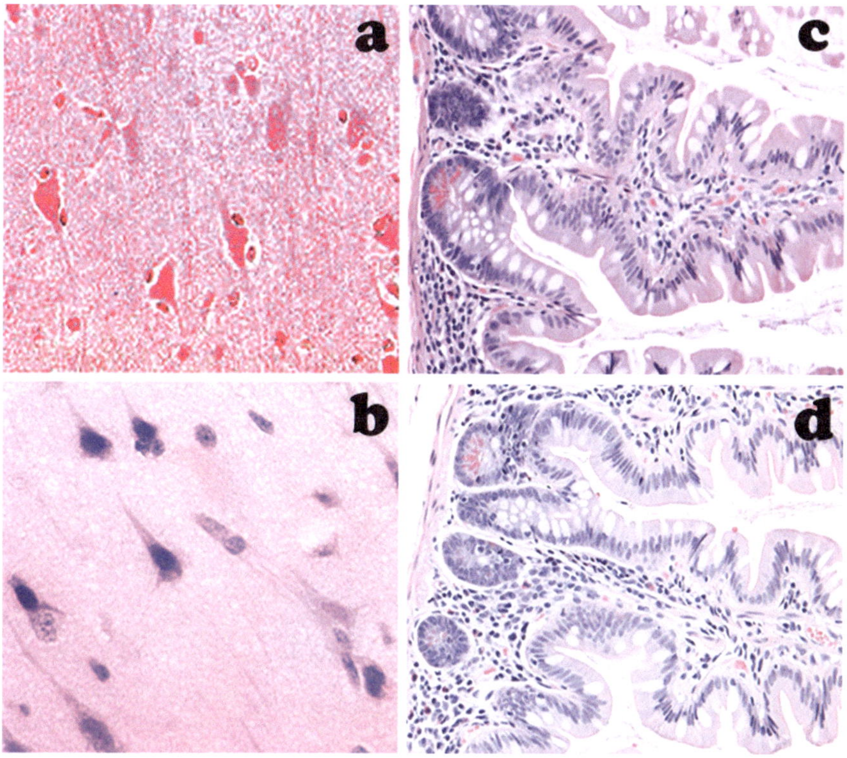

Fig. 1.18 Different qualities of HE stains. Image reproduced by courtesy of Herbert Skip Brown, M. Div., Robert H. Lurie Comprehensive Cancer Center-Northwestern University

3. Differentiation is to differentially extract excess hematin from chromatin and cytoplasm because of over H stain, which is a quantitative procedure and entirely different from "bluing." In Mayer's H stain, there is supposed to be no excess H stain so that differentiation is not necessary, but bluing is still needed (Fig. 1.18b).
4. Color from E stain is also affected by pH: the lower the pH, the more red and the higher the pH, the more pink purple. The standard pH value in E solution is 4.0–4.5. Figure 1.18c, d displays a good H stain because of well-attached blue. However, the stain showed in Fig. 1.18d may suffer a pH value higher than standard.
5. Figure 1.19 compared single H stain (a) with cresyl violet (b) Nissl stain. The image in b shows Nissl stain and eosin counterstained. Note in these two stains, emphasis on cellular organelles is clearly different even though both pictures are stained blue.

1.10 Luxol Fast Blue Myelin Stain

1.10.1 Preparation of Stain Medium

- 0.1 % LFB in 95 % ethanol with 0.5 % glacial acetate acid in solution
- 0.05 % lithium carbonate (Li_2CO_3) in DW
- 1:1 of absolute ethanol: chloroform for frozen sections
- Sigma Eosin Y, Aqueous (certified Eosin Y, 0.5 % (w/v) in water)
- 0.25 % cresyl echt violet with 10 % acetate acid in DW

1.10.2 Luxol Fast Blue Staining Protocol

1. Basic LFB staining
   (a) Bathe frozen sections in 1:1 ratio of pure ethanol and chloroform for 3–5 h at RT.

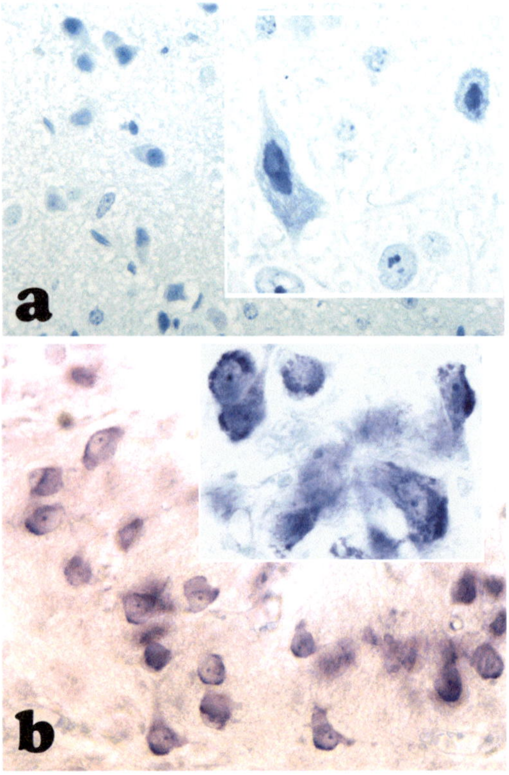

Fig. 1.19 Comparison of single H stain (**a**) with cresyl *violet* Nissl stain (**b**)

Longer time of bathing may be required for frozen sections thicker than 30 μm.

(b) Xylene de-wax for paraffin sections and bathe sections in 100 % ethanol for 2 min twice.

(c) Transfer slides into 95 % alcohol and bathe for 3–5 min, then move into 0.1 % LFB medium.

(d) Incubate slides in 0.1 % LFB staining solution at 56 °C in a thermostat oven overnight. If 1 % LFB working solution is employed, incubate sections at 37 °C or RT overnight or at 56–60 °C for 2–3 h.

(e) Note: Overstained is OK; the key step for satisfactory stain is color differentiation in 0.05 % Li_2CO_3 back and forth (see below). Using a screw-topped plastic, Coplin jar is strongly recommended.

(f) Move the staining jar out of the oven and rinse away the excessive stain with 95 % alcohol.

(g) Rinse shortly in DW and dip slides in 0.05 % Li_2CO_3 for several times or for about 3 s.

(h) Wash shortly in DW and examine the tissue sections under a LM; white matter part of tissue should be in fresh blue and gray matter should be clear; if not, repeat steps f–g until correct color is visualized under a LM examine.

(i) Dehydrate in 70 and 95 % ethanol quickly (shorter than 1 min in each).

(j) Bathe in 100 % ethanol shortly for twice and clarify in xylene twice for 1 min each, then coverslip and seal by either DPX or Permount.

2. LFB stain and (Eosin) fast violet counterstain
   (a) Steps a–f are the same as aforementioned in basic stain.
   (b) After 0.05 % Li_2CO_3 differentiation, move the slides to two sets of 70 % ethanol dip about 10 s in each, and examine the tissue sections under a LM; white matter should be in fresh blue and gray matter should be colorless. Repeat this step until gray matter is clear.
   (c) Optional step: Move slides into Eosin Y solution and stain for 1 min.
   (d) Rinse in DW for 1 min and stain with 0.25 % cresyl echt violet for 1 min.
   (e) Rinse in DW shortly and dehydrate sections in 70, 95 % for 2 min, and then bath with 100 % ethanol twice for 2–3 min each.
   (f) Clarify tissue in xylene twice for 1–2 min each, then coverslip and seal by either DPX or Permount.

   Note: If step c (Eosin) is omitted, it becomes LFB counterstained with cresyl echt violet (Fig. 1.20). If step d is skipped, the stain is counterstained with eosin only. Double counterstain enables an investigator to view more cellular components; in this combined stain (Fig. 1.21), neuronal somata are purple (Nissl bodies), myelinated fibers are blue, and neuropile with another cells is pink (pink glial cells and endothelial cells). During staining, try to avoid using metal jar and tools such as metal forceps; all staining stuff should be plastic or glass.

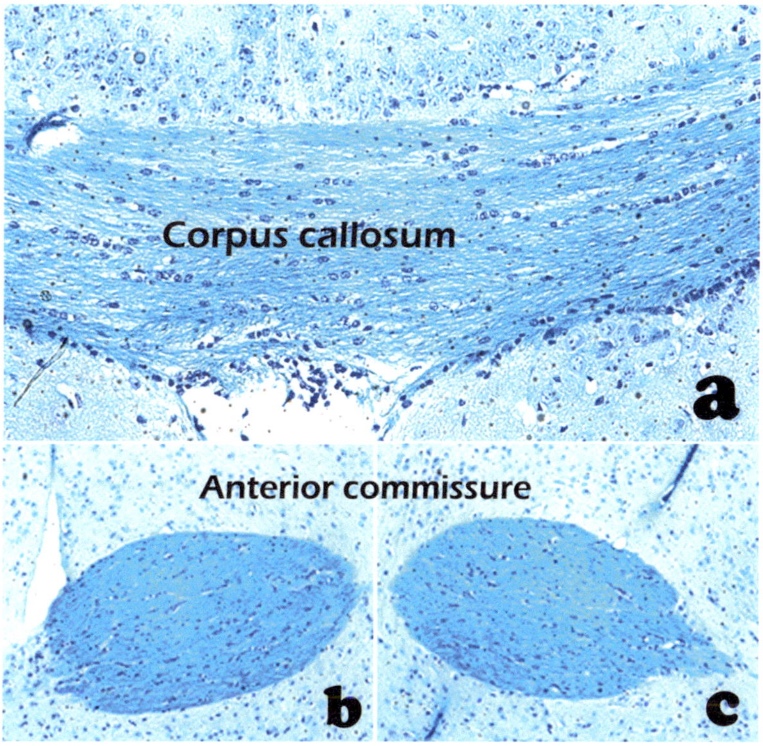

Fig. 1.20 Luxol fast *blue* and cresyl echt *violet*-stained sections

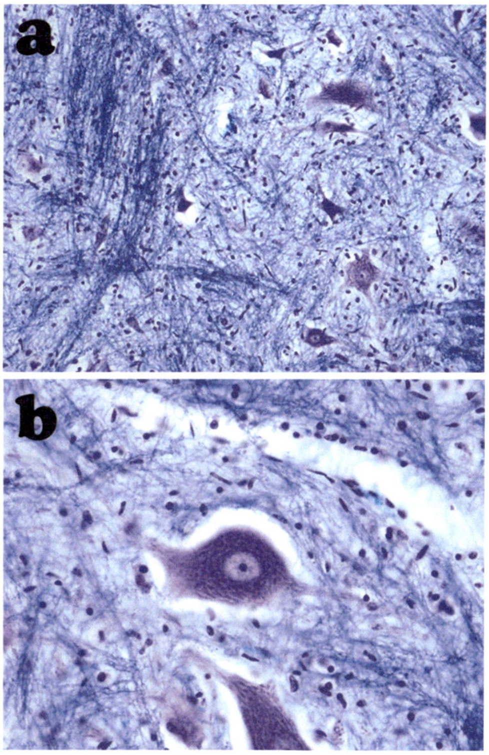

Fig. 1.21 Luxol fast *blue*-, cresyl echt *violet*-, and eosin-stained sections. Image reproduced by courtesy of the Veterinary Pathology Unit, School of Veterinary Sciences, University of Bristol, United Kingdom

References

Bjorklund A (1983) Fluorescence histochemistry of biogenic monoamines. In: Bjoklund A, Hokfelt T (eds) Handbook of neuroanatomy. Elsevier, Amsterdam, pp 50–121

Carriel V, Garzon I, Alaminos M, Campos A (2011) Evaluation of myelin sheath and collagen reorganization pattern in a model of peripheral nerve regeneration using an integrated histochemical approach. Histochem Cell Biol 136:709–717

Gage GJ, Kipke DR, Shain W (2012) Whole animal perfusion fixation for rodents. J Vis Exp (65)pii:3564

Hasegawa R, Takami S, Nishiyama F (2008) Immunoelectron microscopic analysis of the distribution of tyrosine kinase receptor B in olfactory axons. Anat Sci Int 83:186–194

Jones EG (2010) Cajal's debt to Golgi. Brain Res Rev 66:83–91

Kluver H, Barrera E (1953) A method for the combined staining of cells and fibers in the nervous system. J Neuropathol Exp Neurol 12:400–403

Li JL, Wang D, Kaneko T, Shigemoto R, Nomura S, Mizuno N (2000) The relationship between neurokinin-1 receptor and substance P in the medullary dorsal horn: a light and electron microscopic immunohistochemical study in the rat. Neurosci Res 36: 327–334

Li L, Yun SH, Keblesh J, Trommer BL, Xiong H, Radulovic J, Tourtellotte WG (2007) Egr3, a synaptic activity regulated transcription factor that is essential for learning and memory. Mol Cell Neurosci 35: 76–88

Luo P, Haines A, Dessem D (2001) Elucidation of neuronal circuitry: protocol(s) combining intracellular labeling, neuroanatomical tracing and immunocytochemical methodologies. Brain Res Brain Res Protoc 7:222–234

Pinaud R, Jeong JK (2010) Duplex in situ hybridization in the study of gene co-regulation in the vertebrate brain. Methods Mol Biol 611:115–129

Pinaud R, Mello CV, Velho TA, Wynne RD, Tremere LA (2008) Detection of two mRNA species at single-cell resolution by double-fluorescence in situ hybridization. Nat Protoc 3:1370–1379

Pistorio AL, Hendry SH, Wang X (2006) A modified technique for high-resolution staining of myelin. J Neurosci Methods 153:135–146

Rocchietta (1968) Franz Nissl (1860–1919), neuropathologist. JAMA 205:460–461

Soontornniyomkij V, Choi C, Pomakian J, Vinters HV (2010) High-definition characterization of cerebral beta-amyloid angiopathy in Alzheimer's disease. Hum Pathol 41:1601–1608

Steencken AC, Siebert JR, Stelzner DJ (2009) Lack of axonal sprouting of spared propriospinal fibers caudal to spinal contusion injury is attributed to chronic axonopathy. J Neurotrauma 26:2279–2297

Tang ZY, Shu B, Cui XJ, Zhou CJ, Shi Q, Holz J, Wang YJ (2009) Changes of cervical dorsal root ganglia induced by compression injury and decompression procedure: a novel rat model of cervical radiculoneuropathy. J Neurotrauma 26:289–295

Triarhou LC, Del Cerro M (2012) Ramon y Cajal erroneously identified as Camillo Golgi on a souvenir postage stamp. J Hist Neurosci 21:132–138

Wohlrab F, Henoch U (1988) The life and work of Carl Weigert (1845–1904) in Leipzig 1878–1885. Zentralbl Allg Pathol 134:743–751

Yabuta NH, Yasuda K, Nagase Y, Yoshida A, Fukunishi Y, Shigenaga Y (1996) Light microscopic observations of the contacts made between two spindle afferent types and alpha-motoneurons in the cat trigeminal motor nucleus. J Comp Neurol 374:436–450

Yoshida A, Mukai N, Moritani M, Nagase Y, Hirose Y, Honma S, Fukami H, Takagi K, Matsuya T, Shigenaga Y (1999) Physiologic and morphologic properties of motoneurons and spindle afferents innervating the temporal muscle in the cat. J Comp Neurol 406:29–50

Zhang J, Pendlebury WW, Luo P (2003) Synaptic organization of monosynaptic connections from mesencephalic trigeminal nucleus neurons to hypoglossal motoneurons in the rat. Synapse 49:157–169

Zhang J, Luo P, Ro JY, Xiong H (2012) Jaw muscle spindle afferents coordinate multiple orofacial motoneurons via common premotor neurons in rats: an electrophysiological and anatomical study. Brain Res 1489:37–47

Brain Stereotaxic Injection 2

Jingdong Zhang and Huangui Xiong

Abstract

The brain is a precisely organized and compact structure that controls a majority of bodily functions. Precise identification of function related structures is extremely important in neuroscience research. Achievement of this goal would not have been possible without the stereotaxic delivery system. This chapter recalls the history of the stereotaxic instrument and development of system applications. In addition, we use the rat coordinate atlas and stereotaxic instrument as an example and introduce practicable procedures for application of this system in the following sequence: (1) How to read a rat brain atlas with stereotaxic coordinates and determine coordinates for a targeted brain structure. The reference points, bregma, lambda, and interaural line, are interpreted in this section. Identify what parameters in the atlas refer to sagittal, medial–lateral, and anterior–posterior coordinates. In addition, calculate an angled delivery coordinate to avoid damaging any large blood vessel in the brain. (2) How to correctly mount a rat on the instrument and position an electrode or delivery pipette based on coordinates obtained or calculated from the atlas. In this procedure, the head of a rat must first be reliably mounted on the head holder; then actual coordinates for a reference point (bregma, lambda, or interaural line) are required. Theoretically, all should be "0," but actually not.

J. Zhang, M.D., Ph.D. (✉)
Department of Pharmacology and Experimental Neuroscience, University of Nebraska Medical Center, Durham Research Center, Durham Research Plaza, DRC I 8034, Omaha, NE 68198-5880, USA
e-mail: jingdong.zhang@unmc.edu

H. Xiong, M.D., Ph.D.
Department of Pharmacology and Experimental Neuroscience, University of Nebraska Medical Center, Durham Research Center, Durham Research Plaza, DRC I 8034, 45th Street at Emile Street, Omaha, NE 68198-5880, USA
e-mail: hxiong@unmc.edu

H. Xiong and H.E. Gendelman (eds.), *Current Laboratory Methods in Neuroscience Research*, Springer Protocols Handbooks, DOI 10.1007/978-1-4614-8794-4_2,

The actual parameters of a reference point are the number displayed on instrument bars when the tip of the needle aimed at the reference point. (3) The methods of injection include pressure injection, iontophoresis, and dry crystal delivery. The key point for pressure injection is to conduct it quite slowly and keep the micropipette or needle in situ for quite a while after injection. Similarly in iontophoresis delivery, do not withdraw the electrode immediately out of the place after injection. In dry crystal delivery section, a new device for stereotaxic delivery of crystal into deep structures of the brain is introduced. However, online index has not shown any commercial availability of this kind of device.

Keywords
Brain atlas with stereotaxic coordinates • Stereotaxic reference points • Stereotaxic instruments • Stereotaxic positioning of delivery pipette • Pressure or iontophoresis injection

2.1 Introduction

The brain-to-body mass ratio is 1:40 in humans, mouse, and rat (Herschel 1972); but despite its small size, the brain controls activities of the whole body through a complex network of nervous systems. Hence, there are a great number of functional centers in the brain, these areas may communicate with one another to control and modulate different somatic functions such as circulation, respiration, digestion, and locomotion. In order to study neuronal pathways associated with these different functions, investigators created a way through which a specific brain region, such as nucleus or fasciculus, can be targeted. These specific regions can be stimulated and recorded, lesioned, and injected with tracers, or chemicals such as agonists, antagonists or neurotoxic substances. A stereotaxic delivery system is utilized in this method, which consists of a stereotaxic instrument (Fig. 2.1) and a brain atlas with corresponding three-dimensional coordinates (Fig. 2.2). The stereotaxic instrument has a frame baseplate with tooth bar and ear bars to mount animal and a railroad for moving a stereotaxic modulator on which a delivery pipette holder is attached. The holder can hold either an electrode or Hamilton syringe. Identification of the connections of hierarchy centers pertinent to certain function, and the connections between structures related to different functions, would not have been possible without the stereotaxic delivery system.

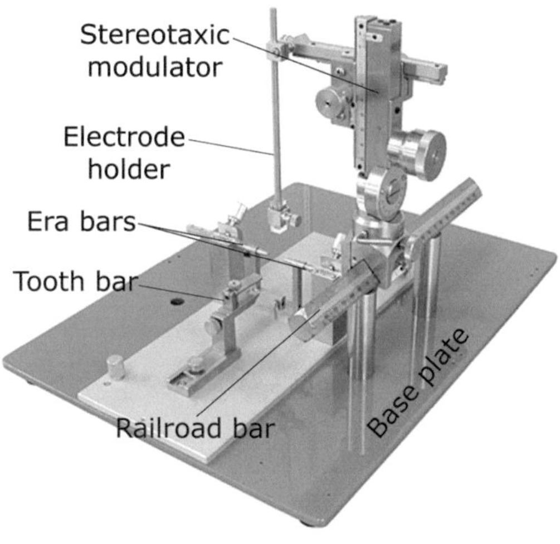

Fig. 2.1 A rat stereotaxic instrument

The first stereotaxic frame was invented by Robert Henry Clarke and published by Horsley and Clarke in 1908 (Horsley and Clarke 1908; Tan and Black 2002). Horsley was a talented neurosurgeon and creative scientist in the field of neurology research in England. This first stereotaxic frame

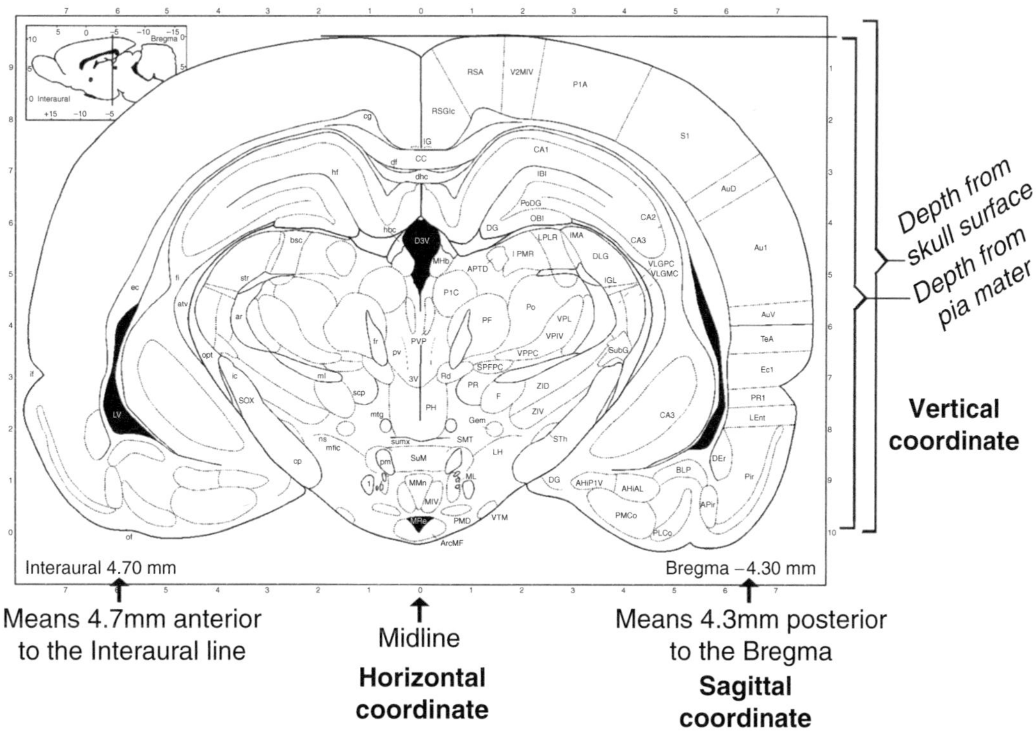

Fig. 2.2 An example page of rat brain atlas from Paxinos and Watson (1998)

was devised for animal research aiming to target, stimulate, and electrolyze the structure in the cerebellum. Spiegel and Wycis (Spiegel et al. 1947) described the first instrument for human use in 1947. Their instrument, which was of the translational type, undoubted stimulated Leksell when, in 1949 (Leksell 1949a, b), he designed the first instrument to be based on the arc-center principle (see pictures of these stereotaxic instruments in "Cyber Museum of Neurosurgery" at http://www.neurosurgery.org/cybermuseum/stereotactichall/92exhibit.html). In the latter half of the century, small mammals became commonly used in biomedical research and stereotaxic instruments for all mammals, including cats, rabbits, rats, and mice, had been produced and successfully applied. In this subchapter, we will use the rat stereotaxic system as an example to introduce the principles and techniques of stereotaxic delivery.

During the last century, the majority of investigators utilized stereotaxic delivery techniques to study anatomy and physiology of neuronal pathways. Since the early 1950s researchers have applied an anterograde Wallerian degeneration method to study specific pathways, by stereotaxic electrolysis of pertinent nuclear neurons and observing the resulting degenerated axons and terminals along the pathway tract (Abercrombie and Johnson 1946; Thomas 1948; Johnson et al. 1950). In parallel, stereotaxically evoking orthodromic action potentials and recording second-order neuronal responses were also widely used in demonstration of functional pathways (Bizzi 1966; Adey and Noda 1973). Later on, many efficacious neuronal tracers came into being and stereotaxic injection of tracers became commonplace in the study of neuronal pathways (Kobbert et al. 2000). However, in recent years of neuroscience studies, a number of novel applications of stereotaxic delivery have been developed. One major new usage is to stereotaxically deliver plasmid vectors designed to carry either pathogenic or therapeutic genes to be transfected into the host cells in the brain

(Liu et al. 2011; Zlokovic and Apuzzo 1997). A second usage, direct stereotaxic delivery of *lentivirus* into targeted brain areas, has been assumed in transfection of either pathogenic or therapeutic genes into host cells (Norgaard Glud et al. 2010; Cetin et al. 2006). Thirdly, catalysts or inhibitors packed in vectors can be stereotaxically injected into the brain and engulfed by host cells in order to modulate some key enzymes in pertinent intracellular pathways (Chu and Etgen 1997; Liu et al. 2012). In addition, stereotaxic delivery of modulatory proteins, peptides, receptor agonists or antagonists, and neurotoxins are still widely in use.

2.2 Rat Brain Atlas with Stereotaxic Coordinates

2.2.1 Stereotaxic Reference Points

1. Bregma. The point of intersection of the coronal and sagittal sutures (Fig. 2.3a, b), which is a connection point of frontal bone and bilateral parietal bones.

   Note: Most of the bregma is regular and situated at the midline; but in a small number of rats, the bregma is irregular as shown by Fig. 2.4, which looks like more than one bregma. In these cases, the reference point is still in the central intersection along the sutures (red arrows and stars in Fig. 2.4).
2. Lambda. The point of intersection of the extending lines of sagittal and lambdoid sutures (Fig. 2.3a, b), which is the central point on the connection of bilateral parietal bones and occipital bone (asterisk in Fig. 2.3a).
3. The interaural line. A horizontal line between the tips of bilateral ear bars. It should be the line between bilateral eardrums when the rat is mounted on the stereotaxic frame correctly.
4. General principle of selection of a correct reference point: If your targeted structure is closer to the bregma, you should use bregma as the reference point; if the aimed structure is closer to the lambda, the interaural line should be selected as the reference point; nonetheless, lambda is the midline point.

2.2.2 Setup of Coordinates for a Targeted Structure

In this subchapter we reference "The Rat Brain in Stereotaxic Coordinates" edited by Paxinos and Watson (1998) as a textbook to elucidate its application. In addition, we use two structures, the oculomotor nucleus (III; red in Fig. 2.5a, b) and the locus coeruleus (LC; blue in Fig. 2.6a, b) as examples to explain the usage of the atlas.

1. Sagittal coordinate. This coordinate reflects structures from anterior to posterior or rostral to caudal in the rat. If the bregma is taken as a reference point (=0), the rat III (Fig. 2.5) is from 6.04 mm posterior to bregma (bregma −6.04) to 7.04 mm posterior to bregma (bregma −7.04). If the interaural line is used as a reference point (=0), the III is from interaural 2.96 to 1.96 (2.96–1.96 mm anterior to the interaural line). For the rat LC (Fig. 2.6), only the interaural line is taken as a reference point, but the principle is the same as when using bregma as a reference point. The rat LC is located rostrocaudally from interaural −0.16 to interaural −1.30 mm. The coordinates are shown from Fig. 44 to Fig. 48 in Rat Brain Atlas edited by Paxinos and Watson (1998).
2. Coronal coordinate. This coordinate index structures from midline towards the bilateral borders of the brain. Both bregma and lambda are situated on the midline of the brain; therefore, the rat III is 0.1–0.5 mm lateral to both bregma and lambda (midline; Fig. 2.5) on both sides. The bilateral rat LCs are situated from 0.9 to 1.4 mm lateral to the midline. The LC itself is 0.2–0.3 mm wide on the coronal plane but shifts from medial to lateral (Fig. 2.6).
3. Vertical coordinate. This is the deepness from the surface of the brain to the targeted structure vertically, in rat it is also referred to as from dorsal to ventral. In an atlas, there are numbers from 0 to 10 (mm) in vertical coordinate (Figs. 2.5 and 2.6); these numbers are given in units of millimeters and signify distance from rat external occipital protuberance to interaural line (Fig. 2.3b). In male Wister rat, weighing approximately 290 g, this distance is about 10 mm

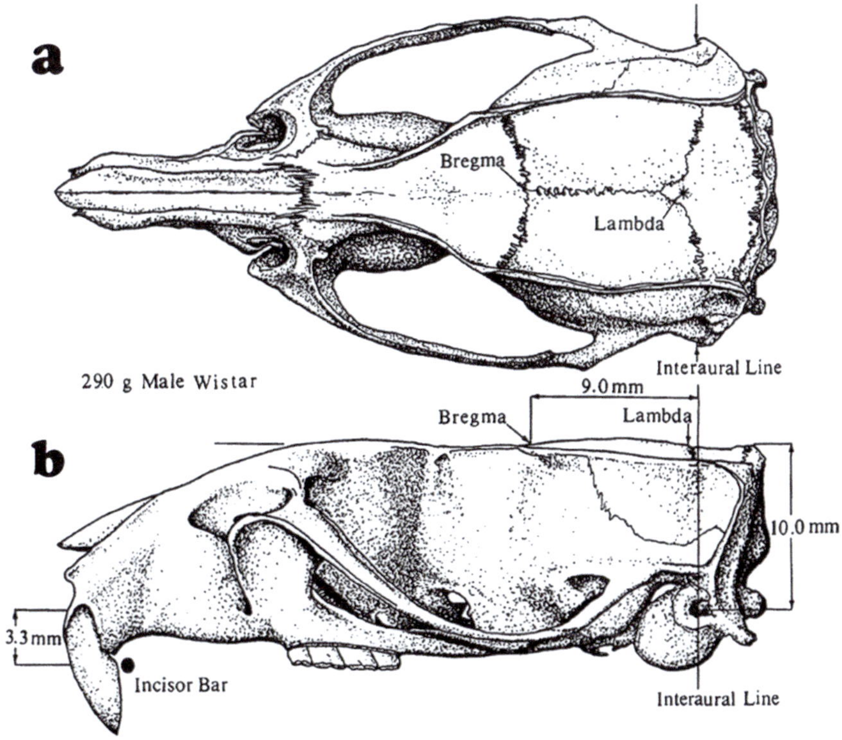

Fig. 2.3 Schematic showing of reference points from rat brain atlas from Paxinos and Watson (1998). (**a**) top view; (**b**) lateral view

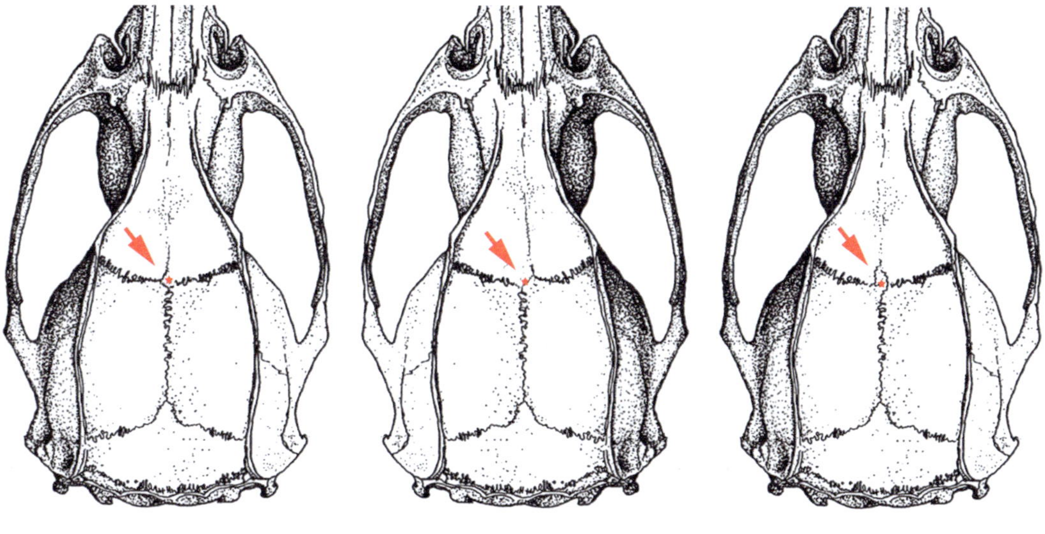

Fig. 2.4 Schematic showing of rat's heteromorphic bregma area and determination of correct points (*red arrows* and *stars*)

(Fig. 2.3b). However, an actual distance from pia mater on the surface of the brain to the targeted structure is commonly applied when stereotaxic injection is performed, because the skull is generally removed by a drill before injection. Therefore, the vertical coordinate for the III is 6.0–6.2 mm from the surface of the brain to center of the III

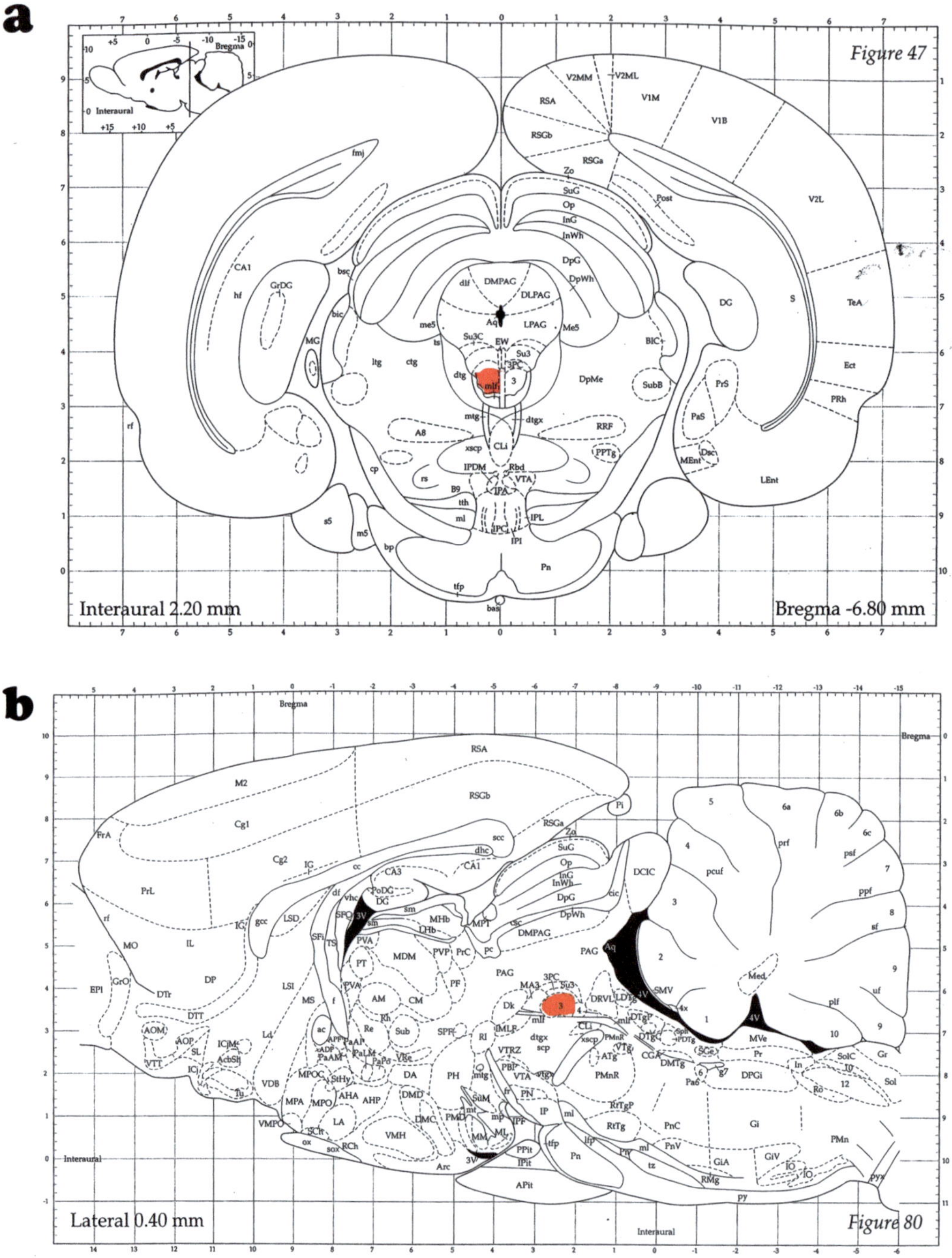

Fig. 2.5 Coordinate parameters for rat oculomotor nucleus on rat brain atlas from Paxinos and Watson (1998), showing an example of how to obtain coordinates from the atlas. (**a**) coronal plane; (**b**) sagittal plane

(Fig. 2.5). While, the vertical coordinate for the LC is 5.5–5.7 mm from surface of the brain to center of the LC on coronal plane (Fig. 2.6).

However, in the actual practice of stereotaxic delivery, you would find the coordinates for the III is feasible but is difficult for the LC because the needle or electrode has to penetrate and pass the

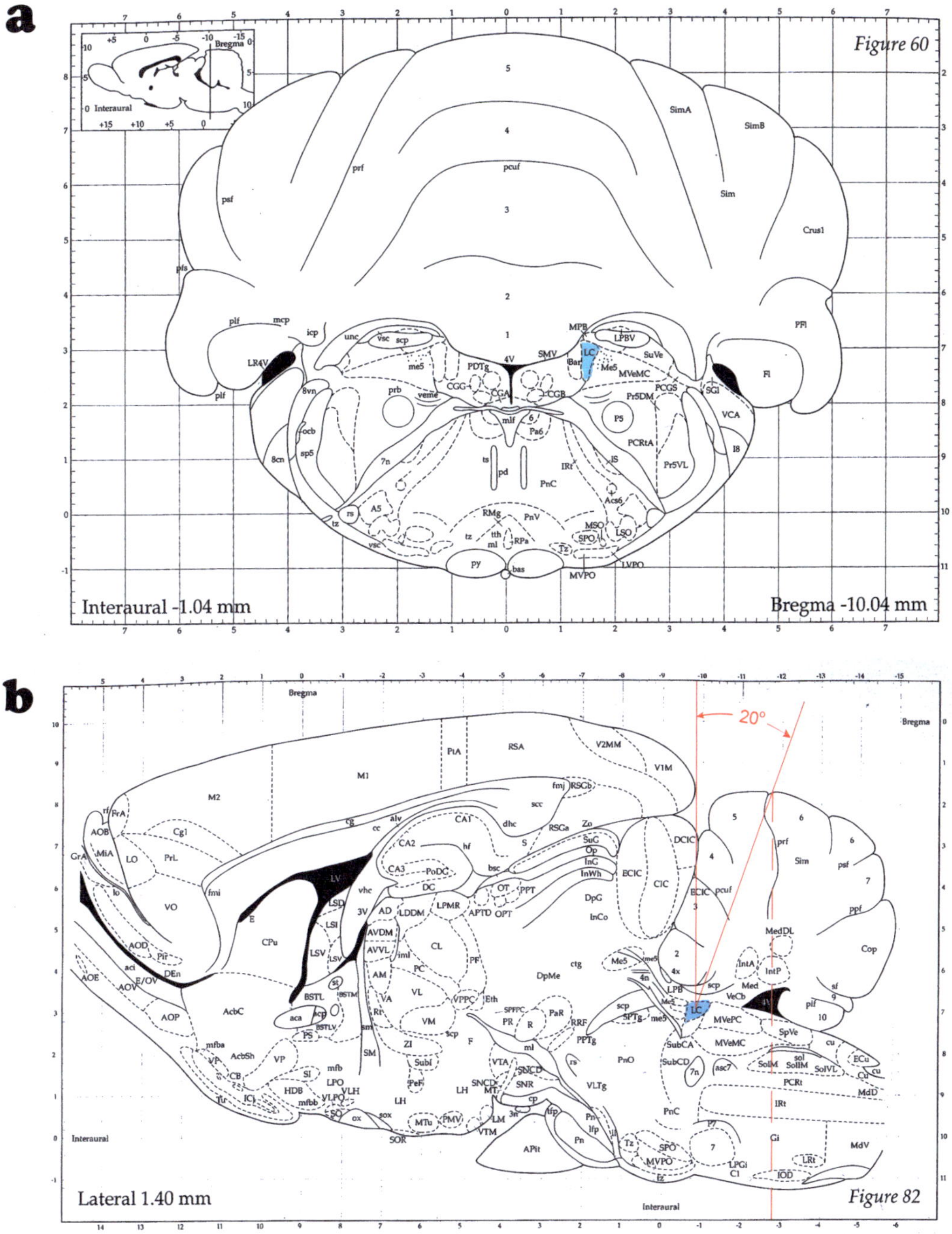

Fig. 2.6 Reformed coordinate parameters for rat locus coeruleus based on rat brain atlas of Paxinos and Watson (1998), showing an example of how to calculate coordinates for angled placement of needle or electrode to avoid hurting large brain blood vessel. (**a**) coronal plane; (**b**) sagittal plane

transverse sinus, a large venous sinus under skull and severe bleeding will occur if broken, which is lethal to the animal. Hence, either an anterior–posterior or a posterior–anterior inclining angle has to be adopted to avoid bleeding. We prefer a 20° posterior–anterior incline as shown in Fig. 2.6b because the distance from surface of the cerebellum is shorter and damaged structures are fewer.

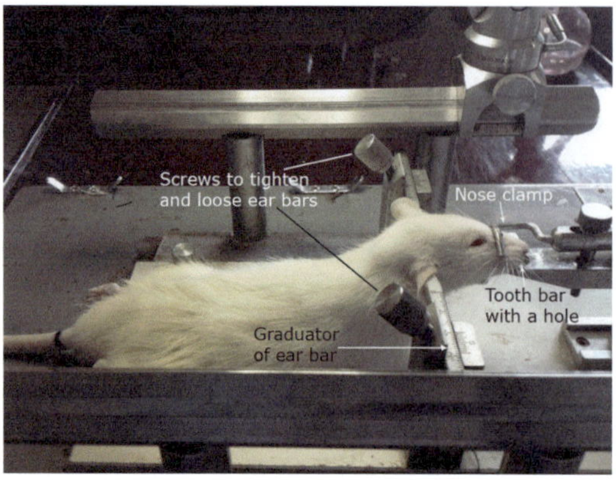

Fig. 2.7 A rat has been correctly mounted on a head holder

In this case (Fig. 2.6b), the coordinates should be 2.8 mm posterior to the interaural line, 1.4 mm lateral to the midline and 5.8 mm from surface of the cerebellum to the LC with a posterior–anterior incline of 20° for electrode or needle.

2.3 Mounting Rat and Positioning Delivery Pipette

2.3.1 Rat Stereotaxic Instrument

1. Rat head holder contains tooth bar and ear bars (Fig. 2.1) standing on the baseplate. The graduations marked on tooth bar show the height you need to adjust according to the atlas you are using. The zero level is at the same height as the interaural line (Fig. 2.3b). Based on the atlas edited by Paxinos and Watson, the tooth bar should be set up in a position of −3.3 mm (Fig. 2.3b). Ear bars should protrude equally into the bony ear canal and fix the head centrally; namely, graduations on ear bar of both sides should be equal (Fig. 2.7). When correctly restrained, all parts of the rat head, with exception of the mandible, have been firmly fixed on the frame (Fig. 2.7).
2. Stereotaxic modulator (Fig. 2.8) is mounted and moved on the anterior–posterior directed railroad bar on the frame base (Figs. 2.1 and 2.8) with a delivery pipette holder attached to (electrode holder in Fig. 2.8). The need for using an electrode or needle is based on the purpose of your stereotaxic delivery. The holder can be maneuvered in two directions by the modulator, medial–lateral and vertical. Combined with the ability of the modulator to move on the anterior–posterior railroad, the electrode or needle can point to any target in the brain. Some modulators have the additional ability to tilt the pipette holder in certain degrees to avoid penetrating large blood vessel as aforementioned (see angle adjustor in Fig. 2.8).
3. Delivery pipette holder. The delivery pipettes are mostly glass micropipette or Hamilton syringes, and different holders are available as accessories to the frame. There is also a hydraulic driving electrode holder attached to modulator, which moves in fine increments to advance the glass pipette or electrode to a certain point. This device is mostly applied in electrophysiology recording and intracellular or extracellular tracer injection after recording. It is also used for delivery of neuroactive substances such as agonists or antagonists in a small area while simultaneously monitoring functional changes of the animal so precise adjustments of the position of the electrode is required.

2.3.2 Mounting Rat and Position Delivery Pipette

1. Fixing the rat head on holder. First, let incisor teeth of the rat bite in the hole on tooth bar, but

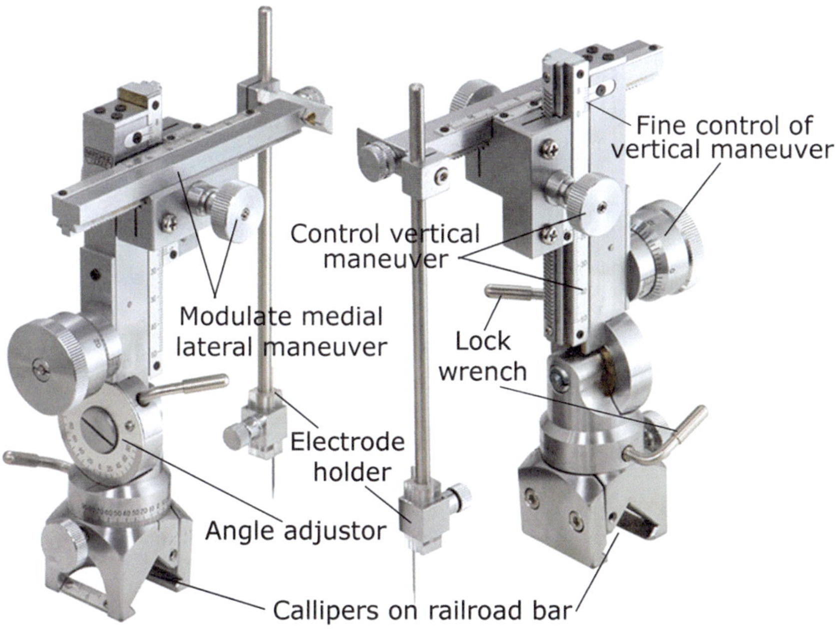

Fig. 2.8 A modulator with electrode holder attached

don't tighten the nose clamp or screws on the baseplate. Place both side ear bars in bar holder and position the rat head in between. Insert ear bars individually into ear canal gently until hearing a gentle crack indicating breaking of the ear drum. This should be accompanied with a blink of eyes, indicating good mounting. Move both sides of ear bar to an equal graduation, tighten the screw for each ear bar, and also tighten the clamp for tooth bar. Then, pull tooth bar rostrally to ensure head is completely immobile and screw the tooth bar firmly on the baseplate (Fig. 2.7).

2. After the skull has been exposed, clear all connective tissue to display bregma and lambda (Fig. 2.9). Move the modulator horizontally (in directions of anterior–posterior and medial–lateral) to aim the tip of electrode at the bregma, then record graduations of both railroad bar and medial–lateral control on the modulator. For example, on horizontal plane, when the tip of pipette exactly aimed at the bregma, the graduation is 9.50 mm anterior to the 0 point (interaural line) on the railroad bar (Fig. 2.10a) and 1.27 mm away from the 0 point (theoretically it should be 0 if the tip of electrode is aligned with 0 point, but this alignment takes time and may not be worthwhile) on the medial–lateral maneuver control (Fig. 2.11a). If the left III of the rat is the targeted area, we already know that horizontal coordinates of the III are about 6.54 mm (center of 6.04 and 7.04) posterior to the bregma and 0.30 mm (center of 0.1 and 0.5) lateral to the midline. Therefore, maneuver the delivery pipette holder 6.54 mm posteri-

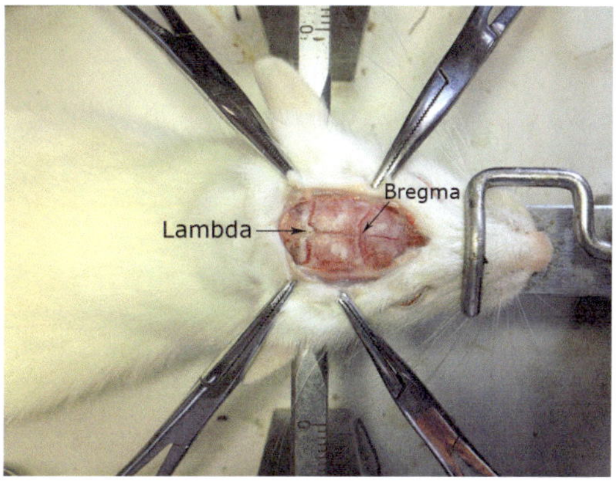

Fig. 2.9 Real appearance of bregma and lambda on a rat skull

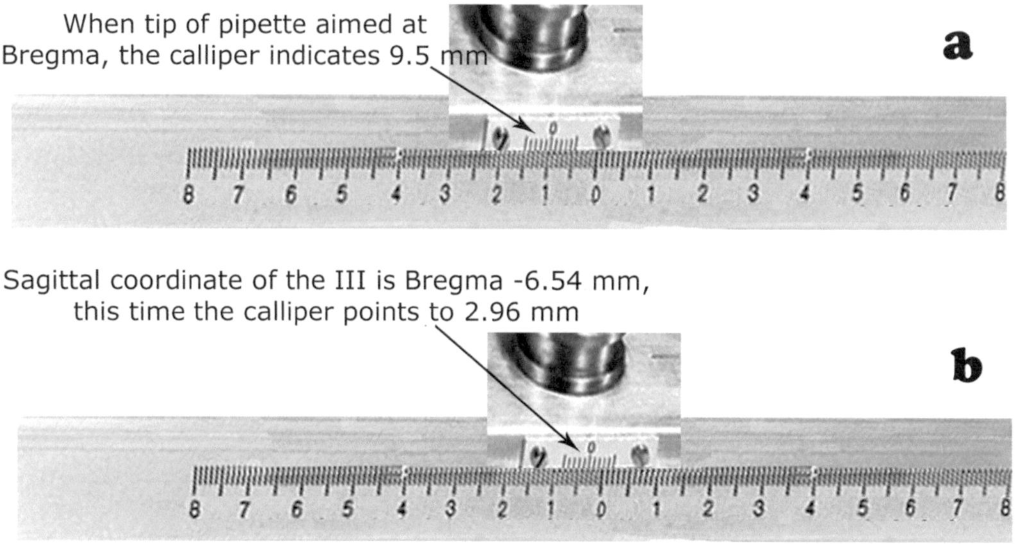

Fig. 2.10 A schematic showing of the actual number of reference point and targeted brain area on sagittal coordinates, namely, coordinates in anterior–posterior direction. (**a**) sagittal coordinates when the tip aimed at the bregma; (**b**) when the tip pointed to the III

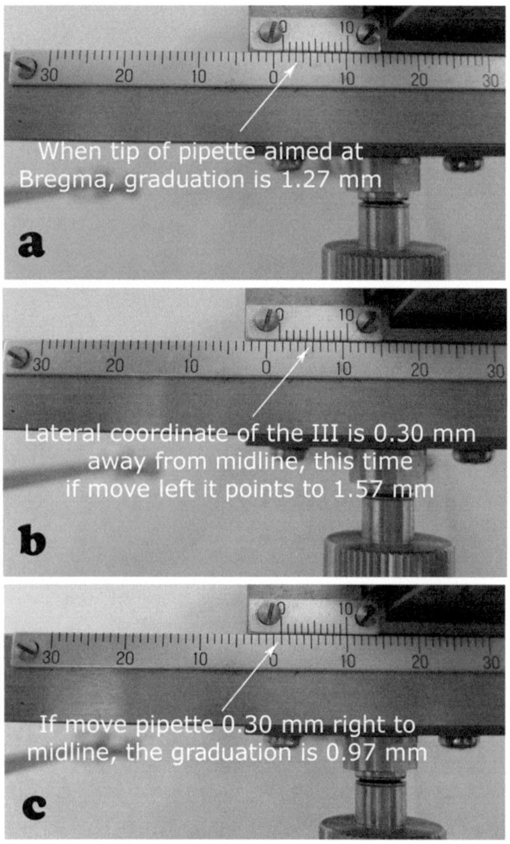

Fig. 2.11 A schematic showing of the actual number of reference point and targeted brain area on coronal coordinates, namely, coordinates in medial–lateral direction. (**a**) coronal coordinates of the bregma; (**b**) the left III; (**c**) the right III

orly; the graduation on the railroad should be 9.50–6.54 = 2.96 mm (Fig. 2.10b); meanwhile, move the holder 0.30 mm left from the midline; the graduation on medial–lateral control of the modulator should be 1.27 + 0.30 = 1.57 mm (Fig. 2.11b). If the injection is aimed at the III in right side, move the holder for 0.30 mm right to the midline, then the graduation on medial–lateral modulator should be 1.27–0.3 = 0.97 mm (Fig. 2.11c).

3. Lower down the pipette until its tip gently touches the surface of the skull (be careful not to break the tip if glass micropipette is used), mark this point by marker pen. A dental drill with a flat drill head is the best tool for opening the skull. Gently drill away the bone without ever pushing the drill down. After a while, only a thin piece of bone is remaining at the bottom of the bony hole, gently peel it away using sharp forceps. The arachnoid mater should be exposed without any bleeding. Set up the fine advance control to 0 point (Fig. 2.12a, b) and gently lower down the pipette holder by rough advance control (Figs. 2.8 and 2.12b, d) until electrode tip touches on the arachnoid mater, immediately maneuver down the pipette by fine advance control into a depth of 6.1 mm (Fig. 2.12c, d), because we already known that

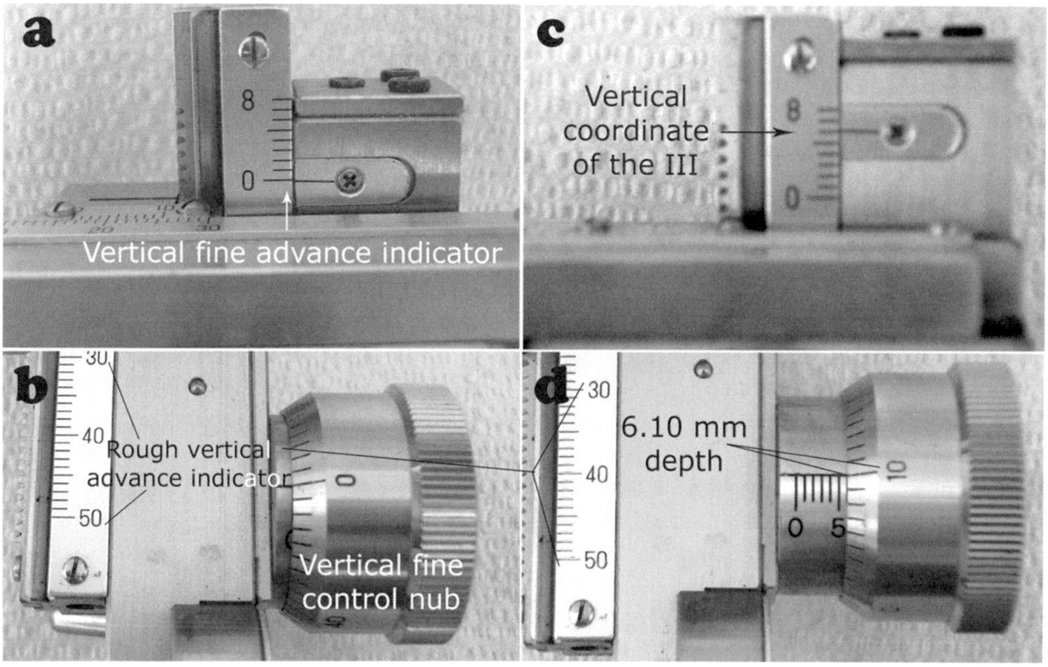

Fig. 2.12 A schematic showing of the actual number of reference point (surface of the brain herein) and targeted brain area on vertical coordinates, namely, coordinates in rat's dorsal–ventral direction. (**a**) and (**b**) fine and rough parameters of the bregma's vertical coordinate; (**c**) and (**d**) fine and rough parameters of the III's vertical coordinate

vertical coordinate of the III is 6.1 mm from surface of the brain. In this case, the III of the rat should have been targeted.

Note: We used a stereotaxic delivery instrument produced by Narishige INC as an example, but there are many companies, such as Kopf or Harvard Apparatus, that manufacture stereotaxic instruments. Each will have a different style of base unit and modulator. Modulators from other companies may not have rough and fine adjustment for moving the delivery pipette holder with the goal of making the instrument more rigid and possibly stable. This will not change the principle of maneuvering the holder. For accessories used in adjusting the angle, each company should have their own. You may select a modulator with angle change function if it is something that will be beneficial. On the other hand, different atlases have different coordinate systems based on the position of the head because the height of the tooth bar setup results in head position change. Nonetheless, the principle of targeting a specific structure is the same.

2.4 Injection

2.4.1 Pressure Injection

A Hamilton syringe was widely used for stereotaxic injection of liquid substances. People commonly used a 1.0 μL syringe to perform stereotaxic injection and the volume of delivered solution was 0.2–0.5 μL in general. Even though the volume was small, it was relatively large for a precise delivery to a specific group of neurons. Later, glass micropipettes have been widely applied for microinjection of a solution commonly connected through silicon tubing to a large syringe or micro pump. Injection of a solution can be driven either by gas pressure or by means of a hydraulic system. The majority of researchers preferred to use the hydraulic delivery system due to a more easily controlled injection volume compared to an air-driven injection. No matter how an injection was delivered, it is worth noting two key steps: one is a slow injection and the other is to keep the needle or pipette in situ for a period after injection. For example, it should take 5–10 min to inject a volume of 0.2 μL into the brain stem, either by Hamilton syringe or with hydraulic driving of pipette injection. After injection, do not move the syringe or pipette immediately; it is best to let it stay in situ for at least 10 min.

2.4.2 Iontophoresis

Most materials can be delivered by iontophoresis, which is more precise and causes less pain comparing with pressure injections. Of course, the current should be small enough to only electrophorese material into tissue instead of electrolyzing the tissue. In cases of intracellular biocytin or Neurobiotin delivery, 5–100 nA current is used to iontophoresis of the tracer. For extracellular delivery, 10–100 μA pulse current is generally assumed. By our experience, administration of a neuronal tracer such as horseradish peroxidase, biotinylated dextran amine, or cholera toxin, 5–50 μA positive current at 1–2 Hz with duration of 100–500 ms is a good range of delivery current. Only a few substances may require negative current for iontophoresis, for example, Pontamine sky blue, since the active elements are negatively charged. Similarly, do not move the syringe or pipette immediately out of the place after injection, it is best to let the delivery pipette stay in situ for at least 10 min either. Use injection of tracers into the rat III as an example to show the correct placement into area III. Indication of a correct placement will be the movement of the ipsilateral eyeball upon initiation of iontophoresis. You may adjust the position of the electrode until you see the strongest eye movement with the lowest current (Zhang et al. 2011), proof you are in the center of your target area—the III. Injection to a motor nucleus is not always the case; however, some regions may be related indirectly to some function. Careful monitoring the animal and keeping a detailed record of any reactions, such as changes in heart or respiration rate etc., and comparing these data with the injection site after brain has been sectioned and stained may be helpful. The sign of correct injection is not always obvious, in which case the only way of adjusting delivery position is by using animals of the same strain, weight, and sex and frequent checking of the injection site and making any necessary corrections for the next injection.

2.4.3 Placing Dry Crystal

The insertion of a crystal dye into targeted brain area, when the target is easy to access, is another method of stereotaxic delivery. The most commonly used crystal delivery in recent years is carbocyanine dyes, including Dil, DiAsp, Dio, and Dia. In most of cases, these dyes are placed as a crystal into the brain of embryonic or newborn rats. However, a device has been created to deliver crystalline carbocyanine to any deep structures as is when dissolved tracers are delivered in adult animals (Marin et al. 2001). In this device, a needle that looks very similar to a Hamilton syringe is capable of holding a tiny crystalline pellet in its tip. During injection, driven by an air pump, a tiny steel rod inside of the needle starts to vibrate and push the crystal into the tissue when the needle has been stereotaxically positioned to a targeted area (Marin et al. 2001). However, online searching has not yet found any company producing this kind of device.

References

Abercrombie M, Johnson ML (1946) Quantitative histology of Wallerian degeneration; nuclear population in rabbit sciatic nerve. J Anat 80:37–50

Adey WR, Noda H (1973) Influence of eye movements on geniculo-striate excitability in the cat. J Physiol 235:805–821

Bizzi E (1966) Changes in the orthodromic and antidromic response of optic tract during the eye movements of sleep. J Neurophysiol 29:861–870

Cetin A, Komai S, Eliava M, Seeburg PH, Osten P (2006) Stereotaxic gene delivery in the rodent brain. Nat Protoc 1:3166–3173

Chu HP, Etgen AM (1997) A potential role of cyclic GMP in the regulation of lordosis behavior of female rats. Horm Behav 32:125–132

Herschel J (1972) A scaled ratio of body weight to brain weight of a comparative index for relative importance of brain size in mammals of widely varying body mass. Psychol Rep 31:84–86

Horsley VA, Clarke RH (1908) The structure and functions of the cerebellum examined by a new method. Brain 31:45–124

Johnson AC, Mc NA, Rossiter RJ (1950) Chemistry of Wallerian degeneration; a review of recent studies. Arch Neurol Psychiatry 64:105–121

Kobbert C, Apps R, Bechmann I, Lanciego JL, Mey J, Thanos S (2000) Current concepts in neuroanatomical tracing. Prog Neurobiol 62:327–351

Leksell L (1949a) A new technique for craniotomy; the osteodural flap. Acta Chir Scand 98:270–272, 272 pl.

Leksell L (1949b) A surgical procedure for atresia of the aqueduct of Sylvius. Acta Psychiatr Neurol 24:559–568

Liu JS, Wang Q, Zhang JB, Kong LJ, Yao SY, Zheng DY, Xu QY (2011) Construction and functional activity of a recombinant vector expressing rat glutamic acid decarboxylase 65. Neurosci Bull 27:430–435

Liu R, Dang W, Jianting M, Su C, Wang H, Chen Y, Tan Q (2012) Citalopram alleviates chronic stress induced depression-like behaviors in rats by activating GSK3beta signaling in dorsal hippocampus. Brain Res 1467:10–17

Marin G, Henny P, Letelier JC, Sentis E, Karten H, Mrosko B, Mpodozis J (2001) A simple method to microinject solid neural tracers into deep structures of the brain. J Neurosci Methods 106:121–129

Norgaard Glud A, Hedegaard C, Nielsen MS, Sorensen JC, Bendixen C, Jensen PH, Larsen K, Bjarkam CR (2010) Direct gene transfer in the Gottingen minipig CNS using stereotaxic lentiviral microinjections. Acta Neurobiol Exp (Wars) 70:308–315

Paxinos G, Watson C (eds) (1998) The rat brain in stereotaxic coordinates, 4th edn. Academic, San Diego

Spiegel EA, Wycis HT, Marks M, Lee AJ (1947) Stereotaxic apparatus for operations on the human brain. Science 106:349–350

Tan TC, Black PM (2002) Sir Victor Horsley (1857–1916): pioneer of neurological surgery. Neurosurgery 50:607–611, discussion 611–602

Thomas GA (1948) Quantitative histology of Wallerian degeneration; nuclear population in two nerves of different fibre spectrum. J Anat 82:135–145

Zhang J, Liang H, Luo P, Xiong H (2011) Unraveling a masticatory—oculomotor neural pathway in rat: Implications for a pathophysiological neural circuit in human? Int J Physiol Pathophysiol Pharmacol 3:280–287

Zlokovic BV, Apuzzo ML (1997) Cellular and molecular neurosurgery: pathways from concept to reality—part II: vector systems and delivery methodologies for gene therapy of the central nervous system. Neurosurgery 40:805–812, discussion 812–803

Neuronal Tract Tracing with Light and Electron Microscopy

3

Jingdong Zhang and Huangui Xiong

Abstract

Neuronal tract tracing technique was developed based on two principles, Wallerian degeneration and axon flow theory, and has been enhanced by knocking in an enhancer or inhibitor to a chain of neurons in studying specific pathways with molecular biological functionality. The tract tracing, including anterograde, retrograde, and transganglion tract tracing, can be achieved through axon transportation of tracers. Based on the visibility of the tracers, they are classified as fluorescent and nonfluorescent tracers. As this chapter focuses on the methods that benefit both light microscopy (LM) and electron microscopy (EM) studies, the protocols of fluorescent dye tract tracing are not discussed. Among the nonfluorescent tracers, horseradish peroxidase (HRP) is a marvelous tracer that can be used in all anterograde, retrograde, and transganglion tract tracing; therefore, the protocol of histochemical visualization of HRP is presented herein. Cholera toxin subunit B (CTB) is also well suited for both anterograde and retrograde labeling. Biotinylated dextran amine (BDA), a predominant anterograde tracer, is broadly used in combination with other tracers for identification of pathway connections or convergent innervations. *Phaseolus vulgaris* Leucoagglutinin (PHA-L) is also a super anterograde tracer that can be applied together with BDA to study convergent projections or combined with any retrograde tracer to explore recipient of the convergent innervations. Fluoro-gold (FG) is a retrograde tracer that can be viewed directly with conventional fluorescent LM or by using anti-FG antibody to modify it and to make any combination as well. The protocols for double labeling either of HRP and BDA, BDA and CTB, or BDA and PHA-L, as well as different combinations of triple labeling with immunostaining of

J. Zhang • H. Xiong (✉)
Department of Pharmacology and Experimental Neuroscience, University of Nebraska Medical Center, Durham Research Center, Durham Research Plaza, DRC I 8034, 45th Street at Emile Street, Omaha, NE 68198-5880, USA
e-mail: jingdong.zhang@unmc.edu; hxiong@unmc.edu

H. Xiong and H.E. Gendelman (eds.), *Current Laboratory Methods in Neuroscience Research*, Springer Protocols Handbooks, DOI 10.1007/978-1-4614-8794-4_3,

neuroactive substances are also presented. These protocols also summarize methods of histochemical and fluorescent illumination, as well as tissue processing, embedding, and immunostaining methods for EM studies. Two major double labeling combinations used in EM observation are delineated: TMB-ST (Tetramethyl benzidine–sodium tungsten) visualization of HRP combined with ABC (Avidin-biotin-complex) histochemical stain and ABC histochemical stain combined with silver-gold pre-embedding method. Each protocol is supplemented with images to illustrate the expected results by following the related protocols.

Keywords
Axonal flew-based neuronal tracer • Anterograde, retrograde, and trans-ganglion tract tracing • Histochemical visualization of tracer labeling • Double or triple labeling • Immuno-electron microscopy

3.1 Introduction

A major goal of neuroscience studies is to explore connections between specific types of neurons and the functions of these connections in the nervous system. Most of what we know today about the pathways and cross pathway connections of neurons has been discovered by using neuronal tract tracing techniques. In combination with stereotaxic injection of tracers, electrophysiology and behavior methods, neuronal tract tracing has made a great contribution to the understanding of brain function by unveiling dramatic neuroanatomical networks. Also, a combination of immunohistochemistry and neuronal tract tracing has delineated the brain maps in terms of chemical neuroanatomy. "The anatomy of memory" by van Strien et al. (2009), illustrates a good example of imaginatively reorganizing known neuronal circuits with functional implication—"memory.". Nonetheless, connections between different areas of the hippocampus and parahippocampus, and regional neuron circuits would not have been demonstrated without the application of neuronal or intracellular tract tracing after function identification (van Strien et al. 2009). From this we can see the important role that neuronal tract tracing work has played in the study of neuroscience. However, in the neuroscience of today, an era of so-called molecular neurosciences, is neuronal tract tracing still useful? Indeed, some new concepts regarding neuronal tract tracing have already emerged; in vivo systemically studies could never be entirely replaced by in vitro studies. How to better combine them is a new area of exploration.

In the early use of neuronal tract tracing, the principle of Wallerian degeneration was applied to the study of anatomic pathways (Abercrombie and Johnson 1946; Thomas 1948; Johnson et al. 1950). The neurons in certain nuclei or functional areas were stereotaxically electrolyzed and their axons and termini were degenerated. This could be revealed by silver stains such as Nauta's or Fink–Heimer stain (Nauta and Gygax 1954; Fink and Heimer 1967). This method could also display degeneration of somata when their axons were electrolyzed or fiber bundles were lanced. Revolutionary methods of tract tracing were achieved after the discovery of the phenomenon of axonal flow (Weiss and Hiscoe 1948). Accordingly, in the early 1970s researchers were attempting to find a compound or compounds that could be up-taken by neurons and transported through axon flow either from somata to axonal terminals or vice versa. Between 1971 and 1972, several groups made a breakthrough and discovered a plant enzyme, horseradish peroxidase (HRP), as a super tracer that can be taken up by both somata and axonal termini (Kristensson and Olsson 1971; LaVail and LaVail 1972) and transported through axons in both directions. Following HRP, cholera toxin subunit B (CTB) was reported to be excellent in both retrograde and anterograde tract tracing (Ruigrok et al. 1995; Angelucci et al. 1996). Moreover, both HRP and CTB can be used in anterograde, retrograde, and transganglionic tract tracing (Zhang et al. 1991; Liu et al. 2004), namely from somata and dendrites to axonal terminals, from terminals to somata and dendrites, and from peripheral processes of the sensory ganglion cells to their central process axonal terminals. In the meantime, autoradiographic anterograde tract tracing by tritium (H^3) conjugated amino acids was used in many labs; however, this method existed only a short time due to the limitation of shipping, usage, and disposal of a radioactive substance. Consequently, many different types of tracers came into being, all relying on axonal transportation.

However, techniques in molecular biology bestowed further novel concepts upon neuronal tract tracing in terms of "genetic dissection of neural circuits" (Wickersham et al. 2007; Luo et al. 2008). These methods, especially "genetic dissection," had been systemically used in invertebrates such as *Elegans* and *Drosophila*, and also attempted in mice. In a system such as the olfactory system, if all cell types are known, you can target any cell to perform "enhancer trap" or "repressor trap," allowing you to gain knowledge about morphology and functional changes of not only those cells but also neurons presynaptic and postsynaptic to them. In mammals, researchers also attempted to apply different modern techniques to tract tracing studies. For example, green fluorescent protein (GFP) is encoded in a deletion-mutant rabies virus (RV), and these RVs are assumed as retrograde tracers to enable live imaging of mammals (Wickersham et al. 2007). Deletion of the gene that encodes glycoprotein, which guides transforming of the RV into other neurons, inhibits transformation of RVs between neurons allowing for retrograde tracing of single neurons that are automatically visible as a consequence of GFP expression. This method enables anatomical study of live animals through either two-photon microscopy and in vivo electrophysiological study of selected neurons, supplemented by histological study after the animal is perfused. On the other hand, tagging synaptic proteins such as synaptophysin with GFP in host cells can be used as functional anterograde tract tracing as well as be used in monitoring dynamic synapses reorganization ex vivo (Rondorf-Klym and Colling 2003; Luo et al. 2008). In mammals, the generation of transgenic animals with a specific group of genetically identifiable neurons is not easy tasks. But one important principle of "genetic dissection" is loss-of-function (LOF) and gain-of-function (GOF), with which investigators can know the contribution of a specific gene to the network function (Luo et al. 2008). Actually, this principle can be applied to mammals by stereotaxic injection of vectors carrying an enhancer or inhibitor gene without in-depth knowledge of the genetics of the targeted neuronal cell type in a pathway or a system. The most innovative tract tracing similar to "genetic dissection" in mammals is knock-in of a viral glycoprotein receptor gene into a specific group of neurons in central nervous system. This can be done by either delivery of a transfecting vectors or by using a receptor gene transgenic mouse and by injecting glycoprotein-coated neurotropic virus to the area containing that specific group of neurons (Beier et al. 2013). When these viruses have been invaginated and transported to somata and dendrites, they will be transformed to the other order of neurons without viral receptor, depending only on their neurotropic nature.

Further, the old method of stereotaxic electrolysis and tracing of degenerated axons has a completely new style of application (Schoene-Bake et al. 2010). By performing lesion surgery on human white matter tracts, researchers can monitor affected tract fibers by diffusion tensor image (DTI) of the magnetic resonance imaging (MRI), instead of Nauta or Fink–Heimer silver stain. In treatment of refractory depression, doctors performed electroconvulsive therapy targeted at specific white matter tract for this kind of depression (Kotowicz 2005). The surgery is based on an idea that targeted white matter tract may conduct brain signals involved in depression-related vicious circle, the extent to which the nerve fibers are lesioned depending on the stimulation intensity, frequency, and duration (Kotowicz 2005). Actually, no one knew the exact range of anatomical connections of those injured axons following the surgery because of no appropriate method to trace the change and range of these axons; but today this goal has been achieved by live neuroimaging (Schoene-Bake et al. 2010). Hence, the concept, methodology, and application of neuronal tract tracing are undergoing development. However, in this chapter, we will focus on neuronal tract tracing using tracers that are transported through axon flow and that can be revealed under both light and electron microscopy (LM, EM).

Another tract tracing technique that will be mentioned, but not covered in detail, is retrograde transsynaptic tract tracing by inactivated pseudorabies virus (PRV) as tracers (Martin and Dolivo 1983; Rouiller et al. 1989). The PRV is a neurotropic virus so they can enter a neuron, duplicate themselves, and spread from soma to terminals presynaptic to this soma. Intracellular tract tracing combined with electrophysiological studies is a large section of tract tracing methodology. Extracellular recording and tracer injection has also resulted in decent labeling (Luo et al. 2006; Zhang et al. 2012) following neurophysiological identification. Electrophysiological recordings will be introduced in another chapter but the staining procedures, commonly for biocytin or neurobiotin, after neurophysiological identification are similar to the Biotinylated dextran amine (BDA) protocols described herein (Luo et al. 2001).

3.2 Classification of Tracers

3.2.1 Nonfluorescent Tracer

1. Anterograde tract tracers
   - HRP, a plant enzyme extracted from horseradish
   - *Phaseolus vulgaris* Leucoagglutinin (PHA-L), a plant lectin; another name of *Phaseolus vulgaris* is common bean, a popular vegetable
   - BDA, dextran amine has been used as neural tracer earlier as both anterograde and retrograde tract tracing; BDA is predominantly anterograde tracer
   - CTB, a cholera toxin protein subunit B secreted from the bacterium *Vibrio cholerae*. This is applied mainly for retrograde tract tracing, but can be used as anterograde tracer as well
2. Retrograde tract tracers
   - HRP
   - Wheat-germ agglutinin (WGA) conjugated HRP (WGA-HRP)
   - CTB
3. Transganglion tract tracer
   - HRP
   - CTB
4. Intracellular (anterograde and retrograde) tracer

   Biocytin. Biotin + L-lysine, biotin is found naturally is Vitamin H, also called Vitamin B_7.

   Neurobiotin. A biotin derivative, *N*-(2-aminoethyl) biotinamide hydrochloride.

3.2.2 Fluorescent Tracers

1. Anterograde tract tracers
   - Rhodamine-isothiocyanate (RITC)
   - Fluorescein-isothiocyanate (FITC)
   - Tetramethylrhodamine-dextran amine, with a trade name Fluoro-Ruby (FR). Carbocyanine dyes such as *Dil, DiAsp, Dio, and Dia*
2. Retrograde tract tracers
   - Fast blue (FB, green)
   - Propidium iodide (PI, orange)
   - Latex beads, either rhodamine-filled (red), or fluorescein-filled (green)
   - Hydroxystilbamidine, with the trade name Fluoro-Gold
   - RITC
   - FITC
   - Carbocyanine dyes such as *Dil, DiAsp, Dio, and Dia*
3. Postmortem (ex vivo) tract tracers
   - Nuclear yellow (NY)
   - Lucifer yellow (LY)

- Carbocyanine dyes such as *Dil*, *DiAsp*, *Dio*, *and Dia*
- Diamidinophenylindole (DAPI)

3.2.3 The Stress on Nonfluorescent Tracers

A positive aspect of a fluorescent tracer is the observation of the histological labeling is simpler.

It involves only injection of tracer, sacrificing the animal, and sectioning brain, and then viewing the labeling under a fluorescent microscope. A disadvantage of a fluorescent tracer is that the labeling lasts only a short time and is not compatible for EM studies. A fluorescent tracer is generally much weaker and fades faster than Alexa Fluor or DyLight due to the fact that tracer compounds are partially lost during cutting and mounting; whereas, Alexa Fluor or DyLight conjugated antibodies have an enlargement function. When Alexa Fluor or DyLight conjugated secondary antibodies are used to visualize nonfluorescent tracers, it is easier to carry out combination staining that results in a stronger signal. This chapter will stress the methods that can benefit both light and electron microscopy studies; thus, the application of fluorescent tracers will not be listed herein with an exception of FG tract tracing.

3.3 HRP Anterograde, Retrograde, and Transganglion Labeling

3.3.1 Delivery and Transportation of HRP

1. Stereotaxic injection of HRP into central nerve system (CNS) as either anterograde or retrograde tract tracer. These studies are mostly performed in rat with two methods for HRP injection: (1) using 1 μL Hamilton syringe (Fig. 3.1a) and injecting 0.2–0.5 μL 20 % HRP in saline or (2) iontophoresis delivery of 10–20 % HRP in saline through a micropipette, using 7–20 μA positive current 7 s on 7 s off for 10 min. Also see Sect. 2.4.
2. Injection of HRP into peripheral nerve trunk as either retrograde or transganglion tract tracer. Using the rat as an example: first, shave and sterilize the skin, then make an incision to uncover muscles, and dissect muscles to expose the nerve trunk (Fig. 3.1b). If the trunk is very thick, such as in the sciatic nerve, a Hamilton syringe can be used to conduct the injection. Otherwise, a thin glass pipette can be used by connecting it to a pressure syringe through silicon tubing (Fig. 3.1b). HRP can be dissolved in saline up to 20–30 %. The injection must be extremely slow, e.g., about 1 μL/10 min. After injection, leave the needle

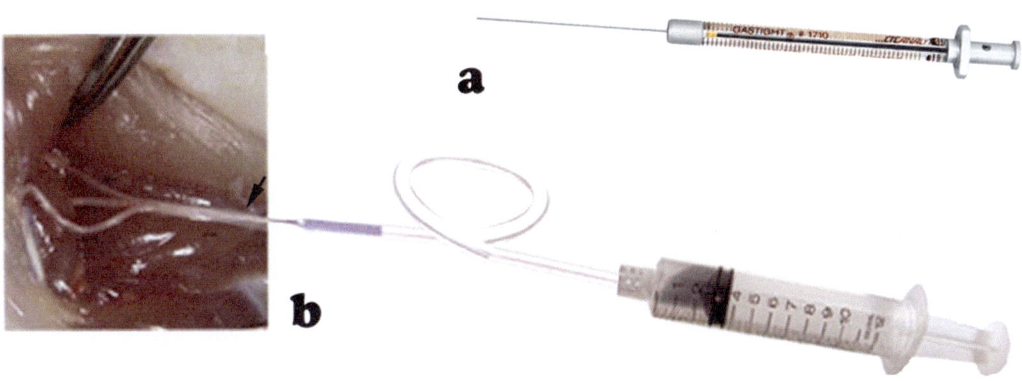

Fig. 3.1 (**a**) Hamilton syringe; (**b**) pressure injection of trace into a nerve trunk through a glass micropipette

or glass pipette in situ about 10–20 min; further, the trunk has to be ligated by a very thin thread and be crushed forcefully with forceps.

3. Injection of HRP into muscles or organ as retrograde tract tracer. In this case, the retrogradely labeled neurons are motoneurons. Injection of HRP into muscles will retrogradely label the motoneurons with their axon terminals distributed in the muscles, but transganglion labeling may be too weak to be visualized. Hence, the best way to generate adequate transganglion labeling is to perform nerve trunk injection and crushing as aforementioned. Similar to the previous method, 20–30 % HRP in saline can be injected into muscles with a Hamilton syringe; alternatively, a crystal of HRP can be placed directly to some organ, such as conjunctiva or taste-goblets.
4. Time course for animal survival is a time including tracer uptake and transport through axon flow. Most neurons capture tracers via endocytosis, phagosome formation, or formation of inclusion bodies are formed. Then, tracer proteins, lectins, or polysaccharides will follow axon flow to be transported. While different tracers have different rates of uptake and transportation, HRP is among the fastest up-taken and transporting tracers (Kobbert et al. 2000). Except for an extremity such as traveling from cortex to anterior horn of the spinal cord in the rat, 48 h is generally sufficient for HRP to travel between any subcortical structure and spinal cord in an adult rat. For example, the distance from principal sensory trigeminal nucleus (Vp) to the ventral lateral thalamus is about 8–9 mm in an adult rat. For this animal, survival of 48 h is adequate (Zhang and Yang 1999).

3.3.2 Histochemical Revealing of HRP-Labeled Terminals

1. Both stereotaxic injection of HRP into the CNS to conduct anterograde tract tracing and nerve trunk injection to carry out transganglion labeling will result in terminal labeling. The best method of revealing these labeled terminals is a reformed (TMB-SNF) Mesulam's TMB (Tetramethyl benzidine) histochemical stain (Mesulam 1978) followed by observation of the projecting terminals under a dark-field illumination equipped LM. Retrograde tracing of HRP-labeled somata and dendrites are also visible using this method; but due to low reaction pH value and because in most cases HRP retrograde labeling will be combined with immunohistochemistry stains or BDA anterograde tracing, this method is not preferred. The method is presented below.
2. Preparing buffer and reagents:
   - 0.2 M acetate buffer (AB; pH 3.3) made by sodium acetate and hydrochloric acid (HCl).
   - For each 100 mL reaction solution, mix 92.5 mL distilled water (DW) with 5 mL 0.2 M pH 3.3 AB and 100 mg sodium nitroferricyanide (SNF).
   - For a 100 mL solution, prepare 5 mg TMB in 2.5 mL absolute ethanol. Keep the solution in the dark by wrapping the container with aluminum foil and placing it in a 37–40 °C oven for 15–20 min. Gently swirl the mixture twice.
   - 5 mL 0.3 % H_2O_2 made from 30 % H_2O_2 and DW.
   - 0.01 M washing solution pH 3.3–3.8.
3. Protocol of TMB-SNF (sodium nitroferricyanide) method:
   (a) Rinse section with washing solution for 3–5 min to acclimate the pH.
   (b) Transfer sections to 0.01 M AB containing SNF and keep shaking.
   (c) Immediately add 2.5 mL ethanol containing 5 mg TMB into the above solution.
   (d) Cover the tray to avoid light and incubate the sections in TMB-SNF solution for 10–15 min and keep shaking.
   (e) Drop 2 mL of the 0.3 % H_2O_2 solution into this 100 mL reaction solution and mix them completely to start histochemistry reaction.

   Note: A grilled strainer seating in a tray (Fig. 3.2) is generally used in this purpose; in this case, lift the strainer containing the sections, drop 0.3 % H_2O_2 into solution and shake

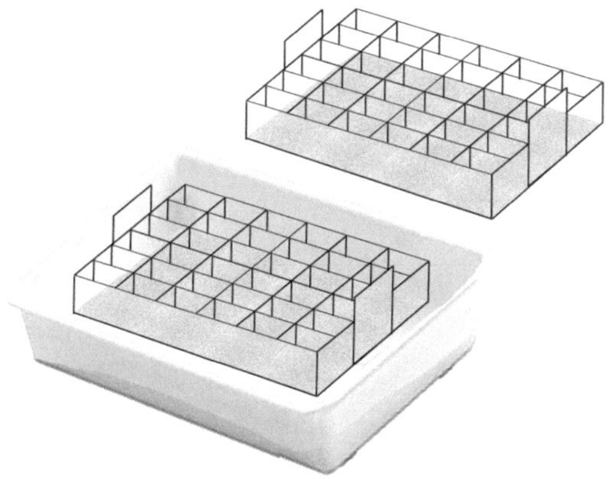

Fig. 3.2 Grilled strainer and tray for holding sections in histochemical reaction

vigorously to mix H_2O_2 evenly with solution before placing strainer with sections back into the solution.

(f) Cover the solution in dark and keep shaking for 15–20 min.

(g) Stop histochemical reaction by dumping reaction solution and exchanging washing AB buffer.

(h) Wash three times with 0.01 M AB (pH 3.3–3.8) for 5 min each time. The sections are now ready to be mounted.

4. The clearest view of HRP transganglion tract labeling is obtained with dark-field observation of TMB-SNF histochemical stain aforementioned. The example result is shown in Fig. 3.3, in which work (Zhang et al. 1991) the author injected 30 % HRP (Toyobo-IC, RZ 3.3) in saline into the rat masseter nerve trunk and the rats were transcardially perfused with 1.5 % paraformaldehyde (PFA) and 1.25 % glutaraldehyde (GA) after 2-day survival.

3.3.3 Histochemical Staining of HRP Retrograde Labeling

1. There are several visualization methods of HRP retrograde labeling: TMB plus sodium tungstate (TMB-ST; pH 5.6–7.4) (Gu et al. 1992); Benzidine dihydrochloride (BDHC) with SNF (pH 4.3–5.0) (Mesulam 1976; De Olmos and Heimer 1977); Diaminobenzidine tetrahydrochloride (DAB; pH 7.4) method (LaVail et al. 1974). Among them, by our experience, the TMB-ST method is the best one for both LM and EM because of its better balance between sensitivity and pH value and easy recognition under EM. Therefore, we introduce only TMB-ST HRP histochemistry stain herein as revealing retrograde HRP labeling, and further relate a method of combination of this stain with anterograde tract tracing for both LM and EM studies.
2. Preparing buffer and reagents:
   - M phosphate buffer (PB) at pH 5.0–5.4 as well as 0.2 M PB at pH 7.4.
   - Prepare 1 % sodium tungsten (ST) in pH 5.0–5.4 PB (1 g for each 100 mL 0.1 M PB), keep in dark.
   - For each 100 mL solution, mix 7 mg TBM in 1 mL acetone and 2 mL absolute ethanol, wrap the container with aluminum foil to keep the solution in dark.
   - 1 N HCl and 1 N NaHO.
   - 5 mL 0.3 % H_2O_2 made from 30 % H_2O_2 and DW.
   - Both 0.1 M PB (pH 5.0–5.4) and PB (pH 7.4) washing solution.
   - For further EM, DAB 25 mg in each 100 mL reaction solution (mainly 0.1 M PB pH 7.4).
   - 20 mg Cobalt chlorate (CoCl) in 100 mL reaction solution.
3. Protocol of TMB-ST (sodium tungsten) method for LM and EM.

(a) Rinse section with PB washing solution (pH 5.4) for 3–5 min to acclimate the pH.

(b) Mix 7 mg TMB in 2.5 mL acetone + ethanol and 1 % ST in 0.1 M PB (pH 5.0–5.4) before using.

(c) The pH value will immediately increase, adjust pH to 6.0–6.4 with 1 N HCl and 1 N NaHO if needed.

(d) Incubate sections in TMB-ST solution, cover the tray to avoid light on solution and keep shaking for about 5 min.

(e) Lift the grilled strainer containing the sections, drop 1–1.5 mL 0.3 % H_2O_2 into this 100 mL reaction solution and mix them completely, then place back the sections to start histochemistry reaction.

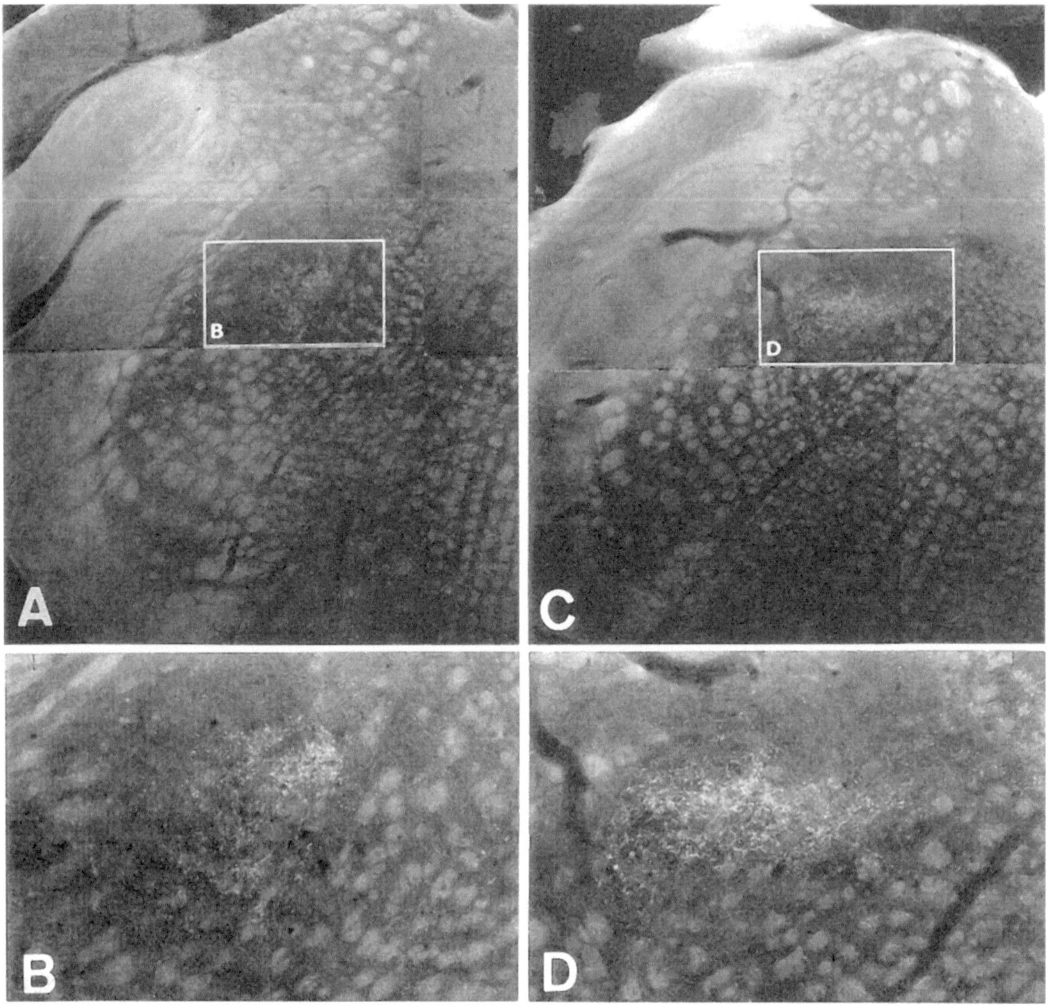

Fig. 3.3 Transganglion HRP-labeled terminals revealed by TMB-SNF histochemical reaction, observed and photographed under dark-field illumination. (**A**) and (**C**), transganglion labeled terminals in the Vodm* and PCRt** within the framed areas. (**B**) and (**D**), framed areas from (**A**) and (**C**), respectively, showing labeled terminals. *dorso-medial part of subnucleus oralis of the spinal trigeminal nucleuss; **parvocellular reticular nucleus

Note: Use the same grilled strainer sitting in a tray (Fig. 3.2) as aforementioned and let 0.3 % H_2O_2 be mixed completely with reaction solution; then place strainer with sections back into the tray.

(f) Protect the solution from light and keep shaking for 10–15 min. After 10–15 min, add another 1.5 mL 0.3 % H_2O_2; shaking for another 15–20 min in the dark, and then add the final 2 mL of 0.3 % H_2O_2.
(g) After 15–20 min, stop the histochemical reaction by pouring off the reaction solution and replacing it with washing PB buffer.
(h) Wash sections three times with 0.1 M PB (pH 5.0–5.4) for 5 min each time. The sections are ready to be mounted for LM.
(i) To continue histochemistry reaction for EM: dissolve 20 mg CoCl in 25 mL DW by stirring and gently heating the solution (keep temperature 40 °C).
(j) Just before using, stir and heat 25 mL DW containing 25 mg DAB until the solution is clear.
(k) Mix 25 mL CoCl, 25 mL DAB and 50 mL 0.2 M PB (pH 7.4). Also before using, prepare 2–3 mL 0.3 % H_2O_2.

(l) Transfer sections mentioned in step h into above reaction solution and bathe for about 2 min.
(m) Add 2 mL 0.3 % H_2O_2 into reaction solution to start reaction.
(n) Examine sections after 3–5 min reaction under LM, labeling should be from green to dark green and finally dark blue.
(o) When desired color is reached, stop reaction by dumping solution and replacing it with 0.1 M washing PB (pH 7.4).
(p) Rinse sections three times for 3 min each time. The sections are ready to be osmicated and EM embedded.

3.4 BDA Anterograde Tract Tracing and Combination

3.4.1 Delivery of BDA and Revealing

1. BDA is predominantly an anterograde tracer used for tract tracing and combined retrograde labeling or immunohistochemistry. It is also broadly used in the study of pathway connections and function. In most cases BDA is delivered to targeted CNS structure by iontophoresis (see Sect. 2.4.2). Ten percent BDA (MW 10,000) in saline is commonly used to conduct iontophoresis with a 5–10 μA positive current at 1–2 Hz and 200–250 ms duration for about 10 min, which yields satisfactory results (Luo et al. 2001; Zhang et al. 2003). The tip diameter of the delivery micropipette is 10–30 μm. Uptake and transportation of BDA by the soma and through axon flow is much slower than HRP. For example, in adult rats the distance from the mesencephalic trigeminal nucleus (Vme) to the hypoglossal nucleus (XII) is approximately 4 mm, and complete uptake and transportation of BDA will take 8–9 days (Zhang et al. 2003). For HRP, 48 h is sufficient to be absorbed by Vme neurons and transported from the Vme to the XII.
2. Revealing of BDA tract tracing with histochemistry staining. Generally the animal is transcardially perfused with 4 % PFA (see Sect. 1.2) about 9 or 10 days after BDA injection. Frozen sections of 30–40 μm thickness are cut and blocked with normal goat serum (NGS) or bovine serum albumin (BSA). But, no antibody is required because biotinylated dextran amine has high affinity to avidin, a compound of the ABC kit—"A" means avidin DH, "B" represents biotinylated HRP, and "C" is the abridge of the complex. Thus, histochemistry revealing of BDA labeling is conducted with ABC kit and HRP-DAB reaction since HRP is the chromogen in ABC.
3. Visualization of BDA labeling by fluorescent dye. Survival time and perfusion of animal are the same, as well as cryostat sectioning (Sect. 3.3.1, step 4). After NGS or BSA blocking, streptavidin conjugated green or red fluorescent dye will be used to reveal BDA labeling instead of ABC kit. As mentioned above, this bacterial avidin will bind biotin compound in BDA with high affinity.

3.4.2 Protocols for Both Histochemistry and Fluorescent Stains

1. ABC kit histochemistry protocol:
   (a) The most broadly used ABC kit is Vector Labs *Elite* standard ABC kit. According to the instructions, add two drops of both reagents A and B in 5 mL buffer (such as 0.05 M PB, pH 7.4). Mixture has to stand for 30 min before use.
   (b) Block section with either 1 % NGS or 1 % BSA in 0.05 M PB, add 1 % Triton X-100 in, for 1 h.
   (c) Rinse the sections in same 0.05 M PB three times, for 5 min each.
   (d) Incubate sections in the primary antibody at 4 °C at least for 24 h.
   (e) Rinse the sections in 0.05 M PB three times for 5 min each.
   (f) Incubate sections in biotinylated secondary antibodies at 4 °C at least overnight.
   (g) Rinse the sections in 0.05 M PB three times for 5 min each.

(h) Bathe sections with ABC solution at RT for at least 5–6 h.
(i) Rinse the sections once with 0.05 M PB (pH 7.4) and twice in 0.05 M Tris–HCl buffer (TB; pH 7.6).
(j) Transfer sections into a mixture of 0.05 % DAB and 0.02–0.025 % NAS (nickel-ammonium sulfate) in 0.05 M TB (pH 7.6).
(k) Add 0.007–0.015 % 30 % H_2O_2 gradually into above solution when incubation has started (about 7–15 μL added to each 100 mL reactive solution). Histochemistry reaction starts when H_2O_2 has been added in solution.
(l) Stop reaction after 20–30 min by removing reactive solution and exchanging with 0.05 M THB (pH 7.6) to wash.
(m) Rinse in 0.05 M TB (pH 7.6) once and twice in 0.05 M PB (pH 7.4). Sections are now ready to be processed for EM plate embedding.

Note: During histochemical reaction, examines a section under LM after total 10 μL of 30 % H_2O_2 has been dropped into the DAB–NAS solution, generally about 10 min after reaction has started. If the labeling is dark enough as well as precise and clear (if everything is correct it is about 15 min after reaction started), the reaction can be stopped; otherwise drop a little bit more H_2O_2 in solution and check the section again until excellent labeling can be observed.

2. Protocol for fluorescent revealing of BDA:
   (a) Block section with either 1 % NGS or 1 % BSA in 0.01 M PBS plus 1 % Triton X-100 for 1 h.
   (b) Rinse the sections in 0.01 M PB (pH 7.4) three times for 5 min each.
   (c) Incubate sections in 1:200 fluorescein or Alexa Flour 568 conjugated streptavidin in 0.01 M PB overnight at RT in the dark.
   (d) Rinse the sections in 0.01 M PB three times for 5 min each in the dark.
   (e) Mount sections with any anti-fading mounting medium in the dark and check result on fluorescent LM.

3.4.3 Combination BDA Anterograde and HRP Retrograde Labeling

1. Application of BDA and HRP double labeling in rats. We have known the time course for uptake and transportation of BDA and HRP is different (see Sect. 3.3.1, step 4 and Sect. 3.4.1, step 1). Therefore, if this double labeling is performed in the same rat to seek connections in a pathway, the HRP should be injected 7 days after the application of BDA, and the animal killed 9 days after BDA and 2 days after HRP injections. To build up a pathway connection, both LM and EM demonstration are indispensable.
2. Protocols of double labeling:
   (a) In this combination, HRP retrograde labeling revealed by TMB-ST method will combine ABC kit visualization of BDA labeling. This combination can be used in both LM and EM studies, with only two major differences. First is the use of a different fixative perfusion: 4 % PFA for LM and 2 % PFA plus 1 % GA for EM. Second is that TX-100 is added in blocking medium for LM but cannot be used for EM; instead, freeze-and-thaw treatment of section before blocking is applied for EM study.
   (b) For LM observation, follow the description in Sect. 3.3.3, step 2 to prepare materials and follow the whole list in Sect. 3.3.3, step 3 to do staining. In HRP and BDA double labeling combination, HRP reaction has to continue to the end as EM work. After TMB-ST HRP reaction, immediately follow the steps in Sect. 3.4.2, step 1 to finish double labeling. Figure 3.4 shows the HRP retrogradely labeled somata and dendrites (a), BDA anterogradely labeled axons and terminals (b), and close apposite of axonal boutons upon soma (c) and dendrite (d).
   (c) For EM study, follow the description in Sect. 3.3.3, step 2 to prepare ingredients and follow the whole list in Sect. 3.3.3, step 3 to perform experiment. After that, follow the

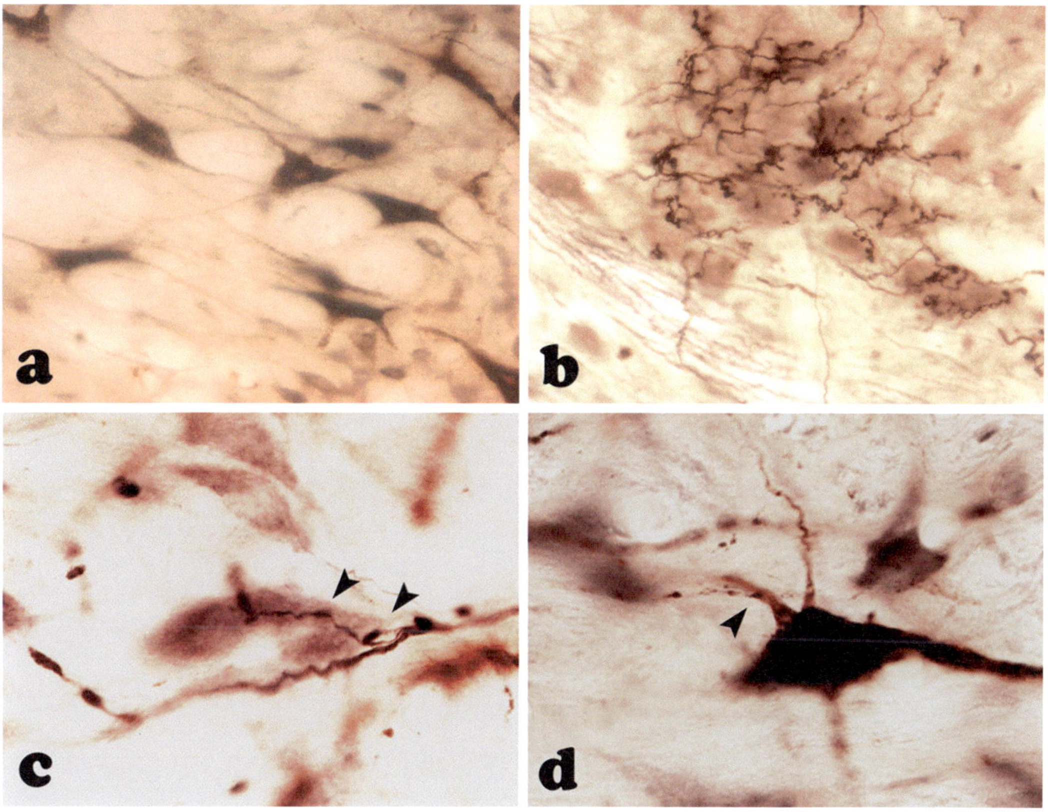

Fig. 3.4 (**a**) HRP retrogradely labeled somata and dendrites revealed with TMB-ST method; (**b**) BDA anterogradely traced axons and terminals revealed by DAB+CoCl histochemical staining of ABC. (**c**) BDA anterogradely labeled terminals closely apposite upon HRP retrogradely traced soma (*arrowheads*); (**d**) contact between BDA positive terminal and HRP-labeled dendrite (*arrowhead*)

steps in Sect. 1.5.2 to carry out a freeze-and-thaw procedure and in Sect. 3.4.2, step 1 to finish the double labeling.

3. Examples of BDA and HRP double labeling: BDA was delivered into a rat Vme with iontophoresis and HRP was injected into ipsilateral tongue of this rat by Hamilton syringe 7 days after BDA delivery. Two days after HRP injection, the animal was euthanized and transcardially perfused. For LM observation, rats were perfused with 4 % PFA; and for EM studies, rats were perfused by 2 % PFA and 1 % GA. Figures 3.5 and 3.6 are cited from this work with permission of the authors (Zhang et al. 2003). Figure 3.5a shows the BDA injection site that was performed by iontophoresis of 10 % BDA in saline. Figure 3.5b, c show BDA-labeled Vme neuronal terminals are closely apposite upon HRP-labeled somata and dendrite (paired arrow/arrowhead pointing contacts between BDA and HRP labeling) of the XII motoneurons. LM work first suggested a potential synaptic connection. Figure 3.6 shows EM verification of the synaptic connections between the labeled boutons and somata or dendrites, which confirmed functional innervation of the Vme neurons through their central processes onto the XII motoneurons.

3.5 CTB Tract Tracing and Combination

3.5.1 Delivery of CTB and Revealing

1. CTB is a 12 kDa small protein with a monosialoganglioside structure. Similar to tetanus toxin C (TTC), which is a trisialoganglioside,

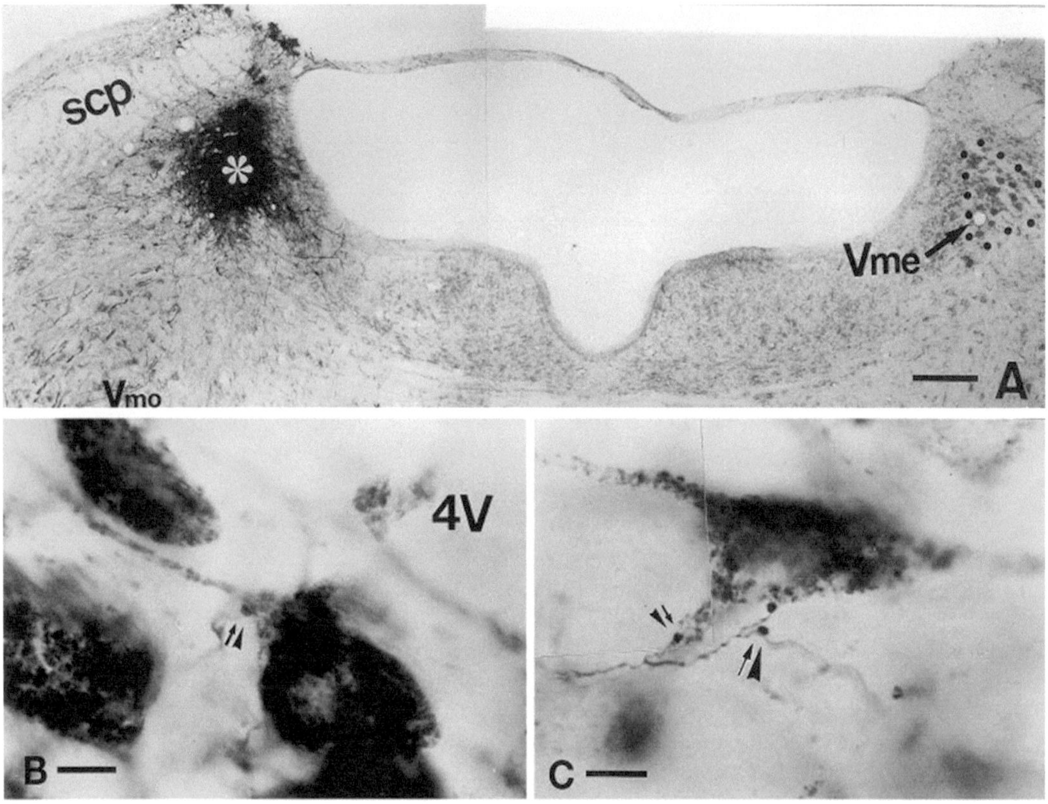

Fig. 3.5 (**a**) BDA injection site in the Vme (*white asterisk*); (**b**, **c**) HRP retrogradely labeled motoneurons in the XII those were closely apposite by BDA-labeled boutons (*arrows/arrowheads*), reflecting Vme neuronal terminals innervate the XII motoneurons

it can be up-taken by axon terminals and transported back to soma and dendrite. TTC is more active and is thought to be capable of transsynaptic tracing so not persistent in a single pathway; therefore, CTB is widely used as retrograde tracer (for review see Kobbert et al. 2000). However, CTB is also assumed as anterograde and transganglionic tract tracer by some investigators (Angelucci et al. 1996; Liu et al. 2004; Pang et al. 2009; Ge et al. 2010), which will be exampled in this chapter. Combination of BDA or immunohistochemistry labeling, CTB is also broadly engaged in the study of pathway connection and function. CTB is mostly delivered by iontophoresis (also see Sect. 2.4.2), and it can be directly injected either into the CNS or peripheral nerve (Zhang et al. 2005, 2012). In these works, 2 % CTB in saline or 0.01 M PBS was iontophorized into CNS with 15 μA positive current at 2 Hz and duration of 200 ms for 10 min and was pressure injected into vagus nerve through micropipette. The tip diameter of micropipette for both central and peripheral injection was 10–30 μm. Uptake and transportation of CTB by axonal terminals is about 5 days for a distance of 2–12 mm (Zhang et al. 2005, 2012).

2. CTB tract tracing can be revealed with either histochemistry staining (Ruigrok et al. 1995) or fluorescent immunostaining (Zhang et al. 2005, 2012). Generally, the animal is transcardially perfused with 4 % PFA (see Sect. 1.2) about 4–6 days after CTB injection. Frozen sections of 30–40 μm thickness are cut; and as immunostain is carried out, the sections are blocked with NGS or BSA. Anti-CTB antibody has to be used in both methods. The fol-

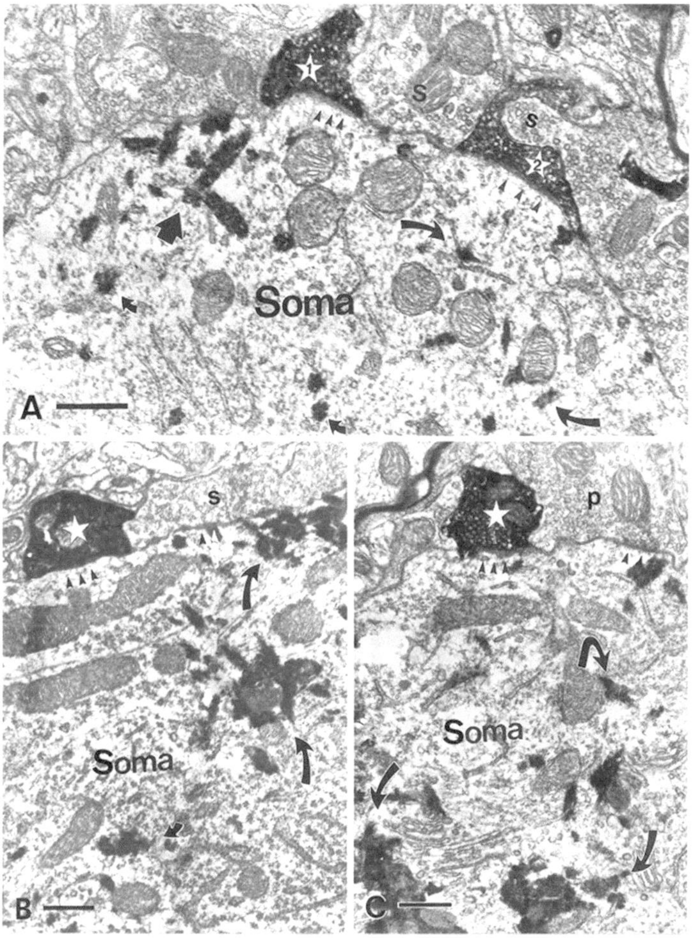

Fig. 3.6 Ultrastructural evidences for functional connections (synapses pointed by small arrowheads in (**a**–**c**)) between BDA-labeled axonal boutons (*stars* in (**a**–**c**)) and HRP retrogradely traced neuron somata (*curved arrows* indicating the HRP labeling in (**a**–**c**))

lowing section will introduce both histochemical and immunofluorescent stains; the former is a modification combined description of Ruigrok et al. for CTB retrograde tract tracing with protocol of Angelucci et al. for CTB anterograde tract tracing (Ruigrok et al. 1995; Angelucci et al. 1996).

3. Protocols for ABC kit histochemical stain of CTB labeling:
   (a) Wash sections a couple of times with 0.01 M PB (pH 7.4) for 3 min each time.

   Note: In protocol of Angelucci et al., the sections are incubated with 0.3 % H_2O_2 in PBS for 20 min, by our experiences it can be omitted if transcardial perfusion was well conducted. Further, after rinse, the sections were also bathed in 0.1 M glycine for 30 min before blocking according to the authors; this step may be helpful in CTB anterograde labeling.

   (b) Incubate sections in 1 % BSA in 0.01 M PB with 1 % TX-100 about 1 h at RT.
   (c) Wash sections with 0.01 M PB three times for 5 min each time.
   (d) Bathe sections in primary anti-CTB antibody, for example, List (List Biological Lab) goat anti-CTB (1:500–5,000) in 0.01 M PB containing 1 % BSA and 1 % TX-100 either overnight at 4 °C or 3–5 h at RT.

(e) Wash sections with 0.01 M PB three times for 10 min each time.
(f) Incubate biotinylated secondary antibodies against primary IgG (or IgM, whatever) in 0.01 M PB containing 1 % BSA and 0.2 % TX-100 for about 3 h at RT. In current case, use biotinylated rabbit anti-goat about 1:200–1,000 (see vendor's recommendation).
(g) Wash sections with 0.01 M PB three times for 10 min each time.
(h) Prepare mixture of A and B reagents as mentioned in Sect. 3.4.2, step 1.
(i) Hatch sections in ABC solution with 1 % BSA and 0.2 % TX-100 for 3–5 h at RT.
(j) Wash sections with 0.01 M PB twice and 0.05 M Tris-HCl (TB, pH 7.6) for 10 min each time (see Sect. 1.5.3, steps 2 and 3).
(k) Transfer sections into 0.05 % DAB and 0.02 % CoCl in 0.05 M TB (pH 7.6).
(l) Acclimate for a couple of minutes and add 5–10 μL (per 100 mL) saturated H_2O_2 to start reaction.
(m) Keep shaking for 20 min, examine sections by LM in between.
(n) Stop reaction by dumping reaction solution and exchange 0.05 M TB (pH 7.6) for rinse.
(o) Rinse section by 0.05 M TB (pH 7.6) for 5 min once and 0.01 M PB (pH 7.4) twice.
(p) The sections can be mounted or continue reaction to perform the other stain.

4. Protocols for immunofluorescent staining of CTB. Combination BDA anterograde tracing with CTB retrograde labeling—BDA-CTB double labeling is presented in this section.

3.5.2 Combination of CTB with BDA by Immunofluorescent Stain

1. Similarly, rats are used as examples for application of CTB retrograde and BDA anterograde tract tracing combination. First, 10 % BDA in saline is injected by iontophoresis into the CNS in the same way as described in Sect. 3.4.1 and allow animal to survive for 10 days. Second, 2 % CTB in saline is either iontophorized into a targeted structure of the CNS or pressure injected into a peripheral nerve trunk or organ in the same rat 5 days after BDA injection. Five days later, the animal is transcardially perfused and sectioned in the same way as aforementioned (see Sect. 3.5.1, step 2).
2. Examples and protocol for double fluorescent labeling:

   Central injection of BDA to the Vme and delivery of CTB into either facial nucleus (VII) or XII of the rats, both through iontophoresis, and then immunofluorescent visualization of the double labeling had been performed after rats were killed and sectioned (Zhang et al. 2012). Figure 3.7 is cited from this study with the permission of the authors and shows the double-labeled contacts between Vme neuronal terminals and premotor neurons of the VII (or XII) under confocal observation.
3. Immunofluorescent double staining protocol assumed in this work (Zhang et al. 2012):
   (a) Incubate sections with 2 % BSA in 0.01 M PB (pH 7.4) with 1 % TX-100 for 1 h at RT.
   (b) Wash sections with 0.01 M PB three times about 5 min each time.
   (c) Incubate sections with 1:1,000 List goat anti-CTB antibody with 1 % BSA and 0.5 % TX-100 at 4 °C overnight or 3–5 h at RT.
   (d) Bathe sections with cocktail of Alexa Fluor 568 conjugated donkey anti-goat (1:200) and Alexa Fluor 488 conjugated streptavidin (1:200–300) for 2–3 h in dark at RT.
   (e) Mount sections with Vectashield anti-fading mounting medium and coverslip.
4. In another example, the BDA injection was also made into the Vme; while the CTB was injected into either vagus nerve trunk (n.X) or a branch of the n.X—recurrent laryngeal nerve. The CTB was applied to nerves 5 days after the BDA injection, then after another 5 days, the animal was transcardially perfused, and the tissue was processed. The protocol is the same as mentioned above (Zhang et al. 2005).

   Note: There are different kinds of antibodies against CTB such as rabbit anti-CTB and

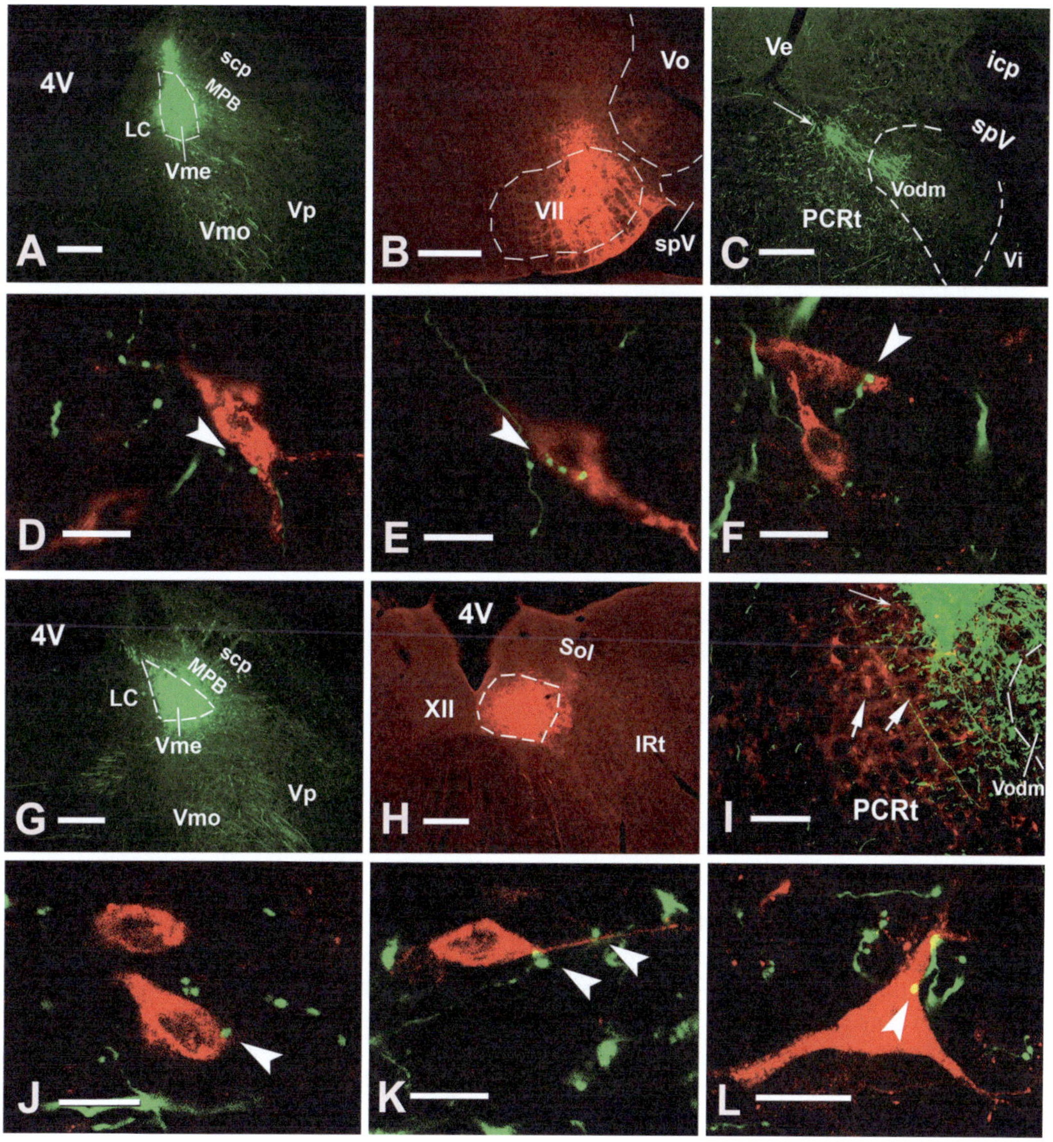

Fig. 3.7 The double labeling contacts between Vme neuronal terminals and premotor neurons of the VII (or XII) under confocal microscopic observation. (**a**, **g**) BDA injection sites in the Vme; (**b**, **h**) CTB injection sites in the VII (**b**) and XII (**h**); (**c**, **i**) BDA-labeled axons and terminals in dorsomedial part of the Vo (Vodm) and adjacent parvocellular reticular nucleus (PCRt); in this area, contacts (*arrowheads* in **d**–**f** and **j**–**l**) between BDA-labeled boutons and CTB-labeled somata or dendrites are observed in high magnification under confocal microscope

mouse anti-CTB antibodies; therefore, if goat anti-CTB is hard to use in combination, more substitutes are available. In the examples mentioned above, BSA was used to block nonspecific binding because primary antibody is hosted from goat; otherwise NGS is commonly used as blocker.

3.5.3 Using CTB as an Anterograde Tract Tracer

1. Since Angelucci et al. (1996) first published their test for using CTB as an anterograde tracer, not many investigators really applied CTB as anterograde tracer. Probably

because visualization of CTB by histochemical stain needs assistance of either PAP or ABC kit and in this circumstance direct employing BDA as anterograde tracer is much simpler if ABC kit is applied to reveal the labeling. In recent year, a group of investigators utilized CTB as anterograde tracer and unveiled the labeling with immunofluorescent stain for LM viewing; further they identified CTB-labeled terminals and synapses formed between CTB positive boutons and postsynaptic dendrites under EM observations (Pang et al. 2009; Ge et al. 2010). In this section, we take these studies as examples to illustrate the application of CTB as anterograde tract tracer.

2. Procedure of experiment:
   The CTB was delivered by pressure injection into different areas [Vp, Vsup (supratrigeminal nucleus), RF (ventral reticular formation of the pons), subnuclei oralis, interpolaris, and caudalis of the spinal trigeminal nucleus (Vo, Vi, and Vc of the Vsp)] to anterograde tracing their projections. The authors stated that anti-CTB did not bind to any endogenous epitopes in the lower brainstem of the rat. This is the reason the authors assumed CTB as an anterograde tract tracer in their studies. The rats were killed 3 or 6 days (6 days for Vp, Vo, Vi, and Vc injection cases; 3 days for Vsup and RF cases) after CTB injections. The aim of the authors was to verify VGluT1 and 2-nergic neurons in above-mentioned areas project to either motor or inter neurons in the trigeminal motor nucleus (Vmo) by triple labeling of CTB, NeuN (Neuron Nuclei) anti-VGluT1 or 2 (Pang et al. 2009; Ge et al. 2010).
3. Protocol of triple labeling includes CTB anterograde tracing:
   (a) Frozen sections of 40 μm were rinsed by 0.01 M PB shortly and bathed in 2 % BSA in 0.01 M PB (pH 7.4) with 0.3 % TX-100 for 1 h at RT.
   (b) Wash sections with 0.01 M PB three times about 5 min each time.
   (c) Incubate sections with cocktail of 1:1,000 List goat anti-CTB, rabbit anti-VGluT1 (1 μg/mL; gift to the authors) or rabbit anti-VGluT2 (1 μg/mL; gift to the authors) and mouse anti-NeuN (1;1,000; Chemicon) with 1 % BSA and 0.2 % TX-100 at RT overnight.
   (d) Wash sections with 0.01 M PB three times about 10 min each time.
   (e) Bathe sections with biotinylated donkey anti-goat (10 μg/mL) in 0.01 M PB for 3–5 h at RT.
   (f) Wash sections with 0.01 M PB three times about 10 min each time.
   (g) Incubate sections with mixture of Alexa Fluor 594 conjugated streptavidin (1:200), Alexa Fluor 488 attached donkey anti-rabbit (1:200) and Alexa Fluor 647 conjugated donkey anti-mouse (1:200) overnight at 4 °C.
   (h) Wash sections with 0.01 M PB three times about 5 min each time.
   (i) Mount sections with Vectashield anti-fading mounting medium and coverslip.
4. Micrographs in Fig. 3.8 exampled CTB injection sites in the Vo, Vi, and Vc (Fig. 3.8a–c), and triple labeling of CTB, anti-VGluT1, and anti-NeuN in the Vmo (Fig. 3.8d, e), as well as labeling of CTB, anti-VGluT2, and anti-NeuN therein (Fig. 3.8f–i). The work had demonstrated VGluT1 and 2-nergic neurons in above-mentioned areas project to either motor or inter neurons in the Vmo. The pictures were cited with the permission of the authors.
5. The part of EM studies on those pathways and concrete protocols will be presented in Sect. 3.8.
6. Transganglionic tract tracing with CTB was reported by Liu et al. about a decade ago (Liu et al. 2004). The authors compared transganglionic labeling results of the CTB with that of WGA-HRP, which showed a difference in regional distribution of the labeled terminals. The authors explain the results as the CTB-labeled myelinated A-fibers project to spinal cord dorsal horn laminae III–V; while WGA-HRP prefer unmyelinated C-fibers and project to the dorsal horn laminae I–II.
   For detailed information, please refer to the original publication (Liu et al. 2004).

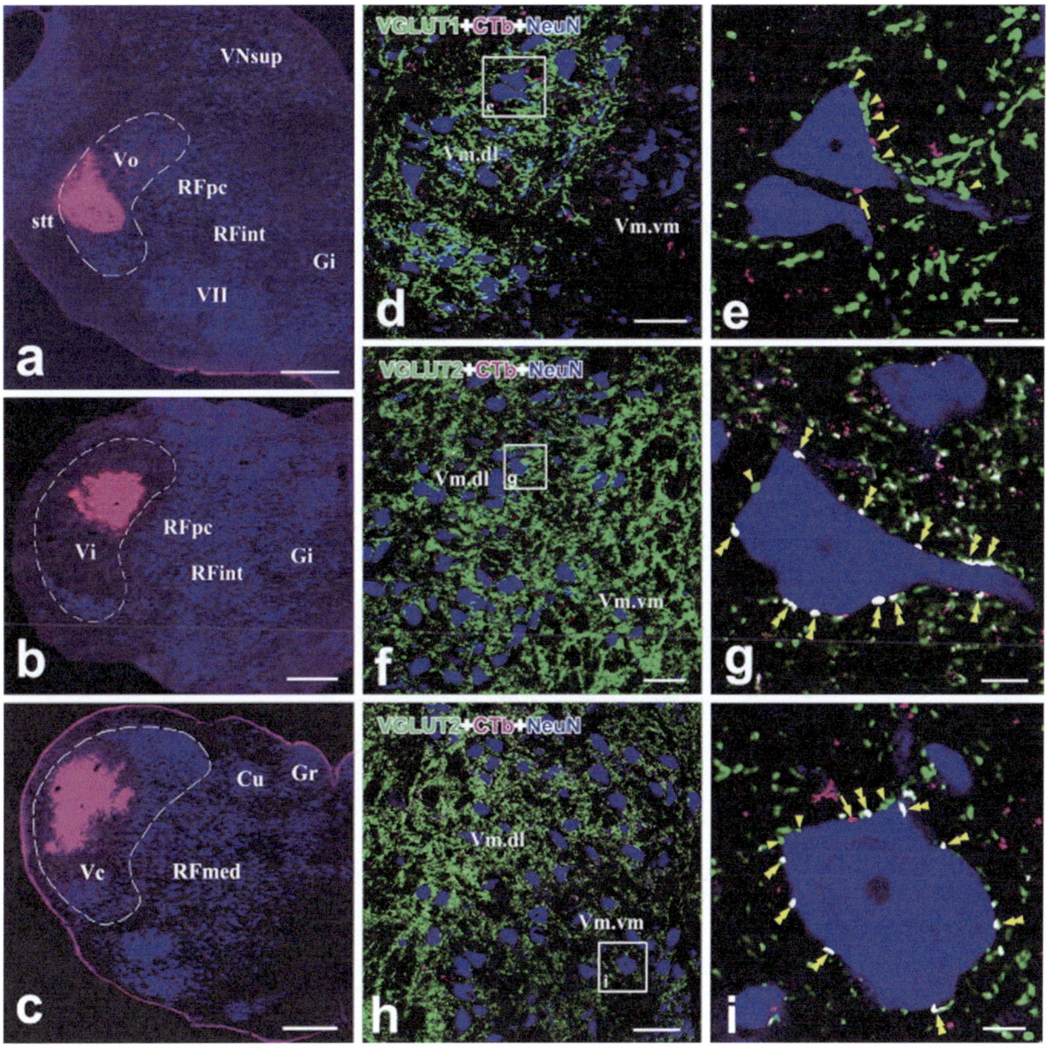

Fig. 3.8 Triple labeling of CTB anterogradely traced terminals, VGluT1 (or 2) positive boutons and NeuN expressed somata and dendrites. (**a–c**) show CTB injection sites in the Vo, Vi, and Vc; (**d**) is an image of triple labeling with VGluT1, CTB, and NeuN; (**f, h**) triple labeling of VGluT2, CTB, and NeuN in the Vmo; (**e, g, i**) high magnification of triple labeling exhibit some CTB traced projections from the Vo, Vi, and Vc are VGluT1 or 2 positive, namely, VGluT1, or 2-nergic neurons in the Vo, Vi, and Vc project to the Vmo; those are tracked by CTB

3.6 PHA-L Anterograde Tract Tracing and Triple Labeling Combination

3.6.1 Delivery of PHA-L and Revealing

1. PHA-L is predominant anterograde tracer employed next to BDA tract tracing and which combined with retrograde labeling or immunohistochemistry is also broadly used in the study of pathway connection and function. Similar to the BDA, PHA-L is delivered mainly to targeted CNS structure by iontophoresis (also see Sect. 2.4.2). The rat is also used as an example to delineate application of the PHA-L. In rats, application of 2.5 % PHA-L in 0.05 M Tris-buffered saline (TBS, pH 7.4) was conducted through iontophoresis by 5 μA positive current for 15–20 min in a pattern of 7 s on and 7 s off. The tip diameter of the delivery micropipette is 10–30 μm, and the result was ideal (Fig. 3.9; Zhang 1998). The

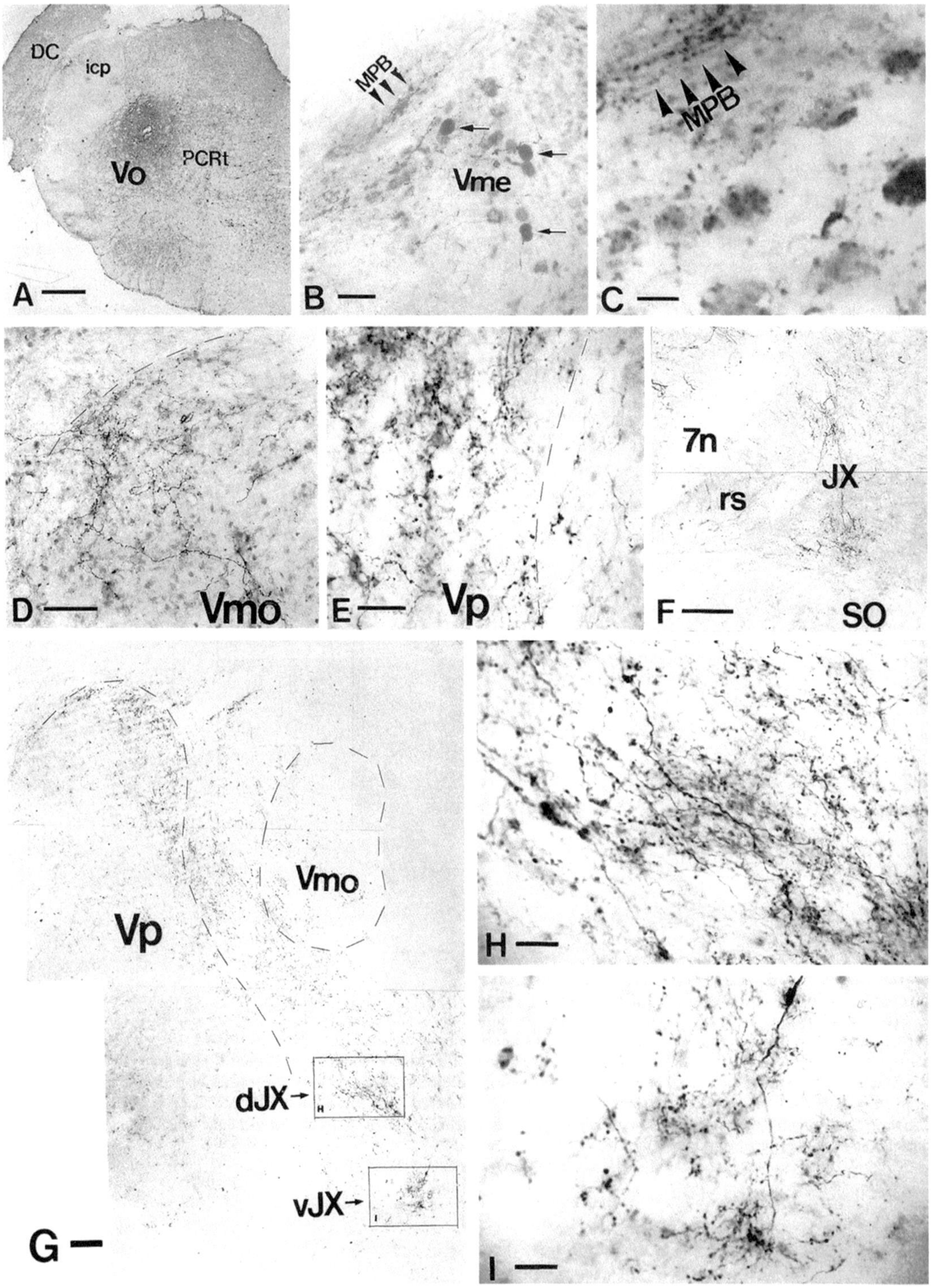

Fig. 3.9 PAP method revealed PHA-L anterograde tract tracing. (**a**) PHA-L injection site in the area of Vodm and PCRt; (**b**–**i**) PHA-L-labeled axons and terminals are distributed in different regions on the coronal planes at the level of the Vme, Vmo, and Vp

survival time of a rat after PHA-L delivery is 5–9 days; for example, in the study aforementioned 7 days total for PHA-L to be up-taken and transported from the Vi and caudal Vo to the level of the Vp, which is about 3 mm in adult rat. Biotinylated PHA-L is also available today and can be directly injected the same way as raw PHA-L, which allows visualization of the labeling directly by ABC kit. However, if PHA-L combined with BDA to explore convergent afferent innervation on to a single targeted area, biotinylated PHA-L and ABC kit revealing of the labeling is not feasible in this case.

2. Revealing of PHA-L or biotinylated PHA-L tract tracing with histochemistry staining. After the animal is transcardially perfused with 4 % PFA (or PFA + GA for EM; see Sect. 1.2) for LM observation frozen sections of 30–40 μm thickness are cut. The sections will be blocked with either NGS or BSA. If biotinylated PHA-L has been applied, ABC kit can be used directly to reveal the labeling (see Sect. 3.4.2, step 1). If raw PHA-L has been iontophorized, either PAP complex or ABC kit can be used in histochemistry revealing of the labeling, which will be delineated in the next section.
3. Protocol for histochemical staining of raw PHA-L labeling:
   (a) Using rabbit anti-PHA-L as a primary antibody and goat anti-rabbit as secondary antibodies as an example; thus, BSA is used as a blocker. Bathe sections in 1–2 % BSA in 0.01 M PBS (pH 7.4) with 1 % TX-100 for 1 h at RT.
   (b) Rinse sections in 0.01 M PBS three times for 5 min in each time.
   (c) Incubate sections with rabbit anti-PHA-L antibody (1:1,000–5,000; Dako. If using other anti-PHA-L, see vendor's recommendation) in 0.01 M PBS with 1 % BSA and 0.2 % TX-100 for 3–5 h at RT (or overnight at 4 °C).
   (d) Wash sections in 0.01 M PBS three times for 10 min in each time.
   (e) For PAP method, bathe sections with goat anti-rabbit (1:100–500; Cappel. If using other anti-rabbit, see vendor's recommendation) in 0.01 M PBS with 0.2 % TX-100 for 3–5 h at RT (or overnight at 4 °C).

       Note: For ABC staining, incubate sections with biotinylated goat anti-rabbit (1:200–500; see vendor's recommendation) in 0.01 M PBS with 0.2 % TX-100 for 3–5 h at RT (or overnight at 4 °C).
   (f) Rinse sections in 0.01 M PBS three times for 10 min in each time.
   (g) For PAP method, incubate sections with rabbit PAP complex (1:200; Cappel) in 0.01 M PBS with 0.2 % TX-100 for 3–5 h at RT (or overnight at 4 °C).

       Note: For ABC staining, incubate sections with ABC mixture (1:50; Vector Labs) in 0.01 M PBS with 0.2 % TX-100 for 3–5 h at RT (or overnight at 4 °C).
   (h) From now on follow the steps listed on Sect. 1.5.3, step 3 from step i to step m including Note.
4. Protocol for histochemical revealing of biotinylated PHA-L. Steps a and b are the same as above mentioned, skip steps c to f, go to step of incubation of sections with ABC mixture (1:50) in 0.01 M PBS with 0.2 % TX-100 for 3–5 h at RT (or overnight at 4 °C). Then, follow the steps listed in Sect. 1.5.3, step 3 from step i to step m including Note.
5. Visualization of PHA-L labeling by immunofluorescent staining. This protocol will be introduced in the next section—combination of PHA-L and BDA.

3.6.2 PHA-L Combined with BDA in Study of Convergent Innervation

1. As mentioned above, rats are also used as examples for description of the method. The rat's survival time for PHA-L and BDA tract tracing is very close, so we applied both tracers to the same rat respectively. In this example, BDA was injected by iontophoresis into the Vme, and PHA-L was delivered simultaneously in a similar way into the medial vestibular nucleus (MVe) in the contralateral side of the same rat. Seven days later, the animal

was transcardially perfused and sectioned the same way as mentioned in Sect. 3.5.1, step 2. The labeling was expected to be located in the III (oculomotor nucleus) ipsilateral to the Vme injection and visualized by immunofluorescent stain also combined with choline acetyltransferase (ChAT) immunostain.

2. Protocol for triple immunofluorescent labeling:
   (a) Frozen sections were cut in 30 μm thickness and were incubated in 1 % BSA in 0.01 M PBS (pH 7.4) with 0.5 % TX-100 for 1 h at RT.
   (b) Wash sections with 0.01 M PB three times about 5 min each time.
   (c) Incubate sections with rabbit anti-PHA-L (1:1,000; Vector Labs or Dako or TSC Biosci.) in 0.01 M PBS (pH 7.4) with 1 % BSA and 0.5 % TX-100 for 3–5 h at RT (or 4 °C overnight).
   (d) Wash sections with 0.01 M PBS three times about 10 min each time.
   (e) Hatch sections with mouse anti-ChAT (1:100; Millipore or Abcam) in 0.01 M PBS with 1 % BSA and 0.5 % TX-100 at RT overnight.
   (f) Wash sections with 0.01 M PBS three times about 10 min each time.
   (g) Bathe sections with cocktail of Alexa Fluor 350 conjugated goat anti-mouse (1:100; Invitrogen), Alexa Fluor 488 conjugated goat anti-rabbit (1:200) and Alexa Fluor 594 conjugated streptavidin (1:200) in 0.01 M PBS at RT for 2–3 h.
   (h) Wash sections with 0.01 M PBS three times about 10 min each time.
   (i) Mount sections with Vectashield antifading mounting medium and coverslip.

   Note: The blue labeling was stained with either AMCA from Jackson Labs or Alexa Fluor 350 from Molecular Probes, both fade relatively faster than other dyes. Therefore, for confocal microscopy, Alexa Fluor 647, 650, or 668 far-red should be used instead of AMCA or Alexa Fluor 350.

3. Figure 3.10 shows the microimages of the injection sites of BDA (Fig. 3.10a) and PHA-L (Fig. 3.10b), and triple labeling of BDA, anti-PHA-L and anti-ChAT in the III (Fig. 3.10c, d) and IV (trochlear nucleus; Fig. 3.10e). The results indicated that ChAT positive motoneurons in the rat III and IV receive convergent innervations from the ipsilateral Vme and contralateral MVe.

3.7 Fluoro-Gold Retrograde Tract Tracing and Triple Labeling Combination

3.7.1 Delivery of FG and Revealing

1. Fluoro-Gold (FG) is a retrograde tract tracer with a broad scope of combination with other tracers or neuroactive substance immunohistochemistry. FG is delivered mainly to targeted CNS structure and can be delivered by either iontophoresis or pressure injection. Similarly, rats are used as examples to describe application of the FG. Generally, researchers use 1–10 % of FG in saline or PB (pH 7.4) for injection. In a recent publication, the authors applied 2 % FG in 0.05 M PB (pH 7.4) through a micropipette in diameter of 15–25 μm and conducted iontophoresis with 4 μA positive current for 20 min in a manner of 7 s on and 7 s off (Dong et al. 2012). The survival time after the FG delivery was 5–7 days; for example, in the study aforementioned 7 days total for FG to be up-taken and traveling a distance about 3–4 mm—from the XII or VII to the level of Vp. The animal was transcardially perfused with 4 % PFA and 15 % (v/v) picric acid 7 days after FG injection.
2. FG is widely used in combination with either fluorescent or nonfluorescent tract tracing, because FG per se can be used as fluorescent tracer and visualized by fluorescent microscope without any histochemical process. Using anti-FG antibody, which can be combined with any kind of immunostain, revealed either histochemical process or other colored immunofluorescent labeling. Commonly, the FG is viewed as orange to red with emission wavelength of 620 nm in physiology conditions, namely at pH 7.2–7.4 environment. When experimental medium has a low pH value, e.g., pH 3.3, the FG exhibits blue with emission wavelength of 440 nm. Hence, the FG is the

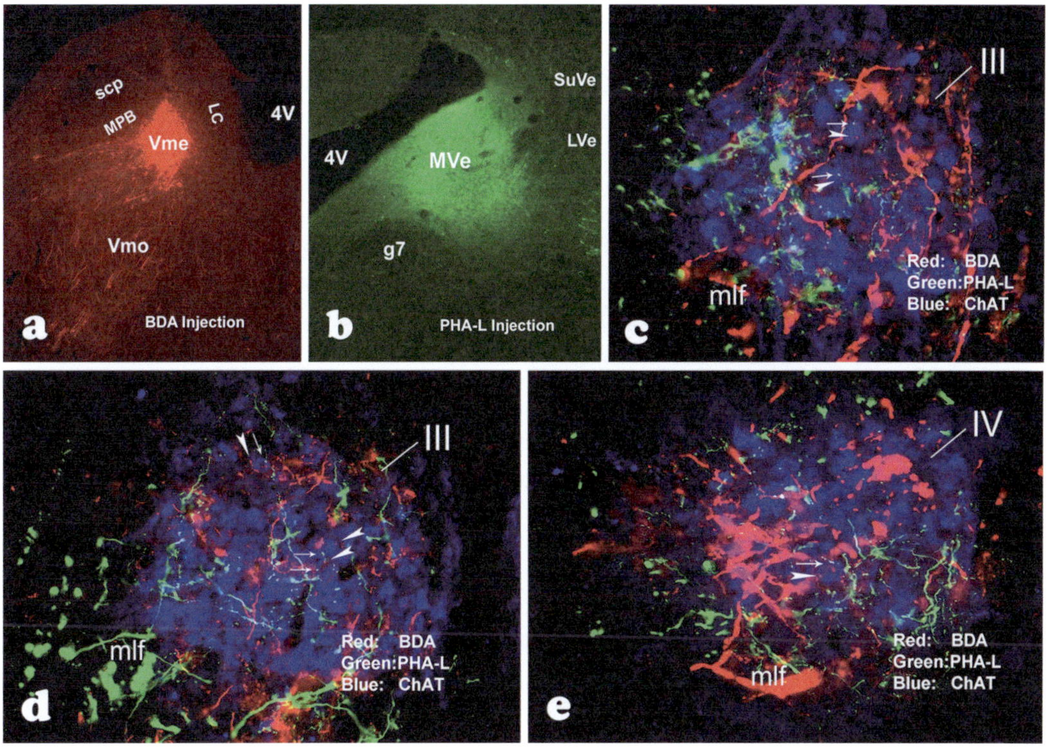

Fig. 3.10 Triple labeling in the rat III and IV by BDA, PHA-L, and ChAT, showing convergent projection from the Vme and MVe onto ChAT positive III or IV motoneurons. (**a**) BDA injection site in the Vme; (**b**) PHA-L injection site in the MVe; (**c–e**) triple labeling of BDA and PHA-L convergent projecting axonal terminals onto ChAT immunostained motoneurons in the III (**c**, **d**) and in the IV (**e**)

best candidate in combination with low pH value TMB-SNF method for HRP histochemistry. This combination can be applied either to discover contact between FG-labeled somadendrite and HRP transganglion traced terminals or to disclose double retrogradely labeled somata. The methods of revealing FG are multiple and in the next section of triple labeling, we will use Dong et al. (2012) publication as an example to introduce a concrete protocol.

3.7.2 Triple Labeling by BDA, FG and Immunostain

1. Application examples. In the above-mentioned article (Dong et al. 2012), the authors assumed two combinations of triple labeling: one was retrograde FG tract tracing carried out with BDA anterograde labeling combined by glutamic acid decarboxylase (GAD) immunofluorescent staining (Fig. 3.11); the other was double labeling the same axonal terminals by both BDA and anti-VGluT2 (vesicular glutamic transporter 2) those were closely apposite upon FG retrogradely labeled somata (Fig. 3.12). In both groups of rats, FG and BDA were delivered to each rat into the VII (or the XII) and the Vc, respectively at the same time. The animals were killed 7 days after both injections. Frozen sections were then cut at 40 μm thickness and triple labeling process of tissue was performed. In both triple-labeling groups, the authors employed anti-FG antibodies to modify the nature emission of the FG. The protocol is summarized below.
2. Protocol of FG, BDA, and GAD triple immunofluorescent labeling:
   (a) Wash sections with 0.05 M PB (pH 7.4) a couple of times.

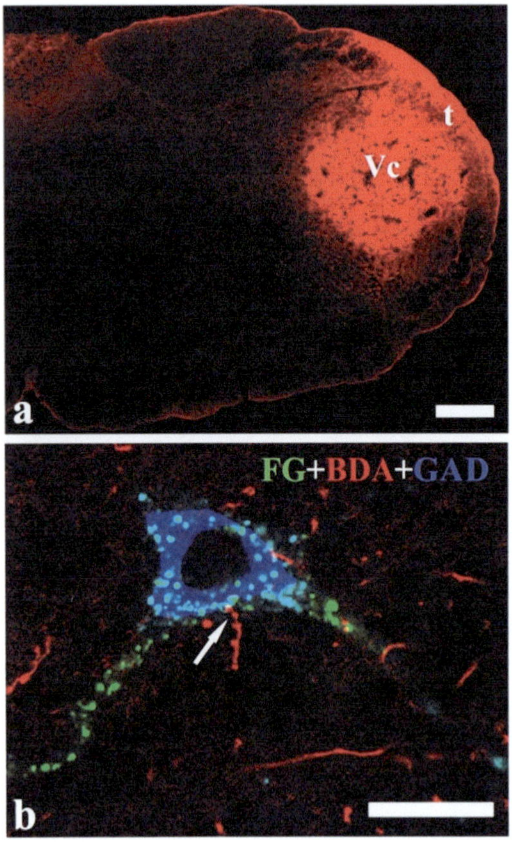

Fig. 3.11 (**a**) BDA injection site in the Vc; (**b**) anterogradely traced BDA axonal terminals ending on a FG retrogradely labeled VII premotor neuron that is also immunostained by anti-GAD

(b) Immunoblotting of sections with 1 % NGS in 0.05 M PB with 0.5 % TX-100 at RT for 1 h.
(c) Wash sections with 0.05 M PB three times for 5 min each time.
(d) Bathe sections in rabbit anti-FG (1;5,000; Chemicom. If other vendors used see their instructions) in 0.05 M PB with 1 % NGS and 0.5 % TX-100 at RT for 3–5 h.
(e) Wash sections with 0.05 M PB three times for 10 min each time.
(f) Incubate sections in mouse anti-GAD (1:500; Millipore. If other vendors, see instructions for dilution) in 0.05 M PB with 1 % NGS and 0.5 % TX-100 at RT overnight.
(g) Wash sections with 0.05 M PB three times for 10 min each time.
(h) Bathe sections in cocktail of Alexa Fluor 488 conjugated donkey anti-rabbit (1:500; Invitrogen. If other vendors, see instruction), Cy5 conjugated donkey anti-mouse (1:500; Jackson Labs or other vendors) and Cy3-labeled avidin D (1:1,000; Jackson Labs or other vendors, see their instruction).
(i) Wash sections with 0.05 M PB three times for 10 min each time and mount sections to view.

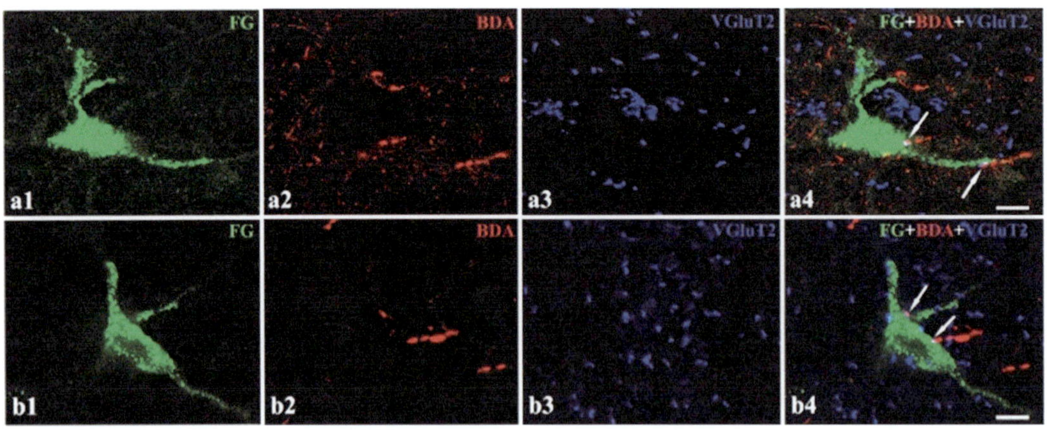

Fig. 3.12 Triple labeling of FG (**a1, b1**), BDA (**a2, b2**), and VGluT2 (**a3, b3**) upon premotor neurons of the VII and XII. In this work, FG was injected into either VII or XII and BDA was delivered to Vc, so the results mean that VGluT2-nergic neurons in the Vc project (traced by BDA) and end onto VII (**a1– a4**) or XII (**b1– b4**) premotor neurons (FG retrogradely labeled)

3. Protocol of FG, BDA, and VGluT2 triple immunofluorescent labeling:
   (a) Steps a–e is the same as mentioned in last section.
   (b) Incubate sections in guinea pig anti-VGluT2 (0.5 μg/mL; gift to authors) in 0.05 M PB with 1 % BSA and 0.5 % TX-100 at RT overnight.
   (c) Wash sections with 0.05 M PB three times for 10 min each time.
   (d) Incubate sections in mixture of Alexa Fluor 488 conjugated donkey anti-rabbit, Alexa Fluor 647 conjugated donkey anti-guinea pig (1:500; Invitrogen or other vendors) and Cy3-labeled avidin D (1:1,000; Jackson Labs or if other vendors, see their recommendation).
   (e) Wash sections with 0.05 M PB three times for 10 min each time and mount sections to view.

   Note: Alexa Flour 647 is far-red and can be visualized under confocal but not in conventional fluorescent LM.

3.8 Double Labeling Immuno-Electron Microscopy

3.8.1 Silver Intensified Immunogold Method

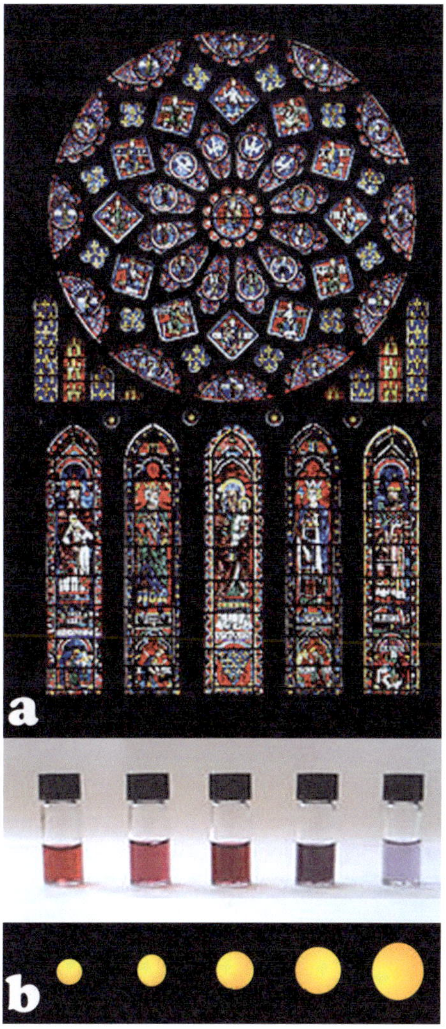

Fig. 3.13 (**a**) Colored glass in the front door of a plaza, those color are stained by colloidal gold; (**b**) different size of gold resulting in different color

1. Development of immunogold method. The word immunogold came from colloidal gold linked to immunoglobulin applied in immunostaining. Colloidal gold has been used in the manufacture of colored glasses for a long time with different sizes of gold particles resulting in different colors (Fig. 3.13). The first publication reported that colloidal gold could be conjugated to antiserum and applied for EM study was in early 1970s by Faulk and Taylor (1971). That was followed by a great number of reports on trying conjugation of colloidal gold to many different proteins in order to visualize them under EM. However, the most applicable approach is to link colloidal gold to secondary antibody IgG and use them just as peroxidase conjugated secondary antibodies in routine immunostain. In the early 1980s, the immunogold methods matured and immunostain was performed on grid after ultrathin sections were cut—so-called post-embedding method (De Mey et al. 1981). In addition to EM, colloidal gold labeling was also used in LM study with silver intensification, which was the earliest silver-gold method applied initially for LM study (De Mey et al. 1981). Even though post-embedding immunogold stain is still used today for immunostain of EM observed tissue,

in this section we will emphasize a simpler and more applicable way of pre-embedding silver-gold stain for EM study. Ideal results were achieved by a group of investigators (Li et al. 2000; Pang et al. 2009; Ge et al. 2010) by using this method; their concrete protocol will be introduced herein.

2. Rough description of a post-embedding immunostain procedure:
   Fixation of tissues. Rats are used as examples. Fixative is 4 % PFA plus 0.2–0.5 % GA in 0.1 M PB (pH 7.2), and transcardial perfusion is conducted as usual. Concentration of GA could be as high as 0.5 % if tract tracing was combined with immunohistochemistry staining of endogenous antigens, or could be lower like 0.075–0.1 % if double immunostain was performed for dual endogenous antigens. Routine vibratome sectioning and LR white embedding was carried out as aforementioned in Sects. 1.5.1 and 1.6.3. Immunostain was performed after ultrathin sections were cut and mounted on the grids.
3. Immunostaining protocols. Among many modifications, an original silver-gold stain is presented herein according to Tapia et al. (1983):
   (a) Bathe grid (face down) in 10 % hydrogen peroxide for 5 min. This "etches" the sections and aids antibody penetration, presumably by permeabilization of the epoxy resin. The grids were then rinsed with DDW.
   (b) Bathe grid in 20 % NGS in 0.05 M TBS (pH 7.2) for 10 min at RT.
   (c) Jet-washing the grid with DDW.
   (d) Bathe the grid with a mixture of the two primary antisera (one rabbit and one guinea pig antiserum) diluted to optimal titers in TBS, for 1 h at RT.
   (e) Subsequently, the grids were "jet-washed" with TBS and with TBS + 0.2 % BSA.
   (f) The grids were placed into droplets of TBS containing 1 % BSA for 5 mm at RT.
   (g) The grids were drained and placed into a mixture of the two second layer antisera (20 nm gold-labeled goat anti-guinea pig IgG and 12 or 40 nm gold-labeled goat anti-rabbit IgG) at optimal titer for 1 h at RT.
   (h) After jet-washing in TBS-BSA, TBS, and DDW, the sections were counterstained in uranyl acetate (saturated in DDW) and lead citrate.
4. Pre-embedding silver-gold method: An entire procedure of animal perfusion, tissue processing, and immunostaining based on Li and her colleagues will be quoted; their published EM microimages will be cited as well with the permission by the authors (Li et al. 2000; Pang et al. 2009; Ge et al. 2010). The rats were transcardially perfused with 4 % PFA and 0.1 % GA in 0.1 M PB (pH 7.4) with 15 % (v/v) picric acid and 50 μm vibratome sections were cut and stored in 0.1 M PB for tissue process and stain.

   Protocol of tissue process and immunostain with single silver-gold method:
   (a) Incubate sections in cryoprotection medium as mentioned in Sect. 1.5.2, steps 2 and 3. In short, 25 % sucrose and 10 % glycerol in 0.05 M PB (pH 7.4).
   (b) Keep shaking for 1 h at RT and prepare LN (also see Sect. 1.5.2).
   (c) Mount sections on a metal mash and quickly freeze and thaw them with LN by inserting and lifting. The sign of good freezing is the sections turn completely white. After freeze-thaw, place sections back to the cryoprotection medium.
   (d) Quickly wash sections with 0.05 M TBS (pH 7.4), keep shaking for 3–5 min.
   (e) Transfer them into 20 % NGS in 0.05 M TBS, incubate them for 1 h at RT.
   (f) Wash sections with 0.05 M TBS three times for 5 min each time.
   (g) Incubate primary antibody in 0.05 M TBS with 2 % NGS for 12 h at RT and then for another 12 h in 4 °C.
   (h) Wash sections with 0.05 M TBS three times for 10 min each time.
   (i) Bathe sections in 1.4 nm gold-particle conjugated secondary antibody against primary IgG (1:100; Nanoprobes) in 0.05 M TBS with 2 % NGS for 12 h at RT.
   (j) Rinse sections with 0.1 M PB (pH 7.4) three times for 10 min each time.
   (k) Post-fixation of sections in 1 % GA in 0.1 M PB for 10 min at RT.

(l) Wash sections with 0.1 M PB once for 10 min and then with DDW three times for 10 min.
(m) Silver intensifying is conducted with HQ silver kit (Nanoprobes) following the instruction in vendor's data sheet. In short: there are reagent solutions A, B, and C in separate bottles; add one drop of "A" into a 1-mL Eppendorf tube and follow by a drop of "B," vortex the tube until A and B are evenly mixed; then add a drop of "C" and mix them by vortex.
(n) Place sections into this A–B–C cocktail for about 5–7 min in the dark and then transfer them to DDW, bring out of dark to examine them under LM. The labeling under LM would show as yellow.
(o) If the colors of the sections are too white, repeat above procedure but use new A–B–C until yellow labeling can be observed under LM.
(p) Wash sections with DDW once for 5 min and then in 0.05 M TBS for 5 min three times. The sections are ready to be processed for EM embedding.

5. Microimages in Fig. 3.14 shows pre-embedding immunogold silver stained ultrathin sections, cited from Li et al. (2000) with the permission of the authors. The structures are clearly labeled by silver-gold particles and tissue preservation is excellent.

3.8.2 Simultaneous Identify Dual Antigens by Silver-Gold and ABC Stains

1. Ultrastructural co-localization of neuroactive substance, such as transmitters or transporters or receptors, is always the significant supplementary evidences for a specific function of the targeted neurons. In this section, ultrastructural double labeling of Neurokinin-1 receptor (NK1R) and Substance P (SP) in rat (Li et al. 2000) is engaged as an example to describe the protocol. The rats were transcardially perfused as usual with PFA + GA + picric acid as aforementioned.
2. Protocol of NK1R and SP double labeling immuno-EM:
   (a) Steps that include cryoprotection, freeze and thaw, and blocking and washing are the same as steps a–f in the last section.
   (b) Continue to step f aforementioned. After washing of sections, incubate them with a mixture of primary antibodies, including rabbit anti-NK1R (1 μg/mL; gift to the authors) and rat anti-SP (1:200; Chemicon), in 0.05 M TBS with 2 % NGS for 12 h at RT and then for another 12 h in 4 °C.
   (c) Wash sections with 0.05 M TBS three times for 10 min each time.
   (d) Bathe sections in 1.4 nm gold-particle attached goat anti-rabbit IgG (1:100; Nanoprobes) and biotinylated donkey anti-rat IgG in 0.05 M TBS with 2 % NGS for 12 h at RT.
   (e) Steps that include washing, post-fixation, silver intensification, and exam sections under a LM are the same as steps 10–15 in the last section.
   (f) Continue to step 15 as aforementioned. Wash sections with DDW once a time for 5 min and then in 0.05 M TBS for 5 min three times.
   (g) Incubate sections in Vector *Elite* ABC cocktail (1:50) in 0.05 M TBS for 3–5 h at RT.
   (h) Wash sections in 0.05 M TBS (pH 7.4) for 5 min once and wash twice in 0.05 M TB (pH 7.6).
   (i) Transfer sections into 0.02 % DAB in 0.05 M TB (pH 7.6) with 0.003 % H_2O_2 to conduct histochemical reaction for about 20–30 min.
   (j) Stop reaction by dumping reaction solution and exchanging fresh 0.05 M TB (pH 7.6).
   (k) Wash them once in 0.05 M TB (pH 7.6) and twice in 0.1 M PB (pH 7.4) for 5 min each. Then the sections are ready to be osmicated and LW embedding.
3. Microimages in Fig. 3.14 exhibit a clear relationship between peroxidase histochemical stained SP-enriched axonal boutons and silver-gold-labeled NK1R-expressed soma

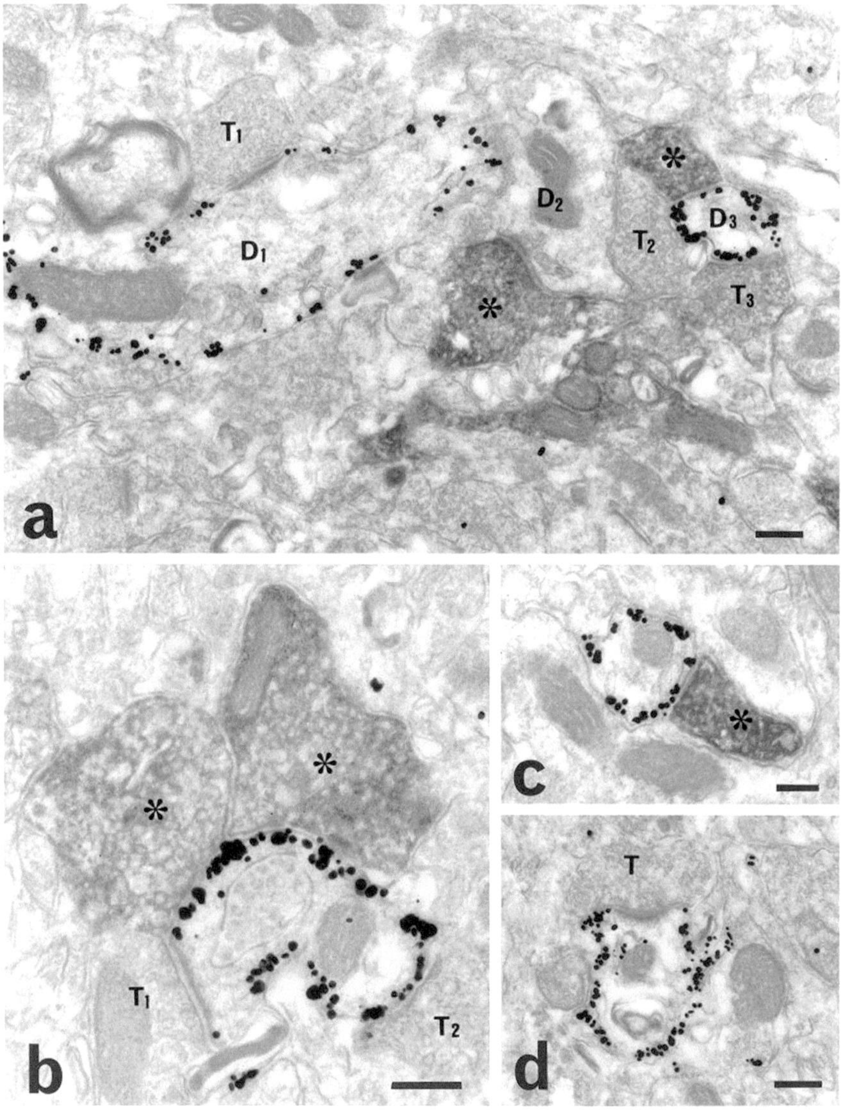

Fig. 3.14 EM microimages showing double labeling of silver-gold particles and peroxidase histochemical staining, they represent two neuroactive substances—NK1R (*black* dots in **a**–**d**) and SP (* in **a**–**c**). The regional and structural specific distribution of the silver-gold particles minimized the possibility of none-specific labeling, reflecting technique sophistication of the authors

and dendrites, cited from Li et al. (2000) with the permission of the authors. The micrographs explicitly displayed that NK1R are distributed in postsynaptic cell membrane and SP-labeled presynaptic boutons terminated on these somata and dendrites, most NK1R labeling are positioned along with the postsynaptic densities, reflecting SP plays a role of direct ligand.

3.8.3 Combination of Silver-Gold with Tract Tracing

1. Two examples of experiment on rats will be tailored herein to interpret the applications of these combinations. One is CTB anterograde tract tracing co-labeled with a neuroactive substance and the other is triple labeling of HRP retrograde tracing, BDA anterograde

labeling and silver-gold revealing of a neuroactive substance. The definite evidence of functional connection finds synaptic formation in the pathway morphologically or recorded orthodromic action potentials on the second-order neurons while stimulus is applied on the first-order neurons as long as they are excitatory. The protocols presented herein are based largely on the works by the same group of investigators aforementioned, and a few of publications will be quoted as original references of the methods (Pang et al. 2009; Ge et al. 2010; Dong et al. 2012); micrographs will be cited to illustrate the results of these protocols with authors' permission.

2. Protocol of tissue process and immunostain for CTB and silver-gold double labeling: The authors delivered 0.25–0.30 μL of 2 % CTB in 0.1 M PB (pH 7.3) as anterograde tracer by pressure injection through a Hamilton syringe, and the rat survives for 6 days after injection. Fixative used in transcardial perfusion is the same as aforementioned, shortly 4 % PFA + 0.1 % GA + 15 % picric acid. For EM work, 50 μm vibratome sections were cut and processed as the follow steps.

   Protocol of tissue processing and staining:

   (a) Steps that include cryoprotection, freeze and thaw, blocking and washing are the same as steps a–f in the Sect. 3.8.1, step 4. The only different is to exchange 20 % NGS with 10 % BSA.
   (b) Continue to step f aforementioned. After washing sections, incubate them with a mixture of primary antibodies, including goat anti-CTB (1:1,000; List Biology) and guinea pig anti-VGluT2 (1 μg/mL; gift to authors), in 0.05 M TBS with 1 % BSA for 12 h at RT and then for another 12 h in 4 °C.
   (c) Wash sections with 0.05 M TBS three times for 10 min each time.
   (d) Bathe sections in 1.4 nm gold-particle conjugated rabbit anti-goat IgG (1:100; Nanoprobes) and biotinylated donkey anti-guinea pig IgG (1:100; Jackson Labs) in 0.05 M TBS with 1 % BSA for 12 h at RT.
   (e) Steps that include washing, post-fixation, silver intensification, and exam sections under a LM are the same as steps j–o in the Sect. 3.8.1, step 4.
   (f) Continue to step o as aforementioned. Wash sections with DDW once a time for 5 min and then in 0.05 M TBS for 5 min three times.
   (g) Steps including Vector *Elite* ABC cocktail incubation, washing, initiation, and termination of DAB reaction are the same as steps g–j in the last section.
   (h) Wash sections once in 0.05 M TB (pH 7.6) and twice in 0.1 M PB (pH 7.4) for 5 min each. The sections are ready to be osmicated and LW embedding.

3. Micrographs in Fig. 3.15 unveil co-localization of CTB and VGluT1 (or VGluT2) in the same axonal boutons those terminate and synapse upon the dendrites in their projecting target region. Those axons were projected from the Vp to the ventral posteromedial thalamic nucleus (VPM), namely VGluT1 (or VGluT2) positive neurons project from the Vp to the VPM traced by CTB (Ge et al. 2010). In the other work, these projections were originated from the Vo to the Vmo, namely VGluT1 (VGluT2)-nergic neurons project from the Vo to the Vmo traced by CTB (Pang et al. 2009). Similarly, the controlling of none-specific binding and tissue preservation in these works are sophisticated. The micrographs are cited with the permission of the authors.

4. Protocol of tissue process and stains for HRP, BDA, and silver-gold triple labeling:

   In this work, 0.2 μL of 10 % BDA in 0.05 M PBS was delivered to the Vc by pressure injection and animal is allowed to survive for 7 days. Then HRP was pressure injected into the VII (or XII) in the same rats; after another 2 days, the rats were transcardially perfused with 4 % PFA and 0.1 % GA in cold 0.1 M PB (pH 7.4; 4 °C). The authors identified co-localization of BDA and VGluT2 in single axonal boutons to demonstrate VGluT2 positive axonal terminals projected from the Vc synapsed on HRP retrogradely labeled VII (or XII) premotor neurons (Dong et al. 2012).

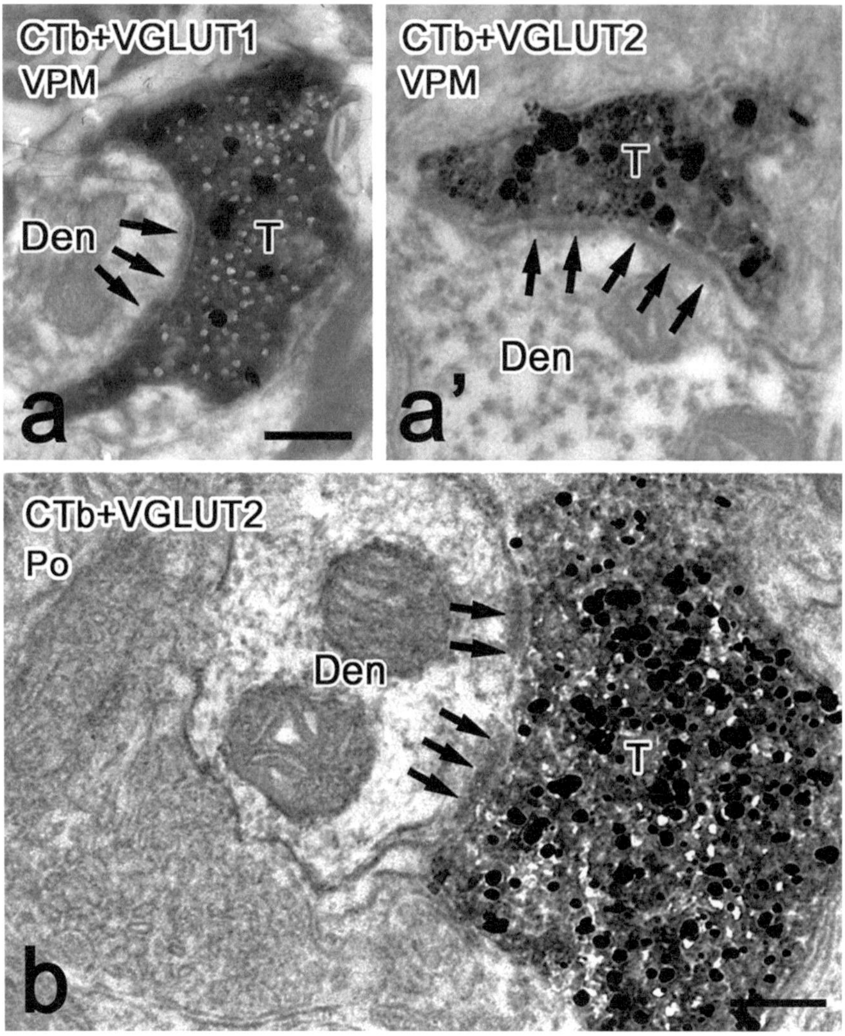

Fig. 3.15 EM micrographs showing double labeling of co-localization of anterograde traced CTB and silver-gold immunostained VGluT1 (**a**) or VGluT2 (**a**' and **b**), CTB labeling represented by peroxidase histochemical staining (**a**–**b**) and VGluT1 (or 2) positive boutons marked by silver-gold particles (**a**–**b**)

Similarly, 50 μm vibratome sections were cut and processed using the follow procedure.

Protocol of tissue processing and staining:

(a) Follow all steps from a to p described in Sect. 3.3.3, step 3.
(b) Follow the steps in Sect. 1.5.2 to carry out a freeze-and-thaw procedure, but use LN directly as mentioned in Sect. 3.8.1, step 4, steps a–c.
(c) Then, follow the steps d–f in Sect. 3.8.1, step 4.
(d) Incubate guinea pig anti-VGluT2 (0.8 μg/mL; gift to the authors) in 0.05 M TBS with 2 % NGS for 12 h at RT and then for another 12 h in 4 °C.
(e) Wash sections with 0.05 M TBS three times for 10 min each time.
(f) Incubate sections in 1.4 nm gold-particle attached goat anti-guinea pig IgG (1:100; Nanoprobes) in 0.05 M TBS with 2 % NGS for 12 h at RT.
(g) Follow steps j–p in Sect. 3.8.1, step 4.
(h) Wash sections with DDW one time for 5 min and then in 0.05 M TBS for three times for 5 min each time.
(i) Follow the steps g–k in Sect. 3.8.2, step 2.

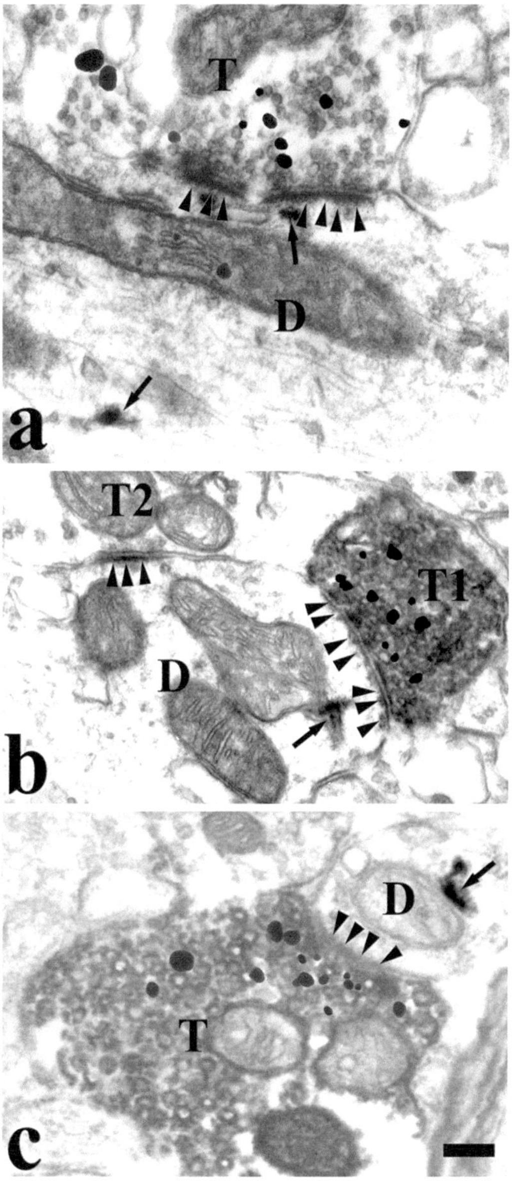

Fig. 3.16 EM microimages showing ultrastructural triple labeling by HRP (*arrows* in (**a–c**)), ABC peroxidase stain (*dark Ts* in (**b**, **c**)) and silver-gold particles. This work means that traced with BDA, the VGluT2-nergic neurons project from the Vc to the premotor neurons of the VII (or XII); those are retrogradely labeled by HRP. This is an example of successful application of HRP, BDA, and anti-neuroactive substance triple labeling at ultrastructural level

(j) The sections are ready to be osmicated and LW embedded.

5. Figure 3.16 shows co-localization of BDA and VGluT2 in the same axonal boutons; those terminate and synapse upon the dendrites are retrogradely labeled with HRP. Thus, the authors traced with BDA that the VGluT2-nergic neurons in the Vc project to premotor neurons of the VII (or XII); those are retrogradely labeled by HRP in rats. The micrographs in this study are cited with the permission of the authors (Dong et al. 2012).

References

Abercrombie M, Johnson ML (1946) Quantitative histology of Wallerian degeneration; nuclear population in rabbit sciatic nerve. J Anat 80:37–50

Angelucci A, Clasca F, Sur M (1996) Anterograde axonal tracing with the subunit B of cholera toxin: a highly sensitive immunohistochemical protocol for revealing fine axonal morphology in adult and neonatal brains. J Neurosci Methods 65:101–112

Beier KT, Borghuis BG, El-Danaf RN, Huberman AD, Demb JB, Cepko CL (2013) Transsynaptic tracing with vesicular stomatitis virus reveals novel retinal circuitry. J Neurosci 33:35–51

De Mey J, Moeremans M, Geuens G, Nuydens R, De Brabander M (1981) High resolution light and electron microscopic localization of tubulin with the IGS (immuno gold staining) method. Cell Biol Int Rep 5:889–899

De Olmos J, Heimer L (1977) Mapping of collateral projections with the HRP-method. Neurosci Lett 6:107–114

Dong YL, Wang W, Li H, Li ZH, Zhang FX, Zhang T, Lu YC, Li JL, Wu SX, Li YQ (2012) Neurochemical properties of the synapses in the pathways of orofacial nociceptive reflexes. PLoS One 7:e34435

Faulk WP, Taylor GM (1971) An immunocolloid method for the electron microscope. Immunochemistry 8: 1081–1083

Fink RP, Heimer L (1967) Two methods for selective silver impregnation of degenerating axons and their synaptic endings in the central nervous system. Brain Res 4:369–374

Ge SN, Ma YF, Hioki H, Wei YY, Kaneko T, Mizuno N, Gao GD, Li JL (2010) Coexpression of VGLUT1 and VGLUT2 in trigeminothalamic projection neurons in the principal sensory trigeminal nucleus of the rat. J Comp Neurol 518:3149–3168

Gu Y, Chen Y, Ye L (1992) Electron microscopical demonstration of horseradish peroxidase by use of tetramethylbenzidine as chromogen and sodium tungstate as stabilizer (TMB-ST method): a tracing method with high sensitivity and well preserved ultrastructural tissue. J Neurosci Methods 42:1–10

Johnson AC, Mc NA, Rossiter RJ (1950) Chemistry of Wallerian degeneration; a review of recent studies. Arch Neurol Psychiatry 64:105–121

Kobbert C, Apps R, Bechmann I, Lanciego JL, Mey J, Thanos S (2000) Current concepts in neuroanatomical tracing. Prog Neurobiol 62:327–351

Kotowicz Z (2005) Gottlieb Burckhardt and Egas Moniz—two beginnings of psychosurgery. Gesnerus 62:77–101
Kristensson K, Olsson Y (1971) Retrograde axonal transport of protein. Brain Res 29:363–365
LaVail JH, LaVail MM (1972) Retrograde axonal transport in the central nervous system. Science 176:1416–1417
LaVail MM, Sidman M, Rausin R, Sidman RL (1974) Discrimination of light intensity by rats with inherited retinal degeneration: a behavioral and cytological study. Vision Res 14:693–702
Li JL, Wang D, Kaneko T, Shigemoto R, Nomura S, Mizuno N (2000) The relationship between neurokinin-1 receptor and substance P in the medullary dorsal horn: a light and electron microscopic immunohistochemical study in the rat. Neurosci Res 36:327–334
Liu Y, Broman J, Edvinsson L (2004) Central projections of sensory innervation of the rat superior sagittal sinus. Neuroscience 129:431–437
Luo P, Haines A, Dessem D (2001) Elucidation of neuronal circuitry: protocol(s) combining intracellular labeling, neuroanatomical tracing and immunocytochemical methodologies. Brain Res Brain Res Protoc 7:222–234
Luo P, Zhang J, Yang R, Pendlebury W (2006) Neuronal circuitry and synaptic organization of trigeminal proprioceptive afferents mediating tongue movement and jaw-tongue coordination via hypoglossal premotor neurons. Eur J Neurosci 23:3269–3283
Luo L, Callaway EM, Svoboda K (2008) Genetic dissection of neural circuits. Neuron 57:634–660
Martin X, Dolivo M (1983) Neuronal and transneuronal tracing in the trigeminal system of the rat using the herpes virus suis. Brain Res 273:253–276
Mesulam MM (1976) The blue reaction product in horseradish peroxidase neurohistochemistry: incubation parameters and visibility. J Histochem Cytochem 24: 1273–1280
Mesulam MM (1978) Tetramethyl benzidine for horseradish peroxidase neurohistochemistry: a non-carcinogenic blue reaction product with superior sensitivity for visualizing neural afferents and efferents. J Histochem Cytochem 26:106–117
Nauta WJ, Gygax PA (1954) Silver impregnation of degenerating axons in the central nervous system: a modified technic. Stain Technol 29:91–93
Pang YW, Ge SN, Nakamura KC, Li JL, Xiong KH, Kaneko T, Mizuno N (2009) Axon terminals expressing vesicular glutamate transporter VGLUT1 or VGLUT2 within the trigeminal motor nucleus of the rat: origins and distribution patterns. J Comp Neurol 512:595–612
Rondorf-Klym LM, Colling J (2003) Quality of life after radical prostatectomy. Oncol Nurs Forum 30:E24–E32
Rouiller EM, Capt M, Dolivo M, De Ribaupierre F (1989) Neuronal organization of the stapedius reflex pathways in the rat: a retrograde HRP and viral transneuronal tracing study. Brain Res 476:21–28
Ruigrok TJ, Teune TM, van der Burg J, Sabel-Goedknegt H (1995) A retrograde double-labeling technique for light microscopy. A combination of axonal transport of cholera toxin B-subunit and a gold-lectin conjugate. J Neurosci Methods 61:127–138
Schoene-Bake JC, Parpaley Y, Weber B, Panksepp J, Hurwitz TA, Coenen VA (2010) Tractographic analysis of historical lesion surgery for depression. Neuropsychopharmacology 35:2553–2563
Tapia FJ, Varndell IM, Probert L, De Mey J, Polak JM (1983) Double immunogold staining method for the simultaneous ultrastructural localization of regulatory peptides. J Histochem Cytochem 31:977–981
Thomas GA (1948) Quantitative histology of Wallerian degeneration; nuclear population in two nerves of different fibre spectrum. J Anat 82:135–145
van Strien NM, Cappaert NL, Witter MP (2009) The anatomy of memory: an interactive overview of the parahippocampal-hippocampal network. Nat Rev Neurosci 10:272–282
Weiss P, Hiscoe HB (1948) Experiments on the mechanism of nerve growth. J Exp Zool 107:315–395
Wickersham IR, Finke S, Conzelmann KK, Callaway EM (2007) Retrograde neuronal tracing with a deletion-mutant rabies virus. Nat Methods 4:47–49
Zhang JD (1998) Projections from dorsomedial part of the subnucleus oralis to the mesencephalic trigeminal neurons innervating the masseter muscle—a PHA-L and HRP double labeling study in the rat. J Hirnforsch 39:55–64
Zhang JD, Yang XL (1999) Projections from subnucleus oralis of the spinal trigeminal nucleus to contralateral thalamus via the relay of juxtatrigeminal nucleus and dorsomedial part of the principal sensory trigeminal nucleus in the rat. J Hirnforsch 39:301–310
Zhang JD, Wang BR, Li HM, Li JS (1991) Projections from neurons innervating the masseter muscle to the subnucleus oralis of the spinal trigeminal nucleus and adjacent lateral reticular formation in the rat. J Hirnforsch 32:641–646
Zhang J, Pendlebury WW, Luo P (2003) Synaptic organization of monosynaptic connections from mesencephalic trigeminal nucleus neurons to hypoglossal motoneurons in the rat. Synapse 49:157–169
Zhang J, Yang R, Pendlebery W, Luo P (2005) Monosynaptic circuitry of trigeminal proprioceptive afferents coordinating jaw movement with visceral and laryngeal activities in rats. Neuroscience 135:497–505
Zhang J, Luo P, Ro JY, Xiong H (2012) Jaw muscle spindle afferents coordinate multiple orofacial motoneurons via common premotor neurons in rats: an electrophysiological and anatomical study. Brain Res 1489:37–47

Stereological Analysis

4

Kristi M. Anderson, Adam M. Szlachetka,
and R. Lee Mosley

Abstract

The use of stereology for quantitative estimates of biological features is becoming commonplace in modern laboratories. Over the past decades, stereology has evolved to keep pace with the ever-changing field of neuroscience. It has moved from its early applications for studying the geologic composition of rocks to an invaluable method to detect both small- and large-scale changes in the complex central nervous system. Recent developments in technology have made the practice highly efficient and accurate while providing a strategy to remove assumptions and investigator biases that lead to inaccurate numerical estimations. This chapter focuses on the use of stereology in neuroscience and provides evidence that validates the results generated via stereological analysis. The reader is introduced to the many applications of stereology in the laboratory including the quantification of neuron populations, measurement of dendrite and axon lengths, analysis of surface area of non-symmetrical shapes, and overall connectivity in the brain. This chapter provides systematic examples for the performance of a stereological study and is presented in a manner allowing the use of this chapter as a manual directly translatable for the neuroscience laboratory.

Keywords

Stereology · Designed-based quantitation · CNS neurons · Fractionators · Disector

4.1 Introduction

Within circumscribed volumes of the central nervous system (CNS), quantitation of cell densities and volumes estimated from histology sections typically has been fraught with inherent errors that have led to under- and overestimation of cell quantities and misinterpretation

The original version of the chapter has been revised. A correction to this chapter can be found at https://doi.org/10.1007/978-1-4614-8794-4_37

K. M.Anderson • A. M.Szlachetka
R. L.Mosley, Ph.D. (✉)
Department of Pharmacology and Experimental Neuroscience, Center for Neurodegenerative Disorders, University of Nebraska Medical Center, Omaha, NE, 68198-5930, USA
e-mail: kristi.anderson@unmc.edu; amszlachetka@unmc.edu; rlmosley@unmc.edu

H. Xiong and H.E. Gendelman (eds.), *Current Laboratory Methods in Neuroscience Research*, Springer Protocols Handbooks, DOI 10.1007/978-1-4614-8794-4_4,

of crucial results. Early corrections of both systematic and endogenous errors were based on assumptions of cell morphology and counting biases, which did not adequately remedy the problem. In contrast, unbiased, design-based stereological methods alleviate many issues and errors associated with quantitative neuromorphological research. Stereological analysis is largely concerned with the three-dimensional interpretation of planar tissue sections that accurately estimates cell parameters, such as number or volume, within a circumscribed three-dimensional volume. Results obtained with design-based stereology are independent of size, shape, spatial orientation, and spatial distribution of the cells investigated; thus, systemic error based on bias is theoretically limited. Therefore, neuroscience methods using design-based stereology are utilized to analyze volumes of brain regions, number of cells (such as neurons and glial cells) within brain regions, cellular compartment volumes (nuclear or perikaryal), length or densities of biological structures (vessels, nerve fibers, or dendrites) within brain regions, cytoarchitecture of brain regions (spatial distribution of cells or arbor density within a region of interest (ROI)), and connectivity (neural network, synapses, and axonal and dendritic processes).

The following chapter attempts to educate the reader with a brief history of stereology and the advances that stereological analysis has provided to biology. We will also discuss how stereology has evolved to become an efficient, reliable, and unbiased method used to estimate cell populations. We will cover quantitations using stereology, as well as potential errors when using stereology. Finally, this chapter will cover the tools and methods utilized in stereological software and how to use them in a way that most benefits the investigator. By consistent adherence to the rules of stereology, quantitative estimates that accurately represent the populations under investigation are efficiently obtainable.

4.2 History of Stereology

Stereology is the study of estimating geometrical quantities in three dimensions through interpretation of two-dimensional images (Sundberg 1984). The geologist Achille Ernest Oscar Joseph Delesse first described stereology when he discovered he could determine a rock's composition using only a single-plane section (Baddeley 2001). Delesse reasoned that if a rock is spatially homogeneous, then any plane section is representative of the entire rock. This rationale worked well in geology but was less applicable in biological science because few biological structures are "homogeneous" in their spatial arrangement. In recent years, stereology has evolved and mathematically compensated for biological structures lacking homogeneity. Use of stereology is now a common and useful tool in biological studies and for some disciplines, such as neuroscience, the standard tool for estimating cell populations such as neurons.

4.2.1 Design-Based Versus Model-Based Stereology

Older models of stereology were "model-based" that relied on a particular model formulated by using known or assumed geometrical properties of the object under study. This form of stereology employed approximations or assumptions about the composition and overall randomness of the subject to make population estimates. After counting the objects of interest, researchers applied formulas to the rough numerical data to correct for these assumptions and obtained what was believed to be an accurate population estimate. The formulas corrected for factors such as width of slice, diameter of cells, distance from top of slice to focal plane, smallest detectable part of the cell, size distribution coefficient, and shape coefficient (Glaser et al. 2007). Incorrect or perfunctory construction of the model resulted in discrepancies in the final population estimate.

Ultimately, errors in quantitation due to incorrect assumptions or formulas warranted a better stereological approach.

Design-based stereology circumvents issues associated with model-based stereology, makes no assumptions about tissues, and requires no preexisting knowledge of size, shape, or orientation of objects and thus is referred to as "assumption- or bias-free." Design-based stereology relies on systematic random sampling (SRS) to ensure counting of objects only once and sampling of the tissue occurs in a fashion that achieves an unbiased estimate of the whole tissue.

4.3 Why Is Stereological Analysis Necessary?

The necessity of stereological analysis for biological systems became apparent when researchers needed to correlate structure and biological function. Stereology provides this ability, while simultaneously alleviates investigator bias, allows detection of small morphological changes, and provides reliable data for statistical analysis and hypothesis testing (Sundberg 1984). Prior to the development of stereology, differences between tissues were qualitatively described using terms such as "large," "small," "many," "few," "absent," or "present" (Schmitz and Hof 2005). These terms may be helpful, but do not reflect subtle changes in tissue composition. They also prevent researchers from statistically testing a quantifiable hypothesis about tissue samples. The need for a quantitative method of measuring samples was paramount for researchers to obtain statistically valid data. Stereology allows researchers to replace qualitative terms with quantitative values and successfully analyze and compare data from differentially treated systems.

4.4 Five General Types of Quantitations Performed with Stereological Analysis

The wide variety of probes that are available for use in stereology provides researchers the freedom to study an array of quantitative questions.

Of interest to most investigators is the number or population density of a specific cell type or any other object. One stereological probe for measuring a population is the Optical Fractionator. This method estimates the total number of cells in a structure and is based on a representative number of cells counted that spans the entire ROI using SRS. Another method, termed the $N_V \times V_{REF}$ method, estimates populations by taking the number of cells counted and divides by the number of counting spaces to obtain a mean cell density. Multiplying the entire volume of the ROI by the mean cell density provides the population estimate. Considering that Optical Fractionator requires no density measurements, it is regarded as an easier method of estimating a population (Glaser et al. 2007).

Length is an easy parameter to understand but a concept that becomes complicated when considering the length of objects that twist and turn. For example, dendrite length is difficult to measure because of constantly changing orientation and tendency to interweave with other dendrites. Using the stereological probe Space Balls, the researcher can estimate the length of tubular structures in thick tissue sections. This technique utilizes virtual spheres (or hemispheres) placed over images acquired at different focal depth intervals. The number of intersections between the spheres (or hemispheres) and the tubular object is then counted and the length estimated (Glaser et al. 2007). (For details on other methods for measuring length, visit www.stereology.info or *Stereology for biological researcher with a focus on neuroscience, Glaser.*)

Surface, in stereology, refers to the area of a shape that is not flat. In contrast, measuring the surface of an area that lies in a single plane is referred to as a profile area or planar surface. Quantitating surface values requires the use of specialized stereological methods, such as the spatial grid probe. Proper and accurate results from this method require a stack of perfectly registered sections (Howard and Sandau 1992). Another method requiring perfectly registered serial sections to measure surface area is the fakir method. This unbiased procedure uses a directly randomized spatial grid rather than sections with randomized directions (Kubinova and Janacek 1998).

Volume is the measure of how an object fills a space. Valid quantitation of volume can be performed using Nucleator, Rotator, or Optical Rotator methods. Obtaining estimates of volume and population at the same time is possible by using the aforementioned methods in conjunction with Optical Fractionator (Glaser et al. 2007).

Connectivity is a geometrical quantity with attributes that assess the way objects are connected to other objects, such as found in neural networks or axonal and dendritic processes. These can be quantitated with a Connectivity Assay, which utilizes the Euler number and Conn Euler counting method as described by Kroustrup and Gunderson (Gundersen et al. 1993; Kroustrup and Gundersen 2001).

4.5 Errors in Cell Counting Committed Without Stereology

Prior to the development of stereological methods, cell populations were estimated using several different approaches. These approaches included maceration (tissue dissolution or softening), biochemistry, and sectional morphometry (Mayhew and Gundersen 1996). Each method proved to have unique errors that rendered them undesirable to researchers. Counting renal glomeruli via maceration was time-consuming and caused damage to the cells. Biochemical methods tried to estimate cell size by using a protein/DNA ratio but were only valid with homogeneous tissues that are exclusively cellular (Mayhew and Gundersen 1996). Biochemical estimates also are disadvantaged when based on DNA since a number of nuclei, especially for cells or syncytia, may have multiple nuclei and yield a higher DNA content/cell that leads to overestimation of populations.

Even with the added advantage of design-based stereological protocols, errors in counting are still an inherent problem associated with achieving an accurate population estimate. Creation of cell fragments by splitting a cell during sectioning whereby cell fragments are counted as single cells (e.g., split cell errors) can lead to overestimation of a population. Moreover, the use of thinner sections and larger cells of interest amplifies the overestimation. In an attempt to correct for overestimation, Abercrombie formulated a novel equation, designated as the Abercrombie correction. He reasoned that the actual number of cells in a population should equal the total count of whole and fragmented cells, corrected by a factor equal to the ratio of the section's thickness, divided by the sum of the section's thickness and the mean height of the cells (Abercrombie 1946). That correction is given by the following equation:

$$\text{Number} = \text{count} \times \left[\frac{(\text{section thickness})\,/}{(\text{section thickness} + \text{cell height})} \right]$$

However, fundamental problems are associated with the value and accuracy of Abercrombie's correction (Williams and Rakic 1988). First, a bias of omission is associated with this method. Potential errors are introduced upon the omission of cells based on size, shape, and orientation. Abercrombie realized his correction was flawed and warned that his equation may result in a 5–10 % overestimation of cell number. This overestimation becomes more serious and less predictable when the size and distribution is skewed or bimodal (Hendry 1976), for instance, the bimodal distribution of dopaminergic neurons along the substantia nigra axis.

A second problem is the difficulty of determining an appropriate value for cell height to include in Abercrombie's correction. Often cell diameter is substituted for cell height. However, these two values are not always equivalent and may result in substantial alterations in the correction ratio. This is especially apparent for oblong cells or in tissues that exhibit uneven shrinkage during processing.

The third problem is that the equation assumes that the knife cleanly cuts the cells in its path. However, if the blade plows the tissue, rather than slicing, cells may be pushed aside and remain uncut, thus resulting in abnormal and uneven distribution of cells. This scenario may yield an underestimation of cell number within the population. Unfortunately, attempting to determine the percentage of cells that are cut and those that are plowed is not an easy task.

The fourth problem addresses the sensitivity of Abercrombie's method to variation in thickness between individual sections. Adjacent paraffin sections can vary up to 50 % in thickness (Graf 1948; Clarke 1968). Frozen sections cut at 30 μm may vary in thickness from 25 to 35 μm due to unstable block temperature and sectioning chatter, i.e., the alternation of thick and thin striations within a section. The more the section thickness varies, the more the counts will vary and sampling error will increase.

The final problem with Abercrombie's method addresses the difficulty of correctly identifying cell fragments at the top and bottom of the sections. Cell fragments mistaken for glial cells or pericytes can result in erroneous inclusion or exclusion from the count.

4.6 Unbiased Counting Methods, Protocols, and Software

As with any other scientific technique, stereology can be susceptible to inaccuracies and bias or systematic error. Minimizing these unwanted factors is essential when trying to acquire "true" data. The development of advanced stereological software and use of a simple system for random sampling has aided in the removal of "sampling biases" associated with stereology.

The first step towards uniform and unbiased counting requires an accurate representative sample. This requires that the entire organ be available for sampling such that the sampling probability is equal between all cells and that probing the population avoids influencing or biasing the result (Geuna 2000; Boyce et al. 2010).

As previously discussed, design-based stereology makes no assumptions about size, shape, or orientation of objects and therefore provides an unbiased estimate of population. However, additional requirements that are needed to minimize bias or systematic error include (1) identical processing conditions between comparison groups to minimize differences due to tissue shrinkage; (2) entire regions of interest need to be accessible, recognizable, and have the same chance of contributing to the stereological sample; (3) sample estimates should be unaffected by loss of structures when cutting the structure or by incomplete staining of the tissue; (4) nonhomogeneous compression of tissue sections along the *z*-axis should be minimized when using methacrylate and paraffin as embedding media; and (5) minimal shrinkage of the sections along the *z*-axis regardless of embedding medium or cutting (Schmitz and Hof 2005).

By using stereology, the investigator reduces the amount of time and effort expended towards quantifying the objects of interest, thus affording a more efficient means of estimating sample populations. Stereology also alleviates most of the problems associated with over- and underestimation of populations by eliminating the need for Abercrombie's correction. Available stereological software includes (in alphabetical order) AID Stereology System (Imaging Research, St. Catharines, ON, Canada), CAST (Visiopharm, Hørsholm, Denmark), Digital Stereology (Kinetic Imaging, Bromborough, UK), Direct Three-Dimensional Counting (3DC-Direct), StereoInvestigator (MicroBrightField, Williston, VT, USA), Stereologer (Systems Planning and Analysis, Alexandria, VA, USA), and Stereology Toolkit Plug-in for NOVA PRIME (Bioquant Image Analysis Corporation, Nashville, TN, USA) (Williams and Rakic 1988; Glaser and Glaser 2000; Schmitz and Hof 2005).

4.6.1 Sampling

The principle of SRS, also referred to as systematic uniform random sampling (SURS), is the sampling methodology utilized in stereology to reduce endogenous variance and error that is associated with nonrandom and other random sampling methods. By using SRS, which incorporates randomization of the initiation section and area of each section sampled, an unbiased estimate of the entire population can be more closely approximated. However, several values must be determined to realize an unbiased estimate. The first value is the section-sampling fraction (ssf) or fraction (*f*). The fraction is defined as 1 over p, where p is the periodicity or section interval chosen by the researcher. For example, if the researcher chooses to sample every fourth section, $p=4$ and ssf = ¼.

The second value required is the area-sampling fraction (asf). This is defined as the area of the unbiased counting frame (B) divided by the area of a grid square (see Sect. 4.6.4) (D). For example, if a counting frame has an area of 8,500 μm^2 and the distance between counting frames was 300 μm in each direction the calculation would be asf $= B/D = 8{,}500\ \mu m^2/(300\ \mu m \times 300\ \mu m) = 0.094$.

The third value to be established is the thickness-sampling fraction (tsf). By taking the height of the unbiased counting frame (H) divided by the average thickness of the tissue section (t), tsf can be calculated by H/t. For example, if the height of the counting frame was 7 μm and the average tissue thickness was 30 μm then tsf $= 7\ \mu m/30\ \mu m = 0.23$.

The result of multiplying $(\text{ssf} \times \text{asf} \times \text{tsf})^{-1}$ yields the value-sampling probability. Using the example numbers above, the equation would be $(¼ \times 0.094 \times 0.23)^{-1} = 185$. By multiplying the number of cells counted by the sampling probability, one can achieve an estimate of the total cell population within a ROI.

Most stereological software packages will perform the required calculations to obtain an estimate that is as accurate as possible. The researcher must first define the variables and the initial parameters such as ssf, asf, and tsf (see Sect. 4.6.10), but calculations are not generally required for routine population estimates.

4.6.2 Counting

Once the ROI has been defined, counting cells in tissue samples using stereological techniques is relatively simple. The software places the grid and counting frames in an unbiased fashion, while the investigator inspects the cells inside the counting frame, clicking the mouse to mark any cell that fulfills the appropriate criteria. Including a cell in the count only occurs if that cell follows the inclusion criteria associated with the counting frame (discussed later) and any additional criteria the researcher has designated for a particular experiment. When considering stereology programs, note that certain computer software programs record the total number of cells counted, the number of cells counted for each tissue section, and the location of each cell counter, while other programs only record the raw number of cells counted.

Prior to beginning a study of cell populations or objects of interest, the investigator must establish a set of criteria for determining whether a cell or object is "countable." This includes selecting identifiable characteristics for each cell to be counted. If the characteristics are not visible, then the cell cannot be included in the final count. Examples of characteristic points include cell top, nucleus top, nucleus, or nucleolus (Glaser et al. 2007; MBFBioscience 2011).

When counting cells with prominent processes or extensions, such as neurons, another set of rules needs to be established. These rules are used to determine whether a neuron is countable if its axon is touching the exclusion line, but the cell body is inside the counting frame. Alternatively, whether to count a neuron when only part of the axon is touching the inclusion line, but the neuronal body itself lies outside the counting frame. A very dense population of neurons in a small area compounds the difficulty of accurately counting because of the difficulty in distinguishing which axons originate from which neuronal bodies. Following a well-defined set of counting rules throughout the entire study is critical to ensure consistent and accurate results.

4.6.3 Coefficient of Error

The coefficient of error or CE is the standard deviation divided by the mean and thus represents the chance for sampling error or the quality of the estimate. Hence, a lower CE indicates less chance for sampling error and a greater chance for an accurate estimate. It is necessary to keep this factor in mind when analyzing stereological data.

For stereological quantitation, the extent of sampling error is unknown; however, several different methods and equations are available to predict the accuracy of a stereological estimate. Common CE methods include those of Gunderson (using smoothness constants (m) of $m=0$ or $m=1$) and Schmitz-Hof. In general, CE values approximating 0.10 are sufficient for reliable sampling errors to estimate populations. Thus, samples that produce consistent CE values greater than 0.10 indicate that the number of cells counted should be increased to provide less chance for sampling error; whereas, consistent CE values less than 0.10 indicate the number of cells counted within a sample can be decreased to allow increased efficiency. Note, if using a section interval of greater than 8, using the Gunderson CE where $m=1$ provides poorer estimates of error, while using $m=0$ yields better estimates of the "true" CE (MBF Bioscience 2011). Adjusting the CE to a desired value using the same variable adjustments to modify the number of cells counted is discussed Sect. 4.6.10.

4.6.4 Fractionator

One of the most common quantitation desired by biological researchers is a quantitative estimate of cell populations. Fractionator or physical fractionator is a stereological method used for counting cells within serial tissue sections and is used in conjunction with the disector probe (see Sect. 4.6.6). Fractionator systematically partitions the outlined ROI into a grid of equal-sized areas (disectors). The disectors are spaced in rows and columns at equal distances from one another (Glaser and Glaser 2000). Random placement of the grid by the software reduces possible sampling bias and thus provides a more accurate sampling of the tissue. The investigator first uses a low magnification lens to identify the ROI and then draws an electronic bit map or contour to encompass and define the entire ROI (Fig. 4.1a). From previous investigator-defined parameters, the software places the disector or grid pattern randomly over the contour (Fig. 4.1b) and within the grid places a counting frame within each grid box (Fig. 4.1c). Finally, the contour-defined ROI, randomly placed grid, and counting frames will be superimposed (Fig. 4.1d). The software records the dimensions of each individual disector and utilizes the asf and ssf to form the final stereological equations that will determine the population estimate.

4.6.5 Optical Fractionator

Optical Fractionator is similar to the fractionator method in that it estimates cell populations that are too large to count exhaustively (Glaser et al. 2007). The main difference is that Optical Fractionator does not use a grid to partition the ROI. Instead, it divides thick sections into several, evenly spaced focal planes and is utilized typically with thicker sections. Optical Fractionator has the advantage of not being affected by tissue shrinkage and does not require rigorous definitions of structural boundaries (Glaser et al. 2007); however, the populations of interest must be amenable to the counting criteria set by each investigator. It should be noted, if antibodies are utilized to visualize the cell markers that define the counting criteria, the tissue section thickness might be limited due to the extent by which the antibodies can permeate the tissue in a timely manner. Optical Fractionator is used primarily in conjunction with the Optical disector probe (see Sect. 4.6.9).

4.6.6 Probes

Stereology uses geometrical shapes to map the outlined ROI and ensure random sampling of the entire ROI but at equal intervals. A probe or grid is typically a set of lines, points, cycloids, or curves that is overlaid on the image of interest to obtain quantitative information (Glaser et al. 2007). Use of different probes in different stereological methods depends on the object of interest and the quantitation desired. Common probes include counting frame, disectors, and Space Balls.

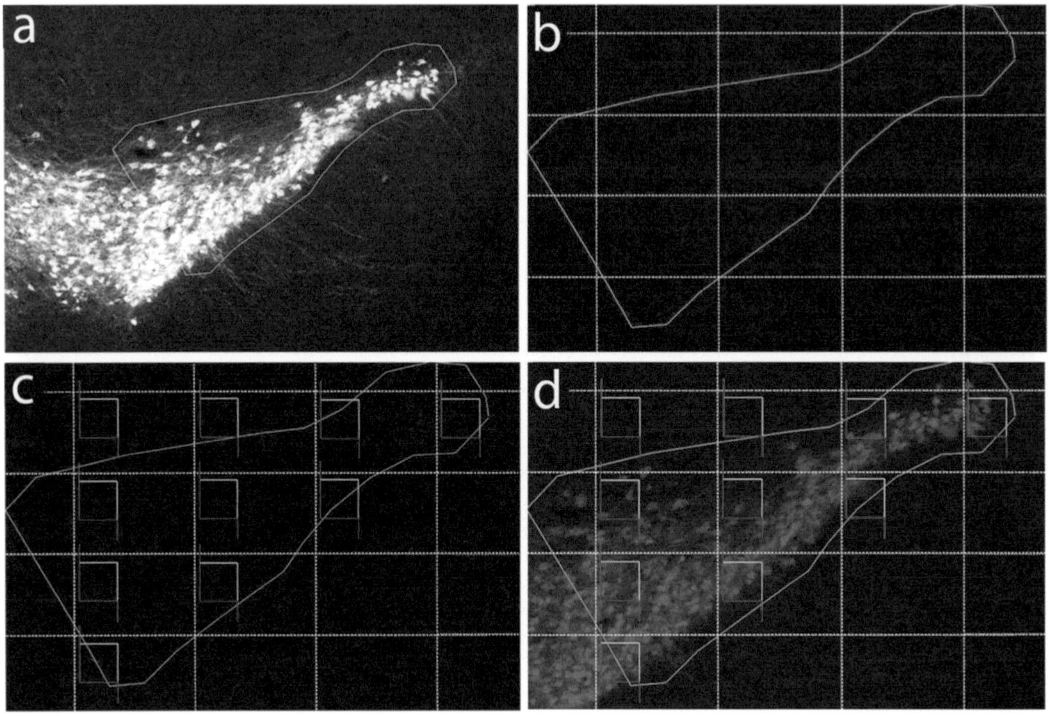

Fig. 4.1 (**a**) Microscopic image of dopaminergic neurons stained for expression of tyrosine hydroxylase (TH) in a mouse ventral midbrain. The *blue* contour or bit map encompasses the ROI or, in this case, the substantia nigra of the midbrain. (**b**) A counting grid is randomly superimposed over the outlined ROI dividing the ROI into equally spaced disectors. (**c**) One counting frame is placed within each of the disectors of the counting grid and superimposed over the ROI. (**d**) Shown is the final merged tissue section, ROI, counting grid, and counting frames

4.6.7 Counting Frame

Stereology superimposes a two-dimensional projection called a counting frame over the ROI to ensure unbiased estimates by preventing repeated counts and overestimates. Each counting frame is placed inside a partitioned segment of the randomly placed probe used in mapping the ROI. Half of the counting frame is outlined in red (exclusion lines), and half of the frame is outlined in green (inclusion lines) (Fig. 4.2a). Objects or cells that meet the appropriate target criteria set by the investigator and do not touch the frame's bounding lines are counted; those that coincide with the green line of the frame are also included in the count; whereas, those coincident with the red line are not counted. Erroneously counting objects that touch the exclusion lines may result in incorrect population estimates. Figure 4.1c shows the placement of a single counting frame inside each partitioned grid box. This placement method provides investigators with a consistently sampled tissue while at the same time limits bias and ensures randomness.

Advanced computer-based stereology utilizes a motorized microscope stage that helps reduce the chance of repeated sampling. After outlining the ROI and random placement of a grid or probe over the ROI, the computer-controlled motorized stage will automatically make very small and precise movements to random counting frames in turn. These small and precise movements are necessitated due to the image acquisition process as discussed in more detail in Sect. 4.7. The software only selects those frames that either coincide with or are within the contoured ROI. Storage of the grid placement orientation and location of each acquired image by the software allows the researcher to revisit any specific location inside an outlined ROI regardless of the

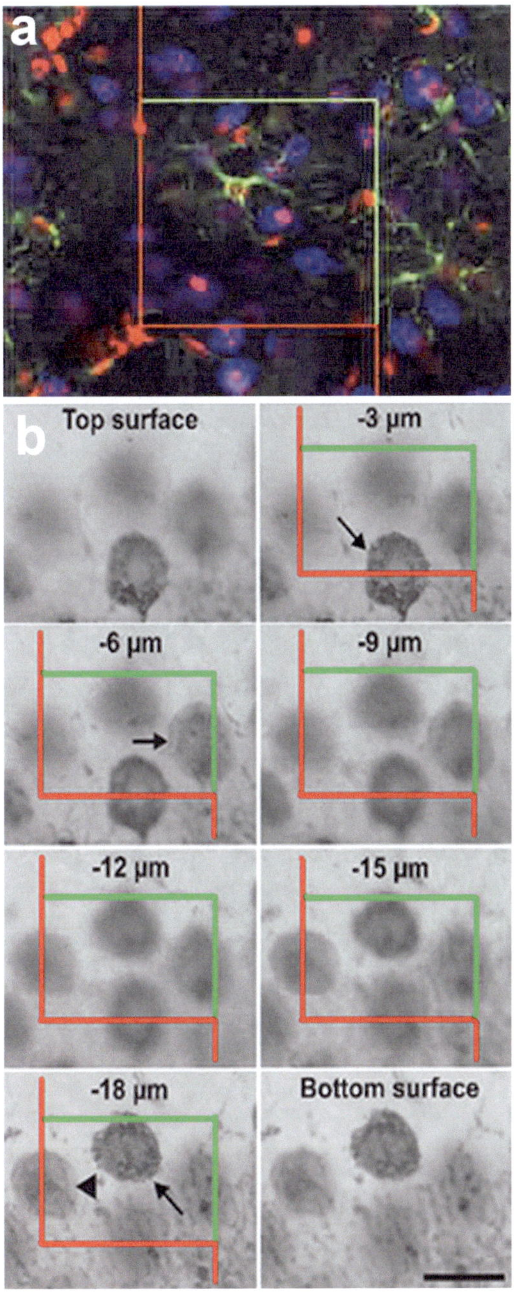

Fig. 4.2 (**a**) A counting frame is superimposed on a merged three-color channel confocal image. The counting frame serves as the visual guide for determining if a cell is included or excluded from the count. (**b**) Tissue section of cells assessed at every 3 µm plane of focus from top surface of the section (*top left*) to the bottom surface (*bottom right*) for a total of six consecutive focal plane images each with a counting frame superimposed on top. Optical Fractionator is utilized to exclude cells (−3 µm, *arrow*; −18 µm, *arrowhead*) or include cells (−6 µm and −18 µm, *arrows*) in different focal planes to ensure a cell is counted once and only once (adapted from MBF Bioscience (2011) www.stereology.info)

magnification (Glaser and Glaser 2000). After counting all cells, the software automatically calculates an estimate of the cell population, coefficient of error, and variance.

4.6.8 Disector

The disector, also referred to as the physical disector, is a stereological probe used to count objects in three dimensions. It uses two parallel, adjacent sections, one denoted as the lookup section and the other called the reference section. The disector requires the investigator to assess objects in two adjacent sections to make a choice about including or excluding the object from the count. If the point of interest is visible in the reference section, but not in the lookup section, then it is included in the count. If the point of interest is contained in both sections, then the object is not counted (Sterio 1984; MBF Bioscience 2011). The only limiting criterion for using disector is that the space between the two sections is small enough that no entire structure of interest can hide between the two sections.

4.6.9 Optical Disector

Optical disector differs from the disector probe in that it utilizes a series of focal planes in a single thick tissue section instead of adjacent tissue sections. For example, a focal plane could be set every 3 µm, and only cells containing the point of interest or cell component (e.g., nucleus) that comes into focus in a single focal plane will be counted. As a result, a cell may come into focus at 3 µm and also be visible at 6 µm, but will be counted only in the 3 µm image. This concept is demonstrated with perikarya (Fig. 4.2b). The focal plane was set to 3 µm where multiple cells are visible at each plane. The cell that comes into focus at −3 µm (marked with an arrow) is not counted for two reasons: (a) the top surface of the cell is clearly visible at −6 µm and (b) the cell touches the exclusion line of the counting frame (see Sect. 4.6.7). At −6 µm, another cell is observed, which is included in the count since it is observed in only this focal plane. At −18 µm, two cells are in focus and should be considered in the count. However, the cell (denoted by the arrow head) that touches the exclusion line is excluded from the count; whereas, the cell denoted with the arrow touches the inclusion line and is therefore included in the count.

4.6.10 Pilot Study

Prior to starting an investigation, researchers must establish the parameters and inclusion criteria to which the objects in all tissues under study must adhere. In addition, regardless of the software utilized, a pilot study should be performed to achieve an appropriate CE. With an appropriate CE, the minimum number of cells and sections necessary for an accurate population estimate can be realized. A pilot study is the best way to ensure accurate results, without spending undue time and effort performing excessive counting. Thus, the pilot study should minimally include a set of tissues similar to those to be examined in the experimental groups and are similarly stained, but not experimentally treated (i.e., negative control), and the ROIs should be drawn as close as possible to those to be examined in the experimental groups. Of note, for tissues wherein neurons define the ROI, diminished densities of neurons in experiments assessing neurodegeneration may affect the ROI in experimental tissues compared to negative controls.

The investigator will need to fine-tune the counting parameters during the pilot study including the section thickness, section increment, counting frame size, and grid size, as well as the minimum number of cells that the researcher needs to count to obtain an adequate CE. As a general starting point in the pilot study, 100–200 "counts" taken across 50 sample areas spanning six to ten sections are thought sufficient to provide a precise estimate for homogeneously distributed cell populations (West et al. 1991; Boyce et al. 2010). However, one must know the distribution of the cell population under investigation. For instance, dopaminergic neurons in the substantia nigra are bimodally distributed along the caudal-to-rostral axis. In addition, a CE target approximating 0.10 is desirable as an estimate of population accuracy. Thus, with pilot studies that count a relatively large number of cells or sections to garner low CEs, counting parameters can be adjusted to achieve optimal values for subsequent experimental studies without excess oversampling. For pilot studies, the counting frame should be big enough to contain, on average, one to six objects or cells. Notably, some researchers choose to rely on a reasonable degree of oversampling and count 500–1,000 cells per subject.

Should the results of the pilot study yield consistently higher or lower counts than the target population, or the CE not be within an optimal range, parameters should be adjusted to optimize the study's estimate. The researcher can increase the number of cells sampled and decrease the CE by decreasing the grid size, increasing the counting frame size, or decreasing the section interval (i.e., increasing the number of sections). Conversely, should the pilot study indicate that tissues have been oversampled (i.e., the CE is exceedingly low), the researcher can decrease the number of cells counted by increasing the grid size, decreasing the counting frame size, or increasing the section interval; all scenarios will ultimately increase the CE. Once parameters of sample size, grid size, and counting frame have been optimized from the pilot study results, the parameters must remain constant for every section within the subsequent experimental study to obtain accurate and valid estimations.

4.6.11 Example Pilot Study

As discussed above, before starting any novel experiment, a pilot study is most important. This will define values for the variables (counting interval, grid size, counting frame size) that will remain unchanged for the duration of the study. A major difference between a pilot study and an experimental study is that during a pilot study, it is best to oversample the sections. Oversampling can be done by manipulating any of the aforementioned variables. Here, we outline the steps taken for a pilot study:

1. Section thickness: If using sections that approximate a certain cell size, the thickness of each section will depend on the height of the cell under study. For example, a dopaminergic neuron is approximately 30 μm thick, thus defining an appropriate thickness for sectioning that is amenable for immunohistochemical processing.
2. Section collection: To begin a pilot study, start by collecting all possible serial sections that may contain the cells of interest.
3. Determine counting frame size: If one is unfamiliar with the ROI and cannot find any guidance in the literature as to the cellular density of the ROI, it is recommended to start by oversampling the ROI. This can be done by making the counting frame relatively large in comparison to the grid box. As a general rule, the counting frame should be small enough to fit into a single field of view and large enough to contain an average of one to six cells (MBF Bioscience 2011).
4. Determine grid box size: The size of the grid will depend on the desired number of cells to be counted. For example, if the target cell count is 1,000 cells over 25 sections and assuming 3 cells counted per counting frame, the grid dimensions would need to be such that there were 13–14 grid boxes per ROI on average. A target cell count of 500–1,000 should provide sufficient oversampling of the ROI.
5. Sampling: Begin by sampling a set of serial sections that contain the entire area of interest and evaluate the resulting population estimate and CE.
   (a) Next, increase the section interval to sample within the sections (e.g., every second, third, or fourth section) and reevaluate the population estimate and CE (Glaser and Glaser 2000). If the increased sampling interval results in a CE less than 0.10, you can continue to increase your sampling interval until the CE approximates 0.10.
   (b) An alternative method, called Resample Oversample, was introduced by StereoInvestigator to help determine the optimum sampling interval by graphically showing population estimates that would result if the sampling interval was increased or decreased (MBF Bioscience 2011).
6. Once the optimal sampling interval has been determined, adjust the asf (as described in Sect. 4.6.1) using the techniques mentioned in the previous section to either increase or decrease the area sample until population estimates are acceptable for the study. For example, begin with a grid size of 245 μm × 240 μm and a counting frame of 120 μm × 100 μm for a total asf of ~20 %. However, it may be more efficient to only sample a smaller percentage. By increasing the grid size to 340 μm × 300 μm and keeping the counting frame size of 120 μm × 100 μm constant, the asf is reduced to ~10 %.

4.6.12 Stereology, Immunohistochemistry, and Immunofluorescence

Computer-based stereology has led to many advances in identifying cells of interest. In addition to histochemical stains and immunohistochemical staining methods relying on oxidative reactions that produce color changes, such as horseradish peroxidase (HRP), fluorescent methods are gaining popularity for detecting cell populations of interest. These include not only immunofluorescence techniques but also tissues that are rendered endogenously fluorescent by constituent or conditional expression of genes encoding fluorescent proteins that have been genetically engineered from one organism to the other, such as transgenic animals expressing genes that encode the green fluorescent protein (GFP) from jellyfish or red fluorescent protein (RFP) from red coral. Immunofluorescence typically relies on the use of primary antibodies that are conjugated to fluorochromes or fluorochrome-conjugated secondary antibodies directed against the primary antibody. Immunofluorescence has the advantage of more readily distinguishing cells of interest from other cells within the tissue. Moreover, compared to immunohistochemistry, immunofluorescence provides greater sensitivity

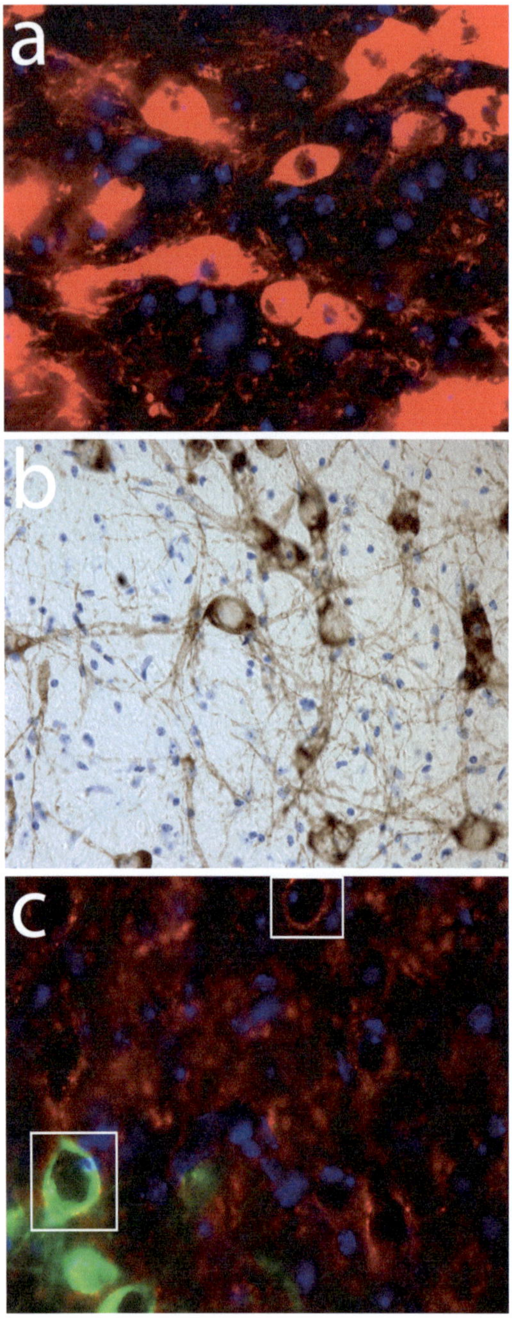

Fig. 4.3 Dopaminergic neurons in the mouse substantia nigra stained with rabbit IgG anti-tyrosine hydroxylase (TH) and visualized with (**a**) Texas Red-conjugated donkey IgG anti-rabbit IgG or (b) HRP-conjugated goat IgG anti-rabbit IgG. Nuclei were visualized with (**a**) 4′,6-diamidino-2-phenylindole (DAPI) or (**b**) thionin-based Nissl stain. (**c**) Multiple fluorescence of mouse ventral midbrain containing the substantia nigra showing dopaminergic and GABAergic neurons from a transgenic mouse expressing green fluorescent protein (GFP) under the control of the TH promoter for dopaminergic neurons and immunostained with rabbit IgG anti-glutamic acid decarboxylase-67 (GAD67) followed by Texas Red-conjugated donkey IgG anti-rabbit IgG for GABAergic neurons. Merged three-color immunofluorescence image showing two distinct populations of neurons in the SNpc, TH+dopaminergic neurons (*green*) and GAD67+ GABAergic neurons (*red, box*) with DAPI+nuclei (*blue*). (**a**–**c**) Magnifications, ×400

for detection of low expression antigens, and depending on the microscopic configuration, can provide greater dynamic ranges. In immunohistochemistry using oxidative reactions for color visualization, the amount of color change can vary between tissue sections due to the time sensitive nature of the reaction, which if developed at different times or in batches can lead to staining variations and bias the cell population estimates within a study. Use of immunofluorescence with either primary or secondary antibody-conjugated fluorochromes can provide more consistent and stable color intensities. Figure 4.3 shows visual conformation of the differences between immunohistochemical and immunofluorescent staining methods. Obvious color and intensity differences among cells are observed in tissues reacted with primary antibody and fluorochrome-conjugated secondary antibody (Fig. 4.3a) and those visualized with HRP-conjugated secondary antibody and subsequent oxidative reaction with chromophore (Fig. 4.3b).

An added advantage of immunofluorescence is that several fluorochromes can be utilized in each section, depending on the capabilities of the microscopic configuration (Fig. 4.3c). This multi-labeling technique utilizing different primary/secondary antibodies conjugated with different fluorochromes allows simultaneous detection of several antigens of interest. The benefits of this feature will be discussed later. Because each fluorochrome is excited at a different wavelength, different emission and absorption filter combinations or filter blocks are required for each color, thus requiring additional equipment to the microscope configuration. If separate colors are to be analyzed in separate channels, a high-sensitivity monochrome camera, while not essential, can be beneficial to reduce the exposure time of each tissue. Minimizing exposure time in immunofluorescence is critical to prevent excessive photo bleaching; an inherent issue associated with immunofluorescence.

When using multiple fluorochromes for labeling, the microscope camera takes multiple images of the same tissue location, each using a different filter combination or block, a process collectively known as multichannel image acquisition. Layering each of the individual images taken by the different filter sets or blocks yields a final image expressing all colors. Imaging software can aid in adjusting the intensity of one or all color channels such that the final image is color-balanced. With multiple colors (or channels), creation of a stepwise protocol for the computer to follow in sequential order must be established by the investigator to acquire each color and also must be tailored for each tissue and immunofluorescent combination. This involves defining the order of acquisition for each color and the exposure time for each color.

The use of multiple fluorochromes is beneficial for the quantitation of more than one cell type simultaneously within each section. To distinguish each cell of interest easily in multi-labeling studies, one must chose primary antibodies derived from different hosts and appropriate secondary antibodies that are extensively absorbed to eliminate cross-reactivity with other species' antibodies and retain affinity only to the primary antibody. Several vendors supply secondary antibodies that are specifically absorbed for multi-labeling studies including Jackson ImmunoResearch (West Grove, PA), Invitrogen (Carlsbad, CA), and Cell Signaling Technologies (Boston, MA). For example, multiple labeling of GABAergic and dopaminergic neurons can be achieved by labeling with rat IgG anti-glutamic acid decarboxylase-67 (GAD67) followed by Texas Red-conjugated donkey IgG anti-rat IgG for GABAergic neurons, and for dopaminergic neurons, rabbit IgG anti-tyrosine hydroxylase (TH) followed by FITC-conjugated goat IgG anti-rabbit IgG can be used. Carefully selecting and testing all antibodies prior to the start of an experiment will ensure minimal cross-reactivity and nonspecific binding.

Additionally, immunofluorescence can detect co-localization of cellular antigens, whereby a single cell, mutually expressing several antigens or markers of interest, can be detected with multiple fluorescent labels. Superimposing the single color images results in a final merged image in which new colors are visualized upon co-localization of antigens. For instance, TH and GAD67 co-expressed by one neuron in the above example would co-localize and be detected as yellow. Using three different fluorochromes in multi-labeling studies can yield images that can produce up to seven different possible colors as shown by the Venn diagram of color combinations for the three initial fluorochromes (Fig. 4.4a). Notably, co-expression of antigens, and thus co-localization of the respective fluorochromes, may not be of equal intensity depending on the antigen, extent of expression, affinity of the antibodies, and fluorochrome/antibody ratio. This can result in a wide spectrum of hues. For example, lime-green, yellow, orange, and orange-red are all possible yellow hues resulting from the co-localization of red and green fluorochromes; whereas, the Venn diagram would lead one to anticipate yellow as the only possible outcome.

In one transgenic strain of mice, expression of the GFP gene from the jellyfish *Aequorea victoria* has been genetically engineered to be driven by the TH promoter; TH being the rate limiting step for dopamine synthesis and expressed by dopaminergic neurons (Fig. 4.4b). Mutually exclusive expression of GFP by dopaminergic neurons was validated by immunofluorescence using a primary rabbit antibody specific for TH and a Texas Red fluorochrome-conjugated secondary antibody specific for rabbit IgG (Fig. 4.4c). The nuclei of all cells were visualized by 4′,6-diamidino-2-phenylindole (DAPI), a membrane permeable fluorescent stain that binds strongly to AT-rich regions in DNA and is used to stain nuclei in fluorescence microscopy (Fig. 4.4d). Counting yellow or orange cells (the result of co-localization of red and green fluorescence) that contain blue nuclei ensured that only the specific cells of interest were included in the final dopaminergic neuron population (Fig. 4.4d). No cells were detected that expressed GFP or stained with Texas Red exclusively.

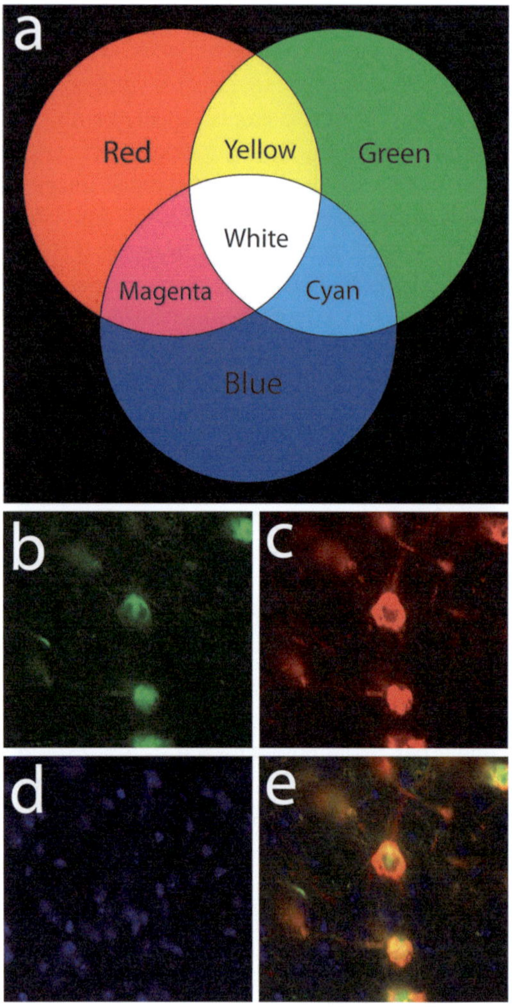

Fig. 4.4 (**a**) Venn diagram of possible color combinations using *red*, *green*, and *blue* fluorochromes in immunofluorescence microscopy. Individual color images from the same tissue section of the substantia nigra from a transgenic mouse that (**b**) expresses GFP under the control of the tyrosine hydroxylase (TH) promoter, (**c**) stained with rabbit IgG anti-TH followed by Texas Red-conjugated donkey IgG anti-rabbit IgG, and (**d**) mounted in media containing DAPI to stain nuclei. (**e**) All three-color images were merged into one image to validate GFP expression by TH+dopaminergic neurons. Magnifications, ×400

Additionally, genetically engineered mice with GFP expression by dopaminergic neurons, driven by the TH promoter, can be used in conjugation with anti-GAD67 antibody to identify multiple neuronal subtypes simultaneously in one tissue section. By first staining the tissues using a rabbit IgG anti-mouse antibody specific for GAD67 and a secondary antibody specific for rabbit IgG and conjugated with Texas Red fluorochrome, GABAergic neurons can be identified as red cells and dopaminergic neurons as green. Figure 4.3c shows both a dopaminergic (green) and GABAergic (red) neuron. The use of transgenic animals, whereby the cells of interest are endogenously fluorescent, affords an added advantage compared to staining with fluorescent antibodies. Penetration of antibodies into a tissue is often limited by the thickness of the section, thus thicker sections are generally not amenable to antibody permeability and may be precluded from stereological analysis by immunohistochemistry or immunofluorescence. However, cells of interest that express a transgenic fluorescent protein lend themselves to ready identification without the necessity of antibodies; thus, thicker sections can be utilized and evaluated with probes such as Optical Fractionator and Optical disector that are specialized for thick section analysis.

Without the aid of motorized stages interfaced with advanced computer-based stereology, immunofluorescence would not be a practical means for estimating cell populations. A pitfall of fluorochrome-based detection systems, particularly those using higher power microscope objectives needed for clear visualization of cells, is the propensity for photochemical destruction of a fluorophore, also known as photobleaching; a problematic characteristic associated with long exposure times as one would encounter with manual counting protocols. Attempting to obtain counts by stereology using immunofluorescence without the aid of an automated system that can quickly acquire images would not be practical due to intensive photobleaching of the stained section. Thus, a high-sensitivity monochrome camera minimizes photobleaching by reducing exposure times to a level of excitation and fluorescence emission that are sufficient for image acquisition without consequence of photobleaching.

4.7 Analysis of Neuronal Tissues

Preparing tissues for use by Optical Fractionator is unlike preparing tissues for other stereological studies. Fractionator and Optical Fractionator require relatively thick sections. The standard recommendation is that sections be no less than 20 μm thick after processing and mounting. Using such thick tissues and immunohistochemistry or immunofluorescence may also require specialized staining methods such as floating the tissues in antibody solutions (free-floating method) to allow antibodies to permeate the tissues from all available sides as opposed to mounted tissues that allow antibody permeation mainly from the open non-mounted face. Optimizing staining methods during the pilot study and staining all sides at the same time with master mixes of reactants will ensure equal staining of all cells and remove a potential source of experimental variation (Glaser et al. 2007).

For stereological assessment, collection of all possible serial sections without loss is also recommended. The necessity for continuous serial sections is due to the requirement for fractionator and Optical Fractionator to count sections at specific intervals. The section interval is determined in the pilot study from the number of sections needed to provide an adequate number of cells sufficient to attain a sampling CE indicative of an accurate population estimate. Choosing a smaller section interval will result in a higher cell count with a smaller CE, and a larger section interval will result in lower cell counts and larger CE.

Clear delineation of the left and right sides of the tissue is often helpful to generate accurate unbiased stereological counts. For instance, in brain tissues wherein total population numbers are required, one hemisphere may be counted for bilateral estimates, thus variations due to sections that may be "cut on the bias" will have little effect on the population estimate. Additionally, for tissues in which an effect or lesion is unilateral, delineating the ipsilaterally affected side is critical to population estimates. To establish the "handedness" of the tissue, introducing a fiducial marker may be helpful as a landmark. This can be achieved with a small notch, needle track, or hole on one side of the tissue prior to sectioning; however, markers should be introduced only in areas of the tissue that definitely will not be required for counting. In tissues such as brain and spinal cord with symmetrical ROIs within each hemisphere or axis and where one hemisphere may be targeted for counting, tissue deformation (e.g., holes, knife chatter, tissue folding) on that side of the section may preclude an accurate sampling, thus necessitating using the identical ROI of the other hemisphere for that section.

Statistical and mathematical principles developed for stereology reduce the introduction of bias into the analysis (Boyce et al. 2010). However, experimental bias or error can still occur during processing of tissues or the quantification process. Definition and strict adherence to a set of parameters and rules minimize introduction of bias during these steps (Coggeshall 2001). These parameters ensure that the entire tissue is available for sampling, each cell is counted only once, and the cells within the population have an equal chance of being sampled.

4.8 Sample Workflow for Analysis with StereoInvestigator Software

Initiation of the quantification process should begin only after tissues have been sectioned, stained, and processed in accordance with the protocols set forth by the investigator. These again should be performed in as uniform manner as possible. When starting the stereological workflow, the first step is to identify the entire ROI. Setting the microscope at low magnification allows visualization of the entire tissue containing the ROI. At this low magnification, the investigator draws a contour encompassing the entire ROI. The software allows complete control over size, shape, and placement of the contour to ensure entire sampling accessibility within the desired region. Notably, a contour that encompasses a slightly larger area than the ROI is considered best.

While the entire ROI is clearly visible at low magnification, single cells are usually difficult to resolve adequately at that magnification. Accurate quantitation requires greater resolution of the objects or cells with higher magnification. The appropriate magnification will depend on the size of the target objects, but a general rule is to use a magnification that allows clear visualization of two entire objects of interest (i.e., resolution).

For each study, the investigator inputs into the software the desired grid and counting frame dimensions that were determined from the pilot study. These parameters should remain unchanged within the same study or series of studies. The software automatically and randomly superimposes a grid using the previously defined dimensions over the outlined ROI. The investigator initiates the software-driven image acquisition. The motorized stage automatically positions the slide such that one image is randomly acquired from inside each partitioned grid box. Depending on the use of a single or multiple fluorochromes, a single image or multiple images will be taken at that grid location, the multiple images layered, and all images from each location stored in a single file for future analysis.

To quantitate the data, the file containing the acquired images is initially opened in the software and while the grid is no longer visible, the images are visible with respect to the contour. As each image is displayed in turn, the counting frame using the pre-defined dimensions is placed over the image. Each object or cell is then counted in accordance with the inclusion criteria and its position recorded by the software. After counting all images, the investigator can view the results, which include, total number of objects counted, number of objects counted in each section, estimated cell population, CE, and areas associated with the ROI, counting frame, and grid box.

4.9 Summary

While the original application of stereology was to assess the granular composition of rock, the applications have advanced greatly over the past decades in biological sciences and have become the golden standard to estimate cell populations in neurological tissues. Contention still exists concerning the usefulness and accuracy provided by stereological analysis; however, most issues involve the introduction or exclusion of investigator bias. The utilization of automated systems has reduced the potential for systematic error by removing unnecessary investigator involvement; however, as in all experimental systems, experimental bias and error still can permeate a study. Adherence to the rules and criteria of inclusion and exclusion must be maintained to achieve accurate and unbiased estimates (Coggeshall 2001). If requirements are simple and clearly defined, compliance should not be an issue; however, vague or nonexistent criteria or rules with which investigators cannot consistently comply will lead to biased and inaccurate quantitative estimates. Careful planning and fine-tuning of the stereological parameters and inclusion criteria during the pilot study and strict adherence to those principles gleaned from the pilot study will diminish bias and experimental errors. Finally, stereology has proven to be a most powerful and useful scientific tool, not only for providing accurate and unbiased quantitative estimates but also for its capacity to diminish the time and effort, and thus increase the efficiency by which those estimates are delivered.

References

Abercrombie M (1946) Estimation of nuclear population from microtome sections. Anat Rec 94:239–247

Baddeley A (2001) Is stereology 'unbiased'? Trends Neurosci 24:375–376, author reply 378–380

Boyce RW, Dorph-Petersen KA, Lyck L, Gundersen HJ (2010) Design-based stereology: introduction to basic concepts and practical approaches for estimation of cell number. Toxicol Pathol 38:1011–1025

Clarke R (1968) A comparative analysis of methods of estimating the size of cell populations from microtome sections. J R Microsc Soc 88:189–203

Coggeshall RE (2001) Commentary of the paper by Benes and Lange. Trends Neurosci 24:376–377, author reply 378–380

Geuna S (2000) Appreciating the difference between design-based and model-based sampling strategies in quantitative morphology of the nervous system. J Comp Neurol 427:333–339

Glaser JR, Glaser EM (2000) Stereology, morphometry, and mapping: the whole is greater than the sum of its parts. J Chem Neuroanat 20:115–126
Glaser J, Greene G, Hendricks S (2007) Stereology for biological research with a focus on neuroscience. MBF Bioscience, Williston, VT
Graf W (1948) The microtome as an error producing factor in quantitative histological investigations. Acta Anat 6:14–44
Gundersen HJ, Boyce RW, Nyengaard JR, Odgaard A (1993) The conneulor: unbiased estimation of connectivity using physical disectors under projection. Bone 14:217–222
Hendry IA (1976) A method to correct adequately for the change in neuronal size when estimating neuronal numbers after nerve growth factor treatment. J Neurocytol 5:337–349
Howard CV, Sandau K (1992) Measuring the surface area of a cell by the method of the spatial grid with a CSLM—a demonstration. J Microsc 165:183–188
Kroustrup JP, Gundersen HJ (2001) Estimating the number of complex particles using the ConnEulor principle. J Microsc 203:314–320
Kubinova L, Janacek J (1998) Estimating surface area by the isotropic fakir method from thick slices cut in an arbitrary direction. J Microsc 191:201–211
Mayhew TM, Gundersen HJ (1996) If you assume, you can make an ass out of u and me: a decade of the disector for stereological counting of particles in 3D space. J Anat 188(Pt 1):1–15
MBF Bioscience (2011) Stereology information for the biological sciences. http://stereology.info. Accessed 21 Oct 2011
Schmitz C, Hof PR (2005) Design-based stereology in neuroscience. Neuroscience 130:813–831
Sterio DC (1984) The unbiased estimation of number and sizes of arbitrary particles using the disector. J Microsc 134:127–136
Sundberg MD (1984) An Introduction to Stereolocial Analysis: Morphometric Techniques for Beginning Biologists, in Tested studies for laboratory teaching, S.E.A. Goldman CA, Hauta PL and Ketchum R, Editor 1984, Association for Biology Laboratory Education Memorial University of Newfoundland 51–72
West MJ, Slomianka L, Gundersen HJ (1991) Unbiased stereological estimation of the total number of neurons in the subdivisions of the rat hippocampus using the optical fractionator. Anat Rec 231: 482–497
Williams RW, Rakic P (1988) Three-dimensional counting: an accurate and direct method to estimate numbers of cells in sectioned material. J Comp Neurol 278:344–352

Part II

In Vitro Preparations

Preparation and Use of Rodent Hippocampal Slices

5

Huangui Xiong and Jianxun Xia

Abstract
Brain slicing is a laboratory technique in the study of the central nervous system. Among the many different brain slices described, the hippocampal slices prepared from rodents have been widely used in electrophysiology and basic neuroscience because of the well-known cytostructures and fiber connections in the hippocampus. To be successful in subsequent experiments, it is essential to prepare the brain slices with minimal damage. This chapter describes the methodologies used in our laboratory in the preparation of hippocampal slices from rodents.

Keywords
Brain slices • Hippocampus • Rodents

5.1 Introduction

Invented and established by McIlwain and his colleagues (McIlwain and Ochs 1952; Li and McIlwain 1957; Yamamoto and McIlwain 1966a, b), an in vitro brain slice preparation became a popular laboratory technique for multidisciplinary studies of brain function (Dingledine 1984). While a variety of brain slice preparations have been employed in neurobiological research (Alger et al. 1984), the hippocampal slice prepared from rodents is most widely used by neurobiologists to study synapses or neural circuits in controlled physiological conditions (Teyler 1980; Schrr et al. 1987). Such an in vitro brain slice preparation, taking the hippocampal slice as an example, offers certain advantages when compared to in vivo study including, but are not limited to, (1) direct visualization of the tissue structure and visual control of the placement of stimulation and recording electrodes which would not be possible in an in vivo system; (2) maintenance of structure integrity needed for studying synaptic transmission and plasticity; (3) mechanical stability for whole-cell patch recordings or sharp electrode intracellular recordings due to lack of interference of heartbeat pulses and respiration;

H. Xiong, M.D., Ph.D. (✉) • J. Xia
Department of Pharmacology and Experimental Neuroscience, University of Nebraska Medical Center, Durham Research Plaza, DRC I 8013, 45th Street at Emile Street, Omaha, NE 68198-5880, USA
e-mail: hxiong@unmc.edu; jxia@unmc.edu

H. Xiong and H.E. Gendelman (eds.), *Current Laboratory Methods in Neuroscience Research*, Springer Protocols Handbooks, DOI 10.1007/978-1-4614-8794-4_5,

(4) easy manipulation and control of extracellular environment such as ionic composition, pH, osmolality, and temperature; (5) no blood–brain barrier, which allows a direct action of drugs, neurotransmitters, and other compounds on neural cells when applied via bath perfusion; (6) maintaining structural connections that are lost in cell cultures and representing a simplified model of the neural circuit in situ; (7) easy extraction from rodent brains; and (8) the slices survive for hours in oxygenated artificial cerebrospinal fluid (ACSF). All of these advantages make rodent hippocampal slices popular in the investigations of neurobiology, especially in the studies of synaptic transmission and plasticity.

In addition to the aforementioned advantages, the brain slice preparation also has its limitations. The most obvious drawback is its lack of input and output connections that are normally present in the intact brain. Further, extraction of the brain and decapitation ischemia/hypoxia can induce neural injury. Importantly, the slicing of the brain can cause damage on both the top and bottom surfaces. Moreover, the life span of a brain slice is limited compared to an in vivo brain tissue.

As brain cells are vulnerable to ischemia/hypoxia and are subject to mechanic damage during the processes of preparation, the slice preparation process is a key determinant for making healthy slices. Although a number of protocols and methods have been described and used in brain slice preparation (Teyler 1980; Schurr et al. 1988; Deng and Xu 2012), this chapter outlines the simple procedures employed in our laboratory for the preparation of rodent hippocampal brain slices.

5.2 Objective

The objective of this chapter is to provide readers with basic methodologies and procedures needed for the preparation of rodent hippocampal brain slices with special targeting for students, laboratory technicians, and other research personnel.

5.3 Equipment

- Beakers (50, 100 mL)
- Dishes (35 mm)
- Whatman filter paper, round (ϕ 90 mm)
- Tissue forceps (Roboz, 4.5 in, RS-8160, teethless)
- Microdissecting forceps (Roboz, R-5135)
- Small animal decapitator (Stoelting, 51330)
- Plastic rodent restraint cone, disposable (Stoelting 51361)
- Stoelting tissue slicer (Model 51425 or 51415)
- Double-edge platinum chrome stainless steel blade (Walgreens)
- Bone rongeur (Roboz, curve)
- Mayo scissors (Roboz, RS-6870)
- Microdissecting scissors (Roboz, straight, sharp)
- Lab spatula, flat
- Lab spatula, with spoon
- Fine and soft paintbrushes
- Scalpel handle (#4) and blade (#22)
- Pasteur pipette (Fisherbrand, ϕ 7 mm)
- Disposable scalpel
- Disposable pipettes
- Metal tray
- Polystyrene cell culture dishes
- Ice and ice bucket
- Parafilm
- Kimwipes
- Gloves, latex
- 75 % alcohol
- Thermometer, traceable expanded-range (VWR)
- Fiber Lite (Model 190, optional)

Equipment used for animal decapitation and brain dissection is shown in Fig. 5.1a, and for hippocampal dissection and brain slice cutting is shown in Fig. 5.1b.

5.4 Methods

5.4.1 Preparation of Artificial Cerebrospinal Fluid

ACSF supplies nutrients to brain slices and removes waste products from brain slice metabolism. It is vital for brain slice survival in an in vitro

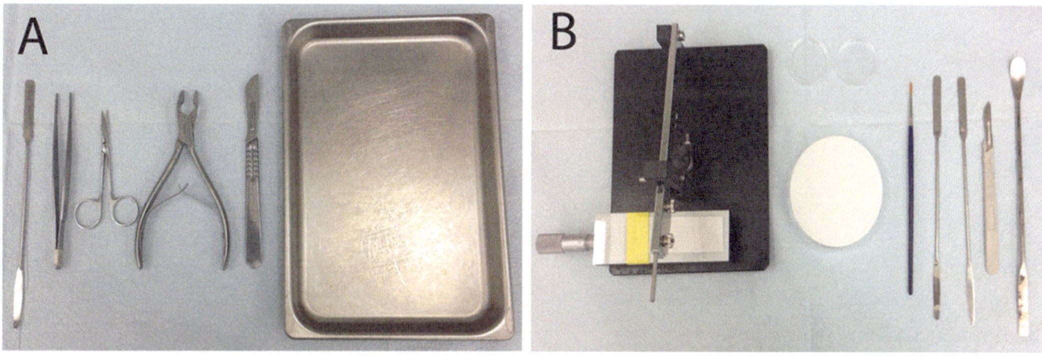

Fig. 5.1 Shows equipment for surgical decapitation and brain isolation (**a**) and hippocampal dissection and slice cutting (**b**)

environment. Chemical composition of ACSF (in mM) is as follows: NaCl 124.0, KCl 3.0, $CaCl_2$ 2.0, $MgCl_2$ 2.0, NaH_2PO_4 1.25, $NaHCO_3$ 26.0, and glucose 10.0.

The exact chemical composition of ACSF varies from laboratory to laboratory, particularly in the concentrations of $CaCl_2$, $NaHCO_3$, and glucose. ACSF is generally used for brain slice cutting, incubation, and bath perfusion during experiments. To minimize neural injury, which may occur during the slice-cutting procedures, some investigators prefer to use a Ca^{2+}-free ACSF or a "cutting" solution. Compared to the ACSF, chemical composition of a "cutting" solution may have (1) reduced NaCl concentration by substitution of sucrose for 50 % of the NaCl or removal of NaCl and replaced with equimolar concentration of sucrose, (2) reduced Ca^{2+} concentration or removal of Ca^{2+} and increase of $MgSO_4$ (or $MgCl_2$), and (3) inclusion of NMDA receptor antagonists or other reagents to protect neural cells from hypoxia/ischemia damage. The chemical compositions of the most frequently used "cutting" solution contain sucrose 230 (or equimolar choline chloride, (Theile et al. 2011), $NaHCO_3$ 26.0, KCl 3.0, NaH_2PO_4 1.25, $CaCl_2$ 0.5, $MgSO_4$ 10.0, and glucose 10.0 equilibrated with 95 % O_2, 5 % CO_2.

It is recommended to make ACSF on experimental day. We prefer to make 2 L fresh ACSF on experimental day using double-distilled water (>18MΩ). As CO_2/HCO_3^- is the interstitial fluid pH buffer system, the pH of the ACSF is adjusted by bubbling with 95 % O_2, 5 % CO_2 using an aquarium bubbler and tubing connected to a 95 % O_2, 5 % CO_2 tank. After bubbling for approximately 20–30 min, the pH of the ACSF is usually around 7.4; otherwise adjust to 7.4 with NaOH.

Alternative way to prepare ACSF is to prepare 10× ACSF stock solution in advance and dilute the 10× ACSF on experimental day. To prevent bacterial growth, it is strongly recommended to omit $NaHCO_3$ and glucose when preparing 10× ACSF and then add these to the ACSF on experimental day after dilution. The 10× stock solution should be stored at 4 °C and is good for use within 1 week.

5.4.2 Prepare Tissue Slicer (Tissue Chopper)

1. Place a blue pad or paper towels on a laboratory bench (if possible, use a lab bench close to a sink). Put the Stoelting tissue slicer (Figs. 5.1 and 5.2) on the blue pad or paper towels. Move the platform (sample stage) forward and backward by dial vernier micrometer to make sure it works properly.
2. Cut a piece of round filter paper (Whatman, ϕ 9 mm) into 2.5×8 cm and place the strip of filter paper on the platform of the tissue slicer. Secure both ends of the filter paper with tape and then moisten the filter paper strip with

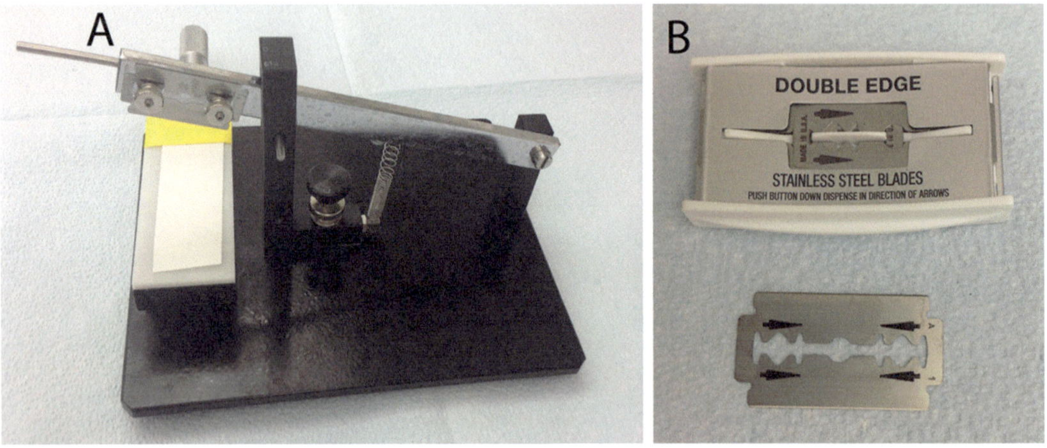

Fig. 5.2 Illustrates Stoelting tissue slicer setting (**a**) and use of double-edge stainless steel blades (**b**) for slice cutting

ACSF. Roll the filter paper from one end to the other using the handle of a fine paintbrush to drive out any air bubble(s) beneath the filter paper and to make it flat without wrinkles.

3. Mount a double-edge stainless steel blade (Fig. 5.2) to the holder in the tissue slicer. Before mounting the blade, wipe one side (one side only) of the blade with a small piece of facial tissue (in a ball shape) moistened with 75 % ethanol for the purpose of surface cleaning (to remove grease, etc). Rinse the blade with distilled water, and then mount it to the tissue slicer with the cleaned side of the blade towards the side with vernier micrometer. Check and make sure the cutting edge of the blade contacts the filter paper evenly when the cutting arm is lowered down. Then, lift the cutting arm and drop it by gravity as usually does during slice cutting. Check and make sure the upper edge of the blade does not shift beyond the top of the arm. Any shift is an indication that the blade is not tightly fastened to the holder and a further fasten is needed.

5.4.3 Surgical Harvest of Hippocampus

1. Place two beakers (50, 100 mL) in a bucket filled with ice. Pour 100 mL of Ca^{2+}-free ACSF or a "cutting" solution into the 100 mL beaker and bubble with 95 % O_2, 5 % CO_2. Check temperature with a digital thermometer, and the temperature usually drops to 0–4 °C after 20–30 min. This oxygenated ice-cold ACSF will be used in subsequent hippocampal dissection and slice cutting. Pour ~50 mL ice-cold and oxygenated ACSF into the 50 mL beaker before animal decapitation.
2. Animal anesthesia and decapitation
   (a) Anesthesia.
   In general, animals are decapitated with a decapitator without anesthesia. In this way, anesthetics are not necessary, and potential influence of anesthetics on brain function can be eliminated. However, decapitation may not be a suitable way for mice and young rats because it may cause brain shock and traumatic injury of hippocampi. Thus, a surgical head dislocation under deep anesthesia is recommended. It is generally believed that anesthesia with either ether or isoflurane has only minor interference on brain function, since these substances can be washed out rapidly in vitro. We prefer to use isoflurane to anesthetize the animals. To avoid side effects of anesthetics, inhalation of CO_2 is also an option to paralyze animals before decapitation (Windhorst and Johansson 1999).
   (b) Decapitation and isolation of hippocampus

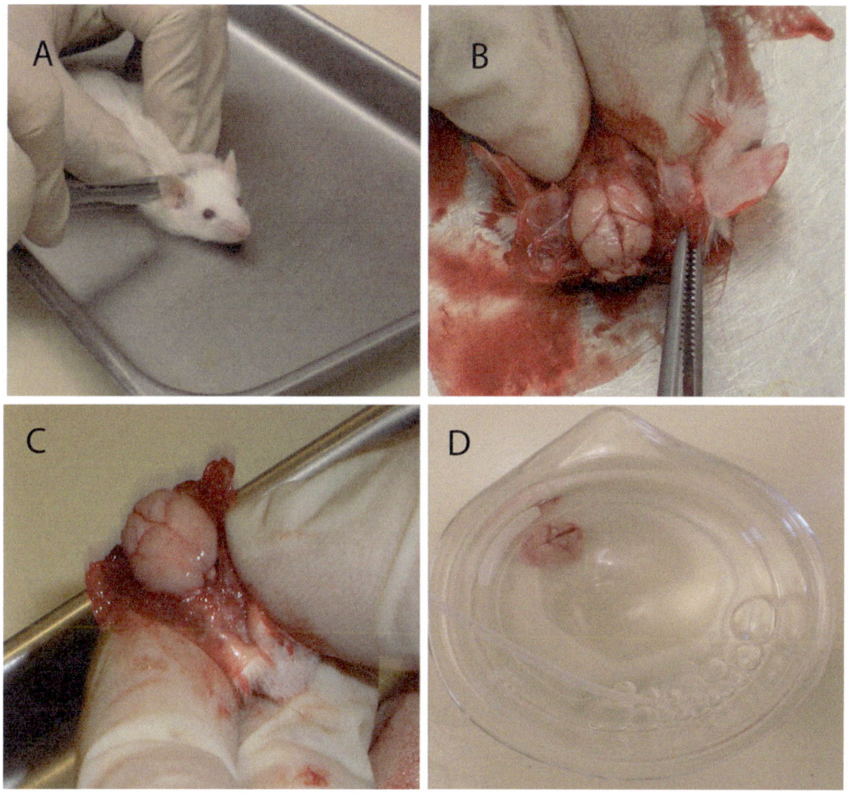

Fig. 5.3 Surgical decapitation (**a**), brain isolation (**b**, **c**), and brain cooldown in ice-cold, oxygenated ACSF (**d**)

- To decapitate, hold the deep anesthetized rodent against the metal tray. Make a midline skin incision in the head–neck junction area and then swiftly decapitate the animal using a scalpel (#22) or surgical scissors (Fig. 5.3a).
- Rinse the wound with ice-cold ACSF and cut the tissue around the foramen magnum off, then put animal head in the metal tray. Hold the animal head firmly against the tray; cut the scalp anterior–posteriorly to expose the skull. In young rats and mice, cut the skull shallowly up the midline using a microdissecting scissors and remove cranial bones with a tissue forceps (Fig. 5.3b). To expose the brain in adult rats, remove the cranial bones using a bone rongeur and a tissue forceps.
- After removal of dura mater, rinse the brain with ice-cold ACSF. Gently slide the spatula with flat end under the brain, flip the brain out of the skull into the 100 mL beaker that contains the ice-cold ACSF, and keep it in ice-cold ACSF for 1 min or so (Fig. 5.3c, d).
- To harvest hippocampi, take the brain out of ACSF using the spatula with spoon place it on the dissecting stage, a bottom-up 9 cm (in diameter) petri dish with a piece of same size filter paper at the top of the dish and moistened with ice-cold ACSF (Fig. 5.4a). After removal of the cerebellum by a surgical cut with a disposable scalpel (Fig. 5.4b), cut the rest of the brain into two halves (hemispheres) along the sagittal fissure (Fig. 4c, d). Transfer half of the brain back into the 50 mL beaker that contains ice-cold ACSF, and leave the

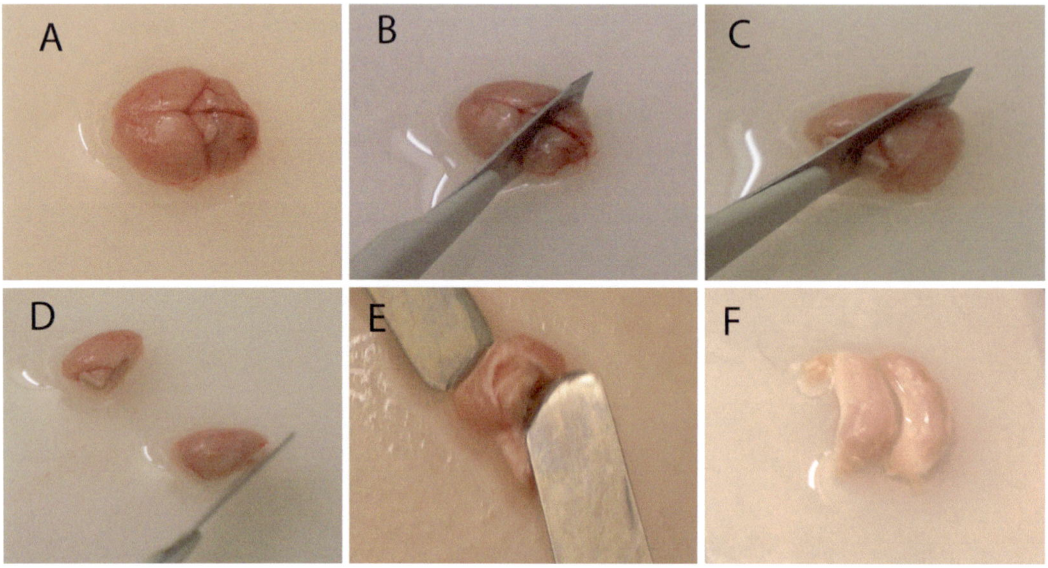

Fig. 5.4 Dissection hippocampi out from two cerebral hemispheres on a dissecting stage (a 9 cm in diameter petri dish with *bottom-up* and a same size ACSF-moistened filter paper on the *top*, (**a–e**)). (**f**) Shows the two hippocampi placed in a 35 mm petri dish (magnification is not the same as others in the figure). For the purpose of picture taking, a circular filter paper was put inside the petri dish and lowered the level of ACSF

other half on the dissecting stage for the isolation of hippocampus.

- Orient the hemisphere medial side up. Isolate the hippocampus gently using two spatulas with flat ends. Hold the hemisphere in place gently with a spatula in one hand, use the spatula in the other hand to flip the hippocampus over, and cut it free of the hemisphere (Fig. 5.4e). Transfer the hippocampus into a 35 mm petri dish filled with ice-cold ACSF. Dissect out the other hippocampus in the same way and place it into the same 35 mm petri dish (Fig. 5.4f).

5.4.4 Preparation of Hippocampal Slices

1. Bring the petri dish which contains the isolated hippocampi close to the tissue slicer, and transfer the hippocampi from the petri dish to the moistened filter paper in the platform of the tissue slicer using the spatula with spoon or using a disposable plastic transfer pipette (cut the end so that a hippocampus can fit inside) to transfer the hippocampi individually.
2. Place the two hippocampi in parallel and perpendicular to the cutting blade. Use soft and fine-tip paintbrush to move the hippocampal tissues if necessary (before use the brush should be wetted in ACSF), and orient the hippocampi so that the side with shadow concavity (also with blood vessels) is faced down (to avoid mechanical tissue damage, do not move the hippocampi with spatula, forceps, etc). Place 1–2 drops of ACSF over the hippocampi using a disposable plastic pipette to avoid the brain tissue drying on the platform. Move the platform forward (from left to the right) by dialing the vernier micrometer to a position so that one side of hippocampi is right under the cutting blade (Fig. 5.5a).
3. To cut hippocampal slices, lift the cutting arm, on which a cutting blade is attached, by its handle with one hand; advance the platform to the right with the other hand by dialing the vernier micrometer clockwise. The desired slice thickness can be achieved by dialing the

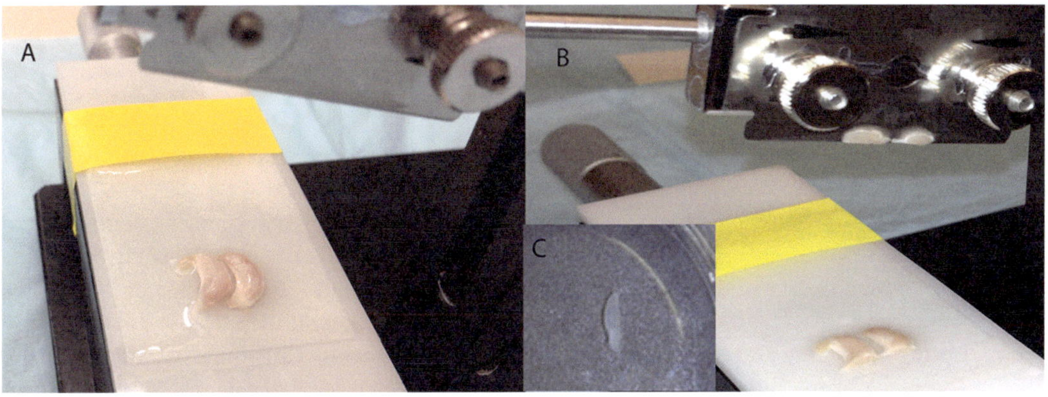

Fig. 5.5 Demonstrates the way of placing hippocampi in cutting stage (**a**) and lift-up of the slices while they adhere to the surface of the blade (**b**) and then removal of the slices from the blade using a fine and soft paintbrush (not shown in the figure). The *inset* in (**b**) shows a hippocampal slice in a petri dish filled with ACSF (**c**)

knob on the vernier micrometer, with a full revolution equaling to 500 μm. Once the wanted slice thickness is set (e.g., 400 μm), simply release the cutting arm and let it fall by gravity (and assisted by the spring) thereby slicing through the tissue at a high velocity. The blade, if correctly installed, will just cut through the filter paper without cutting into the plastic base in the platform. Lift the cutting arm gently and slowly and make sure the slices are adhered to the right-hand side of the razor blade (Fig. 5.5b); then gently remove the two slices off the surface of the blade using a wetted soft fine paintbrush. Place the slices into the other petri dish filled with ice-cold ACSF (Fig. 5.5c). Continue cutting slices using the same procedures until the whole hippocampi is sliced. Special attention should be paid to the process of lifting the cutting arm: lift it slowly and gently to avoid lifting the hippocampal tissues on the left-hand side of the blade. It will ruin the harvest of hippocampal slices if the position of hippocampal tissues is moved during the lift of the cutting arm.

5.4.5 Brain Slice Incubation

Immediately after cutting, the slices are transferred to an incubation chamber filled with oxygenated ACSF using either a fire-polished Pasteur pipette (Fisherbrand, 7 mm in diameter) or a disposable plastic pipette with cut end. The incubation reservoirs vary in size and design depending on an investigator's preference. In our laboratory, we place slices on a plastic petri dish with the bottom being replaced by a fine nylon mesh. The petri dish is fixed to a beaker filled with ACSF in which the solution is bubbled from the bottom. To minimize ischemia-induced cell injury, it is recommended to keep the time from decapitation until placement of slices in the incubation petri dish in no more than 10 min. The slices are regularly incubated in the oxygenated ACSF for at least 1 h at room temperature (22 °C) before recording.

5.5 Potential Problems and Preventive Measures

5.5.1 Common Problems

The most commonly encountered problems in the preparation of hippocampal slices, using the methods described above, occur during the slice-cutting process. One is the slices fail to "stick" to the cutting blade surface when the cutting arm is lifted up. Another is lifting the cutting arm causes the hippocampi position change, a condition that will ruin the slice cutting in most cases.

1. The possible reasons for slices failing to "stick" to the cutting blade are:
   (a) Brain tissue is not completely cut through due to improper setting of the blade or the presence of blood vessels in the concavity side.
   (b) Blade surface should not be cleaned with ethanol.
   (c) The cleaned surface of the blade faces the wrong side.
   (d) Lifting the cutting arm too fast.
   (e) Too much or too little ASCF on the cutting stage.
2. The common reasons for lifting the cutting arm causing the hippocampi position change include:
   (a) Lifting the cutting arm too fast.
   (b) The cutting stage is relatively dry (not enough ACSF in the cutting stage).
   (c) The blade surface facing the hippocampal tissue is not cleaned or not clean enough.
3. To avoid these problems, it is highly recommended to do the following before cutting slices:
   (a) One side of the double-edge platinum chrome stainless blade should be cleaned with a small piece of Kimwipe moistened with 75 % alcohol and then rinsed under a running tap water (just one side, do not clean the other side). The cleaned side of the blade is towards the tissues of hippocampi, and the uncleaned side faces the slices.
   (b) The cutting blade should be mounted appropriately on the mounting slot located on the cutting arm and making sure the cutting edge of the blade is contacting the filter paper evenly when the cutting arm is lowered.
   (c) Remove blood vessels and connective tissues from the concave surface of the hippocampi before putting the hippocampi next to each other on the stage; put hippocampi on the surface with the concave side down.
   (d) Do not put too much or too little ACSF on the cutting stage.
   (e) Lift the cutting arm slowly each time, making sure the hippocampi are not lifted up on one side and the slices "stick" to the blade surface on the other side.

5.5.2 Quick vs. Gentle

During the preparation process, hippocampal slices are subject to both ischemic/hypoxic and mechanic damages. To minimize or avoid such potential damages, it is crucial to cut the hippocampi as quickly (minimize ischemic/hypoxic damage) and as gently (minimize mechanic damage) as possible. The most vulnerable cells in the brain, CA1 neurons of the hippocampus, are widely believed to be fatally injured when ischemia/hypoxia lasts for about 8–10 min. However, our experience indicates that minimization of mechanic damage during the processes of dissecting the brain out of the cranial cavity and isolating the hippocampi from each hemisphere are more important in the preparation of healthy (less or no damage) rodent hippocampal slices than the total time spent on the slice preparation. We have obtained very good and stable whole-cell recordings from the CA1 neurons in the slices, which were prepared carefully (with less damage), though the time spent on the slice preparation was relatively longer (10–15 min) than common expected time (7–8 min). Thus, the key to preparing healthy slices is minimizing mechanic brain tissue damage, that is, do it gently.

5.6 Applications

Hippocampal slicing is the most extensively used brain tissue preparation in the study of the fundamentals of neurophysiology at cellular and synaptic levels due to simple and easy dissecting and cutting procedures and its well-known fiber connections and cytostructures. This preparation can be used to study neuronal biophysical and electrical properties, synaptic transmission, and plasticity in health and diseases. The hippocampal slice preparation is also used as a system to study brain metabolism, pharmacology of drugs in the CNS, and as a model for numerous pathological conditions.

References

Alger BE, Dhanjal SS, Dingledine R, Garthwaite J, Henderson G, King GL, Lipton P, North A, Schartzkroin PA, Sears TA, Segal M, Whittingham TS, Williams J (1984) Brain slice methods. In: Dingledine R (ed) Brain slices. Plenum Press, New York, pp 381–437

Deng P, Xu ZC (2012) Whole-cell patch-clamp recordings on spinal cord slices. Methods Mol Biol 851:65–72

Dingledine R (1984) Brain slices. Plenum Press, New York

Li CL, McIlwain H (1957) Maintenance of resting membrane potentials in slices of mammalian cerebral cortex and other tissues in vitro. J Physiol 139:178–190

McIlwain H, Ochs S (1952) Absence of electrical responses of brain slices on in vitro stimulation. Am J Physiol 171:128–133

Schrr A, Teyler TJ, Tseng MT (1987) Brain slices: fundamentals, applications and implications. S. Kaarger, New York

Schurr A, West CA, Rigor BM (1988) Lactate-supported synaptic function in the rat hippocampal slice preparation. Science 240:1326–1328

Teyler TJ (1980) Brain slice preparation: hippocampus. Brain Res Bull 5:391–403

Theile JW, Morikawa H, Gonzales RA, Morrisett RA (2011) GABAergic transmission modulates ethanol excitation of ventral tegmental area dopamine neurons. Neuroscience 172:94–103

Windhorst U, Johansson H (eds) (1999) Modern techniques in neuroscience research Berlin. Springer, New York

Yamamoto C, McIlwain H (1966a) Potentials evoked in vitro in preparations from the mammalian brain. Nature 210:1055–1056

Yamamoto C, McIlwain H (1966b) Electrical activities in thin sections from the mammalian brain maintained in chemically-defined media in vitro. J Neurochem 13:1333–1343

Single-Cell Neuronal Dissociation for Electrophysiological Studies

6

Yu-Long Li

Abstract

Since the patch-clamp technique was initially applied about 35 years ago, it has been thought to be a powerful method for studying electrical properties of ion channels and cell excitability in the varied excitable cells including peripheral and central neurons. Although we understand that many factors can affect the patch-clamp recording, the viability of the peripheral and central neurons is an essential point for the success of the patch-clamp experiments. In this chapter, I introduce (1) acute dissociation procedures of the peripheral neurons from ganglia and the central neurons from brain (such as equipment, materials, preparation of tissues, enzymatic treatment of the tissues, mechanical agitation of the tissue fragments, and preservation of the isolated cells), (2) explanatory notes for the whole procedure of the acute neuronal isolation, and (3) advantages and disadvantages of the isolated peripheral and central neurons for the electrophysiological studies.

Keywords

Action potential • Cell culture • Cell isolation • Central nervous system • Electrophysiology • Ion channel • Patch clamp • Peripheral nervous system

6.1 Introduction

Neurons are the electrically excitable cells in the central and peripheral nervous systems (CNS and PNS), which include the brain, spinal cord, peripheral ganglia, and sensory organs. Information is processed in a single neuron (the soma, dendrite, and axon) by electrical signals and is subsequently transmitted to other cells through synapses. Examining neuronal ion channels and action potentials is important for understanding the electrophysiological and pathophysiological functions of the CNS and PNS.

Since patch-clamp techniques were first reported to record ion-channel currents (Neher and Sakmann 1976), it has been considered a

Y.-L. Li, M.D., Ph.D. (✉)
Department of Emergency Medicine, University of Nebraska Medical Center, 985850 Nebraska Medical Center, Omaha, NE 68198-5850, USA
e-mail: yulongli@unmc.edu

H. Xiong and H.E. Gendelman (eds.), *Current Laboratory Methods in Neuroscience Research*, Springer Protocols Handbooks, DOI 10.1007/978-1-4614-8794-4_6,

powerful tool to investigate electrophysiological properties of cell membranes in a range of excitable cells that includes neurons. Although established neuronal cell lines (such as human and animal neuroblastomas) may provide valuable information, the extrapolation of such results to primary neuronal cells remains important. The endless debate focuses on whether such transformed cells reflect the genetic, biochemical, and physiological aspects of primary neuronal cells. Therefore, a technique for isolating primary CNS and PNS neurons from adult animals is advantageous for studying the physiological and biochemical neuronal functions. Access to a viable cell is critical for the patch-clamp experiments. This chapter includes (1) a description of the dissociation procedures for PNS and CNS neurons from the ganglia and brain (such as equipment, materials, preparation of tissues, enzymatic treatment of the tissues, mechanical agitation of the tissue fragments, and preservation of the isolated cells) and (2) discussion of the PNS and CNS neuronal electrophysiological studies.

6.2 Procedures for the Isolation of Primary Neurons

6.2.1 Isolation of PNS Neurons

Electrophysiological studies were performed on acutely isolated PNS neurons from adult tissue preparations (Akaike et al. 1978; Ishizuka et al. 1984; Kostyuk et al. 1979; Lamas et al. 1997; Li et al. 2004b, 2008; Oyelese et al. 1995; Tu et al. 2010). Although there are differences in neuronal isolation from ganglia and sensory organs, mechanical and enzymatic dissociation of the neurons is the common, fundamental procedure for the various ganglia and sensory organs. The isolation of adult rat ganglion neurons is an example of the isolation procedure (Li et al. 2008; Tu et al. 2010).

1. Equipment and materials:
   (a) Biological safety cabinet
   (b) Laboratory CO_2 incubator
   (c) Microscope
   (d) Centrifuge
   (e) Shaking water bath
   (f) Sterile filter (0.2 μm)
   (g) Tissue culture plate (96 well) or 35 mm culture dish
   (h) Centrifuge tube
   (i) Disposable Pasteur pipette (5.75 in., 2 mL, borosilicate glass)
   (j) Sterile surgical tools including scissors, retractor, forceps, scalpel, and hemostats
   (k) 75 % ethanol
   (l) Chemicals for Ringer's solution
   (m) Type II trypsin from porcine pancreas and type IV collagenase from Clostridium histolyticum
   (n) Dulbecco's modified Eagle's medium (DMEM)
   (o) F-12 Nutrient Mixture (HAM F-12)
   (p) Fetal bovine serum
   (q) Penicillin-streptomycin liquid
   (r) Insulin-transferrin-selenium-G supplement (ITS)
   (s) Sodium pentobarbital
2. Solution preparation:
   (a) Sterilized Ringer's solution (mM: NaCl 137; $NaHCO_3$ 25; KCl 3; NaH_2PO_4 1.25; $CaCl_2$ 1.2; $MgSO_4$ 1.2; glucose 10; pH 7.2) and Ca^{2+}-/Mg^{2+}-free Ringer's solution (mM: NaCl 137; $NaHCO_3$ 25; KCl 3; NaH_2PO_4 1.25; glucose 10; pH 7.2). After the solution is sterilized by passing it through a sterile filter (0.2 μm), penicillin and streptomycin are added into the solution (100 units/mL penicillin and 100 μg/mL streptomycin in final concentration). Finally, the solution is stored at −80 °C until needed.
   (b) Enzymatic solution A for the first step of digestion: 0.1 % trypsin (type II) and 0.1 % collagenase (type IV) in Ca^{2+}-/Mg^{2+}-free Ringer's solution.
   (c) Enzymatic solution B for the second step of digestion: 0.2 % collagenase (type IV) and 0.5 % bovine serum albumin in Ca^{2+}-/Mg^{2+}-free Ringer's solution.
   (d) Culture medium: DMEM 225 mL, HAM F-12 225 mL, and heat-inactivated fetal bovine serum 50 mL. After the mixed medium is sterilized with sterile filter,

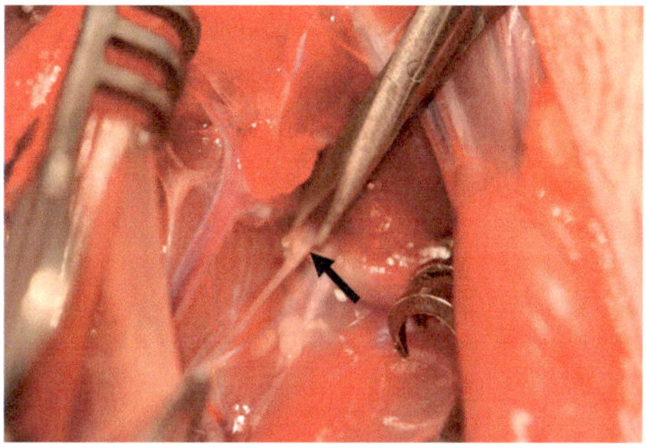

Fig. 6.1 Surgical removal of adult rat nodose ganglia. A *black arrow* indicates *left* nodose ganglion

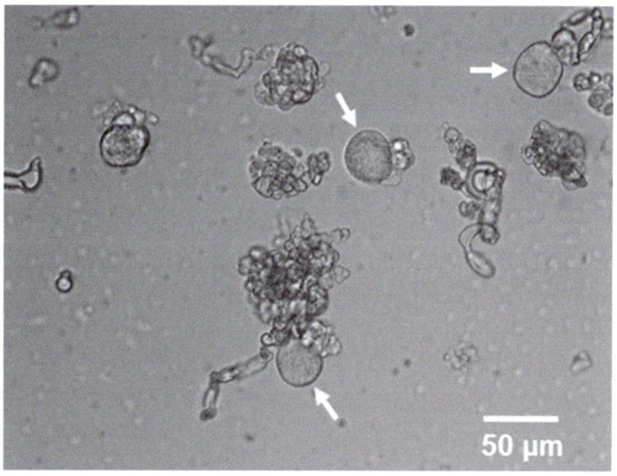

Fig. 6.2 Photomicrograph of primary nodose neurons acutely isolated from adult rats. *Arrows* indicate the spherical neurons with distinct edges and smooth surface

penicillin, streptomycin, and ITS supplement are added in the medium (100 units/mL penicillin, 100 μg/mL streptomycin, and 1 mL/100 mL ITS in final concentration). Store medium at 4 °C until ready for use.

3. Removing rat nodose ganglia.
   The rat is euthanized by intraperitoneal injection of 150 mg/kg sodium pentobarbital. Using sterile technique, a ventral midline incision is made in the neck. The nodose ganglia on both sides are clearly exposed and removed surgically from animal (Fig. 6.1). The removed tissue is placed in ice-cold Ringer's solution and then minced into approximately 30 small pieces with microscissors under the microscope.
4. Isolation of nodose ganglion neurons.
   The minced nodose ganglia are subjected to two-step enzymatic digestion. First, the nodose ganglia are incubated for 30 min at 37 °C in enzymatic solution A and then centrifuged at 150 × *g* for 5 min. After discarding the supernatant, the tissue pellet is dissolved into enzymatic solution B for 30 min incubation at 37 °C. The digested tissue fragments are gently triturated for 1 min every 10 min in the biological safety cabinet during each digestion step. The nodose ganglion neurons are obtained after centrifugation of the digested tissue at 150 × *g* for 5 min. The isolated neurons are then resuspended in 1 mL culture medium, and about 800 neurons in 0.2 mL culture medium are plated into each culture well. Finally, the isolated neurons are cultured at 37 °C in a humidified atmosphere of 95 % air-5 % CO_2 for 4–24 h before the patch-clamp or other experiments.
5. Results.
   Nodose ganglion neurons, immediately following isolation as described above, are shown in Fig. 6.2 and are roughly spherical in shape. Their diameter is about 30–50 μm. Action potential and some ion-channel currents are recorded in the healthy nodose ganglion neurons (Fig. 6.3).
6. Explanatory notes:
   (a) PNS neurons are isolated by a two-step enzymatic digestion protocol. Trypsin and collagenase are selected for the isolation of PNS neurons. However, different digestive enzyme concentrations are used in the isolation of the different PNS neurons for obtaining the isolated healthy neurons. As described above, 0.1 % trypsin and 0.1 % collagenase are used in the isolation of the nodose ganglion neurons, but 0.2 % trypsin and 0.2 % collagenase are used to isolate the glomus cells of the carotid body (Li et al. 2004b; Summers et al. 2002; Tan et al. 2007). For other PNS neurons, a range of 0.05–0.4 % enzymes is used in the preliminary experiments to obtain the optimum concentration for tissue digestion.

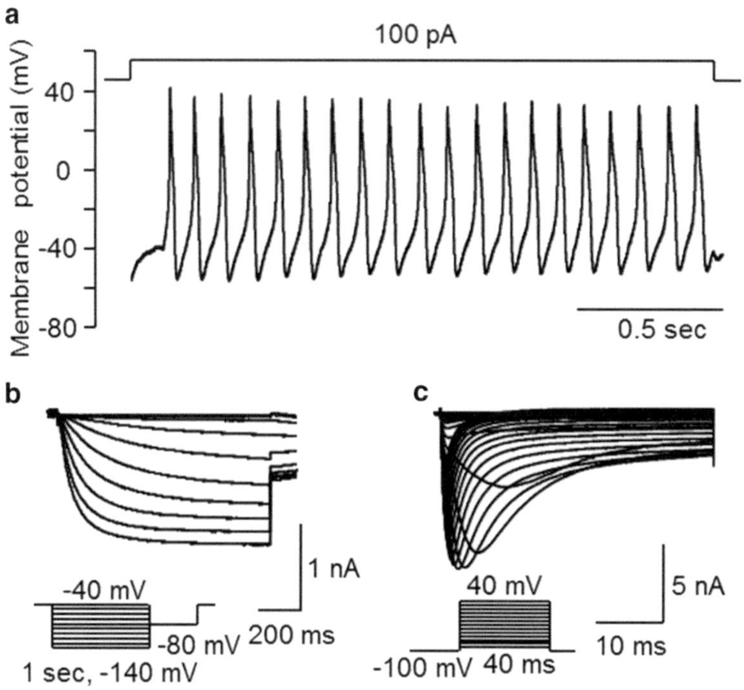

Fig. 6.3 Representative recordings for action potential and ion currents in isolated primary nodose neurons. (**a**) Action potential elicited by current stimulation (100 pA) in current-clamp model. (**b**, **c**) Hyperpolarization-activated cyclic nucleotide-gated current (**b**) and voltage-gated sodium current (**c**) induced by a voltage-step protocol in whole-cell voltage-clamp model

(b) During the process of the tissue digestion, tissue is gently triturated with a sterilized Pasteur pipette (2 mL) to avoid the creation of bubbles. Triturating too vigorously or using a Pasteur pipette with too small or sharp tip can induce cell damage.
(c) Neurons located in the ganglia can innervate a variety of tissues and organs with different cell subpopulations within the ganglia. Therefore, a tracer can be used to label organ-specific neurons in the ganglia. For example, a transported fluorescent dye, DiI (red color), is injected into the adventitia of the aortic arch where it diffuses into the nodose ganglia and labels the aortic baroreceptor neurons (Fig. 6.4) (Tu et al. 2010). Using this technique, organ-specific neurons can be selected for patch-clamp recording or other experiments.

6.2.2 Isolation of CNS Neurons

Neurons from embryonic and neonatal brains are isolated for electrophysiological studies. A critical question is whether the electrophysiological characteristics of these CNS neurons are similar to those present in adult animals. This is especially true for animal models with adult onset or induction of disease (e.g., diabetes, hypertension, and chronic heart failure). Therefore, the ability to successfully isolate neurons from adult brain can aid in investigating CNS neuron functions in physiological and pathophysiological conditions. A detailed procedure for the isolation of adult rat hypothalamic neurons is introduced here (Brewer and Torricelli 2007; Jiang et al. 2009; Kay and Krupa 2001).

1. Equipment and materials:
   (a) Tissue chopper or slicer
   (b) Glass coverslips
   (c) Poly-D-lysine

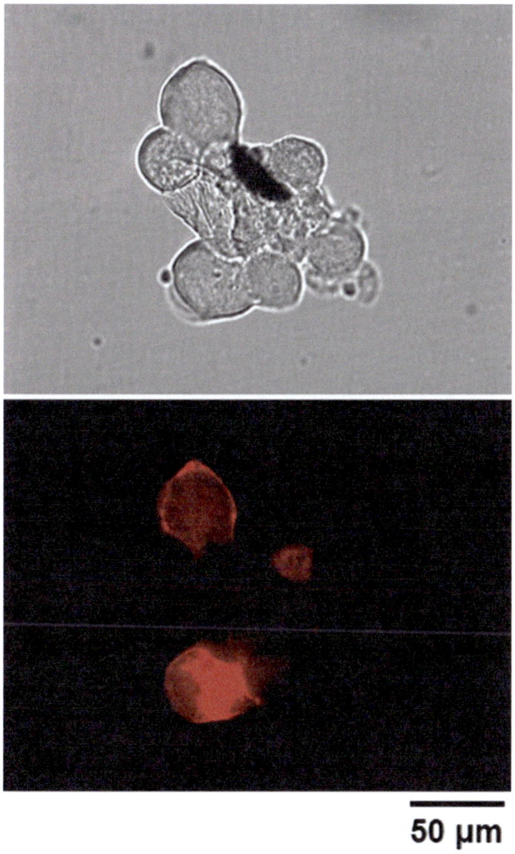

Fig. 6.4 DiI-labeled aortic baroreceptor neurons with *red color* were identified under a Leica fluorescent microscope with appropriate excitation/emission filter. Neurons in the same field were viewed in regular light (*upper panel*) and fluorescent light (*lower panel*) (reprinted from Tu et al. (2010), page 3344, with permission from John Wiley and Sons)

(d) Chemicals for piperazine-*N*, *N′*-bis[2-ethanesulfonic acid] (PIPES) solution and Hank's balanced salt solution (HBSS) with digestive enzyme
(e) Neurobasal A medium and Neurobasal A with B27 supplement
(f) Surgical tools including scissors, rongeurs, forceps, scalpel, micro spatula, and hemostats
(g) Isoflurane
(h) Other equipment and materials are the same as those in the isolation of the PNS neurons

2. Solution preparation:
   (a) PIPES solution (mM): NaCl 120, KCl 5, $CaCl_2$ 1, $MgCl_2$ 1, glucose 25, and PIPES 20 (pH 7.0 adjusted by NaOH).
   (b) HBSS with digestive enzyme (mM): NaCl 137; KCl 5.4; Na_2HPO_4 0.25; KH_2PO_4 0.44; $CaCl_2$ 1.3; $MgSO_4$ 1.0; $NaHCO_3$ 4.2, L-cysteine 5; EDTA 100; and 100 units/mL papain.
3. Removing rat brain tissues:
   (a) A rat is decapitated under deep isoflurane anesthesia. Using sterile technique, the skin over the top of the skull is dissected to expose the skull. The occipital crest (the most caudal extension of the skull visible on the dorsal surface) is broken open to make a hole by rongeurs. The scissors are inserted into spinal canal carefully to cut the calvarium along the midline. The flaps of the skull are removed with rongeurs to expose the brain.
   (b) A sterile micro spatula is used to sever the olfactory bulb and optic nerves. The brain is gently lifted out of the calvarium using the micro spatula and transferred to a dissection dish containing ice-cold PIPES solution.
4. Preparation of brain slices.
   A coronal section of the brain (~1 mm) is cut with a tissue chopper or slicer. From each coronal section containing hypothalamic tissue, a block of the hypothalamus is excised and kept in the ice-cold PIPES solution.
5. Isolation of hypothalamic neurons:
   (a) After the hypothalamic tissue is cut into small pieces, the tissue is incubated with enzyme solution (HBSS containing papain) in a shaking water bath for 30 min at 37 °C.
   (b) Tissue/cell suspensions are then centrifuged at 150×*g* for 5 min. The supernatant is removed and the pellet is resuspended in 2 mL Neurobasal A medium containing trypsin inhibitor, 100 mM bovine serum albumin, and 0.16 mM DNAse and then incubated for 15 min at 37 °C. During this period, the

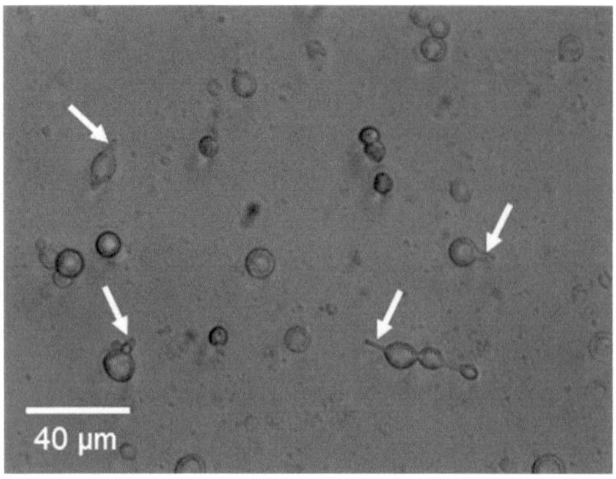

Fig. 6.5 Photomicrograph of acutely isolated rat hypothalamic neurons after 3 h in culture. *Arrows* indicate small process of the hypothalamic neurons

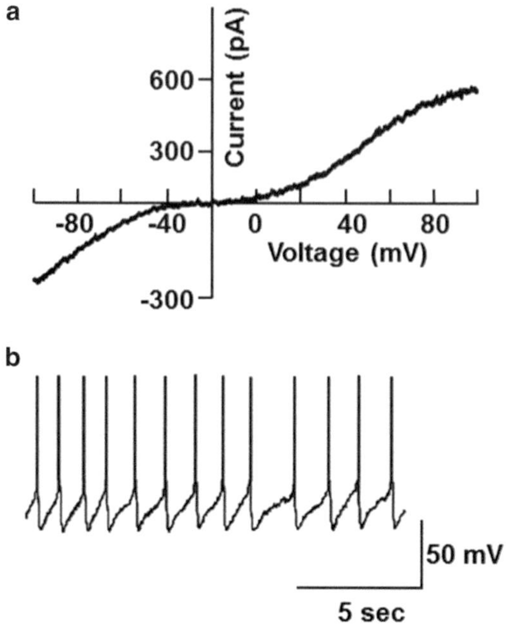

Fig. 6.6 Original recordings for ion currents and action potentials in isolated hypothalamic neurons. (**a**) *Inward* and *outward* ion currents induced by a slow voltage ramp protocol from −100 to +100 mV in whole-cell voltage-clamp model. (**b**) Spontaneous action potentials recorded in current-clamp model

tissue in the solution is mechanically dissociated by trituration (45 s every 3 min) with a fire-polished Pasteur pipette (~0.9 mm diameter, 2 mL). The hypothalamic neurons are collected by centrifugation at 150×*g* for 5 min. Dissociated neurons are suspended in Neurobasal A/B27 medium, and 1×10^4 neurons in 0.5 mL Neurobasal A/B27 medium are plated onto the sterilized glass coverslips precoated with poly-D-lysine. To obtain precoated coverslip, poly-D-lysine (50 μg/mL) is applied on a cleaned 22 mm glass coverslip. Then the coverslip is incubated for 30 min, rinsed three times with sterilized ultrapure water, and allowed to dry about 60 min. The hypothalamic neurons are incubated in Neurobasal A/B27 medium at 37 °C in a humidified atmosphere of 95 % air-5 % CO_2 until the patch-clamp experiments.

6. Results.

   Adult rat hypothalamic neurons typically present as the spherical cells with distinct edges, smooth surface, and strong aureole. A small process appears in some hypothalamic neurons (Fig. 6.5). Representative original recordings for whole inward and outward ion currents and action potential in the isolated hypothalamic neurons are shown in Fig. 6.6.

7. Explanatory notes:
   (a) The present method is suitable to isolate CNS neurons from adult animals from 2 months to 2 years of age (Brewer 1997; Li et al. 2004a). The yield of isolated viable neurons is independent of animal age (Brewer 1997). Therefore, this method provides high quality primary neurons for the study of the electrophysiological characteristics of CNS neurons from adult animals as may be desired for the study of some adult diseases (e.g., diabetes, chronic hypertension, and chronic heart failure).
   (b) The brain must be removed from animals as fast as possible and then kept on ice. Care should be taken to avoid putting excess pressure on the brain to avoid damage to the sensitive CNS tissue.
   (c) Usually experiments using brain slices are performed in solution with 100 % oxygen (Aitken et al. 1995; Lipton et al. 1995). Some studies have reported acute isolation

of CNS neurons from brain slices in oxygen-saturated solution (Huettner and Baughman 1986; Kay and Krupa 2001; Kay and Wong 1986; Kuehl-Kovarik et al. 2003). However, Brewer has found that the number of surviving CNS neurons is not significantly different between use of a solution gassed with 100 % oxygen and one with 21 % ambient oxygen (Brewer 1997). Based on these studies, the solution with regular oxygenation (21 % oxygen) is appropriate for isolating the CNS neurons from adult animals.

(d) There are many commercial digestive enzymes for the acute isolation of the CNS neurons (such as trypsin/proteinase K, protease XIV, protease XXIII, papain, collagenase, dispase, and thermolysin) (Brewer 1997; Jiang et al. 2009; Kay and Krupa 2001; Kuehl-Kovarik et al. 2003; Li et al. 2004a; Ye et al. 2006). Brewer's study demonstrated that there is no significant difference in the yield of the isolated neurons among the different enzyme treatments (Brewer 1997). For prolonged culture (e.g., 4 days), however, viability of isolated CNS neurons is higher with papain treatment compared to other enzyme treatments (Brewer 1997).

(e) To perform molecular genetics experiments (such as gene knock-in and gene knockdown), the CNS neurons must be cultured for extended time periods to ensure viability and functional quality of isolated cells. Neurons are isolated with papain and cultured in Neurobasal A/B27 medium (Brewer 1997). Neurobasal A/B27 medium not only provides nutritive maintenance of neuron viability, but also triggers cell differentiation (Brewer 1997). In addition, avoiding contamination and maintaining well-regulated incubator temperature, humidity, and gas concentration are also important factors for the long-term culture of isolated CNS neurons. After 1-week in culture, healthy neurons with dendritic processes are found on the coverslip (Fig. 6.7).

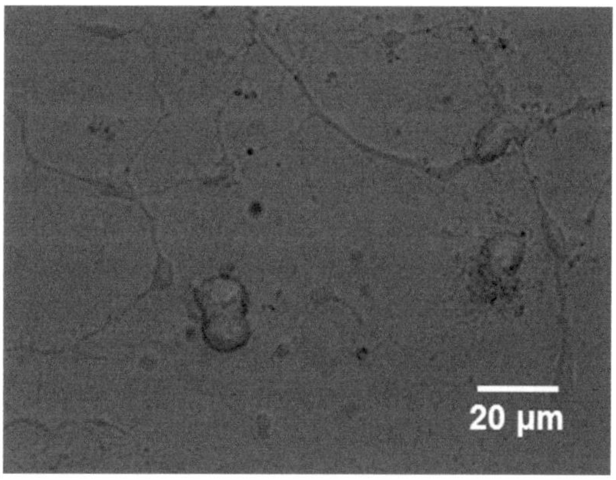

Fig. 6.7 Photomicrograph of acutely isolated rat hypothalamic neurons after 1 week in culture

(f) Exploring cellular and molecular mechanisms of cell death processes can reveal the new therapies for a variety of neurodegenerative diseases. Therefore, rapid and effective measurement of cell viability is essential for calculating cell death. Isolated primary neurons are widely used to assay cell viability. Several detailed protocols for assessment of cell viability have been introduced by some research groups (Aras et al. 2008; Giordano et al. 2011; Stoddart 2011). Regardless of varied assay protocols, preparing healthy primary neurons is a key step for assessment of cell viability.

6.3 Pros and Cons

Isolated primary neurons have several advantages for electrophysiological and biochemical studies. First, concerning in vitro tissue preparations and in vivo animal preparations, it is very difficult to discern the acting sites of a chemical compound treatment. Isolating the primary neurons removes the complicated biophysical and biochemical interactions with neighboring cells, which allows for measurement of the direct effect of the compound treatment on these neurons and avoids the ambiguity of interpreting the cellular target of the treatment compound. Second, isolated primary neurons offer the ability to explore the molecular and cellular mechanisms associated with the cell excitability and the release of transmitters or modulators. Third, using primary neurons, it is easy to precisely change the experimental environment affecting neurons, which is essential for investigating the electrophysiological properties of these cells. Fourth, in long-term culture of isolated primary neurons, time-dependent alteration of morphological characteristics (such as neuritogenesis, connectivity, and toxicity) can be continuously investigated (Brewer 1997). Embryonic and neonatal primary neurons are more appropriate for long-term studies compared to the adult primary neurons because the neurons isolated from embryos and neonatal animals can be more easily isolated and cultured for several weeks (Goslin and Banker 1989; Mattson et al. 1988).

However, some limitations in this method should be considered. First, it is very difficult to investigate the true physiological functions in isolated primary neurons because the cells are devoid of their natural environment. Therefore, care should be taken when extrapolating the data obtained. Second, although one study has shown that the electrophysiological properties of acutely isolated neurons are similar to those recorded in neurons in intact tissues (Leal-Cardoso et al. 1993), the procedure of primary neuron isolation and subsequent cell culture may alter the results obtained through severing axonal fibers and changing protein expression.

6.4 Summary

A detailed method for acute isolation of the PNS and CNS neurons is provided from adult animals. We contend that it is relatively easy to isolate primary neurons from embryos and neonatal animals and these cells may be more suitable for prolonged culture. Combined with other techniques including patch-clamp, single-cell real-time PCR, immunocytochemistry, and gene transfection, successful isolation of primary neurons from adult animals can contribute to our understanding of the physiological and biochemical functions of the brain in health and disease. Furthermore, this technique can aid in the investigation of the effects of neurotransmitters, neuromodulators, and pharmacological compounds on neurons in the physiological and pathophysiological states. However, adjustments of the procedure are necessary for the isolation of primary neurons from different ganglia or different brain regions. In addition, caution is warranted when extrapolating from in vitro findings from isolated primary neurons to possible in vivo neuronal functions and pharmacological effects.

References

Aitken PG, Breese GR, Dudek FF, Edwards F, Espanol MT, Larkman PM, Lipton P, Newman GC, Nowak TS Jr, Panizzon KL (1995) Preparative methods for brain slices: a discussion. J Neurosci Methods 59:139–149

Akaike N, Lee KS, Brown AM (1978) The calcium current of Helix neuron. J Gen Physiol 71:509–531

Aras MA, Hartnett KA, Aizenman E (2008) Assessment of cell viability in primary neuronal cultures. Curr Protoc Neurosci Chapter 7:Unit 7.18.1–7.18.15

Brewer GJ (1997) Isolation and culture of adult rat hippocampal neurons. J Neurosci Methods 71:143–155

Brewer GJ, Torricelli JR (2007) Isolation and culture of adult neurons and neurospheres. Nat Protoc 2: 1490–1498

Giordano G, Hong S, Faustman EM, Costa LG (2011) Measurements of cell death in neuronal and glial cells. Methods Mol Biol 758:171–178

Goslin K, Banker G (1989) Experimental observations on the development of polarity by hippocampal neurons in culture. J Cell Biol 108:1507–1516

Huettner JE, Baughman RW (1986) Primary culture of identified neurons from the visual cortex of postnatal rats. J Neurosci 6:3044–3060

Ishizuka S, Hattori K, Akaike N (1984) Separation of ionic currents in the somatic membrane of frog sensory neurons. J Membr Biol 78:19–28

Jiang N, Shi P, Li H, Lu S, Braseth L, Cuadra AE, Raizada MK, Sumners C (2009) Phosphate-activated glutaminase-containing neurons in the rat paraventricular nucleus express angiotensin type 1 receptors. Hypertension 54:845–851

Kay AR, Krupa DJ (2001) Acute isolation of neurons from the mature mammalian central nervous system. Curr Protoc Neurosci Chapter 6:Unit 6.5.1–6.5.7

Kay AR, Wong RK (1986) Isolation of neurons suitable for patch-clamping from adult mammalian central nervous systems. J Neurosci Methods 16:227–238

Kostyuk PG, Krishtal OA, Pidoplichko VI, Shakhovalov Y (1979) Kinetics of calcium inward current activation. J Gen Physiol 73:675–680

Kuehl-Kovarik MC, Partin KM, Magnusson KR (2003) Acute dissociation for analyses of NMDA receptor

function in cortical neurons during aging. J Neurosci Methods 129:11–17
Lamas JA, Selyanko AA, Brown DA (1997) Effects of a cognition-enhancer, linopirdine (DuP 996), on M-type potassium currents (IK(M)) and some other voltage- and ligand-gated membrane currents in rat sympathetic neurons. Eur J Neurosci 9: 605–616
Leal-Cardoso H, Koschorke GM, Taylor G, Weinreich D (1993) Electrophysiological properties and chemosensitivity of acutely isolated nodose ganglion neurons of the rabbit. J Auton Nerv Syst 45:29–39
Li XM, Li JG, Yang JM, Hu P, Li XW, Wang Y, Qin LN, Gao TM (2004a) An improved method for acute isolation of neurons from the hippocampus of adult rats suitable for patch-clamping study. Sheng Li Xue Bao 56:112–117
Li YL, Sun SY, Overholt JL, Prabhakar NR, Rozanski GJ, Zucker IH, Schultz HD (2004b) Attenuated outward potassium currents in carotid body glomus cells of heart failure rabbit: involvement of nitric oxide. J Physiol 555:219–229
Li YL, Tran TP, Muelleman R, Schultz HD (2008) Blunted excitability of aortic baroreceptor neurons in diabetic rats: involvement of hyperpolarization-activated channel. Cardiovasc Res 79:715–721
Lipton P, Aitken PG, Dudek FE, Eskessen K, Espanol MT, Ferchmin PA, Kelly JB, Kreisman NR, Landfield PW, Larkman PM et al (1995) Making the best of brain slices: comparing preparative methods. J Neurosci Methods 59:151–156
Mattson MP, Dou P, Kater SB (1988) Outgrowth-regulating actions of glutamate in isolated hippocampal pyramidal neurons. J Neurosci 8:2087–2100
Neher E, Sakmann B (1976) Single-channel currents recorded from membrane of denervated frog muscle fibres. Nature 260:799–802
Oyelese AA, Eng DL, Richerson GB, Kocsis JD (1995) Enhancement of GABAA receptor-mediated conductances induced by nerve injury in a subclass of sensory neurons. J Neurophysiol 74:673–683
Stoddart MJ (2011) Cell viability assays: introduction. Methods Mol Biol 740:1–6
Summers BA, Overholt JL, Prabhakar NR (2002) CO(2) and pH independently modulate L-type Ca(2+) current in rabbit carotid body glomus cells. J Neurophysiol 88:604–612
Tan ZY, Lu Y, Whiteis CA, Benson CJ, Chapleau MW, Abboud FM (2007) Acid-sensing ion channels contribute to transduction of extracellular acidosis in rat carotid body glomus cells. Circ Res 101: 1009–1019
Tu H, Zhang L, Tran TP, Muelleman RL, Li YL (2010) Reduced expression and activation of voltage-gated sodium channels contributes to blunted baroreflex sensitivity in heart failure rats. J Neurosci Res 88:3337–3349
Ye JH, Zhang J, Xiao C, Kong JQ (2006) Patch-clamp studies in the CNS illustrate a simple new method for obtaining viable neurons in rat brain slices: glycerol replacement of NaCl protects CNS neurons. J Neurosci Methods 158:251–259

7 Isolation and Culture of Human Neurons, Microglia, and Astrocytes

Li Wu and Santhi Gorantla

Abstract

Neuronal and glial cell cultures are important parts of neuroscience laboratories. Because the human brain is the least accessible organ, experimental studies on brain cell cultures provide valuable information. The ability to experiment on real neurons and glial cells in a controlled environment under specific conditions is essential to understand cellular and molecular mechanisms in the brain during development and in disease. Isolation and culture of neurons and glial cells is tricky but very important techniques for neuroscientists. These cells are very fragile, and primary culture of brain cells can be challenging. This chapter describes step-by-step protocols for the isolation and enrichment of different cell types from human brain tissue, and the optimal conditions needed to achieve viable and healthy neuronal and glial cell cultures. Immunocytochemical staining protocols for the brain cells are described with illustrations.

Keywords

Human • Fetal • Neurons • Astrocytes • Microglia • HIV

7.1 Introduction

Primary neuronal and glial cultures isolated from rodent brains are widely used to study brain function in a controlled in vitro environment. However, certain experimental procedures need human neuronal and glial cells to confirm the cellular mechanisms in human disease and prospective therapies. The human brain is a highly evolved complex organ. It consists of more than ten billion neurons and three times as many glial cells (Pelvig et al. 2008). Neurons carry out brain function, and glial cells provide support for neuronal development and function. The brain is the least accessible organ; therefore, understanding the mechanisms involved in neurodegenerative diseases are delayed. Knowledge acquired from experiments conducted with cells from rodent and invertebrate brains contribute to identifying the underlying mechanisms in brain development and function to some extent. Major obstacles

L. Wu • S. Gorantla, Ph.D. (✉)
Department of Pharmacology and Experimental Neuroscience, University of Nebraska Medical Center, Omaha, NE 68198-5880, USA
e-mail: sgorantla@unmc.edu

H. Xiong and H.E. Gendelman (eds.), *Current Laboratory Methods in Neuroscience Research*, Springer Protocols Handbooks, DOI 10.1007/978-1-4614-8794-4_7,

faced in the study of the human brain are (1) accessibility, (2) availability of the tissue source to isolate normal human brain cells, and (3) reliability of protocols to isolate neuronal and glial cells from human brain tissue. Fetal brain tissue is the primary source of brain cells. However, use of fetal tissue for research purposes has been very limited because of ethical considerations. Nevertheless, research with fetal tissue has made vital contributions in developing vaccines to diseases, such as rubella and polio (Hayflick et al. 1962; Plotkin et al. 1969). Therefore, it is important to develop fetal brain cell isolation and long-term culture techniques with high success due to restricted usage of fetal tissue.

Processes involved in several neurodegenerative diseases such as traumatic brain injury, Alzheimer's, Parkinson's, and Huntington's diseases, and amyotrophic lateral sclerosis (ALS) are poorly understood. Neurons in several brain regions are severely affected in these neurodegenerative disorders, with loss in synapse and compromised neural networks (Gendelman et al. 1994; Petito et al. 1999; Marcello et al. 2012; Picconi et al. 2012; Rumzan et al. 2012). The neuronal loss is associated with neuroinflammatory events involving glial cells (Ghorpade et al. 2005; Kadiu et al. 2005; Mosley et al. 2006; Agostinho et al. 2010; Amor et al. 2010). In vitro studies with human neurons and glial cells play an important role in understanding causes for these diseases and to test new therapeutic approaches. It is important to develop experimental conditions representing disease at the cellular level to understand the function of a single cell type in brain development and disease. Cells obtained from fetal brain contain many different populations including neurons, astrocytes, oligodendrocytes, and microglia. They also contain precursor cells of these different cell types at different stages of differentiation. Primary interest in fetal brain tissue research was for these precursor stem cells as they have potential in clinical applications to regenerate and replace damaged cells in diseases such as Parkinson's and other neurodegenerative diseases (Lindvall et al. 1989; Wu et al. 2002; Kallur et al. 2006; Jordan et al. 2009; Liu et al. 2009). In this chapter, we will describe in detail the isolation and culture of various types of brain cells from human fetal brain tissue. Human fetal brain tissue can be used to generate neuronal, microglia, and astrocyte cultures (Elder and Major 1988; Hassan et al. 1991; Borgmann et al. 2005; Mattson 2005; Jana et al. 2007).

7.2 Materials

7.2.1 Tissue Preparation and Dissociation

- Biosafety cabinet to carry out the cell isolation procedure in aseptic conditions.
- Human fetal brain tissue collected in Hank's balanced salt solution (HBSS), with magnesium and calcium (see Note 2).
- HBSS without (w/o) Mg^{++} and Ca^{++} (see Note 2).
- 100 mm × 20 mm petri dishes.
- 10 and 5 mL pipettes.
- 50 mL conical centrifuge tubes.
- 2.5 % Trypsin (stored at −20 °C as 5 mL aliquots; Sigma, #T4174).
- DNase I (Sigma-Aldrich, St. Louis, MO).
- Fetal bovine serum (FBS) (heat inactivated at 56 °C for 30 min).
- 100- and 70-μm cell strainers.
- Centrifuge (4 °C, 1,500 rpm or 400 × *g*).
- Hemocytometer.
- Trypan blue (Sigma, #T6146).

7.2.2 Cell Culture

- Human neuronal (HN) medium: neurobasal media with 1 % penicillin/streptomycin, 1× B27 serum-free supplement (Invitrogen), and 0.5 mM L-glutamine (Sigma, #G7513).
- Human astrocyte (HA) culture medium: DMEM/F_{12} with HEPES, 2 mM L-glutamine (Invitrogen, #11330-032), 10 % FBS, and 1 % Pen-Strep.
- Human microglial (HM) medium: DMEM with high glucose and phenol red (Gibco, #11965), 10 % FBS, 2 mM L-glutamine, 1 % Pen-Strep, and 1,000 units/mL recombinant

macrophage colony-stimulating factor (MCSF) (R&D Systems, Minneapolis, MN).
- 40-, 70-, and 100-μm cell strainers.
- Disposable plastic syringe, 3 mL.
- Trypan blue.
- Hemocytometer.
- T-75 and T-150 culture flasks (Corning).
- 50 mL conical centrifuge tubes (BD Bioscience, BD #352098).
- Trypsin–EDTA, 0.25 % (Invitrogen).
- Astrocyte freezing media: 70 % DMEM medium, 20 % bovine serum albumin (BSA), and 10 % dimethyl sulfoxide (DMSO).
- Programmed freezing container (Nalgene C1562).
- 96-, 24-, 6-well culture plates (Co-star).
- Cell culture incubator (37 °C and 5 % CO_2).
- Poly-D-lysine (PDL)-coated culture plates and coverslips (can be obtained from neuVitro, El Monte, CA, or coverslips can be coated with PDL following the protocol given below).

7.2.3 Poly-D-Lysine Coating of Tissue Culture Surfaces

- PDL solution, 50 μg/mL of sterile tissue culture grade water
- Tissue culture grade sterile water (autoclaved)
- Microscope coverslips (dimensions 5 and 8 mm to fit 96-well and 24-well plates, respectively)
- Sterile glass beakers (100 mL) and 100 mm glass petri dishes
- 70 % Ethanol
- 70 % Nitric acid
- 96-, 24-, and 6-well culture plates (Co-star)

Comment: All reagents and plastic ware should be sterile and opened only inside the biosafety cabinet.

7.2.4 Immunofluorescence Staining

- Antibodies: Rabbit anti-microtubule-associated protein-2 (MAP-2) (Chemicon, Temecula, CA; #AB5622) and mouse anti-beta III tubulin (Sigma #T3952) for neurons, rabbit anti-ionized calcium-binding adaptor molecule 1 (Iba-1) (1:500; Wako Chemicals) for microglia, and rabbit anti-glial fibrillary acidic protein (GFAP) (Dako, Carpentaria, CA; #M0761) for astrocytes. Alexa Fluor 488 and 594 conjugated anti-rabbit IgG and anti-mouse IgG as secondary antibodies (Invitrogen).
- 4′,6-diamidino-2-phenylindole (DAPI) (Invitrogen).
- Phosphate-buffered saline (PBS) pH 7.4.
- 3 % BSA in PBS.
- 4 % Paraformaldehyde (PFA) in PBS.
- Tween-20.
- Immuno mount or ProLong Gold (Invitrogen).

Comment: Reagents used for immunofluorescence staining do not have to be maintained in sterile conditions.

7.3 Method

7.3.1 Tissue Preparation and Dissociation

Human brain tissue is obtained from elective abortions in full compliance with National Institutes of Health (NIH) guidelines and the institutional ethical codes of conduct. Strict guidelines should be followed with appropriately documented informed consent and no compensation to the patient. Tissue from 12 to 15 weeks of gestational age is generally used because neuronal and glial cell viability in cultures with cells isolated from this age is much greater than younger tissues (Note 1).

(a) The brain tissue is transported in sterile ice-cold HBSS with calcium and magnesium (see Note 2). All procedures are carried out in a biological safety cabinet (BSC) under sterile conditions. Biosafety level-2 (BSL-2) practices, as per NIH guidelines, are required to be followed when handling human specimens. Laboratory coats and gloves must always be worn while working with human material, and the waste must be decontaminated with 50 % bleach. All materials are autoclaved before being discarded.

(b) The first step in isolation of brain cells is cleaning and preparing the tissue for dissociation. Care should be taken to remove meninges, blood vessels, and debris for successful isolation and culture of pure cell populations. Failure in careful removal of meninges and blood vessels may result in fibroblast contamination in neuronal and astroglial cultures.

1. Tissue cleaning and preparation steps 2–9 are common for neuronal, microglial, or astrocyte cell preparation. However, the tissue dissociation methods differ between neuronal, microglial, and astrocyte cell preparations. Enzymatic dissociation method is used to prepare cells for neuronal and microglial cultures, and mechanical dissociation of the tissue is followed for astrocyte culture. This section explains steps involved in enzymatic digestion and Sect. 3.5 describes mechanical dissociation method.
2. Transport brain tissue to laboratory in 50 mL conical tubes on ice, in HBSS with Ca^{++} and Mg^{++} (HBSS). For the cleaning and cell isolation procedure, HBSS without Ca^{++} and Mg^{++} is used (see Note 2).
3. Transfer the tissue contents from the 50 mL conical tube into a 100 mm petri dish. Thoroughly rinse the tissue at least 3–4 times by swirling the tissue in fresh HBSS every time and aspirating the liquid using a 10 mL pipette or with an aspirating pipette fitted to a vacuum line.
4. Remove tissue pieces gently by pipetting the tissue into a 10 mL pipette and place the tissue into a second 100 mm petri dish containing HBSS. Care should be taken not to dissociate the tissue too much. Remove small blood clots, meninges, and other debris with pipette and discard.
5. Transfer cleaned tissue to a 50 mL conical falcon tube. Let the tissue settle to the bottom of the tube for 5 min and using the graduations on the 50 mL conical tube as a guide, estimate the volume of tissue. Add enough HBSS to make up the volume to 2.5 times the tissue volume. Add appropriate amounts of warm trypsin and DNase I to obtain a final concentration of trypsin at 0.25 % and DNase I at 0.1 mg/mL (Note 3).
6. Incubate the tissue at 37 °C for 30 min by placing the 50 mL tube in a water bath. Swirl the tube every 5 min to facilitate complete digestion of the tissue. Alternatively, tissue in trypsin containing HBSS can be transferred into 100 mm petri dish and incubated in a 37 °C incubator.
7. Remove 50 mL tube containing the tissue from water bath. Using a pipette, transfer the tissue to new 50 mL conical tubes containing cold FBS. Final concentration of FBS should be 10 % (see Note 3).
8. Pipette tissue vigorously up and down (triturate) at least 15–20 times with a 5 mL pipette.
9. Add 20 mL of HBSS to each tube and repeat trituration with 5 mL pipette. Most of the tissue will dissociate at this time and the supernatant gets cloudy. Let the undissociated tissue settle down and transfer the cloudy supernatant into new 50 mL conical centrifuge tubes.
10. Resuspend undissociated tissue in 30 mL of fresh HBSS and repeat step 9 to collect more dissociated cells.
11. Pool all the supernatants collected and centrifuge the tubes at 1,500 rpm for 10 min at 4 °C to pellet the cells.
12. Discard the supernatant and suspend the pellet in HBSS. To remove any clumps, filter the cell suspension through a 70 μm cell strainer (Note 4).
13. Count cells using trypan blue and hemocytometer. Determine cell count per mL using the trypan blue cell enumeration protocol from Invitrogen.
14. Repeat step 11 and suspend the cell pellet in appropriate cell culture media.

7.3.2 Coating Cell Culture Surfaces

Neurons require substrates for adherence and growth on culture surfaces. Substrates made of organic polymers are used to coat culture surfaces. A wide range of coating substrates are in use, including, PDL, Poly-L-lysine, laminin, and fibronectin. Here, the protocol for coating surfaces with PDL is defined.

1. Culture plates are coated by treating with PDL overnight. Volume of 50 μg/mL solution of PDL needed to coat 6-, 12-, 24-, and 96-well plates are 1,000, 700, 400, and 200 μL well, respectively. The plates are washed 3–4 times with sterile distilled water and dried in a biosafety cabinet. They can be used immediately or wrapped with plastic wrap and stored at 4 °C until use.
2. Coverslips are also coated to culture neurons. Place the coverslips in a sterile glass beaker and cover them with 100 % ethanol for 2 h at room temperature (RT). Remove ethanol, replace it with 70 % nitric acid and leave them for overnight at RT. Rinse coverslips with water at least six times thoroughly. Coat the coverslips by submerging them in PDL for 3 h at RT or overnight at 4 °C. Rinse them with water 2–3 times and dry them in a laminar hood for 6 h to overnight. Place them in cell culture dishes and they are ready for seeding neurons.

7.3.3 Neuronal Cell Culture

Neurons should be cultured on substrate-coated surfaces for growth and formation of axonal networks. Neuronal cultures from cells isolated from human fetal tissue contain cells in various stages of development. They readily form neuritis in culture and establish synaptic connections (Kerkovich et al. 1999). Axons of human fetal neurons grow slowly compared to rodent cells, and the human neuron cultures can be maintained for relatively longer duration up to 1 month (Mattson 2005).

1. Resuspend the pellet obtained in step 14 of Sect. 3.1 in 5–7 mL HN medium.
2. Filter cell suspension through 40 μm cell strainer to avoid clumps that will lead to astrocytes and microglia in HN culture.
3. Count cells using trypan blue method.
4. Culture cells on PDL-coated surfaces in HN media with seeding density at 0.5×10^5 cells/well in 0.2 mL for 96-well plates, 1.5×10^5 cells/well in 1 mL for 24-well plates, and 5×10^5 cells/well in 1.5 mL for 12-well plates.
5. Change media every 2–3 days. Low L-glutamine in the medium will help to minimize astrocyte contamination (see Note 5).
6. Neurons are ready to use at 12–14 days after initial plating.
7. The purity and integrity of neurons in culture can be determined by immunofluorescence staining described below using antibodies to neuron specific proteins such as MAP-2 (for dendrites) and neurofilament (NF) or beta III tubulin for axons. Figure 7.1 shows immunofluorescence staining of neurons at day 14 in culture.

7.3.4 Microglial Isolation and Culture

Microglia are brain resident macrophages and are phagocytic cells scavenging damaged neurons and infectious agents. Microglial inflammation underlies the pathogenesis of several neurodegenerative diseases. Only 20 % of glial cells in the central nervous system (CNS) are microglia. Isolation and culture of this petite population of CNS cells has been a difficult process (Hassan et al. 1991). Microglial cells are isolated by the principle of preferential adhesion. The isolation method depends on the speed and preference at which different types of cells adhere to the plastic surfaces of culture dishes. Microglial cells are obtained as nonadherent populations released from a confluent monolayer of mixed glial cells over a period of time (Borgmann et al. 2005). Fetal human brain tissue has been the primary source of normal microglial cells, since human adult microglia are more difficult to isolate and culture.

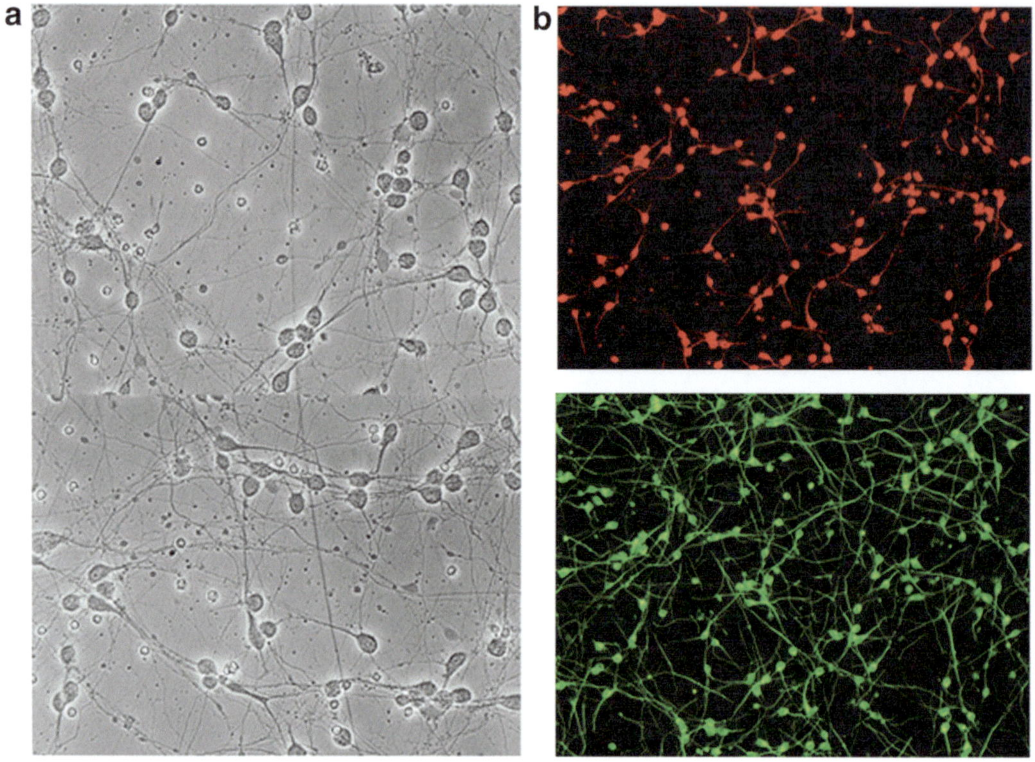

Fig. 7.1 Neuronal cells in culture. (**a**) Light microscopic images of neurons at day 14 in culture. Two fields of view were combined to show the long neurites. Astrocytes are rarely seen. (**b**) Immunofluorescence staining of neurons with antibodies to MAP-2 (*red*) and beta III tubulin (*green*). MAP-2 stains only cell bodies and dendrites. Antibodies to neurofilament or beta III tubulin stain axons. Magnification 20×

1. Pellet obtained in step 14 of Sect. 3.1 is suspended in HM media and seeded in T-75 flasks at 80×10^6 cells per flask in 25 mL.
2. Culture the cells at 37 °C (5 % CO_2). Do not disturb the cells for 7 days. Disturbing cells may hinder attachment (Note 6).
3. On day 7, gently withdraw the entire medium and replace with 25 mL of fresh HM medium.
4. On day 14, gently withdraw the supernatant from the flasks containing floating cells, which are microglia, without disturbing the adhered astrocytes. Collect the supernatants in a fresh tube (see Note 7).
5. Centrifuge at 1,500 rpm at room temperature for 10 min. Decant off as much of the supernatant as possible by pipetting with a 5 mL pipette. Allow the tube to stand for a few minutes. Some medium will still be left in the tube that would accumulate at the bottom. Resuspend the cells in the same medium that has pooled at the bottom.
6. Count the microglial cells which appear as bigger, amoeboid or round, sometimes granulated cells (see Note 8).
7. Seed the cells as a drop of concentrated cell suspension at appropriate density (1×10^6/mL). Cells are plated at 10^5 cells/well in a 96-well plate and 2×10^5 cells/well in a 24-well plate. There is no need to pre-coat culture dishes for microglial cell culture
8. Allow the cells to adhere for 2–4 h. Add fresh microglia medium to the wells to make up the volume to 200 mL/well of a 96-well plate and 500 mL/well of a 24-well plate. Add media from the edge of the well and not in the center of the well.
9. Change medium every 2–3 days by replacing half of the medium with fresh HM medium.

Cells will be ready for experiments after 7 days in culture.

10. Steps 4–9 are repeated again on day 21 and 28 to collect more microglia from the T-75 flasks and culture them for experiments (see Note 7).
11. Purity is tested by immunostaining with antibodies to Iba-1 (Fig. 7.2a) (see Note 9).

Since microglia are the brain resident macrophages, they can be infected with HIV-1. Mononuclear phagocytes (MP; perivascular macrophages and microglia) are the viral reservoirs in the brain. These immune competent cells release inflammatory factors that are responsible for neuronal injury in several neurodegenerative diseases, such as HIV-1 associated neurocognitive disorders (HAND), Alzheimer's disease, Parkinson's disease, and ALS. In HIV-1 infection, MP are viral reservoirs in the brain. They secrete products and toxins upon infection or exposure to HIV-1, which induce neuronal injury. These products include, but are not limited to, the HIV-1 proteins like gp120 and tat, proinflammatory cytokines, arachidonic acid and its metabolites, quinolinic acid, and glutamate (Gendelman et al. 1994; Kaul et al. 2005; Rostasy et al. 2005). If modulation of the inflammation induced by MP can be achieved during disease, the progression and extent of HIV-1-associated cognitive impairment could be significantly slowed (Dou et al. 2003, 2005; Kadiu et al. 2005; Filipovic and Zecevic 2008; Gorantla et al. 2008; Eggert et al. 2009). To reduce MP-induced inflammation, potential candidates for adjunctive therapy are tested on infected microglia or monocyte-derived macrophages (MDM) in vitro (Dou et al. 2003, 2005; Eggert et al. 2009; Huang et al. 2011; Lan et al. 2011, 2012; Zhao et al. 2012). Because of the restricted availability of human microglia, in the majority of studies, MDM have served as models for microglia.

Microglia can be readily infected using a macrophage tropic HIV-1 virus. They are infected at 0.01 MOI (multiplicity of infection) with HIV-1ADA for 4 h and replaced with fresh HM medium. By day 5 post infection, multinucleated giant cells can be clearly observed in culture (Fig. 7.2a). HIV-1 encephalitis is multinucleated giant cell encephalitis with the formation of microglial nodules. Microglial conditioned media (MCM) from infected microglia are collected which contain products responsible for neuronal injury. To test the extent of neuronal injury caused by the factors released by infected microglia, neuronal cultures are treated with 25 % MCM diluted in HN media for 24 h and stained for apoptotic markers (TUNEL assay) or neuronal markers, MAP-2, and beta III tubulin. MCM from HIV-1 infected microglia induce significant dendritic and axonal loss in neuronal cultures when compared to control MCM-treated neurons (Fig. 7.2b). Such in vitro experimental conditions are helpful in discovering potential therapeutic targets and adjunctive therapies to alleviate the inflammatory responses of MP and reduce resulting neuropathology.

7.3.5 Astrocyte Isolation and Culture

Astrocytes play an important role in the survival and guidance of neurons during brain development, maintenance of neuronal plasticity and homeostasis, and also formation and preservation of the blood–brain barrier. Astrocyte cultures are easy to prepare compared to neurons and microglia (Elder and Major 1988; Messam and Major 2000; Ghorpade et al. 2003; Deshpande et al. 2005). They can be isolated from any region of the brain. Astrocytes can reenter the cell cycle and can be subcultured for up to 15 passages. They can be cryopreserved and stored for longer durations.

1. Tissue is cleaned and prepared following steps 2–9 in Sect. 3.1. Last wash is done using HA media.
2. Transfer the tissue into a 100 mm cell strainer placed in a 50 mL conical centrifuge tube. Gently push the tissue by mechanical dissociation through using a sterile piston of a 3 mL plastic disposable syringe (see Note 10).
3. Centrifuge the cell suspension at 1,500 rpm for 10 min in centrifuge at 4 °C and resuspend the pellet in approximately 30 mL of HA media.

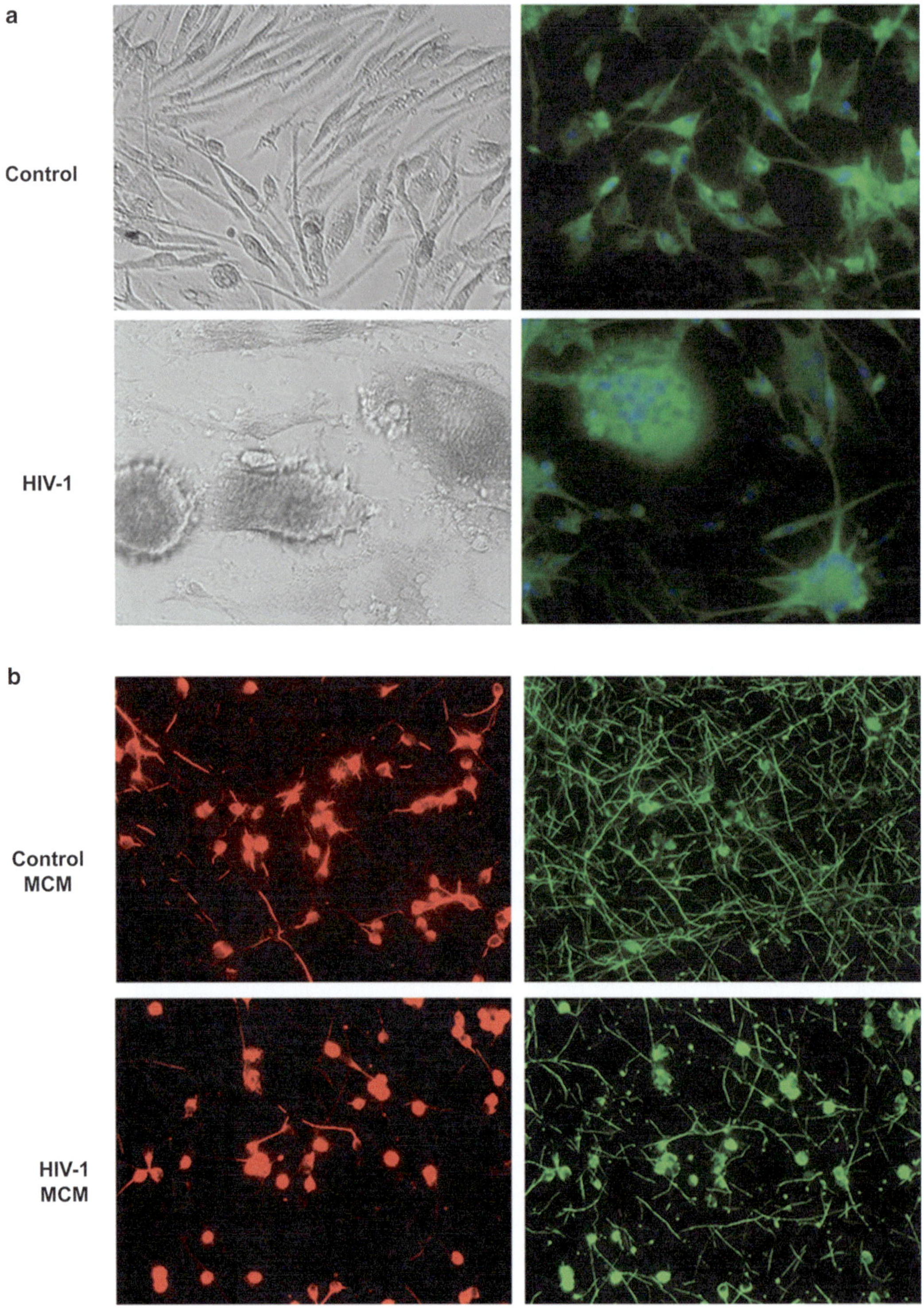

Fig. 7.2 Microglial cells in culture and method to use them to study HIV-1-induced neuropathology. (**a**) Light microscopy of microglia shows bipolar uninfected cells and fused giant cells with HIV-1 infection. Immunofluorescence staining with antibodies to Iba-1 clearly shows the multinucleated giant cells in HIV-1 infection. Magnification 20×. (**b**) Conditioned media from HIV-1-infected microglia affected neuronal survival, stunted dendrites (MAP-2, *red*), and reduced axonal network (beta III tubulin, *green*) compared to MCM from uninfected cells. Magnification 20×

4. Filter the cell suspension again through a 70 mm cell strainer and repeat step 3.
5. Count cells using trypan blue method.
6. Seed T-150 flasks at an optimum density of 50×10^6 cells per flask in 15 mL of HA media, and incubate at 37 °C with 5 % CO_2.
7. Do not disturb the cells for 7 days. Disturbing the cells will interfere with cell attachment to the bottom of the flasks.
8. After 7 days, aspirate the medium with floating cells, and add fresh HA medium (see Note 11).
9. Shake the flask with the hand to detach any loosely bound cells and rinse the culture five times with PBS (see Note 12).
10. Trypsinize the cells with 5 mL Trypsin–EDTA per flask. Incubate the flask at 37 °C for 5 min to detach astrocytes. Add 10 mL of HA medium and harvest the astrocytes (see Note 12).
11. Pellet the cells by centrifugation, resuspend in HA media, count the cells, and reseed the cells at desired density for experiments, 100,000 cells/well in a 24-well plate and 50,000 cells/well in a 96-well plate. Astrocytes can be passaged by plating them at 20 million cells/T-150 flask in 20 mL of HA medium.
12. Astrocytes can be passaged up to 15 times without significant cell death. They can be frozen and stored in liquid nitrogen for further use. For freezing, cells are suspended at 20 million cells/mL of freezing media and placed at −80 °C in a programmed freezing container for 24 h. Cells are transferred to liquid nitrogen for prolonged storage.
13. Cultured astrocytes are immunostained using antibody to GFAP to determine the purity (Fig. 7.3).

7.3.6 Immunofluorescence Staining

1. Remove the media and wash the cells on coverslips or in the cell culture dishes with PBS.
2. Fix the cells by treating them with 4 % PFA or ethanol: acetone (1:1 v/v) solution for 15 min at RT.
3. Wash the cells 2–3 times with PBS.
4. Add blocking solution (3 % BSA in PBS) and incubate for 1 h at RT.
5. Remove the blocking solution and add appropriate antibody diluted in blocking solution with 0.05 % Tween 20. Incubate for 1 h at RT.
6. Wash the cells with washing buffer (PBS with 0.05 % Tween 20) for four times.
7. Add the secondary antibody diluted in blocking buffer with 0.05 % Tween 20 and incubate for 30 min at RT. Wash the cells with washing buffer four times and the final wash with distilled water. Treat the cells with DAPI for 5 min at RT and wash the cells with distilled water. Remove all water and mount them using Immuno mount. Let the Immuno mount dry at RT. Observe the cells and capture the images using a fluorescent microscope such as Nikon Eclipse E800 (Nikon Instruments) with MagnaFire 2.0 software.

7.4 Typical Results

Typically from 10 cm^3 of tissue the yield is ~500 million total brain cells. When total cells obtained are plated in HN medium, at day 12–14 there will be around 10–20 % glial cell contamination. Low glutamine in medium will minimize the astroglial contamination. Alternatively, 5-fluorodeoxyuridine (5FU) can be used in culture to stop the growth of mitotic cells like astroglia. Neurons from fetal brain grow neurites and form refined synaptic connections in culture. Growth of these neurites is slower in human neuronal cultures compared to rodent cell cultures. For this reason, human neuronal cells can be maintained for up to a month in culture (Mattson 2005).

Astrocyte cultures may have oligodendrocyte contamination. After first passage of astrocytes, the oligodendrocytes contamination is reduced, and astrocyte cultures will be ~98 % pure. Shaking the flask before first passage will remove loosely bound neurons and microglia from astrocyte cultures. The presence of FBS in HA medium also inhibits neuronal survival and limits fibroblast growth.

Microglial cell yield is very less compared to neurons and astrocytes. From a 10 cm^3 tissue

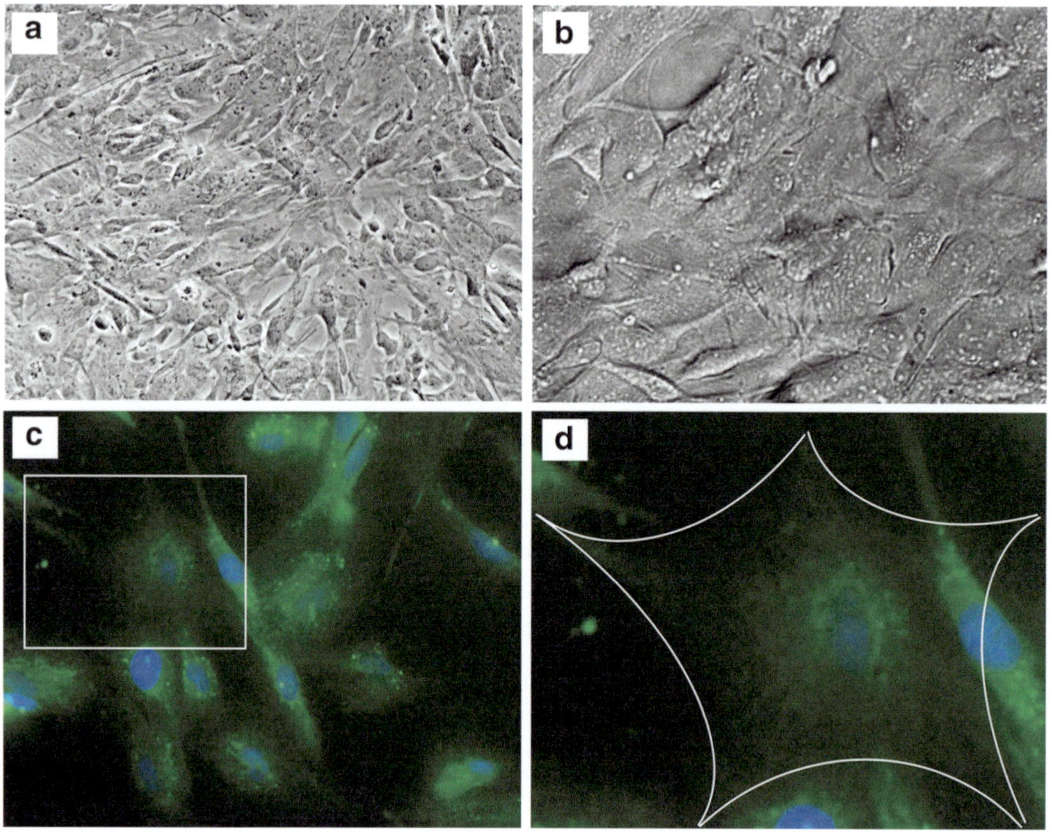

Fig. 7.3 Astrocytes in culture. (**a**, **b**) are light microscopic pictures of astrocytes grown in culture to confluency. Magnification 20× and 40×, respectively. (**c**, **d**) are immunofluorescence images of astrocytes stained using antibody to GFAP. Cells with *star-shaped* (*outlined*) morphology, typical of astrocytes, were observed. (**c**) was at magnification 20× and (**d**) at magnification 60×

sample, the number of microglial cells that are harvested ranges from 20 to 30 million. Microglial cultures are generally >99 % pure and can be infected with HIV-1 as early as day 4 after plating in culture. For β-amyloid treatments to study disease mechanisms in Alzheimer's, microglial cells can be used 2 days after plating (Lan et al. 2011). HIV-1 infection of microglial cells is different from MDM infection with respect to susceptibility, viral replication rate, and the level of inflammation (Huang et al. 2011). Microglial cells are extremely plastic and change their shape according to their activation state. In culture, resting microglia appear as bipolar cells, and activated microglia transform to pancake and amoeboid shape. Infection with HIV-1 induces syncytia and multinucleated giant cell formation.

7.5 Notes

1. Fetal brains collected from the early to middle of first trimester gestation age are suitable for neuronal cultures. Brain tissues obtained from late first trimester to early second trimester gestation age fetuses are suitable for successful yield of viable glial cells.
2. HBSS is an isotonic solution suitable for transport of tissue to the laboratory. Usually transport occurs in HBSS with calcium and magnesium in order to preserve the integrity of cell adhesion molecules. Cell adhesion molecules require calcium and magnesium cofactors to bind properly between cells. Thus, for dissociation purposes HBSS without

calcium, magnesium, and phenol red is used. The lack of calcium and magnesium in the HBSS helps to pry the endogenous calcium and magnesium cofactors from the cell adhesion molecules and allows for easy dissociation of tissue.

3. Trypsin solutions are subject to self-degradation. The stock should be divided into smaller aliquots, frozen at −20 °C, and thawed just prior to use. The tissue is incubated with trypsin to facilitate cell dissociation by hydrolyzing cell adhesion proteins. Incubating cells with trypsin for too long and at high concentrations will cause damage to cellular membranes and lead to cell death. Thus, the optimal concentration of trypsin enzyme and the length of incubation must be determined as the activity of the enzyme can vary from lot to lot. After incubation is complete, trypsin activity can be inhibited by adding serum (FBS) or by adding equimolar concentrations of trypsin inhibitor to maintain a serum-free cellular environment.
4. Filtering the cell suspension will remove clumps and any remaining meninges and blood vessels. This will eliminate the presence of contaminating cells in the culture.
5. Only 50 % of medium is removed and replaced with fresh HN medium. Half exchange of medium is recommended because complete removal of medium may result in drying of neurons, which leads to cell death. Care should be taken not to disturb the neurons while changing the medium. Removing or adding the medium along the walls of the well is recommended. Low glutamine in medium will minimize the astroglial contamination. Alternatively, 5FU or AraC can be used in culture to stop the growth of mitotic cells like astroglia.
6. Do not disturb the flasks at this time at least until day 4 after seeding, not even to observe them under the microscope. Do not change media until 7 days.
7. The days of microglia collection are variable. For example, if the media starts to turn yellow (phenol red indicating an acidic environment), you may need to collect earlier, perhaps on day 12 post-explantation. The caveat is that you need to check the cultures to make sure the microglia are not consuming the growth media and causing a pH change.
8. Count only bigger cells that appear as amoeboid or round and disregard other small and tiny cells. These tiny cells will be washed off during the microglial cell culturing process.
9. Very rarely it is possible to see a cluster of GFAP-positive astrocytes in a microglial culture. Majority of times microglial cultures are ~99 % pure.
10. Mechanical dissociation of the tissue destroys most of neurons and small undifferentiated astroglial cells survive (Ritchie et al. 1988; McCarthy et al. 1998)
11. Astrocytes form monolayer by 7–9 days. Longer period of incubation leads to clustering of astrocytes above the monolayer. This type of clustering will trap the contaminating cells and reduce the purity of astrocyte culture.
12. Shaking of the flask will detach oligodendrocytes from the monolayer of astrocytes. Because astrocytes proliferate faster, further passage of astrocyte culture will eliminate remaining oligodendrocytes.

7.6 Discussion

Cultures derived from the human fetal brain tissue contain diverse cell types and precursor cells at different stages of differentiation. The neuronal cultures from human fetal brain tissue help to understand development of neurons. Studies showed that there were differences in *N*-methyl-D-aspartate (NMDA) and non-NMDA receptors expression pattern and vulnerability to excitotoxicity in culture with respect to the expression of the receptors between human and rodent neurons in culture (Mattson et al. 1991, 1993). Such differences noticed between highly evolved human brain and laboratory animal models exhibit concerns with extrapolating discoveries in rodents to human brain development and disease. This underscores the importance of studying cellular mechanisms and disease processes using human brain cells. Brain development has primarily been studied in invertebrates or rodents. Distinct pattern of gene expression was detected in human fetal cortical areas suggesting evolutionary patterns underlying human-specific neural traits (Lambert et al. 2011). Human-specific histone methylation signatures and epigenetic regulation were also observed using human fetal cortical neurons, which could play an important role in the emergence of human-specific gene expression networks in the brain (Shulha et al. 2012).

In several neurodegenerative diseases, the neuronal loss and functional defects are associated with activation of glial cells (Luessi et al. 2012; Mosley et al. 2006; Reynolds et al. 2007; Agostinho et al. 2010). Glial cells comprise astrocytes, microglia, and oligodendrocytes. Astrocytes are the major glial cells in the brain and carry out important functions, such as structural support, ion homeostasis, regulation of neurotransmitters, and release of neurotrophins (Molofsky et al. 2012; Ransom and Ransom 2012). Astrocyte function is altered in hypoxia, hypoglycemia, and metabolic dysfunction, and they are involved in glutamate-mediated neuronal injury and cell death (Barreto et al. 2011). Microglia are the CNS resident macrophages and primary immune cells of the brain (Streit et al. 1988; Gehrmann et al. 1995; Ousman and Kubes 2012). They are extremely sensitive and react to any small pathological changes in CNS and keep the CNS environment clean by scavenging pathogens and cell debris. Activation of microglial and astrocyte cells result in the release of proinflammatory cytokines and contribute to CNS pathologies (Gendelman et al. 1994; Mosley et al. 2006; Amor et al. 2010). Conditioned media from glial cultures can be used to study the interactions among different cell types of the brain in development and disease and their effect on neuropathology. Cocultures of neurons and glial cells can also be maintained in a controlled environment to follow the cross talk between different cell types. For coculturing purposes, neurons and glial cells obtained from the same donor brain tissue are preferred. Human neurons, astrocytes, oligodendrocytes, and microglia can be isolated successfully from the same fetal tissue (Jana et al. 2007). Human microglia are susceptible to HIV-1 infection and play a major role in neuroinflammation and neuropathology in HAND (Zink et al. 1999; Xiong et al. 2000). Astrocytes, also shown to be infected by HIV-1, serve as reservoirs for HIV genome and contribute to CNS pathology in HAND (Conant et al. 1994; Tornatore et al. 1994a, b; Brack-Werner 1999). For studies on understanding the role of glial cell in HAND, it is important to use human primary glial cell cultures (Lee et al. 1993; Tornatore et al. 1994b; Messam and Major 2000; Ghorpade et al. 2003; Deshpande et al. 2005; Dhar et al. 2006). Neuronal and glial cells are used to study disease mechanisms involved in several neurodegenerative diseases, such as HAND, AD, PD, multiple sclerosis, stroke, and ALS. Either neurons or glia can be treated with amyloid-beta peptide (Aβ), which forms insoluble plaques in the brain. Aβ when added to fetal neurons induces excitotoxic cell injury and death (Mattson 2005; Jana et al. 2007). Conditioned media from microglia treated Aβ is also toxic to neurons (Lan et al. 2011, 2012). Human brain cells obtained from fetal tissue are still developing. To understand underlying pathologies, it is more relevant to study cells isolated from diseased brains obtained at postmortem. However, difficulty in culturing viable cells from adult brains compels the usage of human fetal brain cells. Brain cell cultures are also used to study the direct action of drugs of abuse that are known to affect brain development and function (Buch et al. 2012; Schultz 2011; Cisneros and Ghorpade 2012). Research on human brain cells will help to better understand disease processes observed in small animal models.

References

Agostinho P, Cunha RA, Oliveira C (2010) Neuroinflammation, oxidative stress and the pathogenesis of Alzheimer's disease. Curr Pharm Des 16: 2766–2778

Amor S, Puentes F, Baker D, van der Valk P (2010) Inflammation in neurodegenerative diseases. Immunology 129:154–169

Barreto GE, Gonzalez J, Torres Y, Morales L (2011) Astrocytic-neuronal crosstalk: implications for neuroprotection from brain injury. Neurosci Res 71: 107–113

Borgmann K, Gendelman HE, Ghorpade A (2005) Isolation and HIV-1 infection of primary human microglia from fetal and adult tissue. Methods Mol Biol 304:49–70

Brack-Werner R (1999) Astrocytes: HIV cellular reservoirs and important participants in neuropathogenesis. AIDS 13:1–22

Buch S, Yao H, Guo M, Mori T, Mathias-Costa B, Singh V, Seth P, Wang J, Su TP (2012) Cocaine and HIV-1 interplay in CNS: cellular and molecular mechanisms. Curr HIV Res 10:425–428

Cisneros IE, Ghorpade A (2012) HIV-1, methamphetamine and astrocyte glutamate regulation: combined excitotoxic implications for neuro-AIDS. Curr HIV Res 10:392–406

Conant K, Tornatore C, Atwood W, Meyers K, Traub R, Major EO (1994) In vivo and in vitro infection of the astrocyte by HIV-1. Adv Neuroimmunol 4:287–289

Deshpande M, Zheng J, Borgmann K, Persidsky R, Wu L, Schellpeper C, Ghorpade A (2005) Role of activated astrocytes in neuronal damage: potential links to HIV-1-associated dementia. Neurotox Res 7:183–192

Dhar A, Gardner J, Borgmann K, Wu L, Ghorpade A (2006) Novel role of TGF-beta in differential astrocyte-TIMP-1 regulation: implications for HIV-1-dementia and neuroinflammation. J Neurosci Res 83: 1271–1280

Dou H, Birusingh K, Faraci J, Gorantla S, Poluektova LY, Maggirwar SB, Dewhurst S, Gelbard HA, Gendelman HE (2003) Neuroprotective activities of sodium valproate in a murine model of human immunodeficiency virus-1 encephalitis. J Neurosci 23:9162–9170

Dou H, Ellison B, Bradley J, Kasiyanov A, Poluektova LY, Xiong H, Maggirwar S, Dewhurst S, Gelbard HA, Gendelman HE (2005) Neuroprotective mechanisms of lithium in murine human immunodeficiency virus-1 encephalitis. J Neurosci 25:8375–8385

Eggert D, Dash PK, Serradji N, Dong CZ, Clayette P, Heymans F, Dou H, Gorantla S, Gelbard HA, Poluektova L, Gendelman HE (2009) Development of a platelet-activating factor antagonist for HIV-1 associated neurocognitive disorders. J Neuroimmunol 213:47–59

Elder GA, Major EO (1988) Early appearance of type II astrocytes in developing human fetal brain. Brain Res 470:146–150

Filipovic R, Zecevic N (2008) Neuroprotective role of minocycline in co-cultures of human fetal neurons and microglia. Exp Neurol 211:41–51

Gehrmann J, Matsumoto Y, Kreutzberg GW (1995) Microglia: intrinsic immuneffector cell of the brain. Brain Res Brain Res Rev 20:269–287

Gendelman HE, Lipton SA, Tardieu M, Bukrinsky MI, Nottet HS (1994) The neuropathogenesis of HIV-1 infection. J Leukoc Biol 56:389–398

Ghorpade A, Holter S, Borgmann K, Persidsky R, Wu L (2003) HIV-1 and IL-1 beta regulate Fas ligand expression in human astrocytes through the NF-kappa B pathway. J Neuroimmunol 141:141–149

Ghorpade A, Persidsky Y, Swindells S, Borgmann K, Persidsky R, Holter S, Cotter R, Gendelman HE (2005) Neuroinflammatory responses from microglia recovered from HIV-1-infected and seronegative subjects. J Neuroimmunol 163:145–156

Gorantla S, Liu J, Wang T, Holguin A, Sneller HM, Dou H, Kipnis J, Poluektova L, Gendelman HE (2008) Modulation of innate immunity by copolymer-1 leads to neuroprotection in murine HIV-1 encephalitis. Glia 56:223–232

Hassan NF, Campbell DE, Rifat S, Douglas SD (1991) Isolation and characterization of human fetal brain-derived microglia in in vitro culture. Neuroscience 41:149–158

Hayflick L, Plotkin SA, Norton TW, Koprowski H (1962) Preparation of poliovirus vaccines in a human fetal diploid cell strain. Am J Hyg 75:240–258

Huang Y, Zhao L, Jia B, Wu L, Li Y, Curthoys N, Zheng JC (2011) Glutaminase dysregulation in HIV-1-infected human microglia mediates neurotoxicity: relevant to HIV-1-associated neurocognitive disorders. J Neurosci 31:15195–15204

Jana M, Jana A, Pal U, Pahan K (2007) A simplified method for isolating highly purified neurons, oligodendrocytes, astrocytes, and microglia from the same human fetal brain tissue. Neurochem Res 32:2015–2022

Jordan PM, Ojeda LD, Thonhoff JR, Gao J, Boehning D, Yu Y, Wu P (2009) Generation of spinal motor neurons from human fetal brain-derived neural stem cells: role of basic fibroblast growth factor. J Neurosci Res 87:318–332

Kadiu I, Glanzer JG, Kipnis J, Gendelman HE, Thomas MP (2005) Mononuclear phagocytes in the pathogenesis of neurodegenerative diseases. Neurotox Res 8:25–50

Kallur T, Darsalia V, Lindvall O, Kokaia Z (2006) Human fetal cortical and striatal neural stem cells generate region-specific neurons in vitro and differentiate extensively to neurons after intrastriatal transplantation in neonatal rats. J Neurosci Res 84:1630–1644

Kaul M, Zheng J, Okamoto S, Gendelman HE, Lipton SA (2005) HIV-1 infection and AIDS: consequences for the central nervous system. Cell Death Differ 12(Suppl 1):878–892

Kerkovich DM, Sapp D, Weidenheim K, Brosnan CF, Pfeiffer SE, Yeh HH, Busciglio J (1999) Fetal human cortical neurons grown in culture: morphological differentiation, biochemical correlates and development of electrical activity. Int J Dev Neurosci 17:347–356

Lambert N, Lambot MA, Bilheu A, Albert V, Englert Y, Libert F, Noel JC, Sotiriou C, Holloway AK, Pollard KS, Detours V, Vanderhaeghen P (2011) Genes expressed in specific areas of the human fetal cerebral cortex display distinct patterns of evolution. PLoS One 6:e17753

Lan X, Xu J, Kiyota T, Peng H, Zheng JC, Ikezu T (2011) HIV-1 reduces Abeta-degrading enzymatic activities in primary human mononuclear phagocytes. J Immunol 186:6925–6932

Lan X, Kiyota T, Hanamsagar R, Huang Y, Andrews S, Peng H, Zheng JC, Swindells S, Carlson GA, Ikezu T (2012) The effect of HIV protease inhibitors on amyloid-beta peptide degradation and synthesis in human cells and Alzheimer's disease animal model. J Neuroimmune Pharmacol 7:412–423

Lee SC, Hatch WC, Liu W, Brosnan CF, Dickson DW (1993) Productive infection of human fetal microglia in vitro by HIV-1. Ann N Y Acad Sci 693:314–316

Lindvall O, Rehncrona S, Brundin P, Gustavii B, Astedt B, Widner H, Lindholm T, Bjorklund A, Leenders KL, Rothwell JC, Frackowiak R, Marsden D, Johnels B, Steg G, Freedman R, Hoffer BJ, Seiger A, Bygdeman M, Stromberg I, Olson L (1989) Human fetal dopamine neurons grafted into the striatum in two patients with severe Parkinson's disease. A detailed account of methodology and a 6-month follow-up. Arch Neurol 46:615–631

Liu S, Tian Z, Yin F, Zhao Q, Fan M (2009) Generation of dopaminergic neurons from human fetal mesencephalic progenitors after co-culture with striatal-conditioned media and exposure to lowered oxygen. Brain Res Bull 80:62–68

Luessi F, Siffrin V, Zipp F (2012) Neurodegeneration in multiple sclerosis: novel treatment strategies. Expert Rev Neurother 12:1061–1076; quiz 1077

Marcello E, Epis R, Saraceno C, Di Luca M (2012) Synaptic dysfunction in Alzheimer's disease. Adv Exp Med Biol 970:573–601

Mattson MP (2005) Human fetal brain cell culture. Methods Mol Med 107:163–171

Mattson MP, Rychlik B, You JS, Sisken JE (1991) Sensitivity of cultured human embryonic cerebral cortical neurons to excitatory amino acid-induced calcium influx and neurotoxicity. Brain Res 542:97–106

Mattson MP, Kumar KN, Wang H, Cheng B, Michaelis EK (1993) Basic FGF regulates the expression of a functional 71 kDa NMDA receptor protein that mediates calcium influx and neurotoxicity in hippocampal neurons. J Neurosci 13:4575–4588

McCarthy M, Auger D, He J, Wood C (1998) Cytomegalovirus and human herpesvirus-6 trans-activate the HIV-1 long terminal repeat via multiple response regions in human fetal astrocytes. J Neurovirol 4:495–511

Messam CA, Major EO (2000) Stages of restricted HIV-1 infection in astrocyte cultures derived from human fetal brain tissue. J Neurovirol 6(Suppl 1):S90–94

Molofsky AV, Krencik R, Ullian EM, Tsai HH, Deneen B, Richardson WD, Barres BA, Rowitch DH (2012) Astrocytes and disease: a neurodevelopmental perspective. Genes Dev 26:891–907

Mosley RL, Benner EJ, Kadiu I, Thomas M, Boska MD, Hasan K, Laurie C, Gendelman HE (2006) Neuroinflammation, oxidative stress and the pathogenesis of Parkinson's disease. Clin Neurosci Res 6: 261–281

Ousman SS, Kubes P (2012) Immune surveillance in the central nervous system. Nat Neurosci 15:1096–1101

Pelvig DP, Pakkenberg H, Stark AK, Pakkenberg B (2008) Neocortical glial cell numbers in human brains. Neurobiol Aging 29:1754–1762

Petito CK, Kerza-Kwiatecki AP, Gendelman HE, McCarthy M, Nath A, Podack ER, Shapshak P, Wiley CA (1999) Review: neuronal injury in HIV infection. J Neurovirol 5:327–341

Picconi B, Piccoli G, Calabresi P (2012) Synaptic dysfunction in Parkinson's disease. Adv Exp Med Biol 970:553–572

Plotkin SA, Farquhar JD, Katz M, Buser F (1969) Attenuation of RA 27–3 rubella virus in WI-38 human diploid cells. Am J Dis Child 118:178–185

Ransom BR, Ransom CB (2012) Astrocytes: multitalented stars of the central nervous system. Methods Mol Biol 814:3–7

Reynolds A, Laurie C, Mosley RL, Gendelman HE (2007) Oxidative stress and the pathogenesis of neurodegenerative disorders. Int Rev Neurobiol 82:297–325

Ritchie T, Kim HS, Cole R, deVellis J, Noble EP (1988) Alcohol-induced alterations in phosphoinositide hydrolysis in astrocytes. Alcohol 5:183–187

Rostasy K, Monti L, Lipton SA, Hedreen JC, Gonzalez RG, Navia BA (2005) HIV leucoencephalopathy and TNFalpha expression in neurones. J Neurol Neurosurg Psychiatry 76:960–964

Rumzan R, Chen X, Li YM (2012) Gray matter involvement in patients with multiple sclerosis as shown by magnetic resonance imaging. Chin Med J 125: 2361–2364

Schultz W (2011) Potential vulnerabilities of neuronal reward, risk, and decision mechanisms to addictive drugs. Neuron 69:603–617

Shulha HP, Crisci JL, Reshetov D, Tushir JS, Cheung I, Bharadwaj R, Chou HJ, Houston IB, Peter CJ, Mitchell AC, Yao WD, Myers RH, Chen JF, Preuss TM, Rogaev EI, Jensen JD, Weng Z, Akbarian S (2012) Human-specific histone methylation signatures at transcription start sites in prefrontal neurons. PLoS Biol 10:e1001427

Streit WJ, Graeber MB, Kreutzberg GW (1988) Functional plasticity of microglia: a review. Glia 1:301–307

Tornatore C, Chandra R, Berger JR, Major EO (1994a) HIV-1 infection of subcortical astrocytes in the pediatric central nervous system. Neurology 44:481–487

Tornatore C, Meyers K, Atwood W, Conant K, Major E (1994b) Temporal patterns of human immunodeficiency virus type 1 transcripts in human fetal astrocytes. J Virol 68:93–102

Wu P, Tarasenko YI, Gu Y, Huang LY, Coggeshall RE, Yu Y (2002) Region-specific generation of cholinergic neurons from fetal human neural stem cells grafted in adult rat. Nat Neurosci 5:1271–1278

Xiong H, Zeng YC, Lewis T, Zheng J, Persidsky Y, Gendelman HE (2000) HIV-1 infected mononuclear phagocyte secretory products affect neuronal physiology leading to cellular demise: relevance for HIV-1-associated dementia. J Neurovirol 6(Suppl 1): S14–23

Zhao L, Huang Y, Tian C, Taylor L, Curthoys N, Wang Y, Vernon H, Zheng J (2012) Interferon-alpha regulates glutaminase 1 promoter through STAT1 phosphorylation: relevance to HIV-1 associated neurocognitive disorders. PLoS One 7:e32995

Zink WE, Zheng J, Persidsky Y, Poluektova L, Gendelman HE (1999) The neuropathogenesis of HIV-1 infection. FEMS Immunol Med Microbiol 26:233–241

Isolation and Culture of Neural Stem/Progenitor Cells

8

Hui Peng, Qiang Chen, and Jialin Zheng

Abstract

Neural stem cells, with the capacity of self-renewal and abilities to differentiate into the major cell types of the brain, exist in the developing and adult rodent central nervous system. These cells can be grown in vitro for long periods of time while retaining the capacity to differentiate into neurons, astrocytes, and oligodendrocytes. In this chapter, we describe the steps in detail to isolate and expand neural stem cells in the form of neurospheres from tissue dissections of the embryonic mouse brain. Procedures for the long-term passage of neurospheres, the cryopreservation of neurospheres, and their differentiation are also provided. Although the methodology seems simple, strict adherence to the procedures described is required in order to achieve reliable and consistent results.

Keywords

Neural stem cell • Neural progenitor cells • Neurosphere • Differentiation

8.1 Introduction

Neural stem cells (NSC), which have the capacity to self-renew and differentiate into the major cell types of the brain, exist in the developing and adult central nervous system (CNS) of all mammals (Gage 2000). During development, neuroepithelial germinal cells, described as NSC, proliferate within the ventricular zone and give rise to both neuronal and glial progenitors. In adulthood, new neurons are continually born in predominantly two regions of the brain. The first is the subventricular zone (SVZ), a layer extending along the lateral wall of the lateral ventricle (Doetsch and Scharff 2001), where NSC and progenitors generate new neurons (neuroblasts) and migrate to the olfactory bulb via the rostral migratory stream. The second region is the subgranular zone (SGZ) of the dentate gyrus of the hippocampus (Limke and Rao 2002), a thin cell layer between the granule cell layer and dentate hilus (Seri et al. 2001). Throughout early development, and continuing into adulthood, stem cells function as reservoirs of undifferentiated cell types whose role is to sustain

H. Peng (✉) • Q. Chen • J. Zheng
University of Nebraska Medical Center,
985930 Nebraska Medical Center, Omaha,
NE 68198-5930, USA
e-mail: hpeng@unmc.edu; qche1@unmc.edu;
jzheng@unmc.edu

H. Xiong and H.E. Gendelman (eds.), *Current Laboratory Methods in Neuroscience Research*,
Springer Protocols Handbooks, DOI 10.1007/978-1-4614-8794-4_8,

cell genesis. In the adult, they play an essential homeostatic role by replacing differentiated tissue cells lost by physiological turnover, or lost to injury or disease.

NSC can be cultured as adherent cells, in which they produce large clones containing neurons, glia, and additional stem cells; however, they can also be cultured as floating, multicellular neurospheres. The isolation of NSCs from the CNS through the neurosphere formation assay (NFA) was first described in 1992 (Reynolds and Weiss 1992). Since the 3D structure of neurosphere creates a niche that is more physiologically relevant than 2D culture systems, neurospheres emerged as the assay of choice and became a scheme for isolating and propagating NSC.

8.2 Materials

8.2.1 General Equipment

- Certified laminar flow biosafety level II safety cabinet certified for work
- Routine light microscope and hemocytometer for cell counts
- Inverted microscope with flatfield
- Low-speed centrifuge equipped with biohazard containers
- 37 °C incubators with gas control to maintain >95 % humidity and an atmosphere of 5 % CO_2
- Pipet-Aid

8.2.2 Dissection Equipment

- Dissection microscope
- Large scissors
- Small fine scissors
- Small forceps
- Small fine forceps
- Ultrafine curved scissors
- Spatula

8.2.3 Tissue Culture Equipment

- Flasks: 25–75 cm^2 with 0.2-µm vented filter cap
- 15–50 mL polystyrene conical tubes, sterile
- 40 µM cell strainer
- Culture dishes: 35-mm and 100-mm Petri dishes

8.2.4 Neural Stem Cell Media

- NeuroCult NSC basal medium (mouse, Stemcell Technologies, #05700)
- NeuroCult NSC proliferation supplement (mouse, Stemcell Technologies, #05701)
- Recombinant human epidermal growth factor (EGF, Stemcell Technologies, #02633)
- Recombinant human basic fibroblast growth factor (bFGF, Stemcell Technologies, #02634)

8.2.5 Neural Stem Cell Differentiation Media

- Neurobasal medium (1×) (Life Technologies, #21103049)
- B-27 Serum-free supplements (50×) (Life Technologies, #17504-044)
- L-Glutamine solution (Sigma, #G7513)
- Penicillin- streptomycin 100× solution (10,000 Units) (Life technologies, #15140-122)

8.3 Methods

8.3.1 Establishment of Primary Embryonic Neurosphere Culture

The resident population of NSC in the developing rodent brain peaks prior to embryonic day 12–14 (E12–14) and gradually diminishes due to differentiation into neurons, followed by astrocytes and oligodendrocytes. NSCs exist ubiquitously in the CNS and have been isolated from many regions of embryonic nervous system, i.e., striatum, cortex, spinal cord (SC), thalamus, and ventral mesencephalon of rodent embryos at different stages of development (Davis and Temple 1994; Weiss et al. 1996; Heldmann et al. 2011). Here, we describe how to isolate and expand embryonic NSC and to establish continuous, stem cell lines by means of growth factor stimulation.

1. Tissue isolation setup:
   (a) Set the water bath to 37 °C.
   (b) Disinfect the dissection area with 70 % ethanol.
   (c) Sterilize dissection tools needed to remove brain immediately before use by immersing in gauze-lined beaker filled with 70 % ethanol.
   (d) Add cold, sterile PBS to:
      i. Two 100-mm Petri dishes to hold embryos/tissue
      ii. Several 100-mm Petri dishes to wash embryos/tissue, and to hold mouse embryo head and brains
      iii. Several 35-mm Petri dishes to hold dissected tissues
2. Isolation of embryonic tissue(s):
   Rodents typically mate overnight, are separated the next morning, and checked for the presence of a gestational plug. When a plug is present, it will count as embryonic day zero (E0). Alternatively, pregnant animals at the desired gestational age can be purchased from specialized animal facilities. For establishing embryonic neurosphere cultures, we typically harvest pups at E14–15, sacrificing the mother in accordance with rules dictated by the animal ethics committee. We perform the dissection as quickly as possible (within 2 h), as tissue becomes soft and sticky over time and may be difficult to dissect. Keeping the brain and tissue on ice will help extend viability. Sacrifice of pregnant animals and removal of embryos are performed outside the laminar flow hood. Particular caution should be exercised to avoid microbial contamination.
   (a) Anesthetize pregnant mice by intraperitoneal (i.p.) injection of pentobarbital (120 mg/kg) or other approved anesthetic, and sacrifice by cervical dislocation.
   (b) Lay the female mouse on its back on an absorbent towel, and rinse the abdominal area with 70 % ethanol to sterilize the area.
   (c) Using the large forceps, grasp the skin above the genitalia. With large scissors, cut through the skin and fascia to expose the entire peritoneal cavity sufficiently to view the uteri.
   (d) Using small scissors, open the uterine horns. With small forceps, transfer embryos to a 100-mm dish containing ice-cold PBS. At this point, check the age of the pups and discard those that appear malformed or too small with respect to gestational age.
   (e) Cut off the heads of the pups at the level just below the cervical spinal cord region, with fine forceps; transfer them to a new 100-mm dish containing ice-cold PBS, taking care not to damage the brain.
   (f) Place tissue culture dish under dissecting microscope (under ×10 magnification). Remove the brain by positioning the head dorsal side up and using fine curved forceps, hold it from the caudal side at the ears. Using the coronal cut as an entry point, make a longitudinal cut through the skull along the sagittal suture. Be careful not to damage the brain, make small cuts ensuring that the angle of the blades is as shallow as possible. Using a small wetted curved spatula, peel the skull of each hemisphere outward to expose the brain, scoop the brain using a spatula and place into a Petri dish containing ice-cold PBS.
   (g) At high magnification (×25), dissect out the desired brain region(s) to be used for establishing the culture. Refer to a human or rodent brain atlas for details on how to dissect the specific areas.
   (h) Place tissue from each region into a new 35-mm dish containing ice-cold PBS. Make sure to label each dish with a region-specific code.
3. Dissociation of brain tissue and primary culture:
   (a) Add a total volume of 3 mL of culture medium to 50 mL conical tubes. Using sterile, fire-polished, cotton-plugged glass Pasteur pipettes, transfer tissue pieces into tubes. Before transferring tissue, prewet the pipette with fresh medium to prevent tissue from sticking to the glass wall.
   (b) Dissociate tissue by trituration, until the suspension looks cloudy and only small pieces of tissue are left. Normally, 20–30 strokes are sufficient.

(c) Allow enough time for the undissociated pieces to settle to the bottom of the tube.
(d) Transfer the top cell suspension to a clean, labeled plastic conical 15-mL tube leaving the undissociated tissue behind. Add an appropriate volume of culture medium (depending on volume of tissue) to bring the total volume to 2–4 mL, repeat step 2. During triturating, always avoid foaming and bubbling.
(e) Allow enough time for the undissociated pieces to settle to the bottom of the tube.
(f) Transfer all the volume, except 200–300 μL, to the tube containing the first round of dissociate cells.
(g) Pool the two suspensions you have created, then centrifuge the resulting suspension at 1,000 rpm for 5 min. Discard the supernatant, and then gently resuspend the cells in 2 mL volume of complete NSC medium.
(h) Combine a 10 μL-aliquot of the cell suspension with 90 μL of trypan blue in a microcentrifuge tube, mix, and then transfer 10 μL to a hemocytometer to determine the number of viable cells in the suspension.
(i) For primary cultures, seed cells at a density of 5×10^4 viable cells/cm^2 in growth medium, in the desired tissue culture vessels. Use approximately 6 mL for a 25-cm^2 flask or 10 mL for a 75-cm^2 flask. Note that the cell density for plating primary cells harvested directly from embryonic CNS is higher than that prescribed for subsequent subculturing conditions.
(j) Incubate at 37 °C, in a humidified atmosphere of 5 % CO_2.

Tips: The most important concept of NSC culture is that it represents a selective system in which most of the primary differentiated CNS cells will be eliminated in the existing culture conditions (low cell density, absence of serum, addition of growth factors, and absence of cell adhesion substrate). Conversely, undifferentiated stem cells enter into an active proliferation state. Therefore, upon plating the majority of cells will either die or differentiate; whereas only 0.5–3 % of cells will eventually form primary neurospheres. This occurs depending on the growth factors used, the developmental stage of the original tissue, and the area and the species of origin of the cells.

8.3.2 Culture Propagation: Passaging Neurospheres

Embryonic primary neurospheres should be ready for subculture between 3 and 5 days after initial plating, while passaged neurosphere cultures should be ready for subculture 4–7 days after plating. Nevertheless, one should monitor the cultures each day to ensure that neurospheres are not allowed to grow too large. Typically, a variety of diameters are found in a bulk culture. To determine whether neurospheres are ready to passage, the majority should equal 100–150 μm in diameter. If neurospheres are allowed to grow too large, they become difficult to dissociate and eventually differentiate in situ.

1. Observe the neurosphere cultures under a microscope to determine if the NSC are ready for passage. The average size of neurospheres across the culture should be ~150 μm before passaging. If neurospheres are attached to the culture substrate, forcefully strike the side of the tissue culture flask to dislodge spheres (attempting to minimize vessel movement by applying an equal force with the opposing hand).
2. Remove medium with suspended cells and transfer content of the flask to an appropriate sized sterile conical tube. If some cells remain attached to the substrate, detach them by shooting a stream of media across the attached cells. Wash the flasks out with 5 mL warm basal media (to prevent the cells from being shocked), and add that to the centrifuge tube.
3. Pellet cell suspension by centrifugation at 1,000 rpm for 5 min at room temperature.
4. Remove the supernatant and resuspend the spheres in 1 mL of trypsin–EDTA, incubating at 37 °C in the water bath for 1–3 min. Add 9 mL warm basal media and mix well, but gently. Centrifuge at 1,000 rpm for 5 min.
5. Remove the supernatant down to the actual pellet and resuspend the spheres in 2 mL

culture media. Gently triturate pellet: 25–30× for embryonic rodent cells, 40–50× for adult rodent cells, and up to 150× for human cells. Rinse down sides of tube periodically to dislodge undissociated spheres. Slightly tilt the pipetman, and press tip against the bottom of the tube to generate a fair amount of resistance.

6. Count viable cells by trypan blue exclusion, and seed cells at the appropriated density in culture medium in untreated tissue flasks: 1×10^4 cells/cm^2 for murine cells and 1×10^3 cells/cm^2 for human cells.
7. Subculture when the spheres start to lift off and float in suspension. This will require approx 3–7 days for embryonic murine cells, 5–10 days for adult murine cells, and 7–21 days for embryonic human cells.
8. Depending on both the culture conditions and the region of origin of the cells, the total cell number should increase at each passage: 2–10-fold for embryonic murine cells, 2–5-fold for adult murine cells, and twofold for embryonic human cells.

8.3.3 Cryopreservation and Reestablishment of Neurospheres

Once established, NSC can be effectively expanded to obtain a large number of cells. Cryopreservation of neurospheres at early subculture passages allows for the establishment of a reservoir of homogeneous stock of cells for future experiments.

1. Ensure that freezing jar (Criostep, Nalgene, PBI, #5100-0001) is at room temperature and filled with isopropanol.
2. Make freezing medium (culture basal medium containing 10 % dimethyl sulfoxide).
3. Label cryovials with date, cell type, and passage number.
4. Collect spheres and pellet them by centrifugation at 1,000 rpm for 5 min.
5. Remove the supernatant and resuspend the spheres in 1.5 mL freezing medium.
6. Transfer cells into a labeled 2-mL cryogenic vial.
7. Let cells equilibrate at room temperature for 15 min.
8. Transfer vials into the freezing jar containing isopropyl alcohol. Leave the jar at −80 °C for a minimum for 4 h, to allow a slow and reproducible decrease in temperature (−1 °C/min).
9. Transfer the vials to a liquid nitrogen cryofreezer for long-term storage.
10. To reestablish the neurospheres in tissue culture, remove the cryovial from liquid nitrogen and transfer to 37 °C water bath, and leave until thawed.
11. Slowly add the contents of the cryovial to a 15-mL conical tube containing 10 mL warm culture medium.
12. Centrifuge for 5 min at 1,000 rpm and remove the supernatant.
13. Gently resuspend pellet in fresh medium and plate in flask of appropriate size and incubate at 37 °C with 95 % air and 5 % CO_2.

Tip: Smaller neurospheres will survive the freezing process better than larger neurospheres (in excess of 100 μm in diameter). Therefore, do not let spheres grow too large before harvesting for cryopreservation. Mechanical dissociation of spheres before freezing will increase the number of dead cells, and viability of the culture upon thawing will be very low. Freezing a high concentration of neurospheres will enhance neurosphere reestablishment after freezing. Neurosphere viability can also be enhanced by the addition of 20 % fetal bovine serum to the freezing medium.

8.3.4 Differentiation of Neurosphere Cultures

When growing under the culture conditions described above (in the presence of EGF and/or bFGF), NSC and progenitor cells grow and expand in undifferentiated state and form neurospheres. Cells inside the sphere rarely display a differentiated morphology or antigenic properties typical of neuronal or glia cells. The vast majority of the cells are immunoreactive for nestin, an intermediated filament characteristic of undifferentiated neuroepithelial cells. However, upon the removal of the growth factors and plating stem cell progeny onto a good adhesion substrate (such as poly-D-lysine, laminin, or Matrigel), neurosphere-derived cells are induced to differentiate into neurons, astrocytes, and oligodendrocytes. The procedure described below is a basic one, in which cells are differentiated in serum-free medium, according to the same protocol used to establish enriched neuronal cultures from the embryonic rodent brain. Overall, two methods are described for the differentiation of neurospheres: as whole spheres cultured at low density (typically used to demonstrate individual spheres are multipotent) or as dissociated cells at high density (typically used to determine the relative percentage of differentiated cell types generated). This approach provides the culture conditions to test the effect of specific molecules on the differentiation of stem cells and their neuronal and glial progeny.

1. Differentiation of whole neurospheres:
   (a) Pre-coat glass coverslips by adding a sufficient volume of poly-D-lysine (PDL, 50 μg/mL) to completely cover the glass coverslip for a period of 2 h at 37 °C. Aspirate PDL and immediately rinse three times (10 min each) with sterile PBS (do not allow coverslips or plate to dry). Remove PBS prior to the addition of neurospheres and differentiation medium.
   (b) Once primary or passaged neurospheres reach 150 μm (typically after 4–8 days in vitro), remove medium with suspended cells (neurospheres) and transfer contents of the flask to an appropriate sized sterile conical tube. Spin at 1,000 rpm for 5 min.
   (c) Aspirate essentially 100 % of the growth medium, then gently resuspend (so as not to dissociate any neurospheres) with an appropriate volume of NSC differentiation medium (Neurobasal media + 2 % B-27 supplement). Note that an equal volume of commercially available NSC differentiation medium can also be used here (NeuroCult Differentiation supplement, Stemcell Technologies, #05703).
   (d) Transfer neurosphere suspension to a 60-mm dish (or other sized vessel) to enable the harvesting of individual neurospheres with a disposable plastic pipette.
   (e) Using a sterile disposable plastic pipette, transfer approximately ten neurospheres into individual wells of 24- or 96-well tissue culture plate containing a PDL-coated surface with NSC differentiation medium. Alternatively, commercially available pre-coated chamber slides or culture plates can be employed here.
   (f) After 5–8 days in vitro, individual neurospheres should have attached to the substrate and dispersed to appear as a flattened monolayer of cells.
   (g) Proceed to fix cells with the addition of 4 % paraformaldehyde (in PBS, pH 7.2) for 20 min at room temperature and then process the adherent cells for immunocytochemistry as required.
2. Differentiation of dissociated cells from neurospheres
   (a) Pre-coat glass coverslips by adding a sufficient volume of PDL (50 μg/mL) to completely cover the glass coverslip for a period of 2 h at 37 °C. Aspirate PDL and immediately rinse three times (10 min each) with sterile PBS (do not allow coverslips or plate to dry). Remove PBS prior to the addition of dissociated cells and differentiation medium.
   (b) Once primary or passaged neurospheres reach 150 μm (typically after 4–8 days in vitro), remove medium with suspended cells (neurospheres) and transfer contents of the flask to an appropriate sized sterile conical tube. Spin at 1,000 rpm for 5 min.
   (c) Remove essentially 100 % of the supernatant and resuspend cells using 1 mL of trypsin–EDTA, incubating at room temperature for 1–3 min. Add 9 mL warm basal media and mix well, but gently. Centrifuge at 1,000 rpm for 5 min.
   (d) Remove the supernatant down to the actual pellet and resuspend the spheres in 2 mL differentiation medium. Gently triturate cells until suspension appears milky and no spheres can be seen.

(e) Count viable cells by trypan blue exclusion, and seed cells at the appropriated density in differentiation medium in PDL-coated glass coverslip untreated tissue flasks: 1×10^4 cells/cm^2 for murine cells and 1×10^3 cells/cm^2 for human cells.

(f) Prepare the appropriate cell suspension in complete NSC differentiation media, so as to seed individual wells of 24-well tissue culture plate containing a PDL-coated glass coverslip with 5×10^4 cells. Alternatively, commercially available pre-coated chamber slides can be used here; seed wells at the same density.

(g) After 4–6 days in vitro, neurosphere-derived cells will have differentiated sufficiently. Proceed to fix cells with the addition of 4 % paraformaldehyde (in PBS, pH 7.2) for 20 min at room temperature and then process the adherent cells for immunocytochemistry as required.

3. Immunolabeling to identify differentiated cell types

(a) Fix the cells with 4 % paraformaldehyde (in PBS, pH 7.2) for 20 min at room temperature (0.5 mL/well for 24 well, 0.1 mL/well for 96 wells).

(b) Remove the paraformaldehyde solution using an aspiration system connected to a vacuum pump.

(c) Add PBS (pH 7.2) to the cells and incubate for 5 min. Aspirate PBS using vacuum pump and repeat this washing procedure two more times for a total of three wash steps.

(d) Saturate nonspecific sites and permeabilize with PBS containing 0.1 % Triton and 10 % normal goat serum (NGS) for 60 min at room temperature.

(e) Incubate the cells with primary antibodies (*neuron* (microtubule-associated protein-2 (MAP-2), neuron, β-tubulin type III, and doublecortin), *astrocyte* (glial fibrillary acidic protein (GFAP, S100-β)), and *oligodendrocyte* (myelin basic protein (MBP), O_4, and GalC)) diluted in blocking solution for 1 h at room temperature (or overnight at 4 °C).

(f) Wash cells three times with PBS as step 3.

(g) Incubate the cells with the secondary antibodies (Molecular Probes, Life Technologies, 1:1,000) diluted in blocking buffer for 60 min at room temperature.

(h) Wash cells three times with PBS; include DAPI (1:1,000) in second wash for nuclear counterstain.

(i) Mount on slides using Fluoromount fluorescent mounting media (Sigma, Cal. No. F4680).

(j) Visualize immunostaining under fluorescent microscope using appropriate filters for fluorophore.

8.4 Summary

Owing to a lack of specific and definitive markers, and poorly defined morphological characteristics, NSC are defined based on functional criterion by what they do rather than by what they look like. Therefore, a neural stem cell is defined as an undifferentiated cell that retains the capacity to (a) *proliferate*, (b) *self-renew* (making identical copies of itself), (c) become *multipotent* (generating all the major cell lineages of CNS, including neurons, astrocytes, and oligodendrocytes), and (d) *regenerate tissue after injury* (Jagasia et al. 2006). Identification of a stem cell is achieved when all of these criteria are met. Nevertheless, due to technical or experimental limitations, only some of these criteria may be satisfied in practice. Using the neurosphere formation method to isolate neural stem cells fulfills two central criteria for the definition of a stem cell: self-renewal and multipotency. The propagation of large populations of NSC can provide a source of tissue for many types of applications, from developmental biology to drug screening and cell therapy.

References

Davis AA, Temple S (1994) A self-renewing multipotential stem cell in embryonic rat cerebral cortex. Nature 372:263–266

Doetsch F, Scharff C (2001) Challenges for brain repair: insights from adult neurogenesis in birds and mammals. Brain Behav Evol 58:306–322

Gage FH (2000) Mammalian neural stem cells. Science 287:1433–1438

Heldmann U, Mine Y, Kokaia Z, Ekdahl CT, Lindvall O (2011) Selective depletion of Mac-1-expressing microglia in rat subventricular zone does not alter neurogenic response early after stroke. Exp Neurol 229:391–398

Jagasia R, Song H, Gage FH, Lie DC (2006) New regulators in adult neurogenesis and their potential role for repair. Trends Mol Med 12:400–405

Limke TL, Rao MS (2002) Neural stem cells in aging and disease. J Cell Mol Med 6:475–96

Reynolds BA, Weiss S (1992) Generation of neurons and astrocytes from isolated cells of the adult mammalian central nervous system. Science 255:1707–1710

Seri B, Garcia-Verdugo JM, McEwen BS, Alvarez-Buylla A (2001) Astrocytes give rise to new neurons in the adult mammalian hippocampus. J Neurosci 21:7153–7160

Weiss S, Reynolds BA, Vescovi AL, Morshead C, Craig CG, van der Kooy D (1996) Is there a neural stem cell in the mammalian forebrain? Trends Neurosci 19: 387–393

Isolation of Mitochondria from Brain Tissue and Cells

9

Changhai Tian and Jialin Zheng

Abstract

Mitochondria are one of intracellular organelles acting as the main energy providers of the cell. Increasing evidences have demonstrated that mitochondrial dysfunctions in respiratory chain function, oxidative stress, dynamics (shape, size, fission-fusion, distribution, movement, etc.), calcium-handling capacities, and protein mutation play a central pathogenic role in neurodegenerative diseases such as Parkinson's disease, Huntington's disease, amyotrophic lateral sclerosis, and Alzheimer's disease. Protocols developed for the isolation of pure and intact mitochondria will contribute considerably to the understanding of the pathogenic mechanisms of neurodegenerative disorders.

Keywords

Mitochondria • Brain tissues • Isolation • Reactive oxygen species • Neurodegenerative disorders

9.1 Introduction

Eukaryotic cells contain intracellular membrane-bound structures known as organelles, performing a variety of functions. Mitochondria play a vital role in cellular metabolism and apoptosis (Robertson et al. 2006) and its dysfunction is linked to the pathogenesis of neurodegenerative disorders. Increasing evidence suggests that age-dependent accumulation of somatic mitochondrial DNA (mtDNA) changes is involved in disease progression of neurodegenerative diseases, including Alzheimer's disease (AD), Parkinson's disease (PD), amyotrophic lateral sclerosis (ALS), and Huntington's disease (HD) (Lin and Beal 2006), which is neuronal-specific for each of these degenerative diseases (Reddy and Reddy 2011). Recent findings have greatly expanded our understanding of the role of mitochondria in the pathogenesis of neurodegenerative diseases, including the increased levels of reactive oxygen species (ROS) production, abnormal intracellular calcium levels, and reduced mitochondrial ATP and mutant proteins, such as amyloid beta in

C. Tian (✉) • J. Zheng
Laboratory of Neuroimmunology and Regenerative Therapy, Department of Pharmacology and Experimental Neuroscience, University of Nebraska Medical Center, Omaha, NE 68198-5930, USA
e-mail: ctian@unmc.edu; jzheng@unmc.edu

H. Xiong and H.E. Gendelman (eds.), *Current Laboratory Methods in Neuroscience Research*, Springer Protocols Handbooks, DOI 10.1007/978-1-4614-8794-4_9,

AD, mutant huntingtin in HD, mutant SOD1 in ALS, mutant parkin, mutant DJ-1 and mutant α-synuclein in PD, and frataxin in Friedreich ataxia (FRDA)—which are localized to mitochondrial membranes (Beal 2005; Reddy 2008), which will ultimately contribute to neuron injury by an increased production of free radicals and a low production of cellular ATP. More recent studies on mitochondrial structure in tissues of brains from AD, PD, and HD revealed that mitochondrial fission and fusion also play critical roles in maintaining functional mitochondria, and the dysregulation of mitochondrial dynamics may be the primary cause of mitochondrial dysfunction and neuronal damage (Reddy 2008; Youle and van der Bliek 2012).

To better understand this role, isolated mitochondria from patient tissues or cells can be utilized as a valuable ex vivo model. Using this model, one can analyze the energy metabolism, calcium homeostasis, and ROS generation in vitro, as well as study apoptotic processes by detecting the release of cytochrome c and other apoptotic factors. It is also possible that novel mutations of proteins in mitochondria from patients may be found through the comparisons of mitochondrial proteomics. Finally, isolated mitochondria can be helpful for screening novel disease combating agents, and elucidating their mechanism of action, which will assist in the development and testing of potential effective therapies for neurodegenerative diseases.

In this chapter, our objective is to introduce the basic procedures for the mitochondria isolation from brain tissue and cells using differential centrifugation techniques and detail its many applications in neuroscience research.

9.2 Materials

9.2.1 General Equipment

- Polytron homogenizer (0.1 mL to 2,000 mL sample volume).
- Medium-speed centrifuge with fixed angle rotors.
- Benchtop refrigerated centrifuge (Sigma 1-15K).
- All-glass Dounce homogenizers (Kontes or Wheaton) with a loose pestle and a tight pestle.
- Polycarbonate centrifuge tubes (10–13.5 mL; 50 mL).
- Polypropylene tubes; 1.5 mL microfuge tubes (Eppendorf).
- Medium straight-edged scissors (Fine Science Tools 14 cm straight sharp/blunt or sharp/sharp surgical scissors) and small straight iris scissors.
- Lumbar puncture needle with a blunted tip (between 19 and 23 gauge).
- Medium (200 mm) stainless steel spatula.
- Scalpel with disposable blades (Swann-Morton No. 11 blades).
- Skin graft knife blades.
- Small animal guillotine.
- Disposable plastic and glass Pasteur pipettes.
- Adjustable volumetric pipettes: 100 μL, 1 and 5 mL.

9.2.2 Reagents

- Animals (e.g., Sprague–Dawley or Porton–Wistar rats) are routinely used in this study. The University of Nebraska Medical Center (UNMC) Institutional Animal Use and Care Committee (IACUC) strictly reviews all procedures involving animals. Every effort is made to insure the absolute minimum level of discomfort is maintained at all times. All tissues will be harvested from animals anesthetized prior to decapitation
- Mannitol (Sigma, cat. no. M4125)
- Sucrose (Sigma, cat. no. S9378)
- Tris (Sigma, cat. no. T1503)
- Ethylenediaminetetraacetic acid (EDTA) (Sigma, cat. no. EDS)
- Percoll (GE Healthcare, cat. no. 17089101)
- Bovine serum albumin (BSA), essentially fatty acid-free (Sigma, cat. no. A6003)

9.2.3 Buffers

1. Isolation buffer (MSB) (pH 7.4) (1,000 mL): Weight 38.26 g mannitol (210 mM), 23.96 g

sucrose (70 mM), 0.146 g EDTA (0.5 mM), and 1.21 g Tris (10 mM) and dissolve in approximately 800 mL water in a beaker. Adjust the pH to 7.4 by adding 0.1 M HCl while stirring with a magnetic stirrer. Make up the volume to 1,000 mL by adding water. Finally, weight 2 g BSA (0.2 %) and add to 1,000 mL ready-made isolation buffer, and store this solution at 4 °C. We routinely prepare this solution on the day of mitochondrial isolation.

2. Hypotonic mitochondrial buffer (MB) (pH 7.2) (100 mL): Weight 7.19 g sucrose (210 mM), 1.28 g mannitol (70 mM), 0.029 g EDTA (1 mM), 0.038 g EGTA (1 mM) (ethylene glycol tetraacetic acid), 0.014 g $MgCl_2$ (1.5 mM), and 0.238 g HEPES (10 mM) and dissolve in approximately 90 mL water in a beaker. Adjust the pH to 7.2 by adding KOH (0.5–1 M) while stirring with a magnetic stirrer. Make sure that you use HEPES (acid) and add protease inhibitor cocktail to prevent proteins from degradation.
3. 100 % Percoll solution: 225 mM sucrose, 75 mM mannitol, 1 mM EGTA, and 5 mM HEPES in Percoll. Adjust pH to 7.4 with HCl at 4 °C.

9.3 Isolation and Purification of Mitochondria from Brain Tissue

9.3.1 Procedures

1. All the tools should be pre-cooled at 4 °C and washed with MSB buffer (see Sect. 9.2.3), (4 °C).
2. Decapitate the mouse/rat with an animal guillotine. The procedures should adhere to relevant regulations and guidelines relating to experiments on animals and be approved by the University of Nebraska Medical Center (UNMC) Institutional Animal Use and Care Committee (IACUC).
3. Rapidly remove the brain from the skull. To achieve this, initially use medium-size scissors to carefully make a midline incision of the scalp beginning at the cut edge created by the guillotine and ending in the area between the eyes. Retract the scalp from the surface of brain.
4. Put whole brain into MSB buffer (4 °C) and wash it to remove blood.
5. Transfer whole brain tissue into fresh MSB buffer and wash tissue with MSB for 2–3 times.
6. Homogenize the tissue with an automatic homogenizer at 4 °C three times (600–700 rpm, up and down). If it is unavailable, you can use all-glass Dounce homogenizers (Kontes or Wheaton) with a loose pestle and a tight pestle to replace it, and all operations should be performed on ice.
7. Centrifuge (3,000 rpm) at 4 °C for 10 min, and wash the pellets with MSB buffer for 2–3 times, and then collect all supernatants.
8. Centrifuge supernatants (10,000 rpm) for 10 min; take the pellets (mitochondria fraction).
9. Wash the pellets with MSB buffer for 2–3 times; centrifuge at 10,000 rpm at 4 °C for 10 min.
10. Resuspend the pellets in a small volume of MSB buffer, dependent on experimental requirements.
11. Protein content of mitochondria can be determined by the microbiuret method using BSA as standard (Itzhaki and Gill 1964).
12. The isolated mitochondria can now be used to study mitochondria-mediated apoptosis, ROS generation, etc., or for further purification.

9.3.2 Purification of Mitochondria by Discontinuous Percoll Gradient Centrifugation (see Fig. 9.1) (Kristian 2010)

1. Preparation of the Percoll gradient: 40 % Percoll (1.5 mL), 24 % Percoll (3.7 mL), and 15 % Percoll (3.5 mL). The different concentrations of Percoll solutions will be prepared by diluting 100 % Percoll solution with the isolation buffer (MSB).

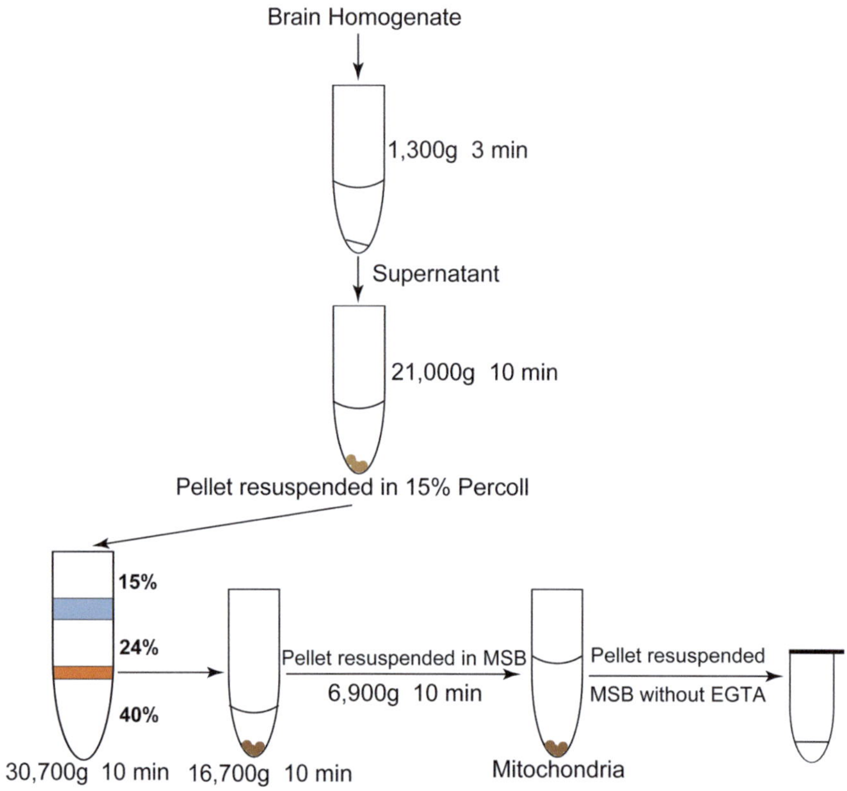

Fig. 9.1 Schematic illustration of mitochondrial purification from brain tissue by using the discontinuous Percoll gradient centrifugation

2. Add 3.7 mL of 24 % Percoll into a 10 mL polycarbonate centrifuge tube, and then collect the 1.5 mL of 40 % Percoll using a disposable plastic pipette and insert the pipette into the 24 % Percoll so the tip touches the bottom of the tube. Slowly introduce the 40 % Percoll solution to the bottom of the tube creating a discontinuous gradient of 24 % Percoll on top of the 40 % Percoll solution. Prepare two tubes with Percoll gradients per one whole rat forebrain (Note: The addition of the 40 % Percoll solution should be sufficiently slow to form a sharp interface between the two different Percoll layers. The tip of the disposable pipette should be leaning against the bottom wall of the centrifugation tube and not held vertically but rather under about 75°).
3. Resuspend the pellet (Sect. 9.3.1, step 9) in 3.5 mL of 15 % Percoll. Using a glass stirring rod, detach the pellet from the bottom of the tube and resuspend the material using a disposable pipette.
4. Layer this material slowly above the 24 % Percoll with a disposable pipette (Note: Make sure that there is a sharp interface between the 24 % Percoll and the Percoll with the resuspended tissue. Begin introduction of the top layer by leaning the tip of the pipette against the tube wall close to the surface of the 24 % Percoll then slowly add the resuspended material on top of this layer).
5. Centrifuge at 30,700 × g at 4 °C for 8 min using slow acceleration (45 s from 0 to 500 rpm followed by normal acceleration) and deceleration (no brakes). This centrifugation will redistribute the tissue material into three major bands.

6. Collect the Percoll solution containing material accumulating at the interface between 40 and 24 % Percoll solution that is enriched by mitochondria and transfer this suspension into a separate 10 mL polycarbonate tube (Note: collect all the material within this band to maximize the mitochondrial yield).
7. Dilute mitochondrial suspension with 2 volume of the isolation buffer (MSB) and pellet the mitochondria by centrifuge at 16,700 × *g* for 10 min at 4 °C.
8. Wash the pellet with isolation buffer to remove Percoll.
9. Resuspend the purified mitochondria in MSB buffer without EDTA.

9.3.3 Most Common Pitfalls and Potential Solutions

1. You can pour off the supernatants, and the loose upper part of the mitochondria pellet will come off as well. Most of the pellet, containing the healthy mitochondria, is dense enough to remain behind. Use a Pasteur pipette to remove the last bit of the liquid. Wipe the inside of the upper part of the tube to remove any fatty material that remains. Any mixing of liquids with the mitochondria supernatants will cause them to uncouple quickly.
2. Use a Teflon rod to homogenize the remaining pellets. It should be a smooth brown paste. Keep the tube on ice when stirring. Transfer the mitochondria to an Eppendorf tube using a 200 μL tip. As mitochondria are easy to adhere to glass surface, we recommend using single-use disposable Pasteur pipettes. Keep all the preparation on ice.
3. When the crude mitochondria preparation is done, mitochondria can be suspended in the same buffer (*without* EGTA and protease inhibitor), and stored on ice.
4. All above steps should be performed without delay; if there must be lapse of time between the preparation and the use, store the mitochondria on ice in pellet form.

9.4 Isolation and Purification of Mitochondria from Cultured Cells

9.4.1 Procedures

1. Harvest cells (2×10^6–1×10^7 cell).
2. Wash the cells with ice-cold PBS (Ca^{2+} free).
3. Resuspend cells with hypotonic mitochondrial buffer (MB) (see Sect. 9.2.3), and keep on ice for 30–60 min with frequent tapping.
4. Wash once with MB and resuspend cells in a minimal volume of MB (with protease inhibitors), which is approximately five times of the volume of the cell pellet.
5. Homogenize for 20–100 strokes with a Dounce homogenizer (trypan blue staining, >30–50 % positive cell). Wash the homogenizer once with minimal buffer (MB).
6. Transfer to Eppendorf tube, which was put on ice before using.
7. Centrifuge at 1,000 × *g* at 4 °C for 5–10 min to remove the nuclei and intact cells. If necessary, you may need to wash the pellet, combine the supernatant, and spin down again to remove the unbroken cells (make sure that you remove the unbroken cell or nuclei by 1,000 × *g* centrifugation before high-speed centrifugation).
8. Collect and spin the supernatant at 10,000 × *g* at 4 °C for 15 min. And collect the pellet containing heavy membrane and mitochondria.
9. Collect supernatant and spin at 100,000 × *g* at 4 °C for 60 min using benchtop ultracentrifuge and specific EP tube and the right rotor to obtain light membrane (ER fraction, pellet) and cytosolic fraction (supernatants) for future use.
10. Lyse the membrane with cell lysis buffer (20 mM Tris–HCl (pH 7.5), 150 mM NaCl, 1 mM Na_2EDTA, 1 mM EGTA, 1 % Triton X-100, 2.5 mM sodium pyrophosphate, 1 mM β-glycerophosphate, 1 mM Na_3VO_4, and 1 μg/mL leupeptin).
11. Adjust the protein concentration and add loading buffer for Western blotting analysis.

9.4.2 Most Common Pitfalls and Solutions

1. Avoid using glassware, which might have excessive Ca^{2+}. Make sure that you check the cell lysis with trypan blue to monitor the homogenization.
2. Keep everything on ice (0-4 °C). Try to use minimal amount of buffer to avoid dilution of the cytosolic fractions.

9.5 Common Experimental Uses of Isolated Mitochondria

9.5.1 Markers and Enzymes Commonly Used to Identify Mitochondria, and Evaluate the Morphology and Function of Mitochondria

When you get mitochondria from cultured cells or tissues, you may determine the distribution of mitochondria in the resulting fractions by detecting the activity of the mitochondrial enzymes, checking mitochondrial protein location, or staining mitochondria with mitochondria-specific fluorescence dye (see Table 9.1).

9.5.2 Electron Microscopy

The mitochondrial samples were fixed overnight with 4 % glutaraldehyde in 0.1 M cacodylate buffer (pH 7.4) at 4 °C, postfixed with 1 % osmic acid in 0.1 M cacodylate buffer, dehydrated, and embedded in LX-112 (Ladd) epoxy resin. Thin sections were observed with a Jeol JEM-1200 EX transmission electron microscope.

Table 9.1 Markers and enzymes used for mitochondrial identification

Enzymes	Succinic dehydrogenase, cytochrome oxidase
Marker proteins	Cytochrome c, VDAC1 and COX IV, etc.
Staining markers	MitoTracker Greeen™ and CMX-rosamine (MitoTracker Red)

Note: *VDAC1* voltage-dependent anion channel; *COX-IV* cytochrome c oxidase subunit IV

9.5.3 Determination of Oxygen Consumption and Respiratory Control Ratio (RCR)

Using a Clark oxygen electrode, recombinant proteins, toxins, and some mitochondria-targeted agents can be used to treat isolated mitochondria, and mitochondrial oxygen consumption rates are monitored at 25 °C in PT-1 buffer (sucrose 250 mM; HEPES 2 mM; KH_2PO_4 0.5 mM; rotenone, 2 M; potassium succinate 4.2 mM).

9.5.4 Mitochondrial Permeability Transition Pore Opening, Mitochondrial Membrane Potential ($\Delta\psi m$), and ROS Determination Ex Vivo

Isolated mitochondria (5 mg protein/mL) are kept in MT buffer containing 250 mM sucrose, 2 mM HEPES, pH 7.4, 0.1 mM EDTA, and 0.1 % fatty acid-free BSA. Permeability transition pore opening can be monitored by the decrease of 90° light scattering at 520 nm at 25 °C in PT-1 buffer containing 250 mM sucrose, 2 mM HEPES, pH 7.4, 0.5 mM KH_2PO_4, 2 mM rotenone, and 4.2 mM potassium succinate, using a Jobin Yvon FluoroMax-2 spectrofluorometer as described (Narita et al. 1998; Tian et al. 2008). To determine the mitochondrial membrane potential ($\Delta\psi m$), the 30 nM rhodamine 123 (Rh123) is added to isolated mitochondria suspension, $\Delta\psi m$ is assessed at 25 °C in PT-1 buffer by measuring the uptake of Rh123 using a spectrofluorometer (Jobin Yvon FluoroMax-2). To evaluate the production of ROS, isolated mitochondria are suspended in PT-1 buffer and incubated with 2′,7′-Dichlorofluorescin-diacetate (DCFH-DA). DCF formation was monitored using a Jobin Yvon FluoroMax-2 spectrofluorometer.

9.5.5 Detection of Cytochrome c Release

Mitochondria are isolated according to the above methods and protein content of isolated mitochondria is determined by the microbiuret method using BSA as a standard. Equal mitochondria fractions from brain tissues are treated with different concentrations of drugs or proteins (e.g., staurosporine and recombinant Bid protein) in PT-1 buffer for 60 min at 25 °C. The samples are then centrifuged at 12,000 × *g* for 15 min at 4 °C. Cytochrome c in the supernatant can be detected by Western blotting using anti-cytochrome c monoclonal antibody and visualized by ECL SuperSignal system (Pierce), and equal protein loading is confirmed by immunodetection of cytochrome c oxidase subunit-IV (COX-IV) or VDAC1 in the mitochondrial pellets. A sample treated with 0.4 μM $CaCl_2$ can be used as a positive control.

Subcellular fractionation is performed following the procedures for mitochondrial isolation from cultured cells. Following step 8 (Sect. 9.4.1), the supernatants collected at 10,000 × *g* at 4 °C for 15 min is subjected to the normalization of protein concentration. The pellet containing mitochondria is lysed with lysis buffer and protein concentrations are determined. Finally, the translocation of cytochrome c from mitochondria will be determined by western blotting analysis with anti-cytochrome c antibody and COX-IV/VDAC1 antibodies in different fractions.

9.6 Summary

Isolated mitochondria can be used in many experimental techniques, including measurement of mitochondrial permeability transition such as cytochrome c release assay, measurement of oxygen consumption and RCR, determination of mitochondrial membrane potential, assay of ROS production and determination of the NAD(P)H redox state. Taken together, isolated mitochondria provide a good model to study the pathologic mechanism of many neurodegenerative disorders. Moreover, the proteomic comparison between purified mitochondria from normal tissue and patient tissue will help verify new mutations of mitochondrial proteins contributing to the pathogenesis of some neurodegeneration.

References

Beal MF (2005) Mitochondria take center stage in aging and neurodegeneration. Ann Neurol 58:495–505

Itzhaki RF, Gill DM (1964) A Micro-Biuret Method for Estimating Proteins. Anal Biochem 9:401–410

Kristian T (2010) Isolation of mitochondria from the CNS. Curr Protoc Neurosci Chapter 7:Unit 7.22

Lin MT, Beal MF (2006) Mitochondrial dysfunction and oxidative stress in neurodegenerative diseases. Nature 443:787–795

Narita M, Shimizu S, Ito T, Chittenden T, Lutz RJ, Matsuda H, Tsujimoto Y (1998) Bax interacts with the permeability transition pore to induce permeability transition and cytochrome c release in isolated mitochondria. Proc Natl Acad Sci U S A 95:14681–14686

Reddy PH (2008) Mitochondrial medicine for aging and neurodegenerative diseases. Neuromolecular Med 10:291–315

Reddy PH, Reddy TP (2011) Mitochondria as a therapeutic target for aging and neurodegenerative diseases. Curr Alzheimer Res 8:393–409

Robertson CL, Soane L, Siegel ZT, Fiskum G (2006) The potential role of mitochondria in pediatric traumatic brain injury. Dev Neurosci 28:432–446

Tian C, Gao P, Zheng Y, Yue W, Wang X, Jin H, Chen Q (2008) Redox status of thioredoxin-1 (TRX1) determines the sensitivity of human liver carcinoma cells (HepG2) to arsenic trioxide-induced cell death. Cell Res 18:458–471

Youle RJ, van der Bliek AM (2012) Mitochondrial fission, fusion, and stress. Science 337:1062–1065

Isolation of Synaptosomes from Archived Brain Tissues

10

Gurudutt Pendyala, James L. Buescher, and Howard S. Fox

Abstract

Synapses in the central nervous system serve as communication points between neurons and are critical regulators of neurotransmission and synaptic plasticity, the latter refers to a process of experience dependent changes in synaptic connectivity, where neurons undergo extensive sculpting and rewiring. Research on understanding the changes at the level of the synapse holds great promise into understanding the biological basis of many neurodegenerative and neuropsychiatric disorders in which brain wiring goes awry. One such approach to understand the changes occurring at the synapse is by isolating synaptosomes. Here, we describe the isolation of synaptosomes from archived human brain tissue using subcellular fractionation, which when combined to high-throughput "omics"-based approaches could yield vital clues into understanding the underlying bases of neurodegeneration.

Keywords

CNS • Neurodegeneration • Synapse • Synaptosomes

10.1 Introduction

The human brain is without a doubt the most complex organ in the human body and consists of multiple specialized cells and structures. While neurons are most associated with central nervous system (CNS) functions, glial cells far outnumber neurons (Williams and Herrup 1988). The neurons in the CNS form a very complex network of interlinking 10^{12} neurons with some 10^{15} synapses (Pocklington et al. 2006). Such synapses play a critical role in regulating neurotransmission and neuroplasticity. The term synapse was coined in 1897 by the English Physiologist Charles Sherrington from the Greek word "synaptein" ("syn": together; "haptein": to fasten or bind). Synapses serve as communication points between two neurons. In general, synapses are comprised of three main components: presynaptic

G. Pendyala (✉) • J.L. Buescher • H.S. Fox
Department of Pharmacology and Experimental Neuroscience, University of Nebraska Medical Center, Omaha, NE 68198-5800, USA
e-mail: gpendyala@unmc.edu; james.buescher@unmc.edu; hfox@unmc.edu

H. Xiong and H.E. Gendelman (eds.), *Current Laboratory Methods in Neuroscience Research*, Springer Protocols Handbooks, DOI 10.1007/978-1-4614-8794-4_10,

terminal, synaptic cleft, and a postsynaptic terminal. Typically, the presynaptic terminal is distinguished from the postsynaptic component by the presence of neurotransmitter-filled vesicles. In response to a depolarization of the presynaptic membrane, the vesicles undergo exocytosis releasing their components into the synaptic cleft and further retrieved back via endocytosis (Sudhof 2004).

The organization and regulation of synaptic connections in the mammalian CNS entail complicated and coordinated molecular and cellular processes that are pivotal in maintaining the integrity of the synapse and for proper neurotransmission. Given the complex circuitry of the CNS, it is thus unequivocal that alterations in the synapse can lead to changes in neurotransmission resulting in aberrations during a CNS disease manifestation. To understand the elaborate neuroadaptive mechanisms and changes associated during a disease state, molecular characterization of synapses warrant further investigation (Pocklington et al. 2006).

Given the complexity of the brain, systematic approaches have been paving the way to understand the normal physiological conditions and functional alterations in a plethora of neurological diseases. One such burgeoning field is the application of mass spectrometry-based proteomics that has been increasingly used in neurosciences (Liao et al. 2009; Bayes and Grant 2009). Diseases relating to the CNS, such as neuropsychiatric and neurodegenerative diseases, often involve multiple interacting proteins and are thus well suited for proteomic analysis (Kim et al. 2004). Apart from allowing a global unbiased identification of molecular markers for disease states, mass spectrometry associated neuroproteomic approaches aid in greater elucidation of the underlying processes.

One key strategy in proteomics is through the performance of quantitative experiments, as many changes in neuronal and synaptic activity are likely to be explained by comparing protein levels and/or activation states between physiological and perturbed states. Two-dimensional gel electrophoresis (2D-GE) has been adapted to quantitative techniques and differentially expressed proteins from such experiments have been identified by mass spectrometry (MS). Furthermore, with the advent of novel quantitative proteomic approaches such as the use of chemical reactions to introduce isotopic tags at specific functional groups on peptides or proteins, such as isotope-coded affinity tags (ICAT) (Gygi et al. 1999) and isobaric tag for relative and absolute quantification (iTRAQ) (Ross et al. 2004), stable isotope labeling by amino acids in culture (SILAC) (Ong et al. 2002; Zhu et al. 2002), and stable isotope labeling in mammals (SILAM) (McClatchy et al. 2008) followed by MS methodologies have greatly enhanced proteomic technologies.

10.1.1 Synaptosomes

Given the large numbers of cell types and structures within the CNS, subcellular fractionation techniques have emerged as key factors in assessing the changes occurring at the synapse. Their analysis is of paramount importance as perturbations in their expression can unravel clues linked to neurodegeneration. One such approach to understand the synapse and gain vital insights into the etiology underlying neurodegeneration is by employing biochemical fractionation in vitro to isolate synaptosomes.

The term "synaptosome" was coined by Whittaker's group in 1964 (Whittaker et al. 1964). Synaptosomes are subcellular membranous structures that are formed during disruption of brain tissues. The shearing forces during homogenization cause the nerve ending to pinch off and subsequently reseal given their high hydrophobic nature. Synaptosomes are comprised of the complete presynaptic terminal including synaptic vesicles and mitochondria along with the postsynaptic membrane (see electron micrograph in Fig. 10.1). Given the above molecular composition necessary for release, uptake, and storage of neurotransmitters, synaptosomes are ideal models to study neurotransmission.

Another important aspect is the isolation of purified synaptosomes feasible by employing subcellular fractionation techniques such as differential sucrose density centrifugation (Gray

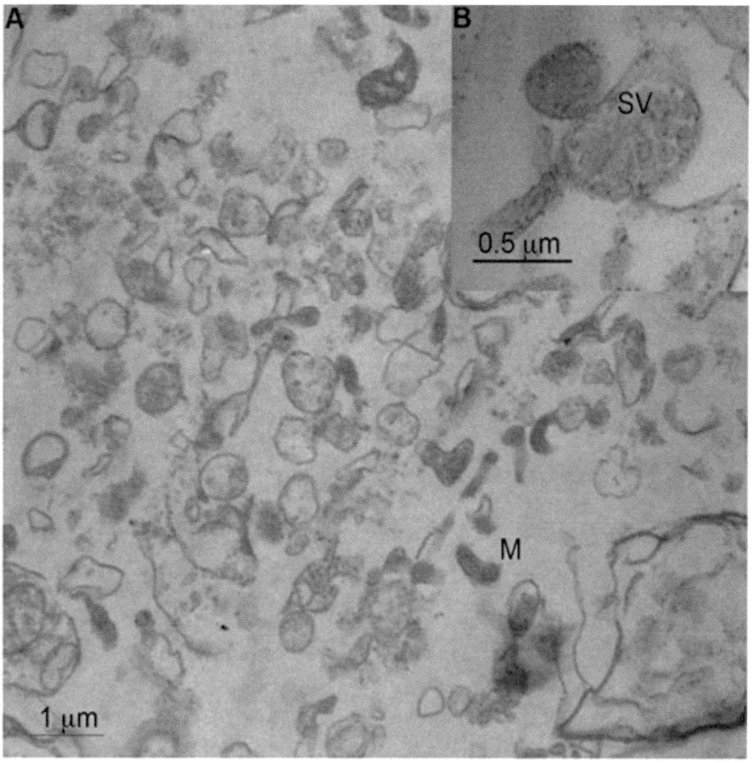

Fig. 10.1 Electron micrograph of a synaptosome fraction isolated from an archived postmortem human brain by sucrose density centrifugation at low magnification (**a**) High magnification (**b**). *M* mitochondria, *SV* synaptic vesicle

and Whittaker 1962; Whittaker et al. 1964), Ficoll/sucrose (Booth and Clark 1978), or Percoll (Nagy and Delgado-Escueta 1984) density gradient centrifugation. While the reagents for formation of the gradient may differ, all concurrent steps involve isolating a purified synaptosomal fraction from its mitochondrial and microsomal contamination.

10.1.2 Neuroproteomics of the Synapse

Advances in technologies have bolstered the field of neuroproteomics with refined tools for analyzing changes occurring in the CNS especially at the synapses in a range of neurological diseases. As described above, the main components of the synapse include the pre- and postsynaptic terminals. Extensive efforts on understanding the changes occurring in these compartments have now started to yield significant clues. Given the limited scope of this book chapter, we present a few exemplars of studies that have employed neuroproteomics to understand the synapse focusing on the synaptosomes. The protein profile of synaptosomes from the rat forebrain has been studied using 2D-GE (Witzmann et al. 2005). Over 900 protein spots were detected by 2D-GE, from which 91 unique proteins were identified; 155 additional proteins were identified through the untargeted studies. Of the total 246 proteins, 61 were proteins with known functions at the synapse. In a separate study, the proteome of synaptosomes isolated from mouse brains was characterized in two separate experiments employing ICAT that identified a total of 1,131 proteins, including representatives of most synaptic structures and functions (Schrimpf et al. 2005). Employing SILAM, another study aimed at analyzing biological changes within the synapses of the cerebellum at four developmental time points (McClatchy et al.

2007). Of the 1,138 quantified proteins, 196 were found to be statistically significant. Some other works on synaptosomal proteomics include analysis of cerebral synaptosomes isolated from a rodent model of Alzheimer's (Yang et al. 2011; Boyd-Kimball et al. 2005), antipsychotic resistance (Ji et al. 2009), and analysis of mouse synaptosome proteome and phosphoproteome (Filiou et al. 2010), to name a few.

While these above mentioned studies have employed an animal model to elucidate the synapse, studies using human brain tissue have been very limited. A recent study reported the identification of 31 candidate protein spots associated with increased immunoproteosome subunits from synaptosomes isolated from the neocortex of HIV-infected decedents with HIV encephalitis (Gelman and Nguyen 2009). Another study assessed the phosphoproteome using synaptosome fractions from human cerebral cortex (DeGiorgis et al. 2005).

10.1.3 Posttranscriptional Regulators at the Synapse

While it is proteins that carry out the greater part of biological events in a cell, including at the synapse, changes at the RNA level may orchestrate a plethora of gene regulatory associated mechanisms. As part of the specialization of the synapse, changes in synaptic protein composition can occur through local translation of mRNA at the synapse. Of course, a prerequisite for local translation is the presence of mRNA (Bramham and Wells 2007) and translation machinery in the vicinity of synapses (Steward and Schuman 2001). A relatively new addition to the mechanisms of translational regulation at the synapse is small noncoding RNAs, in particular microRNAs (miRNAs), which are important posttranscriptional regulators of gene expression in the brain and thus are key players in mechanisms of translational control of neurons (Konecna et al. 2009; Yelamanchili and Fox 2009). In the brain, miRNAs are emerging as important regulators of synaptic plasticity (Chandrasekar and Dreyer 2009; Junn et al. 2009; Khudayberdiev et al. 2009; Laterza et al. 2009; Papagiannakopoulos and Kosik 2009) especially at the synapse (Siegel et al. 2011; Schratt 2009). To date, there have been very limited studies employing a rodent model to characterize the expression of miRNAs at the synapse (Lugli et al. 2008; Smalheiser 2008; Smalheiser and Lugli 2009). Research in this direction is just starting to be recognized, and understanding the potentially huge impact of miRNA-based mechanisms modulating the synaptic proteome will be key to deciphering key mechanisms in many CNS-associated diseases.

As stated above, most of the studies described have employed animal models. Advantages include the relative ease of obtaining fresh tissue versus archived frozen human tissue from brain banks and the use of carefully controlled experiment conditions. However, while animal models are insightful, translational medicine is incomplete without confirming the findings in humans. Furthermore, there are critical aspects of human disease that may be missing from experimentally designed animal models.

Here, we describe the protocol used in our laboratory for isolating synaptosomes from archived human brain tissues using sucrose density centrifugation. It is the discretion of the end user to either isolate protein or RNA; special modifications for maintenance of intact RNA are italicized and depicted in parentheses.

10.2 Materials

(*Note*: For RNA work, it is recommended to autoclave the required glassware prior to starting the experiment. Additionally, usage of RNAse-free microfuge tubes is highly recommended. Spraying of RNAseZap around the workbench also significantly reduces the risk of contamination.)

10.2.1 Materials for Tissue Acquisition

1. Dissection tools: forceps, scalpel blades
2. Dry ice

3. Phosphate-buffered saline (homemade or Invitrogen, Carlsbad, CA, USA, Cat. #A12856-01)

10.2.2 Materials for Fractionation

1. Sucrose (Sigma, St. Louis, MO, USA Cat. #S3089)
2. EDTA (Sigma, Cat. #ED-100g)
3. HEPES (Sigma, Cat. #H3375)
4. Protease Inhibitor cocktail tablets (Roche, Indianapolis, IN, USA Cat. #05 892 791 001)
5. PhosSTOP Phosphatase Inhibitor cocktail tablets (Roche, Cat. #04 906 845 001)
6. *RNAsecure* (Ambion/Life Sciences, Grand Island, NY, USA Cat. #AM7005)
7. *RNAseZap* (Ambion, Cat. #AM9780)
8. *SUPERase In* (Ambion, Cat. #AM2694)
9. Wheaton glass tissue grinder, Potter -ELV (Wheaton, Millville, NJ, USA Cat. #358044)
10. Wheaton Overhead Stirrer (Fisher Scientific, Cat. #22-244-382)
11. 15 mL Falcon Tubes (BD Biosciences, San Jose, CA, USA Cat. #352097)
12. 13.2 mL Thinwall Polyallomer Tubes (Beckman Coulter, Indianapolis, IN, USA Cat. #331372)
13. Microfuge tubes (several manufacturers)
14. SW41 Ti Rotor (Beckman Coulter)
15. Ultracentrifuge (Beckman Coulter)
16. Microfuge (several manufacturers)

10.2.3 Protein Estimation

1. BCA protein assay kit (Thermo Scientific, Hudson, NH, USA Cat. #23225)
2. Bovine serum albumin (BSA), 2 mg/mL (Thermo Scientific, Cat. #23209) (*Note*: BSA is supplied in ampoules as a stock of 2 mg/mL. It is recommended the user perform serial dilution of the stock solution using PBS. Please refer to protein estimation as described elsewhere in the book.)
3. Microplate reader capable of reading 560 nm (Several manufacturers)

10.2.4 RNA

1. Trizol reagent (Invitrogen, Carlsbad, CA, USA Cat. #15596018)
2. Choloroform (Acros Organics, Cat. #610030040)
3. Isopropanol (Acros Organics (Part of Thermo Fisher Scientific), Cat. #327270010)
4. RNA storage solution (Ambion, Cat. #AM7000)
5. Spectrophotometer (Several manufacturers) or NanoDrop 2000 (Thermo scientific) (*Note*: Another option to test the integrity of the RNA sample is by a Bioanalyzer (Agilent Technologies) that assesses the 18S and 28S bands.)

10.3 Methods

10.3.1 Tissue Acquisition

(*Note*: All subsequent preparations are performed at 4 °C unless specified.)

1. Prior to tissue acquisition, place a petri dish on ice.
2. Locate the areas of interest and dissect them carefully using clean forceps and scalpel blades.
3. Weigh the tissue and proceed to fractionation.

10.3.2 Tissue Fractionation

1. Prior to starting the procedure, prepare the sucrose gradients comprising of 0.6, 0.8, and 1.2 M concentrations of sucrose using autoclaved water. (*For RNA preparation, treat sucrose solution with RNAsecure adjusted to 1× concentration followed by addition of SUPERase In at a final concentration of 160 U/mL.*) Prepare the homogenization buffer (0.32 M sucrose, 5 mM HEPES, 0.1 mM EDTA, with the addition of protease and phosphatase inhibitors adjusted to a final concentration of 1× concentration). (*Note*: For RNA, treat homogenization buffer with RNAsecure adjusted to 1× concentration followed by addition of SUPERase In at a final concentration of 160 U/mL.)

2. Using a 5 mL plastic pipette, slowly add 3.3 mL of the 1.2 M sucrose followed by gentle addition of 3.3 mL each of 0.8 M and 0.6 M sucrose solutions into 13.2 mL polyallomer ultracentrifuge tubes. (*Note*: The 1.2 M sucrose is at the bottom followed by the 0.8 M and 0.6 M on the top.) Carefully transfer the tubes to 4° C until further use. (*Note*: It is generally recommended to prepare the gradients prior to the start of the experiment. Also, such casted gradients can be left overnight at 4 °C.)
3. Using a glass homogenizer, homogenize the tissue (at least 250 mg for protein and 1 gm for RNA) with 10 volumes of ice-cold homogenization buffer with 12 strokes moving in an up and down fashion using a Wheaton Overhead Stirrer at 250–300 rpm. Store a 200 μL aliquot of this homogenate for analysis by Western blot or RNA isolation as the starting material to determine purification efficiency. Transfer the rest to a 15 mL Falcon Tubes.
4. Spin at 1,000 × *g* for 10 min at 4° C and collect supernatant. (*Note*: For the ease of working, the supernatant can be collected into clean DNAse- and RNAse-free microfuge tubes.)
5. Spin at 12,000 × *g* for 20 min at 4 °C. Discard the supernatant. The obtained pellet is the crude synaptosomal pellet.
6. Gently resuspend the pellet in the homogenization buffer containing protease and phosphatase inhibitors and overlay carefully on the top of the sucrose gradients. (*Note*: Balance the tubes using the homogenization buffer before placing them in the ultracentrifuge.)
7. Using a SW41 Ti Rotor, spin at 145,000 × *g* for 90 min at 4° C in an ultracentrifuge (*Note* the series of layers obtained, Fig. 10.2).
8. Harvest the synaptosomal band at the interface of 0.8 and 1.2 M sucrose using a 18-guage needle. Pierce gently underneath the synaptosomal band and slowly collect using a syringe.
9. Dilute the obtained purified synaptosomes with 10 volumes of homogenization buffer into a new 13.2 mL ultracentrifuge tube. Perform a wash by centrifuging at 145,000 × *g*, for 30 min at 4° C.
10. Gently discard the supernatant and resuspend the purified synaptosomal pellet in the appropriate buffer (e.g., PBS) for the required downstream application (*see below for RNA*). Estimate the protein concentration using the BCA method and BSA as the standard, following the manufacturer's recommendations. (*Note*: The obtained synaptosomal sample can be stored at −80°C until further use.)

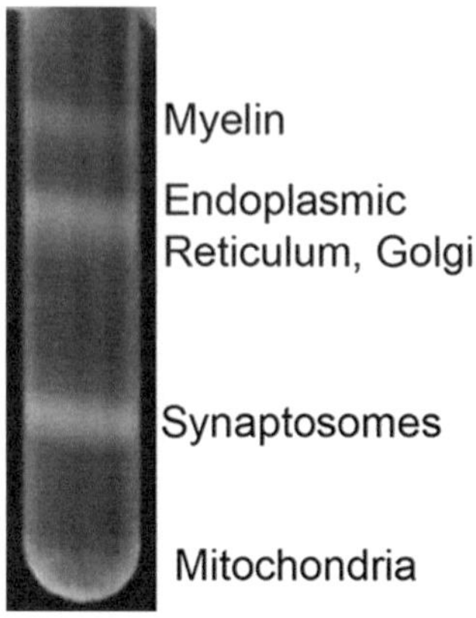

Fig. 10.2 Illustration of layers formed after sucrose gradient centrifugation to isolate synaptosomes

10.3.3 Western Blotting

For protein work, in order to validate the fractionation protocol, western blotting analysis should be performed using the aliquots obtained during the procedure. Antibodies such as synaptosomal associated protein 25-kDa (SNAP-25) and synaptophysin can be used to confirm the enrichment of synaptosomes, and actin or glyceraldehyde 3-phosphate dehydrogenase (GAPDH) for depletion of cytoplasmic components. Utilize the saved aliquot (step 3 of Sect. 10.3.2, with protein quantified by the BCA method as above). Since the quality of the sample is critical, it is recommended to perform this validation prior to the downstream analysis. It is expected that the expression of synaptic markers will be higher, and the cytoplasmic markers lower, in the synaptosomal fraction compared to the homogenate.

Details for western blotting are described elsewhere in this book.

10.3.4 Isolation of RNA

1. Dissolve the isolated synaptosomal pellet in 1 mL of Trizol reagent.
2. Transfer the dissolved pellet into a clean microfuge tube and allow it to sit for 5 min at room temperature.
3. Add 200 μL of chloroform to the sample and vortex vigorously for 15 s. Allow the sample to incubate at room temperature for an additional 2–3 min.
4. Centrifuge at 12,000 × *g* (about 10,000 rpm) in a microfuge at 4 °C for 15 min. Centrifugation separates the sample into three layers: a lower organic layer, an interphase layer containing proteins, and an upper aqueous layer, which contains RNA.
5. Remove the upper aqueous layer using a pipetman (without disturbing the interface) and transfer it into a sterile microfuge tube. If you think you carried over any of the interphase or organic layer, spin the tube in the microfuge for 1 min. This will pellet any sediment and re-separate any layers. Transfer the aqueous phase to a new tube. Add an equal volume of isopropanol to the tube, invert twice then let incubate at room temperature for 10 min. This is the RNA precipitation step.
6. Pellet the RNA by centrifuging the samples at 12,000 × *g* for 10 min. in a microfuge.
7. Centrifuge at 4 °C. A clear to white pellet should result. Pour off the supernatant. Add 1 mL of 70 % ethanol to the pellet to remove any lingering chemicals, invert the tube twice then spin at 7,500 × *g* in a microfuge for 5 min. The pelleted RNA can be dissolved in diethylpyrocarbonate (DEPC)-treated water. Quantify RNA using a spectrophotometer or NanoDrop. Additional quantification includes assessing RNA integrity by Bioanalyzer (*Note*: The obtained RNA needs to be stored at −80 °C until further use).
8. To determine purification efficiency, extract RNA from the material saved in step 3 of Sect. 10.3.2 above. Place 100 μL of the saved aliquot in each of two microfuge tubes, add 1 mL TRIzol, vortex, and allow to sit for 5 min. Continue with steps 3–7 above.

10.4 Summary

In the CNS, synapses are known to be the key structure involved in neurotransmission and neuroplasticity. One ex vivo method to understand the synapse is by isolation of synaptosomes. While it is relatively easier to obtain fresh animal tissue, usage of archived human tissue from the brain banks will significantly enhance in understanding the underlying etiology in an array of neurodegenerative diseases. Here, we describe the isolation of synaptosomes from archived human brain tissue using subcellular fractionation, which when combined to high-throughput "omics"-based approaches could yield vital clues into understanding the underlying bases of neurodegeneration in several neurodegenerative diseases.

References

Bayes A, Grant SG (2009) Neuroproteomics: understanding the molecular organization and complexity of the brain. Nat Rev Neurosci 10:635–646

Booth RF, Clark JB (1978) A rapid method for the preparation of relatively pure metabolically competent synaptosomes from rat brain. Biochem J 176:365–370

Boyd-Kimball D, Castegna A, Sultana R, Poon HF, Petroze R, Lynn BC, Klein JB, Butterfield DA (2005) Proteomic identification of proteins oxidized by Abeta(1–42) in synaptosomes: implications for Alzheimer's disease. Brain Res 1044:206–215

Bramham CR, Wells DG (2007) Dendritic mRNA: transport, translation and function. Nat Rev Neurosci 8:776–789

Chandrasekar V, Dreyer JL (2009) microRNAs miR-124, let-7d and miR-181a regulate cocaine-induced plasticity. Mol Cell Neurosci 42:350–362

DeGiorgis JA, Jaffe H, Moreira JE, Carlotti CG Jr, Leite JP, Pant HC, Dosemeci A (2005) Phosphoproteomic analysis of synaptosomes from human cerebral cortex. J Proteome Res 4:306–315

Filiou MD, Bisle B, Reckow S, Teplytska L, Maccarrone G, Turck CW (2010) Profiling of mouse synaptosome proteome and phosphoproteome by IEF. Electrophoresis 31:1294–1301

Gelman BB, Nguyen TP (2010) Synaptic proteins linked to HIV-1 infection and immunoproteasome induction:

proteomic analysis of human synaptosomes. J Neuroimmune Pharmacol 5(1):92–102
Gray EG, Whittaker VP (1962) The isolation of nerve endings from brain: an electron-microscopic study of cell fragments derived by homogenization and centrifugation. J Anat 96:79–88
Gygi SP, Rist B, Gerber SA, Turecek F, Gelb MH, Aebersold R (1999) Quantitative analysis of complex protein mixtures using isotope-coded affinity tags. Nat Biotechnol 17:994–999
Ji B, Zhang Z, Zhang M, Zhu H, Zhou K, Yang J, Li Y, Sun L, Feng G, Wang Y, He L, Wan C (2009) Differential expression profiling of the synaptosome proteome in a rat model of antipsychotic resistance. Brain Res 1295:170–178
Junn E, Lee KW, Jeong BS, Chan TW, Im JY, Mouradian MM (2009) Repression of alpha-synuclein expression and toxicity by microRNA-7. Proc Natl Acad Sci U S A 106:13052–13057
Khudayberdiev S, Fiore R, Schratt G (2009) MicroRNA as modulators of neuronal responses. Commun Integr Biol 2:411–413
Kim SI, Voshol H, van Oostrum J, Hastings TG, Cascio M, Glucksman MJ (2004) Neuroproteomics: expression profiling of the brain's proteomes in health and disease. Neurochem Res 29:1317–1331
Konecna A, Heraud JE, Schoderboeck L, Raposo AA, Kiebler MA (2009) What are the roles of microRNAs at the mammalian synapse? Neurosci Lett 466:63–68
Laterza OF, Lim L, Garrett-Engele PW, Vlasakova K, Muniappa N, Tanaka WK, Johnson JM, Sina JF, Fare TL, Sistare FD, Glaab WE (2009) Plasma MicroRNAs as sensitive and specific biomarkers of tissue injury. Clin Chem 55:1977–1983
Liao L, McClatchy DB, Yates JR (2009) Shotgun proteomics in neuroscience. Neuron 63:12–26
Lugli G, Torvik VI, Larson J, Smalheiser NR (2008) Expression of microRNAs and their precursors in synaptic fractions of adult mouse forebrain. J Neurochem 106:650–661
McClatchy DB, Yates JR, 3rd (2008) Stable isotope labeling of mammals (SILAM). CSH Protoc 2008:pdb prot4940
McClatchy DB, Liao L, Park SK, Venable JD, Yates JR (2007) Quantification of the synaptosomal proteome of the rat cerebellum during post-natal development. Genome Res 17:1378–1388
Nagy A, Delgado-Escueta AV (1984) Rapid preparation of synaptosomes from mammalian brain using nontoxic isoosmotic gradient material (Percoll). J Neurochem 43:1114–1123
Ong SE, Blagoev B, Kratchmarova I, Kristensen DB, Steen H, Pandey A, Mann M (2002) Stable isotope labeling by amino acids in cell culture, SILAC, as a simple and accurate approach to expression proteomics. Mol Cell Proteomics 1:376–386
Papagiannakopoulos T, Kosik KS (2009) MicroRNA-124: micromanager of neurogenesis. Cell Stem Cell 4:375–376
Pocklington AJ, Armstrong JD, Grant SG (2006) Organization of brain complexity—synapse proteome form and function. Brief Funct Genomic Proteomic 5:66–73
Ross PL, Huang YN, Marchese JN, Williamson B, Parker K, Hattan S, Khainovski N, Pillai S, Dey S, Daniels S, Purkayastha S, Juhasz P, Martin S, Bartlet-Jones M, He F, Jacobson A, Pappin DJ (2004) Multiplexed protein quantitation in Saccharomyces cerevisiae using amine-reactive isobaric tagging reagents. Mol Cell Proteomics 3:1154–1169
Schratt G (2009) microRNAs at the synapse. Nat Rev Neurosci 10:842–849
Schrimpf SP, Meskenaite V, Brunner E, Rutishauser D, Walther P, Eng J, Aebersold R, Sonderegger P (2005) Proteomic analysis of synaptosomes using isotope-coded affinity tags and mass spectrometry. Proteomics 5:2531–2541
Siegel G, Saba R, Schratt G (2011) microRNAs in neurons: manifold regulatory roles at the synapse. Curr Opin Genet Dev 21(4):491–497
Smalheiser NR (2008) Synaptic enrichment of microRNAs in adult mouse forebrain is related to structural features of their precursors. Biol Direct 3:44
Smalheiser NR, Lugli G (2009) microRNA regulation of synaptic plasticity. Neuromolecular Med 11:133–140
Steward O, Schuman EM (2001) Protein synthesis at synaptic sites on dendrites. Annu Rev Neurosci 24:299–325
Sudhof TC (2004) The synaptic vesicle cycle. Annu Rev Neurosci 27:509–547
Whittaker VP, Michaelson IA, Kirkland RJ (1964) The separation of synaptic vesicles from nerve-ending particles ('synaptosomes'). Biochem J 90:293–303
Williams RW and Herrup K (1988) The control of neuron number. Ann Rev Neurosci 11:423–453
Witzmann FA, Arnold RJ, Bai F, Hrncirova P, Kimpel MW, Mechref YS, McBride WJ, Novotny MV, Pedrick NM, Ringham HN, Simon JR (2005) A proteomic survey of rat cerebral cortical synaptosomes. Proteomics 5:2177–2201
Yang H, Qiao H, Tian X (2011) Proteomic analysis of cerebral synaptosomes isolated from rat model of alzheimer"s disease. Indian J Exp Biol 49(2):118–124
Yelamanchili SV, Fox HS (2010) Defining larger roles for "tiny" RNA molecules: role of miRNAs in neurodegeneration research. J Neuroimmune Pharmacol 5(1):63–69
Zhu H, Pan S, Gu S, Bradbury EM, Chen X (2002) Amino acid residue specific stable isotope labeling for quantitative proteomics. Rapid Commun Mass Spectrom 16:2115–2123

Xenopus Oocytes Isolation and Microinjection

11

Huangui Xiong

Abstract

Xenopus oocytes, obtained from sexually mature female *Xenopus laevis*, are widely used as a model system for expression cloning and structure-function studies of ion channels and receptors. The ovarian tissue can be extracted either by a nonfatal surgical removal of parts of the ovary on an anesthetized frog several times during its lifetime or by removal of the ovary from the frog decapitated after deep anesthesia. Then, the *Xenopus* oocytes can be isolated and defolliculated manually or enzymatically or combined. Healthy-looking oocytes in stage V and/or VI are selected for microinjection of RNA or DNA. This chapter describes the methodology and procedures of surgical removal of ovarian tissue, enzymatic isolation and defolliculation of *Xenopus* oocytes, and oocyte microinjection.

Keywords

Xenopus laevis • *Xenopus* oocytes • Oocyte isolation • Defolliculation • Oocyte microinjection

11.1 Introduction

Xenopus laevis (the African clawed frog) oocytes are a popular model system for a wide variety of biological studies. Microinjection of DNA or messenger RNA (mRNA) into the *Xenopus* oocytes leads to the functional expression of the encoded proteins (Gurdon et al. 1971). The proteins expressed in such a new environment can be functionally characterized under well-defined conditions (Gurdon 1974; Dascal 1987; Sigel 1990). This makes oocytes an important expression system for studies of molecular biology and electrophysiology. Such an in vitro expression system is particularly suitable for studying not only the structure and functions of cloned and mutated proteins but also the physiology of neurotransmitter receptors, ion channels, and transporters (Wallingford et al. 2010). In addition, oocytes are

H. Xiong, M.D., Ph.D. (✉)
Department of Pharmacology and Experimental Neuroscience, University of Nebraska Medical Center, Durham Research Center, Durham Research Plaza, DRC I 8034, 45th Street at Emile Street, Omaha, NE 68198-5880, USA
e-mail: hxiong@unmc.edu

H. Xiong and H.E. Gendelman (eds.), *Current Laboratory Methods in Neuroscience Research*, Springer Protocols Handbooks, DOI 10.1007/978-1-4614-8794-4_11,

also used in biochemical analyses of DNA replication and repair (Blow and Laskey 1986) and many other studies in experimental biology.

The *Xenopus* oocytes are classified in stages, from stage I to stage VI, depending on their development stage (Dumont 1972). They are relatively small and immature in stages I–III and become fairly large cells (~1.2–1.3 mm in diameter) in stages V and VI. Beginning at stage IV, oocytes have two poles, the animal pole and the vegetal pole. The animal pole is dark brown in color and the vegetal pole is yellowish. Stage V and VI oocytes are usually used for electrophysiological experiments. As oocytes are surrounded by several layers of structure including vitelline membrane and follicular layer, it is necessary to remove the surrounding structures especially follicle cell layer, because follicle cells possess intrinsic receptors (e.g., angiotensin) and ion channels (delayed-rectifier K channels), which can respond to experimental stimuli and contribute to the oocyte's overall electrophysiological responses (Fraser et al. 1993; Stuhmer and Parekh 1995) .

Since the seminal studies by John Gurdon and his colleagues (Gurdon et al. 1971), *Xenopus* oocytes have been widely used as an expression system because (a) the fully grown oocytes are large in size (~1.3 mm in diameter) which has made them popular cells not only for easy microinjection of DNA, RNA, or membrane impermeable drugs into the cells without compromise to their health, but also for electrophysiological analysis of ion channels and transporters that can be performed by two-electrode voltage clamp recordings; (b) the injected RNA can be faithfully expressed in this expression system; (c) the large size also makes it possible to manually isolate oocyte nuclei to study ion channels at the nucleus (Mak and Foskett 1994, 1998) or remove the oocyte nucleus (enucleation) for examination of the transcription-independent actions; (d) functional analyses can be accomplished within hours to days, as compared to the process of weeks to months that is involved in the isolation of mammalian cells transfected with the cDNA of interest; (e) the oocytes can be surgically extracted several times from a given donor frog; and (f) the handling and care for *X. laevis* and *Xenopus* oocytes are relatively simple, easy, and cheaper in comparison with other transfection and expression systems. The main disadvantages of the oocyte system include the following: (a) there are endogenous ion channels, although few on oocyte membrane that might interfere with the membrane currents conducted by the expressed channels if the currents are small, and (b) post-translational modification for ion channel proteins expressed may be different in the oocyte compared with the native cells. Thus channels may actually function differently in their native environment. As a result, caution must be taken when interpreting experimental data.

11.2 Objective

The objective of this chapter is to provide readers with basic procedures and methodologies needed to extract oocytes surgically from sexually mature female *X. laevis*, particularly targeting graduate students and laboratory technologists.

Fig. 11.1 Female frogs of *Xenopus laevis* (www.xenopus.com)

11.3 Materials

11.3.1 Biological Materials

- Female frogs of *X. laevis* as shown in Fig. 11.1 (also known as African claw-toed frog).

11.3.2 Chemicals, Reagents, and Antibiotics

- Conventional chemicals for making standard oocyte solutions (SOS)
- Pyruvate, sodium salt
- Potassium phosphate, monobasic (KH_2PO_4)
- Potassium phosphate, dibasic (K_2HPO_4)
- HEPES
- Tricaine (3-aminobenzoic acid ethyl ester methanesulfonate salt, $C_9H_{11}NO_2 \times CH_4SO_3$)
- Collagenase type IA (Sigma-Aldrich)
- Gentamicin
- Betadine solution (10 % povidone iodine in water)
- 75 % alcohol

11.3.3 Surgical Equipment

- Straight blunt specimen forceps
- Iris forceps (curved and straight)
- Fine Dumont forceps (No. 5)
- Micro-dissecting scissors (straight, sharp)
- Surgery scissors
- Needle holder
- Suture needles (size 3)
- Monofilament nonabsorbable suture materials (3–0 proline or nylon suture)

11.3.4 Technical Equipment

- Stereo microscope
- Shaker for the tubes
- Fiber Lite (Model 190, optional)
- Pipette for handling oocytes
- Oocyte transfer pipette
- Conical tubes (15 mL)
- Large glass petri or cell culture dishes (100 mm)
- Petri dishes (60 mm)
- Parafilm

11.4 Solutions

11.4.1 Standard Oocyte Solution

- NaCl 100 mM
- KCl 2.0 mM
- $CaCl_2$ 1.8 mM
- $MgCl_2$ 1.0 mM (or slightly less)
- HEPES 5.0 mM

We typically make 20× SOS using double-distilled water (ddH_2O), store at 4 °C, and dilute 20× SOS to 1× SOS with ddH_2O on experimental day (be accurate, therefore use a measuring cylinder).

11.4.2 Supplemented SOS for Oocyte Culture

To make supplemented SOS (sSOS), add 275 mg sodium pyruvate (2.5 mM, tissue culture grade, store at 4 °C) and 1.0 mL of gentamicin (50 μg/mL) to 1,000 mL 1× SOS. Adjust pH to 7.6 with NaOH. Osmolarity 200 ± 20 mOsm. Filter, autoclave, and store at 4 °C. sSOS is used for oocyte incubation/culture.

11.4.3 Defolliculation Solutions

1. Collagenase solution (20 mg/mL)
   Dissolve 400 mg collagenase type 1A in 20 mL of ice cold 1× SOS. Solution is a muddy brown color when dissolved. Aliquot into ~1.5 mL per tube and store at −20 °C. Thaw as needed on experimental day. For each new batch of collagenase, watch oocytes closely as some batches of oocytes are more sensitive to collagenase than others.
2. 0.1 M potassium phosphate solution (pH 6.5)
   Make 200 mL of 0.2 M stock KH_2PO_4 (monobasic) (5.44 g/200 mL) and 100 mL of 0.2 M

stock K_2HPO_4 (dibasic) (3.48 g/100 mL). Mix 118 mL of monobasic and 82 mL of dibasic. Dilute to 400 mL, check pH, and filter.

11.4.4 Tricaine Solution (for Frog Anesthesia)

Make 2 L 0.17 % (w/v) Tricaine solution (3.5 g/2 L ddH_2O).

Note: Store Tricaine (3-aminobenzoic acid ethyl ester methanesulfonate salt) dry powder at −20 °C. Wear gloves when handling and avoid inhalation and touch. Dispose excess Tricaine down the sink followed by water.

The Tricaine solution can also be used to euthanize frogs when needed.

Procedures for euthanizing frogs:

1. Prepare ~4–6 L 0.17 % Tricaine solution in large "*Xenopus* bucket."
2. Place frog(s) in solution for approximately 3.5 h.
3. Remove the frog(s) from solution and place them in plastic bags. Leave in autopsy refrigerator for disposal/incineration.

11.5 *Xenopus* Frog Anesthesia and Surgical Removal of Oocytes

11.5.1 Frog Anesthesia

1. Pour 2 L Tricaine solution into a "Xenopus bucket."
2. Immediately prior to surgery, the frog is immersed in 0.17 % Tricaine solution for 5 min and then cooled by addition of equal volume (or as much as possible) of ice for 10 min.
3. After a total of 15 min anesthesia, check frog for limb reflex. Successful anesthesia is indicated when the frog fails to move when touched on legs or abdomen. The frog can be kept in Tricaine for maximum of 20 min.

11.5.2 Surgical Removal of Oocytes

1. When successful anesthesia is reached, take the frog out of Tricaine solution and lay the frog on ventral up position on a flat bed of ice in a plastic tray; cover the frog's head and legs with ice and wet with water.
2. Illuminate the sight of operation with an optic fiber light source (cold light).
3. Clean the skin of frog abdominal region with Betadine followed by alcohol swab three times.
4. Slightly lift skin up with forceps (with grips) and make a 1 cm incision on the skin with iris scissors. The incision is located on the lower left or right abdomen in parallel with the midline, 05–1.0 cm away from the midline.
5. Make a similar incision in the muscle layer beneath the skin. Care should be taken to prevent cutting anything within the abdomen.
6. Use a blunt forceps to reach into abdomen and gently pull out ovary tissue that contains eggs. It comes out in lobes.
7. Separate and cut off 2–4 ovarian lobes, place them in sSOS, and wait for further isolation processing (Sect. 11.6.1).
8. Suture abdominal muscle and skin separately using a size 3-suture needle and monofilament nonabsorbable suture materials (3-0 proline or nylon suture). The skin sutures will be removed at approximately 10–14 days post-procedure.
9. Place frog in a tripped container filled with 100–200 mL deionized water to keep frog wet but still able to breathe (tilt the container ~15°, keep the frog head in an upright position to avoid drowning). Observe frog every 10–15 min.
10. Once breathing and moving, place frog in recovery tank (filled with 2 L of deionized water). The frog will be kept in recovery tank for at least 12 h for observation before returning to frog colony tank.

Note: A recovery period of 2 weeks is recommended between operations; five operations are the maximum for a given frog.

11.6 Oocyte Isolation and Defolliculation

11.6.1 Oocyte Isolation: Collagenase Treatment

Continuation on Sect. 11.5.2, step 6. All steps done at room temperature:

1. In sSOS, tear apart ovary into small pieces containing 10–25 oocytes/piece. Try to avoid breaking too many oocytes with forceps. The pair of forceps used in tying sutures work well for this step. Change sSOS if it becomes too cloudy. As the solution becomes cloudy, it becomes hypertonic which is not good for the health of oocytes (do this in a glass petri dish).
2. Pour ovary clumps and sSOS into 15 mL conical, screw top tube. Discard supernatant and add fresh sSOS. Ovary clumps filling 1.5–2.0 mL is sufficient for single prep.
3. Rinse ovary clamps several times with sSOS to remove debris and single oocytes. When solution is clear, proceed to step 4. Rinsing is best accomplished using a 1.5 mL Pasteur pipette. Make sure the end of the pipette is not jagged since this would cause damage. Avoid blowing air bubbles into the solution.
4. Remove excess sSOS. Add ~1.5 mL 20 mg/mL collagenase (type 1A) which was previously made in SOS and stored at −20 °C. Screw the cap.
5. Incubate with gentle movement for 30–45 min. Movement at 25–30 rpm on an orbital shaker works well if the tube is laid on its side. The incubation is complete when the solution becomes somewhat cloudy and there is viscous "goo" amongst the ovary.
6. Stop reaction by adding sSOS and quickly decanting solution. Replace solution a second time. The brownness of the collagenase solution should disappear.
7. Continue rinsing the ovary clamps/oocytes. Small white oocytes should appear in the solution as you mix and rinse with a pipette. Smaller oocytes should be discarded unless needed. Proceed when solution is fairly clear and most of the small white oocytes are gone.

11.6.2 Oocyte Defolliculation

Follicular cells are removed by a combination of shifting the pH and mechanical agitation. The defolliculating solution (0.1 M potassium phosphate pH 6.5) is below the pH of sSOS (pH 7.60). A maximum of 10 min at the lower pH is allowed. Shown below are the defolliculation procedures employed in my laboratory:

- Decant sSOS and replace with ~10 mL of defolliculation solution.
- Gently swirl the oocytes around using a Pasteur pipette. Be gentle!
- Change the solution after 1–2 min, then every minute or so until done. Continue swirling and sucking of small oocytes and follicular cells until complete.
- Follicular cells are red and should appear approximately 4 min after the pH shift. You should be able to get most of them off by 7–8 min (10 min maximum) into procedure.
- Stop by switching to sSOS. Rinse a few times and try to make solution clear.

11.6.3 Selecting "Best" Oocytes

1. Once rinsing is complete, decant oocytes and sSOS in to a large glass petri dish (90 mm in diameter).
2. Under a microscope, pick out and discard dead or dying oocytes using a clean pipette. Also discard out any that have a swirling pattern or are discolored.
3. Select the "best" oocytes and transfer to smaller dishes. The best oocytes are generally the largest ones (stage V and/or VI) that have an evenly colored animal pole (Fig. 11.2).

Fig. 11.2 Stage V and VI oocytes (www.xenopus.com)

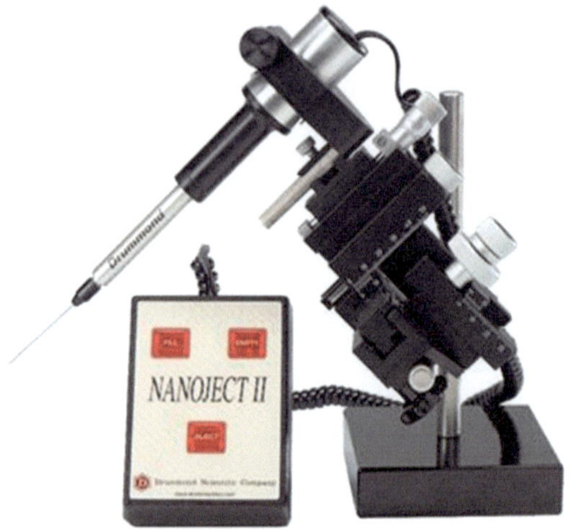

Fig. 11.3 Drummond Nanoject II Nanoliter injector (www.drummondsci.com)

4. The selected oocytes are kept in an incubator at 18 °C until they are ready for microinjection.
5. Change solution daily. If oocytes are dying in large numbers, transfer survivors to clean solution and discard dying oocytes.

11.7 Microinjection of *Xenopus* Oocytes

11.7.1 Injection Equipment

- Stereo microscope
- Fiber Lite (Model 190, optional)
- Right- or left-handed manual manipulator
- Drummond Nanoject II Nanoliter injector (Fig. 11.3)
- Microelectrode puller
- Drummond thin-wall micropipettes

11.7.2 Chemicals, Reagents, and Other

- Desired injection solution (RNA or DNA)
- 0.1 % (v/v) diethylpyrocarbonate (DEPC)-treated water
- Mineral oil (RNase and DNase free)
- 75 % alcohol
- RNaseAway
- 1 mL syringe
- Microfil needle (MF28G-5, WPI)
- Whatman® filter paper, round
- Latex groves

11.7.3 Preparation for Oocyte Microinjection

The injection of oocytes should be performed in a clean, RNase-free laboratory bench area. Experimenter should wear clean lab coat with gloves. Spray RNaseAway in the bench area and wipe gloves with RNaseAway.

1. Setup the Drummond Nanoject II Nanoliter injector.
   (a) Mount and secure the Drummond Nanoject II Nanoliter injector onto the micromanipulator.
   (b) Connect the injector head with control box.
   (c) Plug in the control box to a wall outlet.
2. Fabrication of injection microelectrodes. Using the capillary glass provided (Cat. No. 3-000-203-G/XL) by the Drummond, the microelectrodes are pulled through a Sutter P-1000 puller (or other commercial available pullers). The tip of the injection electrode should be broken to a diameter of 15–20 μm with the edge beveled.

3. Fill a 1 mL sterile disposable syringe with mineral oil.
4. Attach a Microfil needle to the 1 mL syringe and expel the air from the needle.
5. Insert the syringe needle into the injection microelectrode and backfill with approximately 3–4 cm of mineral oil.
6. Place and secure the electrode into the collet of the Drummond Nanoject II Nanoliter injector. Avoid introduction of air bubbles during electrode assembly.
7. Press and hold the EMPTY button to drive mineral oil out of the electrode until an audible beep is heard.
8. Thaw the RNA (or DNA) at room temperature. Spin down the RNA (or DNA) in an Eppendorf microfuge for about 1 min to ensure all the sample is collected at the bottom of the Eppendorf tube. Using a sterile disposable pipette tip, take 2–3 μL of RNA (or DNA) and place it in the middle of a 35 mm petri dish to form a small droplet sphere.
9. Place the petri dish on the platform of a stereo microscope. Place the electrode tip into the sample droplet sphere under the microscope. Press and hold FILL button to fill the injection electrode with the sample (suck the sample into the electrode). Stop filling at any time by letting go of the FILL button and continued filling can be accomplished by pressing the button again. Do not allow air bubbles to form in the electrode; these bubbles can cause inaccurate injection volumes. The filling speed is controlled by #5 dip switch located at side of the control box, with #5 up = 46 nL/s fill rate and #5 down = 23 nL/s fill rate, respectively.

11.7.4 Oocytes Injection

1. Setting injection volumes
The injection volumes and injection speeds are controlled by the positions of the dip switches located on the side of the control box. Switch #1–4 control volume, and switch #5 controls the injection rate and fill rate. Dip switch positions and corresponding volumes are shown in Table 11.1.

Table 11.1 Injection volume is determined by the position of the dip switches

	Dip switches			
Volume (nL)	1	2	3	4
2.3	U	U	U	U
4.6	D	U	U	U
9.2	U	U	U	U
13.8	D	D	U	U
18.4	U	U	D	U
23.0	D	U	D	U
27.6	U	D	D	U
32.2	D	D	D	U
36.8	U	U	U	D
41.4	D	U	U	D
46.0	U	D	U	D
50.6	D	D	U	D
55.2	U	U	D	D
59.8	D	U	D	D
64.4	U	D	D	D
69.0	D	D	D	D

Dip switch #5 controls injection rates; U = fast (46.0 nL/s); D = slow (23.0 nL/s)

2. Oocyte injection
   (a) Put a circular filter paper inside a 35 mm plastic petri dish (to prevent oocytes from rolling away during injection), and fill the dish with 2–3 mL sSOS. Transfer 15–20 selected "best" oocytes (see Sect. 11.6.3) to the petri dish.
   (b) Place the 35 mm petri dish under a dissecting microscope. Move the micromanipulator and position the injection electrode towards the oocytes at approximately 45°. Under ×12 magnification, advance the electrode onto the oocyte and use the electrode tip to dimple the oocyte. Then, use fine adjustment on the micromanipulator to penetrate the oocyte or penetrate the oocyte by a gentle tap on the fine adjustment on the micromanipulator. Once the electrode tip has broken the membrane, the dimple disappears.
   (c) Press the INJECT button on the control box, the amount of RNA (or DNA) in a preset volume is injected into the oocyte.
   (d) Withdraw the electrode from the injected oocyte and inject other oocytes by repeating aforementioned steps (b–d).

(e) Each healthy-looking stage V–VI oocyte is injected with either 36–50 nL water (control) or mRNA/DNA (at concentrations varying from 0 to 1 μg/μL) using the Nanoject II injector.
(f) Transfer the injected oocytes to a large (90 mm in diameter) glass petri dish filled with sSOS.

3. Postinjection care
Oocytes are kept at 18 °C in sSOS for 1–7 days, with changes of sSOS daily. Check oocytes health condition under microscope daily. Pick up unhealthy-looking or dying oocyte(s) and discard.
4. Cautions
(a) It is recommended to perform frog surgery to isolate and select oocytes one day and inject the oocytes in the following day. Some healthy-looking oocytes selected may turn out to be unhealthy or dying oocytes. Thus, those unhealthy and/or dying oocytes should be discarded before the injection.
(b) The vegetal hemisphere is usually the site for injection. Injections made on animal hemisphere may have the potential to damage the nucleus (located in animal hemisphere) and kill the oocyte.
(c) The cytosol may or may not leak out after withdrawal of the electrode. If excessive leakage occurs, this may indicate that the size of the injection electrode tip is too big.
(d) The electrode tip may become clogged after several injections, especially when the tip size of an electrode is relatively small. In this case, no sample solution will be expelled into the oocyte even though the Nanoliter injector is still running when INJECT button on the control box is pressed. A simple way to judge if the electrode is clogged or not is to observe the change in solution level (the interface of solution and mineral oil) in the electrode. Each successful injection will lead to a decrease in the solution level. If the level does not change after injection(s), it suggests the injection electrode is clogged.
(e) There may be oocyte(s) missed (un-injected) during injection. One way to prevent this is to inject one by one from left to the right and row by row, the other way is to transfer a few oocytes to the injection dish each time, to make it easier to remember/recognize which ones are injected and which ones are not.

11.8 Summary

This chapter describes the methodologies and procedures of surgical extraction of oocytes, enzymatic isolation, and defolliculation as well as oocyte microinjection. *Xenopus* oocytes can be obtained from sexually mature female *X. laevis* by surgically removing parts of the ovary. The operation is not fatal and can be performed on a given frog under anesthesia several times (no more than five times) during its lifetime. The isolated oocytes can be injected with RNA or DNA, which can be faithfully expressed in this expression system. The methodologies of surgical extraction of oocytes from frogs with decapitation and manual defolliculation of oocytes are not discussed in this chapter. This other method can be found in Liu and Liu (2006).

References

Blow JJ, Laskey RA (1986) Initiation of DNA replication in nuclei and purified DNA by a cell-free extract of Xenopus eggs. Cell 47:577–587

Dascal N (1987) The use of Xenopus oocytes for the study of ion channels. CRC Crit Rev Biochem 22:317–387

Dumont JN (1972) Oogenesis in Xenopus laevis (Daudin). I. Stages of oocyte development in laboratory maintained animals. J Morphol 136:153–179

Fraser SP, Moon C, Djamgoz MBA (1993) Electrophysiology of Xenopus oocytes: an expression system in molecular neurobiology. In: Wallis DI (ed) Electrophysiology: a practical approach. Oxford University Press, Oxford, pp 65–86

Gurdon JB (ed) (1974) The control of gene expression in animal development. Harvard University Press, Cambridge

Gurdon JB, Lane CD, Woodland HR, Marbaix G (1971) Use of frog eggs and oocytes for the study of messenger RNA and its translation in living cells. Nature 233:177–182

Liu XS, Liu XJ (2006) Oocyte isolation and enucleation. Methods Mol Biol 322:31–41

Mak DO, Foskett JK (1994) Single-channel inositol 1,4,5-trisphosphate receptor currents revealed by patch clamp of isolated Xenopus oocyte nuclei. J Biol Chem 269:29375–29378

Mak DO, Foskett JK (1998) Effects of divalent cations on single-channel conduction properties of Xenopus IP3 receptor. Am J Physiol 275:C179–C188

Sigel E (1990) Use of Xenopus oocytes for the functional expression of plasma membrane proteins. J Membr Biol 117:201–221

Stuhmer W, Parekh AB (1995) Electrophysiological recordings from Xenopus oocytes. In: Sakmann B, Neher E (eds) Single channel recordings. Plenum, New York, pp 341–355

Wallingford JB, Liu KJ, Zheng Y (2010) Xenopus. Curr Biol 20:R263–R264

Part III

Leukocyte Isolation and Application in Neuroscience

Centrifugal Elutriation for Studies of Neuroimmunity

12

Santhi Gorantla, Myhanh Che, and Howard E. Gendelman

Abstract

Centrifugal elutriation is a technique used to fractionate synchronized populations of cells based on size and density. Notably, countercurrent centrifugal elutriation (CCE) can be used for large cell number isolations while minimally interfering with the cells' functional properties. This makes it superior to other cell separation techniques such as adherence and fluorescence cell sorting. Separating cells by large scale CCE isolation for monocytes and lymphocytes from whole blood while technically challenging is critical to recapitulate immune responses operative during disease. This includes central nervous system infections and degenerative and inflammatory disorders. Monocytes differentiate rapidly into macrophage-like cells, which can reflect microglial function. Such cells are the primary innate immune effector cells, the sources of immunoregulatory factors, and those that effect clearance of debris. Perhaps, most importantly, they serve as antigen presenters and the target cells for a range of microbial pathogens that include the human immunodeficiency virus while reacting to misfolded, aggregated, and oxidized proteins such as amyloid beta, alpha synuclein, and superoxide dismutase present in Alzheimer and Parkinson's diseases and amyotrophic lateral sclerosis. They can also readily be examined for their genomic and proteomic structures and functional capacities. Also, it is noteworthy to state that using such unmanipulated cells, reproducible outcomes in an experimental setting reflective of human disease can be performed.

Keywords

Monocytes • Monocyte-derived macrophages • T cells • Leukopak • Countercurrent centrifugal elutriation

S. Gorantla (✉) • M. Che • H.E. Gendelman
Department of Pharmacology and Experimental Neuroscience, University of Nebraska Medical Center, Omaha, NE 68198, USA
e-mail: sgorantla@unmc.edu

H. Xiong and H.E. Gendelman (eds.), *Current Laboratory Methods in Neuroscience Research*, Springer Protocols Handbooks, DOI 10.1007/978-1-4614-8794-4_12,

12.1 Introduction

The brain has an innate immune system regulated through adaptive immunity (Mosley et al. 2012; Ousman and Kubes 2012). The integrity of the blood–brain barrier affects but does not ablate immune responses that play often-pivotal roles in disease pathobiology (3). Immune effector astrocytes and microglia (glial cells) are indigenous to the central nervous system (CNS) while T cells and monocytes traffic in and outside the brain (Rezai-Zadeh et al. 2009; Ransohoff and Brown 2012; Sallusto et al. 2012). During normal condition, these glial cells perform a range of homeostatic functions (Kadiu et al. 2005; Ransohoff and Brown 2012). Brain macrophages and microglia are the primary sentinel cells that affect tissue integrity and homeostasis by engaging in debris clearance or phagocytosis within the brain to clear and destroy bioproducts of cellular metabolism and cell damage and death (Cotter et al. 1999; Prinz et al. 2011). They present antigens that mobilize T cells responses and secrete bioactive molecules, cytokines, and growth factors (Rezai-Zadeh et al. 2009). They are also, notably, critical effectors in the pathobiology of neurological diseases as they are the primary sources of inflammation (Wraith and Nicholson 2012). T cells also serve in immune homeostasis, killing of microbes, clearing aberrant and misfolded proteins, and eliminating cancerous cells. They perform their functions by reacting to specific recall antigens and mobilizing immune responses to clear infection or tumors (Ousman and Kubes 2012; Sallusto et al. 2012). In short, they perform critical surveillance functions and preserve the integrity of the CNS. In this manner immune cells, whether innate or adaptive, are of great interest as we study a variety of neurologic and neurobehavioral diseases. For such studies lymphocytes and monocytes are generally obtained from peripheral blood. Countercurrent centrifugal elutriation (CCE) method is recommended to isolate large populations of nonactivated monocytes and lymphocytes with high purity and unaltered functional integrity (Wahl et al. 1984).

Procurement of a pure homogenous population of cells is more advantageous to understand cell-specific functions than investigating mixed cell populations. There are several methods currently used to isolate monocytes and lymphocytes. Separation methods, like cell adherence or fluorescence cell sorting and magnetic bead separation employing antibodies, result in cell activation and functional changes. Adherence method of cell isolation is commonly used to separate cells such as monocyte/macrophages for their ability to adhere to glass or plastic. This method is fast and easy and does not require expensive equipment or reagents. However, adherence of monocytes induces activation and gene expression (Wahl et al. 2006). Furthermore, the adherent method subsequently requires detachment of cells from the surface either by using physical or chemical procedures, which will greatly affect the cell number and viability.

A desirable method of cell separation is one that yields cell populations with a high degree of purity, viability, and functional integrity. Negative selection methods are recommended to minimize alteration in functional properties. However, negative selection methods like magnetic bead separation and fluorescence cell sorting are not cost effective when large numbers of cells are needed. CCE is a noninvasive and negative selection method for separating large number of cells based on the differences in cell size and sedimentation velocity. It yields nonactivated cells with high purity, since it does not employ antibodies or ligands tagging to the cell surface. CCE has several applications (Siemann et al. 1981; Suzuki 1984; Wahl et al. 1984; Wagner et al. 1990; Uchida et al. 1996; Wahl and Donaldson 2001) and has been applied to separate hematopoietic cells and tumor cells. This method is also useful for separating cells at different stages of the cell division cycle without interfering with cell metabolism (Pandita 2004). CCE is technically difficult compared to other separation techniques and requires Beckman Elutriation System consisting of a Beckman centrifuge equipped with a specialized elutriation rotor. Standard JE-6B rotor is used to separate 2×10^8 cells, and JE-5.0

rotor can separate a larger number of cells (5×10^9 to 10×10^9) (Mason and Weiner 1985; Wahl and Donaldson 2001). Peripheral blood mononuclear cells (PBMC) are placed into the rotor/chamber, and they are subjected to two opposite forces: a centrifugal force maintained by the rotor speed and a centripetal force by a buffer stream that flows in the direction opposite the centrifugal field (Fig. 12.2b). This causes the cells to separate in the chamber with smaller cells of low density accumulating towards the center of the rotor. Cells are sequentially eluted from the rotor based on their size, by incrementally increasing the flow rate of the buffer stream. The smaller cells are the first ones to elute. By balancing centrifugal force against the opposing buffer flow, 6–8 mm sized lymphocytes and 8–10 μm sized monocytes are collected sequentially.

12.2 Materials

1. Biosafety cabinet to carry out the cell isolation procedure in aseptic conditions (*Note 1*).
2. Leukocyte-enriched whole blood (leukopak) ~120 mL. The leukopak should be processed within 6 h of collection for best yield.
3. Lymphocyte separation medium (LSM) or Ficoll–Hypaque.
4. Phosphate buffered saline (PBS), Ca^{2+} and Mg^{2+} free (*Note 2*). PBS can be sterilized either by autoclaving or filtering through 0.2 μm sterile filter.
5. 2 mL aspirating pipettes.
6. 10 mL pipettes.
7. 50 mL conical polypropylene sterile centrifuge tubes.
8. 95 % ethanol.
9. 50 % bleach.
10. 30 mL syringe.
11. Ring stand and holder.
12. Silastic tubing, ¼ × 118 in.
13. Masterflex pump with masterflex silicone tubing (Cole–Parmer Instrument Company, Chicago, IL).
14. J6-MI centrifuge with a JE-5.0 elutriation system (Beckman, Fullerton, CA).
15. Particle count and size analyzer (Beckman Coulter, Fullerton, CA).
16. Isoton diluent, Z-Pak (Beckman Coulter).
17. Dulbecco's minimal essential medium (DMEM) without L-glutamine.
18. Human AB serum (HS), heat inactivated at 56 °C for 30 min.
19. L-glutamine.
20. Gentamicin.
21. Ciprofloxacin.
22. Recombinant human monocyte colony stimulating factor-1 (MCSF).
23. Monocyte medium with MCSF: DMEM, 10 % HS, 2 mM L-glutamine, 50 μg/mL gentamicin, 10 μg ciprofloxacin, 1,000 U/mL rhMCSF-1. Store at 4 °C. Discard after 2 weeks.

Comment: Aliquot reagents 8, 9, 11, and 12 for use in 500 mL bottles of DMEM. Reagents are stored at −20 °C, and freeze–thaw cycles are minimized.

12.3 Methods

12.3.1 Donor Screening and Leukophoresis

Donors are chosen without heart, lung, and liver disease. Those with anemia regardless of cause and women who are pregnant are excluded. Donors undergo a physical examination by a physician before being qualified for leukophoresis. On the day the procedure is performed, donor's hematocrit and platelet count are checked and must be ≥38 % and 150×10^{10}/mL, respectively. Donors are also screened for ABO Rh and antibodies against hepatitis B surface (HBS) or core antigen, hepatitis C virus (HCV), HIV, human T lymphotropic virus (HTLV), cytomegalovirus (CMV), and syphilis as per the institutional regulations.

Leukophoresis is the removal of white blood cells (leukocytes) from the peripheral blood. The process requires access to veins in both arms. Blood is extracted from one arm, and the leukocytes are enriched with small amount of red blood cells (RBC). The remaining blood with the RBC is injected back in to the donor through the vein in the other arm. The procedure usually takes 3–4 h.

12.3.2 Prepare PBMC for Elutriation

All procedures are carried out in a biological safety cabinet (BSC) under sterile conditions (*see Note 1*). Biosafety level-2 (BSL-2) practices are required for handling human specimens given the risks associated with the use and are followed as per National Institutes of Health guidelines, which include laboratory coats and gloves and decontaminating liquid waste with 50 % bleach. All materials are autoclaved before being discarded.

Separation of PBMC and removal of RBC is performed by using a simple method that takes advantage of the differences in their densities. During Ficoll–Hypaque density gradient separation, white blood cells and platelets separate as a buffy coat above the Ficoll–Hypaque layer because they have lower density (Fig. 12.1). RBC and granulocytes have higher density; thus, they collect at the bottom of Ficoll–Hypaque layer. The step-by-step procedure to separate white blood cells containing lymphocytes and monocytes using gradient separation is as follows:

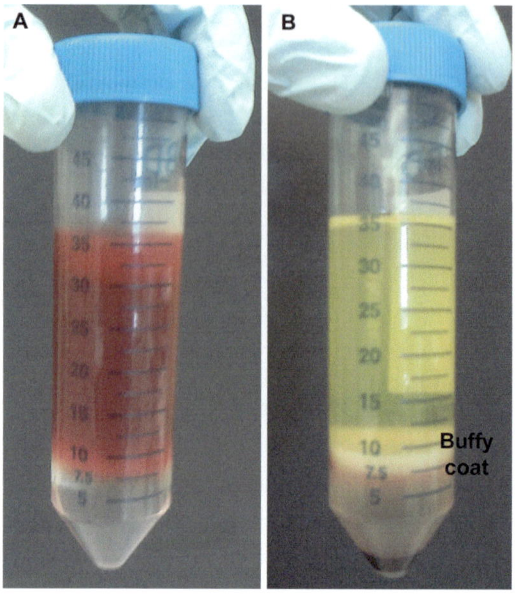

Fig. 12.1 Preparation of PBMC using LSM gradient. Layering of 25 mL whole blood on 10 mL LSM showing the clear interface (**a**). After centrifugation, a buffy coat containing white blood cells is formed at the LSM–blood interface, with clear plasma on top of the buffy coat and RBC pelleted at the bottom of LSM (**b**)

1. In a biosafety cabinet, use sterile scissors to cut the leukopak and transfer the blood sample from the leukopak into a 250 mL centrifuge tube. Dilute the blood sample by adding enough PBS to make up to a total of 250 mL. Pipette up and down to mix gently (*Note 3*).
2. Aliquot 10 mL of LSM into each of the ten 50 mL centrifuge tubes.
3. Layer 25 mL of diluted blood sample carefully onto LSM without disturbing the interface (Fig. 12.1a). To maintain the interface, the tube is held at a 45° angle while layering.
4. Centrifuge at 600 × *g* at room temperature for 20 min without break (*Note 4*).
5. After centrifugation, transfer the tubes from the centrifuge to the biosafety cabinet without disturbing the gradient. RBC settle at the bottom and the PBMC form a buffy coat layer at the LSM and plasma interface (Fig. 12.1b). Aspirate the top layer of diluted plasma and platelets, which is directly above the buffy coat.
6. Collect the buffy coat containing PBMC from each of the gradient tubes and transfer the recovered cells to a fresh 250 mL centrifuge tube.
7. Make the volume up to 250 mL by adding PBS. Mix well by gently pipetting up and down.
8. Centrifuge for 10 min at 400 × *g* at room temperature.
9. Decant supernatant and resuspend the cell pellet in 250 mL of PBS and repeat step 8. The cells are finally suspended in 25 mL of PBS. The sample is ready to be loaded into the elutriation system. At this time the yield of PBMC should be ~5×10^9 to 10×10^9.

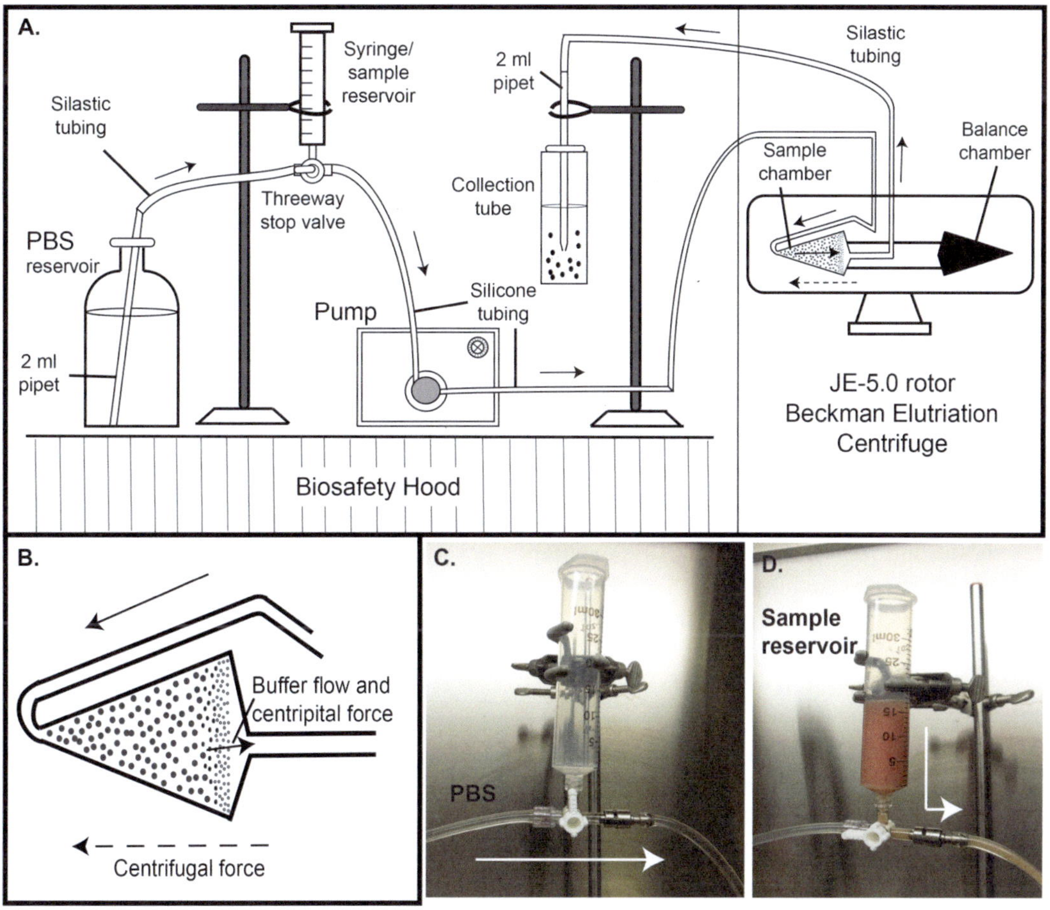

Fig. 12.2 Countercurrent centrifugal elutriation (CCE). Schematic diagram of CCE setup showing the parts to be placed in the biosafety cabinet and the connection to the centrifuge rotor (**a**). *Solid black arrows* show the flow of the liquid. *Dotted arrow* shows the direction of centrifugal force. (**b**) Magnified view of the rotor chamber with cells separated by size. Direction of buffer flow (*solid arrows*) and the direction of centrifugal force (*dotted arrow*) represent the opposite forces the cells are subjected to. (**c**, **d**) Show the position of the three-way stopcock during the PBS flow from the PBS reservoir and cell sample flow from the sample reservoir, respectively

12.3.3 Elutriation Setup

A schematic representation of the elutriation setup is in Fig. 12.2a showing the flow of the system. Parts of the setup that are required to be placed inside the BSC are shown. Assemble the apparatus as shown in the figure.

1. Prepare the elutriation setup before starting to process the leukopak for density gradient separation.
2. Assemble the chambers and the rotor as per the instrument instruction manual.
3. Connect the input tubing and the output tubing to the rotor chamber (*Note 5*).
4. The input tubing is connected through a masterflex peristaltic pump that controls the flow rate. The pump should be connected using silicone tubing and the rest of the system with silastic tubing. To the other end of the input tube, connect a three-way stopcock which is used to control the input of either PBS or cells pumped from two separate reservoirs (the manual provides detailed instructions for the setup). A 30 mL syringe fitted to the stopcock works as the cell sample reservoir. The output tubing goes directly into collection container to collect washes or a sterile bottle while collecting the cells (Fig. 12.2a).

5. The tubings from PBS reservoir and to the collection tube are fitted with 2 mL pipette at the end; the pipettes are inserted into the liquid. Pipettes and syringe are replaced with new sterile ones for every elutriation procedure.
6. Sterilize the chamber and the tubing by pumping 95 % ethanol (~200 mL) through the system by using a flow rate of 80 mL/min (*Note 6*).
7. Fill PBS in both buffer reservoir and cell reservoir and flush the system with PBS to remove air bubbles at 80 mL/min flow rate. Use at least 400 mL of PBS (*Note 7*).
8. Flush PBS through the system by increasing the pump setting to maximum (160 mL/min). This is to ensure that the setup can handle the pressure. Check the system for any air bubbles and avoid them by flushing the system again with more PBS.
10. Load 25 mL of PBMC suspension into the 30 mL syringe (cell reservoir).
11. With PBS flow open (controlled with the stopcock, Fig. 12.2c), turn on the pump at a flow rate of 40 mL/min. Let PBS flow through the system.
12. Turn on the centrifuge, set the temperature to 20 °C, and when the rotor reaches 1,960 rpm (*Notes 9* and *10*), stop PBS flow and let the cells from the cell reservoir enter the system by turning the stopcock. After all the cells are loaded, turn the stopcock back to allow PBS flow. Do not allow air bubbles to enter the tubing. Wait for 10–15 min to allow the cells to separate by size within the rotor chamber (*Note 8*).
13. Turn pump to a flow rate of 45 mL/min and wait for 15 min to remove RBC.
14. Collect aliquots of cell suspension (10 μL) at each speed setting of the pump. Dilute the cell aliquot with 10 mL of isoton diluent. Count and analyze the cells using an electronic counting device (Beckman Coulter cell counter and analyzer), which also gives a profile indicating the cell size. Calibrate the cell analyzer with latex beads of known size before analyzing the cells. Lymphocyte profile will be that of ~6–8 μm diameter cells. Use the cell counter according to the manufacturer's instructions (*Note 9*). Cell number and cell size should be analyzed every time when the flow rate is increased.
15. Turn the pump to 50 mL/min and collect peripheral blood lymphocytes (PBL). Collect 250 mL volume containing lymphocytes.
16. Turn pump up to 55 mL/min to collect more PBL in another 250 mL volume.
17. Count the cells, and if it shows <500 cells/mL, increase the pump flow with increments of +5 mL/min. Each time when the flow rate is increase by 5 mL/min, check the cell number and size using the cell counter. Continue the process until the number of cells with 6–8 μm diameter drops to <500 (*see Note 11*). This drop will be seen when the flow rate is between 65 and 75 mL/min.

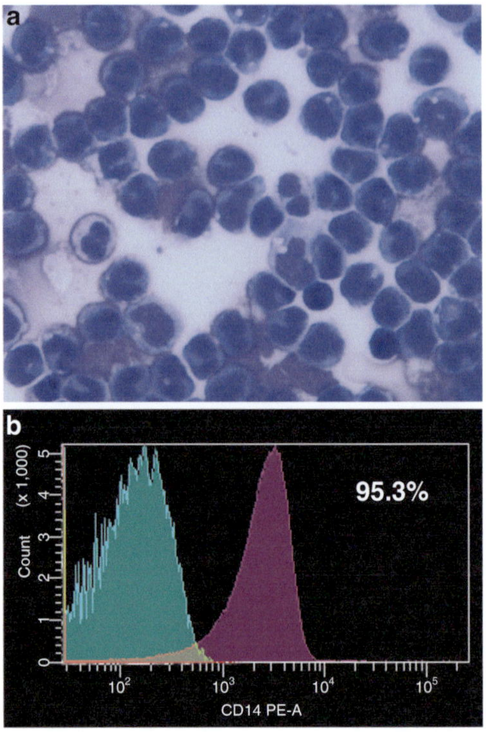

Fig. 12.3 Monocyte isolations show homogeneous cell populations with characteristic morphology, magnification ×40 (**a**). Monocyte fraction was stained for the surface antigen CD14 (**b**). Phycoerythrin (PE)-conjugated mouse IgG against human CD14 was used, and cells were analyzed by flow cytometry. The histogram shows overlap of isotype control (*blue*) and CD14+ monocytes (*purple*). 95.3 % cells in monocyte fraction were positive for the monocyte-specific marker CD14

18. At this time, further increase in flow rate should elute the cells of diameter >8 μm. These cells are mostly monocytes. When a profile of monocytes starts showing (refer to the cell analyzer manual, the profile of cells should be the same as that obtained for cells of 8–10 μm diameter), stop the centrifuge, set the pump to 160 mL/min, and collect all the monocytes. Increased flow rate will drive out all the remaining cells from the chamber.
19. Centrifuge the tubes with PBL or monocytes at 400 × *g* for 10 min at room temperature. Decant the supernatant and resuspend the cells in 25 mL of PBS.
20. Use a small aliquot (10 μL) to count the cells and calculate the total number of cells.
21. To check the purity of PBL and monocyte populations, make a cell smear on a glass slide using a small aliquot from each cell suspension and stain with hematoxylin and eosin (H&E). Observe under the light microscope. Figure 12.3 shows H&E stained monocytes and PBL separated by this procedure. Monocytes have distinct kidney-shaped nucleus and can be easily identified.
22. Purity can also be analyzed by flow cytometry method using fluorescent-tagged antibodies to cell-specific antigens. Figure 12.3b shows flow cytometric analysis of elutriated monocyte population stained for CD14 surface antigen. More than 95 % of cells should be CD14, a monocyte-specific cell surface marker.
23. At the end of elutriation, wash the system with 200 mL distilled water to lyse any cells. Flush the system with 95 % alcohol. The rotor is dissembled and washed thoroughly according to the manufacturer's instructions (*see Note 12*).

12.3.4 Culturing Monocytes

Monocytes obtained by CCE method are >90 % pure and can be used in various cell culture analyses. Monocytes isolated as suspension cells can be plated according to the experimental conditions, while adherent method of monocyte isolation needs cell detachment from the surface for reseeding in culture dishes.

To study monocyte neuroimmune function in disease, monocytes are differentiated into monocyte-derived macrophages (MDM) by culturing in MCSF containing monocyte media (*Note 13*) (Gorantla et al. 2005a).

1. Monocytes are pelleted by centrifuging the cell suspension at 400 × *g* for 8 min. Supernatant is discarded, and the pellet is suspended in monocyte media with MCSF at 2×10^6 cells/mL. An even cell suspension is obtained by gently pipetting up and down.
2. For adherent cultures, cells are plated as follows: 10^5 cells/well of a 96 well plate, 0.5×10^6 cells/well of a 24 well plate, 3×10^6 cells/well of a 6 well plate, and 30×10^6 cells/T75 flask. Cultures are incubated in a 37 °C incubator with 5 % CO_2. Media is half exchanged with fresh monocyte media containing MCSF every other day until 7 days; thereafter, half exchange is done using monocyte media without MCSF. By 7 days in culture, monocytes differentiate into MDM and are ready to use for experimentation.
3. MDM can be maintained in suspension if necessary for certain experimental purposes as scrapping MDM from adherent cultures to obtain suspension cells will result in cell loss, reduced viability, and altered functional integrity. Teflon flasks are used to maintain suspension cultures of monocytes to minimize surface adherence. Cells are cultured at 2×10^6 cells/mL with half media exchanges with fresh monocyte media every other day. While culturing cells in Teflon flask, the cap of the flask should be loosened one thread to allow exchanges of gases in the incubator.

12.4 Typical results

In our laboratory we routinely perform CCE to isolate large number of lymphocytes and monocytes from leukopaks. From mononuclear cells $>5 \times 10^9$, the yield of lymphocytes is ~$1–2.5 \times 10^9$ cells, and the monocytes yield is ~$1–1.5 \times 10^9$ cells. Lymphocytes are collected when eluting cell size is between 6 and 8 μm to minimize contamination with neutrophils, basophils, and eosinophils. The cells eluting with the size ranging from 8 to 10 μm are discarded. Since the monocyte is the largest blood cell, eluting all the remaining cells in the chamber when the cells of 10 μm diameter start to elute results in a homogenous population of monocytes. The elution pattern corresponding to the flow rates should be observed and calibrated during the first few elutriations. This is done by checking cell size and cell number using the cell counter every time the flow rate is increased incrementally by +5 mL/min. The time taken for the whole process of CCE is generally 3–4 h.

Purity of the cells isolated can be tested using flow cytometric method either by antibody to cell-specific surface or by propidium iodide staining (Riccardi and Nicoletti 2006). Monocytes isolated using CCE are usually more than 90 % pure.

12.5 Notes

1. Cell isolation and cell culture procedures must be carried out strictly under sterile conditions in a BSC. All pipettes, centrifuge tubes, and tips must be sterile, and the packages should be opened only inside the BSC. The working area of the BSC should be wiped with 75 % ethanol for sterilization.
2. We used PBS without Ca^{2+} and Mg^{2+} through the cell separation procedure. Some protocols add either 1 % (w/v) bovine serum albumin or 0.02 % (w/v) EDTA to the PBS to avoid cell clumping. Cell clumps will interfere with the cell sedimentation and reduce the purity of isolated cells.
3. To suspend the cells evenly, pipetting up and down gently is suggested. Rapid pipetting results in cell death and less recovery. Also dead cells form clumps, and the presence of clumps will result in reduced purity of monocyte population.
4. Do not forget to switch off the break option or reduce the deceleration speed to the minimum setting before starting the gradient centrifugation. Faster deceleration will disturb the interface and the buffy coat.
5. Tubing, O-ring, and all the connections have to be checked while assembling the elutriation system for the leaks. Leaks occur when chamber gasket is put backwards, chamber screws are not tightened properly, tubing is not connected properly, forget putting O-ring during assembly or nicks, and scratches on the O-ring. Replace worn out tubing and O-ring.
6. Before and after using the elutriation system, it is very important to take proper precautions to clean and maintain the system to avoid clogging and to maintain sterile conditions.
7. Care should be taken not to let the PBS reservoir go empty while pump is on, as air bubbles can enter the system. At least 4 L of PBS are required for the procedure.
8. Air in the system is a major problem, and air bubbles must be completely removed from the system by flushing PBS. Tubes must and the rotor chamber must be thoroughly checked for any air bubbles.
9. The rotor speed is kept constant with gradually increasing the flow rate for best separation of cells. To avoid changes in the rotor speed due to voltage fluctuations and to maintain constant voltage, the centrifuge can be plugged to a voltage stabilizer.
10. The flow rate, rotor speed to elute a particular cell type, is determined by the nomogram given in the instruction manual provided with the elutriation system. In the protocol, the rotor speed is mentioned in rpm, and it is suggested by the nomogram to retain particles of 6 μm or larger at a flow rate of 40 mL/min.
11. Cell analysis should be done fast without time delay, as the pump cannot be stopped during elutriation. Otherwise, there can be loss of cells. Stopping the pump during elutriation will lead to clogging of cells at the entrance of the chamber. This will ruin the elutriation process, since dissembling the rotor and cleaning the chamber is required to remove the clog.

12. After every ten elutriations run 50 % bleach through the system to remove accumulated residual protein. Alterative: soak and clean rotor chamber in Beckman Coulter Cleaning solution 555 after each use. Replace all tubings after every ten elutriation runs.
13. Human AB serum is used to culture monocytes instead of fetal bovine serum to minimize activation and changes in functionality, as monocytes are very sensitive to the environmental factors.

12.6 Discussion/Summary

Monocytes and lymphocytes isolated in large numbers using CCE are used in several applications. The major advantage of CCE is that it is a noninvasive method that does not affect the functional integrity of the isolated cells. Monocytes are differentiated into macrophages to study their innate immune properties in target tissues. Human macrophages are susceptible to HIV-1 infection. They are the viral reservoirs and play a major role in disseminating virus into the brain. In vitro studies on MDM infected with HIV-1 helped to understand their role in HIV-1-induced neuropathology and to unveil several potential therapeutic targets, like innate immune factors that affect neuronal integrity (Meltzer and Gendelman 1992; Epstein and Gendelman 1993; Talley et al. 1995; Nottet et al. 1996; Persidsky et al. 1999; Anderson et al. 2002; Carlson et al. 2004; Kadiu et al. 2005). Treatment of MDM with amyloid beta (Aβ) protein revealed their function in Aβ clearance and the role of inflammation in Alzheimer's disease (Cotter et al. 1999; Lan et al. 2011, 2012).

Human MDM and lymphocytes are used to generate different mouse models of HIV-1 infection (described in Chap. 33). HIV-1-infected MDM is injected into the caudate putamen of immunodeficient severe-combined immune-deficient (SCID) mice to induce HIV-1 encephalitis (HIVE); this model recapitulates the hallmark features of HIV encephalitis such as HIV-1-infected multinucleated giant cells, glial inflammation, and myelin parlor with neuronal loss (Persidsky et al. 1995, 1996). SCID-HIVE mouse model was critical in revealing the central role of HIV-1-infected mononuclear phagocytes in HIV-1 infection-induced encephalitis (Persidsky et al. 1997), and it served as an essential model to test therapeutic strategies to alleviate brain disease (Dou et al. 2003, 2009; Eggert et al. 2009, 2010). The hu-PBL-NOD/SCID-HIVE mouse model was developed by reconstructing NOD/SCID mouse with human PBL and injecting HIV-1-infected MDM into the brains. This model enabled studies on both innate and adaptive immune systems in HIV-1-induced brain disease (Poluektova et al. 2002, 2004). Large number of human PBL and monocytes are needed to generate these mouse models, and moreover, HIV-1-infected MDM should be maintained in suspension cultures in Teflon flasks to inject into the brains of mice. hu-PBL-NOD/SCID mice with systemic reconstitution of human PBL were widely to study the efficacy of antiretroviral therapies (Kanmogne et al. 2012; Roy et al. 2012). Monocytes isolated by CCE can be easily differentiated into dendritic cells, and they have very important clinical application as a therapeutic tool in the treatment of cancers (Van De Velde et al. 2012). Dendritic cell-based vaccination was shown to induce partial protection against HIV-1 infection in hu-PBL-NOD/SCID mice (Gorantla et al. 2005b).

Cell isolation technique like CCE is an attractive method to achieve large number of unmanipulated homogeneous cell populations, which can be used successfully in various experimental studies.

References

Anderson E, Zink W, Xiong H, Gendelman HE (2002) HIV-1-associated dementia: a metabolic encephalopathy perpetrated by virus-infected and immune-competent mononuclear phagocytes. J Acquir Immune Defic Syndr 31(Suppl 2):S43–S54

Carlson KA, Leisman G, Limoges J, Pohlman GD, Horiba M, Buescher J, Gendelman HE, Ikezu T (2004) Molecular characterization of a putative antiretroviral transcriptional factor, OTK18. J Immunol 172:381–391

Cotter RL, Burke WJ, Thomas VS, Potter JF, Zheng J, Gendelman HE (1999) Insights into the neurodegenerative process of Alzheimer's disease: a role for

mononuclear phagocyte-associated inflammation and neurotoxicity. J Leukoc Biol 65:416–427
Dou H, Birusingh K, Faraci J, Gorantla S, Poluektova LY, Maggirwar SB, Dewhurst S, Gelbard HA, Gendelman HE (2003) Neuroprotective activities of sodium valproate in a murine model of human immunodeficiency virus-1 encephalitis. J Neurosci 23:9162–9170
Dou H, Grotepas CB, McMillan JM, Destache CJ, Chaubal M, Werling J, Kipp J, Rabinow B, Gendelman HE (2009) Macrophage delivery of nanoformulated antiretroviral drug to the brain in a murine model of neuroAIDS. J Immunol 183:661–669
Eggert D, Dash PK, Serradji N, Dong CZ, Clayette P, Heymans F, Dou H, Gorantla S, Gelbard HA, Poluektova L, Gendelman HE (2009) Development of a platelet-activating factor antagonist for HIV-1 associated neurocognitive disorders. J Neuroimmunol 213: 47–59
Eggert D, Dash PK, Gorantla S, Dou H, Schifitto G, Maggirwar SB, Dewhurst S, Poluektova L, Gelbard HA, Gendelman HE (2010) Neuroprotective activities of CEP-1347 in models of neuroAIDS. J Immunol 184:746–756
Epstein LG, Gendelman HE (1993) Human immunodeficiency virus type 1 infection of the nervous system: pathogenetic mechanisms. Ann Neurol 33:429–436
Gorantla S, Che M, Gendelman HE (2005a) Isolation, propagation, and HIV-1 infection of monocyte-derived macrophages and recovery of virus from brain and cerebrospinal fluid. Methods Mol Biol 304:35–48
Gorantla S, Santos K, Meyer V, Dewhurst S, Bowers WJ, Federoff HJ, Gendelman HE, Poluektova L (2005b) Human dendritic cells transduced with herpes simplex virus amplicons encoding human immunodeficiency virus type 1 (HIV-1) gp120 elicit adaptive immune responses from human cells engrafted into NOD/SCID mice and confer partial protection against HIV-1 challenge. J Virol 79:2124–2132
Kadiu I, Glanzer JG, Kipnis J, Gendelman HE, Thomas MP (2005) Mononuclear phagocytes in the pathogenesis of neurodegenerative diseases. Neurotox Res 8:25–50
Kanmogne GD, Singh S, Roy U, Liu X, McMillan J, Gorantla S, Balkundi S, Smith N, Alnouti Y, Gautam N, Zhou Y, Poluektova L, Kabanov A, Bronich T, Gendelman HE (2012) Mononuclear phagocyte intercellular crosstalk facilitates transmission of cell-targeted nanoformulated antiretroviral drugs to human brain endothelial cells. Int J Nanomedicine 7:2373–2388
Lan X, Xu J, Kiyota T, Peng H, Zheng JC, Ikezu T (2011) HIV-1 reduces Abeta-degrading enzymatic activities in primary human mononuclear phagocytes. J Immunol 186:6925–6932
Lan X, Kiyota T, Hanamsagar R, Huang Y, Andrews S, Peng H, Zheng JC, Swindells S, Carlson GA, Ikezu T (2012) The effect of HIV protease inhibitors on amyloid-beta peptide degradation and synthesis in human cells and Alzheimer's disease animal model. J Neuroimmune Pharmacol 7:412–423
Mason RR, Weiner RS (1985) Application of the Beckman JE6-B Elutriator System in the isolation of human monocyte subpopulations. Scand J Haematol 34:5–8
Meltzer MS, Gendelman HE (1992) Mononuclear phagocytes as targets, tissue reservoirs, and immunoregulatory cells in human immunodeficiency virus disease. Curr Top Microbiol Immunol 181:239–263
Mosley RL, Hutter-Saunders JA, Stone DK, Gendelman HE (2012) Inflammation and adaptive immunity in Parkinson's disease. Cold Spring Harb Perspect Med 2:a009381
Nottet HS, Persidsky Y, Sasseville VG, Nukuna AN, Bock P, Zhai QH, Sharer LR, McComb RD, Swindells S, Soderland C, Gendelman HE (1996) Mechanisms for the transendothelial migration of HIV-1-infected monocytes into brain. J Immunol 156:1284–1295
Ousman SS, Kubes P (2012) Immune surveillance in the central nervous system. Nat Neurosci 15:1096–1101
Pandita TK (2004) Enrichment of cells in different phases of the cell cycle by centrifugal elutriation. Methods Mol Biol 241:17–21
Persidsky Y, Nottet HS, Sasseville VG, Epstein LG, Gendelman HE (1995) The development of animal model systems for HIV-1 encephalitis and its associated dementia. J Neurovirol 1:229–243
Persidsky Y, Limoges J, McComb R, Bock P, Baldwin T, Tyor W, Patil A, Nottet HS, Epstein L, Gelbard H, Flanagan E, Reinhard J, Pirruccello SJ, Gendelman HE (1996) Human immunodeficiency virus encephalitis in SCID mice. Am J Pathol 149:1027–1053
Persidsky Y, Buttini M, Limoges J, Bock P, Gendelman HE (1997) An analysis of HIV-1-associated inflammatory products in brain tissue of humans and SCID mice with HIV-1 encephalitis. J Neurovirol 3:401–416
Persidsky Y, Ghorpade A, Rasmussen J, Limoges J, Liu XJ, Stins M, Fiala M, Way D, Kim KS, Witte MH, Weinand M, Carhart L, Gendelman HE (1999) Microglial and astrocyte chemokines regulate monocyte migration through the blood–brain barrier in human immunodeficiency virus-1 encephalitis. Am J Pathol 155:1599–1611
Poluektova LY, Munn DH, Persidsky Y, Gendelman HE (2002) Generation of cytotoxic T cells against virus-infected human brain macrophages in a murine model of HIV-1 encephalitis. J Immunol 168:3941–3949
Poluektova L, Gorantla S, Faraci J, Birusingh K, Dou H, Gendelman HE (2004) Neuroregulatory events follow adaptive immune-mediated elimination of HIV-1-infected macrophages: studies in a murine model of viral encephalitis. J Immunol 172:7610–7617
Prinz M, Priller J, Sisodia SS, Ransohoff RM (2011) Heterogeneity of CNS myeloid cells and their roles in neurodegeneration. Nat Neurosci 14:1227–1235
Ransohoff RM, Brown MA (2012) Innate immunity in the central nervous system. J Clin Invest 122:1164–1171
Rezai-Zadeh K, Gate D, Town T (2009) CNS infiltration of peripheral immune cells: D-Day for neurodegenerative disease? J Neuroimmune Pharmacol 4:462–475
Riccardi C, Nicoletti I (2006) Analysis of apoptosis by propidium iodide staining and flow cytometry. Nat Protoc 1:1458–1461
Roy U, McMillan J, Alnouti Y, Gautum N, Smith N, Balkundi S, Dash P, Gorantla S, Martinez-Skinner A, Meza J, Kanmogne G, Swindells S, Cohen SM,

Mosley RL, Poluektova L, Gendelman HE (2012) Pharmacodynamic and antiretroviral activities of combination nanoformulated antiretrovirals in HIV-1-infected human peripheral blood lymphocyte-reconstituted mice. J Infect Dis 206:1577–1588

Sallusto F, Impellizzieri D, Basso C, Laroni A, Uccelli A, Lanzavecchia A, Engelhardt B (2012) T-cell trafficking in the central nervous system. Immunol Rev 248: 216–227

Siemann DW, Lord EM, Keng PC, Wheeler KT (1981) Cell subpopulations dispersed from solid tumours and separated by centrifugal elutriation. Br J Cancer 44: 100–108

Suzuki N (1984) Centrifugal elutriation and characterization of tumor cells from venous blood of tumor-bearing mice: possible relevance to metastasis. Cancer Res 44:3505–3511

Talley AK, Dewhurst S, Perry SW, Dollard SC, Gummuluru S, Fine SM, New D, Epstein LG, Gendelman HE, Gelbard HA (1995) Tumor necrosis factor alpha-induced apoptosis in human neuronal cells: protection by the antioxidant N-acetylcysteine and the genes bcl-2 and crmA. Mol Cell Biol 15:2359–2366

Uchida N, Jerabek L, Weissman IL (1996) Searching for hematopoietic stem cells. II. The heterogeneity of Thy-1.1(lo)Lin(−/lo)Sca-1+ mouse hematopoietic stem cells separated by counterflow centrifugal elutriation. Exp Hematol 24:649–659

Van De Velde AL, Anguille S, Berneman ZN (2012) Immunotherapy in leukaemia. Acta Clin Belg 67: 399–402

Wagner JE, Santos GW, Noga SJ, Rowley SD, Davis J, Vogelsang GB, Farmer ER, Zehnbauer BA, Saral R, Donnenberg AD (1990) Bone marrow graft engineering by counterflow centrifugal elutriation: results of a phase I-II clinical trial. Blood 75:1370–1377

Wahl AF, Donaldson KL (2001) Centrifugal elutriation to obtain synchronous populations of cells. Curr Protoc Cell Biol Chapter 8:Unit 8.5

Wahl SM, Katona IM, Stadler BM, Wilder RL, Helsel WE, Wahl LM (1984) Isolation of human mononuclear cell subsets by counterflow centrifugal elutriation (CCE). II. Functional properties of B-lymphocyte-, T-lymphocyte-, and monocyte-enriched fractions. Cell Immunol 85:384–395

Wahl LM, Wahl SM, Smythies LE, Smith PD (2006) Isolation of human monocyte populations (Chapter 7:Unit 7 6A). In: Coligan JE et al (eds) Current protocols in immunology. Wiley, New York

Wraith DC, Nicholson LB (2012) The adaptive immune system in diseases of the central nervous system. J Clin Invest 122:1172–1179

Flow Cytometry in Neuroscience Research

13

Charles Kuszynski

Abstract

Flow cytometry has been used for numerous applications in immunology, cell biology, and neuroscience research (de Graaf, et al. 2011). With the development of higher sensitivity instrumentation and smaller numbers of cells required, flow cytometry has become an adjunctive methodology for studies of the interface between peripheral immune cells and the nervous system Schwatrz and Kipnis, 2011 (Ref). In this chapter, we will explore the use of flow cytometry to assess and isolate lymphoid cells associated with neuronal injury as well as how this approach would prove helpful in on the identification of neuronal cells. In addition, isolation of cells by other means will be discussed.

Keywords

Flow cytometry neuroscience • Immunological staining

13.1 Introduction/Background

Flow cytometry has been used for numerous applications in immunology, cell biology, and neuroscience research (de Graaf, et al. 2011). With the development of higher sensitivity instrumentation and smaller numbers of cells required, flow cytometry has become an adjunctive methodology for studies of the interface between peripheral immune cells and the nervous system Schwartz and Kipnis 2011. In this chapter, we will explore the use of flow cytometry to assess and isolate lymphoid cells associated with neuronal injury as well as how this approach would prove helpful in on the identification of neuronal cells. In addition, isolation of cells by other means will be discussed.

Flow cytometry is a technique which uses fluorescent probes to identify, quantify, and even isolate cells for further study (de Graaf et al. 2011). Cells can be identified by their surface

C. Kuszynski (✉)
Department of Pathology and Microbiology, University of Nebraska Medical Center, Omaha, NE 68198-5816, USA

Cell Analysis Facility, University of Nebraska Medical Center, Omaha, NE 68198-5816, USA
e-mail: ckuszyns@unmc.edu

H. Xiong and H.E. Gendelman (eds.), *Current Laboratory Methods in Neuroscience Research*, Springer Protocols Handbooks, DOI 10.1007/978-1-4614-8794-4_14,

receptor expression or by cytoplasmic expression of cytokines, chemokines, or intracellular receptors. In addition, cells can be isolated viably and aseptically at high purity based on marker expression. To achieve successful flow cytometry measurements, tissues of origin must be dissociated into a single cell suspension prior to labeling and analysis. Magnetic bead sorting can be used to enrich and/or isolate cells of neuronal origin and facilitate subsequent cell studies.

The normal distribution of lymphoid cells in tissues mimics that found in peripheral blood. In essence, the complete repertoire of white cells may exist in lymphoid tissues. Additionally, specific cells of neuronal origin can be identified in these tissues using flow cytometry. Stem cells, astrocytes, and glial cells have all been identified using specific monoclonal antibodies and are candidates for identification, enumeration, and isolation using flow cytometric techniques (Kraan et al. 2008). A short review of some of the markers expressed mouse and human neural progenitors can be found in de Graaf et al. (2011) and Klassen et al. (2001).

13.2 Objectives

The objectives of this chapter are to provide useful flow cytometry protocols that are adaptable to the identification and isolation of cells involved in neuroscience research and to describe other techniques to augment flow cytometry for these studies.

13.3 Procedures/Methodology

Since the goal of flow cytometry is to assess each cell in a given population one cell at a time, a suspension of single cells is necessary. When cells are isolated from tissue via enzyme digestion, mechanical means, or both, it is often necessary to filter the preparation to remove clumps. This can be done by filtration through nylon mesh. Pore sizes useful for this procedure range from 100 to 40 μM. Filter cap tubes, such as those sold by BD/Falcon™ are useful for this purpose. Loose filters for larger volumes are also sold by BD/Falcon™ and come in various sizes. Catalog numbers for 100 and 70 μm cell strainers are 352360 and 352350, respectively. In addition sheet filter material is available from other sources.

Further, it is imperative that you optimize your disaggregation protocol to prevent cell injury or rupture and the release of DNA which can cause significant clumping. This problem can be alleviated to some extent by using a buffer with EDTA or protease-free DNase. Some cell types and tissues require other digestive enzymes such as collagenase or trypsin. Kraan et al. (2008) provide some assistance with disaggregation protocols.

Once the cells are isolated and washed free of debris and remaining digestive enzymes, the cells are aliquoted into tubes or 96 well round bottom plates for staining. Below is the protocol used in this laboratory, but others exist as web pages from antibody suppliers such as this one from BD Biosciences. http://www.bdbiosciences.com/support /resources/index.jsp.

13.4 Cell Staining Protocols

Cells may be classified into subpopulations on the basis of antigenic determinants on the cell surface. Antibodies may be used to define the subpopulations due to their ability to bind to the surface of viable cells expressing the antigen. When a fluorochrome-conjugated antibody is used to stain the cells, the sample may be enumerated without further manipulation (Sect. 13.4.2). Unconjugated monoclonal antibodies may also be used to mark the cells; however, a second, fluorochrome-conjugated antibody must be used which will bind to the primary antibody (Sect. 13.4.3).

13.4.1 Titration of Antibody

For fluorescently labeled antibodies including directly labeled primary antibodies and fluorescently labeled secondary antibodies, all new lots of conjugated IgG antibodies must be titered to determine the optimum concentrations for use.

For secondary antibodies this is accomplished by testing a known positive monoclonal primary antibody with the fluorochrome conjugated secondary antibody at various dilutions. To accomplish this, work through the following steps:

1. Count and calculate the number of target cells available.
2. Aliquot 2×10^6 cells into one 12×75 mm test tube. Wash cells with PBS containing 1 % protein (BSA or nonimmune serum) by centrifugation at $1{,}000 \times g$ (or at a speed/*g*-force appropriate to pellet the cells) and aspirate the supernatant. Resuspend the cells to a concentration of 2×10^6/mL with PBS + BSA + NaN_3.
3. Aliquot 100 μL of the suspension into each of ten tubes. Arrange the tubes in an ice bath so you have five sets of two. Label 1 tube of the set "Control" and the other with the primary antibody (e.g., "CD38"). Also label them with the dilution of secondary to be used (e.g., 1:20, 1:50, 1:100, 1:200, and 1:300).
4. Pipet 2 μL of control MsIgG (nonspecific antibody of the same isotype as the antibody to be used for specific binding) into each "Control" tube. Pipet 10 μL of your selected primary antibody into the pre-labeled tubes.
5. Incubate on ice for 30 min. While the cells are incubating with the primary antibody, prepare the secondary dilutions using PBS + BSA + NaN_3. As suggested above: 1:20, 1:50, 1:100, 1:200, and 1:300.
6. Add 1 mL of PBS + BSA + NaN_3 to each tube with cells and antibody. Centrifuge at $1{,}000 \times g$ (or at a speed/G-force appropriate to pellet the cells) and aspirate the supernatant. Aspirate supernatant. Pipet 100 μL of the appropriate dilution of secondary antibody into each tube, both control and specific.
7. Incubate on ice for 30 min. Wash as before by centrifugation at $1{,}000 \times g$ (or at a speed/*g-force* appropriate to pellet the cells) and aspirate the supernatant. Then resuspend cell pellet in 0.5 mL of PBS+0.5 % paraformaldehyde. Analyze by flow cytometry. This should be done immediately or with little delay to minimize any variation due to antibody loss or fading.
8. Analyze the data to determine the best binding (fluorescence intensity relative to the control). Use the antibody at the concentration that gives a plateau in mean peak fluorescence as determined by instrument derived values for the fluorescence.

13.4.2 Direct Method

This method utilizes antibodies directly conjugated with fluorochromes.

1. Calculate the total number of cells required for testing the number of antigens in each panel. A minimum of 2×10^5 cells are recommended for adequate analysis of routine surface markers while more cells may be required if rear events are to be examined. Wash cells with PBS containing 1 % protein (BSA or nonimmune serum) by centrifugation at $1{,}000 \times g$ (or at a speed/G-force appropriate to pellet the cells) and aspirate the supernatant. Resuspend the cells to a concentration of 2×10^6/mL (or higher if sorting is to be performed) with PBS + BSA + NaN_3. If you are trying to sort these cells for viability assays or culture growth, do not use sodium azide (NaN_3). Sodium azide paralyzes the membrane to prevent capping of the bound antibodies, but renders the cells nonviable.
2. Vortex the cell suspension well. Pipet 100 μL of the cell suspension into each test tube or well. Again, when looking for rare events, 1×10^7 per well or tube may be necessary.
3. Pipet the appropriate amount of antibody into the tube or well. Be sure to put the tip of the pipet down into the cell suspension to insure that the antibody does not cling to the side of the tube. Change the pipet tip after each antibody or cocktail use. In general, the amounts of antibody to use are often determined by the manufacturers as shown in the example here. Becton-Dickinson; 10 μL, Coulter 5 or 10 μL. These amounts are usually those obtained from the manufacturer's package inserts. It is best to titrate the antibodies you are going to use (see above). In our experience, for research applications, significantly less than the recommended levels give satisfactory results.

In addition, the type of fluorochrome used also affects the fluorescence intensity of the signal.
4. Vortex the tubes or mix the plates and incubate in the dark on ice (4 °C) for 30 min.
5. Pipet 1 mL PBS + BSA + NaN_3 into each tube. Vortex or mix plate. Spin down as before and aspirate supernatant (again, no azide (NaN_3) should be used for cells to be sorted for growth). Staining in plates requires less PBS/BSA per well, but two washes are desirable. Removal of the supernatants from the plates may be accomplished with a flick of the wrist as long cell pellet remains in the well.
6. For most applications, the cells can be fixed prior to analysis. Add 1 mL 0.5 % paraformaldehyde in PBS. Vortex. Analyze by flow cytometry or store samples in the refrigerator for up to 1 week. Do not fix cells to be sorted, these need to be processed immediately and kept on ice to prevent receptor capping. For neuronal cells, storage conditions need to be defined for specific cell types.

13.4.3 Indirect Method

This method is used when the primary antibody is not conjugated to a fluorescent reporter and requires a second fluorescent antibody against the first for labeling.

1. Follow steps 1 and 2 from Sect. 13.4.2 instructions.
2. Pipet the appropriate amount of unconjugated antibody into each tube. Refer to step 3 from above for appropriate volumes of antibody determined empirically by serial dilution and staining as suggested for the titration of antibodies (Sect. 13.4.1).
3. Vortex all tubes and incubate at 4 °C for 30 min.
4. Add 1 mL PBS + BSA + NaN_3 to each tube to wash as spin down as before.
5. Prepare the secondary antibody (e.g., goat anti-mouse Ig-FITC) by diluting with PBS + BSA + NaN_3. Allow 100 μL per tube. Check the label of the antibody for the appropriate dilution (it is usually 1:100, but not necessarily). The secondary antibody may need to be titrated as described above.
6. Pipet 100 μL of diluted secondary antibody into each tube of cells and incubate for 30 min at 4 °C.
7. Wash with 1 mL PBS + BSA + NaN_3 as before. Pipet 1 mL PBS + 0.5 % paraformaldehyde into each tube. Analyze or store at 4 °C for up to 1 week. Fresh cells should be analyzed immediately to prevent capping.

13.4.4 Cytoplasmic Markers

Sometimes proteins and receptors of interest are found in the cytoplasm rather than on the surface of the cell. In order to analyze these markers, the cells must be permeabilized to allow the antibodies to enter the cell and bind to the proteins. The staining procedures for this are similar to those described above, but cells are first treated to make the membrane permeable. Commercial products are available for this purpose. However, most of these have been developed for lymphoid tissue and may need optimization for neuronal tissues. The literature is replete with home brew methods for permeabilizing cells with detergents like saponin, Tween-20, or Triton X-100 (Jamur and Oliver 2010) to open the cell membrane to allow passage of the antibodies into the cytoplasm.

Some commercially available products are Foxp3 Fixation/Permeabilization Concentrate and Diluent™ from eBiosciences and BD Cytofix/Cytoperm™ from BD Biosciences among others.

13.5 Data Analysis

Following data collection, analysis of the data is a critical step in understanding the cell function and hierarchy of the cells. Most flow cytometers store raw data in "list mode" files which adhere to the International Society for the Advancement of Cytometry (ISAC) standard of fluorescent cell sorting (FCS), either FCS2.0 or FCS3.0. This data can be exported to a wide variety of software packages for analysis. Some commercial programs include FACSDiVa® from BD Biosciences, Kaluza® from Beckman-Coulter, Mod-Fit® and

WinList™ from Verity Software House, VenturiOne® from AppliedCytometry, and FlowJo® from Treestar to name a few. A list of free software available on the web can be found at www.cyto.purdue.edu/flowcyt/software/Catalog.htm.

Data imported into the software package can be viewed and analyzed for the binding of antibodies or any other fluorescent molecule used in the experiment. Generally, fluorescence intensity is the measurement of choice. Control samples are compared to experimental samples to determine whether there are any differences. Such differences may include changes in the expression of surface molecules either upward or downward as determined by changes in the mean or median fluorescence. Other measurements may include the percentage of the population expressing the particular molecule. Most software allows the "overlay" of data from different samples to assist in the visualization of the changes.

13.6 Magnetic Bead Sorting

Although not technically a "flow cytometry" technique, magnetic bead sorting can be a useful tool for the enrichment of specific cell types. Commercially available sorting devices include MACS® from Miltenyi Biotech, https://www.miltenyibiotec.com/en/support/resources.aspx and Dynabeads® from Life Technologies (Neurauter et al. 2007).

Cell suspensions are prepared in a manner similar to flow cytometry except that the antibodies used are coupled to iron particles instead of fluorochromes. These particles render the cells magnetic when they are loaded onto a column under magnetic flux. The cells which have the antibody/iron particle on the surface are trapped in the magnetic field of the column and retained while the unlabeled fraction is eluted from the column. When the magnetic field is removed, the trapped cells are eluted. The eluted cells are highly purified. These cells, either the negative or positive fraction, can then be used for flow cytometry of experimental studies. Further, enriching for rare cells significantly reduces the time required to flow sort rare populations and often increases the yield.

13.7 Flow Cytometric Neuronal Research Studies

In a study to assess the expression of beta-amyloid protein on cells from Alzheimer's patients, Jung et al. (1996) used flow cytometry to assess the level of this protein on the surface of freshly isolated mouse brain cells and noted that on archival tissue, this protein was rapidly shed and was only found in the cytoplasm.

Bilsland et al. (2006) developed a method to identify cycling cells in the hippocampus of mice. This study utilized the uptake of BrDU by cells in vivo followed by isolation of the cells from the hippocampus and assessing the amount of BrDU incorporated into the dentate gyrus and the subventricular zone using flow cytometry and antibodies to BrDU on fixed tissues.

On a different note, Stevens et al. (2002) utilized flow cytometry to assess infiltrating inflammatory cells into ischemic areas of the cerebellum. They found that flow cytometry could be used to rapidly and accurately assess the specific types and numbers of immune cells which were infiltrating the ischemic area. These cells expressed markers of activated T cells, such as CD11b. There was a pattern of macrophage and neutrophil infiltration, which is thought to activate pro-inflammatory pathways resulting in cell death at the ischemic site.

In a similar study, Badie et al. (2000) examined the experimental use of Dexamethasone as a mechanism to reduce or eliminate the damage caused by the inflammatory reaction. He noted, using flow cytometry, the common infiltrating cells, lymphocytes, and microglia were inhibited at low doses of Dexamethasone and suggested this as a potential clinical strategy for the prevention of brain tumors.

McLarena et al. (2001) evaluated neural stem cells using flow cytometry. She and her colleagues treated rat neurospheres with cytokines and following separation into single cells, fixed and stained the cells to examine Nestin, beta-tubulin, and GFAP expression in these cells. After staining, cells were analyzed by flow cytometry, and the data demonstrated that cells could be categorized into specific cell subsets by scatter parameters and specific expression of cytoplasmic markers (GFAP for astrocytes and beta-tubulin for neurons). Nestin expression (which is specific to neural stem cells) varied between subsets of differentiated cells from neurospheres. The expression of Nestin and GFAP during self-renewal and differentiation has been described in stem cells of the central nervous system and in retinal tissues (Kawaguchi et al. 2001; Bhattacharya et al. 2002).

The synaptosome has recently been examined using flow cytometric techniques. This method produced better results than that observed using microscopic techniques because the vesicles were not degraded by the fixation process and were easily identified and characterized using scatter parameters and SNAP 25 along with GFAP (Gylys et al. 2000).

Flow cytometry was used by Behbahani et al. (2005) to assess the toxic effects of tumor necrosis factor-alpha and other stressor chemicals on primary cultured neurons. Using Annexin V and propidium iodide to assess apoptosis and cell death, they were able to compare populations of necrotic, early apoptotic, late apoptotic, and viable cells following treatments. This technique successfully demonstrated that flow cytometry could be used as a screening technique for assessing treatments of neurodegenerative diseases.

Kipnis et al. (2012) and Schwartz and Kipnis (2011) have studied the cognitive relationship between the brain and immunity through the identification of specific T cells that have precognitive properties. By using an immune-deficient mouse model, they demonstrated that both T cells from the immune compartment as well as innate immunity including microglia and infiltrating monocytes are responsible for differences observed in brain activity and behavior between control and immune-deficient mice.

In our laboratory, we have used flow cytometry in similar ways. We have examined the infiltration of immune cells into the brains of Parkinson's patients (Mosley et al. 2012). Using flow cytometry and magnetic bead separations, we identified and quantified activated T cells expressing surface CD25, CD127, and CD45RA and RO (Fig. 13.1).

13.8 Common Problems/Pitfalls

The use of flow cytometry for the analysis of any type of cell preparation depends greatly on whether a single cell suspension is obtained. This is critical for identifying each and every cell that passes through the instrument. Secondly, the selection of antibodies specific for the cell type to be analyzed is critically important. Many cell types express the same markers so that you must select one or more very specific antibodies to unique surface markers to clearly identify the desired cell population. Third, fixation and permeabilization protocols may need to be modified, in terms of concentration of the agent, time of incubation, and buffering conditions.

It should be noted that with modern flow cytometers, multiparametric studies of 5–30 independent markers can be detected using 1 tube, provided appropriate single marker controls are provided. Core facilities at most institutions will be able to assist you with experimental design and antibody/fluorochrome selections.

13.9 Discussion/Summary

In general, flow cytometry techniques can readily be developed for examining neuronal cells. Although flow cytometry is thought to be a technique for examining immune cells, recent work has

Fig. 13.1 Memory and naïve phenotypes of CD4 regulatory T cells and effector T cells. (**a**) Forward and side scatter dot plot demonstrates the size and granularity of lymphocytes, monocytes, and granulocytes. (**b**) CD4+ T cells can be identified by low side scatter (size) and high expression of CD4. (**c**) Regulatory T cells (Treg) can be identified as CD25+ CD127–, while effector T cells are identified as CD25+ CD127+. (**d**) Memory T cells (Q1) are identified as CD45RO+ CD45RA–, while naïve T cells (Q4) are identified as CD45RO–CD45RA+ in both the Treg and Teff populations of CD4+ T cells. (**e**) Population hierarchy demonstrates the percentages of CD4+ T cells that are Treg and Teff and the percentages of Treg and Teff that are of the memory or naïve phenotype

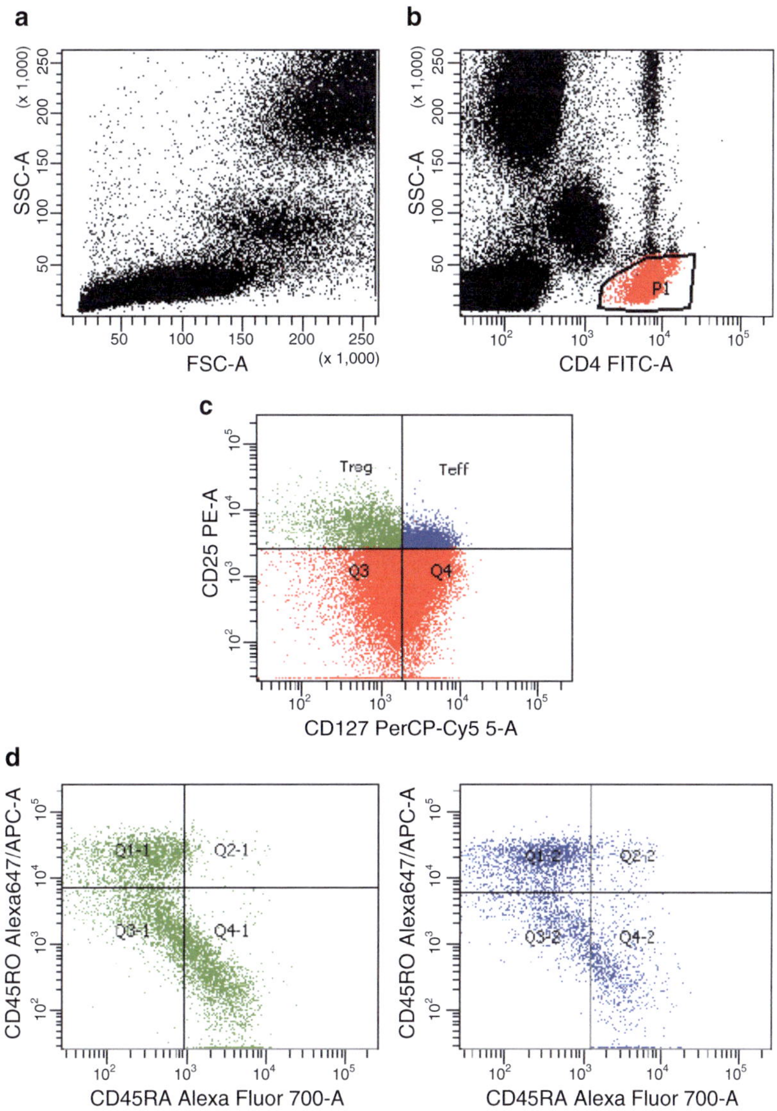

e

Well: 309-2

Population	#Events	%Parent	%Total
All Events	694,042	####	100.0
P1	101,245	14.6	14.6
Treg	9,491	9.4	1.4
Q1-1	5,145	54.2	0.7
Q2-1	171	1.8	0.0
Q3-1	1,915	20.2	0.3
Q4-1	2,260	23.8	0.3
Teff	6,747	6.7	1.0
Q1-2	4,766	70.6	0.7
Q2-2	181	2.7	0.0
Q3-2	1,006	14.9	0.1
Q4-2	794	11.8	0.1
Q3	36,385	35.9	5.2
Q4	48,622	48.0	7.0

shown it to be an easily adaptable technique to analyze almost any single cell suspension including neuronal populations. The use of flow cytometry to assess the interactions of immune cells with neuronal cells has also been well documented here.

With the appropriate antibodies, cells from neuronal tissues can be identified, quantified, and sorted using flow cytometric techniques, and in doing so, the speed and accuracy of the experimental measurements will also be enhanced.

References

Badie B, Schartner JM, Paul J, Bartley BA, Jl V, Preston K (2000) Dexamethasone-induced abolition of the inflammatory response in an experimental glioma model: a flow cytometry study. J Neurosurg 93:634–639

Behbahani H, Rickle A, Concha H, Ankarcrona M, Winblad B, Cowburn RF (2005) Flow cytometry as a method for studying effects of stressors on primary rat neurons. J Neurosci Res 82:432–441

Bhattacharya S, Jackson JD, Das AV, Thoreson WB, Kuszynski C, James J, Joshi S, Ahmad I (2002) Direct identification and enrichment of retinal stem cells/progenitors by Hoechst 33342 dye efflux assay. Invest Opthalmol Vis Sci 44(6):2764–2773

Bilsland JG, Haldon C, Goddard J, Oliver K, Murray F, Wheeldon A, Cumberbatch J, McAllister G, Munoz-Sanjuan I (2006) A rapid method for the quantification of mouse hippocampal neurogenesis *in vivo* by flow cytometry: validation with conventional and enhanced immunohistochemical methods. J Neurosci Methods 157:54–63

de Graaf MT, de Jongste AHC, Kraan J, Boonstra JG, Sillevis Smitt PAE, Gratama JW (2011) Flow cytometric characterization of cerebrospinal fluid cells. Cytometry B Clin Cytom 80B:271–281

Gylys KH, Fein JA, Cole GM (2000) Quantitative characterization of crude synaptosomal fraction (P-2) components by flow cytometry. J Neurosci Res 61:186–192

Jamur MC, Oliver C (2010) Permeabilization of cell membranes. Methods Mol Biol 588:63–66

Jung SS, Nalbantoglu J, Cashman NR (1996) Alzheimer's beta-amyloid precursor protein is expressed on the surface of immediately ex vivo brain cells: a flow cytometric study. J Neurosci Res 46:336–348

Kawaguchi A, Miyata T, Sawamoto K, Takashita N, Murayama A, Akamatsu W, Ogawa M, Okabe M, Tano Y, Goldman S, Okano H (2001) Nestin-EGFP transgenic mice: visualization of the self-renewal and multipotency of CNS stem cells. Mol Cell Neurosci 17:259–273

Kipnis J, Gadani S, Derecki NC (2012 Sep) Pro-cognitive properties of T cells. Nat Rev Immunol 12(9):663–669. doi:10.1038/nri3280. Epub 2012 Aug 20, PubMed PMID: 22903149

Klassen H, Schwartz MR, Bailey A, Young M (2001) Surface markers expressed by multipotent human and mouse neural progenitor cells include tetraspanins and non-protein epitopes. Neurosci Lett 312:180–182

Kraan J, Gratama JW, Haioun C, Orfao A, Plonquet A, Porwit A, Quijano S, Stetler-Stevenson M, Subira D, Wilson W (2008) Flow cytometric immunophenotyping of cerebrospinal fluid. Curr Protoc Cytom Chapter 6:Unit 6.25

McLarena FH, Svendsenb CN, Van der Meidec P, Jolya E (2001) Analysis of neural stem cells by flow cytometry: cellular differentiation modifies patterns of MHC expression. J Neuroimmunol 112:35–46

Mosley RL, Hutter-Saunders JA, Stone DK, Gendelman HE (2012) Inflammation and adaptive immunity in Parkinson's disease. Cold Spring Harb Perspect Med 2(1):a009381, PMID:22315722

Neurauter AA et al (2007) Cell isolation and expansion using Dynabeads®. Adv Biochem Eng Biotechnol 106:41–73

Schwartz M, Kipnis JA (2011) A conceptual revolution in the relationships between the brain and immunity. Brain Behav Immun 25(5):817–819, PubMed PMID: 21187141; PubMed Central PMCID: PMC3074045

Stevens SL, Bao J, Hollis J, Lessov NS, Clark WM, Stenzel-Poorea MP (2002) The use of flow cytometry to evaluate temporal changes in inflammatory cells following focal cerebral ischemia in mice. Brain Res 932:110–119

Part IV

Standard Laboratory Nucleic Acid and Protein Detections

14 Western Blotting Technique in Biomedical Research

Jianuo Liu, James Haorah, and Huangui Xiong

Abstract

Western Blot is the most important and powerful technique frequently used in laboratory research. It is used to identify specific proteins in biological samples isolated from cells or tissues. Similar to the Southern Blot for DNA and Northern Blot for RNA, the Western Blot procedures rely upon three key elements: the separation of protein mixtures by size utilizing sodium dodecyl sulfate-polyacrylamide gel electrophoresis, the efficient transfer of separated proteins from the gel to a nitrocellulose or polyvinylidene difluoride membrane, and identification of a target protein via conjugation with specific primary and enzyme-labeled secondary antibodies. This specificity of the antibody–antigen interaction enables a target protein to be identified in the midst of a complex protein mixture. Once detected, an appropriate substrate is then added to the enzyme, and together, they produce a detectable band visible on a blotting membrane, X-ray file, or imaging system. Western Blot is rapid, and simple, enabling easy to interpret, unique, and unambiguous results. Along with other immunoassays, Western Blot is routinely used in research and clinical settings.

Keywords

Western Blot • Proteins • Sodium dodecyl sulfate-polyacrylamide gel electrophoresis • Nitrocellulose membrane • Polyvinylidene difluoride membrane • Primary antibody • Secondary antibody

14.1 Introduction

Western Blot is an analytical technique used to detect specific proteins in biological samples. The detection of a specific protein is achieved by transferring the electrophoretically separated

J. Liu (✉) • J. Haorah • H. Xiong
Department of Pharmacology and Experimental Neuroscience, University of Nebraska Medical Center, Durham Research Center, Durham Research Plaza, DRC I 8034, 45th Street at Emile Street, Omaha, NE 68198-5880, USA
e-mail: jnliu@unmc.edu; jhaorah@unmc.edu; hxiong@unmc.edu

H. Xiong and H.E. Gendelman (eds.), *Current Laboratory Methods in Neuroscience Research*, Springer Protocols Handbooks, DOI 10.1007/978-1-4614-8794-4_14,

proteins from the gel onto a nitrocellulose or polyvinylidene difluoride membrane. The transferred proteins of different molecular size that bind to the membrane are then identified by conjugating with specific primary and labeled secondary antibodies. Thus, the Western Blotting technique is used for separation, identification, and localization of proteins in biomedical research for diagnosis of many human diseases. Since Western Blotting can detect only matured functional protein levels, this technique is different from Northern (detection of RNA) and Southern (detection of DNA) Blotting techniques.

This technique was first introduced and developed by Towbin et al. (1979) in 1979. The name "Western Blot" is a play on the name Southern Blot, which is named after its inventor Edwin Southern (1975). Western Blot is also referred to as protein immunoblotting because an antibody is used to detect its antigen (LeGendre 1990). The specificity of the antibody–antigen interaction enables a target protein to be identified in the midst of a complex protein mixture. Similar to the Southern Blot for DNA, Western Blot is a powerful and important tool for the immunodetection of proteins following electrophoresis, particular for proteins of low abundance. Currently, Western Blot is a routine tool widely used in the fields of molecular biology, biochemistry, immunogenetics, and other molecular biological disciplines. This technique also serves as an effective diagnostic tool in the investigation of CNS disease (Gajovic et al. 2010; Torian et al. 2011; Zoller et al. 1993). For example, diagnosis of HIV infection is based on an initial ELISA test to detect HIV-1 antibody, followed by Western Blot as a complimentary test along with HIV-1 protein size determinations (Masciotra et al. 2000, 2011; Pappaioanou et al. 1993).

In this chapter, we discuss the entire process of Western Blot procedures, including sample preparation, gel electrophoresis, protein transfer from gel to membrane, immunostaining with a probe, and protein visualization. In addition to the diagnostic application of Western Blot technique in human diseases, this chapter also discusses the troubleshooting techniques in Western Blot procedures.

14.2 Sample Preparation

14.2.1 Lysis Buffers

An important part of Western Blot is sample preparation. Cells and tissues need to be lysed in order to release proteins from within the cell membrane, which are able to migrate individually through a separating gel. Thus, choosing the correct cell or tissue lysis buffer is a critical step for efficient protein extraction. Although lysis buffer preparations vary, only few compliment Western Blot experiments. The main difference is in their ability to solubilize proteins. The lysis buffers which contain sodium dodecyl sulfate (SDS) and other ionic detergents are considered to be most efficient and, therefore, most likely to give the highest yield. The following shows commonly used cell or tissue lysis buffers.

RIPA Buffer (Radio Immunoprecipitation Assay Buffer)

- 150 mM NaCl (sodium chloride)
- 1.0 % NP-40 or 0.1 % Triton X-100
- 0.5 % sodium deoxycholate
- 0.1 % SDS
- 50 mM Tris–HCl pH 8.0

RIPA buffer is useful for cell lysates and membrane-bound proteins. It enables efficient cell lysis and protein solubilization while avoiding protein degradation and interference with protein's immunoreactivity and biological activity. RIPA buffer also gives low background, but will disrupt protein–protein interaction and may be problematic for immunoprecipitations and molecular pull–down assays.

Nonidet-P40 (NP-40) buffer

- 150 mM NaCl
- 1.0 % Nonidet-P40 (NP-40)
- 50 mM Tris–HCl pH 8.0

NP-40 buffer can solubilize many soluble cytoplasmic, membrane bound, or whole cell proteins, but is best used when a lower denaturing level is preferred. However, NP-40 buffer is not suitable for nuclear proteins as it is not strong enough to lyse the nuclear membrane. Lastly, the pH of NP-40 buffer greatly affects the efficiency of extraction.

Tris-Triton Buffer

- 10 mM Tris, pH 7.4
- 100 mM NaCl
- 1 mM EDTA
- 1 mM EGTA
- 1 % Triton X-100
- 10 % glycerol
- 0.1 % SDS
- 0.5 % deoxycholate

Tris-Triton buffer is the ideal buffer with detergent required to release membrane or cytoskeleton-bound proteins.

Table 14.1 Protein location and lysis buffer choice

Protein location	Buffer recommended
Whole cell	NP-40 or RIPA
Cytoplasmic (soluble)	Tris–HCl
Cytoplasmic (cytoskeletal bound)	Tris-Triton
Membrane bound	NP-40, RIPA, or Triton X-100
Nuclear	RIPA or nuclear enrichment fraction protocol
Mitochondria	RIPA with spatial protocol

14.2.2 Protein Condition and Lysis Buffer Choice

Most commercial antibodies recognize reduced and denatured proteins and, therefore, should be used under reducing and denaturing conditions. Lysing cells or tissues with denaturing detergent will keep the proteins in a denatured state. When an antibody is able to recognize the denatured form of a target protein, RIPA buffer is the best choice. However, some antibodies will only recognize a protein in its native state, non-denatured form and is unrecognizable when extracted with a denaturing detergent such as SDS, deoxycholate, or some mild denaturing detergent such as Triton X-100 and NP-40. In this case, lysis buffers without detergent or relatively mild nonionic detergents should be used.

14.2.3 Protein Location and Lysis Buffer Choice

Proteins that are in a subcellular location can be enriched in a lysate of subcellular fraction. Isolation of subcellular fractions allow for more efficient identification and study of proteins of interest. Currently, many scientific subcellular protein fractionation kits and methods enable stepwise separation and extraction of cytoplasmic, membrane, nuclear soluble, chromatin-bound, and cytoskeletal proteins from mammalian-cultured cells. The commonly used lysis buffers for extractions of subcellular proteins are given in Table 14.1.

14.2.4 Protease and Phosphatase Inhibitors

Cells contain many different types of proteases. Once cells are disrupted, proteases are released and can quickly degrade the protein. This can drastically reduce the yield of the protein during the isolation procedure. To prevent proteolysis, dephosphorylation, and degradation of proteins, protease and phosphatase appropriate inhibitors are incorporated to fresh lysis buffer. Currently, various ready to use cocktails of inhibitors are available on the market, but one can also easily make an appropriate protein inhibitor cocktail (Table 14.2).

14.2.5 Procedure of Protein Extraction

Protein Extraction from Cell Culture: General Protocol

1. Place the cell culture dish on ice and remove the culture media.
2. Wash the cells two times with ice-cold PBS and rock plates several times. Completely remove each wash. Insufficient washing will contaminate lysate with media components.
3. Add ice-cold lysis buffer to chilled, washed cells and scrape adherent cells off the dish using a cold plastic cell scraper; gently trans-

Table 14.2 Protein inhibitors

Inhibitor	Protease/phosphatase inhibitor	Effective concentration	Stock solution (store at −20 °C)
Aprotinin	Trypsin, chymotrysin, plasmin	1–2 μg/mL	10 mg/mL in water or 0.01 M HEPES pH 8.0
Leupeptin	Lysosomal, trypsin, papain	5–10 μg/mL	10 mg/mL in water
Pepstatin A	Aspartic proteases	1 μg/mL	1 mg/mL in methanol
PMSF	Serine, cysteine proteases	1 mM	10 mM in ethanol
EDTA	Metalloproteases that require Mg^{++} and Mn^{++}	5 mM	0.5 M in water, pH 8.0
EGTA	Metalloproteases that require Ca^{++}	1 mM	0.5 M in water, pH 8.0
Na fluoride	Serine–threonine phosphatases	5–10 mM	Dilute in H_2O
Na orthovanadate	Tyrosine phosphatase	1 mM	Dilute in H_2O

fer the cell suspension into a prechilled microcentrifuge tube (commonly, 1 mL lysis buffer is added to 10^7 cells/100 mm dish or 150 cm^2 flask; 0.5 mL per 5×0^6 cells/60 mm dish or 75 cm^2 flask.).

4. Cells growing in suspension should be concentrated by centrifugation and then resuspended in appropriate ice-cold lysis buffer at a concentration of 10×10^6 cells/mL.
5. Gently rock the suspension on either a rocker in a cold room or simply provide constant agitation on ice for 30 min to lyse cells.
6. Centrifuge the lysate at 12,000 × *g* for 15 min at 4 °C. Centrifuge force and time vary depending on cell type.
7. Gently remove the tubes from the centrifuge and place on ice. Immediately transfer the supernatant to a fresh tube and store it on ice or at −70 °C, depending on the sensitivity of the target antigen to freezing and thawing.

Protein Extraction from Tissues

1. Dissect the tissue of interest with sterile tools. To prevent degradation by proteases, all procedures must be done with prechilled reagents on ice and as quickly as possible.
2. Place the tissues in an Eppendorf tube and immerse in liquid nitrogen to "snap freeze." Store samples at −80 °C for later use or keep on ice for immediate homogenization.
3. Add lysis buffer rapidly to the tube containing frozen tissue, homogenize the tissues with electric homogenizer, 2–5 strokes, and rinse the blade twice with the same amount of lysis buffer. The volume of lysis buffer must be determined in relation to the amount of tissue present. For a ~5 mg piece of tissue, add 300 μL lysis buffer for homogenization, then rinse the blade twice with another 2 × 300 μL lysis buffer. Protein extract should not be too diluted to avoid loss of protein. Optimal protein concentration is 1–5 mg/mL.
4. Incubate with constant agitation of lysate for 2 h at 4 °C.
5. Centrifuge at 12,000 × *g* for 20 min at 4 °C using microcentrifuge.
6. Transfer supernatant to a fresh tube cautiously. Do not take any lipid from the surface layer or any precipitate from the bottom, which may interfere with Western Blot.

14.2.6 Quantification of Protein Concentration in Extracts

Testing relative differences in protein levels between different samples using the Western Blot analysis is accurate if equivalent amounts of total protein are loaded. Therefore, quantification of protein concentration is an important step to a successful Western Blot.

Protein concentration can be assessed by a Bradford assay (Bradford 1976; Kruger 1994), Lowry assay (Lowry et al. 1951) or a BCA assay (Smith et al. 1985; Wiechelman et al. 1988). Bovine serum albumin (BSA) is frequently used as a protein standard (Bradford 1976; Noble and

Bailey 2009). After determining the protein concentrations, the samples are ready to load onto the gels or preserved at −20 or −80 °C for later use.

14.2.7 Preparation of Samples for Loading into Gels

As mentioned above, most antibodies typically recognize reduced or denature protein samples because these antibodies only identify a small portion of the protein epitope, which may reside within the 3D conformation of the protein. To enable antibody access to this portion, it is necessary to unfold protein by denaturing it. The common method is that protein lysates are mixed with a loading buffer and then boiled at 95–100 °C for 5 min.

The standard loading buffer is called 2× Laemmli buffer (Laemmli 1970). It contains the anionic denaturing detergent SDS that can reduce the protein to a primary linear structure and coats it with a negative charge. Loading buffer also contains β-mercaptoethanol or dithiothreitol (DTT) that reduces disulphide bridges in proteins. 2× Laemmli buffer is mixed in a 1:1 ratio with the protein sample.

Alternatively, an antibody may only recognize a native and nonreduced sample. If this is the case, SDS, β-mercaptoethanol, and DTT must be left out of the loading buffer, and Western Blot will be run in nonreducing and native conditions.

2× Laemmli buffer:

- 4 % SDS
- 10 % 2-mercaptoethanol
- 20 % glycerol
- 0.004 % bromophenol blue
- 0.125 M Tris–HCl
- Adjust the pH to 6.8

14.3 Gel Electrophoresis

After denaturing, the protein is ready to be separated using gel electrophoresis, which can be one-dimension (1D) or two-dimension (2D). One-dimensional gel electrophoresis is routinely used for protein separation, while 2D is typically used for fingerprinting. Here we focus on 1D electrophoresis.

14.3.1 Sodium Dodecyl Sulfate-Polyacrylamide Gel Electrophoresis

Almost all analytical electrophoresis of proteins are carried out in polyacrylamide gels under conditions that ensure dissociation of proteins into their individual polypeptide subunits while minimizing aggregation (Ornstein 1964; Davis 1964). This procedure is abbreviated as SDS-PAGE (sodium dodecyl sulfate-polyacrylamide gel electrophoresis). Protein separation within a gel is in reference to polypeptide size, independent of its sequence. The effective range of separation depends on the concentration of polyacrylamide. Thus, carefully choosing the percentage of gel is very important as it determines the rate of migration and degree of separation from one protein to the others. Table 14.3 shows materials and reagents for SDS-PAGE preparation.

14.3.2 Preparation of Running Samples

1. Molecular Weight (MW) Markers: MW marker is a mixture of several recombinant proteins with different molecular weights. It reveals protein size and also monitors the progress of electrophoresis.
2. Positive Controls: When running SDS-PAGE, a positive control can monitor the working protocol correctly in order to identify the target protein. It is always better to use a positive control while performing a new experiment because it gives confidence and rationale explanation to the outcome of the experimental procedures.
3. Load Controls: Loading controls must be run along with the samples. This will verify that the gel lanes are evenly loaded, especially when a comparison is made between the levels

Table 14.3 Materials and effective range of separation of SDS-PAGE

	Gel percentage (%)			
	7 %	10 %	12 %	15 %
Protein size (kDa)	35–90	15–100	10–70	12–45
H_2O	15.3 mL	12.3 mL	10.2 mL	7.2 mL
1.5 M Tris–HCl pH8.0	7.5 mL	7.5 mL	7.5 mL	7.5 mL
20 % (w/v) SDS	0.15 mL	0.15 mL	0.15 mL	0.15 mL
Acrylamide/Bis-acrylamide (30 %/0.8 % w/v)	6.9 mL	9.9 mL	12.0 mL	15.0 mL
10 % APS	0.15 mL	0.15 mL	0.15 mL	0.15 mL
TEMED	0.02 mL	0.02 mL	0.02 mL	0.02 mL

of protein expression from one experimental condition to the other. Loading controls are often used to standardize the results from these studies. They are also used to check for even transfer from gel to membrane. Commonly used loading controls include β-actin (MW: 42 kDa), GAPDH (MW: 30–40 kDa), and tubulin (MW: 55 kDa).

4. Loading Samples: Special gel loading tips are used to load samples into each well, and one lane is reserved for a MW maker. Commonly 10–40 μg of protein can be loaded per mini-gel well. Never overfill wells; it may result in poor data if samples spill into adjacent wells, resulting in poor resolution bands.

14.3.3 Standard Running Buffer (1× Tris-Glycine-SDS Running Buffer)

The electrophoresis apparatus is set up with the cathode buffer covering the gel in the negative electrode chamber, and the anode buffer in the lower positive electrode chamber. Tris-glycine-SDS (TGS) is used for both the anode buffer and the cathode buffer. Thus, it is the most commonly used buffer for SDS-PAGE gel preparations. Run voltage between 100 and 150 V for vertical gel electrophoresis units.

1× Tris-glycine-SDS running buffer:

- 25 mM Tris
- 192 mM Glycine
- 0.1 % SDS

14.3.4 Procedure for Running SDS-PAGE

- Pour 1× SDS-PAGE running buffer into the Western Blot tank and position the acrylamide gel into the gel holder assembly and then immerse into the tank.
- Fill the inner chamber (between the two gels) with 1× SDS-PAGE running buffer.
- Carefully load the MW makers and equal amounts of protein samples into designated wells of the gel.
- Place the lid on the tank and plug it into power source.
- Run the apparatus at 100 V for 1–2 h allowing the samples to separate. Continue running the apparatus until the loading buffer incorporated into each sample has reached the bottom of the gel (Fig. 14.1).

14.4 Transfer Protein

Following electrophoresis, the separated protein will be transferred from the gel matrix to a membrane, referred to as the "blot." The process of transferring protein can be achieved by (a) simple diffusion, (b) vacuum blotting, and (c) electrophoretic elution or electrotransfer (Towbin et al. 1979; Peferoen et al. 1982; Kurien and

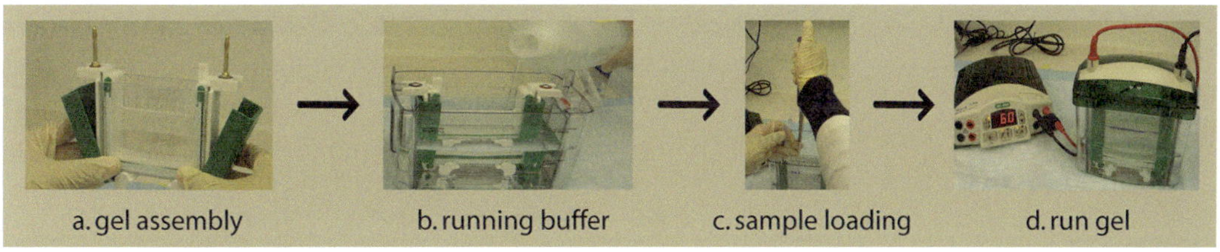

Fig. 14.1 Flowchart of SDS-PAGE (**a**) Place the gel cassette in the electrode assembly. (**b**) Position the gel assembly into the tank and apply the running buffer. (**c**) Pipet the samples into the wells. (**d**) Run gel at 60–100 V

Scofield 1997). Electrophoretic transfer is the most commonly used transfer method for protein because of its speed and transfer efficiency. The efficient transfer depends on the nature of the gel, the molecular mass of the proteins being transferred, and the membrane used (Kurien and Scofield 2002, 2006).

14.4.1 Membranes

The two common membrane types used for Western Blot are nitrocellulose (Burnette 1981) and polyvinylidene difluoride (Matsudaira 1987).

1. Nitrocellulose membrane: Nitrocellulose (NC) membrane is one of the first types of membranes used for Western Blot. The advantages of NC membrane include the following: (1) it is easily wetted with high adsorption capacity for protein and nucleic acids; (2) nonspecific binding can be blocked with BSA, gelatin, or other surface passivating agents that are not effective on nylon or other membrane types; and (3) it has lower "burn through" in Western Blot electrotransfer, allowing longer time and complete transfer from the gel, thus improving sensitivity.
2. Polyvinylidene difluoride membrane: Polyvinylidene difluoride (PVDF) membrane was first developed for protein blotting by Millipore Corporation in 1985. Compared with NC, the PVDF membrane offers higher protein adsorption capacity and retention than any other membrane. Also, protein immobilized on PVDF is not easily removed. The strength of PVDF membranes is an asset when one is interested in stripping and re-probing for further immunodetection. However, PVDF requires careful pretreatment which includes soaking the membrane in methanol for 1–2 min, followed by rinsing with deionized water and incubation in ice-cold transfer buffer for 5 min.

14.4.2 Electrophoretic Transfer

Protein transfer can be done in wet (tank transfer) or semidry transfer conditions. Both are based on the same principles, but differ only in the mechanical devices used to hold the gel/membrane stack along with the application of the electrical field.

1. Wet transfer: Wet transfer is the traditional technique where the gel and membrane are sandwiched between a sponge and filter paper with the gel/membrane stack completely immersed in the buffer reservoir as current is applied. It is important that the membrane is adjacent to the positive electrode and the gel adjacent to the negative electrode. The negatively charged proteins move towards the positively charged electrode; the membrane binds to the proteins and prevents them from continuing on. Wet transfer systems are typically run at a constant voltage. Mixing the buffer during transfer keeps the current relatively constant.
2. Semidry transfer: Instead of a buffer reservoir, in a semidry transfer, the gel–membrane sandwich is soaked in transfer buffer and then placed directly between positive and negative electrodes (cathode and anode respectively). The advantage of semidry transfer is that it is generally faster (15–45 min) and efficient.

Semidry transfer has a low buffer capacity and thus is inappropriate for prolonged transfers. Semidry transfer is typically run at a constant current; the voltage normally increases during the transfer period.

14.4.3 Transfer Buffer

A standard transfer buffer is Towbin system buffer (Towbin et al. 1979). Generally, transfer buffer without SDS is better; however, for proteins that have a tendency to precipitate or have a molecular weight larger than 80 kDa, SDS should be included at a final concentration of 0.1 %. Methanol is typically present in the transfer buffer and aids in stripping SDS from proteins and increases the binding capacity of NC for proteins (Gershoni and Palade 1982, 1983). For general protein transfer, methanol is suggested to a final concentration of 20 %.

1× Standard transfer buffer (Towbin system buffer)

- 25 mM Tris
- 192 mM Glycine
- 20 % Methanol

14.4.4 Procedure of Protein Transfer

Procedure for Wet Transfer

1. Fill the ice-cooling unit with water and store at −20 °C until ready to use.
2. Prepare the transfer buffer and store in refrigerator, using buffer at 4 °C will improve heat dissipation.
3. Wearing gloves, cut the membrane and filter paper to the size of gel and mark the membranes with pencil to ensure they can be identified during the testing.
4. Membrane equilibration depends on the procedure and type of the membranes:
   (a) For nitrocellulose membrane: Wet the membrane with deionized water at least 5 min and equilibrate in transfer buffer until use.
   (b) For PVDF membrane: Wet the membrane 1–2 min with methanol followed by rinsing with deionized water, and incubate in cold transfer buffer until use.
5. Equilibrate the gel and soak the filter paper and fiber pads in transfer buffer for 10–60 min depending on gel thickness.
6. Carefully set up gel sandwich as follows: from cathode side (−) to anode (+) side.
   (a) Place the cassette on a clean large glass container containing transfer buffer. The cathode side is lying in the buffer, and anode side is folded out.
   (b) Place one pre-wetted fiber pad on the top of the cathode side.
   (c) Place a piece of pre-wetted filter paper on the fiber pad.
   (d) Place the equilibrated gel on the top of the filter paper.
   (e) Place the pre-wetted membrane on the gel.
   (f) Place a piece of pre-wetted filter on the membrane.
   (g) Place the last fiber pad on the filter paper.
   (h) It is important to press gel/membrane stack into the buffer and squeeze out any bubbles that might be trapped in the gel sandwich. A glass tube may be used to gently roll out air bubbles.
7. Close the sandwich by folding in anode side and locking it in place.
8. Insert the gel sandwich cassette into the transfer module and place in tank.
9. Put the frozen ice-cooling unit in tank and completely fill the tank with buffer.
10. Connect the tank to power supply and start the transfer. Follow the manufacturer's recommendations for current strength. Transfers are generally complete in 60–90 min (Fig. 14.2).

Procedure for Semidry Transfer

1. Prepare the transfer buffer (see wet transfer).
2. Following electrophoresis, equilibration of gels, precut membrane, and filter paper (see wet transfer).
3. Assembly of the unit for standard transfers:
   (a) Carefully remove the cover and the stainless steel cathode assembly.

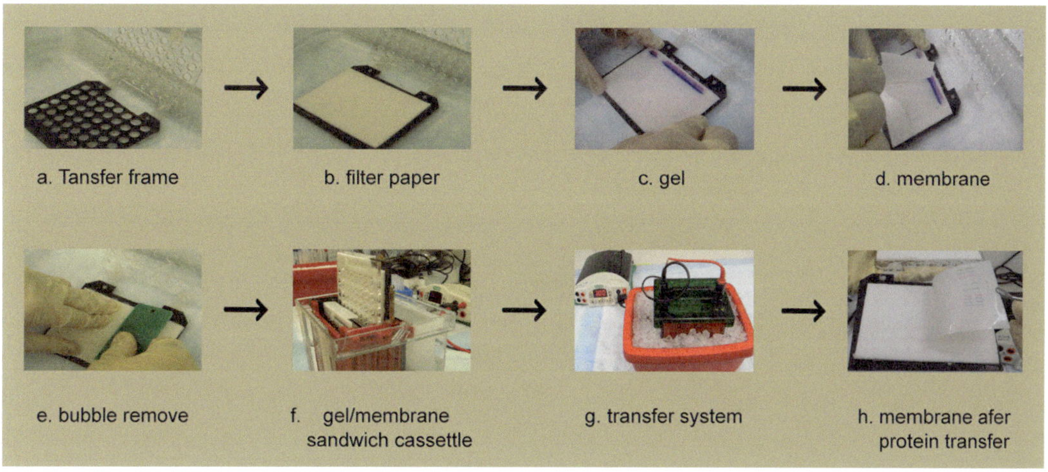

Fig. 14.2 Flowchart of wet transfer. (**a**) Transfer cassette: the *black frame* is the cathode side, and *white frame* is the anode side. (**b**) Place pre-wetted filter paper on the cathode side. (**c**) Lay the equilibrated gel on the top of the filter paper. (**d**) Place a pre-wetted membrane on the gel. (**e**) Apply pre-wetted filter paper on the top of the membrane and remove the air bubbles trapped within the gel sandwich. (**f**) Insert gel/sandwich cassettes into transfer module. (**g**) Fill transfer buffer into the tank and connect the power supply and then start the transfer. (**h**) After transfer, protein MW makers will show on the membrane

(b) Place two presoaked pieces of filter paper onto the platinum anode and remove the air bubbles over the surface of the filter paper.
(c) Lay the pre-wetted membrane on top of the filter paper and roll out the air bubbles.
(d) Carefully place the equilibrated gel on top of the membrane, making sure no portion of the gel is outside the membrane otherwise the transfer will be incomplete. Remove all air bubbles by gently rolling with a glass tube.
(e) Place the other piece of filter paper on top of the gel removing air bubbles completely.
(f) If more than one full-size gel is to be transferred, place a pre-equilibrated membrane on top of the filter paper stack and then place the gel on top of the membrane and filter paper on top of the gel. Thoroughly roll out air bubbles in each layer.
(g) Carefully connect the cathode onto the stack. Press to engage the latches without disturbing the filter paper stack.
(h) Place the safety cover on the unit. Plug the unit into power supply. Transfer mini gels for 15–30 min at 10–15 V, and large gels can be transferred for 30–60 min at 15–25 V (Fig. 14.3).

14.5 Blocking Binding Site on the Membrane

Membranes have a high capacity to bind proteins including antibodies and target proteins; therefore, it is necessary to coat the membrane with excess protein to prevent interactions between the membrane and the antibody used for detection of the target protein. Two blocking solutions are commonly used: nonfat-dried milk or BSA (Cohn fraction V). Dried milk is the most efficient and least expensive. It is easy to use and compatible with all immunological detection systems. Dried milk is not recommended for studies of phosphor-proteins because milk contains casein, a phosphor-protein can cause high background. Blocking the membrane with BSA causes some antibodies to give a stronger signal for unknown reasons.

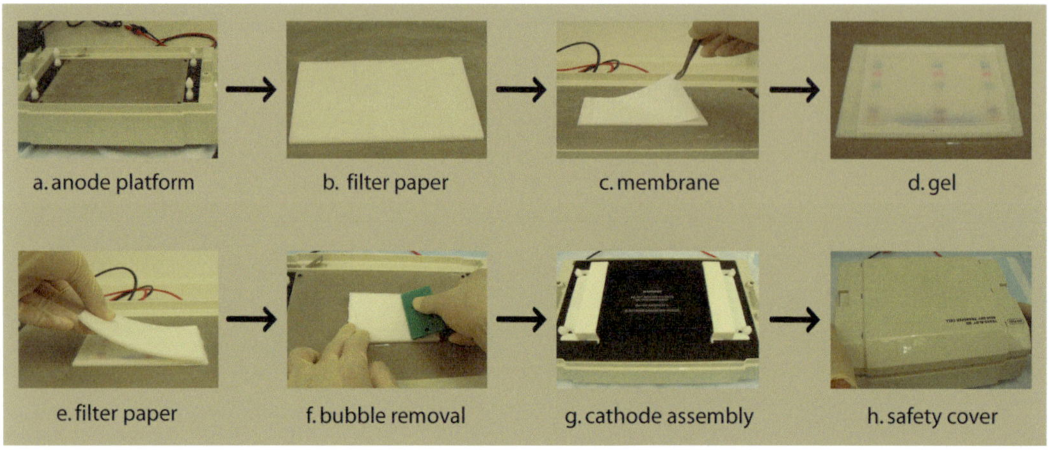

Fig. 14.3 Flowchart of semidry transfer. (**a**) Semidry transfer module: anode platform. (**b**) Place presoaked extra thick filter paper on the anode platform. (**c**) Apply a pre-wetted membrane on top of the filter paper. (**d**) Place the equilibrated gel on top of the membrane. (**e**) Place presoaked extra thick filter paper. (**f**) Remove the air bubbles from between each layer. (**g**) Carefully place the cathode onto the stack. (**h**) Place the safety cover on the unit; plug the unit into the power supply and start to transfer

14.5.1 Blocking Solution

1. 5 % (w/v) nonfat-dried milk:
   - 5.0 g Nonfat-dried milk
   - 100 mL Tris-buffered saline (TBS)
2. 3–5 % (w/v) BSA
   - 3.0–5.0 g BSA
   - 100 mL TBS

It is to be noted that (1) if the background of nonspecific binding of probes is high, adding Tween 20 to TBS buffer to a final concentration of 0.05 % (TBS-T) is recommended. (2) Blocking solution must be well mixed and filtered. Failure to filter can lead to "spotting" where tiny dark grains will contaminate the blot during development.

14.5.2 Procedure of Blocking

1. Place the membrane in Western Blot chamber box.
2. Add blocking solution to immerse the membrane.
3. Incubate the membrane for 1 h at room temperature or overnight at 4 °C with gentle agitation on a platform shaker.
4. After blocking, the membrane is rinsed once with TBS-T for 5 s.

14.6 Antibody Detection

To detect a target protein, the membrane is probed with a modified antibody in two steps: an unlabeled antibody (primary antibody) specific to the target protein is first incubated with membrane followed by incubation with second antibody, an anti-immunoglobulin. The advantage of these steps allows primary antibody to bind the antigen on the membrane, and binding labeled anti-immunoglobulin to the primary antibody–antigen complex enhances the immunoreactive signal.

14.6.1 Primary Antibody

Primary antibody is the specific detection tool that binds the protein directly and forms an antibody–protein complex with the protein of interest. Primary antibodies can be polyclonal or monoclonal antibodies. Normally, both polyclonal and monoclonal antibodies work well. Polyclonal antibodies are less expensive and often have a high affinity for the antigen. In contrast, monoclonal antibodies are valued for their specificity, purity, and consistency that result in a lower background.

14.6.2 Secondary Antibody

In general, the primary antibody that recognizes the target protein is not directly detectable. Therefore, tagged secondary antibodies are used for aiding detection of the target antigen (indirect detection). The choice of secondary antibody depends on the species of animal in which the primary antibody was raised (the host species). For example, if the primary antibody is a mouse monoclonal antibody, the secondary antibody must be an anti-mouse immunoglobulin raised from a non-mouse host.

In Western Blot, secondary antibodies are commonly coupled with different labels. These labels include biotin, fluorescein, rhodamine, horseradish peroxidase (HRP), and alkaline phosphatase. Thus, the choice of secondary antibody also depends upon the type of label that is desired. Enzymatic labels are used most commonly and consistently produce excellent results. Alkaline phosphatase (AP) and HRP are the two enzymes extensively used in labeling.

14.6.3 Procedure of Antibody Detection

1. Dilute the primary antibody in TBS-T with 5 % nonfat-dried milk or 5 % BSA at the suggested dilution or optimize the dilution according to the result. Too much antibody will result in nonspecific binding.
2. Incubate the membrane with diluted primary antibody with gentle agitation.
   (a) Incubation time can be a few hours or overnight depending on the binding affinity of the antibody for the protein and abundance of the protein.
   (b) Incubation temperature is preferably at 4 °C, especially when incubating overnight.
   (c) Agitation of the antibody is recommended to ensure complete antibody expose to the membrane and to prevent uneven binding.
3. Wash the membrane four times in TBS-T with agitation, 10 min each time to remove residual primary antibody.
4. Incubate the membrane with secondary antibody diluted in TBS-T at suggested dilution or optimize the dilution according to the results for 1–2 h at room temperature with agitating.
   (a) The secondary antibody can be diluted in blocking buffer, but it may reduce specific signal.
   (b) Secondary antibodies can be HRP conjugated or AP conjugated. Using HRP-conjugated secondary antibodies are preferred because AP-conjugated secondary antibodies are not sensitive enough.
5. Wash the membrane four times with TBS-T for 10 min each and then rinse in TBS once.

14.7 Enzyme Substrates

The appropriate substrate choice for antibody signal development depends on the enzyme label. The performance of a particular substrate may vary dramatically because of the concentration and purity of the substrate and by other additives and buffer components that are part of the formulation. Thus, for HRP-conjugated antibody signal development, ECL and ECL-plus are the traditional kits used.

14.7.1 Procedure of ECL Detection

1. Mix the ECL detection reagent at 1:1 ratio, for example, 1.5 mL reagent 1 and 1.5 mL reagent 2.
2. Apply the mixed ECL reagent to the surface of the membrane and incubate for 5 min.
3. Drain off the ECL reagent and cover the membrane in transparent plastic wrap.

14.8 Development

There are several methods for capturing data generated from chemiluminescence Western Blot, including X-ray film development and digital images.

Digital images are a new generation technique in which electronic data is captured with digital cameras and transforms the signal into a digital image for rapid analysis with software.

As the technologies improve and equipment prices decline, traditional manual film development is still wildly used. Most of the data obtained is still captured on X-ray film. Commonly, it is necessary to expose several films for different time periods to obtain the proper balance between signal and background. An over expressed film is not suitable for determination of the relative amount of protein.

14.9 Stripping and Re-probing

Stripping and re-probing is useful; and it can investigate more than one protein on the same blot instead of running multiple gels. This process has the advantage of saving samples, materials, and time. A big difference between NC and PVDF membranes is their ability to support stripping antibodies off and reusing the membrane for subsequent antibody probes. PVDF allows for easier stripping and more versatile tests before the antigenic protein gets depleted from the membranes.

14.9.1 Standard Stripping Buffer

- 15 g glycine
- 1 g SDS
- 10 mL Tween 20
- Adjust pH to 2.2
- Bring volume up to 1 L with ultrapure water

14.9.2 Procedure of Stripping and Re-probing

- Wash membrane in TBS-T.
- Incubate the membrane in stripping buffer for 30 min at room temperature.
- Wash membrane in TBS-T three times for 10 min per wash.
- Ready for blocking stage and follow normal protocol for Western Blots.

14.10 Results

Western Blot is a powerful tool to identify the specific protein of interest. The SDS-PAGE provides information on the mass, charge, purity, and presence of a proteins allowing visualization by staining of Coomassie Brilliant Blue or silver staining. After staining, the separated proteins will appear as distinct bands within the gel. MW marker is used to calibrate the gel and determine the approximate molecular mass of unknown proteins by comparing the distance traveled relative to the MW marker. Furthermore, SDS-PAGE can be processed to transfer the proteins onto the surface of a membrane, where the proteins become tightly attached. After immunoblotting, the HRP enzyme conjugated on the secondary antibody converts a luminol substrate to a light-releasing substance. This signal is detected as immunoreactive bands on film. The relative intensity of these immunoreactive bands is quantified as arbitrary units, which are then corrected by the arbitrary data of the internal control standard marker protein and presented as final corrected results.

14.11 Summary

The antigen–antibody interaction can detect a specific protein in the midst of a complex protein mixture. Western Blots allow us to identify the specific protein with the molecular weight and measure relative amounts of the protein present in different conditions. To summarize, (1) the protein samples are prepared from tissues or cell cultures in a lysis buffer that protects the protein of interest from degradation. (2) Isolated protein is separated by gel electrophoresis and subsequently transferred onto a membrane (NC or PVDF). (3) Protein of interest is probed by a primary antibody, which recognizes a specific target protein within a complex enzyme-labeled antibody. (4) Identification of these targeted proteins is achieved by capturing the immunoreactive bands using X-ray film or digital imaging. The intensity of these immunoreactive bands is quantified as arbitrary units using appropriate software, and the arbitrary data is then interpreted as findings of the experiments.

14.12 Western Blot Troubleshooting

1. High background
   (a) Possible Cause: BSA used for blocking.
   Solution: Blocking solution containing BSA may cause high background on the

membrane. Try to add SDS to reduce the background or use a different blocker, such as nonfat-dried milk.

(b) Possible Cause: Antibody concentration may be too high.
Solution: Optimize primary and secondary antibody concentration or incubate for longer but in more dilute antibody.

(c) Possible Cause: Cross-reaction between blocking agent and antibody.
Solution: Milk is usually contaminated with IgG and will cross-react with anti-goat secondary antibodies. Milk also contains casein which is a phosphor-protein, thus phosphor-specific antibody will detect the casein present in the milk and increases the background. Add a mild detergent such as Tween 20 to the washing buffer. Switch blocking solution by using BSA reagent instead of milk.

(d) Possible Cause: Insufficient wash.
Solution: Increase the number of washes and buffer volume.

(e) Possible Cause: Background of membrane.
Solution: NC membrane is considered to give less background than PVDF. Always use Tween 20 (0.1–0.2 % final concentration) during the secondary incubation step. Do not allow the membrane to dry.

2. Poor transfer quality

(a) Possible Cause: Excess methanol in the transfer buffer.
Solution: Although methanol aids in protein binding to PVDF or NC membrane, too much methanol in the transfer buffer decreases the transfer efficiency of protein from the gel to membrane. Reduce the concentration of methanol; balance is needed.

(c) Possible Cause: Air bubbles between the gel and membrane.
Solution: Air bubbles are not conductive; they disturb the protein transfer. Thus, using a pipette to roll over the gel/membrane sandwich removes air bubbles out.

(e) Possible Cause: Insufficient transfer time and current.
Solution: Large molecular weight proteins may be difficult to elute from gel to membrane. Therefore, increase the transfer current and prolong the transfer duration.

3. Weak or no signal

(a) Possible Cause: Insufficient antibody used.
Solution: (1) Primary antibody may offer a low affinity. Increase the amount of antibody and extend incubation time or try different source of antibody. (2) Primary antibody does not recognize the protein in species being tested. Run positive control. (3) Primary and secondary antibodies are not compatible. Use secondary antibody that was raised against the species in which the primary was raised.

(c) Possible Cause: Insufficient antigen load.
Solution: Load more protein on the gel. It is best to use 20–30 μg protein per lane. Try using the narrowest possible well size to concentrate antigen. Use protease inhibitors to avoid protein degradation.

(e) Possible Cause: The protein of interest is not abundantly present in the cell culture and tissue.
Solution: Use an enrichment subfraction to concentrate protein of interest.

(f) Possible Cause: Too much detergent present and excessive washing; signal being washed away.
Solution: Decrease Tween 20 concentration in diluted antibodies and wash buffer. Do not over wash the membrane.

(g) Possible Cause: Poor transfer of protein to membrane.
Solution: (1) Check the transfer with a reversible stain such as Ponceau S. (2) Use pre-stained MW marker to monitor transfer and stain the gel after transfer to make sure no protein is retained in the gel.

4. Nonspecific or multiple bands

(a) Possible Cause: Antibody concentrations too high.
Solution: High concentrations of antibody will bind nonspecifically. (1) Decrease the concentration of antibody used. (2) Reduce incubation time. (3) Increase Tween 20 in diluted antibodies. Add SDS in diluted secondary antibodies.

(b) Possible Cause: The antibody has not purified.

Solution: Try to use affinity-purified antibody. This will help to reduce nonspecific binding.

(c) Possible Cause: Insufficient blocking.

Solution: Blocker used may affect background bands. Use a blocking peptide to differentiate between specific and nonspecific bands. Only specific bands should be blocked. If this is the case, try a different blocker, an optimal blocking reagent.

5. White bands on black blot

(a) Possible Cause: Use too much primary and/or secondary antibody.

Solution: Dilute the antibody more.

6. Smile effect of bands

(a) Possible Cause: Protein migration was too fast and/or too hot.

Solution: Reduce running current to slow down the migratory speed and run the gel in a cold room or on ice.

References

Bradford MM (1976) A rapid and sensitive method for the quantitation of microgram quantities of protein utilizing the principle of protein-dye binding. Anal Biochem 72:248–254

Burnette WN (1981) "Western blotting": electrophoretic transfer of proteins from sodium dodecyl sulfate–polyacrylamide gels to unmodified nitrocellulose and radiographic detection with antibody and radioiodinated protein A. Anal Biochem 112:195–203

Davis BJ (1964) Disc electrophoresis. II. Method and application to human serum proteins. Ann N Y Acad Sci 121:404–427

Gajovic O, Todorovic Z, Nesic L, Lazic Z (2010) [Lyme borreliosis—diagnostic difficulties in interpreting serological results]. Med Pregl 63:839–843

Gershoni JM, Palade GE (1982) Electrophoretic transfer of proteins from sodium dodecyl sulfate-polyacrylamide gels to a positively charged membrane filter. Anal Biochem 124:396–405

Gershoni JM, Palade GE (1983) Protein blotting: principles and applications. Anal Biochem 131:1–15

Kruger NJ (1994) The Bradford method for protein quantitation. Methods Mol Biol 32:9–15

Kurien BT, Scofield RH (1997) Multiple immunoblots after non-electrophoretic bidirectional transfer of a single SDS-PAGE gel with multiple antigens. J Immunol Methods 205:91–94

Kurien BT, Scofield RH (2002) Heat-mediated, ultra-rapid electrophoretic transfer of high and low molecular weight proteins to nitrocellulose membranes. J Immunol Methods 266:127–133

Kurien BT, Scofield RH (2006) Western blotting. Methods 38:283–293

Laemmli UK (1970) Cleavage of structural proteins during the assembly of the head of bacteriophage T4. Nature 227:680–685

LeGendre N (1990) Immobilon-P transfer membrane: applications and utility in protein biochemical analysis. Biotechniques 9:788–805

Lowry OH, Rosebrough NJ, Farr AL, Randall RJ (1951) Protein measurement with the Folin phenol reagent. J Biol Chem 193:265–275

Masciotra S, Livellara B, Belloso W, Clara L, Tanuri A, Ramos AC, Baggs J, Lal R, Pieniazek D (2000) Evidence of a high frequency of HIV-1 subtype F infections in a heterosexual population in Buenos Aires, Argentina. AIDS Res Hum Retroviruses 16:1007–1014

Masciotra S, McDougal JS, Feldman J, Sprinkle P, Wesolowski L, Owen SM (2011) Evaluation of an alternative HIV diagnostic algorithm using specimens from seroconversion panels and persons with established HIV infections. J Clin Virol 52(Suppl 1): S17–22

Matsudaira P (1987) Sequence from picomole quantities of proteins electroblotted onto polyvinylidene difluoride membranes. J Biol Chem 262:10035–10038

Noble JE, Bailey MJ (2009) Quantitation of protein. Methods Enzymol 463:73–95

Ornstein L (1964) Disc electrophoresis. I. Background and theory. Ann N Y Acad Sci 121:321–349

Pappaioanou M, Kashamuka M, Behets F, Mbala S, Biyela K, Davachi F, George JR, Green TA, Dondero TJ, Heyward WL et al (1993) Accurate detection of maternal antibodies to HIV in newborn whole blood dried on filter paper. AIDS 7:483–488

Peferoen M, Fransen P, De Loof A (1982) Influence of procaine HCl on larval development, adult lifespan and acid phosphatase activity in Musca domestica L. Arch Int Physiol Biochim 90:309–315

Smith PK, Krohn RI, Hermanson GT, Mallia AK, Gartner FH, Provenzano MD, Fujimoto EK, Goeke NM, Olson BJ, Klenk DC (1985) Measurement of protein using bicinchoninic acid. Anal Biochem 150:76–85

Southern EM (1975) Detection of specific sequences among DNA fragments separated by gel electrophoresis. J Mol Biol 98:503–517

Torian LV, Forgione LA, Punsalang AE, Pirillo RE, Oleszko WR (2011) Comparison of Multispot EIA with Western blot for confirmatory serodiagnosis of HIV. J Clin Virol 52(Suppl 1):S41–4

Towbin H, Staehelin T, Gordon J (1979) Electrophoretic transfer of proteins from polyacrylamide gels to nitrocellulose sheets: procedure and some applications. Proc Natl Acad Sci U S A 76:4350–4354

Wiechelman KJ, Braun RD, Fitzpatrick JD (1988) Investigation of the bicinchoninic acid protein assay: identification of the groups responsible for color formation. Anal Biochem 175:231–237

Zoller L, Cremer J, Faulde M (1993) Western blot as a tool in the diagnosis of Lyme borreliosis. Electrophoresis 14:937–944

15 Polymerase Chain Reaction (PCR) and Real-Time PCR

Georgette D. Kanmogne

Abstract

The PCR technique, developed over 3 decades ago, has revolutionized molecular biology. PCR applications encompass several fields, including genetics, molecular cloning, diagnosis of infectious and hereditary diseases, and forensic sciences. Improvements of the PCR technique have resulted in its expansion to include variants such as reverse-transcription PCR, real-time PCR, and real-time reverse-transcription PCR. This chapter provides an overview of current PCR and PCR variants methods.

Keywords

PCR • Quantitative PCR • mRNA • Complementary DNA • Reverse transcription

15.1 Introduction

The polymerase chain reaction (PCR) is an enzymatic amplification of DNA sequences mediated by primers and deoxynucleoside triphosphates (dNTPs) and catalyzed by DNA polymerase. The reaction can generate millions of copies of the original DNA template. Since its development almost 3 decades ago (Saiki et al. 1985, 1986; Mullis et al. 1986), the PCR technique has been improved and widely expanded to include variants such as reverse-transcription PCR, real-time PCR, and real-time reverse-transcription PCR. These PCR-based procedures are now commonly used molecular biology techniques in many laboratories (Bartlett and Stirling 2003; Hue-Roye and Vege 2008); and their applications encompass several fields, including genetics, molecular cloning, diagnosis of infectious and hereditary diseases, and forensic sciences (Gause and Adamovicz 1994; Ishmael and Stellato 2008).

15.2 Polymerase Chain Reaction

15.2.1 Reagents

DNA template: 1–100 ng (up to 500 ng for genomic DNA)

G.D. Kanmogne, Ph.D., M.P.H. (✉)
Department of Pharmacology and Experimental Neuroscience, University of Nebraska Medical Center, 985800 Nebraska Medical Center, Omaha, NE, USA
e-mail: gkanmogne@unmc.edu

H. Xiong and H.E. Gendelman (eds.), *Current Laboratory Methods in Neuroscience Research*, Springer Protocols Handbooks, DOI 10.1007/978-1-4614-8794-4_15,

Primers: Oligonucleotides that define the sequence to be amplified, optimal length 18–22 base pairs (bp) (generally 15–30 bp with about 50 % GC content). Final concentration, 0.1–0.5 μM for each primer

dNTPs: Working concentration, 200 μM each for dATP, dGTP, dCTP, and dTTP

10× PCR buffer: Contains 100 mM Tris–HCl, pH 8.3, and 500 mM KCl

Magnesium chloride: Stock concentration: 25 mM; final working concentration, 1.5–4 mM

Enzyme: Taq DNA polymerase

Final reagent concentration

PCR mix (for 25 μL reaction volume)

Reagents	Volume per sample	Final concentration
Ultrapure autoclaved water	#	1×
10× PCR buffer	2.5 μL	1× (10 mM Tris–HCl, 50 mM KCl)
25 mM $MgCl_2$	1–4 μL	1–4 mM
dCTP (10 mM)	0.5 μL	200 μM
dGTP (10 mM)	0.5 μL	200 μM
dATP (10 mM)	0.5 μL	200 μM
dTTP (10 mM)	0.5 μL	200 μM
Forward primer	#	0.2–1 μM
Reverse primer	#	0.2–1 μM
DNA template	#	100–500 ng
AmpliTaq DNA polymerase (5 U/μL)	0.125 μL	0.625 U
Total	25 μL	

#Calculate the volume of primers and DNA template necessary to obtain the desired final concentration, and add ultrapure water to bring the total reaction volume to 25 μL. Use equal amount of DNA for all samples.

Notes

- The optimal $MgCl_2$ concentration can vary and should be determined empirically for each DNA template and primers.
- The optimal primer concentration can vary and should be determined empirically for each primer pair.
- To make reaction conditions more uniform and minimize pipetting losses, make a reaction mix consisting of PCR buffer, $MgCl_2$, dNTPs, primers, and enzyme for all samples, aliquot them into labeled tubes, and add each DNA template into its corresponding tubes.

15.2.2 DNA Amplification Steps

1. *Initialization* (94–98 °C for 1–10 min): This step is required for Hot Start PCR.
2. *Denaturation* (94–98 °C for 20 s to 1 min): Heating disrupts hydrogen bonds that link complementary bases, thus separating double-stranded DNA into separate single strands.
3. *Annealing* (37–65 °C for 30 s to 1 min): During this step, the primers anneal to the single-strand DNA, the polymerase enzyme binds to the primer-DNA template and starts DNA synthesis.
4. *Extension/elongation* (72–76 °C for 30 s to 2 min): During this step, DNA polymerase adds dNTPs in the 5′–3′ direction to synthesize a new complementary DNA strand for each template DNA strand.
5. *Final elongation* (70–76 °C for 5–15 min): This step is performed at the end of the PCR cycle to ensure complete elongation of any remaining DNA strands.
6. *Hold* (4 °C): This step maintains the samples at an appropriate temperature for short-term storage.

   Example of PCR cycle:

95 °C, 1 min	
95 °C, 30 s 55 °C, 45 s 72 °C, 30 s	×35
72 °C, 5 min	
4 °C, hold	

Note: The optimal annealing temperature is dependent on primer's melting temperature and should be determined empirically for each primer pair. The optimal number of cycle is dependent on the target gene, quality and quantity of DNA samples, and should be determined empirically for each target gene.

15.2.3 Detection of PCR Products

PCR products are generally detected by agarose gel electrophoresis using Tris–borate–EDTA (TBE) buffer or Tris–acetate–EDTA (TAE) buffer.

Make a stock of 5× or 10× TBE buffer or a stock of 50× TAE buffer and store at room temperature. Dilute to 1× working concentration just before use.

Buffer Composition

1× TBE: 90 mM Tris, 90 mM boric acid, and 2 mM EDTA, pH 8.3.

1× TAE buffer: 40 mM Tris, 20 mM acetic acid, and 1 mM EDTA, pH 8.0.

Preparation of the Agarose Gel

- Mix the desired amount of agarose with the buffer (TBE or TAE) in an Erlenmeyer flask. Lower agarose concentrations (0.5–1 %) provide better resolution for higher MW DNA (above 500 bp), while higher agarose concentrations (1–2 %) give better resolution for lower MW DNA (below 500 bp).
- Melt by heating the agarose solution in a microwave for 1–3 min, using oven gloves; mix the solution by gently swirling to ensure a complete and homogeneous melting of the agarose.
- Let the agarose solution cool down to about 50 °C, add ethidium bromide (EtBr) (0.2–0.5 μg/mL), and mix by gentle swirling. EtBr is a dye that intercalates between base pairs of double-stranded DNA and fluoresces when exposed to UV light.
- Pour the gel in the casting tray containing a comb (the casting tray must be prepared in advance by blocking the ends with tape or with the gel caster walls, to avoid leaking). Two combs can be used, depending on the number of samples. Let the gel set (solidify) at room temperature; this can take about 1 h.

Electrophoresis

- Assemble the electrophoresis apparatus as directed by the instrument manufacturer.
- Remove the comb and the gel casting gates (or tapes if tapes were used for gel casting).
- Put the gel tray in a leveled electrophoresis tank, making sure the part of the gel with comb is on the cathode (black electrode) side of the tank, because DNA is negatively charged and will migrate toward the anode (red electrode).
- Fill the electrophoresis chamber with electrophoresis buffer (1× TAE or 1× TBE) and make sure the gel is submerged.
- Mix each DNA sample with loading buffer containing a tracking dye, bromophenol blue or xylene cyanol dyes (e.g., Invitrogen TrackIt Cyan/Orange loading buffer (Cat. No: 10482-028) or BlueJuice Gel Loading Buffer (Cat. No: 10816-015)), according to the manufacturer's instructions. The final sample volume will depend of the size and thickness of the comb.
- Pipette DNA samples and load them in the wells carefully to avoid diffusion and dispersion of samples in the electrophoresis buffer.
- Load a molecular weight marker in at least one well.
- Put the lid on the electrophoresis box; plug the wires to the power supply (red wire on the red plug and black wire on the black plug).
- Turn the power supply on to apply current (about 70 V, 40 mA). Run the gel for 30 min to 1 h, or until the tracking front dye is about half way or two-third down the gel.
- Turn off the power supply, unplug the wires, and remove the lid from the electrophoresis box. Remove the gel tray; visualize the DNA on the gel using a UV transilluminator, and take photograph of the gel (For example see Fig. 15.1). Use lab safety procedures and take necessary precautions to avoid skin exposure to chemicals and UV light.
- Clean up all equipment used and discard the gel in appropriate biohazard containers.

Notes

1. For each PCR experiment, include a no-DNA negative control (a sample that has all reagents in the same final concentrations used in all other samples, but no DNA template), to control for cross-contamination or reagent

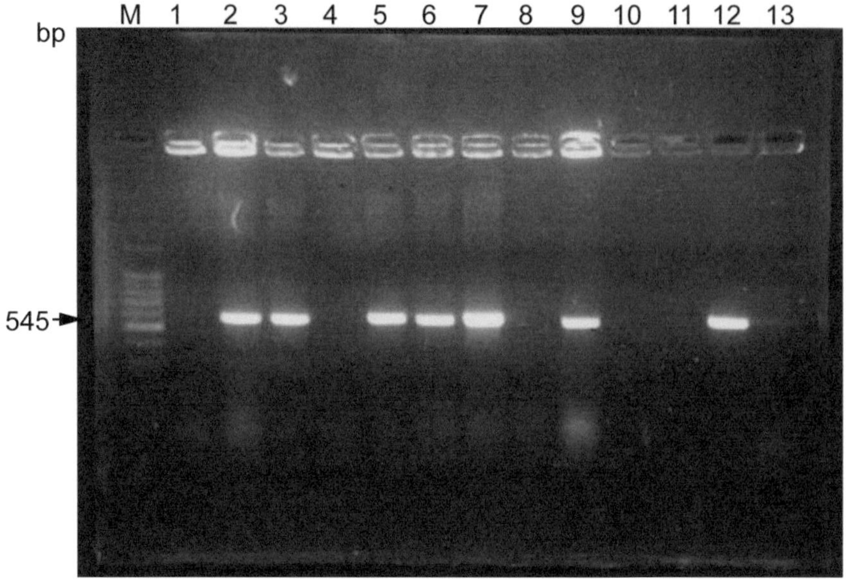

Fig. 15.1 Gel electrophoresis of amplified plasmid DNA samples from HIV-infected patients. PCR targeting an HIV-1 envelope sequence of 545 bp show amplification for samples in lanes 2, 3, 5, 6, 7, 9, and 12. No amplicon was detected for samples in lanes 1, 4, 8, 10, 11, and 13

contamination. Also include a positive control if available.

2. The volume of agarose depends on the size of the casting tray.
3. Agarose concentration will depend on the experiments and molecular weight (MW) of the target DNA. Lower agarose concentrations (0.5–1 %) provide better resolution for higher MW DNA (above 500 bp), while higher agarose concentrations (1–2 %) give better resolution for lower MW DNA (below 500 bp).
4. EtBr staining can also be done after electrophoresis, by incubating the gel for about 30 min in an EtBr solution (0.5 μg/mL), followed by about 30 min destaining in distilled water to remove excess EtBr and reduce background fluorescence.
5. Gel loading buffer can be prepared in the lab. Example: 6× blue/orange loading dye is composed of 0.4 % orange G; 0.03 % bromophenol blue; 0.03 % xylene cyanol; 15 % Ficoll 400; 10 mM Tris–HCl (pH 7.5); and 50 mM EDTA, pH 8.0.

15.3 Variants of PCR

15.3.1 Hot Start PCR

Hot Start PCR is the starting of the PCR enzymatic reaction after the initial denaturation phase. The Hot Start reduces primer-dimer formation, primer misalignment, and nonspecific amplification, which results in increased PCR specificity and sensitivity and increased PCR amplification yields (D'Aquila et al. 1991; Chou et al. 1992). Hot Start PCR was previously performed by adding DNA polymerase after the initial denaturation step, which involved additional sample manipulation (Chou et al. 1992). Currently, enzymes specific for Hot Start PCR are available from several commercial sources (e.g., JumpStart™ Taq or REDTaq DNA Polymerase (Sigma, St. Louis, MO); GoTaq® Hot Start Polymerase (Promega, Madison, WI); AmpliTaq Gold® or Platinum® Taq DNA Polymerase (Life Technologies, Foster City, CA). These Hot Start enzymes are often bound to neutralizing antibodies which block Taq polymerase activity at room temperature, and the polymerase activity is restored during the first PCR denaturation step.

15.3.2 Nested PCR

Nested PCR is a two-step PCR procedure which uses two pairs of primers. The first pair of primers (outer primers) is used in the first PCR step to amplify the target sequence. An aliquot of the first PCR product (10–20 % of the reaction volume) is then used as DNA template in a second PCR step with inner (nested) primers that bind inside the first amplicon to amplify a second DNA target sequence located inside the first PCR product, which is shorter than the first sequence. This results in increased PCR specificity because if no target sequence is present initially and none is amplified during the first PCR, it is unlikely that any amplification will occur during the second PCR. Nested PCR is also useful for amplifying long DNA sequences with increased specificity and sensitivity.

15.3.3 Reverse-Transcription PCR

This technique is used for the detection and semiquantification of RNA expression (Clementi et al. 1995; Salomon 1995; Cale et al. 1998; Halford et al. 1999). Using the reverse transcriptase enzyme and oligo(dT) primers or random hexamers, RNA in target samples is reverse transcribed into complementary DNA (cDNA). An aliquot of the cDNA is then used as template, with primers targeting the gene of interest, in a PCR, to amplify the sequence corresponding to the gene of interest (Clementi et al. 1995; Salomon 1995; Cale et al. 1998; Halford et al. 1999). Amplified cDNA can be detected by agarose gel electrophoresis or phosphorimaging.

15.3.3.1 Quantification of RT-PCR Products

Reverse-transcription PCR (RT-PCR) products can also be quantified by simultaneous amplification of the target RNA with an internal control (e.g., a transcript with known uniform expression across samples). Amplicons in each sample are then normalized to the sample internal control, which allows for comparisons between experimental samples (Gause and Adamovicz 1994; Clementi et al. 1995). Target RNA can also be simultaneously amplified with known amounts of an unrelated transcript. After PCR, the amplification of the known amounts of the unrelated transcript is used to generate a standard curve which is used to estimate transcript levels in other samples (Tsai and Wiltbank 1996; Halford et al. 1999).

Note: Reverse transcriptase and reverse transcription kits are available from several commercial sources, and each reagent often comes with the manufacturer's recommended reverse transcription protocol.

15.4 Real-Time PCR or Quantitative PCR

Real-time PCR is a sensitive technique used for absolute and real-time quantification of nucleic acids (Arya et al. 2005; Kubista et al. 2006). It uses the same principle as conventional PCR (denaturation, annealing, and extension) but also uses a fluorescent DNA-binding dye (e.g., SYBR Green) or probes (e.g., TaqMan probes) that emit a fluorescent signal during amplification, thus allowing detection and quantification of PCR products in real time (Lie and Petropoulos 1998; Johansson 2006). Real-time PCR can enable quantification of small amounts of mRNA, DNA, or cDNA in a biological sample (Lie and Petropoulos 1998; Ginzinger 2002; Johansson 2006).

15.4.1 Types of Real-Time PCR

There are currently three major methods of gene detection and quantification by real-time PCR (VanGuilder et al. 2008; Whitman and Dunbar 2008):

1. *Real-time PCR using DNA-binding dyes* (e.g., SYBR Green, SYTO 9, EVA Green, LC Green 1): The PCR is performed in the presence of a

fluorescent dye which intercalates within a double-strand DNA and emits fluorescence. During PCR, an increase in the number of amplicons (double-strand DNA) is associated with an increase in the amount of fluorescence, which can be measured in real time. This method is cheaper than other quantitative PCR (qPCR) methods. The fluorescent dyes used are not sequence specific, so they intercalate with any double-strand DNA, including mismatched strands, primer dimers, and fragments from nonspecific amplification. Thus, when using DNA-binding dyes method, it is important to ensure that there is no contaminant in DNA or RNA samples, primers are specific, and PCR conditions (e.g., melting temperature) are optimal to ensure no primer-dimer formation and no sequence mispriming.

2. *Real-time PCR using fluorescent primers* (e.g., Amplifluor® system or LUX™ Fluorogenic Primers): For these experiments, a fluorophore is attached directly to a target-specific primer. Thus, each time the primer is used to synthesize a new DNA strand (during annealing and elongation), there is increase in fluorescence. The fluorophore-labeled primer usually has a hairpin structure with a fluorophore attached near the 3′ end and a quencher near the 5′ end (Amplifluor® system) (Lie and Petropoulos 1998). When the primer is not bound to DNA, the hairpin structure suppresses the fluorescence. When the primer is bound to the target DNA (following annealing and elongation) and becomes linear during amplification, fluorescence is emitted (Johansson 2006; Quellhorst and Rulli 2008).
3. *Real-time PCR using fluorescent probes* (e.g., TaqMan® probes): This procedure is very specific and very sensitive. It uses two target-specific primers (forward and reverse primers) and a probe (Lie and Petropoulos 1998; McChlery and Clarke 2003; Quellhorst and Rulli 2008). A fluorophore (e.g., FAM) is covalently attached to the 5′ end of the probe, while a second fluorescent dye (quencher; e.g., TAMRA) is covalently attached to the probe 3′ end. For intact probes, both fluorophore and quencher are in close proximity and the florescence activity of the fluorophore is inhibited by the quencher (Quellhorst and Rulli 2008). During amplification, the 5′–3′ exonuclease activity of the Taq DNA polymerase cleaves the probe which releases the fluorophore, which emits a fluorescence signal that can be quantified in real time (Lie and Petropoulos 1998; McChlery and Clarke 2003; Johansson 2006). The probe is generally complementary to a sequence within the target DNA to be amplified, which increases reaction specificity.

15.4.2 RNA Quantification by Real-Time PCR

Quantitative reverse-transcription PCR (qRT-PCR) is a powerful method for absolute quantification of RNA or mRNA (Tsai and Wiltbank 1996; Bustin 2000; Bustin et al. 2005). Total RNA or mRNA is reverse transcribed into cDNA using the reverse transcriptase enzyme, random hexamers, or oligo(dT) primers; and the resulting cDNA is used for real-time PCR (Lutfalla and Uze 2006). This allows RNA or mRNA quantification in real time (Bustin 2000; Bustin et al. 2005; Schefe et al. 2006). qRT-PCR is generally a two-step procedure, but one-step qRT-PCR can also be performed.

1. *One-step qRT-PCR*: For these experiments, both reverse transcription and qPCR are performed in one step in a single tube containing RNA, reverse transcriptase, and other reverse transcription reagents (a target-specific primer, buffer, dNTPs), as well as Taq DNA polymerase and other qPCR reagents (primers, probes, buffers, dNTPs, and $MgCl_2$). The reaction-cycling conditions are set up so that reverse transcription is performed first, followed by qPCR (Invitrogen 2008; Quellhorst and Rulli 2008). In principle, Taq DNA polymerase is inactive during the reverse transcription step, and as soon as the cDNA amplification step starts, the high temperature (first denaturation step) deactivates reverse transcriptase while activating Taq DNA polymerase.

   One-step qRT-PCR offers the convenience of lower number of experimental steps; however, there are potential problems. For example, the fact that primers and probes are present during reverse transcription can increase the risk of mispriming and formation of primer dimers (Chou et al. 1992). Also, most reverse transcription reactions are optimized individually, and most DNA/cDNA amplification reactions are also optimized individually based on primers, annealing temperatures, and other PCR conditions (Cale et al. 1998). In one-step PCR, with both reverse transcriptase buffer and Taq DNA polymerase buffers mixed, it is difficult to get optimal reaction for both reverse transcription and qPCR. Lastly, in one-step qRT-PCR, all cDNAs generated are used for one qPCR experiment, this can be problematic when dealing with limited samples or limited biological materials.

2. *Two-step qRT-PCR*: Here, RNA is first used in a reverse transcription experiment, using reverse transcriptase, random hexamers or oligo(dT) primers, and dTNPs, as described above (RT-PCR). Ten to 20 % of the cDNA generated is then used as template for qPCR (Invitrogen 2008; Quellhorst and Rulli 2008). Although two-step qRT-PCR involves more time and manipulation steps, it has several advantages. The cDNA generated can be stored and used in multiple qPCR experiments to analyze several genes; this is an important attribute when dealing with limited biological samples. Two-step qRT-PCR can be more sensitive because reverse transcription and qPCR are performed separately, and each of those experiments can be performed in more optimal conditions.

15.4.3 Controls for qRT-PCR and qPCR

To determine amplification specificity and efficiency and ensure that there is no contamination, controls should be included in each PCR. These controls include:

- *Negative control 1*: To check for genomic DNA contamination. During the reverse transcription stage, include a tube containing the same reagents in the same final concentrations as in other tubes, but not with reverse transcriptase. Without this enzyme, there should be no cDNA generated. Thus, if there is any DNA amplification, it will be contaminating genomic DNA. DNase I is often added to samples during reverse transcription to diminish the risk of genomic DNA contamination in RNA samples.
- *Negative control 2*: To check for contamination or cross-contamination. For qPCR, include a sample containing the same reagents in the same final concentrations as in other tubes, but no RNA or cDNA.
- *Positive control*: When the experimental goal is to demonstrate the presence or absence of a particular gene, a sample known to express that gene of interest can be used as positive control.

These controls can also be used in conventional PCR and RT-PCR.

15.4.4 qRT-PCR and qPCR Normalization and Quality Control

The following can be done to ensure standardization and optimal reaction amplification:

- Make sure nucleic acids are not degraded and check for RNA (or DNA) purity and quality: Good quality RNA has an absorbance A_{260}/A_{280} ratio between 1.8 and 2.0 (Wilfinger et al. 1997), while a good quality DNA has an A_{260}/A_{280} ratio between 1.7 and 2.0 (Promega 2008).
- For RT-PCR and qRT-PCR, quantify RNA from all samples and use similar amounts of RNA for all samples.
- For qPCR, measure DNA concentration in all samples and use similar amounts of DNA for all samples.
- Make sure there is no sequence complementarity between the two primers or between primers and probes. This avoids annealing between primers and probe and the formation of primer dimers.
- For optimal amplification, primers' length should be 18–28 bp, with about 50 % GC content, and should not have long sequences of repeated nucleotides (Invitrogen 2008).
- Avoid pipetting errors, which can introduce discordance in sample amounts.

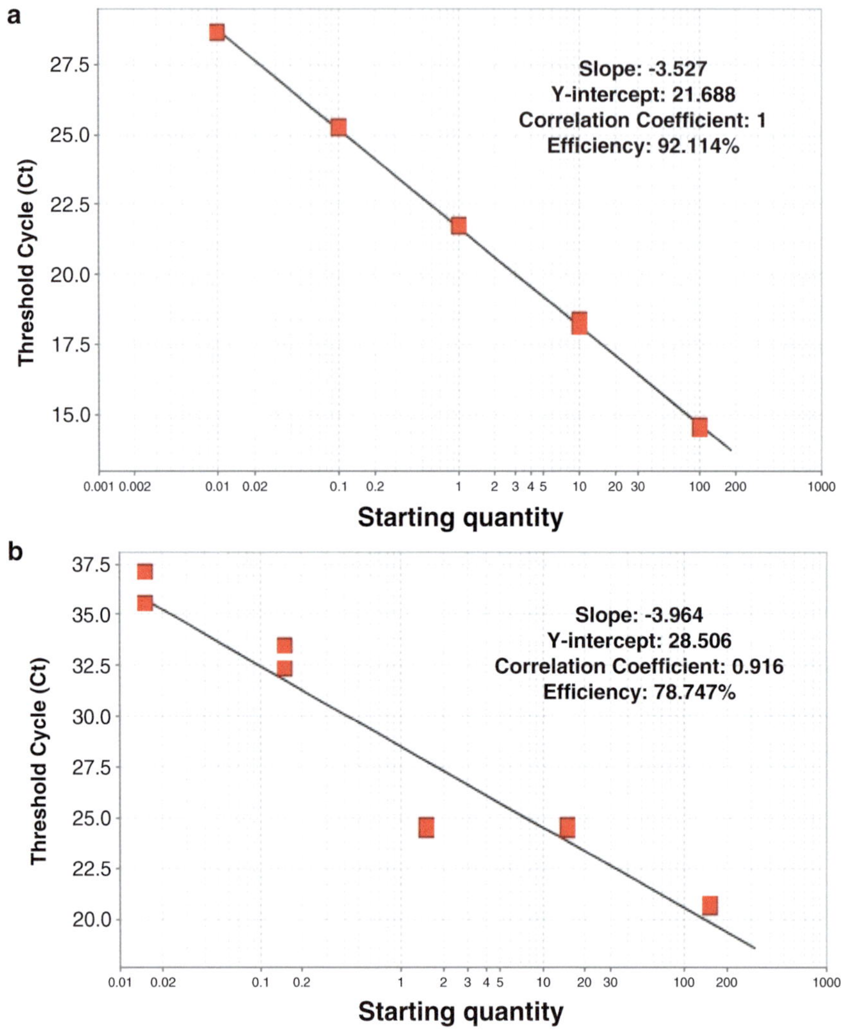

Fig. 15.2 Standard curves generated using serial dilutions of human brain microvascular endothelial cells RNA (RNA concentrations: 0.01, 0.1, 1, 10, and 100 ng/mL). Following reverse transcription, cDNA samples were used in real-time PCR with primers for human GAPDH. Experiment *A* standard curve shows a more efficient amplification reaction and better fit of the data, compared to experiment *B* standard curve. Replicate samples on this second standard curve *B* also show larger standard deviations, possibly due to pipetting errors

- Before using the real-time PCR instrument, make sure it is calibrated as instructed by the manufacturer.
- Run each sample in duplicate or triplicate to assess reproducibility.
- Choose a reference gene or housekeeping gene that is abundant and consistently present in all samples (e.g., GAPDH or beta-actin) and not affected by sample treatment.
- Do qRT-PCR or qPCR of the target gene simultaneously with the reference gene, and normalize all results to that of the reference gene.
- For each qPCR experiment, make a serial dilution of a reference sample to generate a standard curve. This will help determine amplification efficiency and sensitivity, as well as potential pipetting errors (Fig. 15.2).

15.5 Comparison of PCR and Real-Time PCR

- PCR and reverse-transcription PCR are mostly qualitative; the final amount of product obtained following amplification is not related to the initial amount of DNA or mRNA/cDNA used in the reaction.
- With PCR, amplification products are detected at the end-point stage, and accurate quantification is not possible at this stage. Thus, PCR is not quantitative (at best, it can be semiquantitative). On the contrary, real-time PCR enables kinetic detection (quantitative) of amplification products; and exponential phase plots show quantity of amplicons over time (Hue-Roye and Vege 2008).
- In real-time PCR, there is simultaneous gene amplification and detection, which allows reaction monitoring and quantification in real time, during each cycle.
- Following PCR, additional steps (agarose gel electrophoresis and image analysis) are required to detect amplification products; while for real-time PCR, no post-PCR manipulations are required after amplification.
- PCR has lower sensitivity and a short dynamic range (>2 logs) while real-time PCR is more sensitive, with a very wide dynamic linear ranges (from 10 copies to 10^9 copies) (Quellhorst and Rulli 2008).

Acknowledgment Dr. Georgette Kanmogne's research is supported by grants from the National Institute of Mental Health grants 1RO1 MH081780 and 1RO1 MH094160.

References

Arya M, Shergill IS, Williamson M, Gommersall L, Arya N, Patel HR (2005) Basic principles of real-time quantitative PCR. Expert Rev Mol Diagn 5:209–219

Bartlett JM, Stirling D (2003) A short history of the polymerase chain reaction. Methods Mol Biol 226:3–6

Bustin SA (2000) Absolute quantification of mRNA using real-time reverse transcription polymerase chain reaction assays. J Mol Endocrinol 25:169–193

Bustin SA, Benes V, Nolan T, Pfaffl MW (2005) Quantitative real-time RT-PCR—a perspective. J Mol Endocrinol 34:597–601

Cale JM, Shaw CE, Bird IM (1998) Optimization of a reverse transcription-polymerase chain reaction (RT-PCR) mass assay for low-abundance mRNA. Methods Mol Biol 105:351–371

Chou Q, Russell M, Birch DE, Raymond J, Bloch W (1992) Prevention of pre-PCR mis-priming and primer dimerization improves low-copy-number amplifications. Nucleic Acids Res 20:1717–1723

Clementi M, Menzo S, Manzin A, Bagnarelli P (1995) Quantitative molecular methods in virology. Arch Virol 140:1523–1539

D'Aquila RT, Bechtel LJ, Videler JA, Eron JJ, Gorczyca P, Kaplan JC (1991) Maximizing sensitivity and specificity of PCR by pre-amplification heating. Nucleic Acids Res 19:3749

Gause WC, Adamovicz J (1994) The use of the PCR to quantitate gene expression. PCR Methods Appl 3:S123–S135

Ginzinger DG (2002) Gene quantification using real-time quantitative PCR: an emerging technology hits the mainstream. Exp Hematol 30:503–512

Halford WP, Falco VC, Gebhardt BM, Carr DJ (1999) The inherent quantitative capacity of the reverse transcription-polymerase chain reaction. Anal Biochem 266:181–191

Hue-Roye K, Vege S (2008) Principles of PCR-based assays. Immunohematology 24:170–175

Invitrogen (2008) Real-time PCR: from theory to practice. www.invitrogencom

Ishmael FT, Stellato C (2008) Principles and applications of polymerase chain reaction: basic science for the practicing physician. Ann Allergy Asthma Immunol 101:437–443

Johansson MK (2006) Choosing reporter-quencher pairs for efficient quenching through formation of intramolecular dimers. Methods Mol Biol 335:17–29

Kubista M, Andrade JM, Bengtsson M, Forootan A, Jonak J, Lind K, Sindelka R, Sjoback R, Sjogreen B, Strombom L, Stahlberg A, Zoric N (2006) The real-time polymerase chain reaction. Mol Aspects Med 27:95–125

Lie YS, Petropoulos CJ (1998) Advances in quantitative PCR technology: 5′ nuclease assays. Curr Opin Biotechnol 9:43–48

Lutfalla G, Uze G (2006) Performing quantitative reverse-transcribed polymerase chain reaction experiments. Methods Enzymol 410:386–400

McChlery SM, Clarke SC (2003) The use of hydrolysis and hairpin probes in real-time PCR. Mol Biotechnol 25:267–274

Mullis K, Faloona F, Scharf S, Saiki R, Horn G, Erlich H (1986) Specific enzymatic amplification of DNA in vitro: the polymerase chain reaction. Cold Spring Harb Symp Quant Biol 51(Pt 1):263–273

Promega (2008) Genomic DNA Purification Instructor's Manual. www.promegacom

Quellhorst G, Rulli, S. (2008) A systematic guideline for developing the best real-time PCR primers. What we have learned from designing assays for more than 14,000 genes. SABiosciences www.SABiosciencescom

Saiki RK, Bugawan TL, Horn GT, Mullis KB, Erlich HA (1986) Analysis of enzymatically amplified beta-globin and HLA-DQ alpha DNA with allele-specific oligonucleotide probes. Nature 324: 163–166

Saiki RK, Scharf S, Faloona F, Mullis KB, Horn GT, Erlich HA, Arnheim N (1985) Enzymatic amplification of beta-globin genomic sequences and restriction site analysis for diagnosis of sickle cell anemia. Science 230:1350–1354

Salomon RN (1995) Introduction to reverse transcription polymerase chain reaction. Diagn Mol Pathol 4:2–3

Schefe JH, Lehmann KE, Buschmann IR, Unger T, Funke-Kaiser H (2006) Quantitative real-time RT-PCR data analysis: current concepts and the novel "gene expression's CT difference" formula. J Mol Med (Berl) 84:901–910

Tsai SJ, Wiltbank MC (1996) Quantification of mRNA using competitive RT-PCR with standard-curve methodology. Biotechniques 21:862–866

VanGuilder HD, Vrana KE, Freeman WM (2008) Twenty-five years of quantitative PCR for gene expression analysis. Biotechniques 44:619–626

Whitman DF, Dunbar SA (2008) Real-time polymerase chain reaction detection methods. Recent Pat DNA Gene Seq 2:20–26

Wilfinger WW, Mackey K, Chomczynski P (1997) Effect of pH and ionic strength on the spectrophotometric assessment of nucleic acid purity. Biotechniques 22(474–476):478–481

Part V

Nanomedicine

Nanoformulations

16

Xin-Ming Liu and Tatiana K. Bronich

Abstract

Existing diagnostic and therapeutic strategies for central nervous system (CNS) disorders are restricted to access the CNS. Meanwhile, the blood–brain barrier (BBB) hinders the delivery of many potentially important diagnostic and therapeutic agents to the brain. To resolve these problems, various nanoformulation technologies were developed to improve the delivery of diagnostic and therapeutic agents into CNS. In this chapter, we discussed the manufacture, characteristics, and potential problems of widely used nanoformulations including liposomes, polymeric micelles, and nanosuspensions. We expect that this chapter would provide novel nanoformulation tools for the basic and clinical research of neuroscientists.

Keywords

Nanoformulations • Liposomes • Micelles • Nanosuspensions

16.1 Introduction

The advancement of neuroscience research can have a major impact on understanding the pathological mechanisms of peripheral and central nervous system (CNS) disorders and the development of new therapeutic modalities, especially for disorders of the CNS (Orlacchio et al. 2010). CNS disorders include, but are not limited to, viral and bacterial neuroinfections and associated encephalitis, neuroinflammatory diseases such as multiple sclerosis, neurodegenerative diseases that include Alzheimer's and Parkinson's diseases (AD and PD), and brain tumors. The causes of CNS disorders are complex and may be associated with various factors such as aging, various peripheral diseases or infectious agents, genetic alterations, problems of the immune system, brain or spinal cord injuries, and exposure to environmental

X.-M. Liu, Ph.D. (✉) • T.K. Bronich, Ph.D. (✉)
Department of Pharmacology and Experimental Neuroscience, Center for Drug Delivery and Nanomedicine, University of Nebraska Medical Center, Omaha, NE 68198, USA

Departments of Pharmaceutical Sciences, Center for Drug Delivery and Nanomedicine, University of Nebraska Medical Center, Omaha, NE 68198, USA
e-mail: xliu@unmc.edu; tbronich@unmc.edu

H. Xiong and H.E. Gendelman (eds.), *Current Laboratory Methods in Neuroscience Research*, Springer Protocols Handbooks, DOI 10.1007/978-1-4614-8794-4_16,

toxins (Mayeux 2003). Neurological diseases are among the most destructive and costly public health problems. Unfortunately, the existing therapeutic options provide only limited therapeutic efficacy and face considerable challenges. This is in part due to the extremely heterogeneous cellular and molecular environment of the brain and the restricted anatomical access of the CNS. Furthermore, the blood–brain barrier (BBB) can hinder the delivery of many potentially important diagnostic and therapeutic agents to the brain (Loch-Neckel and Koepp 2010). To date, delivery of therapeutic molecules into the brain often involves highly invasive techniques. New technologies are desperately needed to resolve these challenges and improve the diagnosis and treatment of CNS disorders. Drug nanoformulations, the pharmaceutical application of nanotechnology, provide revolutionary tools for biomedical research, most importantly for cancer research. These technologies can play a significant role in improving delivery of therapeutic or diagnostic agents into the CNS and have the potential to offer novel therapeutic modalities that do not exist today (Haney et al. 2011).

In order to provide nanoformulation tools for basic and clinical neuroscientists, we have selected widely studied nanoformulations for discussion in this chapter. These nanoformulations include liposomes, polymeric micelles, and nanosuspensions. We will also discuss the use of a bioconjugation approach for modifying the surface of such constructs to target them to specific cell types. The background and potential applications of each type of nanoformulation will be discussed. Detailed preparation procedures, typical results, and potential problems will also be described.

16.2 Liposomes

16.2.1 Objective

The poor BBB penetration of many anticancer drugs requires the use of high doses of conventional systemic chemotherapeutics for brain tumor treatment (Groothuis 2000). The resulting side effects are therefore the limiting factors impeding the clinical application of these formulations. A delivery system that is selective for a tumor target or BBB penetration might maximize the therapeutic efficacy of anticancer drugs and minimize systemic exposure and resulting side effects (Krauze et al. 2007). Liposomes, one of the first carriers developed for drug delivery, represent nanoscale-sized spheres consisting of one or more lipid bilayers surrounding a central aqueous core. Drug molecules can be incorporated into the aqueous compartment (hydrophilic substances) or lipid bilayers (lipophilic substances). Their circulation longevity can be improved by grafting polyethylene glycol (PEG) onto the liposome surface. To target PEGylated liposomes to the disease site, they can be further vectorized with specific ligands (Torchilin 2005). As an example, we provide here the protocols for the preparation of long-circulating liposomes modified with folic acid as the targeting ligand and carrying doxorubicin as a model drug.

16.2.2 Coupling of Folic Acid to DSPE-PEG$_{2000}$-Amine Lipid

1. Add 0.5 g of folic acid into 10 mL of anhydrous dimethyl sulfoxide (DMSO), and stir the mixture for 4–6 h or until the folic acid is completely dissolved.
2. Add 0.25 mL of anhydrous triethylamine and 0.26 g of *N*-hydroxysuccinimide (NHS) to the reaction solution. Then slowly add 0.47 g of *N,N′*-dicyclohexylcarbodiimide (DCC).
3. Stir the reaction mixture overnight at room temperature in the dark.
4. Filter the reaction mixture using a Buchner funnel to remove the insoluble by-product, dicyclohexylurea (DCU).
5. Add the filtrate dropwise into 100 mL of diethyl ether to precipitate folate-NHS ester.
6. Filter the product using a Buchner funnel, and redissolve the powder into anhydrous DMSO for the next reaction or storage at 4 °C.
7. Add 50 mg 1,2-distearoyl-sn-glycero-3-phosphoethanolamine-*N*-[amino (polyethylene glycol)-2000] (DSPE-PEG$_{2000}$-amine) into a solution of 20 mg folate-NHS ester in 1 mL of anhydrous DMSO.

8. Allow the reaction to proceed at room temperature for about 4 h. Monitor the reaction by placing 2 μl of the reaction mixture on a silica gel thin layer chromatography (TLC) plate, immersing the plate in a 5 % solution of ninhydrin in ethanol, and heating the plate to over 100 °C. The disappearance of a ninhydrin positive DSPE-PEG_{2000}-amine spot on the TLC plate will confirm successful conjugation.
9. To remove unconjugated folic acid, purify the final product on a Sephadex LH-20 column using 100 % methanol as the eluent.
10. Confirm the structure of the final product by ^{1}H NMR.

16.2.3 Preparation and Characterization of Doxorubicin-Loaded Liposomes by Extrusion and Remote Loading Method (Gabizon et al. 2003)

1. Dissolve 50 mg of total lipid mixture of dipalmitoylphosphatidylcholine (DPPC)/cholesterol/DSPE-PEG_{2000}/DSPE-PEG_{2000}-folate at 80:20:4.5:0.5 M ratios (optimized formulation) in 5 mL chloroform/methanol (2:1 v/v) in a round bottom flask.
2. Evaporate the organic solvents under vacuum in a rotary evaporator to form a thin lipid film on the wall of flask; remove traces of the solvents by placing the flask in a vacuum desiccator for 4 h.
3. Add 2 mL of 250 mM ammonium sulfate into the flask containing the thin lipid film, and rotate at 120 rpm in a rotary evaporator at 60 °C for 15–20 min for hydration.
4. Extrude the solution through 400, 200, and 100 nm polycarbonate membranes (11 times for each of the membranes) at 55–60 °C using an extruder (Avanti Polar Lipids, USA), then keep the sample at room temperature for at least 2 h in the dark under a nitrogen atmosphere.
5. Equilibrate a Sephadex G50 column with phosphate buffered saline (PBS), load the liposomal dispersion on the column, and elute the column with PBS at a flow rate of 1 mL/min to replace the ammonium sulfate outside the liposomes with PBS and produce an ammonium sulfate concentration gradient between the inner aqueous space of liposomes and surrounding media to facilitate the loading of doxorubicin inside the liposomes.
6. Transfer the liposome fraction into a round bottom flask, preheat to 60 °C, and then add 5 mg doxorubicin hydrochloride into the preheated suspension. The round bottom flask is rotated using a rotary evaporator at 100 rpm for 1 h at 60 °C for the remote loading of doxorubicin. Reserve an aliquot (50 μl) of the lipid–drug mixture for determination of the doxorubicin loading efficiency.
7. Remove unencapsulated doxorubicin using a Sephadex G50 column. For this purpose, Sephadex G50 bulk material is suspended in water and filled in a GE Healthcare, HR 10/30 column. After equilibrating the column with PBS, the sample is separated on the column with PBS as eluting buffer at a flow rate of 1 mL/min.
8. Quantify doxorubicin in the initial doxorubicin–lipid mixture and in the final formulation by absorbance at 495 nm using a UV/Vis spectrophotometer. Prepare a calibration curve of standard solutions of doxorubicin (0–150 nmol) in the presence of Triton X-100 (final concentration of 1 % v/v). Prepare assay samples in the same manner. Lipid content can be quantified using standard phosphate assay (Fan et al. 2012). The percentage of doxorubicin loading is determined as: % loading = $[(DOX/L)_{\text{formulation}} \times 100]/(DOX/L)_{\text{initial}}$, where DOX and *L* are the concentrations of doxorubicin and lipids in the corresponding samples.
9. The size, particle size distribution, and zeta-potential of resulting liposomes can be determined by dynamic light scattering (DLS, e.g., using Nano-Zetasizer, Malvern).

16.2.4 Preparation of Liposomes by Sonication Method

1. Prepare multilamellar lipid vesicles (MLV) as per steps 1–3 of Sect. 16.2.3.
2. Transfer the MLV dispersion into a glass vial, and place it in a 0 °C water bath.

3. Immerse the sonicator probe into the sample, and adjust it to at least 1 cm (0.4 in.) above the bottom of the vial.
4. Flush the vial with nitrogen, and keep it purged with nitrogen during the sonication process.
5. Switch on the Cole Parmer ultrasonic processor (Vernon Hills, IL), and set the sonication power at about 50 % amplitude output.
6. Set the sonication time to about 10 min with 30 s on and 30 s off (depending on the sample volume and concentration), and sonicate until the dispersion turns from milky to opalescent.
7. Keep the sample at room temperature in the dark under nitrogen atmosphere for 4 h.
8. Centrifuge the sample at 12,000 rpm for 10–15 min to remove large liposomes and titanium particles shed from sonicator.
9. Perform steps 5–9 of Sect. 16.2.3 to prepare and characterize doxorubicin-loaded liposomes.

16.2.5 Typical Results

1. Typically, liposomes prepared using the sonication method have an average size of 60–80 nm (Gast et al. 1982); this is a good method to produce small unilamellar vesicles (SUV). The extrusion method is widely used to prepare large unilamellar vesicles (LUV) of diameters from 100 nm to a few hundred nm depending on the pore size of the polycarbonate membranes used.
2. For weakly basic drugs like doxorubicin, high drug loading capacity can be achieved through remote loading, which is a commonly used method based on the generation of a transmembrane pH gradient as the driving force for the accumulation of weak bases inside the liposomes. In this case, more than 95 % doxorubicin is encapsulated into liposomes with high drug-to-lipid ratios.

16.2.6 Notes

1. Ethyl-dimethyl-aminopropylcarbodiimide (EDC) and *N*-hydroxysuccinimide (NHS) coupling is widely used for the conjugation of targeting ligands onto the liposome surface. It should be noted that the conjugation efficiency varies for different ligands and reaction solvents. We recommend performing preliminary experiments to optimize conjugation conditions such as the lipid/ligand ratio, reagent concentrations and ratios, solvent, pH, and temperature.
2. For the preparation of lipid film, it is important to produce as thin a film as possible. Large lumps or thick films are difficult to hydrate.
3. During extrusion of liposomes through membranes, it is critical to extrude very slowly to avoid rupture of the membrane. Also, extrusion should be done above the phase transition temperature of the lipid (liquid crystalline state).
4. It should be pointed out that vigorous sonication may destroy some agents. SUV prepared by sonication are usually not very stable. They tend to fuse together to form larger vesicles due to the high curvature energy of the lipid bilayer. It is recommended to keep the sample at room temperature for a few hours in the dark and then remove the small amount of MLV fused by centrifugation.
5. The ammonium/pH gradient resulting from buffer exchange is only stable during the loading period. To achieve high loading, buffer exchange and remote loading should be performed in a single working process.
6. For the removal of unencapsulated doxorubicin from liposomes, it can be seen that most of the encapsulated doxorubicin (red color) will be eluted with the liposomal fraction first, the retention time of which is normally 6–10 min. Small amount of unencapsulated doxorubicin can be seen in the column as faint red color.
7. For encapsulating nonbasic hydrophobic drugs into liposomes, drugs and lipids should be dissolved together and the drug/lipid film prepared for hydration following extrusion or sonication. For encapsulating nonbasic, hydrophilic drugs into liposomes, drugs should be dissolved into the hydration buffer for extrusion or sonication.

16.2.7 Discussion

Liposomes are widely used as delivery vehicles for biological studies and pharmaceutical applications because they have a unique ability to simultaneously encapsulate and deliver lipid-soluble (in the bilayer) as well as water-soluble (in the core) agents (Maruyama 2011). PEGylation of liposomes to minimize their clearance by the mononuclear phagocytic system (MPS) has been applied to significantly extend the circulation time of liposomes (Immordino et al. 2006). As delivery vehicles, long-circulating liposomes have the abilities to prevent enzymatic degradation and rapid clearance, alter biodistribution, and reduce the toxicities of encapsulated agents. Several liposome formulations are currently on the market for the treatment of different diseases, such as Doxil and Caelyx (PEGylated liposomal doxorubicin) and DaunoXome (non PEGylated liposomal daunorubicin) (Gaitanis and Staal 2010; Abraham et al. 2005). To further enhance drug delivery to the desired site, great efforts have been made over the years to develop targeted liposomes by immobilization of a targeting ligand (peptide, protein, or oligosaccharide) on the surface of liposomes. Studies have also focused on conjugation of highly specific monoclonal antibodies for specific cellular targeting and delivery. Several immunoliposome constructs were successfully used to deliver drugs and DNA to the brain (Xu et al. 2002). Indeed, liposomes have emerged as versatile and efficient delivery systems for various applications, including neuroscience research (Modi et al. 2010).

16.3 Micelles

16.3.1 Objective

Polymeric micelles are nanosized assemblies of amphiphilic block copolymers with a diameter of tens of nanometers and a unique core–shell structure. The inner core of the micelles serves as a nanocontainer for hydrophobic or amphiphilic drug molecules. The shell stabilizes the micelles in dispersion and masks the drug from interactions with serum proteins and untargeted cells. Here, we outline the preparation of polymeric micelles based on Pluronic block copolymers (also known as "Poloxamers"), the triblock copolymers of PEG-*b*-poly(propylene glycol)-*b*-PEG. The non-nucleoside reverse transcriptase inhibitor (NNRTI) efavirenz (EFV) is selected as the model drug for incorporation into the Poloxamer 407 micelles.

16.3.2 Preparation of Drug-Loaded Polymeric Micelles by Solid Dispersion Method

1. Dissolve Poloxamer 407 (10 mg/mL) and EFV (1.5 mg/mL) in a common organic solvent, such as ethanol, methanol, acetonitrile, acetone, dichloromethane, or chloroform; transfer the solution to a round bottom flask, and stir for 0.5 h at 37 °C.
2. Remove the organic solvents under vacuum in a rotary evaporator at 45 °C to form a thin waxlike film on the wall of the flask; remove traces of the solvents by placing the flask in a vacuum desiccator for 4 h.
3. Hydrate the dried film with water while stirring until micelles containing drug are formed.
4. Remove the unincorporated drug by centrifugation at 1,000 × *g* for 10 min or filtration of the micelle dispersion through a 0.2 μm polycarbonate membrane filter.
5. Determine EFV concentration in the micelles using HPLC: YMC Octyl C8 column (Waters Inc., Milford, MA) with a C8 guard cartridge, mobile phase consisting of 48 % acetonitrile/52 % 25 mM KH_2PO_4 (pH 4.15) with a flow rate of 0.4 mL/min and UV/Vis detection at 212 nm.
6. The size, particle size distribution, and zeta-potential of EFV-loaded micelles can be determined by DLS at a concentration of approximately 1 mg micelles/mL.

16.3.3 Preparation of Drug-Loaded Polymeric Micelles by Micelle Extraction Method

1. Disperse Poloxamer 407 (10 mg/mL) into water with stirring until micelles are formed.
2. Dissolve EFV in a common water-miscible organic solvent, such as dimethylformamide, tetrahydrofuran, acetonitrile, acetone, ethanol, or methanol (EFV concentration >20 mg/mL), and then add the EFV solution into the micelle dispersion with agitation (the final concentration of EFV is about 1.5 mg/mL).
3. Place the drug-micellar solution into dialysis tubing with 3,000 Da cutoff, and remove organic solvent by dialysis against 2 L of distilled water, changing water at intervals of 2–4 h for 24 h.
4. As an alternative strategy to steps 2–3, add EFV powder (1.5 mg/mL) directly into the micelle solution, and agitate the mixture for 2–5 days.
5. Follow steps 4–6 in Sect. 16.3.2.

16.3.4 Preparation of Drug-Loaded Polymeric Micelles by Solvent Evaporation Technique

1. Dissolve Poloxamer 407 (10 mg/mL) and EFV (1.5 mg/mL) in a common water-miscible, volatile, organic solvent, such as ethanol, methanol, acetone, or acetonitrile, and stir 0.5 h at 37 °C.
2. Add or inject Poloxamer 407 and EFV mixture dropwise into ten volumes of water while vigorously stirring, giving a final polymer concentration of 3.3 mg/mL. Micelles form immediately.
3. Stir the micelle dispersion for 2 h, and then remove the remaining organic solvent in a rotary evaporator at reduced pressure.
4. Follow steps 4–6 in Sect. 16.3.2.

16.3.5 Typical Results

1. The EFV loading capacity (defined as the weight percent of the drug incorporated into the polymeric micelles relative to the weight of the drug-loaded micelles) in Poloxamer 407 micelles is about 15 %.
2. The EFV-loaded Poloxamer 407 micelles display particle sizes between 25 and 35 nm. The size of polymeric micelles usually varies from 10 to 100 nm.

16.3.6 Notes

1. The loading capacity of polymeric micelles is influenced by several factors including the structure of both the core-forming block and drug, molecular characteristics of the copolymer such as composition and molecular weight, and the solution temperature. The polymer–drug miscibility is one of the most important parameters to govern the drug loading.
2. The micelle loading capacity for hydrophobic drugs can be optimized by blending block copolymers with different chain lengths (Oh et al. 2004).
3. To maintain the colloidal stability of micelles, we recommend formulating micelles in pure water and then reconstituting micelles in PBS or other desired media.
4. For the preparation of micelles, volatile solvents are mainly used in the solid dispersion method (such as methanol, tetrahydrofuran, dichloromethane, or chloroform) and evaporation method (water-miscible solvents such as acetone, methanol, or tetrahydrofuran) to facilitate removal of the solvents. The selection of organic solvent and the ratio of water to solvent employed have been shown to greatly affect the physical properties and drug loading of the micelles (Kohori et al. 2002).

16.3.7 Discussion

As a result of the capability to load lipophilic molecules into the hydrophobic core, polymer micelles are promising nanocarrier systems for drug delivery. Important properties of the polymer micelles as drug carriers such as particle size, stability, loading capacity, and release kinetics of drugs can be modulated by the structure and composition of the block copolymers. Beyond solubilizing hydrophobic drugs, polymeric micelles can enhance permeability across physiological barriers, overcome drug resistance (Kabanov et al. 2002; Batrakova and Kabanov 2008), and deliver their payload to specific tissues passively or actively through the enhanced permeability and retention (EPR) effect (Maeda et al. 2000) or via surface modification with specific ligands for brain targeting and delivery (Shao et al. 2010). Furthermore, multiple drugs and imaging agents can be incorporated into a single micelle allowing both diagnosis and therapy (Ferrari 2005). During the last decade several micellar formulations of antitumor drugs have entered into clinical trials. The results from these clinical studies demonstrated that the polymeric micelle formulations reduced the toxicity associated with these drugs, which, in turn, resulted in a higher therapeutic effect. Overall, this strategy has potential for developing novel modalities for the delivery of various drugs to the brain (Gilmore et al. 2008).

16.4 Nanosuspensions

16.4.1 Objective

Current antiretroviral therapy (ART) has greatly reduced disease morbidity and mortality in people infected with human immunodeficiency virus (HIV). However, lifelong, daily administration of ART inevitably leads to drug resistance and increased risk of toxicity. In addition, the limited or failed delivery of ART agents into viral tissue sanctuaries such as the CNS and genitalia is reflected by varying degrees of development of cognitive and motor disorders as well as accelerated disease progression. The development of ART nanosuspensions (nanoARTs) has proven to be a promising strategy to provide long-acting, well-tolerated, less toxic therapy, capable of targeting viral tissue sanctuaries, especially for CNS infection (Kanmogne et al. 2012; Kadiu et al. 2011; Nowacek et al. 2010; Nowacek et al. 2011). This section describes how to develop and characterize nanoARTs. The protease inhibitor atazanavir (ATV) is selected as a model drug.

16.4.2 Preparation of ATV Nanosuspension by Wet-Milling Using the NETZSCH MicroSeries Wet Mill (Balkundi et al. 2010)

1. Mix ATV crystals (1 g) and polymeric surfactant, Poloxamer 407 (0.5 g), with HEPES buffer (10 mM, pH 7.8, 100 mL) using a T-18 Ultra-turrax mixer until completely dispersed.
2. Turn on the chiller and the compressor.
3. Turn on the control unit, rotate the grinding tank, and load the milling media (zirconium ceramic beads, 0.8 mm, 50 mL) into the grinding chamber. Clean up any beads in the threads of the grinding chamber to avoid leakage from the mechanical seal.
4. Seal and secure the milling chamber with appropriate screen mesh size and lid assembly.
5. Lower the grinding chamber to the horizontal position, and put the outlet of the product tube into the collecting vessel.
6. Add the ATV/Poloxamer 407 suspension into the product inlet, and start the pump using the control unit. The flow rate can be varied from 50 to 150 mL/min.
7. Turn the agitator on and adjust the speed with the control unit (620–4,320 rpm). Monitor the temperature through the temperature gauge on the grinding chamber, and avoid overheating in the case heat-sensitive compounds.
8. At various time points collect an aliquot of the suspension, monitor the size and particle size distribution of the ATV nanosuspension using DLS until the desired size of the ATV nanosuspension is obtained.
9. Collect the sample, turn off the pump, loosen the nuts on the lid assembly, and remove all beads from the chamber.
10. Centrifuge the resulting ATV nanosuspension at 10,000 rpm for 30 min, collect the pellets, and resuspend them in the same amount of fresh surfactant solution.
11. Determine ATV concentration in the nanosuspension using HPLC: YMC Octyl C8 column (Waters Inc., Milford, MA) with a

C8 guard cartridge, mobile phase consisting of 48 % acetonitrile/52 % 25 mM KH_2PO_4 (pH 4.15) with a flow rate of 0.4 mL/min and UV/Vis detection at 212 nm.

16.4.3 Preparation of ATV Nanosuspension by High-Pressure Homogenization Using Avestin EmulsiFlex C5 Homogenizer (Balkundi et al. 2010)

1. Transfer suspension of ATV and Poloxamer 407 (see step 1 in Sect. 16.4.2) into the vessel of the homogenizer and begin recirculation.
2. Turn the chiller on and start homogenization.
3. Increase the pressure gradually to 20,000 ± 2,000 psi, continue to homogenize, and monitor the size of ATV suspension by DLS until the desired particle size is obtained.
4. Collect the resulting ATV nanosuspension and then switch off the homogenizer.
5. Centrifuge the resulting ATV nanosuspension at 10,000 rpm for 30 min, collect the pellets, and resuspend into the same volume of fresh surfactant solution.
6. Determine ATV concentration (see step 11 in Sect. 16.4.2).

16.4.4 Typical Results

By using either the wet-milling or high-pressure homogenization method, ATV nanosuspensions can be prepared with an average size of 300 nm, polydispersity index (PDI) of 0.2, and zeta-potential of −20 mV when Poloxamer 407 is used as the surfactant. These technologies also yield high ATV loading capacity (about 75 %). The typical morphology of the ATV particles in nano-suspension is thin rods with smooth edges regardless of preparation method.

16.4.5 Notes

1. Typically, multiple homogenization cycles are required to obtain the desired particle size. The number of required cycles depends on the hardness of the drug, desired mean particle size, and required homogeneity of the product. It is anticipated that the higher the number of homogenization cycles, the smaller the particle size obtained. The optimum number of homogenization cycles can be determined by analyzing the particle size and polydispersity index of the drug after each cycle.
2. For wet-milling, the screen mesh size should be at least half the size of the beads to avoid the product from clogging the filter. Do not start milling until the outlet pressure reaches around 100 psi. The milling speed can be regulated from 620 to 4,320 rpm. It can be adjusted during the milling process through monitoring the milling temperature to avoid overheating in the case of heat-sensitive compounds.
3. A problem associated with wet-milling technology is the possible erosion of the milling material and the subsequent contamination during the milling process. However, for some drugs with high hardness, it may be difficult to prepare nanosuspensions using homogenization, even using more than 20 cycles. In that case, wet-milling is still a preferred method; but the type and amount of milling media, the drug concentration and volume, and the milling time should be carefully considered to avoid erosion of milling media. The severity of this problem has been reduced to a great extent with the advent of polystyrene resin-based milling medium (Patravale et al. 2004).
4. The type and amount of surfactant/stabilizer has a pronounced effect on the physical stability and in vivo behavior of nanosuspensions. In some cases, a mixture of stabilizers is required to obtain a stable nanosuspension.

16.4.6 Discussion

A large proportion of newly developed drug candidates are water insoluble and therefore poorly bioavailable. Formulating them into nanosuspensions has emerged as a promising strategy to resolve these problems. Wet-milling and high-pressure homogenization, the two major top-down technologies to disintegrate coarse material into nanosized colloidal dispersions of pure solid drug crystals coated and stabilized with surfactants, are generally used to prepare nanosuspensions (Junghanns and Muller 2008; Van Eerdenbrugh et al. 2008). Nanosuspensions provide several major advantages compared to other nanoparticulate carriers, such as very high drug loading (more than 70 %), increased physical and chemical stability of the drug, reduced amount of pharmaceutical excipients and associated volume of administered drug, and avoidance of and removal of toxic organic solvents. Recent studies using crystalline nanoART suspensions have demonstrated sustained release and improved pharmacokinetics and efficacy in animal models of HIV infection (Dou et al. 2009; Kanmogne et al. 2012).

References

Abraham SA, Waterhouse DN, Mayer LD, Cullis PR, Madden TD, Bally MB (2005) The liposomal formulation of doxorubicin. Methods Enzymol 391:71–97

Balkundi S, Nowacek AS, Roy U, Martinez-Skinner A, McMillan J, Gendelman HE (2010) Methods development for blood borne macrophage carriage of nanoformulated antiretroviral drugs. J Vis Exp (46): 2460

Batrakova EV, Kabanov AV (2008) Pluronic block copolymers: evolution of drug delivery concept from inert nanocarriers to biological response modifiers. J Control Release 130:98–106

Dou H, Grotepas CB, McMillan JM, Destache CJ, Chaubal M, Werling J, Kipp J, Rabinow B, Gendelman HE (2009) Macrophage delivery of nanoformulated antiretroviral drug to the brain in a murine model of neuroAIDS. J Immunol 183:661–669

Fan J, Zhang Y, Chuang-Smith ON, Frank KL, Guenther BD, Kern M, Schlievert PM, Herzberg MC (2012) Ecto-5′-nucleotidase: a candidate virulence factor in Streptococcus sanguinis experimental endocarditis. PLoS One 7(6):e38059

Ferrari M (2005) Cancer nanotechnology: opportunities and challenges. Nat Rev Cancer 5:161–171

Gabizon A, Horowitz AT, Goren D, Tzemach D, Shmeeda H, Zalipsky S (2003) In vivo fate of folate-targeted polyethylene-glycol liposomes in tumor-bearing mice. Clin Cancer Res 9:6551–6559

Gaitanis A, Staal S (2010) Liposomal doxorubicin and nab-paclitaxel: nanoparticle cancer chemotherapy in current clinical use. Methods Mol Biol 624:385–392

Gast K, Zirwer D, Ladhoff AM, Schreiber J, Koelsch R, Kretschmer K, Lasch J (1982) Auto-oxidation-induced fusion of lipid vesicles. Biochim Biophys Acta 686:99–109

Gilmore JL, Yi X, Quan L, Kabanov AV (2008) Novel nanomaterials for clinical neuroscience. J Neuroimmune Pharmacol 3:83–94

Groothuis DR (2000) The blood–brain and blood-tumor barriers: a review of strategies for increasing drug delivery. Neuro Oncol 2:45–59

Haney MJ, Zhao Y, Li S, Higginbotham SM, Booth SL, Han HY, Vetro JA, Mosley RL, Kabanov AV, Gendelman HE, Batrakova EV (2011) Cell-mediated transfer of catalase nanoparticles from macrophages to brain endothelial, glial and neuronal cells. Nanomedicine (Lond) 6:1215–1230

Immordino ML, Dosio F, Cattel L (2006) Stealth liposomes: review of the basic science, rationale, and clinical applications, existing and potential. Int J Nanomedicine 1:297–315

Junghanns JU, Muller RH (2008) Nanocrystal technology, drug delivery and clinical applications. Int J Nanomedicine 3:295–309

Kabanov AV, Batrakova EV, Alakhov VY (2002) Pluronic block copolymers for overcoming drug resistance in cancer. Adv Drug Deliv Rev 54:759–779

Kadiu I, Nowacek A, McMillan J, Gendelman HE (2011) Macrophage endocytic trafficking of antiretroviral nanoparticles. Nanomedicine (Lond) 6:975–994

Kanmogne GD, Singh S, Roy U, Liu XM, McMillan J, Gorantla S, Balkundi S, Smith N, Alnouti Y, Gautum N, Zhou Y, Poluektova L, Kabanov A, Bronich T, Gendelman HE (2012) Mononuclear phagocyte intercellular crosstalk facilitates transmission of cell targeted nanoformulated antiretroviral drugs to human brain endothelial cells. Int J Nanomedicine 7:2373–2388

Kohori F, Yokoyama M, Sakai K, Okano T (2002) Process design for efficient and controlled drug incorporation into polymeric micelle carrier systems. J Control Release 78:155–163

Krauze MT, Noble CO, Kawaguchi T, Drummond D, Kirpotin DB, Yamashita Y, Kullberg E, Forsayeth J, Park JW, Bankiewicz KS (2007) Convection-enhanced delivery of nanoliposomal CPT-11 (irinotecan) and PEGylated liposomal doxorubicin (Doxil) in rodent intracranial brain tumor xenografts. Neuro Oncol 9:393–403

Loch-Neckel G, Koepp J (2010) The blood–brain barrier and drug delivery in the central nervous system. Rev Neurol 51:165–174

Maeda H, Wu J, Sawa T, Matsumura Y, Hori K (2000) Tumor vascular permeability and the EPR effect in macromolecular therapeutics: a review. J Control Release 65:271–284

Maruyama K (2011) Intracellular targeting delivery of liposomal drugs to solid tumors based on EPR effects. Adv Drug Deliv Rev 63:161–169

Mayeux R (2003) Epidemiology of neurodegeneration. Annu Rev Neurosci 26:81–104

Modi G, Pillay V, Choonara YE (2010) Advances in the treatment of neurodegenerative disorders employing nanotechnology. Ann N Y Acad Sci 1184:154–172

Nowacek AS, McMillan J, Miller R, Anderson A, Rabinow B, Gendelman HE (2010) Nanoformulated antiretroviral drug combinations extend drug release and antiretroviral responses in HIV-1-infected macrophages: implications for neuro AIDS therapeutics. J Neuroimmune Pharmacol 5: 592–601

Nowacek AS, Balkundi S, McMillan J, Roy U, Martinez-Skinner A, Mosley RL, Kanmogne G, Kabanov AV, Bronich T, Gendelman HE (2011) Analyses of nanoformulated antiretroviral drug charge, size, shape and content for uptake, drug release and antiviral activities in human monocyte-derived macrophages. J Control Release 150:204–211

Oh KT, Bronich TK, Kabanov AV (2004) Micellar formulations for drug delivery based on mixtures of hydrophobic and hydrophilic Pluronic block copolymers. J Control Release 94:411–422

Orlacchio A, Bernardi G, Orlacchio A, Martino S (2010) Stem cells: an overview of the current status of therapies for central and peripheral nervous system diseases. Curr Med Chem 17:595–608

Patravale VB, Date AA, Kulkarni RM (2004) Nanosuspensions: a promising drug delivery strategy. J Pharm Pharmacol 56:827–840

Shao K, Huang R, Li J, Han L, Ye L, Lou J, Jiang C (2010) Angiopep-2 modified PE-PEG based polymeric micelles for amphotericin B delivery targeted to the brain. J Control Release 147:118–126

Torchilin VP (2005) Recent advances with liposomes as pharmaceutical carriers. Nat Rev Drug Discov 4:145–160

Van Eerdenbrugh B, Van den Mooter G, Augustijns P (2008) Top-down production of drug nanocrystals: nanosuspension stabilization, miniaturization and transformation into solid products. Int J Pharm 364:64–75

Xu L, Huang CC, Huang W, Tang WH, Rait A, Yin YZ, Cruz I, Xiang LM, Pirollo KF, Chang EH (2002) Systemic tumor-targeted gene delivery by anti-transferrin receptor scFv-immunoliposomes. Mol Cancer Ther 1:337–346

17

Neuronanomedicine

JoEllyn M. McMillan, Xin-Ming Liu,
and Howard E. Gendelman

Abstract

The need for the development of nanomedicines for the treatment of human immunodeficiency viral (HIV) infections is urgent and immediate. To this end, our laboratories have devised the means to utilize mononuclear phagocytes as nanoparticle carriers. Drug nanocarriers are specifically relevant to the central nervous system as they can target sites of inflammation including those contained within a diseased brain and enhance therapeutic efficacy. Nonetheless and despite the promise for HIV infections, nanomedicines have yet to make their mark. This, we posit, will change as translation of nanoformulation technology for clinical therapeutic applications shows its complete potential. This chapter reviews the promise of this technology and outlines in clear form the methods for manufacturing and testing the particles. Such works will have broad applicability beyond viral infections of the nervous system and are particularly relevant for this book series.

Keywords

Nanoparticles • HIV • Antiretroviral therapeutics • NanoART • Cell-based drug delivery • Macrophages

J.M. McMillan, Ph.D
Department of Pharmacology and Experimental Neuroscience, University of Nebraska Medical Center, 985800 Nebraska Medical Center, Omaha, NE, USA
e-mail: jmmcmillan@unmc.edu

X.-M. Liu • H.E. Gendelman, M.D. (✉)
Department of Pharmacology and Experimental Neuroscience and Center for Drug Delivery and Nanomedicine, University of Nebraska Medical Center, 985880 Nebraska Medical Center, Omaha, NE, USA

Department of Pharmaceutical Sciences, and Center for Drug Delivery and Nanomedicine, University of Nebraska Medical Center, 985880 Nebraska Medical Center, Omaha, NE, USA
e-mail: xliu@unmc.edu; hegendel@unmc.edu

H. Xiong and H.E. Gendelman (eds.), *Current Laboratory Methods in Neuroscience Research*, Springer Protocols Handbooks, DOI 10.1007/978-1-4614-8794-4_17,

17.1 Introduction

Eradication of the human immunodeficiency virus (HIV) will require delivery of both antiretroviral and immune modulatory agents to reservoir sites of infection that include the gut, lymphoid, and nervous system tissues (Palmer et al. 2011). One means to achieve this goal is through targeted nanoformulated antiretroviral therapy (nanoART) (Kim et al. 2010; Batrakova et al. 2011; McMillan et al. 2011; Mahajan et al. 2012; Re et al. 2012). Such inventions have led to a near "revolution" in brain targeted therapies for a broad range of disorders that include infection, trauma, stroke, misfolded aggregated protein-linked disorders, environmental toxins, and metabolic dysregulation (de Vries et al. 1997, 2012; Rosenberg 1997; Frank-Cannon et al. 2009; Graves and Vernino 2012; Kanwar et al. 2012). For HIV specifically, we reasoned that this could be facilitated by taking advantage of the natural functions of mononuclear phagocytes (MP; monocyte, macrophages, and dendritic cells) for phagocytosis, cell-to-cell communication, and exchange and trafficking throughout body tissues (Mosser 2003; Mosser and Edwards 2008). Thus, our laboratories developed a MP nanoART carriage system that uses cells as drug depots and a means to move therapy from blood to tissue sites of active viral growth (Dou et al. 2006, 2009; Nowacek et al. 2009, 2011; Kadiu et al. 2011; Dash et al. 2012; Gautam et al. 2013). The carriage system is facilitated by known comorbid inflammation that occurs as a consequence of chronic viral infection (Doitsh et al. 2010; d'Ettorre et al. 2011; Nischang et al. 2012). Indeed, sustained immune responses against the virus provide a target for attraction of the drug carrying MP (Dash et al. 2012; Roy et al. 2012; Martinez-Skinner et al. 2013). In this way the cell can seek and attenuate viral infection. The protective consequences of this therapeutic approach can also yield secondary and sustained phenotype switches for the brain's glial (microglia and astrocytes) network with deactivation of innate immunity and decreased release of a broad range of neurotoxic inflammatory mediators that include excitotoxins, pro- and anti-inflammatory cytokines, and chemokines. These ultimately improve brain homeostasis and notably can reduce cognitive and motor symptoms reflective of an ongoing HIV-associated metabolic encephalopathy and yield parallel improvements in microbial clearance for phagocytosis, intracellular killing, and facilitated antigen presentation as well as repair of any damage seen in the blood–brain barrier (BBB) with known extravasation of toxic proteins and other immune products (de Vries et al. 1997; Rosenberg 1997; O'Callaghan et al. 2008; Kanwar et al. 2012). While complex in mechanism, each of these components in disease provides opportunities for therapeutic intervention.

The nanoformulation procedures outlined below provide the pathways for how our laboratories developed the polymer-encased drugs, tested these in in vitro blood-borne macrophage-based assays, and then translated the laboratory test results to pharmacokinetic and biodistribution investigations. The results also provide the student and researcher an opportunity to use the same or similar platforms to design other MP carriage systems for brain disease that seek to take advantage of the carrying and storage capacity of these unique cells for improvements in disease outcomes of the nervous system.

17.2 Procedures

17.2.1 NanoART Preparation

1. Preparation of nanoART suspensions by high-pressure homogenization is described in Chap. 16 and summarized below.
2. Weigh 1 g hydrophobic drug crystals (e.g., free base atazanavir) and 0.5 g polymer surfactant (poloxamer 407 (P407)) and mix with 100 mL HEPES buffer (10 mM, pH 7.8) using a T-18 Ultraturrax mixer (IKA® Works Inc, Wilmington, NC) to obtain complete dispersion.
3. Transfer the suspension to the homogenizing vessel (Avestin C3 homogenizer; Avestin Inc, Ottawa, ON), making sure that the chiller is on before beginning recirculation.
4. Increase the pressure gradually to 20,000 ± 2,000 psi and continue to homogenize

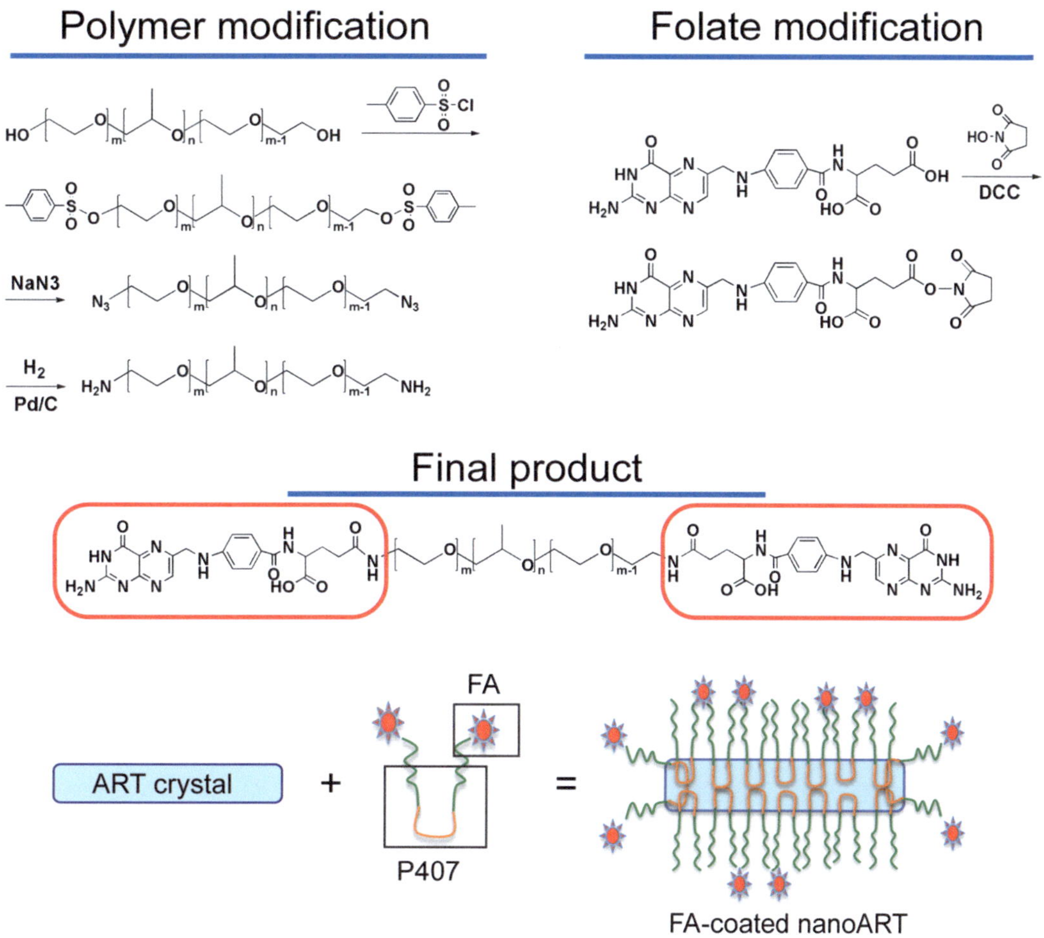

Fig. 17.1 Synthesis of folate-coated nanoART. Activated poloxamer 407 (P407) and folic acid (FA) are combined to form FA-modified P407 (FA-P407). Each polymer chain is modified with two FA moieties (shown in *red boxes*). FA-P407 is used in combination with P407 to coat antiretroviral (ART) drug crystals by high-pressure homogenization yielding FA-coated nanoART

until the target particle size is achieved, as determined by dynamic light scattering (DLS). This usually takes between 45 and 60 min.

5. Transfer the homogenized suspension to 50 mL centrifuge tubes and centrifuge at 10,000 × *g* for 30 min. Collect the pellets and resuspend them with an equal volume of fresh 0.2 % (*w*/*v*) aqueous solution of P407. Determine particle size, charge, and polydispersity by DLS.
6. Determine drug concentration of the samples using reversed phase high-performance liquid chromatography with UV/visible detection (HPLC-UV/Vis) (Nowacek et al. 2009) and/or ultraperformance liquid chromatography–tandem mass spectrometry (UPLC-MS/MS) (Huang et al. 2011).

17.2.2 Preparation of Folate-Targeted NanoART

1. Coupling of folic acid to amine-terminated P407 is similar to the method described in Chap. 16, except replacing 1,2-distearoyl-phosphatidylethanolamine-methyl-polyethyleneglycol conjugate-2000 ($DSPE\text{-}PEG_{2000}$) -amine with amine-terminated P407 (Fig. 17.1).
2. To prepare folic acid-modified nanoART, weigh 1.0 g crystalline drug, 0.2 g folic acid-modified P407 (FA-P407), and 0.3 g P407, and mix with HEPES buffer (10 mM, pH 7.8) using an Ultraturrax mixer until completely dispersed.

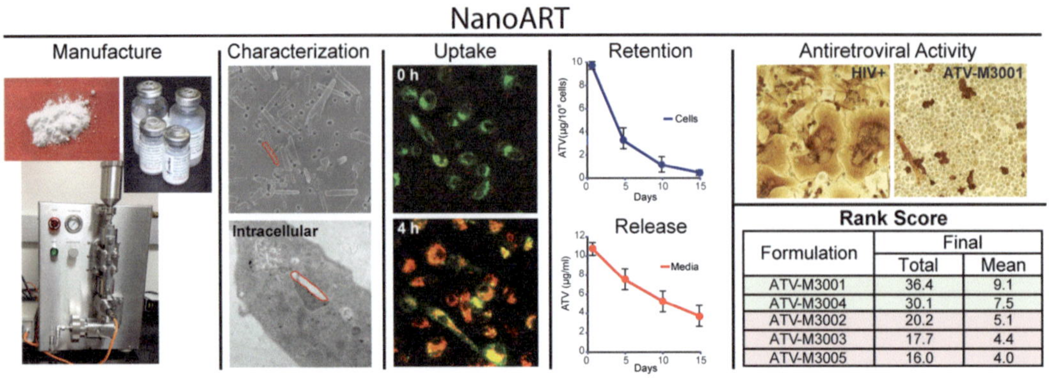

Fig. 17.2 Schematic for manufacture, characterization, cell-based screening, and scoring of nanoART. Crystalline antiretroviral drug is coated with polymer surfactant by high-pressure homogenization and characterized by size, polydispersity, zeta potential, and shape. NanoART is screened in cell-based assays for macrophage uptake, retention, release, and antiretroviral efficacy. The formulations are rank scored for selection of those suitable for animal testing as described by Nowacek et al. (2011)

3. Homogenize using an Avestin C3 homogenizer as described in Sect. 17.2.1 for nanoART preparation.
4. Following centrifugation at 10,000 × *g* for 30 min, resuspend the nanoparticle pellet in an equal volume of fresh 0.2 % P407 aqueous solution.
5. Determine drug content as described in Sect. 17.2.1.

17.2.3 Cellular Uptake of NanoART

1. *Cell culture*. Culture human peripheral blood monocytes in Dulbecco's Modified Eagles Medium (DMEM) containing 10 % heat-inactivated human serum, 1 % glutamine, 50 μg/mL gentamicin, 10 μg/mL ciprofloxacin, and 1,000 U/mL recombinant human macrophage colony-stimulating factor (MCSF) at a density of 1.5×10^6 cells/mL on 12- or 6-well tissue culture plates in a 37 °C, humidified, 5 % CO_2 incubator. Culture cells for 7 days to allow differentiation of monocytes into macrophages (MDM). Change one-half the medium with fresh medium on day 5.
2. *NanoART treatment*. Add nanoART to fresh culture medium (without MCSF) for desired drug concentration. Vortex to obtain homogeneous nanoART dispersion. Add treatment medium to triplicate culture wells.
3. *NanoART uptake*. The procedure for cell testing is summarized in Fig. 17.2. For determining cell uptake of nanoART, at desired timepoint(s) remove treatment medium from triplicate wells and rinse three times with phosphate buffered saline (PBS). Scrape cells into 1 mL of fresh PBS and place in 1.7 mL microcentrifuge tube. Centrifuge cells at 1,000 × *g* for 10 min at 4 °C. Remove medium and add 200 μL of HPLC-grade methanol. Lyse cells by briefly sonicating with probe sonicator at 20 % amplitude. Store samples at −80 °C until drug analysis.

17.2.4 Cellular Retention and Release of NanoART

1. At a selected timepoint, usually determined as the time when >80 % uptake has occurred, remove cell treatment medium and wash cells three times with PBS. Add fresh medium without MCSF and nanoART.
2. On desired day(s) from each treatment well, collect 1 mL of culture medium and corresponding cells as described for cell uptake (Sect. 17.2.3, Step 3).

3. Process and store cells as described in Sect. 17.2.3.
4. Process medium by adding 1 mL of 100 % HPLC-grade methanol to 150 μL medium in 1.7 mL microcentrifuge tube. Vortex for 10 s. Centrifuge samples at 20,000 × *g* for 10 min. Transfer supernatant to clean microcentrifuge tube. Evaporate sample without heat using a SpeedVac® (Savant, Thermo Fisher Scientific, Waltham, MA). This may take up to 4 h. Store dried samples at −80 °C until drug analysis.

17.2.5 In Vitro Antiretroviral Efficacy of NanoART

1. At a selected timepoint, as determined in cell retention experiments (Sect. 17.2.4), remove treatment medium from cells and wash three times with PBS. Add fresh medium without MCSF and nanoART.
2. On desired days, usually days 1, 5, 10, and 15, remove medium and replace with fresh medium containing HIV-1_{ADA} at a multiplicity of infection of 0.01 viral particles/cell. After 24 h remove viral infection medium and replace with fresh medium without virus.
3. Culture cells for an additional 10 days with one-half medium exchange every other day, or as needed to keep cells alive. At 10 days after infection, collect 3 × 10 μL of medium from each well and place in a 96-well plate. Store at −80 °C for assay of HIV reverse transcriptase (RT) activity.
4. Wash corresponding cells three times with 1 mL PBS. Add 4 % paraformaldehyde to each well and leave overnight at 4 °C. The following morning rinse cells with PBS and store at 4 °C in PBS until ready to stain.
5. *RT assay*. To each 10 μL sample from Step 3, add 10 μL of 100 mM Tris–HCl (pH 7.9), 300 mM KCl, 10 mM dithiothreitol (DTT), and 0.1 % nonyl phenoxylpolyethoxylethanol-40 (NP-40) in water. Incubate for 15 min at 37 °C. Add 25 μL of 50 mM Tris–HCl (pH 7.9), 150 mM KCl, 5 mM DTT, 15 mM $MgCl_2$, 0.05 % NP-40, 10 μg/mL poly(A), 0.25 U/mL oligo $d(T)_{12-18}$, and 10 μCi/mL ^{3}H-TTP in water and incubate for 18 h at 37 °C. Add 50 μL ice-cold 10 % trichloroacetic acid to each sample and harvest samples onto individual glass microfiber filters. Determine incorporation of ^{3}H-TTP by β-scintillation spectroscopy.
6. *HIV-1 p24 staining*. Remove PBS from fixed cells from Step 4 and add 10 % bovine serum albumin (BSA) with 1 % Triton X-100. Incubate for 30 min at room temperature. Add mouse anti-human p24 antibody (M0857, Dako, Carpinteria, CA), diluted 1:100 in 10 % BSA with 1 % Triton X-100. Incubate for 2–3 h at room temperature. Wash with PBS for 15 min at room temperature. Add anti-mouse secondary antibody (K4000, Dako EnVision + System HRP labeled polymer anti-mouse) for 30–45 min at room temperature. Wash with PBS for 15 min at room temperature. Add 300–400 μL diaminobenzidine (D4293-SET, Sigma, St. Louis, MO; 1 tablet in 5 mL deionized water) until brown color develops. Remove and add 500 μL hematoxylin stain for 60 s. Wash with PBS three times for 5 min. Image using bright-field microscopy.

17.2.6 Pharmacokinetics and Biodistribution of NanoART

1. Lyophilize nanoART or targeted nanoART suspension from Sect. 17.2.1 or 17.2.2. Determine drug content (wt%) using HPLC-UV/Vis and UPLC-MS/MS.
2. Determine how much drug will be needed for the study based upon drug dose (mg drug/kg body weight) and number of animals for each treatment.
3. Weigh out appropriate amount of nanoART to provide calculated amount of drug for study.
4. Suspend nanoART in enough PBS to deliver drug dose in a final injection volume of no greater than 50 μL per 25 g mouse. Vortex injection suspension to obtain a homogeneous nanoART dispersion. Save 10 μL of suspension for drug quantitation by HPLC-UV/Vis and UPLC-MS/MS.

5. Record weight of each mouse prior to nanoART treatment. Administer dose of nanoART by intramuscular or subcutaneous injection using a tuberculin syringe with a 27G needle.
6. Collect 100 μL blood from individual mice for up to 21 days into heparin-coated microcentrifuge tubes by serial cheek puncture using a 0.5 mm Goldenrod animal lancet (MEDIpoint, Inc., Mineola, NY). To obtain plasma, centrifuge tubes at 1,500 × *g* for 10 min at 4 °C and remove plasma supernatant to clean tube. Store at −80 °C until drug quantitation by UPLC-MS/MS.
7. At the end of the study anesthetize mice using an approved anesthesia method. Collect blood by cardiac puncture using a 1 cm^3 syringe with 27G needle into a heparin-coated tube and prepare plasma as in Step 6. To remove residual blood from tissues, cannulate the right atrium using a 27G butterfly needle and, using a peristaltic pump, flush the animal with PBS. Collect and weigh all tissues. Quick-freeze tissues in liquid nitrogen immediately upon collection and store at −80 °C until drug analysis by UPLC-MS/MS.

17.3 Notes

1. Preparation of crystalline nanoART is a viable method for only hydrophobic drugs. Because nanoART preparation requires the drug to be insoluble in an aqueous buffer, the method is unsuitable for preparation of nanoformulations of hydrophilic drugs.
2. These procedures describe use of poloxamers for preparation of nanoART. Other polymer surfactants can also be used, such as DSPE-mPEG2000, polyvinyl alcohol (PVA), and sodium dodecyl sulfate (SDS) (Nowacek et al. 2011). These polymers will impart a negative zeta potential to the nanoparticles. Addition of 1,2-dioleoyloxy-3-trimethylammoniumpropane (DOTAP) to the polymer coating will provide a positive zeta potential. The use of biocompatible and biodegradable polymers is recommended for development of nanoART for translational studies.
3. These procedures describe the addition of folic acid as a targeting ligand. Other targeting ligands, such as mannose, *N*-formylmethionyl-leucyl-phenylalanine (fMLP), HIV-gp120, and transferrin, may be used to target specific leukocyte and T-cell receptors. Addition of a fluorescence probe onto the polymer coating is possible and provides a tool for determining subcellular distribution (Kadiu et al. 2011; Puligujja et al. 2013) and particle uptake by macrophages in vivo (Gautam et al. 2013).
4. Quantitation of drug content of nanoART suspensions and lyophilized nanoART by reversed phase HPLC with UV/visible detection is described. The type of HPLC column, mobile phase, and wavelength for detection are dependent on the drug being formulated (Nowacek et al. 2011). HPLC-UV/Vis quantitation is usually adequate for measuring drug in nanosuspensions and cells. Improved sensitivity for detection and quantitation of drug in in vivo samples is achieved using UPLC-MS/MS (Huang et al. 2011).
5. These procedures describe testing of nanoART using human monocyte-derived macrophages. Other cell types, including human or animal immortalized macrophage and T-cell lines, lymphoblastoid cells, endothelial cells, and neuronal cells, may be used (Dou et al. 2006; Kanmogne et al. 2012). For testing targeted formulations, the choice of test cells is of importance.
6. Pharmacokinetic and biodistribution testing in mice is described. Pharmacokinetic analyses may be conducted using other animal species, including nonhuman primates (Gautam et al. 2013).
7. Pharmacokinetic and biodistribution studies in mice provide key information for design of dosing regimens and route of administration for pharmacodynamic and antiviral efficacy studies in animal models of HIV infection, including humanized mice and nonhuman primates (Dash et al. 2012; Roy et al. 2012; Epstein et al. 2013).

17.4 Discussion

Nanoformulations of antiretroviral drugs for cell-based delivery can be easily prepared using high-pressure homogenization. Crystalline nanoformulations are suitable for hydrophobic drugs; however, other nanoformulation strategies can be used for formulating hydrophilic drugs. As described in this chapter, suitability of the formulations for cell-based delivery, including cell uptake, retention, and antiviral efficacy, can be screened in relevant cell cultures, such as MDM. Additional experimental systems can be used to determine trafficking of nanoparticles to non-degrading intracellular compartments for storage and release (Kadiu et al. 2011). The effect of nanoART carriage on cell function can be assessed using proteomics, cytokine arrays, functional assays, and toxicity tests (Bressani et al. 2011; Kanmogne et al. 2012; Martinez-Skinner et al. 2013). Actively targeting nanoART to specific cell receptors can enhance cell uptake, retention, and antiviral efficacy. In vivo pharmacokinetics and pharmacodynamics assessments will provide necessary confirmation of extended drug delivery, antiviral efficacy, and targeting to viral reservoir sites such as the CNS (Dash et al. 2012; Kanmogne et al. 2012; Roy et al. 2012; Epstein et al. 2013; Puligujja et al. 2013). These studies highlight in vitro and in vivo preclinical studies necessary for ultimate translation of nanoART to clinically relevant therapy (Fig. 17.3).

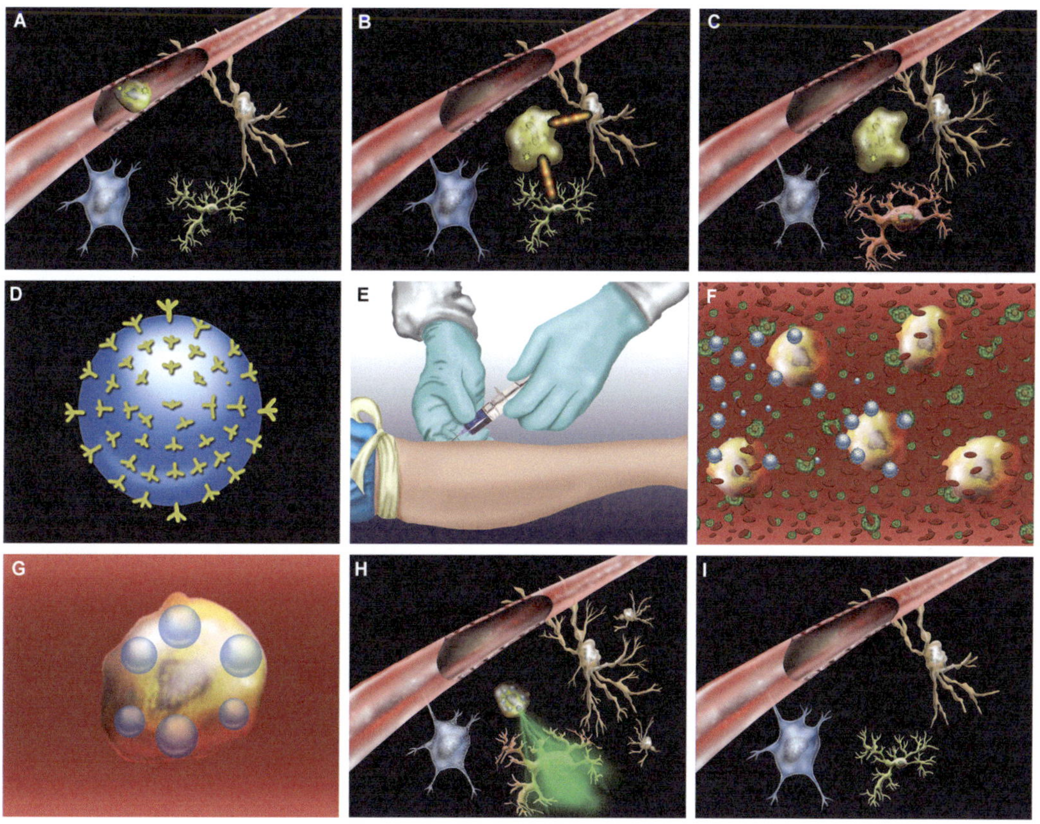

Fig. 17.3 Paradigm for targeted nanoART treatment of HIV-1-associated neurodegenerative disorders. (**a**) Circulating monocytes infected with HIV-1 enter the CNS parenchyma and (**b**) infect microglia and astrocytes and release cytokines, chemokines, excitotoxins, and reactive oxygen species that lead to (**c**) neuronal cell damage. (**d**) Folate-coated nanoART (FA-nanoART) is (**e**) administered to patients and (**f**) enters circulating monocytes by receptor-mediated endocytosis. (**g**) FA-nanoART is stored in recycling endosomes in the cells. (**h**) FA-nanoART-loaded cells enter the CNS parenchyma where drug is released to inhibit HIV replication, reduce microglial and astrocyte activation, and (**i**) restore neuronal integrity

References

Batrakova EV, Gendelman HE, Kabanov AV (2011) Cell-mediated drug delivery. Expert Opin Drug Deliv 8:415–433

Bressani RF, Nowacek AS, Singh S, Balkundi S, Rabinow B, McMillan J, Gendelman HE, Kanmogne GD (2011) Pharmacotoxicology of monocyte-macrophage nanoformulated antiretroviral drug uptake and carriage. Nanotoxicology 5:592–605

d'Ettorre G, Paiardini M, Ceccarelli G, Silvestri G, Vullo V (2011) HIV-associated immune activation: from bench to bedside. AIDS Res Hum Retroviruses 27:355–364

Dash PK, Gendelman HE, Roy U, Balkundi S, Alnouti Y, Mosley RL, Gelbard HA, McMillan J, Gorantla S, Poluektova LY (2012) Long-acting nanoformulated antiretroviral therapy elicits potent antiretroviral and neuroprotective responses in HIV-1-infected humanized mice. AIDS 26:2135–2144

de Vries HE, Kuiper J, de Boer AG, Van Berkel TJ, Breimer DD (1997) The blood–brain barrier in neuroinflammatory diseases. Pharmacol Rev 49:143–155

de Vries HE, Kooij G, Frenkel D, Georgopoulos S, Monsonego A, Janigro D (2012) Inflammatory events at blood–brain barrier in neuroinflammatory and neurodegenerative disorders: implications for clinical disease. Epilepsia 53(Suppl 6):45–52

Doitsh G, Cavrois M, Lassen KG, Zepeda O, Yang Z, Santiago ML, Hebbeler AM, Greene WC (2010) Abortive HIV infection mediates CD4 T cell depletion and inflammation in human lymphoid tissue. Cell 143:789–801

Dou H, Destache CJ, Morehead JR, Mosley RL, Boska MD, Kingsley J, Gorantla S, Poluektova L, Nelson JA, Chaubal M, Werling J, Kipp J, Rabinow BE, Gendelman HE (2006) Development of a macrophage-based nanoparticle platform for antiretroviral drug delivery. Blood 108:2827–2835

Dou H, Grotepas CB, McMillan JM, Destache CJ, Chaubal M, Werling J, Kipp J, Rabinow B, Gendelman HE (2009) Macrophage delivery of nanoformulated antiretroviral drug to the brain in a murine model of neuroAIDS. J Immunol 183:661–669

Epstein AA, Narayanasamy P, Dash PK, High R, Bathena SP, Gorantla S, Poluektova LY, Alnouti Y, Gendelman HE, Boska MD (2013) Combinatorial assessments of brain tissue metabolomics and histopathology in rodent models of human immunodeficiency virus infection. J Neuroimmune Pharmacol, In Press

Frank-Cannon TC, Alto LT, McAlpine FE, Tansey MG (2009) Does neuroinflammation fan the flame in neurodegenerative diseases? Mol Neurodegener 4:47

Gautam N, Roy U, Balkundi S, Puligujja P, Guo D, Smith N, Liu XM, Lamberty B, Morsey B, Fox HS, McMillan J, Gendelman HE, Alnouti Y (2013) Preclinical pharmacokinetics and tissue distribution of long-acting nanoformulated antiretroviral therapy. Antimicrob Agents Chemother 57(7):3110–3120

Graves D, Vernino S (2012) Immunotherapies in neurologic disorders. Med Clin North Am 96:497–523, x

Huang J, Gautam N, Bathena SP, Roy U, McMillan J, Gendelman HE, Alnouti Y (2011) UPLC-MS/MS quantification of nanoformulated ritonavir, indinavir, atazanavir, and efavirenz in mouse serum and tissues. J Chromatogr B Analyt Technol Biomed Life Sci 879:2332–2338

Kadiu I, Nowacek A, McMillan J, Gendelman HE (2011) Macrophage endocytic trafficking of antiretroviral nanoparticles. Nanomedicine (Lond) 6:975–994

Kanmogne GD, Singh S, Roy U, Liu X, McMillan J, Gorantla S, Balkundi S, Smith N, Alnouti Y, Gautam N, Zhou Y, Poluektova L, Kabanov A, Bronich T, Gendelman HE (2012) Mononuclear phagocyte intercellular crosstalk facilitates transmission of cell-targeted nanoformulated antiretroviral drugs to human brain endothelial cells. Int J Nanomedicine 7:2373–2388

Kanwar JR, Sriramoju B, Kanwar RK (2012) Neurological disorders and therapeutics targeted to surmount the blood–brain barrier. Int J Nanomedicine 7:3259–3278

Kim BY, Rutka JT, Chan WC (2010) Nanomedicine. N Engl J Med 363:2434–2443

Mahajan SD, Aalinkeel R, Law WC, Reynolds JL, Nair BB, Sykes DE, Yong KT, Roy I, Prasad PN, Schwartz SA (2012) Anti-HIV-1 nanotherapeutics: promises and challenges for the future. Int J Nanomedicine 7:5301–5314

Martinez-Skinner AL, Veerubhotla RS, Liu H, Xiong H, Yu F, McMillan JM, Gendelman HE (2013) Functional proteome of macrophage carried nanoformulated antiretroviral therapy demonstrates enhanced particle carrying capacity. J Proteome Res 12:2282–2294

McMillan J, Batrakova E, Gendelman HE (2011) Cell delivery of therapeutic nanoparticles. Prog Mol Biol Transl Sci 104:563–601

Mosser DM (2003) The many faces of macrophage activation. J Leukoc Biol 73:209–212

Mosser DM, Edwards JP (2008) Exploring the full spectrum of macrophage activation. Nat Rev Immunol 8:958–969

Nischang M, Sutmuller R, Gers-Huber G, Audige A, Li D, Rochat MA, Baenziger S, Hofer U, Schlaepfer E, Regenass S, Amssoms K, Stoops B, Van Cauwenberge A, Boden D, Kraus G, Speck RF (2012) Humanized mice recapitulate key features of HIV-1 infection: a novel concept using long-acting anti-retroviral drugs for treating HIV-1. PLoS One 7:e38853

Nowacek AS, Miller RL, McMillan J, Kanmogne G, Kanmogne M, Mosley RL, Ma Z, Graham S, Chaubal M, Werling J, Rabinow B, Dou H, Gendelman HE (2009) NanoART synthesis, characterization, uptake, release and toxicology for human monocyte-macrophage drug delivery. Nanomedicine (Lond) 4:903–917

Nowacek AS, Balkundi S, McMillan J, Roy U, Martinez-Skinner A, Mosley RL, Kanmogne G, Kabanov AV, Bronich T, Gendelman HE (2011) Analyses of nanoformulated antiretroviral drug charge, size, shape and content for uptake, drug release and antiviral activities in human monocyte-derived macrophages. J Control Release 150:204–211

O'Callaghan JP, Sriram K, Miller DB (2008) Defining "neuroinflammation". Ann N Y Acad Sci 1139:318–330

Palmer S, Josefsson L, Coffin JM (2011) HIV reservoirs and the possibility of a cure for HIV infection. J Intern Med 270:550–560

Puligujja P, McMillan J, Kendrick L, Li T, Balkundi S, Smith N, Veerubhotla RS, Edagwa BJ, Kabanov AV, Bronich T, Gendelman HE, Liu XM (2013) Macrophage folate receptor-targeted antiretroviral therapy facilitates drug entry, retention, antiretroviral activities and biodistribution for reduction of human immunodeficiency virus infections. Nanomedicine, In Press

Re F, Gregori M, Masserini M (2012) Nanotechnology for neurodegenerative disorders. Nanomedicine 8(Suppl 1):S51–S58

Rosenberg GA (1997) Neuroinflammatory disease. IBC meeting on neuroinflammatory disease: research and treatment strategies. London, UK, 17 and 18 September 1996. Mol Med Today 3:12–13

Roy U, McMillan J, Alnouti Y, Gautum N, Smith N, Balkundi S, Dash P, Gorantla S, Martinez-Skinner A, Meza J, Kanmogne G, Swindells S, Cohen SM, Mosley RL, Poluektova L, Gendelman HE (2012) Pharmacodynamic and antiretroviral activities of combination nanoformulated antiretrovirals in HIV-1-infected human peripheral blood lymphocyte-reconstituted mice. J Infect Dis 206:1577–1588

Part VI

Bioimaging

Confocal Imaging of Nerve Cells

18

You Zhou

Abstract

Confocal laser scanning microscopy has been widely utilized for real-time, cytochemical, or immunofluorescence image analysis in many studies using various types of biological samples. Recently developed fluorophores and fluorescent proteins with higher fluorescence intensities in narrower ranges of excitation and emission wavelengths give researchers more choices for selections of fluorescent tags and labeling combinations. This chapter provides brief introduction of confocal microscopy, autofluorescence with focus on aging brains, signal and noise ratio, and common fluorophore selections/combinations. In addition, this paper describes general procedures for confocal image analysis using live cells or fixed samples, using examples for (1) real-time imaging analysis of intracellular ROS response of Neuro-2A cells to specific treatments and (2) triple-labeling immunofluorescence confocal microscopy using mouse brain sections.

Keywords

Autofluorescence • Confocal laser scanning microscopy • Fluorophores and fluorescent proteins selection • Immunofluorescence labeling • Nerve cells • Real-time imaging • Signal-noise ratio

18.1 Introduction

18.1.1 Brief Summary of Current Confocal Imaging Technology

The confocal imaging technique was pioneered by Marvin Minsky at Harvard University in 1955, and later he patented the confocal principles in 1957. However, confocal microscopy did not become a widely used, powerful tool for imaging

Y. Zhou, Ph.D. (✉)
Center for Biotechnology and School of Veterinary Medicine & Biomedical Sciences, University of Nebraska-Lincoln, 1901 Vine Street, Lincoln, NE, 68588-0665, USA
e-mail: yzhou2@unl.edu

H. Xiong and H.E. Gendelman (eds.), *Current Laboratory Methods in Neuroscience Research*, Springer Protocols Handbooks, DOI 10.1007/978-1-4614-8794-4_18,

acquisition and analysis until improvement of data computing technology and computer speed in late 1980s. A confocal laser scanning microscope (often just called a confocal microscope) is an integrated computer-operated optical imaging system, using laser beams as the light source and photomultiplier tubes (PMT) as detectors to acquire a single high-resolution in-focus image or a series of images by "optical sectioning" through live or fixed specimens (Carter 1999; Paddock 1999a; Tsien et al. 2006; Murray et al. 2007; Swedlow 2012). Controlling the focal plane and depth of focus field is the key advantage of confocal over conventional microscopes and allows minimization of background noise or out-of-focus information, serial optical images of a given focal plane through a thick sample, and 3D image reconstruction analysis. A commonly used confocal system is either an upright or an inverted fluorescence microscope equipped with 2–5 lasers which provides different laser spectral lines ranging from ~350 nm (UV) to ~650 nm (far-red) wavelengths. The advanced confocal system can provide single, double, multi-laser-line excitation (either sequentially or simultaneously), to obtain a single, dual, or multichannel simultaneous display from a specimen, respectively. The more recent models of confocal systems with tunable lasers and filter wheels, or Acousto-Optical Tunable Filter (AOTF) which enables high-speed changes in laser beam wavelength and intensity, allow faster automated process (such as "sequential scan" or "line scan"; see more explanation below) and more excitation laser lines and emission spectral displays (Paddock 1999a; Garini et al. 2006; Svoboda and Yasuda 2006; Swedlow 2012).

This chapter will focus on the principles and practical protocols of live-cell image analysis and immunofluorescence-labeling confocal microscopy using fixed samples [for further readings, see (Lippincott-Schwartz and Patterson 2003; Giepmans et al. 2006; Tsien et al. 2006; Selever et al. 2011; Herberich et al. 2012; Swedlow 2012)]. For live-cell imaging, most studies use cultured cells transfected with the gene(s) of interest tagged with a fluorescent protein, such as the "green fluorescence protein" (GFP) or the "red fluorescence protein" (RFP), with or without loading cells with another fluorescent marker for specific cellular components (such as the nucleus, membrane, Golgi, mitochondria, and lysosomes). This type of live-cell imaging confocal microscopy allows scientists to carry out real-time study of genes or proteins of interest for localization analysis and protein trafficking in response to external stimuli, such as drug treatment, oxidation stress, viral infection, or physical damage. After real-time image collection, these samples can then be fixed and processed for immunohistochemical studies. Immunofluorescence-labeling confocal microscopy experiments are usually designed for localization analysis of 1–3 specific proteins or molecules, using specific primary antibodies from different hosts and corresponding secondary antibodies conjugated with different fluorescent dyes with desired excitation and emission wavelengths (see Table 18.1 for examples of dye combinations). Direct labeling fluorescence confocal microscopy can also be applied to a localization study by using fluorescent-dye-conjugated primary antibodies, especially when the primary antibodies available are from the same host and same type, such as mouse monoclonal antibodies. The immunofluorescence-labeled brain sections can be counterstained (i.e., for the nucleus) for better orientation or identification of regions and cell types (see second part of this chapter for detailed procedures).

18.1.2 General Considerations for Confocal Laser Scanning Microscopy

Numerous publications have shown data obtained by means of live-cell imaging or immunofluorescence confocal microscopy using primary cell cultures or stabilized cell lines. While most images in recent publications are obtained by or with the assistance of imaging specialists, it is important for a researcher to understand the basic microscopic and imaging concepts (including magnification, brightness, resolution, and fluorescence) and advanced optical technology associated with a confocal system, such as pinhole and optical sec-

Table 18.1 Common fluorophore selections and fluorescent labeling combinations

Excitation laser lines (range)	405 nm (395–420 nm)	488 nm (480–515 nm)	543/568 nm (540–580 nm)	633/647 nm (630–650 nm)
Emission range/(color)	400–450 nm (violet-blue)	500–550 nm (green-yellow)	580–620 nm (orange-red)	640–700 nm (red-deep red)
Fluorophores conjugates and fluorescent protein tags[a]				
Alexa Fluor/cyanine (Cy) dyes	Alexa Fluor-405	Alexa Fluor-488, Cy2	Alexa Fluor-546, Cy3	Alexa Fluor-633, Cy5
Enhanced fluorescent proteins	TaqBFP, EBFP	EGFP, mVenus	ERFP, mCherry	mKate2
Non-permeant nuclear counterstains for labeling fixed and permeabilized cells				
DNA/RNA (for permeabilized cells)	SYTOX Blue, DAPI	SYTOX Green	SYTOX Orange, PI	SYTOX Red
Cell-permeant fluorescent markers for cell components using live/unfixed cells[b]				
DNA/RNA	Hoechst, SYTO 40 blue	SYTO 16 Green	SYTO 85 orange	SYTO 62 far-red
Mitochondria		MitoTracker Green FM	MitoTracker Red FM	MitoTracker Deep Red FM
Lysosomes/late endosomes (weaker)	LysoTracker Blue DND-22	LysoTracker Green DND-26	LysoTracker Red DND-99	
Endoplasmic reticulum/ Golgi		ER-Tracker Green	ER-Tracker Red	
Golgi with weak ER labeling		BODIPY FL C5 ceramide-BSA	BODIPY TR C5 ceramide-BSA	
Plasma/lipid membranes		DiOC6, DiOC18(3)	CellMask Orange, DiI	CellMask Deep Red, DiD

[a]The mKate2 should be excited with 568 nm laser line or with 543 nm laser line, which has a weaker excitation intensity

[b]Cells loaded with the permeant dyes can be fixed with non-alcohol-containing agents for additional labeling (see their specific information before fixation)

tion thickness (Paddock 1999b). In addition, several practical factors, as briefly discussed below and by others (Tsien et al. 2006; Matsuda et al. 2011; Herberich et al. 2012), should be considered before starting an experiment using fluorescence technology and confocal microscopy.

1. Autofluorescence and signal-noise ratio.

   Autofluorescence is named for "self-fluorescence," the natural fluorescence arising from endogenous molecules/components, which have an excitation range of 330–600 nm and an emission range of 480–620 nm, including reduced pyridine nucleotides (such as mitochondrial NADH); oxidized flavins (also abundant in mitochondria); chlorophylls in chloroplasts of the plant cells (Brehmer et al. 2004); and lipofuscin and neuromelanins (aging pigments), which accumulate with age in the cytoplasm of nerve cells (such as neurons, astrocytes, and microglia) and other types of cells (Koistinaho et al. 1986; Elleder and Borovansky 2001; Double et al. 2008; Eichhoff et al. 2008; Zecca et al. 2008; Herberich et al. 2012). Autofluorescence can be induced or enhanced by fixation with aldehyde-containing fixatives, such as paraformaldehyde, formaldehyde, and glutaraldehyde (Beisker et al. 1987). Autofluorescent noise in mammalian cells can be generated using excitation in a range of 350–550 nm (UV to blue/green light), with an emission range of 440 nm to ~600 nm (blue/green to orange/red). As shown in Fig. 18.1, the autofluorescent noise can be seen in both green and red channels when using single 405 nm (panels a1 and a2) or 488 nm (panels b1 and b2) laser line for excitation; however, the noise can be seen in only the red channel (panels c1 and c2) when using 543 nm laser line for excitation. Furthermore, this autofluorescent noise is

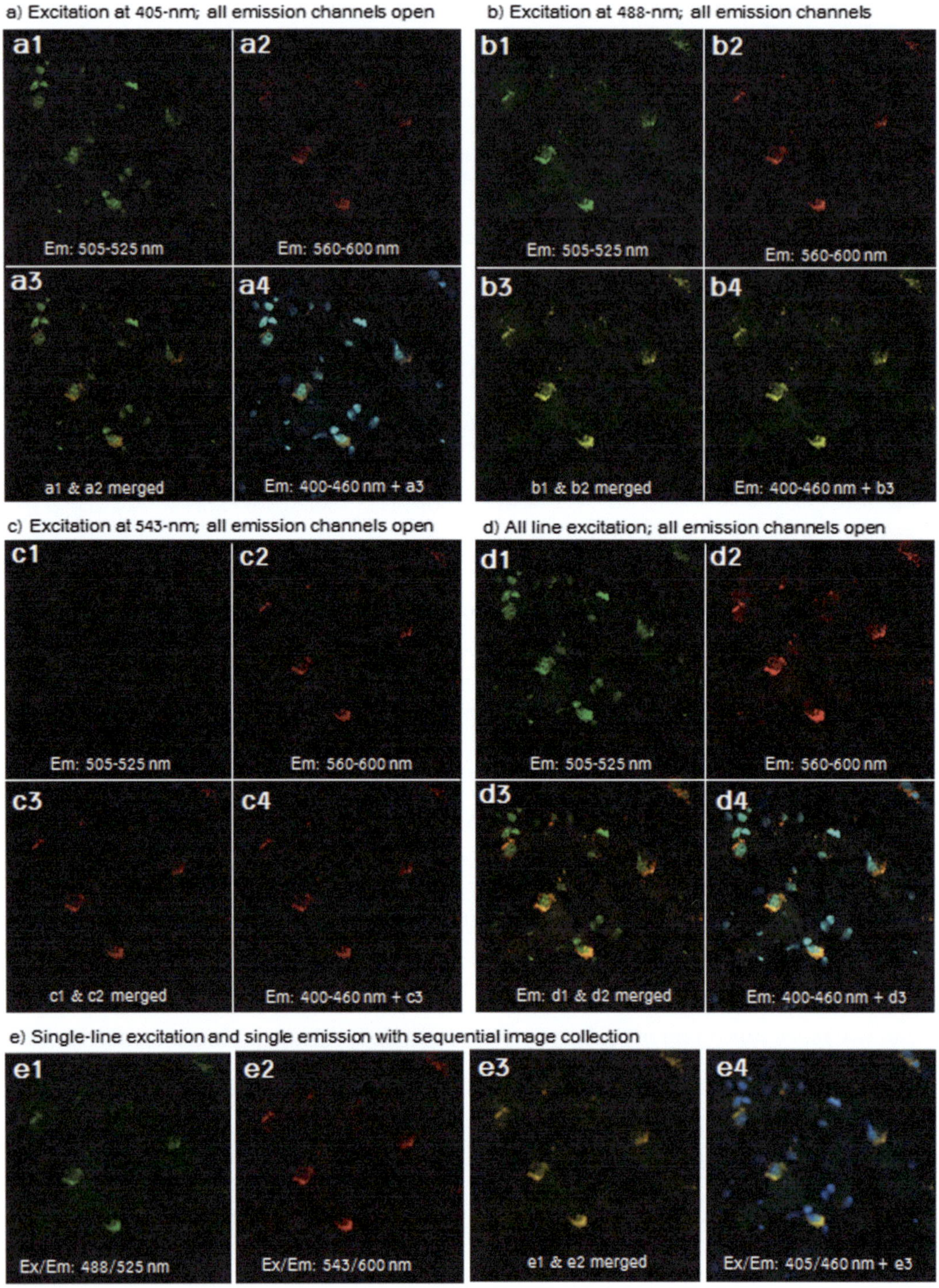

Fig. 18.1 Confocal images collected from the same region of an aging human brain sample showing the autofluorescence of "aging pigments" in the nerve cells. Five sets of images ((**a–e**) four-panel displays) were collected using different imaging modes but under the same settings (such as gain, *black* level, and laser power). The autofluorescence of the aging pigments can be seen using each of three excitation laser lines (Ex) in all three emission channels (Em), even when using sequential image collection program (*panels* **e1–e4**). *Panel 1* images (i.e., **a1**, **b1**, **c1**, **d1**, **e1**) are collected using 505–525 nm Em filter (shown in *green color*), *panel 2* images are from 560 to 600 nm Em filter (*red*), *panel 3* shows merged images of *panels 1* and *2* (the age pigments shown in *yellow/orange color*), and *panel 4* images are same as the *panel 3* but overlaid with blue fluorescence from DAPI-stained nuclei. Note that the DAPI fluorescence "crossover" or "bleed-through" in the 505–525 Em channel can be seen in both (**a**) and (**d**) (all Em channels open) but not in (**e**) (sequential mode), when using 405 nm excitation

stronger when using all three laser lines simultaneously for excitation, as seen in panels d1 and d2. Even when using the sequential scanning mode, which minimizes the fluorescence "bleed-through" or spectral overlap with a higher-cutoff emission filter, the autofluorescence from the aging pigments is still seen in both the green and red emission channels (panels e1 and e2). Therefore, careful controls and additional procedures to increase the signal-noise ratio should be considered, especially when the targeted signals of interest are relatively weak and likely associated with punctuated structures. In localization studies using immunofluorescence confocal microscopy of nerve cells, especially those in fixed brain tissues, one should first consider to use a fluorophore emitting in far-red (640–700 nm) for the target and use a fluorophore emitting in the ~500–530 nm (green) to 580–610 nm (red) range to label a well-known marker for co-localization, as discussed in more detail below. Finally, an alternate way to reduce autofluorescence noise is to pretreat samples with certain types of blocking agents, such as $CuSO_4$ and Sudan Black B (Schnell et al. 1999; Yao et al. 2003; Viegas et al. 2007).

2. Selection of commonly used fluorophores and specific markers.

   There are many fluorophores and their conjugates commercially available. Some of them have a broad range of excitation and/or emission, such as DAPI and PI (propidium iodide), a fluorophore for nucleic acid stain that has an emission range of 560–700 nm. These types of dyes are suitable for initial image analysis, but are not best for co-localization experiments, especially when the signal of interest is weak. Several series of fluorophores (such as cyanine dyes: Cy2, Cy3, and Cy5; Alexa dyes: Alexa 405, 488, 568, or 633/647) and constructs with fluorescent protein tags, such as TagBFP, EGFP, mVenus, ERFP, mCherry, and mKate2 (See Table 18.1), have been widely utilized for single or combination of labeling using corresponding excitation laser lines and emission filter sets of common confocal systems (Tsien et al. 2006; Rizzo et al. 2009; Kremers et al. 2011; Lam et al. 2012). In general, one should always try to select a combination of fluorophores with minimum overlap of spectrum for excitation and emission (Tsien et al. 2006; Lam et al. 2012). For example, when studying the expression of a GFP-tagged protein or using an Alexa Fluor-488 or Cy2 conjugate, it is best to use a far-red dye such as Alexa Fluor-633 or Cy5 conjugates for co-localization. If the expression of the GFP-tagged gene/protein appears to be associated with the mitochondrial, lysosomal, endoplasmic reticulum, or Golgi structure, it is quicker to obtain results by means of live-cell imaging analysis using a cell permeable fluorescent marker with red/far-red emission, such as the MitoTracker, LysoTracker, or ER-Tracker, respectively. Table 18.1 shows the most commonly used dye combinations that are suitable for confocal systems.

3. Fixation, permeabilization, and targeted labeling.

   There are three types of commonly used fixatives based on the known chemical-reactive mechanisms: cross-linking agents (e.g., aldehydes-paraformaldehyde, formaldehyde, and glutaraldehyde), protein denaturing/precipitating agents (e.g., acetic acids, ethanol, methanol, and acetone), and oxidizing agents (e.g., osmium tetroxide). For immunofluorescence microscopy, glutaraldehyde fixation should be avoided since it can induce high autofluorescence noise and may reduce antigen-antibody binding efficiency due to its fast and strong cross-linking property. Alternatively, initial fixation with paraformaldehyde followed by treatment with ethanol (prechilled at −20 °C) or a mild detergent (such as Triton X-100) has been widely used for fixation and permeabilization for immunofluorescence-labeling image analysis. Quick fixation using ethanol, methanol, or even acetone has been used for staining nucleic acids or cytoskeleton bundles, and for fixing frozen sections. Nevertheless, alcohol-only quick fixation should be avoided when studying molecules that are cytosolic, compartmentalized/membrane-bound, and tagged with fluorescent protein (BFP, GFP YFP, or RFP), or lipoproteins, especially when seeking detailed localization information of a protein of interest and its translocation in response to a treatment (Chidlow et al. 2011; Matsuda et al. 2011).

18.2 Materials

1. Phosphate-buffered saline (PBS; 0.1 M).

   Prepared from 10× PBS stock solution, dilute 1–10 with distilled water.

   - 10× PBS (pH 7.2): 10.9 g Na_2HPO_4, 3.2 g NaH_2PO_4, 90 g NaCl, and 1 L distilled water.

2. Paraformaldehyde (PFA; 4 %, *v*/*v*) in PBS.

   The 4 % PFA solution should be prepared freshly from specially packaged PFA stock solution (16 % formaldehyde, methanol-free) or directly from PFA powders. It is much easier and relatively inexpensive to prepare from the stock solutions than from PFA powders. Use the PFA stock solution (10 mL vials sealed with inert gas, 16 % prepared from PFA powder) made by Electron Microscopy Sciences (EMS; Cat# 15710; 10×10 mL), by adding the 10 mL PFA stock solution into 26 mL distilled water and 4 mL 10× PBS stock solution to make the final concentration of 4 % PFA in PBS.

   - *Note*: *If aliquots are made from fresh-made 4 % PFA, they can be stored at −20°C. It works fine if the aliquots are thawed only once (i.e., do not refreeze the 4 % PFA).*

3. Formaldehyde (3.7 %) or 10 % formalin in PBS.

   Prepared from the commonly used 37–40 % formaldehyde solution (methanol-free) or from a commercial formalin (37 % formaldehyde) stock solution with 6–15 % methanol. Dilute the stock solution 1–10 with PBS before use.

   - *Note*: *The commercial formalin fixative contains methanol which can minimize cell shrinkage and make membranes permeable for small compound to enter the cells. However, to determine whether a protein is localized to the cell surface (and not just underneath or inside the surface plasma membrane), which also includes the transmembrane proteins, one should use PFA or methanol-free formaldehyde (not formalin) to fix the cells, followed by the immunofluorescence labeling and washes without detergents (i.e., no Tween-20 or Triton X-100). Samples fixed in 3.7 % formaldehyde at room temperature can be stored in the same fixative at 4 °C for months.*

4. Cold (−20 °C) ethanol/methanol/acetone.

   Industrial (~95 %) ethanol prechilled at −20 °C can be used for fixation and permeabilization.

   - *Note*: *Ethanol/methanol/acetone should not be used as the primary fixatives for preserving cellular or intracellular structures, as they can remove lipid or lipo-molecules and precipitate proteins and induce possible translocation of the water-soluble cytoplasmic proteins. However, cold ethanol/methanol (−20 °C) can be used preparation of frozen sections of samples fixed with PFA/formalin before air-drying the tissue sections.*

5. Triton X-100 (0.05 %).

   Triton X-100 is a nonionic surfactant commonly used in laboratories for permeabilization of the membranes of eukaryotic cells under unfixed or lightly fixed (such as short fixation by 4 % PFA) conditions. Prepare 0.05 % (*v*/*v*) of Triton X-100 in PBS before use.

   - *Note*: *Triton X-100 should not be used when studying a membrane protein (especially for quantitative image analysis using confocal microscopy) since Triton X-100 treatment can solubilize and remove membrane proteins, causing a reduced signal, even in lightly fixed cells.*

6. Tween-20 (0.05 %).

   Tween-20 is also a nonionic detergent commonly used in immunochemical-labeling experiments to reduce nonspecific antibody binding. Prepare the 0.05 % (*v*/*v*) Tween-20 in PBS before use.

   - *Note*: *It is generally considered as a milder detergent than Triton X-100 and can also solubilize proteins found mostly on the membrane surface (i.e., non-transmembrane). Therefore, Tween-20 should not be used if the protein or antigen of interest is on the surface of the plasma membrane.*

7. Bovine serum albumin (BSA; Fraction V; 3 and 1 % in PBS).

   The dry BSA stock should be stored at −20 °C for long-term storage. Prepare 3 % (*w*/*v*) solution in PBS for blocking. Use a 15 or 50 mL conical tube to make a 3 % BSA in PBS solution by vortexing at top speed with a benchtop vortex for 2–3 min. Make 1 % BSA-PBS-Tween-20 from the blocking solution before diluting antibodies.

   - *Note*: *BSA is expensive*; *therefore*, *one should make just enough 3% BSA-PBS solution needed for blocking. The BSA blocking solution should be freshly made before use and not be reused due to the increase in nonspecific fluorescent background.*

8. Other materials.

   (a) 35 mm cover glass-bottom culture dishes are designed for use with 60× or 100× lens for high-resolution, live-cell confocal real-time imaging and for higher resolution under controlled temperature and CO_2 conditions. If temperature or CO_2 condition is not a factor, the HEPES-buffered culture medium should be used to maintain a consistent pH.

   (b) Coverslips for most oil-immersion lenses should be type no. 1 (1 oz) and nonfluorescent cover glass if using 60× or 100× oil lens for fluorescence microscopy.

   (c) Water-based with stable pH mounting medium should be used for mounting of the cells or tissue sections labeled with fluorescent dyes and antibody conjugates. A commonly used mounting medium is Fluoro-Gel (EMS).

18.3 Methods/Procedures

18.3.1 Live-Cell and Time-Lapse Confocal Imaging Analysis of Cultured Neuronal Cells

For more than 2 decades, time-lapse confocal microscopy has been widely utilized to investigate dynamic changes of a molecule of interest, protein trafficking and interactions, or cellular response to external stimuli, in a given time frame in many different types of cells (Zhou et al. 1995; Pawley 2002; Chen et al. 2007; Waters 2007; Malinouski et al. 2011; Lam et al. 2012). Confocal images can be collected from cells in pH-balanced medium (HEPES-buffered MEM or PBS) under room temperature for short period of time (<15 min) or in a regular culture medium such as MEM under a humidified chamber with controlled CO_2 (5 %) and temperature (37 °C). Specific fluorescent proteins, dyes, or markers can be used as an indicator of cell response or for labeling cell organelles in vivo for real-time confocal imaging analysis (Zhou et al. 1995). This section will describe the procedures for live-cell imaging and in vitro analysis of neuronal cell responses to H_2O_2 treatment, using one of the commonly used neuronal cell lines, neuroblastoma (Neuro-2A) cells.

1. The Neuro-2A cells were cultured in a 35 mm cover glass-bottom dish in order to fit into a specially designed confocal chamber and to save dyes and other treatment materials.
2. Preload the cells with 2′,7′-dichlorofluorescein-diacetate (DCFH-DA, Molecular Probes), a fluorescence indicator for ROS levels, in 500 μL of serum-free medium (5 μM final concentration) for ~20 min in the cell culture incubator.

   Note: *Take the fluorescent dyes or other treatment materials mentioned below from the stock solution, add it into 500 μL medium (serum-free for dye loading), and mix well before adding into the culture dish for preloading. This also applies when adding the medium containing treatment materials such as* H_2O_2 *(also see notes below).*
3. After the DCFH-DA preloading incubation, replace the dye medium with new warm medium containing serum to remove the unloaded DCFH-DA. Change the culture medium again (use 500 μL or other desired volume) before placing the dish into the special chamber with 37 °C and 5 % CO_2 for confocal microscopy.
4. Find the cells of interest under phase contrast or transmitted light of the microscope before confocal imaging.

Note: Do not use mercury lamp/UV light during this experiment since the cells preloaded with DCFH-DA are very sensitive to UV exposure, which can increase intracellular ROS levels.

5. Test and set up the confocal image collection conditions. The following are optimal settings on an Olympus FV500-IX80 confocal system for this experiment:
   (a) Excitation laser line: 488 nm
   (b) Emission filter: 505–522 nm
   (c) Laser power: no more than 15 %
   (d) Confocal apertures (pinhole) size: ~150 μm (see note below)
   (e) PMT tuning: PMT voltage, ~500 V; gain, 2–3×; and offset at 2 %
   (f) Scan speed at ~0.5 s/frame (512×512)

   Note: The size of the pinhole opening is key factor controlling the thickness of a focal plane (a plane of focus within a specimen). The smaller pinhole size yields the thinner focal plane and lower noise. Smaller pinhole size should be considered for collecting Z-optical images (see more discussions in Sect. 18.3, step 2), especially when collecting the Z-series of optical sections for 3D image reconstruction (Murray et al. 2007). *However, for live-cell imaging in a time course study when the cells have a fast, dynamic response/change, such as the experiment described here, the pinhole size should not be too small in order to obtain signals from a cluster of whole cells.*
6. Collect a time series of optical images from the same group of cells without any treatment as a set of control data for dye quenching rate and response to a given number of laser scans of the cell loaded with the fluorescent dye.
7. Find another region of cells and add 500 μL of the same medium containing 2× concentration of treatment materials, for example, 20 μM H_2O_2 or 20 μM dexamethasone (DEX, a potent synthetic member of the glucocorticoid class of steroids), into the dish that contains 500 μL of medium (final concentration for DEX or H_2O_2 is 10 μM). Mix the solution well by gently rocking the dish.
8. Using the same time course as the one set up for the control data, collect a series of optical images immediately after adding the DEX or H_2O_2 solution.
9. Image collection, analysis, and results.

   Representative results from this example are shown in Fig. 18.2 and indicate that the DEX induced free radicals in mainly mitochondria-like structures, while H_2O_2 caused much higher ROS levels in entire cells.

18.3.2 Multi-labeling Immunofluorescence Confocal Microscopy

Many studies on neuronal cell responses to treatment utilize indirect immunofluorescence-labeling confocal microscopy using fixed neuronal cells or tissues, including neuronal cell lines or paraffin/frozen sections of nervous tissues such as brain or spinal cord sections (Svoboda and Yasuda 2006; Selever et al. 2011; Dunaevsky 2012). This section will focus on immune-labeling and confocal image collection procedures using mouse brain sections (Peng et al. 2008; Zhang et al. 2008). In general, frozen section preparation is a relatively simpler and quicker method than the paraffin-embedding/section method and may also yield better immunolabeling results when studying certain antigens that are sensitive to cross-linking fixative, temperature, and a series of dehydration steps. However, paraffin methods can provide thinner sections and better morphology than frozen sections.

1. Section collection and pre-labeling steps.

   Both paraffin and frozen sections should be collected onto poly-L-lysine or other material coated slides for better adhesion of the brain sections so they can stay on the slides after multiple wash steps in the labeling procedure.
   (a) Paraffin sections of brain samples, which should be fixed by perfusion with 4 % paraformaldehyde (see Sect. 18.2, step 2 above) before paraffin embedding, will need to go through the conventional deparaffin steps (two times in xylene and two times in 100 % ethanol, 5 min each time). After the

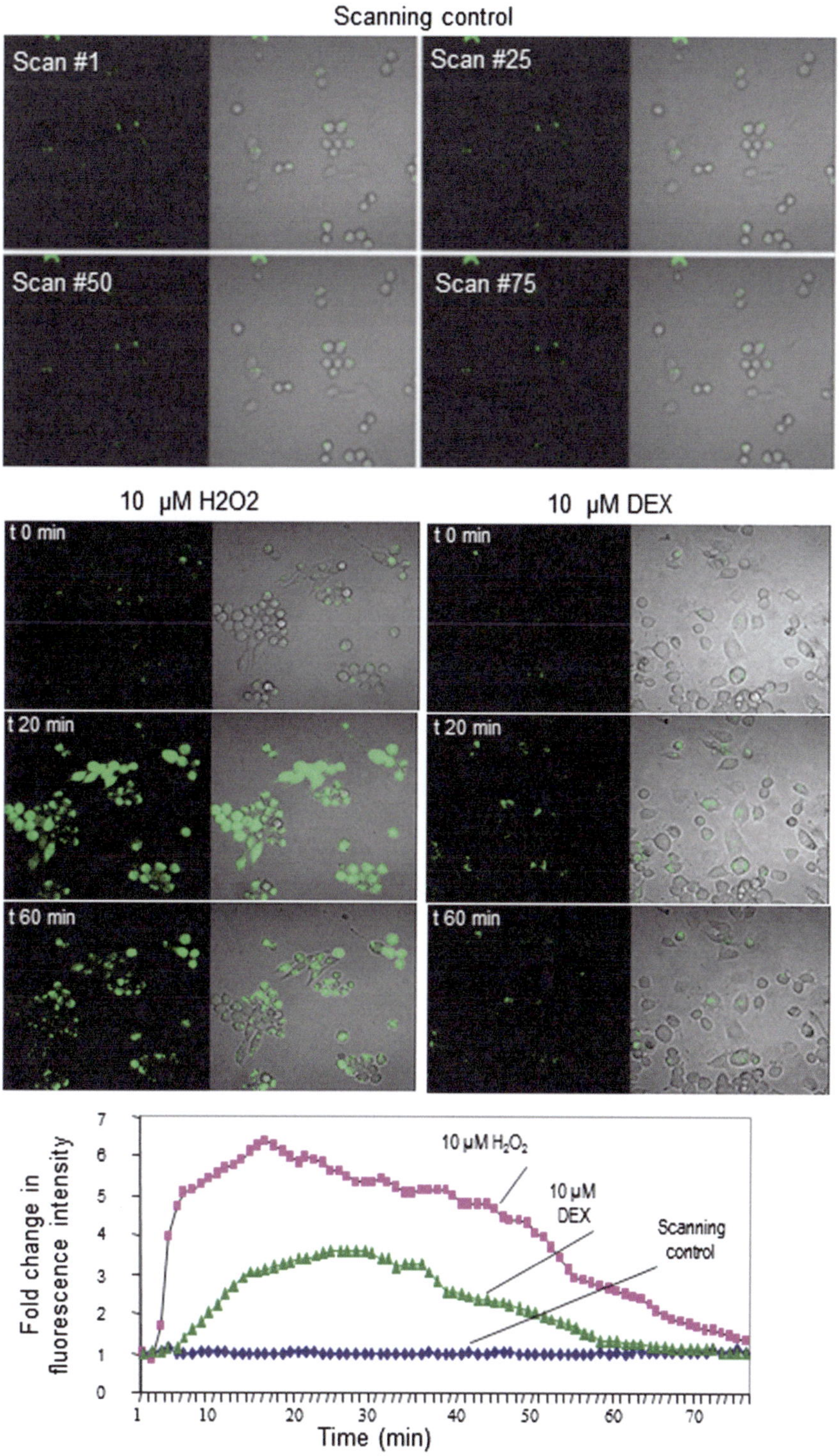

Fig. 18.2 Representative optical images from a scan-control series (the *top set*) and time series of live Neuro-2A cells showing ROS changes as indicated by DCF fluorescence in response to H_2O_2 or DEX treatment. The dynamic changes in fluorescence intensity were presented (the *bottom panel*) as fold changes from the baseline fluorescence (before treatment) averaged from the same field of 20–40 cells. The control series were collected under same confocal settings from a different dish of cells without treatment

ethanol step, allow the slides to air-dry (<5 min) and label the slide using a pencil.

(b) Frozen sections of paraformaldehyde-fixed brain (usually ~10 μm thick but can be up to 30 μm thick if needed) should be collected onto the coated glass slides and placed immediately into a slide rack on dry ice (at this stage, brain slides can be stored in a sealed slide box at −20 °C for months before the next step). Submerge the sections into cold (−20 °C) ethanol or methanol for 5 min, followed by a 5-min air-dry. Label the slide using pencil at this stage.

Note: *6–8 serial sections can be collected onto one slide to save process time and antibodies.*

2. After rehydration in PBS for 5 min (0.05–0.1 % Triton X-100 can be used here for additional permeabilization, if needed), block the brain samples (either from paraffin or frozen section) with PBS containing 0.05 % Tween-20 (PBST) and 3 % BSA for 1 h.

   Note: *After this step, avoid drying the sections throughout the labeling process.*
3. Incubate the brain sections with ~150 μL (for 6–8 brain sections) of PBST plus 1 % BSA and primary antibodies (dilute the antibodies as recommended by the provider/company, usually 1:100–200), and cover the samples with a sheet of 26×65 mm of parafilm, 1 h in a humidified chamber. In this example, two antibodies were used: rabbit anti-MAP2 (microtubule-associated protein-2) and monoclonal anti-mitochondrial glutaminase, an enzyme that localizes to the mitochondrial membrane and catalyzes the conversion of glutamine to glutamate (Shapiro et al. 1991; Tsien et al. 2006; Tian et al. 2012).
4. Wash 3×10 min using PBST in a slide tray while rocking gently.
5. Add secondary antibody solutions diluted in PBST plus 1 % BSA and incubate for 1 h (similar to step 3 above). In this example, a mixture of Alexa Fluor-488-conjugated Donkey anti-rabbit IgG and Alexa Fluor-633-conjugated Donkey anti-mouse IgG was used, 1:500 dilution.
6. Wash the samples in PBST containing SYTOX Blue (for 405 nm laser line), Hoechst, or DAPI for 15 min, followed by 2×5 min washes in PBS.
7. Mount the brain samples in Gel mount or other mounting media. Let the sample mounting media polymerize at room temp (under a foil to block the light) for 40 min to 1 h, before storing them in a slide folder in a refrigerator.

   Note: *For control samples, pre-immune serum can be used under the same conditions or no primary antibodies after blocking step.*
8. Image collection, analysis, and results.

   Optical images in this example were collected with an Olympus FV500-IX81 confocal laser scanning microscopy imaging system, using multiline excitation, sequential image collection, and simultaneous multichannel display procedures. More specifically, three excitation (Ex) laser lines and emission (Em) filters (Ex 405 nm/Em 420 nm, Ex 488 nm/Em 520 nm, and Ex 633 nm/Em 650 nm) were used for a Z-series (0.5 μm z-steps) of sequential image collections for the nucleus structure (DAPI, blue fluorescence), microtubules (Alexa Fluor 488, green), and mitochondria (Alexa Fluor 633, deep red), respectively, using a 60× oil lens.

Figure 18.3 shows two sets of confocal images collected from different Z-step (0.5 μm) of optical sections (Z-section #6 and Z-section #24) and a set of expended focusing images including all 44 Z-step optical images. Each set of six-panel images shows the nucleus (blue), microtubules (green), and mitochondrial glutaminase expressed within mitochondria (red) obtained from the same focal plane. More detailed information on nerve cells and their spatial relationship can be obtained when all Z-optical data are projected into one plane of image (extended focus image). The results from this example indicate that neurons have a higher level/amount of expression of mitochondrial glutaminase than other types of nerve cells in the mouse brain. This series of optical images can be further used for 3D image reconstruction and analysis (Paddock 1999a; Murray et al. 2007).

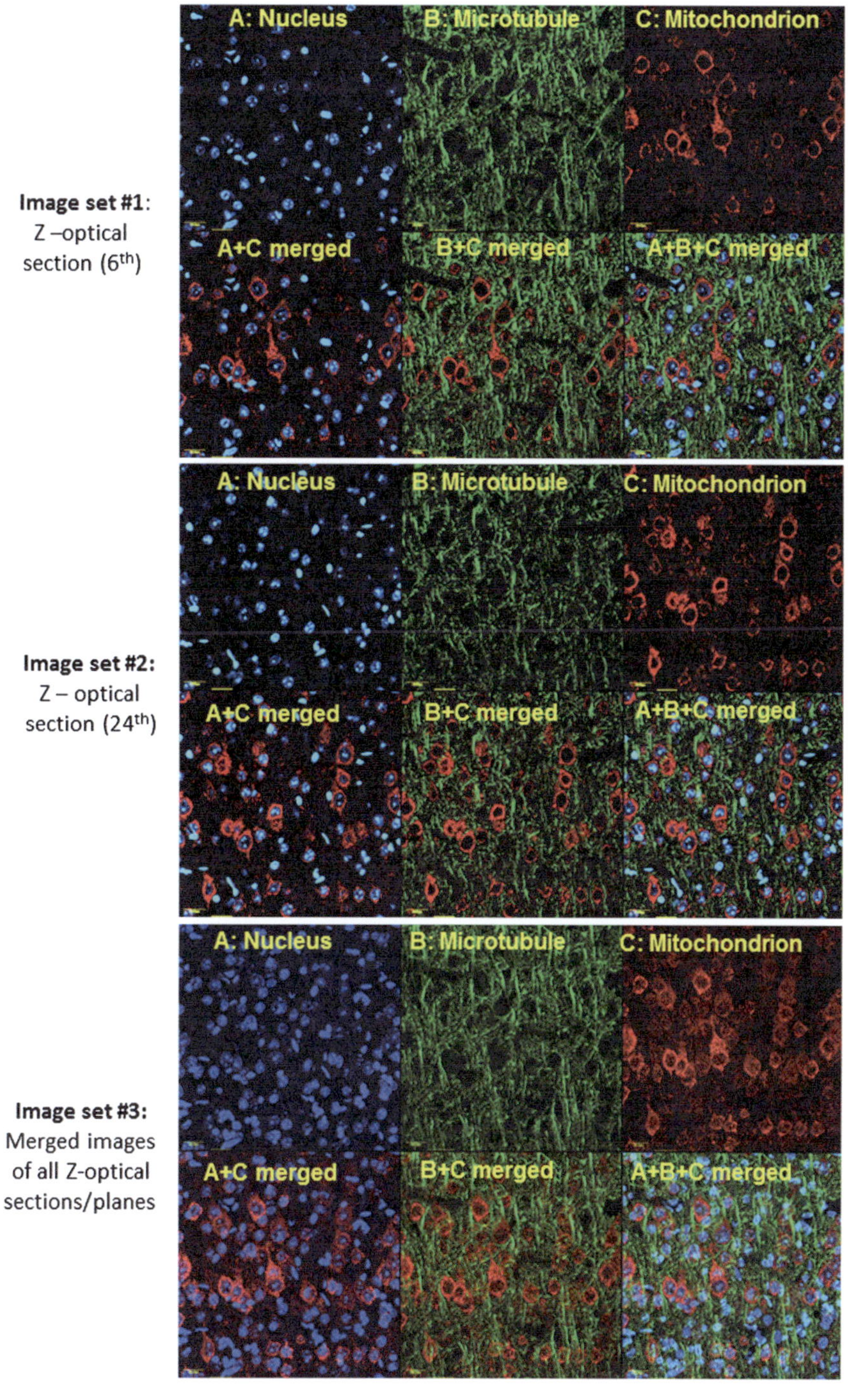

Fig. 18.3 Confocal images of a triple-labeled brain sample demonstrating the difference in detected contents between single focal planes and extended focus images (merged all 44 Z-optical images). The cell components (the DAPI-stained nuclei and neuronal marker labeled microtubules and mitochondria) shown in single color panel (*blue, green*, and *red color*) or color overlaid panels are indicated by the text at the top of each of the images, respectively

References

Beisker W, Dolbeare F, Gray JW (1987) An improved immunocytochemical procedure for high-sensitivity detection of incorporated bromodeoxyuridine. Cytometry 8:235–239

Brehmer A, Blaser B, Seitz G, Schrodl F, Neuhuber W (2004) Pattern of lipofuscin pigmentation in nitrergic and non-nitrergic, neurofilament immunoreactive myenteric neuron types of human small intestine. Histochem Cell Biol 121:13–20

Carter D (1999) Practical considerations for collecting confocal images. Methods Mol Biol 122:35–57

Chen KC, Zhou Y, Zhang W, Lou MF (2007) Control of PDGF-induced reactive oxygen species (ROS) generation and signal transduction in human lens epithelial cells. Mol Vis 13:374–387

Chidlow G, Daymon M, Wood JP, Casson RJ (2011) Localization of a wide-ranging panel of antigens in the rat retina by immunohistochemistry: comparison of Davidson's solution and formalin as fixatives. J Histochem Cytochem 59:884–898

Double KL, Dedov VN, Fedorow H, Kettle E, Halliday GM, Garner B, Brunk UT (2008) The comparative biology of neuromelanin and lipofuscin in the human brain. Cell Mol Life Sci 65:1669–1682

Dunaevsky A (2012) Neuron-glial interactions in the developing cerebellum. Microsc Microanal 18:742–744

Eichhoff G, Busche MA, Garaschuk O (2008) In vivo calcium imaging of the aging and diseased brain. Eur J Nucl Med Mol Imaging 35(Suppl 1):S99–S106

Elleder M, Borovansky J (2001) Autofluorescence of melanins induced by ultraviolet radiation and near ultraviolet light. A histochemical and biochemical study. Histochem J 33:273–281

Garini Y, Young IT, McNamara G (2006) Spectral imaging: principles and applications. Cytometry A 69:735–747

Giepmans BN, Adams SR, Ellisman MH, Tsien RY (2006) The fluorescent toolbox for assessing protein location and function. Science 312:217–224

Herberich G, Windoffer R, Leube RE, Aach T (2012) Signal and noise modeling in confocal laser scanning fluorescence microscopy. Med Image Comput Comput Assist Interv 15:381–388

Koistinaho J, Sorvaniemi M, Alho H, Hervonen A (1986) Microspectrofluorometric quantitation of autofluorescent lipopigment in the human sympathetic ganglia. Mech Ageing Dev 37:79–89

Kremers GJ, Gilbert SG, Cranfill PJ, Davidson MW, Piston DW (2011) Fluorescent proteins at a glance. J Cell Sci 124:157–160

Lam AJ, St-Pierre F, Gong Y, Marshall JD, Cranfill PJ, Baird MA, McKeown MR, Wiedenmann J, Davidson MW, Schnitzer MJ, Tsien RY, Lin MZ (2012) Improving FRET dynamic range with bright green and red fluorescent proteins. Nat Methods 9:1005–1012

Lippincott-Schwartz J, Patterson GH (2003) Development and use of fluorescent protein markers in living cells. Science 300:87–91

Malinouski M, Zhou Y, Belousov VV, Hatfield DL, Gladyshev VN (2011) Hydrogen peroxide probes directed to different cellular compartments. PLoS One 6:e14564

Matsuda Y, Fujii T, Suzuki T, Yamahatsu K, Kawahara K, Teduka K, Kawamoto Y, Yamamoto T, Ishiwata T, Naito Z (2011) Comparison of fixation methods for preservation of morphology, RNAs, and proteins from paraffin-embedded human cancer cell-implanted mouse models. J Histochem Cytochem 59:68–75

Murray JM, Appleton PL, Swedlow JR, Waters JC (2007) Evaluating performance in three-dimensional fluorescence microscopy. J Microsc 228:390–405

Paddock SW (1999a) An introduction to confocal imaging. Methods Mol Biol 122:1–34

Paddock SW (1999b) Confocal laser scanning microscopy. Biotechniques 27:992–996, 998–1002, 1004

Pawley JB (2002) Limitations on optical sectioning in live-cell confocal microscopy. Scanning 24:241–246

Peng H, Whitney N, Wu Y, Tian C, Dou H, Zhou Y, Zheng J (2008) HIV-1-infected and/or immune-activated macrophage-secreted TNF-alpha affects human fetal cortical neural progenitor cell proliferation and differentiation. Glia 56:903–916

Rizzo MA, Davidson MW, Piston DW (2009) Fluorescent protein tracking and detection: applications using fluorescent proteins in living cells. Cold Spring Harb Protoc 2009:pdb top64

Schnell SA, Staines WA, Wessendorf MW (1999) Reduction of lipofuscin-like autofluorescence in fluorescently labeled tissue. J Histochem Cytochem 47:719–730

Selever J, Kong JQ, Arenkiel BR (2011) A rapid approach to high-resolution fluorescence imaging in semi-thick brain slices. J Vis Exp Jul 26(53):2807

Shapiro RA, Farrell L, Srinivasan M, Curthoys NP (1991) Isolation, characterization, and in vitro expression of a cDNA that encodes the kidney isoenzyme of the mitochondrial glutaminase. J Biol Chem 266:18792–18796

Svoboda K, Yasuda R (2006) Principles of two-photon excitation microscopy and its applications to neuroscience. Neuron 50:823–839

Swedlow JR (2012) Innovation in biological microscopy: current status and future directions. Bioessays 34:333–340

Tian C, Sun L, Jia B, Ma K, Curthoys N, Ding J, Zheng J (2012) Mitochondrial glutaminase release contributes to glutamate-mediated neurotoxicity during human immunodeficiency virus-1 infection. J Neuroimmune Pharmacol 7:619–628

Tsien RY, Ernst L, Waggoner A (2006) Chapter 16: Fluorophores for confocal microscopy: photophysics and photochemistry, 3rd edn. Springer, New York

Viegas MS, Martins TC, Seco F, do Carmo A (2007) An improved and cost-effective methodology for the reduction of autofluorescence in direct immunofluorescence studies on formalin-fixed paraffin-embedded tissues. Eur J Histochem 51:59–66

Waters JC (2007) Live-cell fluorescence imaging. Methods Cell Biol 81:115–140

Yao PJ, O'Herron TM, Coleman PD (2003) Immunohistochemical characterization of clathrin assembly protein AP180 and synaptophysin in human brain. Neurobiol Aging 24:173–178

Zecca L, Bellei C, Costi P, Albertini A, Monzani E, Casella L, Gallorini M, Bergamaschi L, Moscatelli A, Turro NJ, Eisner M, Crippa PR, Ito S, Wakamatsu K, Bush WD, Ward WC, Simon JD, Zucca FA (2008) New melanic pigments in the human brain that accumulate in aging and block environmental toxic metals. Proc Natl Acad Sci U S A 105:17567–17572

Zhang Y, Zhou Y, Schweizer U, Savaskan NE, Hua D, Kipnis J, Hatfield DL, Gladyshev VN (2008) Comparative analysis of selenocysteine machinery and selenoproteome gene expression in mouse brain identifies neurons as key functional sites of selenium in mammals. J Biol Chem 283:2427–2438

Zhou Y, Marcus EM, Haugland RP, Opas M (1995) Use of a new fluorescent probe, seminaphthofluorescein-calcein, for determination of intracellular pH by simultaneous dual-emission imaging laser scanning confocal microscopy. J Cell Physiol 164:9–16

Magnetic Resonance Imaging 19

Yutong Liu

Abstract

Magnetic resonance imaging (MRI) is an essential tool in neuroscience investigations employing laboratory animals. In this chapter, the basic physics of nuclear magnetic resonance (NMR) signals are briefly described, followed by the principle of MRI formation and pulse sequences. MRI instrumentation is discussed including the necessary accessories for animal imaging. The focus of this chapter is the introduction of a variety of MRI techniques and their applications in neuroscience. The physiologic basis of MRI contrasts based on proton density and relaxation times (T_1, T_2, and T_2^*) in tissue is first described, followed by their applications in neuroscience study. Advanced techniques, including diffusion and perfusion MRI, are then introduced. In diffusion MRI, diffusion-weighted imaging (DWI) is first described followed by diffusion tensor imaging (DTI). The physics, physiologic basis, and applications in neuroscience of these techniques are addressed respectively. There are two techniques in perfusion MRI: bolus tracking and arterial spin labeling. Both are introduced with a focus on bolus tracking technique. Functional MRI (fMRI) and its applications are presented. The stimulations used in fMRI are described, and resting state fMRI is also discussed. The possible pitfalls in fMRI applications are also addressed. The applications of exogenous contrast agents in MRI are also discussed, especially manganese-enhanced MRI (MEMRI) and in vivo cell tracking using superparamagnetic iron oxide (SPIO) particles.

Keywords

MRI • Relaxation time • Pulse sequence • Diffusion-weighted imaging (DWI) • Diffusion tensor imaging (DTI) • Perfusion MRI • Functional MRI (fMRI) • Manganese-enhanced MRI (MEMRI) • Superparamagnetic iron oxide (SPIO)

Y. Liu, Ph.D. (✉)
Departments of Radiology and Pharmacology and Experimental Neuroscience, University of Nebraska Medical Center, 981045 Nebraska Medical Center, Omaha, NE, 68198-1045, USA
e-mail: yutongliu@unmc.edu

H. Xiong and H.E. Gendelman (eds.), *Current Laboratory Methods in Neuroscience Research*, Springer Protocols Handbooks, DOI 10.1007/978-1-4614-8794-4_19,

19.1 Introduction

Magnetic resonance imaging (MRI) is an essential arm of clinical diagnostic imaging. Future innovations in diagnostics and therapeutic developments rely on it as a tool in neuroscience investigations employing laboratory animals (Dijkhuizen and Nicolay 2003; Lythgoe et al. 2003; Van Bruggen and Roberts 2003; Dijkhuizen 2006). MRI has been used in the studies of a variety of neurological disease models of stroke (Liu et al. 2005, 2007a, b, c; Gao et al. 2006; Justicia et al. 2008; Wang et al. 2008; Uppal et al. 2010; Baskerville et al. 2011; Guo et al. 2011; Nagaraja et al. 2011), Parkinson's disease (Podell et al. 2003; Kondoh et al. 2005; Pelled et al. 2005, 2007; Boska et al. 2007; Strome and Doudet 2007), Alzheimer's disease (Borthakur et al. 2006; Braakman et al. 2006; Dhenain et al. 2006; Benveniste et al. 2007; Faber et al. 2007; Falangola et al. 2007; Jack et al. 2007; Poduslo et al. 2007; Smith et al. 2007; Strome and Doudet 2007; Alexander et al. 2008; Lau et al. 2008; Luo et al. 2008; Ramakrishnan et al. 2008; Wengenack et al. 2008; Winkeler et al. 2008), and NeuroAIDS (Boska et al. 2004; Nelson et al. 2005; Liu et al. 2008; Dash et al. 2011), among others. Specially designed MRI systems for laboratory animals are now commercially available. Software tools for MRI acquisition and post-processing are widely accessible. Private and public funding has supported the development of MRI-based brain atlases and databases.

MRI can provide noninvasive three-dimensional information of organ systems with high anatomical resolution. It offers a variety of controls over image contrast. In this chapter, the basic physics of nuclear magnetic resonance (NMR) signals are described, followed by the principle of MRI formation and pulse sequences. MRI contrasts and advanced techniques including diffusion and perfusion MRI will be described. Functional MRI (fMRI), manganese-enhanced MRI (MEMRI), cell tracking MRI will also be discussed.

19.2 Basic Principle of MRI

MRI signals are generated by the hydrogen nuclei (protons) of water molecules within an animal or a tissue sample while the subject is placed inside a strong (main) magnetic field. The protons, acting as small magnets, align along the direction of this field. Transmitting a radiofrequency (RF) field tips the protons away from their initial directions. When the RF is turned off, the protons precess about the main magnetic field, and the precessing protons produce an RF signal. Spatial information is encoded into the RF signal using gradient magnetic fields. These impose a linear variation in all three dimensions in the main magnetic field. As a result of the variation, the precessional frequencies and phases of the protons are affected by their spatial locations. The RF signal is detected using an RF coil (probe) and digitized. A mathematical operation, called inverse Fourier transformation, is then performed to convert the signal into an image.

19.2.1 Spins and Net Magnetization Vector

The percentage of water within the body changes with age and body composition. Water typically comprises about 70 % of the human body and 85 % of the brain. Water is also the most abundant compound within most laboratory animals such as rats, mice, and rabbits. The hydrogen nucleus (or proton) in the water molecule possesses a quantum mechanical property termed "spin." The spinning proton or spin (spin and proton are synonymous herein) produces a magnetic dipole similar to a tiny bar magnet (Fig. 19.1a, b) with a tiny magnetic moment (shown using a vector in Fig. 19.1a). In a strong external magnetic field (we call this field B_0 thereafter, in the unit of Tesla), the spins are distributed into two energy levels: alignment with (parallel to) B_0 at a low energy level and alignment against (antiparallel to) B_0 at a slightly higher energy level. A slight majority of spins exist in the low energy state

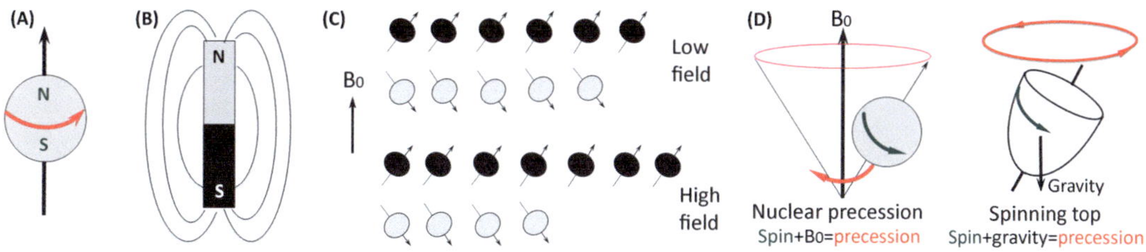

Fig. 19.1 The hydrogen nucleus (or proton) in the water molecule possesses a quantum mechanical property termed "spin." The spinning proton or spin (**a**) produces a magnetic dipole like a tiny bar magnet (**b**) with a tiny magnetic moment (shown using a vector in (**a**)). (**c**) In a strong external magnetic field B_0, the spins are distributed into two energy levels: alignment with the (parallel to) B_0 at a low energy level (colored in *black*) and alignment against (antiparallel to) B_0 at a slightly higher energy level (colored in *gray*). A slight majority of spins exist in the low energy state. The population difference between the two energy levels is determined by the external magnetic field strength. Spins experience a torque from the external magnetic field B_0 that causes precession. The spin precession can be analogized to a spinning *top* wobbling due to the force of gravity (**d**)

Table 19.1 Larmor frequencies in typical clinical and laboratory magnetic fields

B_0 (Tesla)	ω_0 (MHz)
1.0	42.58
1.5	63.86
3.0	127.74
4.7	200.13
7.0	298.06
9.4	400.25
14.0	596.12

(Fig. 19.1c). The population difference between the two energy levels is determined by the external magnetic field strength (Fig. 19.1c). The magnitude of the MRI signal is proportional to the difference in populations. Therefore, higher signal-to-noise ratio can be achieved using a stronger main magnet. The protons also experience a torque from the external magnetic field that causes precession. The proton precession can be analogized to a spinning top wobbling due to the force of gravity (Fig. 19.1d). The precessional frequency is called the Larmor frequency and depends on the strength of the external magnetic field B_0. Table 19.1 lists the proton Larmor frequencies in several typical magnetic fields used in clinical medicine and in laboratory.

The large number of individual protons precessing in the parallel and antiparallel directions can be represented by two cones (Fig. 19.2a). By convention, the external magnetic field B_0 is directed parallel to the z-axis of a three-dimensional Cartesian coordinate system. The x- and y-axes are perpendicular to the z direction (Fig. 19.2a). The proton's magnetic moment precesses about the z-axis at the Larmor frequency. Each magnetic moment is represented as a vector with magnitude and direction (Fig. 19.2a). The net magnetic moment (named as the net magnetization vector, M) is equal to the vector sum of all the individual magnetic moments (Fig. 19.2b). The magnetization vector is said to be at the equilibrium state pointing to the $+z$ direction with the magnitude of M_0. Apparently at the equilibrium state, the z component (also known as the longitudinal magnetization, M_z) is maximal and is denoted as M_0 — the equilibrium magnetization (i.e., $M_z = M_0$, and $M_x = M_y = 0$). The vector sum of the components in x- and y-axis is in the xy plane and is known as the transverse magnetization (i.e., $M_{xy} = M_x + M_y$). If the magnetization vector is away from the equilibrium state such as in Fig. 19.2c, then $M_z < M_0$ and $M_{xy} > 0$.

19.2.2 Radiofrequency (RF) Pulses

At equilibrium, the transverse magnetization is zero ($M_{xy} = 0$). The MRI signal is produced by the xy component of the net magnetization vector. In order to obtain MRI signal, the net magnetization

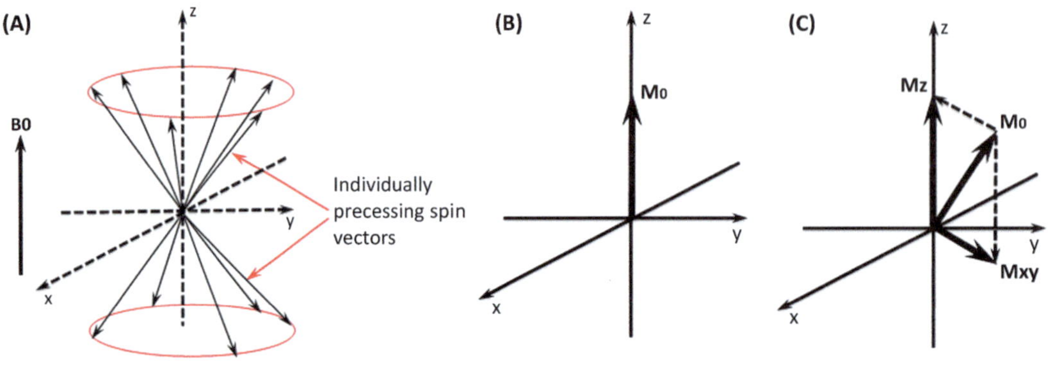

Fig. 19.2 (**a**) The large number of individual protons precessing in the parallel and antiparallel directions can be represented by two *cones*. (**b**) The net magnetization vector M_0 equals to the vector sum of all the individual magnetic moments. At the equilibrium state, the z component $M_z = M_0$, and $M_x = M_y = 0$. The components in x- and y-axis can be combined together as the component in xy plane, i.e., $M_{xy} = M_x + M_y$. (**c**) If the magnetization vector is away from the equilibrium state, then $M_z < M_0$ and $M_{xy} > 0$

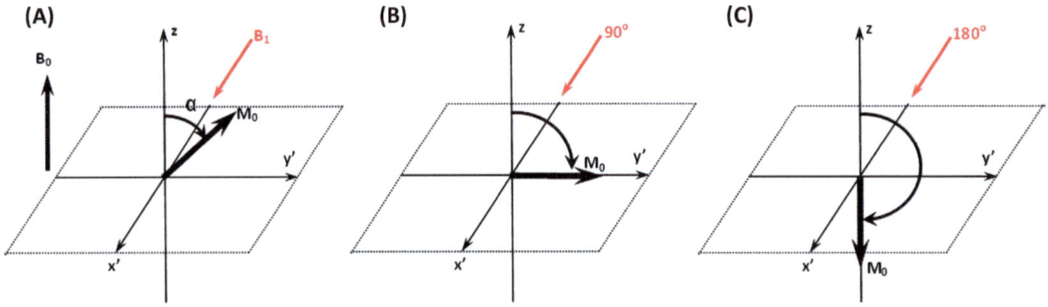

Fig. 19.3 (**a**) An RF pulse B_1 tips the magnetization vector M_0 away from the equilibrium state and rotates it about the direction of B_1. The flip angle α is defined as the angle through which the magnetization vector is rotated by B_1. (**b**) A 90° pulse results in a flip angle of 90° and the maximum value of the M_{xy} component and thereby the maximum MRI signal. (**c**) A 180° pulse converts M_z into $-M_z$, producing no transverse magnetization and thereby no signal

vector has to be tipped away from the equilibrium state and into the transverse plane as illustrated in Fig. 19.3 by applying certain energy to the system. The energy is supplied by an oscillating electromagnetic field termed as B_1. B_1 oscillates at the RF regime in the electromagnetic spectrum, thereby B_1 is also known as RF field. Because the B_1 field is applied only for a period of duration, it is usually referred to as an RF pulse. When the frequency of B_1 matches to the precessional frequency of the protons, i.e., the Larmor frequency, a condition of *resonance* is produced. The net magnetization M_z is excited under the resonance condition and rotates around the axis of the applied B_1 field. The flip angle α is defined as the angle through which M_z is rotated by B_1 (Fig. 19.3a), and depends on the power and duration of the RF pulse. A flip angle of 90° results in the maximum value of the M_{xy} component and thereby the maximum MRI signal (Fig. 19.3b). The B_1 field that produces a flip angle of 90° is termed as a 90° pulse. A 180° pulse convert M_z into $-M_z$, producing no transverse magnetization, and thereby no signal (Fig. 19.3c).

19.2.3 T_1 and T_2 Relaxation

A 90° RF pulse rotates the longitudinal magnetization M_z from its equilibrium state M_0 to the xy plane resulting in $M_z = 0$, and maximum possible transverse magnetization M_{xy} for a given sample

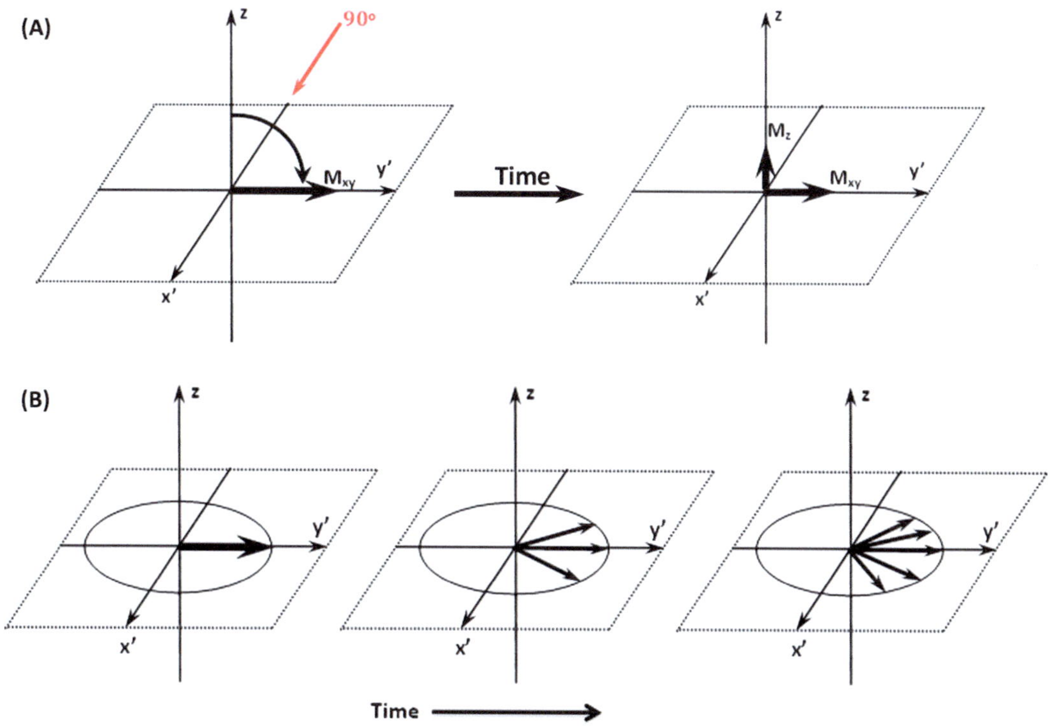

Fig. 19.4 (a) 90° RF pulse rotates the longitudinal magnetization M_z from its equilibrium state M_0 to the *xy* plane resulting in $M_z=0$, and maximum possible transverse magnetization M_{xy} for a given sample volume. After the RF is turned off, M_z and M_{xy} return to their equilibrium values over time, i.e., the spins realign with B_0 and the M_z vector begins to grow in magnitude to M_0 while the magnitude of M_{xy} decreases. (b) After the 90° pulse is off, M_{xy} rotates about B_0 at the Larmor frequency. Due to the spin–spin interaction, each spin experiences a slight different local magnetic field strength, whereby the individual spins precess at different frequencies. Some spins travel faster and some slower, resulting in a loss of phase coherence. This process is called dephasing. Because M_{xy} is the vector of all individual magnetic moments, the dephasing results in the decay of M_{xy}

volume. After the RF is turned off, M_z and M_{xy} must return to their equilibrium values over time, i.e., the spins realign with B_0 and the M_z vector begins to grow in magnitude to M_0 while the magnitude of M_{xy} decreases (Fig. 19.4a). The T_1 relaxation time is the time needed to recover 63 % of the longitudinal magnetization. T_1 relaxation results from the transfer of energy from the excited spins to the surrounding lattice. This energy transfer between spins and lattice is called spin–lattice interaction; thus, T_1 relaxation is also termed as spin–lattice relaxation. The inverse of relaxation time is called the spin–lattice or longitudinal relaxation ratio ($R_1=1/T_1$). Different tissues may have different T_1 values, and this difference is the basis of T_1-weighted (T_1-wt) contrast on MR images.

After the 90° pulse is off, M_{xy} rotates about B_0 at the Larmor frequency. Due to the interaction among individual spins (that is called spin–spin interaction), each spin experiences a slightly different local magnetic field strength, whereby the individual spins precess at different frequencies. Some spins travel faster and some slower, resulting in a loss of phase coherence. This process is called dephasing. Because M_{xy} is the vector of all individual magnetic moments, the dephasing results in the decay of M_{xy} (Fig. 19.4b). The M_{xy} decay is termed as T_2 relaxation. The elapsed time between the maximum M_{xy} and 37 % of the maximum level is the T_2 relaxation time. T_2 relaxation is also called spin–spin relaxation. The inverse of T_2 relaxation time is called spin–spin or transverse relaxation ratio ($R_2=1/T_2$). As in the case of T_1

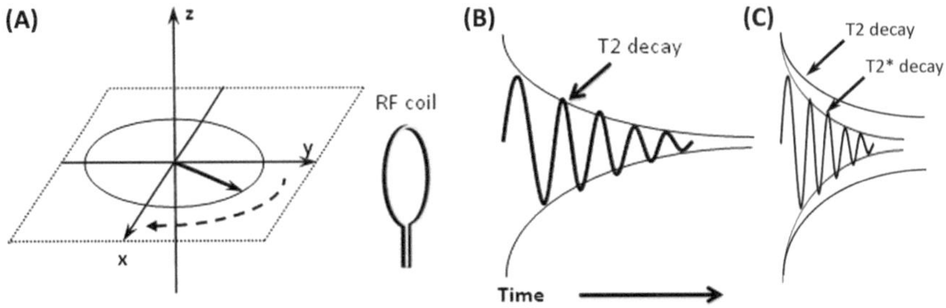

Fig. 19.5 (**a**) M_{xy} rotates about B_0 at the Larmor frequency. The alternating magnetization produces (by magnetic induction) a damped sinusoidal electrical signal known as the free induction decay (FID) signal in an RF receiver coil. (**b**) The FID signal oscillates at the Larmor frequency and decays with time due to the T_2 relaxation. T_2 is the decay time resulting from intrinsic magnetic properties of the sample. T_2^* is the decay time that results from both intrinsic and extrinsic magnetic inhomogeneities. T_2^* is usually much shorter than T_2. (**c**) Because of the T_2^* effect, the FID decays faster than when only T_2 relaxation is considered

relaxation, different tissues may have different T_2 values, and the difference can be used to generate T_2-weighted (T_2-wt) signal in MRI. In both humans and laboratory animals, T_2 is usually much shorter than T_1.

As described in the previous paragraph, after the 90° pulse is turned off, the M_{xy} rotates about B_0 at the Larmor frequency. The alternating magnetization produces (by magnetic induction) a damped sinusoidal electrical signal known as the free induction decay (FID) signal in an RF receiver coil. The FID signal oscillates at the Larmor frequency and decays with time due to the T_2 relaxation (Fig. 19.5a, b).

In reality, the loss in phase coherence occurs more rapidly than from the spin–spin interactions by themselves due to the presence of extrinsic magnetic inhomogeneities such as imperfect main magnetic field or susceptibility contrast agents that are present in tissues. The overall relaxation time is designated as T_2^* when both intrinsic spin–spin interactions and the extrinsic magnetic inhomogeneities are considered. T_2^* is usually much shorter than T_2. Therefore, because of the T_2^* effect, the FID decays faster than when only T_2 relaxation is considered as shown in Fig. 19.5c.

The contents discussed in this section can be found in many books (Cho et al. 1993; Haacke 1999; Bushberg 2002; Webb 2003; Berry and Bulpitt 2009) and online resources (Hornak).

19.3 Magnetic Resonance Imaging

The MR signal described so far is from the whole sample volume and simply the sum of individual signals of each spin in the volume. To generate an MR image, spatial information has to be acquired to distinguish the positions of the signal in a three-dimensional object (the sample). The spatial information is encoded in the MR signal using magnetic field gradients. Magnetic field gradients are superimposed upon the homogeneous main magnetic field B_0 and change the main magnetic field in a controlled and predictable pattern so the field is no longer homogeneous. In the coordinate system used in all MRI scanners, the main magnetic field B_0 direction is designated as the *z* direction. In horizontal MRI systems (i.e., B_0 is oriented in the horizontal direction), by convention, the *y*-axis corresponds to the vertical direction and the *x*-axis from right to left. Three independent gradients in *x*, *y*, and *z* directions are used to provide three-dimensional spatial information for MRI. It needs to be noted that the magnetic field directions of the three gradients are all in *z* direction (B_0 direction), and it is the field strength *change* that is in either *x*, *y*, or *z* direction.

As shown in Fig. 19.6, when a magnetic field gradient is turned on, it adds or subtracts from the main field about a point in space called the

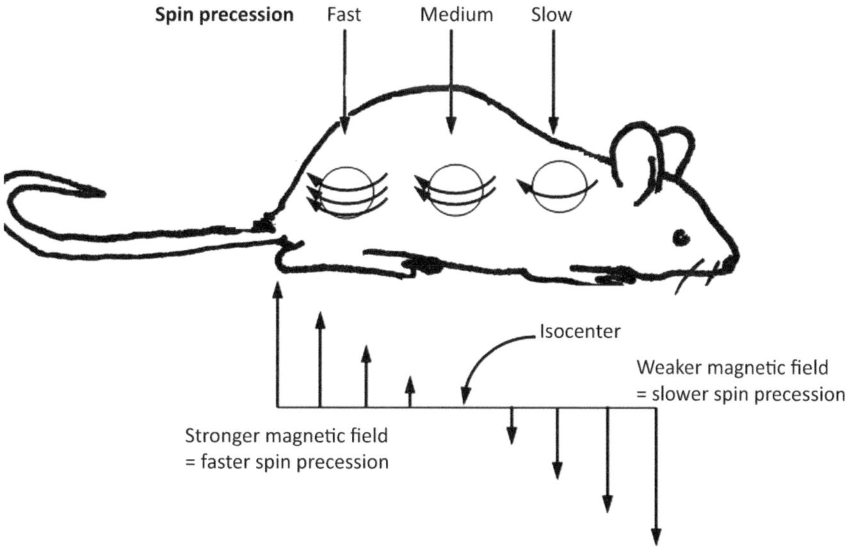

Fig. 19.6 When a magnetic field gradient is turned on, it adds or subtracts from the main field about a point in space called the isocenter ($x=0$, $y=0$, $z=0$), where no change in the main magnetic field exists, thus the magnetic field is simply B_0. The magnetic field linearly increases (or decreases) away from the isocenter. As the magnetic field increases or decreases with position, the precessional frequency increases or decreases with position as well

isocenter ($x=0$, $y=0$, $z=0$), where no change in the main magnetic field exists, thus the magnetic field is simply B_0. The magnetic field linearly increases (or decreases) away from the isocenter. As the magnetic field increases or decreases with position, the precessional frequency increases or decreases with position as well. In a gradient field, position is directly and linearly proportional to frequency (Fig. 19.6). Similar expressions of the spatial dependence of the precessional frequencies can be obtained for the x and y gradients.

The process of image formation can be broken down into three components: slice selection, frequency encoding, and phase encoding. Each component is accomplished by a single gradient or a combination of the x, y, and z gradients.

19.3.1 Spatial Encoding

Most MRI studies acquire a series of slices through the subject to be imaged. To acquire the MR signal from a slice, a slice selection gradient (G_{ss}) is applied resulting in protons precessing at different frequencies along the slice selection direction.

Slice selection gradient can be applied along any axis of the body. The slice selection direction in conjunction with the animal's position in the magnet dictates the orientation—coronal, axial, or sagittal—of the image. For an animal in the "head first prone" position, the three orientations correspond to slice selection in the z, the y, or the x direction, respectively.

When the slice of tissue to be imaged has been defined using the slice selection gradient, another gradient is applied to further localize the MRI signal to columns within the slice. The gradient encodes the columns using different frequencies; thus, it is called frequency-encoding gradient (G_{freq}), also known as the readout gradient. The frequency-encoding gradient is applied in a direction perpendicular to the slice selection gradient.

Position of the protons in the third dimension is determined by assigning the protons different precessing phases using a phase-encoding gradient (G_{phase}). The phase-encoding gradient is applied along the third perpendicular axis.

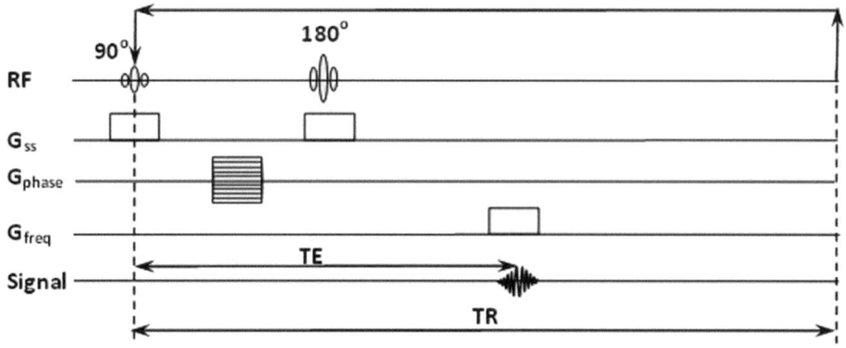

Fig. 19.7 A simplified yet complete pulse sequence. In the *top line*, the 90° pulse is followed by the 180° pulse with the time interval of TE/2. The sequence is repeated a number of times (depending on the number of phase-encoding steps) at an interval of TR (time of repetition). The next line shows the G_{ss}. It is applied at the time of the 90° and 180° RF pulses, so that only the spins in that selected slice are excited and contribute to the signal. The G_{phase} is applied once per repetition, with different strength on each occasion. The sequence diagram indicates the multiple gradients by showing several of the gradients at once. The frequency-encoding gradient (G_{freq}) that is timed out coincides with the time of the echo. The *last line* shows when the signal is detected at the time of TE after the 90° pulse

The phase-encoding gradient has to be applied a number of times with a different gradient strength each time to generate an image.

The result of a single measurement made with both phase and frequency encodings is that each spatial position along the phase-encoding direction is represented by a unique phase and each position along the frequency-encoding direction is represented by a unique frequency. These unique combinations of frequency and phase encode the coordinates of each pixel.

19.3.2 Pulse Sequences

An MR image is generated by running an imaging pulse sequence using an MRI system. A pulse sequence is the timing, order, polarity, and repetition of the RF pulses and spatial encoding gradients. It is represented by individual lines correlated in time showing RF pulses, magnetic gradients, and MR signal acquisition as well.

Figure 19.7 shows the simplified yet complete drawing of a pulse sequence. It has five lines: one for the RF pulses, one for each of the three gradients, and one for the measured signal. In the top line, the 90° pulse is followed by the 180° pulse. The sequence is repeated a number of times (depending on the number of phase-encoding steps) at an interval of TR (time of repetition). TR is the time between two 90 excitation pulses. The next line shows the G_{ss} using a rectangle, and its width is the duration of the gradient, and its height is the strength. It is applied at the time of the 90° and 180° RF pulses. The G_{phase} is applied once per repetition, with different strength on each occasion. The sequence diagram indicates the multiple gradients by showing several of the gradients at once. The last line shows the signal acquisition. Usually instead of acquiring the FID signal, the dephased spins as shown in Fig. 19.5b are refocused using certain methods, and the signal acquired results from the refocused spins. The signal is called an echo. The time from the 90° pulse to the center of the echo is called time to echo or TE. TR and TE are two important parameters in this sequence, and by adjusting them, different physiology or pathology can be emphasized as described in details in the following paragraphs. The frequency-encoding gradient (G_{freq}) that is timed out coincides with the time of the echo as shown in Fig. 19.7.

Detailed explanation of the spatial encoding in MRI and pulse sequences can be found in several books (Cho et al. 1993; Haacke 1999; Bushberg 2002; Webb 2003; Berry and Bulpitt 2009) and online resources (Hornak).

19.3.3 Instrumentation

An MRI system includes the following major components: the main magnet, the gradient system, the RF coil, the control circuits, and the computer system as shown in the schematic representation in Fig. 19.8a. The main magnet produces a strong and static field (B_0). Most magnets are of the superconducting type. A superconducting magnet is an electromagnet made of superconducting wire. The wire is immersed in low-temperature liquid helium to generate the superconducting condition. A current is provided to flow in the superconducting wire to generate the main magnetic field. Inside the main magnet are the three gradient coils for MRI spatial encoding. Within the gradient coils is the RF coil. The RF coil produces the B_1 magnetic field necessary to rotate the spins. The RF coil also detects the signal from the spins within the body. The gradient coils and the RF coil are controlled by a computer through the control circuits. The MRI signal detected by the RF coil is transferred through the digitizer to the computer and processed to generate an image. For animal imaging, some other accessories are necessary such as an anesthesia system and a vital monitoring system. It has to be noted that these systems need to be MRI compatible, i.e., they are MR safe and will not significantly affect the quality of MRI, nor will their operation be significantly affected being in the MR environment. Figure 19.8b is the picture of a typical small animal MRI scanner. Figure 19.8c shows a mouse that is ready to be put into the scanner. The mouse is on a holder, and its head is under an RF coil. A tube is attached to provide inhalational anesthetic to the mouse.

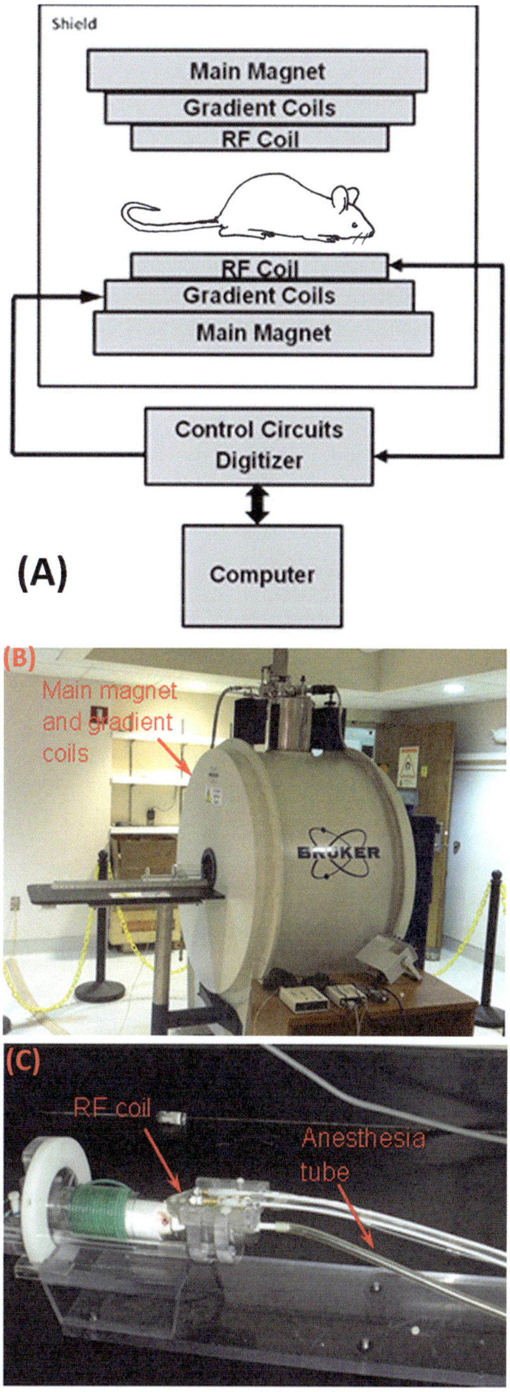

Fig. 19.8 (**a**) Schematic of the MRI system. (**b**) A typical small animal MRI scanner. (**c**) A mouse to be put into the scanner. The mouse is on a holder, and its head is under an RF coil. A tube is attached to provide inhalational anesthetic to the mouse

19.4 MRI Applications for Experimental Neuroscience

19.4.1 Proton Density and Relaxation Times

1. Physiological Basis.

   As it has been described earlier, MRI signals are generated by the hydrogen nuclei

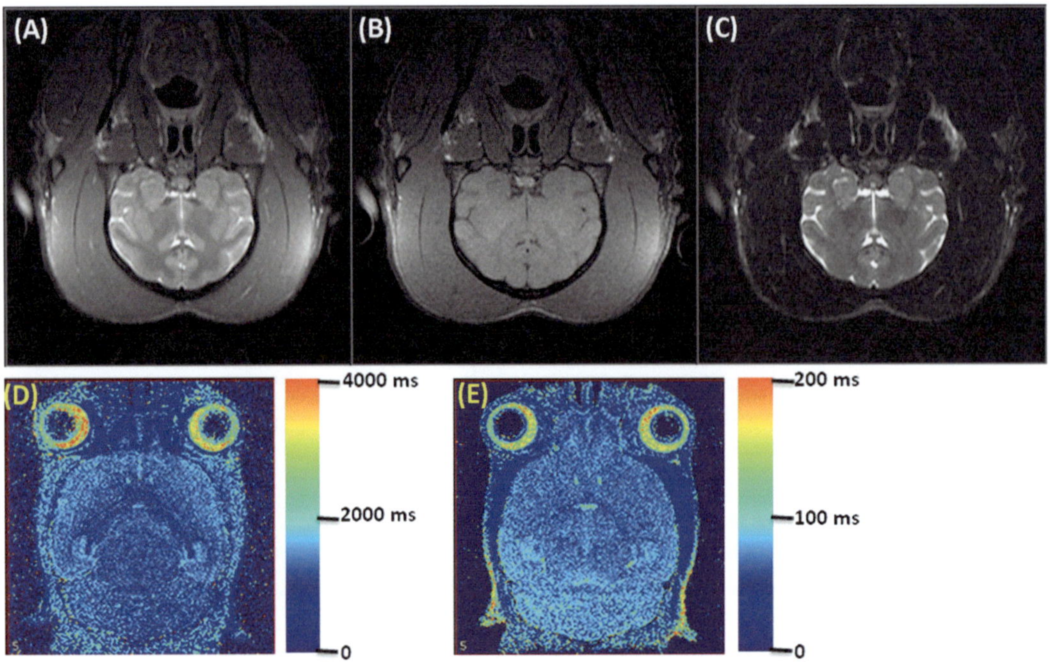

Fig. 19.9 (**a**) PD-wt image of a monkey brain. CSF is *bright* on the PD-wt image because it contains more water molecules compared to other tissues. (**b**) T_1-wt image. CSF is *dark*, because it contains large amount of free water molecules but not much macromolecular content, resulting in long T_1 relaxation time. (**c**) T_2-wt image. CSF is *bright* because the long T_2 relaxation time resulted from the high mobility of CSF water. (**d**) T_1 map of a mouse brain (in axial direction). (**e**) T_2 map of the same mouse brain

(protons) of water molecules. The proton density (PD, also called spin density (SD)) is one of the primary factors influencing the signal intensity. The relaxation times T_1, T_2, and T_2^* are other factors. An MR image contains combined influences of proton density and relaxation times. By selecting appropriate image sequences and scan parameters, it is possible to accentuate on one factor. The MRI emphasizing the tissue contrast or appearance dependent on PD in the tissue and minimizing the effects of T_1 and T_2 relaxations is called PD-weighted (PD-wt) MRI. Similarly, the MRI maximizing T_1 or T_2 effect and minimizing other effects is called T_1- or T_2-wt MRI, respectively.

Apparently PD-wt MRI is a reflection of the abundance of water molecules in a tissue; high proton density generates hyperintense signal on PD-wt images. For example, in Fig. 19.9a, in which a PD-wt image of a monkey brain is shown, cerebrospinal fluid (CSF) is bright on the PD-wt image because it contains more water molecules compared to other tissues.

The physiological basis of T_1- and T_2-wt MRI is more complex. Both T_1- and T_2-wt images can be viewed as reflecting related features of the physical microenvironment of the water molecules in tissue. T_1 relaxation is determined by the energy transfer between excited (by RF) protons (spins) and their surroundings (lattice). This is why it is also called spin–lattice relaxation. In the microenvironment with high macromolecular content (e.g., proteins, lipids), the energy transfer occurs fast and thus T_1 is short. If free water molecules are abundant in a tissue, the energy transfer occurs in a much slower fashion, and thus T_1 is long. On a T_1-wt image, tissue with long T_1 relaxation time generates *hypointense* signal. It can be seen that in Fig. 19.9b, CSF in the monkey brain is dark on the T_1-wt image, because it contains large amount of free water molecules but not much macromolecular content resulting in long T_1 relaxation time.

T_2 relaxation is also called spin–spin relaxation. In spin–spin relaxation, there is no energy transfer. Instead, it is caused by the interaction of neighboring water molecule protons. In a tissue in which water molecules move slowly, the interaction of protons is prolonged resulting in fast signal loss thus short T_2 relaxation time. Whereas in a highly mobile milieu, the interaction of protons is only transient, leading to slow signal loss and long T_2 relaxation time. On a T_2-wt image (Fig. 19.9c), CSF is bright because of the long T_2 relaxation time resulted from the high mobility of CSF water.

A specific weighting is achieved on an image by varying the MRI sequence parameters including TR and TE. To appropriately set these parameters, the relaxation values of the tissue have to be known beforehand using T_1/T_2 measurement sequences. The relaxation times are calculated from the data acquired using the sequences. The T_1 and T_2 values of a mouse brain are shown in Fig. 19.9d, e, respectively. The picture-like presentation of the relaxation times is called relaxation time map.

T_1 and T_2 can be considered as intrinsic properties of a tissue. On the other hand, T_2^* relaxation results from the extrinsic inhomogeneity of the main magnetic field B_0. Due to the T_2^* effect, signal loss is faster than when only T_2 relaxation is considered. At the interface of neighboring tissues with different magnetic susceptibility, e.g., between tissue and air or between soft tissue and bone, local magnetic field is spatially varied. The protons in this region then precess at different frequencies, leading to fast dephasing and thus undesirable signal loss at the tissue interface. However, the magnetic susceptibility sensitivity can also be exploited as an MRI contrast. T_2^* relaxation is the image contrast basis for many MRI applications, such as perfusion MRI, and fMRI. We will discuss these applications in following sections.

2. Applications.

   As shown in Fig. 19.9a–c, MRI can provide anatomical information of brain structures. Different anatomical structures can be accentuated using MRI weightings. For example, the ventricular system can be better visualized using PD- or T_2-wt MRI than on T_1-wt MRI, because CSF is bright on PD- and T_2-wt images.

   Some neurological diseases cause morphological changes in whole brain or certain structures. The changes can be revealed using MRI. In fact, MRI has widely been used for morphological studies in neuroscience (Ullmann et al. 2010; Bock et al. 2006; Wengenack et al. 2008; Aggarwal et al. 2009; Van der Linden et al. 2009; Chuang et al. 2011; Dazai et al. 2011; Lebenberg et al. 2011). Many types of pathology, such as tumors, edema, and hemorrhage, are associated with changes in water content and relaxation rates (Pirko and Johnson 2008; Durukan and Tatlisumak 2009; Brandt 2011). Pathological conditions may be visualized using MRI by emphasizing the altered PD or relaxation times. The change of relaxation times resulted from pathology can also be viewed using relaxation time maps.

   However, the proton density and relaxation changes are not universal, and pathology can be either light, dark, or isointense with normal tissue on either T_1- or T_2-wt images, depending upon the nature of the cellular abnormality. For example, the infarct area in the chronic (48 h after ischemia induction) stroked rat brain is bright on T_2-wt image (Fig. 19.10a), while appears to be normal on T_1-wt image (Fig. 19.10b). Moreover, the proton density and relaxation changes are not specific, and there are many other mechanisms by which the MRI signal intensity may be modulated.

19.4.2 Diffusion MRI

1. Diffusion-Weighted Imaging (DWI).
   (a) Physiological Basis.

      Diffusion is the random (or pseudo-random) motion of water molecules due

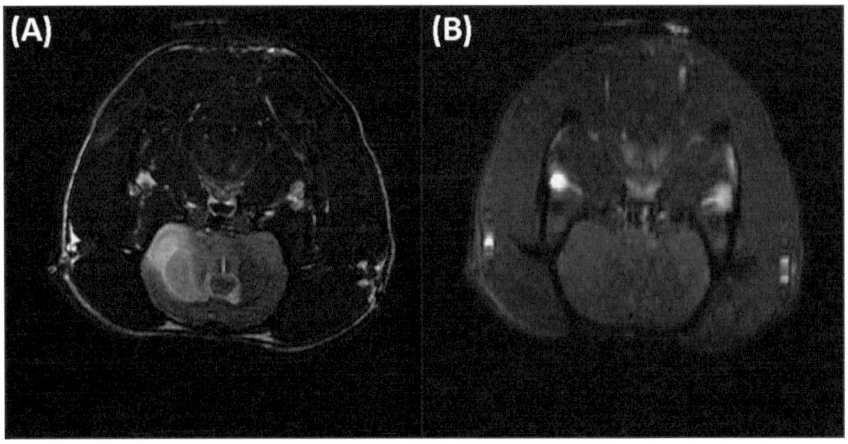

Fig. 19.10 (**a**) T_2-wt image of a stroked rat brain at chronic stage. The infarct area is *bright*. (**b**) T_1-wt image of the same brain. The infarct area appears to be normal

Fig. 19.11 Diffusion is the random (or pseudo-random) motion of water molecules due to nonzero temperature. For reference, the diffusion coefficient of water molecules is 2.2×10^{-5} cm^2/s at room temp. That is, on average a water molecule moves and covers a patch of area 0.000022 cm^2 every second

to nonzero temperature. For reference, the diffusion coefficient of water molecules is 2.2×10^{-5} cm^2/s at room temperature. That is, on average a water molecule moves and covers a patch of area 0.000022 cm^2 every second (Fig. 19.11). Water diffusion depends on the microscopic structural environment in the tissue, and measurement of it can provide valuable insights into the regional functional and structural changes occurring in disease states. The greatest diffusion rate occurs in free water with no boundaries. In brain tissue, the boundaries of cell membranes restrict the motion of water to a degree which depends upon the mean free path of the water molecules. For example, in cerebral ventricles, water diffusion is relatively unimpeded and CSF has a high diffusion coefficient. Gray matter (GM) and white matter (WM) have lower diffusion coefficient than CSF because of their complex tissue structure.

The water diffusion can be detected or measured using the diffusion-weighted imaging (DWI) technology. DWI is sensitized to the random molecular motion of water in tissue by applying magnetic field gradients (diffusion gradients) in the RF pulse sequence. In a DWI sequence, the diffusion weighting is determined by a parameter called "*b*-value," which is in the unit of second per square millimeter (s/mm^2). High "*b*-value" generates high diffusion weighting, and no diffusion weighting is generated when $b=0$. On a diffusion-weighted image, the tissue that contains high diffusing water generates hypointense signal. A map of apparent diffusion coefficient (ADC) of water molecules can be calculated from the diffusion-weighted image.

(b) Applications.

The use of DWI on neurological studies has attracted a lot attention (Carano et al. 2003). It has been shown that a wide range of neuropathology causes DWI signal changes. One of the most successful applications is in the stroke study. T_1- and T_2-wt MRI failed to detect the ischemic

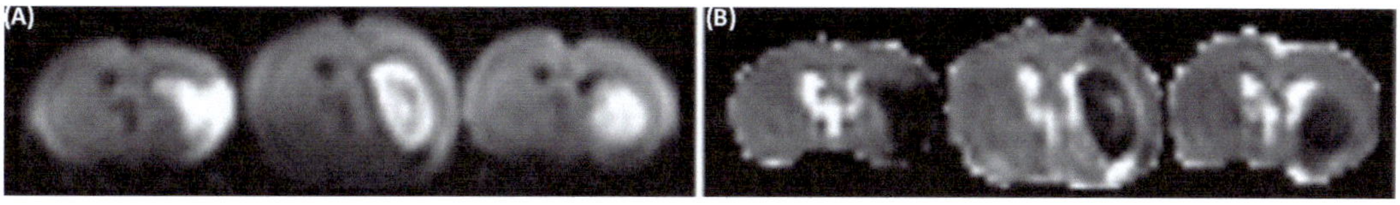

Fig. 19.12 DWI (**a**) and the ADC map (**b**) of a stroked rat brain. The ischemic region has elevated DWI signal intensity indicating decreased water diffusion in this region. ADC map calculated from the DWI shows decreased ADC in the same region

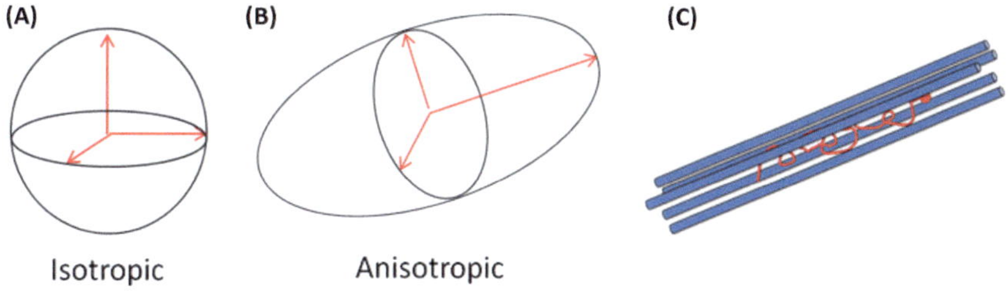

Fig. 19.13 In a uniform environment, diffusion is isotropic in all directions and can be represented by a *sphere* (**a**). If water molecule movement is restricted in certain directions, the diffusion becomes anisotropic, represented by an *ellipsoid* (**b**). (**c**) In fiber-like cell structures, such as white matter tracts, the diffusion is relatively free along the long axis of the fiber tract, but restricted in the other two dimensions

lesion in acute stroke. On the other hand, the lesion can be detected using DWI. Studies have shown that DWI can reveal the immediate temporal changes in ADC that occur upon induction of ischemia (Liu et al. 2007c). Figure 19.12 shows the DWI and the ADC map of a stroke model using rats. The ischemic region has elevated DWI signal intensity indicating decreased water diffusion in this region. ADC map calculated from the DWI shows decreased ADC in the same region.

The underlying pathology of the ADC change during ischemia remains unclear. Several theories exist that try to explain the observation. One of them is the cell swelling theory. This theory assumes that water diffusion is slower inside cells than in the extracellular space (Moseley et al. 1990). The disruption of blood supply in stroke induces cell swelling (cellular edema). Water molecules then spend more time diffusing in swollen cells, and thus decreasing ADC. Another theory assumes that the changes in cell membrane permeability may contribute to the ADC reduction (Szafer et al. 1995). A loss of active intracellular water transport with energy failure may be another cause of the decreased water diffusion.

2. Diffusion Tensor Imaging (DTI).
   (a) Physiological Basis.

   DTI is an extension of DWI. Diffusion is a three-dimensional process. In a uniform environment, it is isotropic in all directions and can be represented by a sphere (Fig. 19.13a). If water molecule movement is restricted in certain directions, the diffusion becomes anisotropic, represented by an ellipsoid (Fig. 19.13b). For example, in fiber-like cell structures, such as white matter tracts, the diffusion is relatively free along the long axis of the fiber tract, but restricted in the other two dimensions. The diffusion in cellular structures is described mathematically by a tensor. A tensor is a 3×3 matrix. The diffusion tensor can be measured using DTI with diffusion gradients in appropriate strength (Basser et al. 1994a, b). After a series of mathematical manipulations, the axes of the diffusion ellipsoid

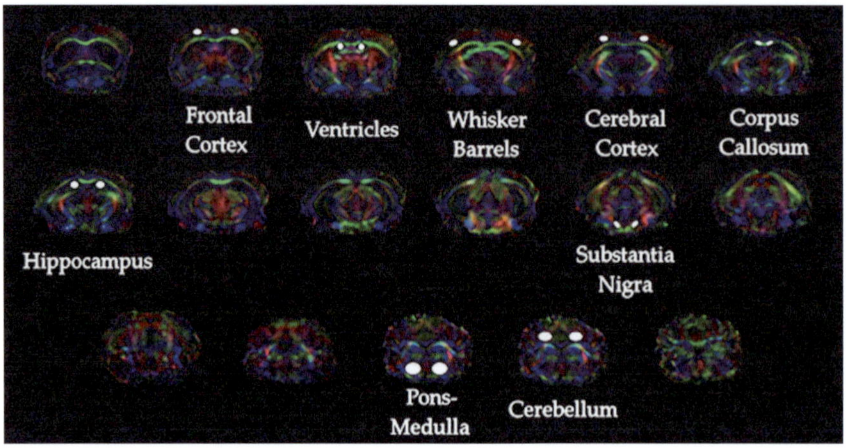

Fig. 19.14 Color-encoded direction of the principle eigenvector of the diffusion tensor in a mouse brain. The brightness was scaled by the FA value

and the diffusion magnitudes along the axes can be calculated.

Determination of the diffusion axes and magnitudes enables the computation of diffusion anisotropy in tissues. Several indices have been proposed to represent the diffusion anisotropy and the fractional anisotropy (FA) is a commonly used one (Basser and Pierpaoli 1996). FA is a scalar quantity from 0 to 1. High FA value indicates that water diffusion is more likely anisotropic. FA provides information for identifying and quantifying anisotropic tissue structures.

(b) Applications.

Apparently the principal axis (the axis with the maximum diffusion magnitude) of the diffusion ellipsoid points to the preferred diffusion direction. It is reasonable to think that for WM, the favorite diffusion direction is along the WM tracts, and thus the principal axis is parallel to fiber tract orientation. Several methods have been proposed to display the principal axis, one of which is the so-called color-encoding technique. In the color-encoding technique, the three components along x, y, and z directions of the principal axis are encoded with the primary colors (red, x component; green, y component; and blue, z component) and the brightness is scaled by an anisotropy index such as FA. Figure 19.14 shows an example of this method on multiple image slices of a mouse brain where the brightness was scaled by the FA value.

Information on diffusion anisotropy in brain tissue can be used for modeling neuronal tracts. The principal axis is used to connect similar voxels and is the basis for DTI-based fiber tracking. Many fiber tracking algorithms (Lori et al. 2002; Dubois et al. 2006; Jian et al. 2007; Price et al. 2007; Chen and Song 2008; Burgel et al. 2009; Zalesky and Fornito 2009; Jiao et al. 2011; Mishra et al. 2011) have been developed to reconstruct fiber architecture and confirm connections among different regions of the brain of human and animals. The application of fiber tracking allows us to determine neuronal connectivity of different regions of the brain. Despite certain limitations, such as relatively poor spatial resolution as compared with axon dimensions and reconstruction difficulties at fiber branching or merging sites, DTI provides a unique means to study integrity of fiber bundles in particular neurologic disorders.

If a disease damages the microstructure of a tissue, most likely the water diffusion anisotropy in the tissue is altered.

This assumption has rationalized the study of brain tissue using DTI in a variety of neurological diseases including, but not limited to, stroke, multiple sclerosis (MS), Alzheimer's disease (AD), neuroAIDS, Krabbe's disease, Alexander's disease, and Wallerian degeneration (Carano et al. 2003). More or less, diffusion anisotropy changes were observed in these studies. However, because of the lack of complete understanding of the mechanisms underlying the changes, the findings are still in controversy.

19.4.3 Perfusion MRI

Cerebral perfusion is crucial for brain functions. It is represented by the amount of blood flow (in the unit of mL g^{-1} s^{-1}) in the capillary bed to a given region of the brain (flow per unit volume of tissue). Two distinct perfusion MRI techniques have been developed, differing with regard to use of exogenous and endogenous MRI contrast agents. The first technique, known as bolus tracking MRI, requires injection of a bolus of MRI contrast agents and tracks the passage of the bolus in the tissue. The other technique is called arterial spin labeling (ASL) in that it uses RF pulses to label moving spins in flowing blood. MRI has been used to assess perfusion as a noninvasive technique. Perfusion MRI has been used to demonstrate regions of tissue at risk of further damage from stroke and blood flow abnormalities and highlight regions of hyperperfusion in tumors.

1. Bolus Tracking MRI.

   The contrast agent used in the bolus tracking MRI is a paramagnetic substance (e.g., gadolinium (Gd)), which has a small magnetic susceptibility. Gd, usually in a chelated form, causes fast spin dephasing (i.e., decrease in T_2^*) during its passage through the tissue (Villringer et al. 1988). Thus, this technique is also called dynamic susceptibility contrast (DSC) MRI. In a typical DSC MRI, a series of baseline T_2^*-wt MRI is first acquired. Gd is then injected into a vein or an artery. The animal is being continuously scanned using T_2^*-wt MRI for another period of time until the Gd has passed through the tissue.

   Figure 19.15 shows several typical images among a series T_2^*-wt MRI in a rat stroke study. The ischemia was created by occluding the middle cerebral artery (MCA). The signal drop and recovery can be observed in the contralateral hemisphere (right side from the reader's point of view). The signal change in the ischemic hemisphere is mild compared to the other side. Schematic plots of signal versus time in the normal tissue and occluded region are shown in Fig. 19.15. The occluded region has a perfusion deficit with lower signal drop, delayed entering and washout, and prolonged pass-through. The relative contrast concentration change with time can be calculated from the MRI signal as shown in Fig. 19.15. Several parameters can be derived from the concentration curve to describe the perfusion in tissue. The parameters usually include the relative cerebral blood volume (CBV or rCBV) that is the area under the concentration curve; mean transit time (MTT) of the bolus through the tissue; and time to peak (TTP), which is the time it takes for the contrast agent to reach the maximum concentration. The cerebral blood flow can be calculated by dividing CBV by MTT. The perfusion parameters are usually calculated pixel by pixel in the MR images, and maps of the parameters are generated for further analysis. For example, the maps of CBV and TTP on the rat brain in the stroke study are shown in Fig. 19.15. The regions with lower CBV and prolonged TTP are identified.

2. ASL.

   Bolus tracking MRI is kind of invasive due to the injection of the contrast agent. ASL is a noninvasive technique by labeling the spins of the intravascular arterial water using one or two RF pulses (Dixon et al. 1986; Kim 1995). Another advantage of ASL over the bolus tracking MRI is that ASL can be repeated many times in a single study, which makes it appealing for the continuous monitoring of the perfusion. In a typical ASL, RF pulses first

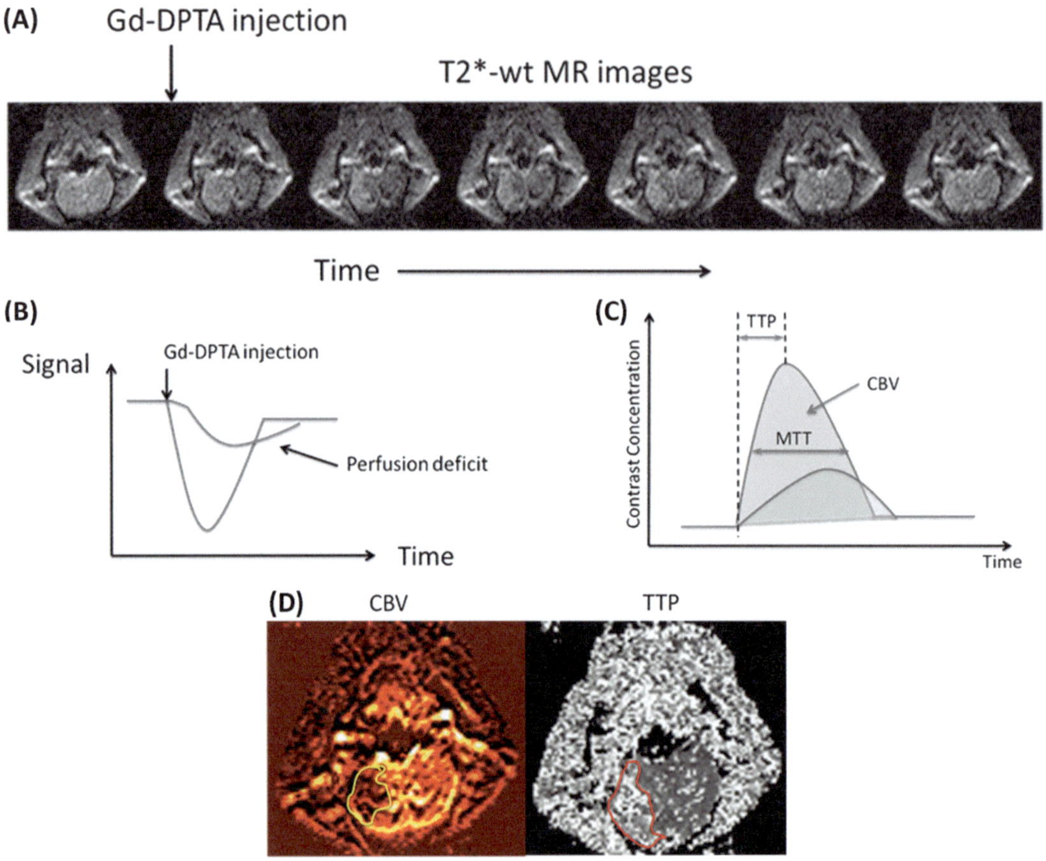

Fig. 19.15 A perfusion MRI in a rat stroke study. (**a**) A series T_2^*-wt MRI pre- and post-Gd injection. (**b**) T_2^*-wt signal drop in the contralateral hemisphere and ischemic tissue. (**c**) Contrast concentration change with time and perfusion parameters measured from the concentration *curve*. (**d**) CBV and TTP maps generated from the perfusion study. CBV: cerebral blood volume, MTT: mean transient time, TTP: Time to peak

label the spins in the arterial blood just upstream of the tissue of interest by inverting the longitudinal magnetization Mz. The spins then perfuse into the tissue of interesting, and thus the signal in the tissue is modified by the labeled spins. A second unlabeled acquisition serves as a reference to then calculate the perfusion images.

19.4.4 Functional MRI

1. Physiological basis.

   During the last decade, mapping of cerebral activity has become a major field in MRI research (Van Bruggen and Roberts 2003). Standard fMRI techniques measure hemodynamic responses to neural activity in a functioning brain during cognitive, perceptual, sensory, or motor processes or during a pharmacologic challenge (Van Bruggen and Roberts 2003). Blood oxygenation level-dependent (BOLD) contrast is mostly used as implications of the neural activity. BOLD MRI is based upon the magnetic property of blood, which is dependent upon the oxygenation state of hemoglobin (Ogawa et al. 1990). Deoxygenated hemoglobin (Hb) is paramagnetic, whereas oxyhemoglobin (HbO_2) is diamagnetic. Deoxygenation results in increased local magnetic susceptibility differences between intra- and extravascular compartments and thereby causes signal loss on T_2-wt and T_2^*-wt MR images (Fig. 19.16). This mechanism is essentially similar to T_2^*-shortening by exogenous contrast agents, be it that BOLD MRI makes use of an endogenous contrast agent (i.e., deoxyhemoglobin). When a brain region is activated by stimulation, the oxygenated blood supply increases in the region. The MRI signal then increases due to the oversupply of oxygenated blood in the capillaries.

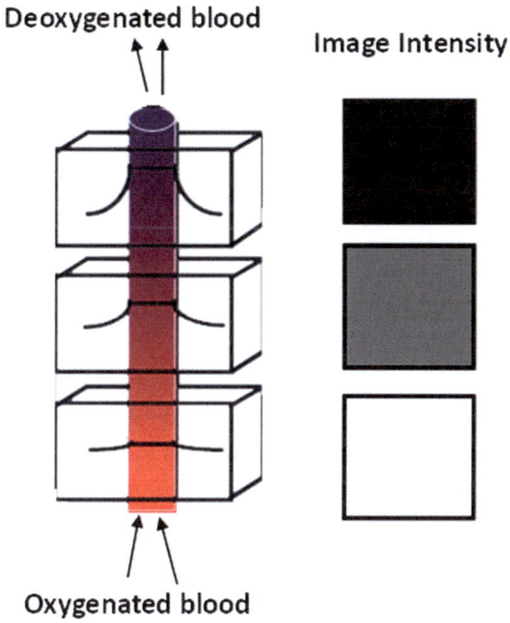

Fig. 19.16 Oxyhemoglobin (HbO_2) is diamagnetic. Deoxygenated hemoglobin (Hb) is paramagnetic. Deoxygenation results in increased local magnetic susceptibility differences between intra- and extravascular compartments and thereby causes signal loss on T_2-wt and T_2^*-wt MR images

2. Applications.

In small animal fMRI, usually a sequence of stimulus is applied on the whisker or front/hind paws during image acquisition. Figure 19.17 shows a typical fMRI paradigm, in which the stimulus lasts 16 s and is repeated four times. The interval between the stimuli is also 16 s. It has to be noted that fMRI does not measure neuronal activity directly, but instead measures the hemodynamic response to the stimulation. And the signal from hemodynamic response is further modulated by physiological and environmental noise. Usually an fMRI paradigm needs to be applied several times for reliable data collection, and statistical methods are used for data analysis.

Recently resting state fMRI (rsfMRI) has been used in animal studies. In rsfMRI, the BOLD signal is detected when an animal is not stimulated but in the resting state. The resting state approach is useful to explore the brain functional organization. Using rsfMRI, a number of function networks have been identified in healthy subjects (Biswal 2012; Rosazza and Minati 2011).

Performing fMRI in awake conditions is possible but usually requires time-consuming and challenging training to acclimate animals to the environment in the scanner. Usually the animals are anesthetized for fMRI studies. It has been shown that the hemodynamics and functional activity in brains are sensitive to anesthetics and level of sedation. Many anesthetics reduce or suppress the coupling of functional activity and hemodynamic response. Several anesthetics have been tested for functional

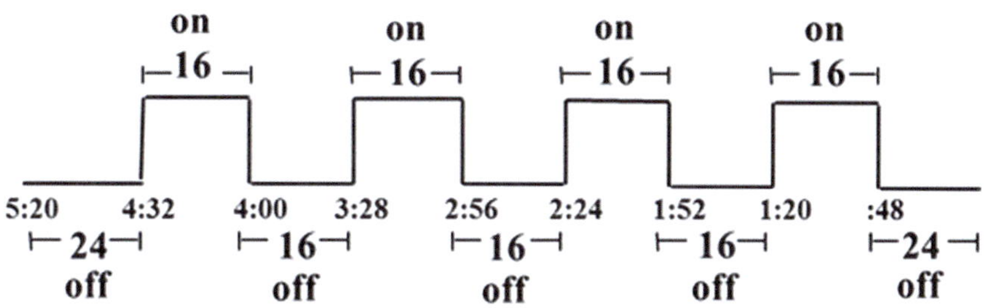

Fig. 19.17 A typical fMRI stimulation paradigm. The stimulus lasts 16 s and is repeated four times. The interval between the stimuli is also 16 s

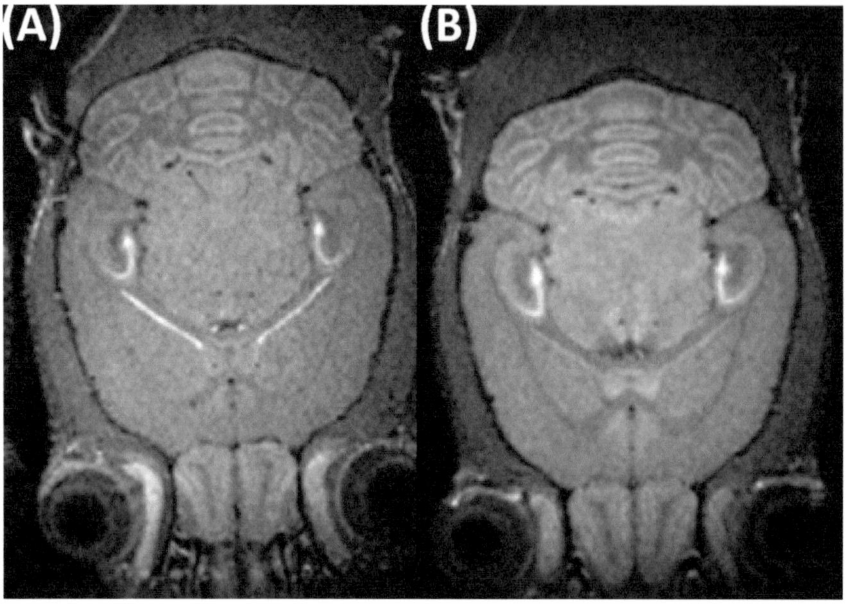

Fig. 19.18 MEMRI enables the visualization of some detailed neurostructures including the layers in cerebral and cerebellar cortex, olfactory bulbs, and sub-hippocampal structures. (**a**) A mouse injected with $MnCl_2$ at the dose of 120 mg/kg and (**b**) at the dose 240 mg/kg

studies including isoflurane, urethane, α-chloralose, medetomidine, and propofol. Results from these studies are controversial, and so far no standard anesthesia protocol exists for fMRI (Masamoto and Kanno 2012).

19.4.5 Manganese-Enhanced MRI

1. Physiologic Basis.

   This technique relies on the ability of manganese ions (Mn^{2+}) to substitute for calcium ions (Ca^{2+}). The influx of Ca^{2+} into cells is required for the release of neurotransmitters during functional activation. Paramagnetic Mn^{2+} has van der Waals radii similar to those of Ca^{2+} and can therefore enter cells through voltage-gated Ca^{2+} channels. Cells that have increased uptake of Ca^{2+} during brain stimulation will selectively accumulate Mn^{2+}. This, in turn, leads to a decrease in T_1 relaxation time that is readily detectable using T_1-wt imaging.
2. Applications.

   MEMRI enables the visualization of some detailed neurostructures including the layers in cerebral and cerebellar cortex, olfactory bulbs, and sub-hippocampal structures as shown in Fig. 19.18. For the neurostructure visualization and analysis, systematic administration of Mn^{2+} (usually in the form of $MnCl_2$) is used by intravenous or intraperitoneal injection. MRI is acquired about 24 h after the manganese administration.

   Using MEMRI, investigators have been able to detect activated brain regions in animals (Aoki et al. 2002). This technique has high sensitivity and exquisite spatial resolution compared to BOLD fMRI. Manganese can trace axonal pathways as well since it is taken up by active cells and then transported down the axonal projection to a connected area (Pautler et al. 1998). To apply MEMRI in fiber tracking, Mn^{2+} needs to be injected intracranially into the region of interest. The major drawback of MEMRI is the neurotoxicity of Mn^{2+}. A method called fractional administration was recently developed to inject a small amount of $MnCl_2$ daily for several days to minimize the side effects (Grunecker et al. 2011).

19.4.6 Cell Tracking MRI

The prospect to trace specific cells and follow their distribution in vivo allows monitoring of transplanted cells for therapeutic purposes. Labeling of cells with MR contrast agent provides a tool to detect and track cells with MRI (Norman et al. 1992; Bulte et al. 2002a). Several methods for cellular incorporation of MR contrast agent exist, as has been reviewed by Bulte et al. (2002b). One approach makes use of phagocytosis of intravascularly administered superparamagnetic iron oxide (SPIO) particles (Laurent et al. 2010). SPIOs are T_2-shortening particles. Macrophage activity in brain disorders that involve inflammation can be detected using SPIO MRI. In recent years, stem cell therapy has proven to be a promising means to improve neurologic function in various brain pathologies and has excellent perspectives to be clinically effective on a large scale. With MRI, the distribution and survival of stem cells labeled with gadolinium complex, SPIO particles can be monitored.

References

Aggarwal M, Zhang J, Miller MI, Sidman RL, Mori S (2009) Magnetic resonance imaging and micro-computed tomography combined atlas of developing and adult mouse brains for stereotaxic surgery. Neuroscience 162:1339–1350

Alexander GE, Chen K, Aschenbrenner M, Merkley TL, Santerre-Lemmon LE, Shamy JL, Skaggs WE, Buonocore MH, Rapp PR, Barnes CA (2008) Age-related regional network of magnetic resonance imaging gray matter in the rhesus macaque. J Neurosci 28:2710–2718

Aoki I, Tanaka C, Takegami T, Ebisu T, Umeda M, Fukunaga M, Fukuda K, Silva AC, Koretsky AP, Naruse S (2002) Dynamic activity-induced manganese-dependent contrast magnetic resonance imaging (DAIM MRI). Magn Reson Med 48: 927–933

Baskerville TA, Deuchar GA, McCabe C, Robertson CA, Holmes WM, Santosh C, Macrae IM (2011) Influence of 100% and 40% oxygen on penumbral blood flow, oxygen level, and T2*-weighted MRI in a rat stroke model. J Cereb Blood Flow Metab 31:1799–1806

Basser PJ, Pierpaoli C (1996) Microstructural and physiological features of tissues elucidated by quantitative-diffusion-tensor MRI. J Magn Reson B 111: 209–219

Basser PJ, Mattiello J, LeBihan D (1994a) MR diffusion tensor spectroscopy and imaging. Biophys J 66: 259–267

Basser PJ, Mattiello J, LeBihan D (1994b) Estimation of the effective self-diffusion tensor from the NMR spin echo. J Magn Reson B 103:247–254

Benveniste H, Ma Y, Dhawan J, Gifford A, Smith SD, Feinstein I, Du C, Grant SC, Hof PR (2007) Anatomical and functional phenotyping of mice models of Alzheimer's disease by MR microscopy. Ann N Y Acad Sci 1097:12–29

Berry E, Bulpitt AJ (2009) Fundamentals of MRI: an interactive learning approach. CRC, Boca Raton, FL

Biswal BB (2012) Resting state fMRI: a personal history. Neuroimage 62:938–944

Bock NA, Kovacevic N, Lipina TV, Roder JC, Ackerman SL, Henkelman RM (2006) In vivo magnetic resonance imaging and semiautomated image analysis extend the brain phenotype for cdf/cdf mice. J Neurosci 26:4455–4459

Borthakur A, Gur T, Wheaton AJ, Corbo M, Trojanowski JQ, Lee VM, Reddy R (2006) In vivo measurement of plaque burden in a mouse model of Alzheimer's disease. J Magn Reson Imaging 24:1011–1017

Boska MD, Mosley RL, Nawab M, Nelson JA, Zelivyanskaya M, Poluektova L, Uberti M, Dou H, Lewis TB, Gendelman HE (2004) Advances in neuroimaging for HIV-1 associated neurological dysfunction: clues to the diagnosis, pathogenesis and therapeutic monitoring. Curr HIV Res 2:61–78

Boska MD, Hasan KM, Kibuule D, Banerjee R, McIntyre E, Nelson JA, Hahn T, Gendelman HE, Mosley RL (2007) Quantitative diffusion tensor imaging detects dopaminergic neuronal degeneration in a murine model of Parkinson's disease. Neurobiol Dis 26:590–596

Braakman N, Matysik J, van Duinen SG, Verbeek F, Schliebs R, de Groot HJ, Alia A (2006) Longitudinal assessment of Alzheimer's beta-amyloid plaque development in transgenic mice monitored by in vivo magnetic resonance microimaging. J Magn Reson Imaging 24:530–536

Brandt CT (2011) Experimental studies of pneumococcal meningitis. Dan Med Bull 57:B4119

Bulte JW, Duncan ID, Frank JA (2002a) In vivo magnetic resonance tracking of magnetically labeled cells after transplantation. J Cereb Blood Flow Metab 22: 899–907

Bulte JW, Zhang SC, van Gelderen P, Herynek V, Jordan EK, Janssen CH, Duncan ID, Frank JA (2002b) Magnetically labeled glial cells as cellular MR contrast agents. Acad Radiol 9(Suppl 1):S148–S150

Burgel U, Madler B, Honey CR, Thron A, Gilsbach J, Coenen VA (2009) Fiber tracking with distinct software tools results in a clear diversity in anatomical fiber tract portrayal. Cent Eur Neurosurg 70:27–35

Bushberg JT (2002) The essential physics of medical imaging, 2nd edn. Lippincott Williams & Wilkins, Philadelphia

Carano RAD, Van Bruggen N, De Crespigny A (2003) MRI measurement of cerebral water diffusion and its application to experimental research. In: Van Bruggen N, Roberts T (eds) Biomedical imaging in experimental neuroscience. CRC, New York, pp 55–92

Chen B, Song AW (2008) Diffusion tensor imaging fiber tracking with local tissue property sensitivity: phantom and in vivo validation. Magn Reson Imaging 26:103–108

Cho ZH, Jones JP, Singh M (1993) Foundations of medical imaging. Wiley, New York

Chuang N, Mori S, Yamamoto A, Jiang H, Ye X, Xu X, Richards LJ, Nathans J, Miller MI, Toga AW, Sidman RL, Zhang J (2011) An MRI-based atlas and database of the developing mouse brain. Neuroimage 54:80–89

Dash PK, Gorantla S, Gendelman HE, Knibbe J, Casale GP, Makarov E, Epstein AA, Gelbard HA, Boska MD, Poluektova LY (2011) Loss of neuronal integrity during progressive HIV-1 infection of humanized mice. J Neurosci 31:3148–3157

Dazai J, Spring S, Cahill LS, Henkelman RM (2011) Multiple-mouse neuroanatomical magnetic resonance imaging. J Vis Exp 2011(48): 2497

Dhenain M, Delatour B, Walczak C, Volk A (2006) Passive staining: a novel ex vivo MRI protocol to detect amyloid deposits in mouse models of Alzheimer's disease. Magn Reson Med 55:687–693

Dijkhuizen RM (2006) Application of magnetic resonance imaging to study pathophysiology in brain disease models. Methods Mol Med 124:251–278

Dijkhuizen RM, Nicolay K (2003) Magnetic resonance imaging in experimental models of brain disorders. J Cereb Blood Flow Metab 23:1383–1402

Dixon WT, Du LN, Faul DD, Gado M, Rossnick S (1986) Projection angiograms of blood labeled by adiabatic fast passage. Magn Reson Med 3:454–462

Dubois J, Hertz-Pannier L, Dehaene-Lambertz G, Cointepas Y, Le Bihan D (2006) Assessment of the early organization and maturation of infants' cerebral white matter fiber bundles: a feasibility study using quantitative diffusion tensor imaging and tractography. Neuroimage 30:1121–1132

Durukan A, Tatlisumak T (2009) Ischemic stroke in mice and rats. Methods Mol Biol 573:95–114

Faber C, Zahneisen B, Tippmann F, Schroeder A, Fahrenholz F (2007) Gradient-echo and CRAZED imaging for minute detection of Alzheimer plaques in an APPV717I×ADAM10-dn mouse model. Magn Reson Med 57:696–703

Falangola MF, Dyakin VV, Lee SP, Bogart A, Babb JS, Duff K, Nixon R, Helpern JA (2007) Quantitative MRI reveals aging-associated T2 changes in mouse models of Alzheimer's disease. NMR Biomed 20:343–351

Gao H, Liu Y, Lu S, Xiang B, Wang C (2006) A reversible middle cerebral artery occlusion model using intraluminal balloon technique in monkeys. J Stroke Cerebrovasc Dis 15:202–208

Grunecker B, Kaltwasser SF, Peterse Y, Samann PG, Schmidt MV, Wotjak CT, Czisch M (2011) Fractionated manganese injections: effects on MRI contrast enhancement and physiological measures in C57BL/6 mice. NMR Biomed 23:913–921

Guo J, Zheng HB, Duan JC, He L, Chen N, Gong QY, Tang HH, Li HX, Wang L, Cheng JQ (2011) Diffusion tensor MRI for the assessment of cerebral ischemia/reperfusion injury in the penumbra of non-human primate stroke model. Neurol Res 33:108–112

Haacke EM (1999) Magnetic resonance imaging: physical principles and sequence design. Wiley, New York

Hornak JP The Basics of MRI. http://www.cis.rit.edu/htbooks/mri/

Jack CR Jr, Marjanska M, Wengenack TM, Reyes DA, Curran GL, Lin J, Preboske GM, Poduslo JF, Garwood M (2007) Magnetic resonance imaging of Alzheimer's pathology in the brains of living transgenic mice: a new tool in Alzheimer's disease research. Neuroscientist 13:38–48

Jian B, Vemuri BC, Ozarslan E, Carney PR, Mareci TH (2007) A novel tensor distribution model for the diffusion-weighted MR signal. Neuroimage 37: 164–176

Jiao F, Phillips JM, Stinstra J, Krger J, Varma R, Hsu E (2011) Metrics for uncertainty analysis and visualization of diffusion tensor images. Lect Notes Comput Sci 6326(2010):179–190

Justicia C, Ramos-Cabrer P, Hoehn M (2008) MRI detection of secondary damage after stroke: chronic iron accumulation in the thalamus of the rat brain. Stroke 39:1541–1547

Kim SG (1995) Quantification of relative cerebral blood flow change by flow-sensitive alternating inversion recovery (FAIR) technique: application to functional mapping. Magn Reson Med 34:293–301

Kondoh T, Bannai M, Nishino H, Torii K (2005) 6-Hydroxydopamine-induced lesions in a rat model of hemi-Parkinson's disease monitored by magnetic resonance imaging. Exp Neurol 192:194–202

Lau JC, Lerch JP, Sled JG, Henkelman RM, Evans AC, Bedell BJ (2008) Longitudinal neuroanatomical changes determined by deformation-based morphometry in a mouse model of Alzheimer's disease. Neuroimage 42:19–27

Laurent S, Bridot JL, Elst LV, Muller RN (2010) Magnetic iron oxide nanoparticles for biomedical applications. Future Med Chem 2:427–449

Lebenberg J, Herard AS, Dubois A, Dauguet J, Frouin V, Dhenain M, Hantraye P, Delzescaux T (2011) Validation of MRI-based 3D digital atlas registration with histological and autoradiographic volumes: an anatomofunctional transgenic mouse brain imaging study. Neuroimage 51:1037–1046

Liu Y, D'Arceuil H, He J, Duggan M, Seri S, Hashiguchi Y, Nakatani A, Gonzalez RG, Pryor J, de Crespigny A (2005) Dynamic susceptibility contrast perfusion imaging of cerebral ischemia in nonhuman primates: comparison of Gd-DTPA and NMS60. J Magn Reson Imaging 22:461–466

Liu Y, Karonen JO, Nuutinen J, Vanninen E, Kuikka JT, Vanninen RL (2007a) Crossed cerebellar diaschisis in acute ischemic stroke: a study with serial SPECT and MRI. J Cereb Blood Flow Metab 27: 1724–1732

Liu Y, D'Arceuil H, He J, Duggan M, Gonzalez G, Pryor J, de Crespigny A (2007b) MRI of spontaneous fluctuations after acute cerebral ischemia in nonhuman primates. J Magn Reson Imaging 26:1112–1116

Liu Y, D'Arceuil HE, Westmoreland S, He J, Duggan M, Gonzalez RG, Pryor J, de Crespigny AJ (2007c) Serial diffusion tensor MRI after transient and permanent cerebral ischemia in nonhuman primates. Stroke 38:138–145

Liu Y, Uberti MG, Dou H, Banerjee R, Grotepas CB, Stone DK, Rabinow BE, Gendelman HE, Boska MD (2008) Ingress of blood-borne macrophages across the blood–brain barrier in murine HIV-1 encephalitis. J Neuroimmunol 200:41–52

Lori NF, Akbudak E, Shimony JS, Cull TS, Snyder AZ, Guillory RK, Conturo TE (2002) Diffusion tensor fiber tracking of human brain connectivity: acquisition methods, reliability analysis and biological results. NMR Biomed 15:494–515

Luo F, Seifert TR, Edalji R, Loebbert RW, Hradil VP, Harlan J, Schmidt M, Nimmrich V, Cox BF, Fox GB (2008) Non-invasive characterization of beta-amyloid(1–40) vasoactivity by functional magnetic resonance imaging in mice. Neuroscience 155:263–269

Lythgoe MF, Sibson NR, Harris NG (2003) Neuroimaging of animal models of brain disease. Br Med Bull 65:235–257

Masamoto K, Kanno I (2012) Anesthesia and the quantitative evaluation of neurovascular coupling. J Cereb Blood Flow Metab 32:1233–1247

Mishra A, Anderson AW, Wu X, Gore JC, Ding Z (2011) An improved Bayesian tensor regularization and sampling algorithm to track neuronal fiber pathways in the language circuit. Med Phys 37:4274–4287

Moseley ME, Kucharczyk J, Mintorovitch J, Cohen Y, Kurhanewicz J, Derugin N, Asgari H, Norman D (1990) Diffusion-weighted MR imaging of acute stroke: correlation with T2-weighted and magnetic susceptibility-enhanced MR imaging in cats. AJNR Am J Neuroradiol 11:423–429

Nagaraja TN, Ewing JR, Karki K, Jacobs PE, Divine GW, Fenstermacher JD, Patlak CS, Knight RA (2011) MRI and quantitative autoradiographic studies following bolus injections of unlabeled and (14)C-labeled gadolinium-diethylenetriaminepentaacetic acid in a rat model of stroke yield similar distribution volumes and blood-to-brain influx rate constants. NMR Biomed 24:547–558

Nelson JA, Dou H, Ellison B, Uberti M, Xiong H, Anderson E, Mellon M, Gelbard HA, Boska M, Gendelman HE (2005) Coregistration of quantitative proton magnetic resonance spectroscopic imaging with neuropathological and neurophysiological analyses defines the extent of neuronal impairments in murine human immunodeficiency virus type-1 encephalitis. J Neurosci Res 80:562–575

Norman AB, Thomas SR, Pratt RG, Lu SY, Norgren RB (1992) Magnetic resonance imaging of neural transplants in rat brain using a superparamagnetic contrast agent. Brain Res 594:279–283

Ogawa S, Lee TM, Kay AR, Tank DW (1990) Brain magnetic resonance imaging with contrast dependent on blood oxygenation. Proc Natl Acad Sci U S A 87: 9868–9872

Pautler RG, Silva AC, Koretsky AP (1998) In vivo neuronal tract tracing using manganese-enhanced magnetic resonance imaging. Magn Reson Med 40:740–748

Pelled G, Bergman H, Ben-Hur T, Goelman G (2005) Reduced basal activity and increased functional homogeneity in sensorimotor and striatum of a Parkinson's disease rat model: a functional MRI study. Eur J Neurosci 21:2227–2232

Pelled G, Bergman H, Ben-Hur T, Goelman G (2007) Manganese-enhanced MRI in a rat model of Parkinson's disease. J Magn Reson Imaging 26: 863–870

Pirko I, Johnson AJ (2008) Neuroimaging of demyelination and remyelination models. Curr Top Microbiol Immunol 318:241–266

Podell M, Hadjiconstantinou M, Smith MA, Neff NH (2003) Proton magnetic resonance imaging and spectroscopy identify metabolic changes in the striatum in the MPTP feline model of parkinsonism. Exp Neurol 179:159–166

Poduslo JF, Ramakrishnan M, Holasek SS, Ramirez-Alvarado M, Kandimalla KK, Gilles EJ, Curran GL, Wengenack TM (2007) In vivo targeting of antibody fragments to the nervous system for Alzheimer's disease immunotherapy and molecular imaging of amyloid plaques. J Neurochem 102:420–433

Price G, Cercignani M, Parker GJ, Altmann DR, Barnes TR, Barker GJ, Joyce EM, Ron MA (2007) Abnormal brain connectivity in first-episode psychosis: a diffusion MRI tractography study of the corpus callosum. Neuroimage 35:458–466

Ramakrishnan M, Wengenack TM, Kandimalla KK, Curran GL, Gilles EJ, Ramirez-Alvarado M, Lin J, Garwood M, Jack CR Jr, Poduslo JF (2008) Selective contrast enhancement of individual Alzheimer's disease amyloid plaques using a polyamine and Gd-DOTA conjugated antibody fragment against fibrillar Abeta42 for magnetic resonance molecular imaging. Pharm Res 25:1861–1872

Rosazza C, Minati L (2011) Resting-state brain networks: literature review and clinical applications. Neurol Sci 32:773–785

Smith KD, Kallhoff V, Zheng H, Pautler RG (2007) In vivo axonal transport rates decrease in a mouse model of Alzheimer's disease. Neuroimage 35:1401–1408

Strome EM, Doudet DJ (2007) Animal models of neurodegenerative disease: insights from in vivo imaging studies. Mol Imaging Biol 9:186–195

Szafer A, Zhong J, Anderson AW, Gore JC (1995) Diffusion-weighted imaging in tissues: theoretical models. NMR Biomed 8:289–296

Ullmann JF, Cowin G, Kurniawan ND, Collin SP (2010) Magnetic resonance histology of the adult zebrafish brain: optimization of fixation and gadolinium contrast enhancement. NMR Biomed 23:341–346

Uppal R, Ay I, Dai G, Kim YR, Sorensen AG, Caravan P (2010) Molecular MRI of intracranial thrombus in a rat ischemic stroke model. Stroke 41:1271–1277

Van Bruggen N, Roberts T (2003) Biomedical imaging in experimental neuroscience. CRC, Boca Raton, FL

Van der Linden A, Van Meir V, Boumans T, Poirier C, Balthazart J (2009) MRI in small brains displaying extensive plasticity. Trends Neurosci 32:257–266

Villringer A, Rosen BR, Belliveau JW, Ackerman JL, Lauffer RB, Buxton RB, Chao YS, Wedeen VJ, Brady TJ (1988) Dynamic imaging with lanthanide chelates in normal brain: contrast due to magnetic susceptibility effects. Magn Reson Med 6:164–174

Wang HH, Menezes NM, Zhu MW, Ay H, Koroshetz WJ, Aronen HJ, Karonen JO, Liu Y, Nuutinen J, Wald LL, Sorensen AG (2008) Physiological noise in MR images: an indicator of the tissue response to ischemia? J Magn Reson Imaging 27:866–871

Webb AR (2003) Introduction to biomedical imaging. Wiley, Hoboken, NJ

Wengenack TM, Jack CR Jr, Garwood M, Poduslo JF (2008) MR microimaging of amyloid plaques in Alzheimer's disease transgenic mice. Eur J Nucl Med Mol Imaging 35(Suppl 1):S82–S88

Winkeler A, Waerzeggers Y, Klose A, Monfared P, Thomas AV, Schubert M, Heneka MT, Jacobs AH (2008) Imaging noradrenergic influence on amyloid pathology in mouse models of Alzheimer's disease. Eur J Nucl Med Mol Imaging 35(Suppl 1):S107–S113

Zalesky A, Fornito A (2009) A DTI-derived measure of cortico-cortical connectivity. IEEE Trans Med Imaging 28:1023–1036

X-Ray, Positron Emission, and Single Photon Emission Tomographic Bioimaging

20

Katherine A. Estes, Jacob C. Peterson, Adam M. Szlachetka, and R. Lee Mosley

Abstract

Computer-aided tomography is commonly used for biomedical imaging in medicine and research. Most systems utilize some form of energized probe to either visualize tissues and/or localize a labeled compound. Common to all forms of tomographic imaging is the acquisition of cross-sectional images in a 360° rotation. Computer-interfaced software is used to reconstruct two-dimensional images into a three-dimensional representation. This allows selected two-dimensional images to be analyzed in any axis. In computed tomography (CT), also called computer-aided tomography or computerized axial tomography, X-rays produced by a cathode ray tube provide an external source of radiation that passes through the subject and are collected by a detector array. Density variations of different organs, tissues, and structures allow different amounts of radiation to penetrate and reach the detector to create an anatomical image.

In emission computed tomography, radionuclide probes are administered to the subject and provide an internal source of radiation. Positron emission tomography (PET) is used to detect radionuclides that emit positrons (β^+ particles), and single photon emission computed tomography (SPECT) detects radionuclides that emit gamma (γ) rays. The radionuclide probes are conjugated to diagnostic or therapeutic agents that ultimately target specific tissues. CT used in conjunction with PET or SPECT generate images that are co-registered, resulting in merged images wherein anatomical regions of interest can be defined and analyzed to detect the location and intensity of radionuclide probes. Such methods are used for diagnostic and therapeutic purposes as well as to monitor disease progression.

K.A. Estes • J.C. Peterson • A.M. Szlachetka
R.L. Mosley, Ph.D. (✉)
Department of Pharmacology and Experimental Neuroscience, Center for Neurodegenerative Disorders, University of Nebraska Medical Center, Omaha, NE 68198-5930, USA
e-mail: rlmosley@unmc.edu

H. Xiong and H.E. Gendelman (eds.), *Current Laboratory Methods in Neuroscience Research*, Springer Protocols Handbooks, DOI 10.1007/978-1-4614-8794-4_20,

Keywords
Bioimaging • Emission computed tomography • CT • PET • SPECT • Tomography • Radioisotopes • Digital image reconstruction

20.1 X-Ray CT

CT was first introduced in 1972 to improve image quality over traditional X-rays. In conventional X-ray imaging, the X-ray beam passes through the subject and is collected by a relatively large two-dimensional detector surface. This detector has traditionally been energy-sensitive silver bromide/iodide-coated plastic film; but more recently, film is being replaced with reusable detectors or screens that capture X-ray images and convert those signals to digital images. In addition to the obvious cost benefits, digital technology also provides greater dynamic range of signal than film. In conventional X-ray imaging, X-ray scatter leads to high background and decreasing image contrast. Superimposition of three-dimensional structures onto a two-dimensional surface makes interpretation of images logistically more difficult. The limited X-ray beam of CT scanners greatly decreases scatter and increases contrast, which provide image data for multiple two-dimensional slices of the subject such that the three dimensions are not compressed into a single two-dimensional image, thus greatly improving image quality (Mahesh 2002).

20.1.1 CT Instrument Design

The basic components of a CT scanner include a patient bed, X-ray source, and X-ray detector with the latter two components placed on a rotating gantry at 180° opposition (Fig. 20.1). Although the basic design has remained unchanged, CT scanners have vastly improved with progressing technology over time. The development of modern CT scanners began in 1967 by the work of Godfrey Hounsfield at EMI Corporation in England. Hounsfield postulated that many X-ray images of a subject taken at many positions around the subject could be reconstructed into three-dimensional images of soft tissues, leading to the first CT scanner built by EMI in 1971. The revolutionary images from this head-only scanner stimulated the rapid development of more advanced CT scanners.

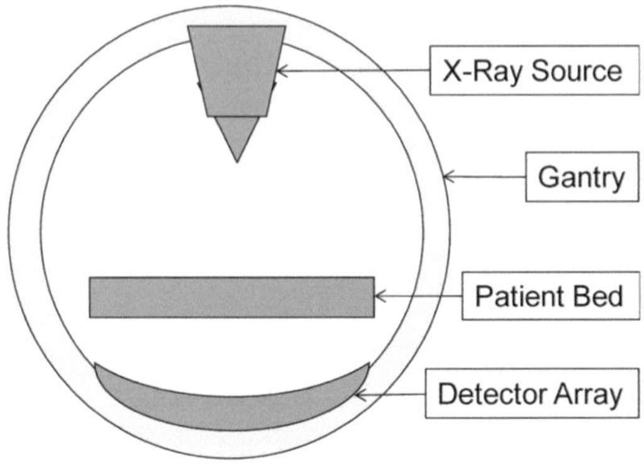

Fig. 20.1 Basic design of a CT instrument. CT scanners are composed of a patient bed, an X-ray generating source, usually a high-voltage cathode ray tube, and an X-ray detector that is situated at 180° opposition on a rotating gantry

First-generation scanners consisted of an X-ray source that emitted a pencil-sized beam and was paired with a detector on the other side of the subject. The X-ray source was a cathode ray tube (CRT), a glass vacuum tube that operated at high voltage to discharge electrons at high velocity that bombard the anode, resulting in X-ray emission. The X-ray tube and detector moved together in parallel to scan across the width of the subject, from which the measurements were acquired to produce a view or exposure. The source and detector rotated around the subject, typically at 1° intervals, and scanned across the subject at each interval. This process of translation and rotation was repeated to obtain many views over 180°, and the combined views were then reconstructed to comprise one transverse section or slice of the subject. Scans took 5–6 min, which introduced movement errors, thus good images of the body trunk were acquired with difficulty (Goldman 2007).

Second-generation scanners were developed in the mid-1970s and used multiple X-ray beams coupled to multiple detectors, allowing for simultaneous acquisition of many images, thus reducing scan time. For example, 30 detectors reduce the number of movements and also the scan time by a factor of 30, making trunk and whole-body scans possible. Nevertheless, the mechanical complexity of the rapid translational and rotational movement and realignment of the many scanner components became a major limitation on reducing scan time (Mahesh 2002; Goldman 2007).

The first third-generation scanners appeared in late 1975. Third-generation scanners utilized an X-ray tube that produced a fan beam covering the entire width of the subject. Measurements were acquired by a semicircular array of detectors, each receiving a small portion of the fan beam. This eliminates the translational movement, with the X-ray tube and the detector array rotating in tandem around the subject. Fourth-generation scanners have an entire ring of detectors, and only the X-ray tube rotates; however, only a fraction of the detectors are utilized at one time. Technological advances improving third-generation scanners and the high cost of detectors have virtually eliminated the use of fourth-generation scanners (Goldman 2007).

A major improvement in CT design came in the early 1990s with the use of slip-ring technology. Slip rings are electromechanical devices consisting of contactors and brushes that transmit electricity from stationary components to rotating components. Previously, after each full rotation during a scan, data acquisition stopped while the cables were unwound for the next rotation. Slip rings allow data from detectors to be transmitted wirelessly, making continuous rotation possible. Thus, interscan delays are limited to the movement of the patient bed to the next slice position (Mahesh 2002; Goldman 2007).

Another major development around 1990 was helical (also called spiral) CT scanning. The patient bed moves smoothly as slices are continuously acquired, creating a spiral path relative to the patient. The movement of the bed per rotation divided by the slice thickness is defined as the pitch. Adjusting the pitch is a trade-off between coverage and accuracy; greater pitch provides more coverage, but lower resolution for the same scan period (Goldman 2007). Since projections are not all in a single plane, new reconstruction algorithms were needed. Interpolation methods to generate projections in a single plane were developed allowing data to be reconstructed into three-dimensional radiographs as well as allowing the display of one slice image at a particular point (Mahesh 2002).

The latest major advance in CT technology was the introduction of multi-slice scanners in the 1990s. Multi-slice scanners have rows of detector arrays for acquisition of several slices at the same time, allowing more coverage to be achieved with each rotation. This greatly reduces the time necessary for whole-body scan while increasing resolution of the images. Faster scanning reduces subject motion artifacts and thinner sections improve resolution. Thus, multi-slice helical CT decreases scan time while yielding high-resolution, near-isotropic images (Rydberg et al. 2000; Mahesh 2002; Flohr et al. 2005; Goldman 2008).

Ultrafast electron-beam CT (EBCT) technology has been developed for applications requiring especially high-speed acquisition, such as imaging the heart (Goldman 2007). An electron beam from the cathode is focused and deflected to strike a point on the anode, producing a very fine X-ray beam which can scan across the anode very quickly. In EBCT scans, the X-ray tube and detectors do not move; instead the electron beam moves. Another method of scanning the heart is cardiac-gated CT, in which image acquisition is coupled to the heart rate so that movement blur is minimized (Flohr et al. 2005).

Under development is flat panel volume CT (VCT) whereby a wide-angle X-ray tube produces a cone beam. The detector array is a large panel such that imaging the volume of interest takes only one rotation. Flat panel VCT is useful for dynamic (4D) imaging that is most beneficial during cardiac surgical procedures (Gupta et al. 2008).

20.1.2 CT Acquisition and Image Reconstruction

X-ray tube and scanner design have advanced simultaneously given that sources with rotating anodes, higher power, and higher heat capacities were required (Mahesh 2002). X-ray tubes are typically operated at voltages of 120 eV to 120 keV, depending on the thickness and density of the tissues. Collimators are used to restrict and shape the X-ray beam, and filters are used to adjust the intensities of penetrating X-rays. As the ionizing radiation passes through the subject, the attenuated energy impinges the detector and the energy is translated into electrical current proportional to the intensity of the energy. Early scanners used detectors made of sodium iodide or cesium iodide crystals coupled to photomultiplier tubes (PMTs). The energy of the X-ray is absorbed by the crystal, causing it to emit photons, which are captured by the PMTs and converted into electrical current. These crystals sometimes have continued fluorescence called afterglow and were later replaced by other crystals such as bismuth germinate and calcium tungstate to alleviate afterglow. Third-generation scanners began using gas detector arrays. These are detector arrays with chambers containing pressurized xenon gas separated by long, thin electrically charged tungsten plates. When radiation strikes the detector, the gas becomes ionized, producing an electrical current. Solid-state detectors, now used in most advanced CT scanners, contain scintillation materials (most commonly ceramic with a rare-earth element such as yttrium or gadolinium) coupled to photodiodes to convert the light to electrical current (Mahesh 2002; Goldman 2007).

CT image reconstruction involves analyses that divide a slice of the subject into small, three-dimensional boxes called voxels. Each voxel will have been traversed by many rays yielding many measurements of the attenuated beams. Since each measurement is related to the sum of the attenuation values from all of the voxels in one X-ray path and each voxel is included in many measurements, it is possible to computationally estimate the attenuation value for each voxel (Mahesh 2002; Goldman 2007). Another computational method of assessing the data is to generate a sinogram whereby the acquired data of a cross-sectional image is transformed by the Radon transformation function, a function closely related to the Fourier transformation function (Bazañez-Borgert 2006; Goldman 2007). Two fundamentally different types of computational processes have been developed for tomographic analyses; iterative and analytical. The iterative approach is called the algebraic reconstruction technique (ART) and was developed first; however, since measurements are typically "noisy" (high levels of random error), ART estimates are very time consuming, lie within some tolerance range of the noisy data, and give rise to blurred images. One analytical approach involving back projection was then developed; however, this also produced blurred images. To better resolve the images, functional analysis using convolution, a mathematical operation combining two functions to obtain a third function or filter function, can be applied to each view before back projection (Mahesh 2002; Goldman 2007). This reconstruction algorithm design is called filtered back projection (FBP) and utilizes Fourier analyses in its computational processes (Bazañez-Borgert 2006). With the continuing increase in computational power and the advances in scanner technology, more advanced reconstruction algorithms continue to be developed. For example, adaptive statistical iterative reconstruction (ASIR) combines FBP with iteratively reconstructed images, allowing the use of a lower radiation dose to obtain an image of comparable quality to using FBP analysis alone (Flohr et al. 2005; Fleischmann and Boas 2011). Model-based iterative reconstruction (MBIR) is an iterative technique that includes parameters to model the geometry of the scanner while also lowering the radiation dose needed for quality images (Fleischmann and Boas 2011).

In the end of the analyses, computed attenuation values (μ) are eventually converted into CT numbers or Hounsfield units (HU), which are based on the attenuation in water (μ_{water}), by the following equation:

$$\text{CT number}\,(\text{HU}) = \frac{m - m_{\text{water}}}{m_{\text{water}}} \times 1{,}000$$

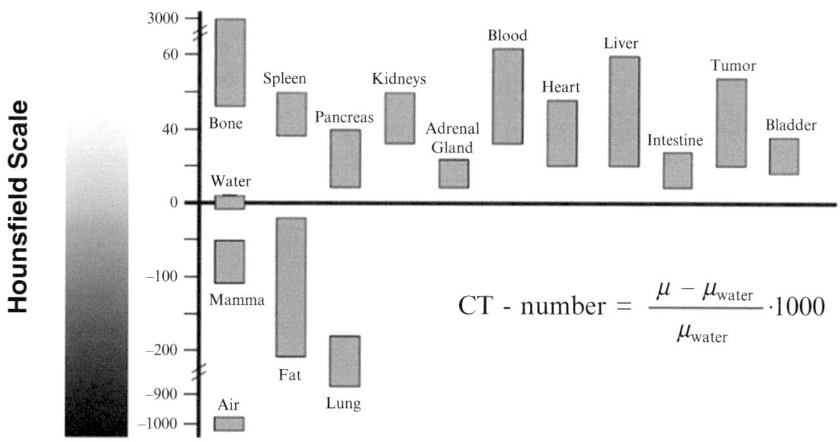

Fig. 20.2 The Hounsfield scale. The Hounsfield unit (HU) is a linear transformation of the attenuation coefficient of a substance's radiodensity relative to that of water at standard pressure and temperature (STP). The density of water is defined as zero HUs, while the radiodensity of air is −1,000 HUs (adapted from Bazañez-Borgert 2006)

Subsequently, CT numbers are represented on the Hounsfield scale by pixels, the two-dimensional equivalent of voxels, with densities that fall within a gray scale range of 4,000 HU (Fig. 20.2). The range of the scale determines the intensity and contrast of image produced; therefore, a large range shows most major structures encompassing a broad range of densities. To visualize fine structure in tissues of more similar density, a small range of Hounsfield numbers can be chosen (Bazañez-Borgert 2006). Separate two-dimensional images of the three-dimensional volume or voxel can be viewed, or alternatively a three-dimensional image can be generated from reconstructed two-dimensional exposures.

20.2 Emission Computed Tomography

Emission computed tomography (ECT) comprises several noninvasive imaging techniques which are similar in principle to CT, but different due to their source of radiation. While CT uses external X-rays, ECT uses probes introduced into the subject as an internal source of radiation. These specific probes are conjugated to radioisotopes as sources of radiation, which results from the radioactive decay of the nuclide. Isotopes of an element have the same atomic number (number of protons), but different mass numbers (thus different numbers of neutrons). Radioisotopes have unstable nuclei and undergo radioactive decay, yielding α-, β-, γ-, and/or X-ray emissions, depending on the nuclide and the type of decay. Unfortunately, very few radioisotopes are suitable as probes for in vivo imaging. A suitable probe should have the following characteristics: (1) it accumulates to a higher concentration in target (organ/tissue) than nontarget regions; (2) it typically possesses a high binding affinity suitable to allow the completion of a study within a reasonable time; (3) a low dose provides quality images with as low as reasonably achievable (ALARA) radiation exposure; (4) preparation is simple, convenient, fast, and cost-effective; and (5) it affords little interference with normal physiology.

In ECT, the subject is administered radionuclide probes that differentially bind to cells or tissues. The probes must bind targets with high sensitivity and specificity in order to achieve a higher signal-to-noise ratio. Targets can be measured qualitatively or quantitatively and the results used to assess pathophysiological processes. Examples include characterization of tumors and tumor burden (Bybel et al. 2006), examination of dopamine transporters and dopaminergic receptors in Parkinson's disease (Trott and El Fakhri 2008), and monitoring increased

Table 20.1 Common β^+ emitting radionuclides for PET

Isotope	Energy max (MeV)	Half-life	Production
^{11}C	0.959	20.4 min	Cyclotron
^{13}N	1.197	9.96 min	Cyclotron
^{15}O	1.738	2.07 min	Cyclotron
^{18}F	0.65	109.8 min	Cyclotron
^{68}Ga	1.899	68 min	Cyclotron
^{124}I	1.532, 2.135	4 days	Reactor

Modified from Badawi (1999)

expression of peripheral benzodiazepine receptors by microglia in neuroinflammation (Doorduin et al. 2008). Cellular processes can also be analyzed, such as the migration of peripheral blood mononuclear cells to identify cryptic inflammation (Meikle et al. 2005) and gene incorporation (Acton and Zhou 2005).

The two types of ECT are positron emission tomography (PET), which uses positron (β^+) particle-emitting radioisotopes, and single photon emission computed tomography (SPECT), which uses gamma (γ) photon-emitting radionuclides.

20.2.1 Positron Emission Tomography

PET is the most sensitive ECT imaging method for analyzing physiological processes in vivo (Bazañez-Borgert 2006). PET probes are pharmaceuticals or substrates that are involved in metabolic processes or bind a particular tissue/cell type of interest. These probes are conjugated with radionuclides that emit positrons (β^+ particles, antielectrons, or antimatter). When a positron interacts with matter (subject tissues) and encounters a low-energy electron, those particles annihilate each other to produce a low-energy γ-photon. The emitted γ-photons are detected by a PET scanner and provide an image rendition that allows subsequent quantitation of intra-tissue probe levels (Townsend 2004).

20.2.1.1 PET Probes

Table 20.1 lists the positron-emitting radionuclides useful for PET, their energies, half-lives, and instrumentation needed for their production. Most radionuclides for PET probes are produced in a cyclotron (Bazañez-Borgert 2006) and generally have relatively short half-lives. This property is considered favorable for achieving radiation exposures that are as low as reasonable (i.e., within ALARA guidelines), however, this very property precludes shipping the probe over long distances. Thus, use of PET usually requires that the isotope be prepared within a short distance of the patient and detecting instrumentation (PET scanner). This usually necessitates that a cyclotron be within close proximity to the study site. The majority of PET applications utilize ^{18}F due to several advantageous properties such as longer half-life, which allows for metabolic and quantitative studies, and lower energy level of the positron (Table 20.1) which minimizes subject radiation exposure and results in a shorter distance to annihilation; the latter improves image quality (Townsend 2004; Basu et al. 2011). As an example, ^{18}F-labeled-deoxyglucose (FDG) is widely used in oncology, since glucose uptake tends to be increased in cancer cells compared to surrounding tissues (Warburg 1956). FDG uptake parallels normal glucose uptake, but is retained in cells and not metabolized further (Bybel et al. 2006). Bone-scanning procedures utilize ^{18}F in the form of NaF (Townsend 2004), while L-6-[^{18}F]fluoro-3,4-dihydroxyphenylalanine (^{18}F-DOPA) is used for studies of dopaminergic neuron uptake in Parkinson's disease (Pavese et al. 2011). For more information, a compendium of probes for PET and CT can be found in the *Molecular Imaging and Contrast Agent Database* (*MICAD*) on the NCBI Bookshelf (http://www.ncbi.nlm.nih.gov/books/NBK5330/).

20.2.1.2 PET Detectors

When a positron (β^+) emitted from a PET probe collides with an electron (β^-) in the subject (usually within 1 mm), both particles are annihilated and produce two equal energy (511 keV) γ-photons (Fig. 20.3a). These γ-photons radiate oppositely from the point of impact at a 180° angle to form a line of response (LOR). To capture these events, most PET scanner detectors are arranged in parallel rings surrounding the subject or may have a two-head camera system that rotates around the subject (Robilotta 2004). Electronic coincidence detectors report only coincidence events, those events that are detected within a certain time interval, usually a few nanoseconds, and

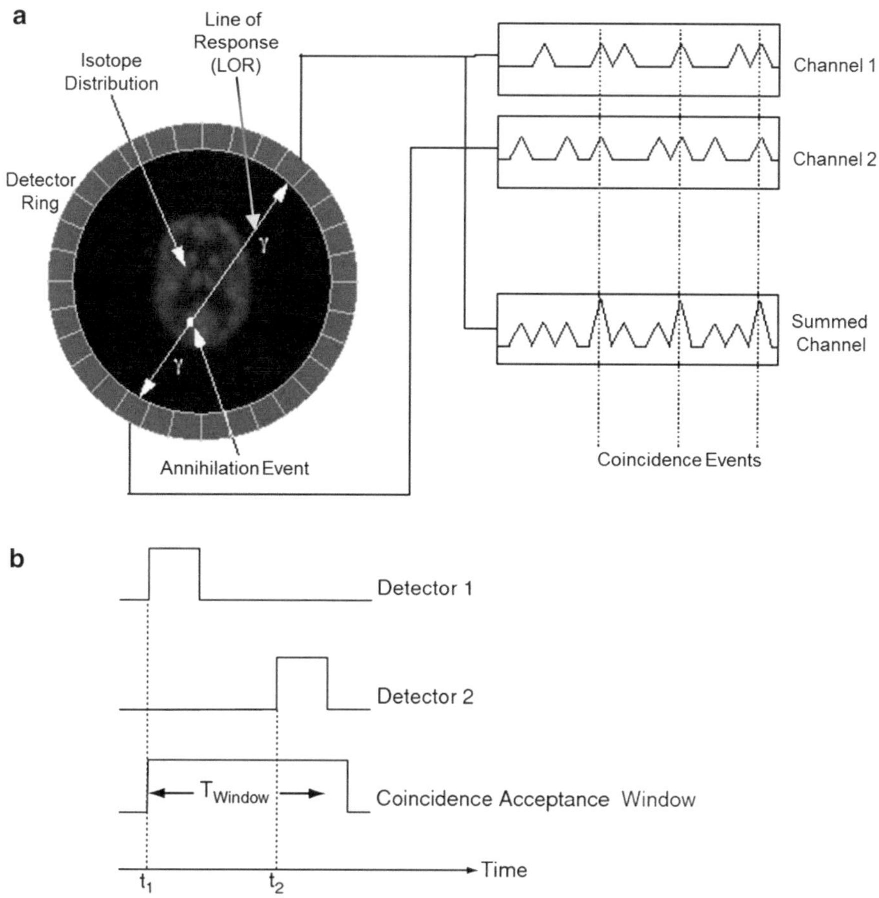

Fig. 20.3 Schematic of coincidence detection of PET probes. (**a**) The PET detector ring is comprised of many individual detectors situated such that paired detectors are 180° in opposition. The annihilation of a positron (β^+) produces two photons of equal energy that traverse in opposite directions, which form the basis of coincidence detection. Both photons strike opposing detectors within nanoseconds of each other and produce electronic pulses in both channel 1 and channel 2. These pulses are then combined in coincidence circuitry, and if the pulses fall within a short time window (T_{window}), they are deemed coincident and are counted (adapted from Badawi 1999). (**b**) Schematic of coincidence timing within a short time window, whereby an incident γ-photon induces in detector 1 a scintillation signal at time t_1 which triggers a coincidence acceptance window with a temporal width (T_{window}) within which a second signal must coincide. If a second detection event occurs in another detector at time t_2 whereby the time differential is less than the acceptance window, then the signals are coincident and the signal is counted. If the time differential is greater than the temporal window, then the event is ignored (adapted from McParland 2010)

occur in opposite directions (Lewellen 2008). This process is termed electronic collimation and provides increased sensitivity and accuracy of the point source whence the annihilation event occurred (Bazañez-Borgert 2006). Use of time-of-flight (TOF) PET detection, wherein the arrival time difference between the two photons is measured (or coincidence time resolution), constrains the two-photon emission point to a particular segment along the LOR and affords greater spatial accuracy of the annihilation event, thus leading to better resolution of the image (Fig. 20.3b) (Moses 2007).

Early PET scanners used scintillation crystals such as thallium-doped sodium iodide, NaI(Tl), coupled to PMTs. However, this one-on-one coupling limited spatial resolution due to the size of the PMTs. Block designs were developed using arrays of crystals viewed by arrays of PMTs. Many other scintillators are currently used, including bismuth germinate (BGO), germanium orthosilicate (GSO), and more commonly lutetium

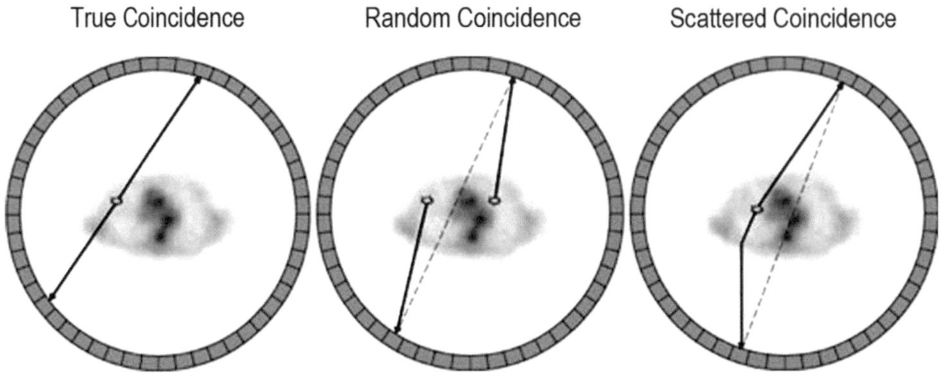

Fig. 20.4 Possible types of PET-associated coincidence detection events. With PET detection, the detector generates a pulse with each incident γ-photon and if the pulse falls within the temporal acceptance window, the pulse is deemed to be coincident and is counted. Coincident events fall into three categories: true coincidence (*left*), random coincidence (*middle*), and scattered coincidence (*right*) (modified from Bengel et al. 2009)

oxyorthosilicate (LSO) or lutetium yttrium orthosilicate (LYSO). The brighter scintillators constructed with lanthanum chloride and cerium-doped lanthanum chloride, among others, are used in TOF PET scanners, which measure coincidence intervals to reduce statistical noise (Moses 2007; Lewellen 2008). PMTs have improved and now have multiple dynode chains within the same vacuum tube. Alternatives to PMTs that have been developed include avalanche photodiodes and silicon PIN diodes; the latter composed of an intrinsic (I) semiconductor region between p (P)- and n (N)-type semiconductors. These offer one-to-one coupling and greater space efficiency (Lewellen 2010).

Originally, PET scanners had lead or tungsten separators between detector rings to shield the detectors from scattered photons. However, this arrangement limited reconstruction to two-dimensional image sets and decreased the efficiency of radiation detection. Scanners with retractable septa were developed in the 1990s to allow the generation of two-dimensional or fully three-dimensional images (Townsend 2004).

20.2.1.3 PET Acquisition and Image Reconstruction

PET image reconstruction involves compilation of coincidence data from many LORs and performing various algorithms to generate an image. Coincidence events can occur as true annihilation events or as unrelated annihilation photons detected within the coincidence time window superimposing low-frequency noise over true signal events. Noise could be due to scatter of the photons (scatter), either inside or outside of the patient, or to random coincidence events (randoms) via various unknown mechanisms (Fig. 20.4). True events occur when both photons from an annihilation event are detected in coincidence. Scattered coincidence occurs when one or sometimes both photons have undergone Compton scattering whereby the incident photon energizes an electron. Scattered coincidence events add background to the true coincidence distribution, decrease contrast, and lead to overestimation of the count. Random coincidence occurs when two γ-photons, not derived from the same annihilation event, are incident within the temporal pulse acceptance window and are counted. As with scatter, inclusion of random coincident events will add to the background noise and lead to overestimation of the count and isotope concentration. Attempts to increase the net true events and reduce unrelated annihilation events have been made by narrowing the coincidence time windows. BGO detectors use a typical 12 ns coincidence time window; whereas GSO and NaI detectors reduce the coincidence window to 8 ns, and LSO-based systems can reduce the window to 6 ns. Photon scatter also can attenuate true coincidence measurements; therefore, scatter events not only can simulate true coincidence events but also replace them.

However, the amount of attenuation can be determined from a transmission scan, whereby a scan is performed using a radiation source that is external to the subject and produces attenuation patterns similar to those obtained from sources within the subject. Resulting transmission scans can then be used to filter the within-subject scan. Other methods involve storage of all coincidence events (trues, scatter, and randoms) and also all single non-coincident events detected during data acquisition. By using mathematical formulas, corrections due to scatter, randoms, and attenuation can be easily sorted, filtered, and subtracted prior to reconstruction (Townsend 2004; Bazañez-Borgert 2006). Although simple and efficient, this method requires large memory capacities for storage and exceedingly long processing times. One faster and more common correction method estimates randoms and scatter with a delayed coincidence window method, whereby all coincidence events processed into the delayed window are known to be totally unrelated and consequently used to estimate the level of randoms and scatter from the true signals.

Three-dimensional reconstruction algorithms involve filtering and back projection (FBP). Faster rebinning methods such as Fourier rebinning (FORE) and single-slice rebinning (SSRB) of three-dimensional data sets into two-dimensional sets also have been developed. This allows the use of many algorithms available for two-dimensional reconstruction, such as ordered subset expectation maximization (OSEM) (Townsend 2004; Basu et al. 2011) and the maximum likelihood expectation maximization (MLEM) algorithm (Robilotta 2004). Additionally, four-dimensional reconstruction methods and algorithms have been developed to correct for blurring due to subject movement during the scan, such as cardiology PET techniques (Tang et al. 2010; van Elmpt et al. 2011).

20.2.1.4 PET/CT

PET imaging alone often lacks anatomical structure references, making it difficult to interpret images relative to subject landmarks (Basu et al. 2011). Therefore, to provide critical anatomical landmarks to PET images, CT images are superimposed or co-registered with PET images to provide greater ability for spatial recognition within the PET image. The first integrated PET/CT scanner was developed in the late 1990s (Beyer et al. 2000). Images for PET and CT scanners are not acquired simultaneously since the CT and PET detection arrays are normally located such they cannot operate in tandem. Therefore, the CT image is rapidly acquired first, and then the longer PET scan is performed. The CT scan can also serve as a transmission scan since it is an external radiation source (see Sect. 20.2.1.3), whereby attenuation correction can be determined and utilized for reconstruction and analysis of the PET scan, thus eliminating the need for a transmission scan using the same isotope (Basu et al. 2011).

PET/CT is used for diagnosis, staging, and restaging in oncology, especially for lymphoma, melanoma, and cancers of the lung, breast, esophagus, head, neck, and colon. It is used in cardiology for assessing myocardial viability. Applications in neurology include diagnosis of Alzheimer's disease, indicated by regions of decreased glucose metabolism, and the localization of the focus site of epileptic seizures for surgical treatment (Bybel et al. 2006). Dynamic, quantitative PET/CT can be used for kinetic studies for clinical or research purposes (Seo et al. 2008b).

20.2.2 Single Photon Emission Computed Tomography and SPECT Probes

SPECT is another noninvasive ECT imaging modality. Similar to PET, it is based on administration (usually by parenteral routes) and internalization of nanomolar to picomolar amounts of tracer molecules (less than concentrations required for a pharmacological response) that are labeled with a radioactive nuclide (Accorsi 2008). In contrast to PET, which utilizes nuclides which emit positrons, SPECT employs radionuclides that emit γ-radiation that generally results from radioactive decay known as isomeric transition, the transition of an unstable isotope to a more stable ground state with the ensuing loss of energy (Bazañez-Borgert 2006). Gamma-decay emits photons within a spectrum of γ-radiation that possess energies of 35–700 keV depending on the

Table 20.2 Common γ-emitting radionuclides for SPECT

Isotope	Max γ energy(keV)	Half-life	Production
^{99m}Tc	140	6.0 h	^{99}Mo generator
^{201}Tl	70–80, 135, 167	73 h	Cyclotron
^{67}Ga	93.5, 184.5, 296, 388	78 h	Cyclotron
^{123}I	159, 285	13.3 h	Cyclotron
^{125}I	20–35 (27 keV X-ray)	59 days	Reactor
^{131}I	284, 364	8 days	Reactor
^{111}In	23, 171, 245	67.5 h	Reactor
^{153}Sm	103 (also β^-)	46.3 h	Reactor

nuclide. Unlike PET, where annihilation of a positron emits two photons of equal energies in opposite directions, nuclides used in SPECT produce a single, univector photon per decay event (hence, single photon emission). Because of the nature of the photon, several issues of detection relative to instrumentation and sensitivity arise that are not encountered in PET imaging (see Sect. 20.2.2.1). Although PET sensitivity is generally several orders of magnitude higher than SPECT, radionuclides used in SPECT (Table 20.2) generally have longer half-lives, allowing for studies of longer duration (Rahmim and Zaidi 2008). Moreover, the relatively longer half-lives of SPECT isotopes permit transport over longer distances, thus sites of isotope and probe manufacture (e.g., cyclotrons or reactors) do not need to be in close proximity to study sites.

For probe manufacture, peptides and antibodies can be labeled with technetium (Tc) or iodine (I) relatively easily; therefore probes are readily available for SPECT studies, and dual-isotope imaging is possible with most systems (Meikle et al. 2005). The most commonly used isotope in SPECT is ^{99m}Tc (Robilotta 2004), with ^{99m} Tc-abeled hexamethylpropyleneamine oxime (^{99m}Tc-HMPAO) being commonly used for perfusion studies (Accorsi 2008). Together with correction factors for attenuation, SPECT analyses can yield quantitative information (Rosenthal et al. 1995; Meikle et al. 2005) with a general expected resolution of at least 0.8 mm (Funk et al. 2006). A database of SPECT probes can also be found in the *MICAD* on the NCBI Bookshelf (http://www.ncbi.nlm.nih.gov/books/NBK5330/).

20.2.2.1 SPECT Imaging Systems

SPECT imaging systems detect γ-emissions in a manner necessary for an accurate image of the source to be obtained. A SPECT imaging system consists of one or more cameras, which in turn consist of a detector or detector array and at least one collimator for each camera. Most systems have interchangeable collimators which allow modification of sensitivity and resolution. The collimator permits only photons traveling along a particular path to reach the detector, thus providing directional information about the path. The detector(s) is comprised of photon-sensitive materials that convert the photon energy into an electronic signal and send it to a computer. This allows for the detection of γ-emissions, whose source can then be spatially derived and used as a noninvasive, structural investigation of a subject.

Non-scattered photons from a radiation source will exit the subject in a straight line or LOR. The path of a photon must be known in order to pinpoint the location of its emission source (probe). To accomplish this, collimators serve as filters, allowing the passage of γ-rays with specific trajectories. A collimator placed between the subject and the detector is permeable only to photons with specific LORs, thus allowing the specific location of the radiation source to be determined (Fig. 20.5). Without a collimator, divergent LORs from a source would strike the detector in many different places with divergence proportional to the distances and angles by which the photons traverse (Fig. 20.5a). However with a collimator in place, divergent photons are blocked and only those with a LOR aligned with the collimator's channel are allowed to enter the camera and be detected (Fig. 20.5b). To block photons with divergent LORs, collimators are made of dense materials, usually tungsten or lead. Thus, blockage of undesirable photons and passage of only a small percentage of photons to the detectors result in the loss of relative sensitivity compared to the PET systems, which do not require collimators to track the LOR.

Collimators also vary greatly in design depending on the type of imaging that is to be performed (Accorsi 2008) (Fig. 20.6a). Parallel-hole collimators, most commonly used in clinical SPECT systems, consist of parallel channels arranged in a regular three-dimensional lattice (Madsen 2007). The projected image resulting from photons reaching the detector through parallel channels will represent the actual size of the source object.

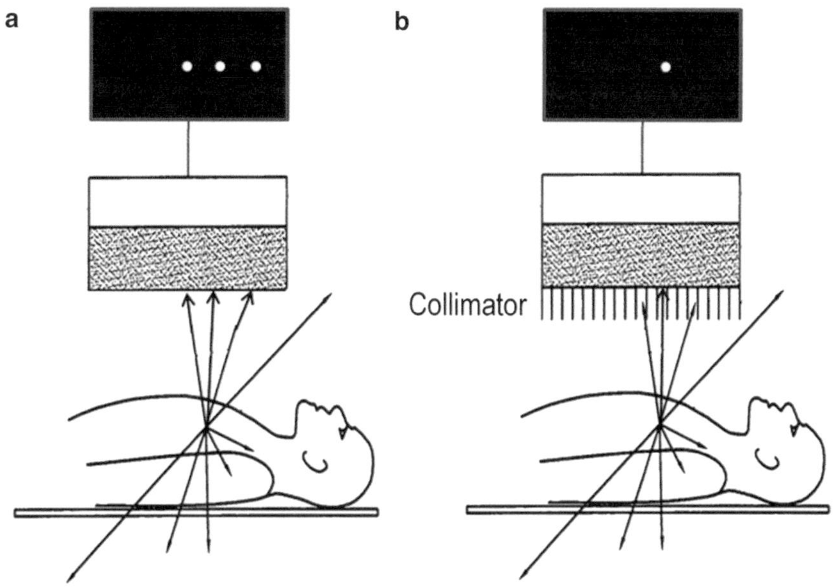

Fig. 20.5 Necessity of collimation in single photon emission computed tomography (SPECT) analysis. (**a**) When no collimator is used, multidirectional single photons from a single point source impinge on the detector and are scattered over the range of trajectory angles emitting from the source. Thus, no clear trajectory can be established from whence the photons originated, and the image appears blurred due not only to detection of the multitude of signals from one point source (shown here as three signals) but also to the exponential number of signals that are overlapping from the other point sources in the body. (**b**) A collimator uses channels or pinholes that allow single photons on a parallel or near-parallel trajectory to enter the detector. The other photons impinge the channel wall or are severely restricted due to the minute pinhole. Thus only a small percentage of photons are utilized to track the trajectory back to the point source such that a clear image of the point source can be resolved (shown here as one signal) (modified from Shwartz and Ohana 2003)

The channels are of a finite width and length, thus allowing photons with near-parallel LORs to be accepted through the collimator to the detector (Fig. 20.6b). Wider or shorter channels allow more photons to pass and will increase sensitivity with the improvement of counting statistics, but at the expense of increased noise (Fig. 20.6c). In contrast, longer or narrower channels limit the number of photons allowed to the detector with lower sensitivity and counting statistics, but with diminished noise to signal ratio (Fig. 20.6d). Compared to the photon acceptance angle of shorter or wider channels (Fig. 20.6c), longer or narrower channels diminish this angle that defines more compacted LORs and increases the spatial definition leading to better resolution (Fig. 20.6d). Thus, for SPECT analysis, resolution must be balanced with limits of detection efficiency. If high resolution is not required, high sensitivity will minimize the subject's radiation exposure (Accorsi 2008). On the other hand, increased resolution may be a key parameter to better discriminate spatial location within the organs or tissues of small animals.

With parallel collimators, the photons from the object are projected along parallel lines to the detector and the projection is the same size as the object. With converging collimators, the channels converge along the "focal line" to a "focal point" beyond the object (Fig. 20.6a, e). This distance from the focal point to the collimator is the focal length of the collimator. Projection with a converging collimator results in magnification of the object onto the detector and the extent of magnification increases as the distance from the source to the focal locus decreases (Accorsi 2008), but since magnification varies with distance, some image distortion will occur (Bazañez-Borgert 2006).

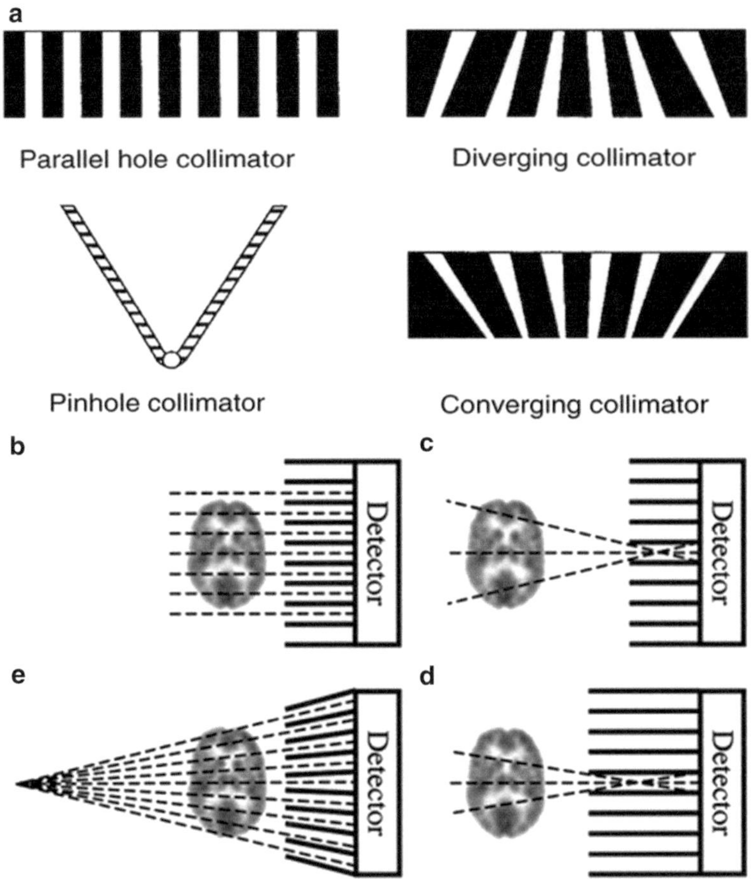

Fig. 20.6 SPECT collimator designs. (**a**) To accomplish different goals, several different types of SPECT collimators are employed including parallel-hole collimators, diverging and converging collimators, and pinhole collimators (adapted from Saha 2010). (**b–d**) With parallel-hole collimators, images of the subject are projected onto the detector without magnification, and the projections do not use the entire detectors' FOV. Incident γ-photons reach the detector only by traveling along LORs (*dashed lines*) that are parallel to the collimator channels. (**c**, **d**) The width and length of the collimator channel allows the LORs that are nearly parallel to the collimator channel also to reach the detector. Therefore, channels that are shorter and wider (**c**) increase the acceptance angle of the γ-photon LORs with enhanced sensitivity, but diminished resolution; whereas longer and narrower channels (**d**) decrease the acceptance angle with diminished sensitivity, but enhanced resolution. (**e**) Convergent collimators permit the γ-photon LORs to converge to a focal point beyond the subject, and when using the entire FOV of the detector affords a magnified projection onto the detector (adapted from Accorsi 2008)

In addition to the benefit of a magnified image and despite a blurred image, resolution is improved compared to parallel collimators due to the fact that the object's proportions are maintained by the magnification geometry. Moreover, sensitivity is increased due to the principle that as the source moves away from the collimator, each channel collects fewer photons because of the increasing distance; however, more channels are utilized in a converging collimator and more photons pass. The trade-off with converging collimators is that the field of view (FOV) is diminished compared to parallel collimators. However, for small animal or single organ (e.g., brain) evaluations, a large FOV may be unnecessary. Examples of converging collimators include cone-beam collimators, fan beam collimators, and astigmatic collimators.

Diverging collimators have channels that diverge from the detector face (Fig. 20.6a), but have a line of focus that converges on a point behind the collimator and detector. Photons reaching the detector through divergent collimator geometry will produce a smaller, non-inverted image of the source, thus the projection of the object is minified. This increases the FOV and the useful image area becomes larger; however, resolution and sensitivity are generally diminished. Increases in FOV and useful area make this collimator particularly useful with smaller detectors or for imaging larger organs or larger animals (Bazañez-Borgert 2006).

A pinhole collimator (Fig. 20.6a) primarily consists of a pinhole aperture bored into the end of a cone made of a high-density metal such as lead, tungsten, or platinum. An inverted image of the source is projected onto the detector, and the image is magnified if the distance from the source to the aperture is less than the length of the cone (Bazañez-Borgert 2006). With pinhole collimators, resolution and sensitivity are maximal when the collimator is in close proximity to the subject, not at a point further removed as resolution and sensitivity diminish with increased distance from the subject. However, greater sensitivity and resolution are afforded close to the subject, but at the expense of a smaller FOV. Because a limited FOV is typically sufficient, the added benefit of increased resolution makes pinhole collimators most useful for small animal imaging (MacDonald et al. 2001; Cao et al. 2005; Meikle et al. 2005; Funk et al. 2006; Accorsi 2008). Since sensitivity suffers from the smaller pinhole aperture, systems have been developed to increase sensitivity (Cao et al. 2005; Funk et al. 2006). These systems usually include multiple detectors with pinhole collimators surrounding the animal to obtain nonoverlapping views and multiple-pinhole collimators that project overlapping views onto the detector (Fig. 20.7) (Meikle et al. 2005; Engdahl et al. 2007). Increasing the number of pinholes decreases the effective acceptance angle, the focal length, the viewing area, and the magnification (Table 20.3) (Engdahl et al. 2007). More importantly, with increased pinhole number, the sensitivity increases but at the expense of resolution.

Photons passing through the collimator interact with the detector activating the detector's photon sensing circuitry. This entire system is often referred to as a gamma camera since it basically requires γ-photons to produce an image. Most gamma camera detectors are composed of a scintillator and an electronic light sensor (ELS) (Peterson and Furenlid 2011).

SPECT detector scintillators are made of dense materials, usually inorganic salts doped with small amounts of rare-earth elements, such as thallium (Tl) or cerium (Ce). These materials are noted for their property not only to readily absorb photonic energy but to reemit that energy in the form of light or a "scintillation." The most common type of gamma camera is also called an Anger camera, developed by Hal Anger (Anger 1958). Earlier Anger-type cameras use a large-area, thin sodium iodide, thallium-doped [NaI(Tl)] crystal coupled to a photomultiplier (PMT) array. Other types of scintillators used in gamma cameras include bismuth orthogermanate (BGO) and cesium iodide laced with thallium [CsI(Tl)]. The detector scintillator may be constructed as large blocks or in arrays of small smaller blocks or pixels (Madsen 2007; Accorsi 2008; Peterson and Furenlid 2011). When incident γ-photons reach the detector material, the ionizing energy of the photon is absorbed by the scintillator, whereby the ionizing energy is converted to radiant or luminescent energy which is reemitted as light or scintillation. The amount of light produced is proportional to the energy of the radiation (Bazañez-Borgert 2006), which is subsequently processed by an ELS.

Two types of ELSs commonly used in gamma cameras are PMTs and photodiodes. PMTs offer the advantage of high amplification gain, but have disadvantages of relatively low quantum efficiency, drifting gain, and cathode failure and are rather bulky in size. Newer PMTs include position sensitive PMTs (PS-PMTs) in which the position of the incident γ-photon on the PMT can be estimated from the electron distribution on a set of orthogonally arranged anodes. The newest generation of PS-PMTs provides larger anodes in array formations, thus are referred to as multi-anode PMTs (MA-PMTs). An alternative light to electrical energy conversion process is achieved with photodiodes. Photodiodes operate by absorption of γ-photons and generate a flow of current in an external circuit proportional to the incident

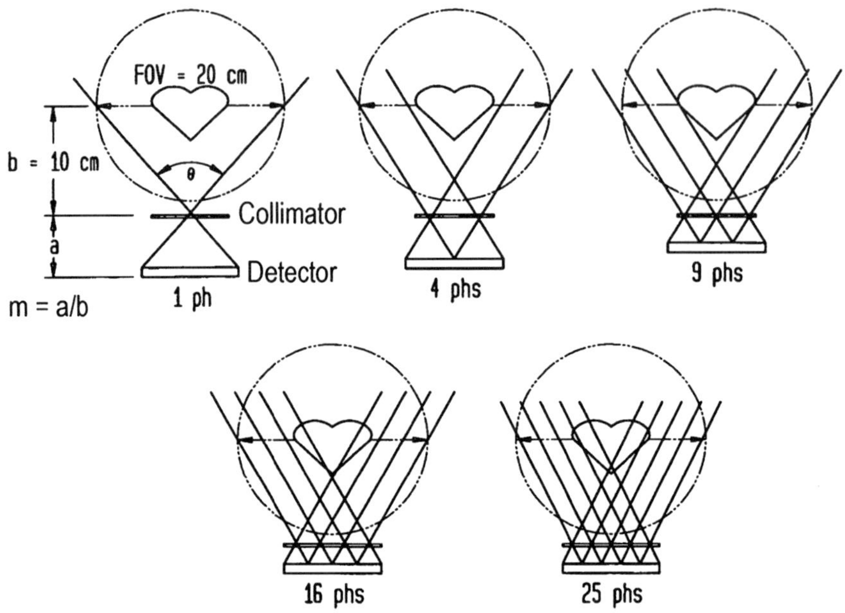

Fig. 20.7 Multiple-pinhole collimators. Examples of pinhole collimators with 1, 4, 9, 16, and 25 pinholes with diameters of 1 mm and an object to collimator distance fixed at 10 cm (*b*). For a one pinhole collimator with a focal length of 5 cm (*a*), imaging of a 20×20 cm^2 field of view with an acceptance angle of 90° provides a total circular viewing area of πr^2 (3.14×10^2=314 cm^2). With a magnification ($m=a/b$) of 0.5, the image is captured onto a 10 cm×10 cm detector with a spatial resolution of 3.6 mm and relative sensitivity of 1.0. With increasing numbers of pinholes, the entire area of the detector is utilized by segregated and nonoverlapping acceptance angles, but the acceptance angles and focal lengths diminish. The number of pinholes determines the balance of acceptance angle, focal length, viewing area, magnification, resolution, and sensitivity (Table 20.3) (modified from Engdahl et al. 2007)

Table 20.3 Effects of multiple-pinhole collimation on resolution and sensitivity

Number of pinholes	Acceptance angle (°)	Focal length (cm)	Detector area (cm^2)	Magnification	Spatial resolution (mm)	Relative sensitivity
1	90	5.0	100	0.50	3.6	1.0
4	73	3.4	25	0.34	4.9	4.6
9	67	2.5	11	0.25	6.4	8.6
16	64	2.0	6	0.20	7.8	12.1
25	62	1.7	4	0.17	9.1	14.4

Adapted from Engdahl et al. (2007)

energy of the photon. These devices are smaller and more efficient and can be individually coupled to scintillator crystals; however, the lack of signal amplification and capacitance of the electrodes limits their signal-to-noise ratio (Peterson and Furenlid 2011). Silicon drift detectors (SDDs) provide a current of electrons to a small anode with a lower capacitance that offers a higher efficiency for conversion of light to electrical energy while producing a lower signal-to-noise ratio. SSDs can cover a large active area and thus be used as tiles to cover a larger detection area. Avalanche photodiodes (APDs) operate at potentials that yield higher gain utilization and increase sensitivity. Charge-coupled devices (CCDs) and complimentary metal-oxide-semiconductors (CMOSs), which provide high noise immunity and low static power consumption, are currently being utilized in SPECT detectors (Madsen 2007; Peterson and Furenlid 2011).

As an alternative to scintillator technology, some gamma cameras use solid-state components that serve as semiconductor radiation detectors. These convert gamma rays directly into electric signal and have the potential for greater spatial and energy resolution, important for small animal and multi-isotope imaging, respectively (Accorsi 2008; Peterson and Furenlid 2011). The most common semiconductor for SPECT solid-state detectors is cadmium/zinc/telluride (CZT) (Madsen 2007; Accorsi 2008; Peterson and Furenlid 2011).

20.2.2.2 SPECT Acquisition and Image Reconstruction

In SPECT, many two-dimensional images are acquired as the gamma camera rotates around the subject. Systems often have two to four gamma cameras to speed data acquisition, keeping in mind that most of γ-photons will be undetected due to the use and physical configuration of the collimator. For typical SPECT studies wherein the entire ROI is contained within the FOV, thus not requiring axial scanning, only rotational movement is needed, usually rotating every 3–6° for each exposure. Circular or elliptical orbits are typically used depending on the system; however, as resolution is highest when closest to the collimator, most advanced scanners use elliptical orbits that are adjusted based on the proximity to the subject (Accorsi 2008). The projections are initially processed to generate sonograms or image representations of the raw data to be used later for tomographic reconstruction (Robilotta 2004). Eventually, all individual two-dimensional images acquired are reconstructed into a single three-dimensional tomogram.

Reconstruction utilizes two basic methods: FBP or iteration. In FBP, algorithms are commonly used in SPECT image reconstruction, and a filter function is necessary due to the blurring inherent to the back projection process. Projection data are weighted in a fashion determined by the filter selected and then projected along straight lines back into the reconstruction FOV (Accorsi 2008). This provides a fast reconstruction of the data, but does not readily compensate for inherent noise found in the SPECT data path. Also, image degradation along the data path is due principally to the finite resolution of the gamma camera and requires further filtering of noise. Over time, many different filter functions have been developed for FBP processes and typically are chosen to optimize signal-to-noise ratios and resolution. Three-dimensional filtering or volume smoothing is a method which filters in both dimensions of two-dimensional images and is used to further improve image quality. Another method of reconstruction uses iterative algorithms, which is analogous to choosing a number and having a colleague guess the number by providing information as being too high or too low with each guess or iteration. Similar to this analogy, this reconstruction algorithm uses an initial estimate of the image assuming all pixels have the same values as those that will be analyzed (forward projections). The estimated forward projections are then converged and compared with the actual acquired projections and are updated based on the actual projections. This process is repeated until no iterative update provides further clarification of the converged images and your colleague guesses the number. Iterative reconstruction not only has the advantage of adequately handling noise and other aberrations but is able to model other abnormalities of SPECT such as the high-energy γ-photon penetration of the collimator and heavy metal shielding of the detector. However, this comes at the expense of longer reconstruction times and computer capacity. Additionally, the number of iterations is critical; too few iterations will result in partial reconstruction, while too many iterations may yield a noisy, illegible tomogram. Since noise is inherent in the SPECT data path, thus limiting a total exact reconstruction of the object being imaged, several iterative algorithms have been devised to maximize the most likely activity that represents the γ-photon source. The most common of these techniques is MLEM algorithm. A more widely used technique to improve the speed of reconstruction is the OSEM algorithm, whereby comparisons and updates are based on a smaller number or sampling of projections that are chosen to maximize the input of new information at each iteration. SPECT reconstruction often utilizes MLEM and OSEM algorithms in conjunction with filter functions to obtain better resolution (Groch and Erwin 2000; MacDonald et al. 2001; Robilotta 2004). In dynamic SPECT analysis, wherein the biodistribution of the radiolabeled probe is constantly changing, multiple detectors and modifiable reconstruction algorithms are required to compensate for the changing radioactivity distribution during the course of the scan (Madsen 2007).

20.2.2.3 Multiple Isotope SPECT

A unique capability of SPECT, but not PET, is the ability to image two probes simultaneously by using isotopes with gamma emissions of different energies. This permits different metabolic processes or tissues to be analyzed during the same scan. Most commonly, the isotopes used are ^{99m}Tc with ^{201}Tl or ^{123}I. When two isotopes are used simultaneously, it is necessary to optimize the energy windows used for imaging, since many isotopes have overlapping energies or cross talk (Accorsi 2008; Trott and El Fakhri 2008). ^{111}In also has a low energy emission and can be used with a higher-energy emitter (Wagenaar et al. 2006). ^{125}I is another possibility for dual-isotope studies, used with either ^{99m}Tc or ^{123}I. Since ^{125}I has lower energy emissions, cross talk is less of a problem, although a more common concern is down-scatter, the detection of photons emitted by the higher-energy isotope in the energy window used for imaging the lower-energy isotope (Meikle et al. 2005). For instance, with dual-isotope SPECT using ^{99m}Tc- and ^{201}Tl-labeled probes, greater than 10 % of the total detected isotope is due to down-scatter caused by X-rays emitted from incident ^{201}Tl photons on the lead shielding. With ^{99m}Tc/^{153}Gd isotope combinations, a fraction of the detected ^{153}Gd isotope is from down-scatter that originates from ^{99m}Tc photons which only partially deposit their energy in the detector crystal (i.e., due to crystal interactions). Therefore, the use of high-energy isotopes requires high-resolution detectors that can adequately separate the different energy peaks of the isotopes (Wagenaar et al. 2006; Accorsi 2008; Trott and El Fakhri 2008). Gamma cameras that utilize CZT detectors typically have improved energy resolution compared to SPECT instruments that use NaI(Tl) crystals and yield higher contrast images with reduced cross talk. Not only does dual-isotope imaging hold great potential for diagnostic uses, but may be critical for research applications such following biodistribution of two drugs simultaneously or discerning differential distribution of drug and excipient (Wagenaar et al. 2006).

20.2.2.4 SPECT/CT

SPECT/CT, like PET/CT, has the advantage of providing the anatomical reference for nuclear imaging, and the CT also provides the added benefit of improved attenuation correction (Madsen 2007). While the original SPECT/CT design used only one detector for acquiring SPECT and CT images simultaneously, this proved unfeasible for clinical applications for several reasons including low CT resolution, small FOV, and lengthy scan times. Modern SPECT/CT systems have a similar arrangement to PET/CT scanners, with the two systems separated by about 50 cm (Bybel et al. 2008; Seo et al. 2008a). Small animal SPECT/CT scanners are able to position complete SPECT and CT systems around the same FOV, but scans are still acquired sequentially (Wagenaar et al. 2006; Seo et al. 2008a). SPECT/CT has been especially important for attenuation correction in myocardial perfusion studies, where attenuation could be construed as vascular deficiencies. In general, CT image co-registration with SPECT imaging greatly improves anatomical localization, and SPECT imaging is useful to differentiate malignancies from abnormalities or infection on CT images (Bybel et al. 2008).

20.3 Experimental Considerations for Animal Studies

The first step in planning an imaging experiment is to develop or obtain a labeled probe for use in the appropriate animal model. A database of bioimaging probes and media can be found in the *MICAD* (http://www.ncbi.nlm.nih.gov/books/NBK5330/). Examples of probes include, (1*R*)-2beta-carbomethoxy-3beta-(4-[^{123}I]-iodophenyl) tropane ([^{123}I]-β-CIT), which binds to the dopamine transporter and has been shown to be a useful in assessing surviving neurons in Parkinson's disease (Asenbaum et al. 1997; Gendelman et al. 2003). In cancer studies, molecules that bind to the tumor, such as tumor antigen-associated antibodies, can be radiolabeled and utilized as probes (Baranowska-Kortylewicz et al. 2005). One radiopharmaceutical agent, [^{111}In]oxyquinoline (oxine) is routinely used in the clinic to label autologous leukocytes. Labeled leukocytes are then injected back into the patient and followed by SPECT to find areas of inflammation. Similarly, this strategy can be employed in immune cell trafficking in animals used for studies of autoimmune or neuroinflammation (Gong et al. 2011).

Invariably, the proper amount of probe to be utilized must be determined. This will depend on the specific activity, the amount of probe utilized, and the type of radionuclide. If the probe is a pharmaceutical, dosing information is necessary, and if the probe is a marker, maximum signal without toxicity is desired. For small animals such as mouse or rat, activities in the range of 0.5–2.0 mCi per animal are typically good starting activities for initial pilot studies.

Biodistribution and elimination kinetics of the probe are often important to ascertain when probe detection is optimal during a tomographic study (i.e., maximal signal-to-noise ratio). However, the time required for data acquisition during a tomographic study may preclude that method for determining kinetics. On the other hand, planar studies, wherein the distribution of the probe can be monitored at intervals of seconds or minutes, may prove more beneficial. Typically for planar studies, the animal is anesthetized and placed prone under the γ-camera that is fitted with a parallel-hole collimator to allow for higher sensitivity and shorter acquisition times, but at the expense of spatial resolution. The probe is administered to the subject and static planar images are acquired at predetermined time intervals. Image analysis may show an initial period of systemic biodistribution when the activity of the total probe is present and the bound probe cannot be discerned. Eventually, the unbound probe is eliminated from the systemic circulation (washout) and the target-bound probe is retained for a period of time. The kinetics of biodistribution, washout, and target retention are idiosyncratic for each probe and target pair, and must be determined experimentally. Once the probe's kinetics and optimal time of detection have been determined, then tomographic studies comparing probe binding under control and test conditions within the animal can commence.

In a study of different polymer effects on a ligand targeting the bone growth plates, the bone binding moiety, Tyr-D-Asp_8, was labeled with ^{125}I and conjugated to polymers of *N*-(2-hydroxypropyl) methacrylamide of varying molecular masses (Wang et al. 2006). Planar analysis indicated that by 24 h, the biodistribution, washout, and target retention of [^{125}I]-Tyr-d-Asp_8 provided optimal detection of accumulated ligand in the skeleton (Fig. 20.8, Planar images). However, the 24 kDa copolymer formulation of *N*-(2-hydroxypropyl) methacrylamide of [^{125}I]-Tyr-D-Asp_8 (P-D-Asp_8, 24 kDa) accumulated mostly in the kidneys and was eliminated quickly, whereas the 46 kDa co-polymer of [^{125}I]-Tyr-D-Asp_8 (P-D-Asp_8, 46 kDa) showed high accumulation throughout the peripheral tissues, particularly the liver and bladder as well as the skeleton. Together, these data demonstrate that while the parental ligand was not changed, the formulations of the ligand had significant and virtual opposite effects on the behavior of the probe. Also note that compared to the planar image of [^{125}I]-Tyr-D-Asp_8 binding of the growth plates in the knee, the tomographic study performed with a pinhole collimator showed greater resolution with clear separation of the femoral and tibia growth plates (Fig. 20.8, Tomography).

For studies whereby (1) the signal of the bound probe is sufficiently active and (2) does not change appreciably in concentration or activity over the time necessary to acquire an image and (3) the spatial location of the probe (e.g., organ, tissue, cell, or tumor) is desired, then tomographic SPECT imaging should be utilized. Typically tomographic imaging is performed with a pinhole collimator consisting of one to five pinholes to provide better resolution, but with lower sensitivity than with planar collimators; therefore, longer

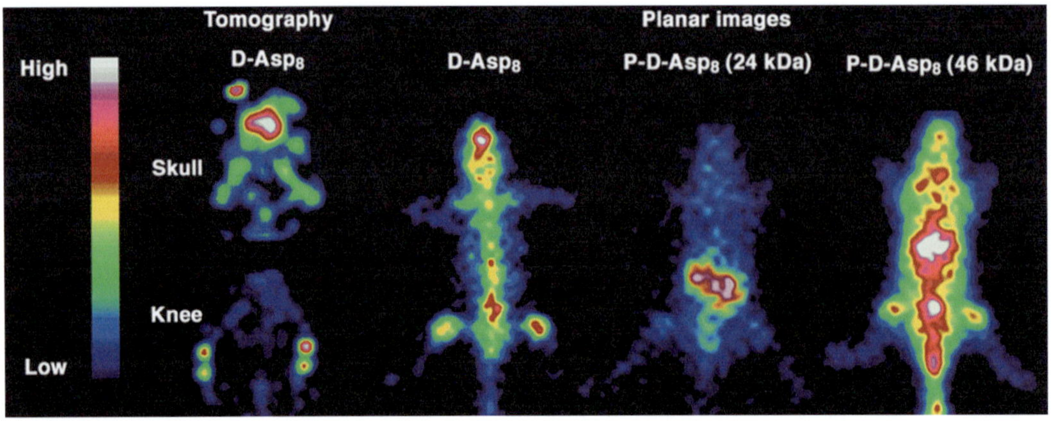

Fig. 20.8 Planar and tomographic SPECT images. BALB/c mice were injected i.v. with 10–20 μCi of [^{125}I]-labeled-Tyr-d-Asp$_8$ and P-D-Asp$_8$ conjugated copolymers of 24 and 46 kDa molecular weight. SPECT scans of mice were performed 24 h post-administration (reprinted with permission from Wang et al. 2006, copyright 2006, American Chemical Society)

acquisition times are necessary and typically range from 15 to 60 s/exposure or 16 to 64 min/64 exposure acquisition. The probe is administered to the subject and scans can be performed at intervals of hours, days or even weeks, depending on the half-life of the isotope and the affinity of the probe.

Probe affinity for the target and the isotope utilized may also influence imaging strategy. For instance, 2beta-carbomethoxy-3beta-(4-iodophenyl)tropane (β-CIT) is a ligand for the dopamine transporter (DAT, SLC6A3). [^{123}I]-β-CIT is a radiopharmaceutical (^{123}I half-life = 13 h) and is used to image the extent of dopaminergic loss in the striatum of Parkinson's disease (PD) patients and animal models of PD (Gendelman et al. 2003). Under clinical conditions, a 3–5 mCi dose of [^{123}I]-β-CIT is administered intravenously, SPECT images are acquired 3–6 h later, and by 24 h, most of the bound β-CIT is released and eliminated. In the 1-methyl-4-phenyl-1,2,3,6-tetrahydropyridine (MPTP) model of PD, administration of 500 μCi of [^{123}I]-β-CIT results in robust labeling of the striatum in the control mouse; however, lack of β-CIT signal in the striatum of the MPTP-treated mouse portends the loss of DAT-expressing dopaminergic termini (Fig. 20.9a) (Gendelman et al. 2003). Administration of 500 μCi ^{125}I-β-CIT (^{125}I half-life = 60 days) instead of [^{123}I]-β-CIT to animals revealed that while β-CIT is released from striatal DAT and eliminated, the probe is retained in the eyes for up to 6 days post-administration (Fig. 20.9b, c). These data demonstrate that the same probe may have differential elimination kinetics or affinities for its target in different tissues and suggest that utilizing different isotopes with longer half-lives may be required to detect differential properties of ligand-target interactions.

With SPECT imaging alone, spatial localization of probe activity within defined tissues and organs is difficult in the absence of anatomical landmarks. Thus, imaging methods by which those anatomical reference points can be merged with SPECT images are required. One strategy is to co-register MRI and SPECT, which requires separate lengthy scans and alignment of fiducial markers common to both scanned images. The combination of CT and SPECT within the same instrument, and software that co-registers both images, offers the anatomical reference points necessary for quantitative SPECT imaging of defined tissues and organs. Typically, a CT scan consisting of 1,024 images taken over a 5 min continuous 360° rotation ("fly mode") is performed prior to SPECT image acquisition. Once the SPECT images are acquired, reconstruction software merges the CT and SPECT images into a single three-dimensional representation from which two-dimensional slices can be obtained. If quantitation is desired, a region of interest that encompasses the organ or tissue is generated electronically within the reconstructed tomo-

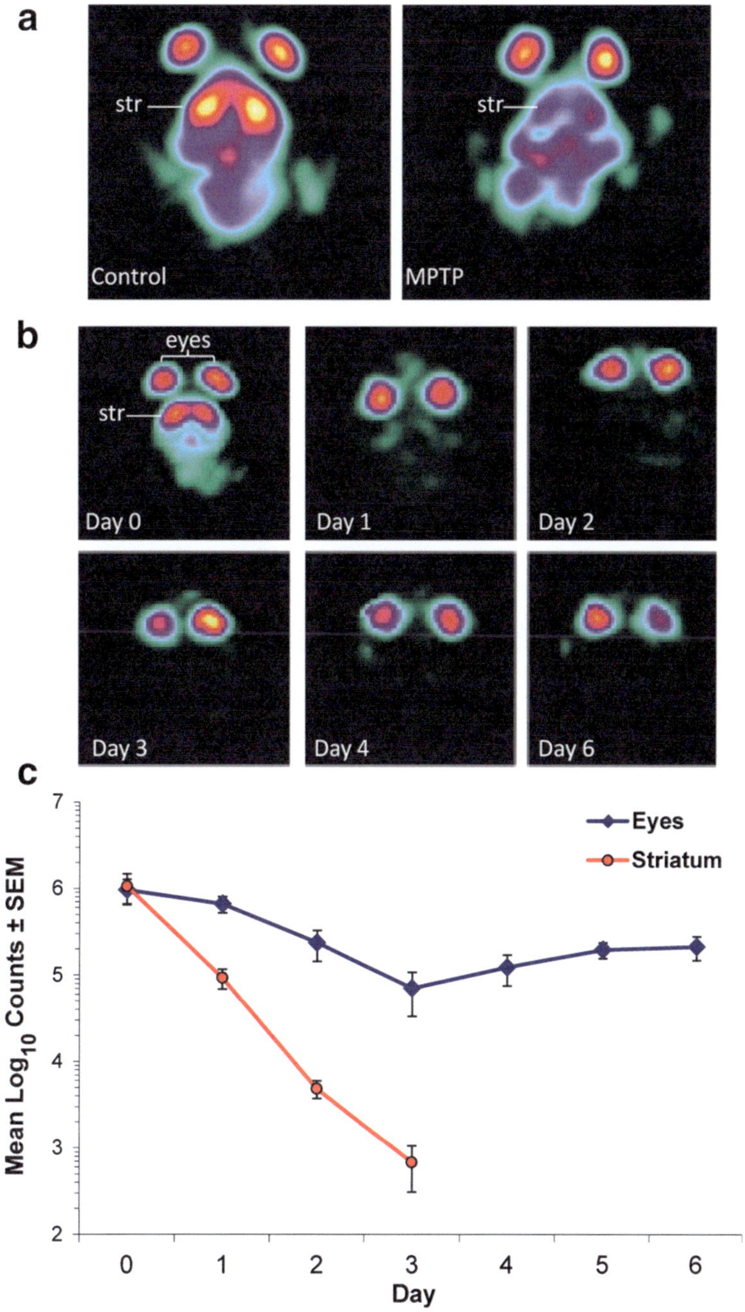

Fig. 20.9 SPECT analysis of β-CIT binding to dopamine transporter (DAT). C57BL/6J mice were administered 4 s.c. injections of PBS (Control) or 1-methyl-4-phenyl-1,2,3,6-tetrahydropyridine (MPTP). Each dose was given every 2 h at 18 mg free base/kg body weight. Seven days after MPTP intoxication, mice were injected i.v. with (**a**) 500 μCi [^{123}I]-β-CIT into both control and MPTP mice, or (**b, c**) 500 μCi [^{125}I]-β-CIT into PBS mice. Striatal binding of β-CIT to the DAT was assessed by SPECT at 1 h post-administration (**a**, **b**) or every day thereafter until day 6 (**b**). Increased intensity of binding is depicted by shift from *blue* to *red* to *white*. (**b, c**) Regions of interest were electronically drawn to encompass the striatum and eyes, and the degree of radioactivity was determined by digital image analysis. Data are the mean $\log_{10}$ counts (±SEM) for five animals. Linear regression analysis indicated loss of activity within the striatum was strongly time dependent ($r^2=0.990$, $P=0.0027$), whereas loss of activity from eyes was less time dependent ($r^2=0.416$, $P=0.118$). Comparison of the regression lines indicated they were significantly different ($P=0.0002$)

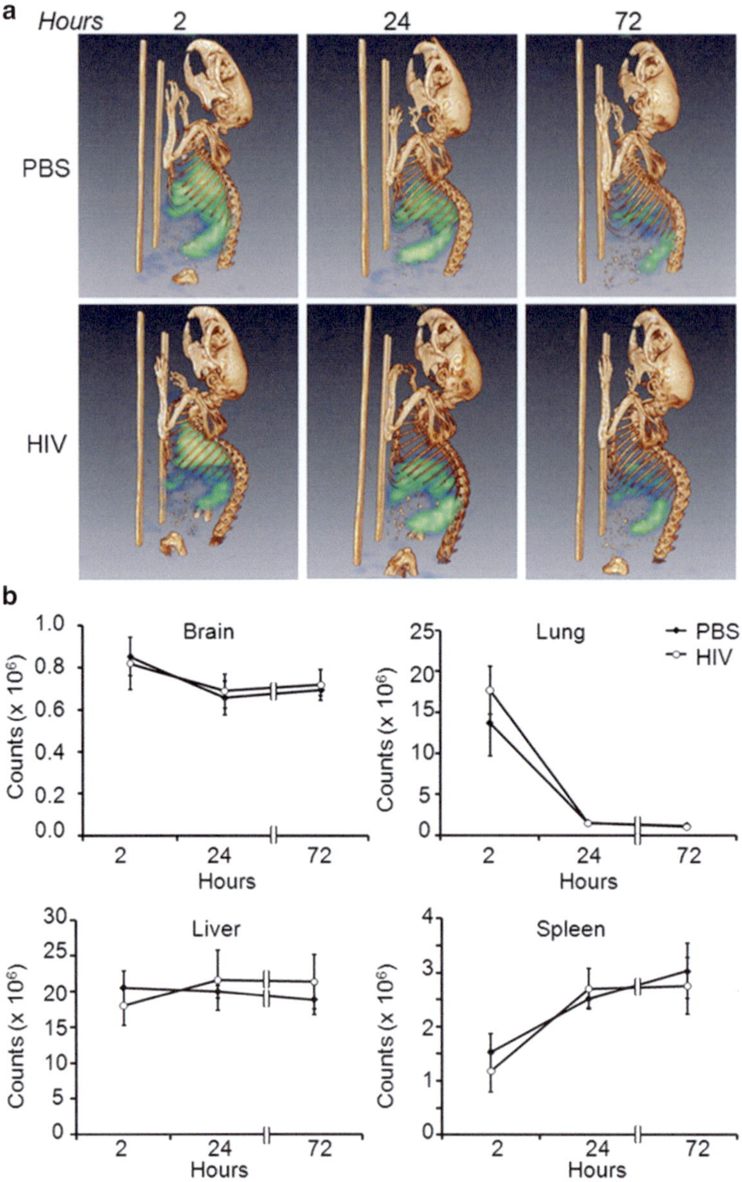

Fig. 20.10 CT/SPECT distribution of Treg. CD4⁺CD25⁺FoxP3-GFP⁺ Treg were labeled with ^{111}In oxine, and 20×10^6 labeled Treg were adoptively transferred to HIVE mice or sham-injected control mice. Distribution of ^{111}In-labeled Treg was evaluated by computed tomography/single photon emission computed tomography (CT/SPECT) analysis at 2, 24, and 72 h post-transfer using a FLEX Triumph CT/SPECT pre-clinical imaging system (Gamma Medica-Ideas, Northridge, CA). (**a**) Representative CT/SPECT images of ^{111}In-labeled Treg in PBS sham control mice or HIVE mice acquired at 2, 24, and 72 h. Intensities of radioactive signals are depicted as pseudo-colored *blue-green* signals. (**b**) Three-dimensional electronic bit maps were drawn encompassing brain, lung, liver, or spleen and total radioactive counts of ^{111}In-labeled Treg distributions in mice were calculated by digital image analysis using VIVID software (Gamma Medica-Ideas). Counts at each time point were decay corrected to time of injection. Relative radioactive counts as a function of pixel intensities were determined and presented as the means ± SEM for three mice per group. No statistical differences in ^{111}In-labeled Treg intensities were detected between HIVE mice and PBS controls at any time point. In lungs of both HIVE and sham control mice, significant diminutions of ^{111}In-labeled Treg counts were observed at 24 and 72 h compared with 2 h post-transfer ($P < 0.001$). No other tissues exhibited significant increases in radioactive counts at any times compared with 2 h post-transfer, although intensities trended to increase at 24 h ($P < 0.1$) and 72 h ($P < 0.07$) in both HIVE and PBS sham controls (modified from Gong et al. 2011)

graph, and the volume and the intensity of radioactivity for that region is determined by the analysis software. An example of this strategy is provided in a study assessing trafficking of T cells to the brain in a mouse model of HIV-1 encephalitis (HIVE) (Gong et al. 2011). C57BLJ/6 mice were injected intracranially (i.c.) with PBS as control or HIV-1-infected bone marrow-derived macrophages. Regulatory T cells (Tregs) were labeled with [^{111}In]-oxyquinoline (indium oxine, GE Healthcare, Arlington Heights, IL) at a dose of 100 μCi per 20×10^6 cells/mL RPMI 1640 supplemented with 10 mM HEPES for 45 min at 37 °C. [^{111}In]-labeled Tregs were adoptively transferred to HIV-infected and PBS control mice via tail vein injection 24 h after establishment of HIVE. Mice were imaged at 2, 24, and 72 h after adoptive transfer. SPECT/CT images show that adoptively transferred Tregs largely accumulate first in the lungs and liver and over time leave the lungs, remain in the liver, and migrate to the spleen (Fig. 20.10a). Quantitation of the radioactivity in the different organs, including the brain, was achieved with digital image analysis software (VIVID) to define regions of interest that encompass each organ within the CT images and calculate the volume and the relative radioactivity of that region (Fig. 20.10b). While radioactivity was detected in the brain, no difference was discerned between HIV-infected and control animals (Gong et al. 2011).

20.4 Summary

As bioimaging technology has advanced, computer-aided tomographic examinations utilizing X-ray, positron, and photon emissions have become the gold standard for clinical monitoring of disease progression and therapeutic outcomes in many neurodegenerative disorders. While animal models of these disorders allow greater capacities to probe associated mechanisms, the effects of altering these mechanisms ultimately should be measurable by tools utilized in the clinic. Thus, for longitudinal studies in animal models of disease that are monitored principally by bioimaging, inclusion of noninvasive bioimaging measurements would represent a most valuable component of translational research strategies.

References

Accorsi R (2008) Brain single-photon emission CT physics principles. Am J Neuroradiol 29:1247–1256

Acton PD, Zhou R (2005) Imaging reporter genes for cell tracking with PET and SPECT. Q J Nucl Med Mol Imaging 49:349–360

Anger HO (1958) Scintillation camera. Rev Sci Instrum 29:27–33

Asenbaum S, Brucke T, Pirker W, Podreka I, Angelberger P, Wenger S, Wober C, Muller C, Deecke L (1997) Imaging of dopamine transporters with iodine-123-beta-CIT and SPECT in Parkinson's disease. J Nucl Med 38:1–6

Badawi R (1999) Introduction to PET physics. http://depts.washington.edu/nucmed/IRL/pet_intro/intro_src/section2.html

Baranowska-Kortylewicz J, Abe M, Pietras K, Kortylewicz ZP, Kurizaki T, Nearman J, Paulsson J, Mosley RL, Enke CA, Ostman A (2005) Effect of platelet-derived growth factor receptor-beta inhibition with STI571 on radioimmunotherapy. Cancer Res 65:7824–7831

Basu S, Kwee TC, Surti S, Akin EA, Yoo D, Alavi A (2011) Fundamentals of PET and PET/CT imaging. Ann N Y Acad Sci 1228:1–18

Bazañez-Borgert M (2006) Basics of SPECT, PET and PET/CT imaging. http://www14.informatik.tu-muenchen.de/konferenzen/Jass06/courses/6/files/Bazanez_Borgert_paper.pdf

Bengel FM, Camici P, Lamare F (2009) Nuclear cardiology (PET and SPECT): basic principles. In: Zamorano JL, Bax JJ, Rademakers FE, Knuuti J (eds) The ESC textbook of cardiovascular imaging. Springer, London, pp 73–88

Beyer T, Townsend DW, Brun T, Kinahan PE, Charron M, Roddy R, Jerin J, Young J, Byars L, Nutt R (2000) A combined PET/CT scanner for clinical oncology. J Nucl Med 41:1369–1379

Bybel B, Brunken RC, Shah SN, Wu G, Turbiner E, Neumann DR (2006) PET and PET/CT imaging: what clinicians need to know. Cleve Clin J Med 73:1075–1087

Bybel B, Brunken RC, DiFilippo FP, Neumann DR, Wu G, Cerqueira MD (2008) SPECT/CT imaging: clinical utility of an emerging technology. Radiographics 28:1097–1113

Cao Z, Bal G, Accorsi R, Acton PD (2005) Optimal number of pinholes in multi-pinhole SPECT for mouse brain imaging—a simulation study. Phys Med Biol 50:4609–4624

Doorduin J, de Vries EF, Dierckx RA, Klein HC (2008) PET imaging of the peripheral benzodiazepine receptor: monitoring disease progression and therapy response in neurodegenerative disorders. Curr Pharm Des 14:3297–3315

Engdahl JC, Joung J, Chowdhury S (2007) Multi-pinhole collimation for nuclear medical imaging, Siemens Medical Solutions USA, Inc. US Patent 7,166,846

Fleischmann D, Boas FE (2011) Computed tomography—old ideas and new technology. Eur Radiol 21:510–517

Flohr TG, Schaller S, Stierstorfer K, Bruder H, Ohnesorge BM, Schoepf UJ (2005) Multi-detector row CT systems and image-reconstruction techniques. Radiology 235:756–773

Funk T, Despres P, Barber WC, Shah KS, Hasegawa BH (2006) A multipinhole small animal SPECT system with submillimeter spatial resolution. Med Phys 33:1259–1268

Gendelman HE, Destache CJ, Zelivyanskaya ML, Nelson JA, Boska MD, Biskup TM, McCarthy MK, Carlson KA, Nemachek C, Benner EJ, Mosley RL (2003) Neuroimaging and proteomic tracking of neurodegeneration in MPTP-treated mice. Ann N Y Acad Sci 991:319–321

Goldman LW (2007) Principles of CT and CT technology. J Nucl Med Technol 35:115–128, quiz 129-130

Goldman LW (2008) Principles of CT: multislice CT. J Nucl Med Technol 36:57–68, quiz 75-56

Gong N, Liu J, Reynolds AD, Gorantla S, Mosley RL, Gendelman HE (2011) Brain ingress of regulatory T cells in a murine model of HIV-1 encephalitis. J Neuroimmunol 230:33–41

Groch MW, Erwin WD (2000) SPECT in the year 2000: basic principles. J Nucl Med Technol 28:233–244

Gupta R, Cheung AC, Bartling SH, Lisauskas J, Grasruck M, Leidecker C, Schmidt B, Flohr T, Brady TJ (2008) Flat-panel volume CT: fundamental principles, technology, and applications. Radiographics 28:2009–2022

Lewellen TK (2008) Recent developments in PET detector technology. Phys Med Biol 53:R287–R317

Lewellen TK (2010) The challenge of detector designs for PET. AJR Am J Roentgenol 195:301–309

MacDonald LR, Patt BE, Iwanczyk JS, Tsui BMW, Wang Y, Frey EC, Wessell DE, Acton PD, Kung HF (2001) Pinhole SPECT of mice using the LumaGEM gamma camera. IEEE Trans Nucl Sci 48:830–836

Madsen MT (2007) Recent advances in SPECT imaging. J Nucl Med 48:661–673

Mahesh M (2002) Search for isotropic resolution in CT from conventional through multiple-row detector. Radiographics 22:949–962

McParland BJ (2010) The biodistribution (II): human. In: Medicine N (ed) Radiation dosimetry. Springer, London, pp 533–574

Meikle SR, Kench P, Kassiou M, Banati RB (2005) Small animal SPECT and its place in the matrix of molecular imaging technologies. Phys Med Biol 50:R45–R61

Moses WW (2007) Recent advances and future advances in time-of-flight PET. Nucl Instrum Methods Phys Res, Sect A 580:919–924

Pavese N, Rivero-Bosch M, Lewis SJ, Whone AL, Brooks DJ (2011) Progression of monoaminergic dysfunction in Parkinson's disease: a longitudinal 18F-dopa PET study. Neuroimage 56:1463–1468

Peterson TE, Furenlid LR (2011) SPECT detectors: the anger camera and beyond. Phys Med Biol 56: R145–R182

Rahmim A, Zaidi H (2008) PET versus SPECT: strengths, limitations and challenges. Nucl Med Commun 29:193–207

Robilotta C (2004) Emission tomography: SPECT and PET. Computación Sistemas 7:167–174

Rosenthal MS, Cullom J, Hawkins W, Moore SC, Tsui BM, Yester M (1995) Quantitative SPECT imaging: a review and recommendations by the Focus Committee of the Society of Nuclear Medicine Computer and Instrumentation Council. J Nucl Med 36:1489–1513

Rydberg J, Buckwalter KA, Caldemeyer KS, Phillips MD, Conces DJ Jr, Aisen AM, Persohn SA, Kopecky KK (2000) Multisection CT: scanning techniques and clinical applications. Radiographics 20:1787–1806

Saha GB (2010) Instruments for radiation detection and measurements. In: Saha GB (ed) Fundamentals of nuclear pharmacy, 6th edn. Springer, New York, NY, pp 33–48

Seo Y, Mari C, Hasegawa BH (2008a) Technological development and advances in single-photon emission computed tomography/computed tomography. Semin Nucl Med 38:177–198

Seo Y, Teo BK, Hadi M, Schreck C, Bacharach SL, Hasegawa BH (2008b) Quantitative accuracy of PET/CT for image-based kinetic analysis. Med Phys 35:3086–3089

Shwartz SC, Ohana I (2003) SPECT gamma camera, USA, U.C.G. Technologies. US Patent US20030208117

Tang J, Kuwabara H, Wong DF, Rahmim A (2010) Direct 4D reconstruction of parametric images incorporating anato-functional joint entropy. Phys Med Biol 55:4261–4272

Townsend DW (2004) Physical principles and technology of clinical PET imaging. Ann Acad Med Singapore 33:133–145

Trott CM, El Fakhri G (2008) Sequential and simultaneous dual-isotope brain SPECT: comparison with PET for estimation and discrimination tasks in early Parkinson disease. Med Phys 35:3343–3353

van Elmpt W, Hamill J, Jones J, De Ruysscher D, Lambin P, Ollers M (2011) Optimal gating compared to 3D and 4D PET reconstruction for characterization of lung tumours. Eur J Nucl Med Mol Imaging 38: 843–855

Wagenaar DJ, Zhang J, Kazules T, Vandehei T, Bolle E, Chowdhury S, Parnham K, Patt BE (2006) In vivo dual-isotope SPECT imaging with improved energy resolution. In: Conference Record IEEE Nuclear Science Symposium, San Diego, CA, pp 3821–3826

Wang D, Sima M, Mosley RL, Davda JP, Tietze N, Miller SC, Gwilt PR, Kopeckova P, Kopecek J (2006) Pharmacokinetic and biodistribution studies of a bone-targeting drug delivery system based on N-(2-hydroxypropyl)methacrylamide copolymers. Mol Pharm 3:717–725

Warburg O (1956) On the origin of cancer cells. Science 123:309–314

Noninvasive Neurophysiological Imaging with Magnetoencephalography

21

Tony W. Wilson

Abstract

Magnetoencephalography (MEG) is a noninvasive neurophysiological recording technique that is primarily utilized in human studies of system-level brain function. Although the seminal MEG measurements occurred more than 40 years ago, the precision of the instruments and the analytical sophistication of the field have dramatically increased over the past decade. At present, MEG is the only noninvasive high-resolution neurophysiological imaging technique and the only functional brain imaging method to offer both high temporal (<1 ms) and spatial (2–5 mm) resolution. The current chapter will provide a brief historical introduction to MEG and functional magnetic resonance imaging (fMRI), which is the most common method of functional brain imaging. Description of the physical and physiological bases of the signals measured in noninvasive functional imaging will follow, with an emphasis on the neuromagnetic signals quantified in human MEG measurements. An introduction to the most common MEG analysis methods will be presented; and thereafter, several examples of MEG applications will be discussed to illustrate the type of questions often pursued in MEG research and the general areas of study where MEG measurements are making an impact. The chapter will conclude by presenting some new applications for MEG-based functional brain imaging.

Keywords

Electrophysiology • Oscillations • MEG • Gamma • Beta • Neuroimaging • Synchronization • Desynchronization • Synchrony • Cortex

T.W. Wilson, Ph.D. (✉)
Department of Pharmacology and Experimental Neuroscience, University of Nebraska Medical Center (UNMC), Omaha, NE USA

Center for Magnetoencephalography, University of Nebraska Medical Center, 988422 Nebraska Medical Center, Omaha, NE 68198, USA

Department of Neurological Sciences, University of Nebraska Medical Center (UNMC), Omaha, NE USA

Munroe-Meyer Institute for Genetics and Rehabilitation, University of Nebraska Medical Center (UNMC), Omaha, NE USA
e-mail: tony.w.wilson@gmail.com

H. Xiong and H.E. Gendelman (eds.), *Current Laboratory Methods in Neuroscience Research*, Springer Protocols Handbooks, DOI 10.1007/978-1-4614-8794-4_21,

21.1 Introduction

Noninvasive functional neuroimaging is one of the most widely used tools in cognitive neuroscience, a field of research that aims to illuminate the biological correlates of cognition and the relationship between the brain and the mind. While there are many different methods and modalities for deriving functional brain images, functional magnetic resonance imaging (fMRI) is by far the most common. The technique was pioneered by Ogawa and colleagues who first demonstrated that local differences in blood-oxygen level could serve as signal contrast (Ogawa et al. 1990). The blood-oxygen-level-dependent (BOLD) signal was born, and shortly thereafter three groups independently and simultaneously reported BOLD signal changes in humans (Bandettini et al. 1992; Kwong et al. 1992; Ogawa et al. 1992). This development made functional neuroimaging research feasible on a geographically large scale, as MRI scanners were already available in hospitals and universities around the globe and could be quickly and easily retrofitted to enable such functional measurements. Additionally, the technique did not require the introduction of radioactive isotopes, thus providing a venue for the repetition of functional brain measurements on individual people. Overnight fMRI became a mainstay in cognitive neuroscience, and almost immediately assumed its current role as the most widely used functional brain imaging modality.

Meanwhile, another noninvasive technique for studying human brain function was taking a monumental step forward, as the world's first whole-head magnetoencephalography (MEG) system was being installed and tested at Helsinki University of Technology during the early 1990s (Ahonen et al. 1993; Hämäläinen et al. 1993). Previous MEG systems could only measure a relatively limited area of the human cortex at a given moment (e.g., one quadrant, Fig. 21.1), which precluded MEG being applicable for many cognitive neuroscience paradigms. MEG detects and quantifies minute magnetic fields that naturally accompany the tiny electrical currents that underlie brain function and was pioneered by David Cohen (1968) who first recorded the magnetic signature of occipital alpha rhythm changes (i.e., synchronization) during eyes closed compared with eyes open. The system installed at Helsinki was equipped with 122 planar gradiometer-type sensors positioned quasi-equidistantly above the scalp and provided the first glimpse of whole-cortex, real-time human physiology on the sub-centimeter scale. The utility of MEG's combined spatial (<5 mm) and temporal precision (<1 ms) was quickly recognized; however, hefty costs and technical complications prevented the rapid growth seen with fMRI, and to some degree these issues continue to stymie the widespread application of MEG techniques. That said, many major medical centers are currently equipped with MEG, and systems with 200–300 cortical sensors are relatively standard (Fig. 21.2).

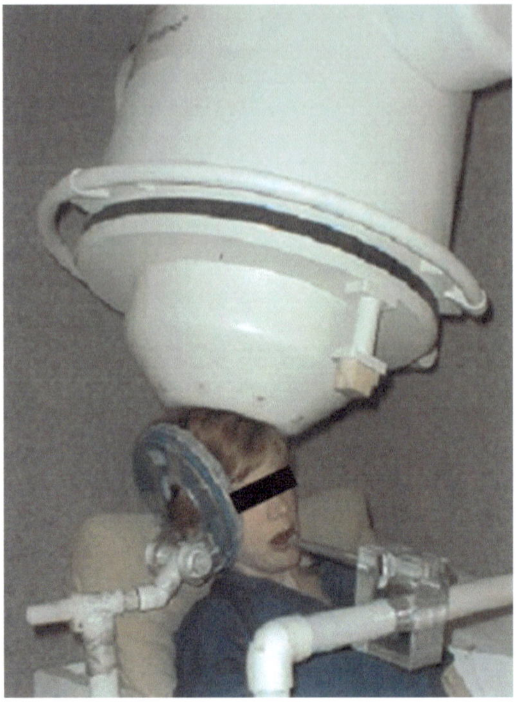

Fig. 21.1 MEG system with a 37-sensor array positioned over a portion of the left hemisphere

Through these methods and a large group of related technologies, such as positron-emission tomography (PET), high-density electroencephalography (EEG), transcranial magnetic stimulation (TMS), functional near-infrared spectroscopy

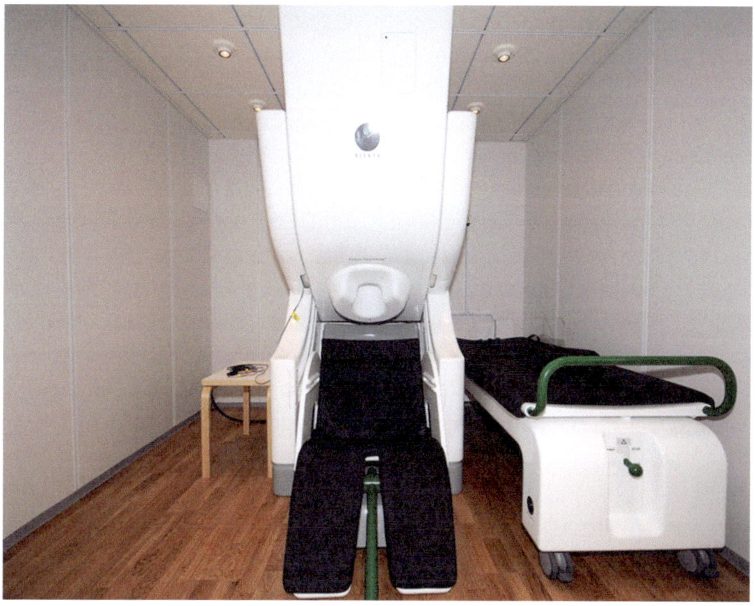

Fig. 21.2 A modern whole-head MEG system equipped with 306 cortical sensors

(fNIRS), and many other imaging methods, knowledge concerning the biological underpinnings of human cognitive functions has increased dramatically in the past 20 years and continues to rapidly evolve. While the scope of this chapter will be largely limited to the technique of MEG, with some minor discussions of EEG and fMRI, all functional imaging modalities have their own set of strengths and weaknesses that position them as ideal for particular applications. Thus, to truly appreciate the capacity of these research tools and the current status of this research area, further reading is essential (excellent reviews include Humm et al. 2003; Huettel et al. 2008; Hansen et al. 2010).

21.2 Physical and Physiological Basis of Signals

Imaging modalities are defined by the signal each quantifies, as well as the advantages and disadvantages of the particular signal's relationship with the underlying neurophysiology. Variation in these signals reliably correlates with task parameters; and through this correlation, certain neural regions can be linked with certain cognitive functions. This section briefly reviews the physical basis of signals underlying common applications of fMRI and MEG and subsequently connects these physical signals to their neurophysiological correlates.

21.2.1 The Physical Signals of Functional Imaging

Functional Magnetic Resonance Imaging

The fMRI signal is often criticized because it is only an indirect measure of neural activity. Unlike PET-based imaging, the BOLD-fMRI signal does not require the introduction of extraneous substances, but instead harnesses a paradoxical phenomenon of neurobiology. Under normal conditions, neural cells acquire energy through the oxidative metabolism of glucose. However, during brief periods of intense activity, neural cells apparently resort to the more rapid, although less energetically efficient, anaerobic glycolysis (for a well-developed cellular model, see Magistretti and Pellerin 1999; Aubert et al. 2007). That is, the brain uses highly inefficient anaerobic means to produce energy during periods of heightened activity. BOLD-fMRI takes advantage of this phenomenon as a contrast agent for the signal (for review, see Raichle and Mintun 2006). Basically, the procedure relies on the oxygenation-dependent magnetic susceptibility of hemoglobin. Deoxygenated hemoglobin is paramagnetic, whereas oxygenated hemoglobin is slightly diamagnetic, so the signal intensity in MRI voxels containing more deoxygenated blood is attenuated relative to those containing oxygenated blood (Bandettini and Ungerleider 2001; Heeger and Rees 2002). Thus, the ratio of oxygenated-to-deoxygenated hemoglobin provides contrast for the local oxygen content of blood.

The signal change is usually very small, rarely exceeding 5 % of the total MRI signal. Consequently, experiments typically consist of many trials or at least several blocks, where the subject is repeatedly stimulated (e.g., with a flashing light) or is instructed to repeat the behavior of interest numerous times. In short, when activity increases in a neural area, blood flow also increases to supply the necessary metabolic elements (e.g., glucose); however, oxygen consumption is not equivalent to glucose utilization, which causes a local buildup of oxygenated relative to deoxygenated hemoglobin, and in doing so creates signal contrast changes that correlate with neuronal activity. Thus, in BOLD-fMRI, brain activation is deduced from patterns of increased/decreased oxygenated-to-deoxygenated hemoglobin. The spatial resolution of the activation maps depends on multiple variables, including magnetic field strength and many other data processing parameters, but in most applications it is on the order 3 mm. The ultimate temporal resolution is arguably constrained by blood flow rate, but 1–3 s is attainable in event-related designs. There are many great in-depth reviews of fMRI (e.g., Huettel et al. 2008).

Electroencephalography

The physical signals measured by MEG and EEG are direct reflections of ongoing neural physiology and thus are superficially more straightforward than those yoked to hemodynamic parameters. First, it is helpful to consider the case of EEG, which quantifies electrical voltage fluctuations that occur across the scalp. These small fluctuations reflect very complex patterns of ionic currents in the tissue surrounding populations of active neural cells. Scalp measurements are possible because the intervening tissues (i.e., scalp, skull, CSF, and brain) passively conduct the electrical currents, although with significant distortion due to nonuniform conductivity properties. Of course, these voltage changes are (by definition) referenced to something, and there are many different and valid referencing techniques in EEG applications; the most appropriate reference is often contingent on the ultimate goal of the experiment. In typical applications, conductive gel is applied to each electrode to keep impedances low, and all electrodes are positioned in a quasi-standard configuration across the subject's head. In principle, EEG provides a continuous record of overall brain activity precise enough to detect minute voltage fluctuations on areas of the scalp. However, in practice, participants generally complete many repetitions of the same stimulation or behavior (i.e., trials) during a given experiment. The ongoing record of brain activity is then time-domain averaged across the trials by aligning them to the occurrence of an external event (e.g., stimulus onset). This attenuates all voltage changes that are not time-locked and phase-locked to stimulus onset (e.g., spontaneous cortical rhythms, EEG system noise) and thereby improves the signal-noise ratio (SNR) of the recorded data. Thus, in many neuroscience applications, the EEG record is reduced to a single event-related potential (ERP) per experimental condition.

Under optimal circumstances, the ERP will contain all time-locked and phase-locked voltage changes that occurred at each electrode across the time course of each trial. The temporal resolution of the ERP depends on the sampling rate of the original recording (typically, 250–1,000 Hz, or 1–4 ms) and, to a lesser extent, the parameters of any filters applied to the data. The spatial resolution is often 2–3 cm due to the conductivity problems mentioned above (i.e., the skull is a major insulator and all intervening tissues contribute to the signal smearing); however, this problem can

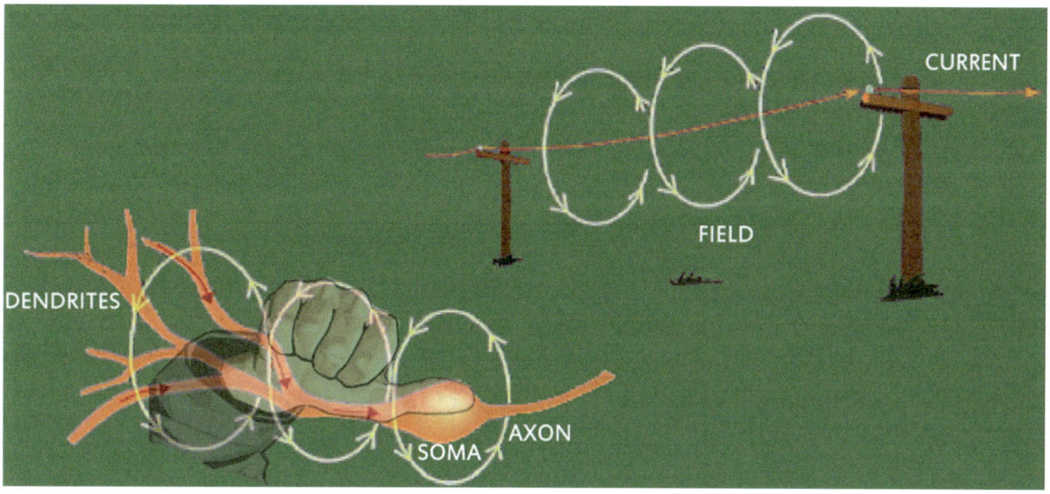

Fig. 21.3 The magnetic fields produced by neuronal currents follow the same electromagnetic principles (e.g., right-hand rule) as the fields that are more common in our environment

be ameliorated through higher spatial sampling frequencies (i.e., more electrodes) and more sophisticated head models. Currently, electrode sensor nets of 256 are common, as is the use of multilayer head models, which account for the inhomogeneous conductivity profiles of each of the intervening tissues.

Magnetoencephalography

The physical signal quantified in MEG comprises the summated magnetic fields produced by ionic currents within active neural areas. Just as electric current passing through power lines produces magnetic fields, ionic currents within brain tissue also produce magnetic fields according to the same physical principles (e.g., the right-hand rule, Fig. 21.3). However, neuromagnetic fields are ultra minute (10^{-15} T) relative to all other magnetic fields that are typically encountered (e.g., the earth's static field is about 10^{-5} T); thus, MEG recordings are normally carried out in magnetically shielded rooms (MSRs) using special sensors and/or noise-cancellation software. The most common MEG sensors are magnetometers, axial gradiometers, and planar gradiometers. Magnetometers measure the zero gradient perpendicular to the coil (i.e., the field directly) and hence are highly sensitive to both brain activity and environmental noise. First-order axial gradiometers measure first- and higher-order radial gradients by utilizing two coils (i.e., a pickup coil and a reference coil) wound in opposition to one another; thus, the measurement is the difference between the two coils, and accordingly fields with uniform spatial gradients (e.g., fields from strong distant sources) are self-canceling. Planar gradiometers are similar to their axial counterparts, except the measured gradients are tangential to the scalp surface. Since field strength decreases with the square of the distance from the source current, gradiometer class sensors are less sensitive than magnetometers to ambient environmental noise and to sources deep in the brain (i.e., gradiometers are "near-sighted"). Regardless of type, each sensor is coupled to a superconducting quantum interference device (SQUID), which acts as a low-noise magnetic flux-to-voltage converter. All sensors and SQUIDs are housed within a cryogenic dewar and operate at the superconducting temperature of liquid helium.

In summary, the magnetic flux emanating from electrically active local brain tissues induces tiny currents in each of a small set of sensors. The magnetic fields associated with each these induced currents impinge on their SQUID, which proportionally "converts" the flux into a voltage that can be amplified and recorded by conventional electronics. It should be noted that each MEG sensor is associated with a dedicated SQUID. Overall, MEG provides a continuous record of magnetic flux patterns above the scalp. However, as with EEG, this record is typically averaged within condition to form some type of event-related field (ERF). The reasons for this are analogous to those described for EEG. As discussed later, such averaging can take many forms (e.g., time domain, frequency domain, time-frequency domain) depending on the experimental goals of the investigators. Like EEG, the temporal resolution of MEG is linked to the sampling rate of the original recording (1 kHz is common, thus 1 ms), but the two techniques do differ greatly in spatial precision. Intervening tissues (e.g., skull) distort electrical currents, but do not generally affect magnetic fields; therefore, these tissues can be considered transparent in MEG and this significantly enhances the spatial precision of the technique. MEG systems with ~250 or more sensors are currently available, and under optimal circumstances these instruments are spatially accurate within 2–5 mm (Leahy et al. 1998; Wagner et al. 1997). Excellent reviews of MEG techniques and instrumentation are available (Lu and Kaufman 2003; Hansen et al. 2010).

21.2.2 Physiological Correlates of Functional Imaging Signals

The utility of any functional imaging modality is ultimately determined by the relationship of the measured signal to the underlying neurophysiology. Research in this area has generally relied on invasive recordings coupled with the imaging modality of interest; thus, where appropriate, short introductions to the neurophysiological activity gauged by invasive techniques will be provided.

Functional Magnetic Resonance Imaging

Early studies used single-unit recordings in monkeys and fMRI in humans to evaluate the relationship between BOLD variation and neural activity (e.g., see Heeger et al. 2000; Rees et al. 2000). In single-unit recordings, a microelectrode is placed near a neuron's soma (or axon), and the measured mean extracellular field potential (mEFP) indicates the spike traffic of the neuron and normally several of its neighbors. The spiking activity of individual neurons is then discerned analytically after the recording. Typically, recordings are taken over a reasonable span of time, and the electrodes are moved periodically as to provide data on more neurons. These early studies were informative in many ways, but more definitive evidence came from seminal work using simultaneous multiunit recordings and fMRI in ten anesthetized monkeys (Logothetis et al. 2001). Multiunit recordings are similar to single-unit recordings, except the electrode tip is placed further from spike-generating sources as to prevent action potentials from dominating the signal. The output measure of the two techniques is the same (i.e., mEFP), but in the case of multiunit recordings the mEFP is often filtered to provide two distinct indices of neural activity. Typically, the mEFP is high-pass filtered at 300 Hz to yield multiunit spiking activity (MUA), and low-pass filtered below 300 Hz to obtain local field potential (LFP). The MUA indicates the weighted sum of action potentials from all neurons within a sphere of <300 μm radius (electrode tip at center), while the LFP reflects slower oscillations such as synaptic potentials and dendritic potentials. In short, the MUA represents the output of a cortical area, and the LFP indicates the input and all local intracortical processing of the same cortical area (including excitatory and inhibitory interneurons; see Logothetis 2003 for a review). Logothetis et al. indicated both MUA and LFP strongly correlate with BOLD signal measurements in primary visual cortices. However, LFP was significantly more correlated with the BOLD signal than was MUA, explaining 7.6 % more variance. According to Logothetis et al., MUA is only a "fortuitous" predictor of the BOLD signal, its power being mostly a by-product of MUA's normal correlation with the LFP (i.e., input to a cortical area often, but not always, leads to output in the form of action potentials).

Logothetis et al. concluded that the BOLD signal most closely reflects synaptic activity, excitatory and inhibitory interneuronal activity, and the effects of modulatory neurotransmitter input (e.g., dopamine) on intracortical processing. The key results of Logothetis et al. (2001) were recently replicated in a group of awake behaving monkeys (Goense and Logothetis 2008), which further substantiates this important study.

Electroencephalography and Magnetoencephalography

In contrast to the hemodynamic studies, research on the physiological basis of EEG and MEG signals have focused on which aspects of neural activity are "not" reflected in the measurement, and how intervening tissues (e.g., skull) affect spatial precision. The magnetic fields detected in MEG arise from a subset of the same currents measured in EEG. First, it is useful to distinguish between two types of neuronal activity: (a) fast depolarizations of the membrane that result in action potentials and are mediated by voltage-gated sodium and potassium channels and (b) slower postsynaptic potentials (dendritic currents) that are mediated by several neurotransmitter systems and their interactions with specific receptors, ionic channels, and second messenger systems. These slower postsynaptic potentials are the primary source of the fields measured in MEG and are generally segregated into two main types, excitatory and inhibitory postsynaptic potentials (EPSPs and IPSPs), based on the transmitter involved. Unlike action potentials, these EPSPs and IPSPs are often extended in time (hundreds of milliseconds) allowing thousands of neighboring cells to have active postsynaptic potentials at overlapping time periods, which clearly promotes the local summation necessary to produce fields that are measurable from a distance (i.e., with a MEG sensor). Briefly, we know that there is no accumulation of charge anywhere in the medium from the basic biophysics; thus, any currents flowing into a given cell must be compensated by opposite flowing currents elsewhere in the cell. In the case of an EPSP, when excitatory neurotransmitter binds to a postsynaptic cell, positive ions flow into the soma and apical dendrites of the cell and this transmembrane current causes an active sink at the level of the excitatory synapse in the extracellular space (i.e., a current sink). Consequently, there will be multiple current sources along the soma-dendritic membrane that function to leak current and thereby complete the circuit; thus, charge is conserved by a continuous flow of current around the cell. Since every neuron receives inputs from many other neurons, numerous currents are simultaneously present causing a distribution of sources and sinks that is very complex. The intracellular currents are referred to as primary currents, whereas the extracellular currents are termed secondary or volume currents. Generally, only primary currents create a magnetic field measurable from a distance (i.e., with a MEG sensor), as those associated with volume currents are often self-canceling. However, the local geometry also plays a paramount role as, based on the physical principles alone, the cellular architecture can promote field summation (e.g., cortical pyramidal neurons arranged in palisades with optimally angled and elongated dendritic arbors) or local attenuation (e.g., the cellular arrangements within many subcortical nuclei). Empirical support comes from a series of studies using guinea pig hippocampal slices and a custom-made micro-SQUID device, which showed that primary currents in parallel-oriented pyramidal cells were the dominant source of MEG signals in the mammalian brain (Okada et al. 1997). Finally, in regard to spikes (action potentials), they have been historically understood as only minor or negligible contributors to the MEG signal, and a recent modeling study provided additional support to this view (Murakami and Okada 2006). The duration of an individual spike is the main factor limiting their contribution, as cortical areas are rarely synchronized enough to have ~10,000 action potentials occurring at overlapping time periods, which would be required to attain the necessary summation. Therefore, the fields measured in MEG arise mostly from primary currents associated with postsynaptic (dendritic) activity, with EPSP/IPSPs in cortical pyramidal cells being the optimal MEG signal producers. In contrast to

MEG, EEG readily detects both primary and secondary currents. Furthermore, since all cells are immersed in a conductive saline solution (i.e., extracellular matrix) and the skull has very high electrical resistance, secondary currents flow through the orbits of the eyes and any other skull openings into the scalp, where they then contribute to the EEG measurement (for a review, see Lopes da Silva and van Rotterdam 2005; Lopes da Silva 2010).

However, it is important to recognize that many of these basic signal generation issues remain unresolved. For instance, the necessary volume of cortex that must be synchronously active to produce a detectable signal is unknown. Also, what is the impact of synchrony? Perhaps surprisingly small cortical volumes can produce detectable signals given ample synchronicity in their intrinsic local networks. Another important issue involves action potentials, as they are commonly assumed to contribute negligibly (or not at all) to the MEG signal. This assumption is grounded in the idea of necessary temporal summation (i.e., spikes are too brief an occurrence and too asynchronous for the temporal summation necessary to create a detectable signal), which currently lacks direct empirical support. Lastly, the sensitivity of MEG to the magnetic fields of tangential versus radial (superficial) currents, and deep currents (e.g., basal ganglia) remains unclear. Intuitively, it seems doubtful that purely radial currents would exist in practice, given the average size of a detectable cortical patch and the unique vantage of each MEG sensor in today's whole-head high-density systems. This issue has been assessed using whole-head 151-sensor MEG and simulation data involving the human cortex (Hillebrand and Barnes 2002), which suggested that only thin strips (~2 mm wide) near the very crest of gyri were poorly resolvable with MEG; overall, these areas constituted less than 5 % of the entire cortical mantle. Hillebrand and Barnes concluded that source depth was the ultimate limiting factor in source resolvability, which is sensible given the known effect of distance on field strength. Of course, this study by no means concludes the debate and MEG studies showing activity in deep brain areas such as basal ganglia structures and amygdala are relatively common.

21.3 Introduction to the Methods of MEG

21.3.1 Experimental Design and Basic MEG Signal Processing

As alluded to earlier, MEG experiments, like most physiological experiments, comprise many "trials" where the external stimulus or other event of interest is repeated. As an example, consider an experiment where the goal is to map in time and space (spatiotemporal) the brain's response to a brief instance of mechanical stimulation. In such an experiment, the participant might sit comfortably in the MSR with their head within the MEG helmet, as a single stimulation is delivered to their left index finger pad every 2–4 s for a total of about 6 min. In this hypothetical example, we will use a MEG system equipped with 248 axial gradiometer type sensors and sample all sensors in parallel at a rate of 1 kHz. Overall, this would generate about ~120 trials of mechanical stimulation to the index finger. From previous studies, we know that this type of stimulation elicits activity about 50 ms after stimulus onset in the contralateral postcentral gyrus in what is likely the finger representation in primary somatosensory cortex. We also know that other brain regions become active slightly later and that neuronal activity will have largely dissipated by 400 ms. Thus, in this example, it would make sense to define an analysis period (or epoch) that included at least 400 ms of data after the stimulation onset. Of course, we want to know how the stimulus changed brain activity (i.e., neurons in this cortical area were not silently at rest before stimulus onset) so we will also need to include a baseline period, and the time period just preceding the stimulus onset is as good as any. For this example, we will arbitrarily choose a 200 ms baseline period, which means that the overall analysis epoch will be of 600 ms duration with stimulus onset being defined as 0 ms (i.e., baseline, −200 to 0 ms; active period, 0 to 400 ms). In practice, investigators often utilize longer analysis epochs than are necessary in their initial analysis of a dataset, because they rarely know the temporal extent (i.e., overall duration) of neural activity that their experiment will elicit.

To review, we have ~120 trials of mechanical stimulation and for analysis each trial has been defined as a 600 ms epoch (−200 to 400 ms). Since the MEG signal was sampled at 1 kHz at each of the 248 sensors, we conveniently have 600 data points per epoch at each of the 248 sites. The SNR is quite low in raw MEG recordings and data from each of the ~120 trials would appear to be predominantly noise. This noise is partly due to the subject (e.g., eyeblinks, heartbeats), environment (e.g., nearby cars, vibrations), and the instrumentation (e.g., SQUID noise), but a significant portion is actually of neural origin. That is, the brain's ongoing spontaneous oscillations, which are not directly related to the stimulation, function as a major noise source when the goal is to extract responses specific to the stimulation. To improve the SNR, one can average the MEG data across trials (per sensor). Such signal averaging can be done in many different ways, but the most common is referred to as time-domain averaging and consists of averaging amplitude values time point by time point with respect to the stimulus onset. This would result in the ~120 trials being reduced to a single trial (per sensor), which still consists of 600 data points (i.e., 600 ms). Provided that the noise is temporally uncorrelated, such averaging reduces the noise as $1/\sqrt{N}$, where N is the number of trials. Beyond the time domain, MEG signals are sometimes averaged in the frequency domain (after Fourier analyses) and/or the time-frequency domain. These methods will be briefly discussed later in the chapter, but suffice to say that the best approach to signal averaging depends on the specific goals of the experiment. Finally, before moving forward, it is important to recognize that investigators generally peruse the raw signals before averaging to identify and discard any epochs that contain eyeblinks or other events that cause large artifacts in the MEG. These bad epochs are excluded from the average, as they would actually add noise to the final average. In some studies (e.g., children), the number of bad epochs can be large and this is important to consider in the initial task design.

21.3.2 Sensor Space and Source Space in MEG

In most MEG experiments, it is critical to identify the neural regions that generated the observed responses, in short, to move from "sensor space" to "source space." To conceptualize this issue, consider the spatial relationship between the 248 MEG sensors (from the example above) and the underlying brain. The sensors are arranged in a fixed array that will remain constant in regard to intersensor distances and orientations. However, brain structure will clearly vary from subject to subject in regard to size, shape, and intrinsic gyrus/sulcus patterns (these patterns alone will significantly affect the MEG signals that are observed outside of the head). Furthermore, the distance from each MEG sensor to the brain surface will vary tremendously from person to person and within the same person from experiment to experiment. For example, consider how measurements from each of the 248 sensors would be affected if the subject simply turned their head to the far left, raised their chin slightly, or slouched down in the MEG chair. Thus, using patterns in the sensor array to decipher which brain areas generated a particular response is tricky and generally not possible, except maybe on a gross level (e.g., frontal region, occipital regions).

Coregistration and Head Modeling

Several major analysis steps are necessary to move from sensor to source space. The technical details of these steps have consumed entire volumes of literature and are mostly beyond the scope of this chapter. Thus, only the high points will be covered here to introduce the problems and the current crop of solutions. First, each participant generally undergoes a high-resolution structural MRI scan shortly after his/her MEG session. This MRI data generally contains enough information to reconstruct the individual's skin surface, which is used to co-register the structural

MRI data with their functional MEG data using a series of structural landmarks (i.e., fiducials) that were recorded in the MEG array's spatial reference frame prior to (and during) the MEG session. Once the structural and functional data are in the same physical space, the next hurdle is to define the conducting volume. In MEG, the most common head model continues to be a single-shell spherically symmetric conductor model, which possesses a number of very attractive elements for MEG (Sarvas 1987; Hämäläinen and Sarvas 1989). These spherical models are generally aligned to approximate the brain surface. As an aside, more realistic head modeling such as the multi-shell boundary element method, where tissues with distinct conductivity profiles are each represented by a separate layer (e.g., scalp, skull, CSF) and conductivity is assumed to be homogenous with the layer, is a mainstay in EEG but is less frequently employed in MEG as it is less of a necessity.

Forward and Inverse Problems

No other area of MEG methods has as much technical detail or dedicated literature as source localization, but we will skip much of this detail in the interest of promoting a conceptual and pragmatic basic understanding of the method. First, the ultimate goal of MEG source localization is to identify the neural regions that have generated the measured MEG signals. Within this goal, two modeling problems can be distinguished. The first can be called the forward problem, which basically consists of predicting the magnetic fields that will exist at the sensor level given a known source model. In other words, the forward problem addresses the question of "what the resulting distribution of MEG sensor measurements would be given a neural source of known location, orientation, and amplitude." This problem has a unique solution and is solvable through conventional physics and mathematics. On the other hand, the second modeling problem, the inverse problem, addresses the question of "what location, orientation, and amplitude of neural source(s) would create the observed distribution of MEG sensor measurements" (i.e., identifying active neural areas from the surface measurements). As recognized by Helmholtz more than a century ago, the general inverse problem is ill-posed. In other words, it lacks a unique solution as there are an infinite number of possible source configurations that could equally well explain the electromagnetic fields observed outside of a volume conductor. It is important to realize that the inverse problem is not a problem specific to MEG; in general, it exists in many fields (e.g., astronomy, computer vision, geophysics). While the inverse problem may first appear to be a total blockade, bringing a wide array of contextual information into the equation makes the problem substantially more tractable, and in MEG there is plenty of a priori information that can be used to reduce the scale of the problem. For instance, as noted earlier, the MEG signal is primarily generated by pyramidal cells in the cortical gray matter. Thus, limiting MEG source space to the cortex would be a reasonable constraint. It would significantly reduce the overall source volume (i.e., number of locations where signal generators could possibly be located). Many other assumptions and constraints are obviously possible and even sensible to implement in particular applications. Moreover, there are different conceptual approaches to the MEG inverse modeling problem. Only the two broadest categories will be described here with individual examples of each approach later in the chapter.

Localizing and Imaging Signal Generators

The most fundamental difference between these two approaches can be reduced to the source model. Essentially, the *dipole-localization method* assumes that the observed MEG signals are generated by a group of active brain regions and that the location of these brain regions can be recovered or estimated from the data. The actual source model can comprise one or many elements (i.e., equivalent-current dipoles—ECDs), each of which represents a spatially distinct active brain region. In contrast, the *imaging method* does not attempt to estimate the location of sources from the data; instead, it begins with a dense source model that comprises 2,000–5,000 individual elements spaced across the source volume (e.g., the cortical gray matter). The primary aim of the imaging approach is then to estimate the amplitude distribution of all neural currents at the physiological scale of MEG signals. Thus, unlike the regionally discrete nature of the dipole-localization method, brain activity appears more continuous in the imaging method with neural areas of maximum activity surrounded by regions with slightly less activity (i.e., activity gradients). However, it is important to recognize that the continuous appearance of these MEG "images" is dictated by the model and not necessarily the data, although such an appearance is in agreement with the patterns of cortical activation found using other methods (e.g., electrocorticography—ECoG, fMRI, fNIRS). Finally, it is vital to appreciate that each of these approaches have their strengths and weaknesses and that there are particular applications where each approach is clearly superior to the other.

While there are examples of inverse modeling methods that perfectly fit each of these fundamentally different approaches, in reality, most inverse methods would be considered a slight nuance of one approach or even a near middle of the road option. Several of these nuance methods will be demonstrated in the applications section (below), and at that time a few more technical details will be presented so that the strengths and weaknesses of the particular method, in the context of a given experiment, can be more fully appreciated. Readers wanting a more in-depth understanding of MEG methods are encouraged to peruse the excellent reviews that are available (Hämäläinen et al. 1993; Mosher et al. 1999; Baillet et al. 2001; Baillet 2010; Parkkonen 2010). Lastly, it should be noted that *many other* preprocessing steps that go into mainstream MEG analyses were completely skipped for brevity. It's also worth noting that *all* of the processes described above, including inverse modeling, are typically done on each subject's data set in isolation; the output of these analyses are then entered into an appropriate (multi-subject) statistical model.

21.4 Applications of MEG: Some Examples

MEG data from basic sensory, cognitive, and motor paradigms will be presented below to illustrate the type of questions that are often pursued in MEG research and to demonstrate some of the methods in action. In each example, data from a single subject will be focused on and the section will close by presenting the overall group effect.

21.4.1 Somatosensory Stimulation

To begin, the hypothetical experiment presented initially in the MEG Methods section will be continued, except this time with real data. The goal of this experiment was to spatiotemporally map the brain's response to an air-puff that was applied to the pad of the left index finger. Participants were seated inside the MSR with their head resting in the MEG helmet as a series of air-puffs were presented at a rate of one puff every 2–4 s. These data were subjected to a noise rejection algorithm, divided into 600 ms epochs (−200 to 400 ms), inspected for artifacts (bad epochs were discarded), and coregistered with structural MRI.

In this first example, a classic version of the dipole-localization approach to identifying neural generators will be utilized. First, the artifact-free epochs were time-domain averaged and low-pass filtered at 40 Hz. These sensor-level data were then projected onto a two-dimensional plane, so that time periods with dipolar field patterns could be readily identified. Essentially, cortical activations often produce magnetic field distributions that are dipolar in nature and can be modeled with an ECD. An ECD is a reasonable approximation of current flowing within a limited cortical structure being recorded from a distance of several centimeters (Hämäläinen et al. 1993), and each ECD is meant to indicate the three-dimensional location, orientation, and amplitude of the center of a cortical current. Thus, dipolar field patterns in the two-dimensional plots (i.e., contour plots) are informative because

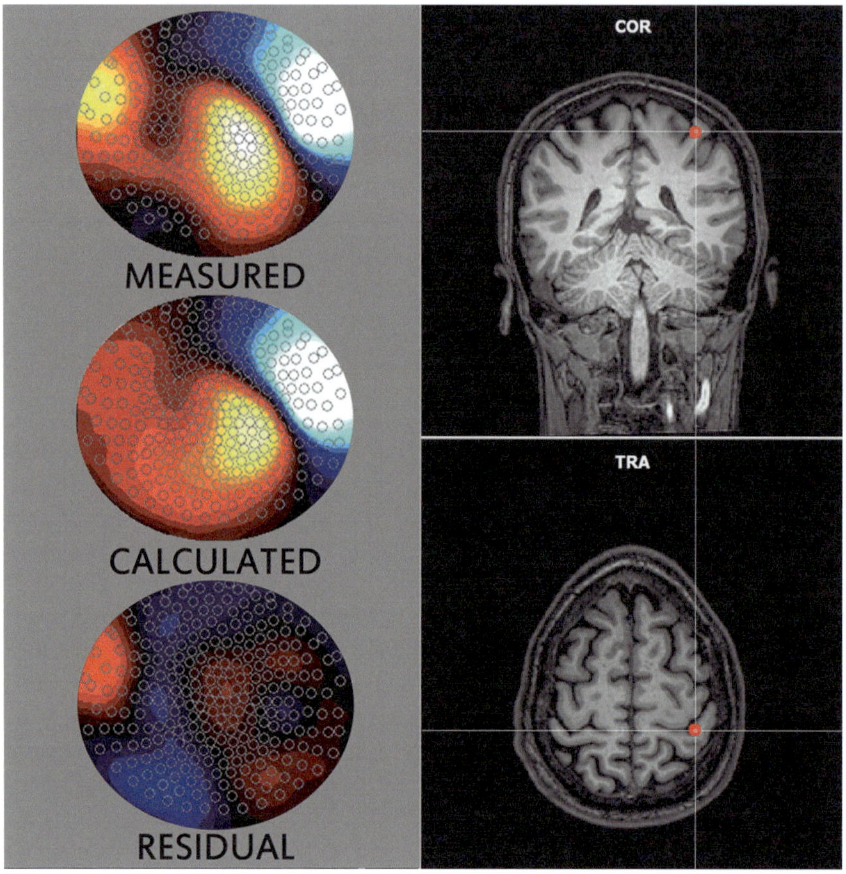

Fig. 21.4 The two-dimensional contour plots used in dipole localization are shown to the *left* with the *top* plot showing the measured MEG signal amplitudes across the array at 54 ms, the *middle* plot showing the calculated fields produced by the optimal ECD, and the *bottom* plot depicting the residual (difference between the measured and calculated fields). To the *right*, the optimal dipole-localization results are overlaid on sagittal and coronal MRI slices (*left*=*left*)

they generally indicate the local presence of an active cortical area. In the current subject, only one clear dipolar pattern was identified, and the peak amplitude of this response was reached at ~54 ms poststimulation. Thus, magnetic field measurements at 54 ms (from all sensors) were modeled as a single ECD using an optimization algorithm. Briefly, the location, amplitude, and orientation of the ECD were derived through an iterative procedure in which all three variables were allowed to vary until the square of the difference between the measured data and the computed (forward) solution of the ECD was minimized. In other words, the best dipole parameters were found, in a least-squares sense, by iteratively changing each of the ECD parameters and recomputing the forward solution (per iteration) until the residual variance had reached a putative minimum. The results of this analysis are shown in Fig. 21.4 and indicated that mechanical stimulation to the left index finger activated the right postcentral gyrus and that this activity peaked at 54 ms with an amplitude of ~14 nÅ. The dipole orientation is also informative, but less so from a neuroscience perspective. From the sensor-level waveforms and contour plots, one can also conclude that activity in this area began slightly before (~48 ms) and had largely dissipated by 67 ms. Thus, not surprisingly, this dipolar field pattern likely reflected the primary somatosensory response to the mechanical stimulation.

For didactic purposes, a second dataset from this same experiment will now be evaluated using a spatial filtering approach, which does not fit neatly into either of the categories of inverse methods discussed earlier. Spatial filtering or beamforming was initially developed in the fields of sonar and radar for localizing the origin of signals. Like the imaging methods, beamforming begins with a well-defined and dense source model (e.g., ~2,500 dipoles spaced equidistantly) that varies slightly between different studies. The idea is that a spatial filter for each point in the model is constructed and that these filters pass activity from a given point while minimizing activity in all other source points. The output is a three-dimensional map reflecting activity during the task (van Veen et al. 1997; Vrba and Robinson 2001; Hillebrand et al. 2005). Beginning with the artifact-free single trial epochs (un-averaged), a time-frequency analysis was performed on each epoch and the results were averaged across all trials. This type of signal averaging transforms the magnetic field measurements from each sensor into a representation that reflects the spectral power at a given frequency per unit time for that sensor (Fig. 21.7). The time-frequency plots for this task showed robust power in the 12–26 Hz range between 40 and 120 ms (12–26 Hz reflects the frequency at which activity in the target neural population is oscillating). This time-frequency window and a window of equal duration from the baseline period were evaluated using a beamformer. In principle, the beamformer operator generates a spatial filter that passes signals without attenuation from a region of interest, while suppressing activity in all other brain areas. The filter properties arise from the forward solution (lead field) for each location on a volumetric grid specified by input voxel space and from the MEG data covariance matrix. Basically, for each target voxel, a set of beamformer weights is determined which amounts to each MEG sensor being allocated a sensitivity weighting for activity in the particular voxel. This set of beamformer weights is the spatial filter unique to the target voxel, and this procedure is iterated until such a filter is computed for each voxel in the brain. Activity in each target voxel is then determined by multiplying the signal in each MEG sensor with its corresponding weight and summing the weighted sensor signals. This can be done independently and sequentially for each voxel to produce a volumetric map of source activity with relatively high spatial resolution (~5 mm). The output of this beamformer analysis is shown in Fig. 21.5. As can be discerned, the beamforming analysis indicates that several brain areas including the bilateral postcentral gyri and central sulcus were activated by the air-puff stimulation in this participant. Further probing of the data indicated that activity was initially detected in the contralateral (to stimulation, right) postcentral gyrus and central sulcus at ~50 ms. Activity then spread to the ipsilateral postcentral gyrus at ~75 ms and thereafter shifted to a more medial aspect of the right postcentral gyrus (Fig. 21.5). Interestingly, when this same analysis was conducted on a larger sample of subjects (~15), the neural areas that exhibited statistically significant activation included those described above but also the cerebellar cortices ipsilateral to stimulation (see Wilson et al. 2009).

21.4.2 Response Inhibition

MEG is also a valuable tool for imaging the brain regions that serve more complex cognitive processes. In this example, data from a response-inhibition experiment will be examined to illustrate a new method and application. The participant of interest was diagnosed with attention-deficit/hyperactivity disorder (ADHD) and was being treated with a stimulant-based medication, which is the most common type of pharmacotherapy for ADHD. The experiment involved a rapid-presentation stop-signal motor task. This task requires speeded motor responses, pressing of either a left or right button on a magnetically silent pad, according to the direction of "go" arrows (pointing left or right) that appear on a visual display in front of the subject. However, on 20 % of trials an up-facing arrow appeared shortly after the "go" stimulus (i.e., left/right arrow), signaling participants to withhold the motor response. These no-go trials (20 %) were of primary interest

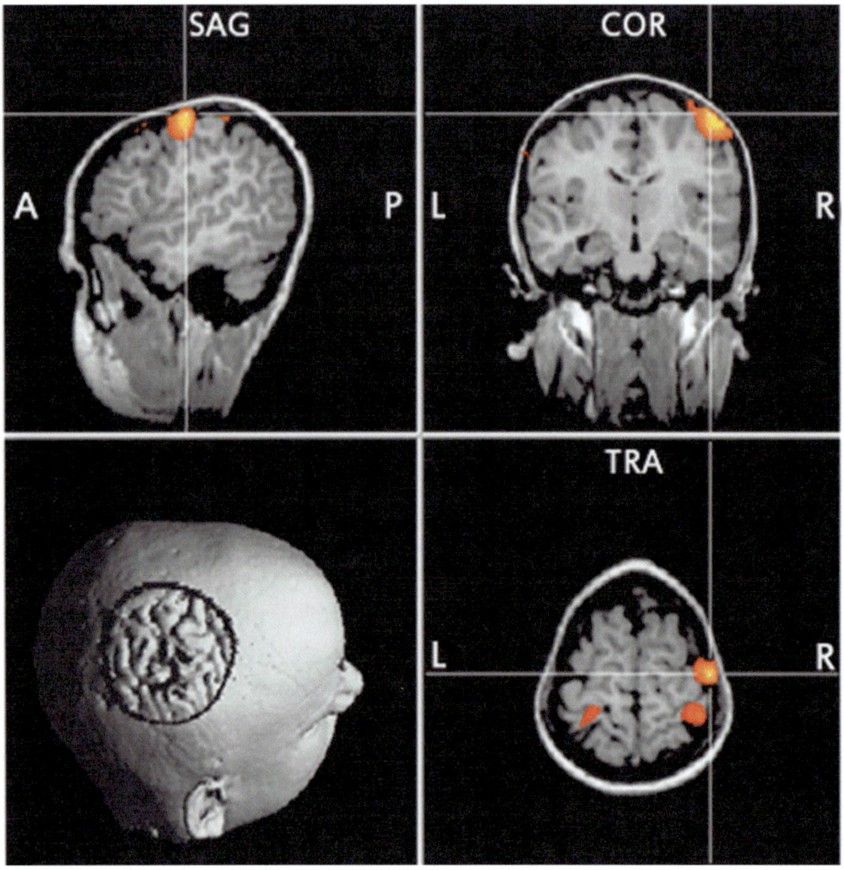

Fig. 21.5 Stimulation of the left index finger elicited activity in the right postcentral gyrus, right central sulcus, and the left postcentral gyrus. These single-subject maps reflect neuronal activity in the 12–26 Hz frequency band that was imaged using a spatial filtering approach

in the MEG analysis because the participant was actively inhibiting a prepotent motor response. The participant completed the task two times, before and after their daily medication, and during the task brain responses were recorded with a 306-sensor MEG system. For this example, only data from the post-medication run will be analyzed. All MEG data were subjected to a noise rejection algorithm, inspected for artifacts, divided into 2,000 ms epochs (−800 to 1,200 ms, with 0 ms defined as the onset of the up-facing arrow), and coregistered with structural MRI.

These data will be examined using a cortically constrained variant of minimum-norm estimation (MNE; Hämäläinen and Ilmoniemi 1994), which is a classic example of the imaging approach to the inverse modeling problem. The source model consisted of the brain's cortical gray matter, which was decimated into ~3,200 points (sources). In MNE, the amplitudes of all cortical sources are estimated in parallel, with the solution having the lowest total amplitude (across the model) being the preferred solution. This minimum amplitude criterion functions as a prior for regularizing the inverse modeling problem. Initially, the artifact-free single trials from the post-medication run were averaged in the time domain. The entire sensor-level time series was then transformed into source space using cortically constrained MNE. The results indicated robust activation 150–400 ms after the up-facing arrow in the left dorsolateral prefrontal cortices, bilateral supplementary motor area (SMA), and the left precentral gyrus (Fig. 21.6). Thus, one

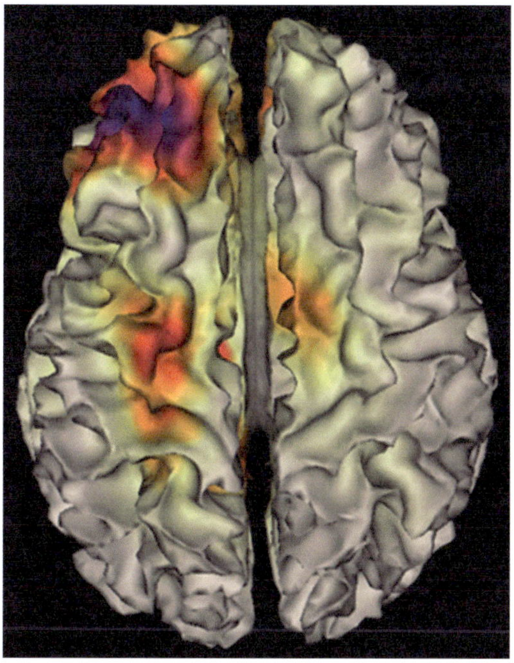

Fig. 21.6 Brain areas activated as a participant attempts to inhibit a prepotent motor response. Activation was strongest in the left dorsolateral prefrontal cortices. Results were generated using cortically constrained MNE

can conclude from this analysis that these three brain regions are involved in inhibiting a prepotent response (with the right hand) and that the first 400 ms after the "no-go" stimulus are critical for successfully withholding the response. The group results (12 subjects) were virtually identical to Fig. 21.6, except the statistically significant areas of activation were slightly larger. Finally, it is important to recognize that, for brevity, only the 150–400 ms time period was discussed. There was also activity during other time windows; for example, the primary motor and somatosensory areas were active a few hundred milliseconds before the onset of the up arrow.

21.4.3 Basic Motor Control

As discussed previously, superior temporal resolution is a major methodological advantage for MEG over comparable imaging techniques. This advantage will be exploited in the current example by sequencing the brain areas involved in basic motor control and performance. The experiment required a single flexion-extension movement of the right index finger every ~6 s, in reference to an external visual pacing device (i.e., a clock-like image). During task performance, participants were seated inside the MSR with their head resting inside the MEG helmet. To track finger movements, accelerometer chips were attached to the right index finger and were sampled along with the high-density MEG array at 1 kHz. All MEG data were subjected to a global noise filter, inspected for artifacts, divided into 6,000 ms epochs (−3,000 to 3,000 ms, with 0 ms defined as the movement onset), and coregistered with structural MRI.

In this example, the beamforming approach will be used for source imaging (described previously). First, the artifact-free single trials (−3,000 to 3,000 ms) were transformed into the time-frequency domain using complex demodulation and averaged across all trials. This type of signal averaging transforms the raw field measures from each sensor into a representation of the spectral power at a given frequency per unit time. The time-frequency plot of a sensor located in the vicinity of the sensorimotor cortices is shown as an example in Fig. 21.7. The resulting plots showed robust decreases in neuronal activity at 20–32 Hz before movement onset, strong increases in high-frequency gamma activity (74–86 Hz) that coincided with the movement onset, and finally increases in activity at 14–26 Hz after the movement terminated. Each of these time-frequency windows was imaged along with a baseline window of equal duration and bandwidth taken from the −2,000 to −1,400 ms baseline period. The beamforming results indicated that the SMA, left postcentral gyrus, and the right cerebellar cortices were activated before movement onset (Fig. 21.8). As the movement commenced, gamma activity was detected in the left precentral gyrus (primary motor cortex) and the SMA, and following the movement offset 14–26 Hz activation was present in the anterior and inferior aspects of the left precentral gyrus (premotor area) and bilateral cerebellar cortices. The group level results were very similar and

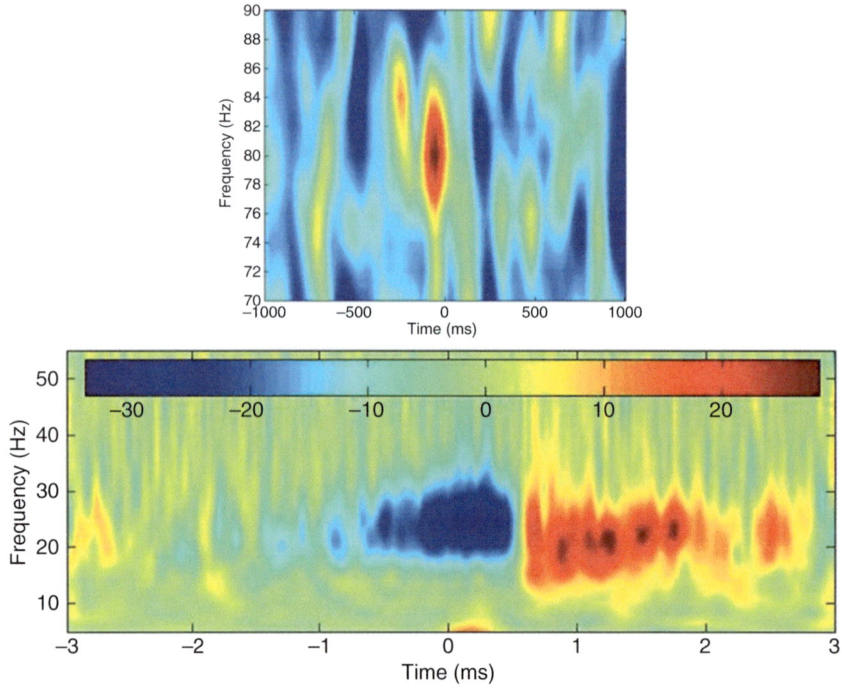

Fig. 21.7 Time-frequency representation of sensor near sensorimotor cortices. *Bottom* plot shows data for lower frequencies (5–45 Hz) and *upper* plot shows that for higher frequencies (70–90 Hz). In each plot, frequency is shown on the *y*-axis in (Hz) and time is on the *x*-axis in (ms). *Red* depicts increases in spectral power relative to the baseline period (−2,000 to −1,400 ms) and *blue* reflects decreases in power. Movement onset was defined as 0 ms. See text for additional details

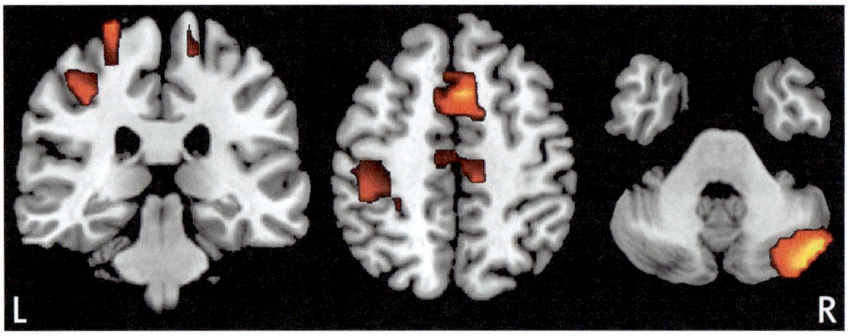

Fig. 21.8 The brain areas activated before movement onset are shown in *red-to-yellow* intensity and included the supplementary motor area (SMA), left postcentral gyrus, and right cerebellum

Fig. 21.9 shows the peak brain activation per region during each phase of the movement (see also Wilson et al. 2010, 2011). Thus, based on these analyses, one can conclude that the right cerebellum, bilateral SMA, and left postcentral gyrus were strongly activated during pre-movement planning stages, whereas the primary motor cortex and SMA were active during the instantiation of movement, and finally the left premotor area and cerebellum were active after movement and perhaps have a role in resetting motor areas for future movements.

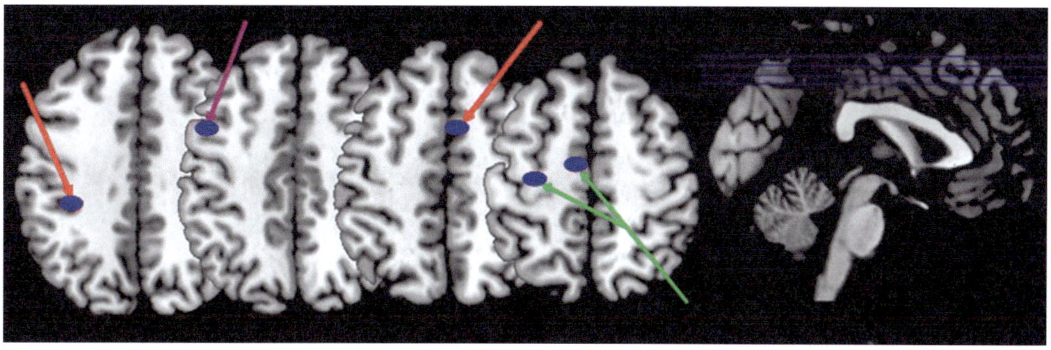

Fig. 21.9 Regional activation peaks (group level) per phase of right index finger movements, based on MEG beamforming analysis. *Red arrows* indicate pre-movement activity; *magenta* shows post-movement activity; and *green arrows* reflect activation that coincided with movement onset. From *left* to *right*, *blue ovals* indicate left postcentral gyrus (pre-movement), left premotor area (post-movement), right SMA (pre-movement), and left precentral gyrus and SMA (during)

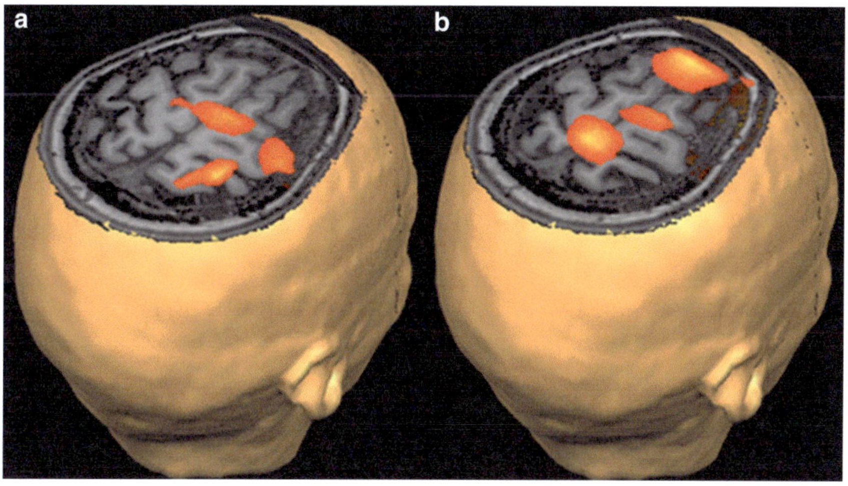

Fig. 21.10 MEG study of brain plasticity. (**a**) Cortical regions that were activated during leg movements in a participant before therapy (i.e., at baseline). (**b**) Cortical regions activated after several weeks of physical therapy (same task and participant). Note that activation in lateral motor areas dissipated after therapy and that activity in medial cortices was enhanced. Healthy participants generally show activation restricted to medial areas of the sensorimotor cortices

21.5 Current and Future Directions for MEG-Based Functional Imaging

The use of MEG has grown rapidly over the past decade and especially so in the past few years. This is true for the clinical uses of MEG in epilepsy and tumor patients, as well as in purely research settings. Some of the hottest areas for MEG are in translational neuroscience. The search for MEG-based biomarkers of disease is especially busy, with a plethora of new articles appearing on autism, schizophrenia, and most other major psychiatric disorders. MEG research teams are also actively pursuing biomarkers for Parkinson's disease, Alzheimer's, and moderate (non-lesional) traumatic brain injury. Using MEG to monitor therapeutic programs is a growing area as well, with several groups using the method to chart the evolution of functional brain alterations (plasticity) across treatment sessions (Fig. 21.10). Finally, the use of MEG in areas of pharmacology (pharmaco-MEG) has seen rapid expansion. MEG is the only high-resolution functional imaging modality that is a direct measure of neurophysiology, which positions it as ideal for evaluating the systems-level mechanisms that underlie the action of drugs. As you may recall, the physical signals of comparable imaging modalities (e.g., fMRI) are vascular-based and thus are indirect indices of neuronal activity. In some cases, this means that drug-induced modulations of the vasculature (e.g., vasodilation) will bias the signal, irrespective of whether the drug actually changed neuronal activity. Since MEG is not sensitive to such biases, it is quickly becoming the method of choice for understanding how new and existing drugs affect brain function in a number of psychiatric and neurological disorders. In the future, MEG may see action in the early stages of drug development to ascertain where drugs are primarily acting. This is especially likely for disorders where certain brain areas are known to underlie key aspects of the symptomatology. Thus, MEG will likely appear in many new venues in the coming years, as it continues to benefit from the scientific community's growing focus on translational research.

References

Ahonen AI, Hamalainen MS, Kajola MJ, Knuutila JET, Laine PP, Lounasmaa OV, Parkkonen LT, Simola JT, Tesche CD (1993) 122-channel SQUID instrument for investigating the magnetic signals from the human brain. Physica Scripta T49:198–205

Aubert A, Pellerin L, Magistretti PJ, Costalat R (2007) A coherent neurobiological framework for functional neuroimaging provided by a model integrating compartmentalized energy metabolism. Proc Natl Acad Sci U S A 104:4188–4193

Baillet S (2010) The dowser in the fields: searching for MEG sources. In: Hansen PC, Kringelbach ML, Salmelin R (eds) MEG: an introduction to the methods. Oxford University Press, New York, pp 83–123

Baillet S, Mosher JC, Leahy RM (2001) Electromagnetic brain mapping. IEEE Signal Process Mag 18(6):14–30

Bandettini PA, Ungerleider LG (2001) From neuron to BOLD: new connections. Nat Neurosci 4:864–866

Bandettini PA, Wong EC, Hinks RS, Tikofsky RS, Hyde JS (1992) Time course EPI of human brain function during task activation. Magn Reson Med 25:390–397

Cohen DS (1968) Magnetoencephalography: evidence of magnetic fields produced by alpha rhythm currents. Science 161:784–786

Goense JB, Logothetis NK (2008) Neurophysiology of the BOLD fMRI signal in awake monkeys. Curr Biol 18:631–640

Hämäläinen MS, Ilmoniemi RJ (1994) Interpreting magnetic fields of the brain: minimum norm estimates. Med Biol Eng Comput 32(1):35–42

Hämäläinen MS, Sarvas J (1989) Realistic conductivity geometry model of the human head for interpretation of neuromagnetic data. IEEE Trans Biomed Eng 36(2):165–171

Hämäläinen M, Hari R, Ilmoniemi RJ, Knuutila J, Lounasmaa OV (1993) Magnetoencephalography—theory, instrumentation, and applications to noninvasive studies of the working human brain. Rev Mod Phys 65(2):413–495

Hansen PC, Kringelbach ML, Salmelin R (eds) (2010) MEG: an introduction to the methods. Oxford University Press, New York

Heeger DJ, Rees D (2002) What does fMRI tell us about neuronal activity? Nat Rev Neurosci 3:142–151

Heeger DJ, Huk AC, Geisler WS, Albrecht DG (2000) Spikes versus BOLD: what does neuroimaging tell us about neuronal activity? Nat Neurosci 3:631–633

Hillebrand A, Barnes GR (2002) A quantitative assessment of the sensitivity of whole-head MEG to activity in the adult human cortex. Neuroimage 16:638–650

Hillebrand A, Singh KD, Holliday IE, Furlong PL, Barnes GR (2005) A new approach to neuroimaging with magnetoencephalography. Hum Brain Mapp 25:199–211

Huettel SA, Song AW, McCarthy G (2008) Functional magnetic resonance imaging, 2nd edn. Sinauer, Sunderland, MA

Humm JL, Rosenfeld A, Del Guerra A (2003) From PET detectors to PET scanners. Eur J Nucl Med Mol Imaging 30:1574–1597

Kwong KK, Belliveau JW, Chesler DA, Goldberg IE, Weisskoff RM, Poncelet BP et al (1992) Dynamic magnetic resonance imaging of human brain activity during primary sensory stimulation. Proc Natl Acad Sci U S A 89:5675–5679

Leahy RM, Mosher JC, Spencer ME, Huang MX, Lewine JD (1998) A study of dipole localization accuracy for MEG and EEG using a human skull phantom. Clin Neurophysiol 107:159–173

Logothetis NK (2003) The underpinnings of the BOLD functional magnetic resonance imaging signal. J Neurosci 23:3963–3971

Logothetis NK, Pauls J, Augath M, Trinath T, Oeltermann A (2001) Neurophysiological investigation of the basis of the fMRI signal. Nature 412:150–157

Lopes da Silva FH (2010) Electrophysiological basis of MEG signals. In: Hansen PC, Kringelbach ML, Salmelin R (eds) MEG: an introduction to the methods. Oxford University Press, New York, pp 1–23

Lopes da Silva FH, van Rotterdam A (2005) Biophysical aspects of EEG and magnetoencephalographic generation. In: Niedermeyer E, Lopes da Silva FH (eds)

Electroencephalography, basic principles, clinical applications and related fields, 5th edn. Lippincott Williams & Wilkins, Philadelphia, pp 1165–1198
Lu ZL, Kaufman L (eds) (2003) Magnetic source imaging of the human brain. Lawrence Erlbaum Associates, Mahwah, NJ
Magistretti PJ, Pellerin L (1999) Astrocytes couple synaptic activity to glucose utilization in the brain. News Physiol Sci 14:177–182
Mosher JC, Baillet S, Leahy RM (1999) EEG source localization and imaging using multiple signal classification approaches. J Clin Neurophysiol 16(3):225–238
Murakami S, Okada Y (2006) Contributions of principal neocortical neurons to magnetoencephalography and electroencephalography signals. J Physiol 575:925–936
Ogawa S, Lee TM, Kay AR, Tank DW (1990) Brain magnetic resonance imaging with contrast dependent on blood oxygenation. Proc Natl Acad Sci U S A 87:9868–9872
Ogawa S, Tank DW, Menon R, Ellermann JM, Kim SG, Merkle H et al (1992) Intrinsic signal changes accompanying sensory stimulation: functional brain mapping with magnetic resonance imaging. Proc Natl Acad Sci U S A 89:5951–5955
Okada YC, Wu J, Kyuhou S (1997) Genesis of MEG signals in a mammalian CNS structure. Clin Neurophysiol 103:474–485
Parkkonen L (2010) Instrumentation and data processing. In: Hansen PC, Kringelbach ML, Salmelin R (eds) MEG: an introduction to the methods. Oxford University Press, New York, pp 24–64
Raichle ME, Mintun MA (2006) Brain work and brain imaging. Annu Rev Neurosci 29:449–476
Rees G, Friston K, Koch C (2000) A direct quantitative relationship between the functional properties of human and macaque V5. Nat Neurosci 3:716–723
Sarvas J (1987) Basic mathematical and electromagnetic concepts of the biomagnetic inverse problem. Phys Med Biol 32(1):11–22
van Veen BD, van Drongelen W, Yuchtman M, Suzuki A (1997) Localization of brain electrical activity via linearly constrained minimum variance spatial filtering. IEEE Trans Biomed Eng 44:867–880
Vrba J, Robinson S (2001) Signal processing in magnetoencephalography. Methods 25:249–271
Wagner H, Eiselt M, Zwienger U (1997) Exactness of source analysis of biomagnetic signals of epileptiform spikes by the method of spatial filtering: a computer simulation. Med Biol Eng Comput 35:708–714
Wilson TW, Slason E, Hernandez OO, Asherin RM, Reite ML, Teale PD, Rojas DC (2009) Aberrant high frequency desynchronization of cerebellar cortices in early-onset psychosis. Psychiatry Res 174:47–56
Wilson TW, Slason E, Asherin RM, Kronberg E, Reite ML, Teale PD, Rojas DC (2010) An extended motor network generates beta and gamma oscillatory perturbations during development. Brain Cogn 73:75–84
Wilson TW, Slason E, Asherin RM, Kronberg E, Teale PD, Reite ML, Rojas DC (2011) Abnormal gamma and beta MEG activity during finger movements in early-onset psychosis. Dev Neuropsychol 36:596–613

Multiphoton Brain Imaging

22

Anna Dunaevsky

Abstract

Imaging approaches have revolutionized neuroscience research. The discovery and development of dyes, fluorescent proteins, light-activated proteins (optogenetics), as well as transgenic animals with subset of cells expressing fluorescent proteins allows scientists to probe the structure and function of neurons with ever-increasing specificity. The ability to perform such investigations in intact animals affords linkages between cellular form and function to the dynamics of neural networks, animal behavior, and disease. Two-photon laser scanning microscopy (TPLSM) is amongst the best approaches available for high-resolution imaging of neurons and other cells that provides spatiotemporal information in the intact brain not feasible by other methods.

The objective of this chapter is to address the following questions:

1. Why is there a need for TPLSM for in vivo imaging?
2. How does TPLSM work?
3. What are its common applications?
4. How to perform in vivo imaging of the brain using TPLSM?

Keywords

Multiphoton • Confocal • GFP • In vivo • Cranial window

22.1 Introduction and Background

22.1.1 The Demand for In Vivo Imaging Techniques

Most living preparations pose several optical challenges and are generally quite a hostile environment for microscopists. The main problem is the thickness of the tissue. The interaction of

A. Dunaevsky (✉)
Department of Developmental Neuroscience, Munroe Meyer Institute for Genetics and Rehabilitation, University of Nebraska Medical Center, 985960 Nebraska Medical Center, Omaha, NE 68198-5960, USA
e-mail: adunaevsky@unmc.edu

H. Xiong and H.E. Gendelman (eds.), *Current Laboratory Methods in Neuroscience Research*, Springer Protocols Handbooks, DOI 10.1007/978-1-4614-8794-4_22,

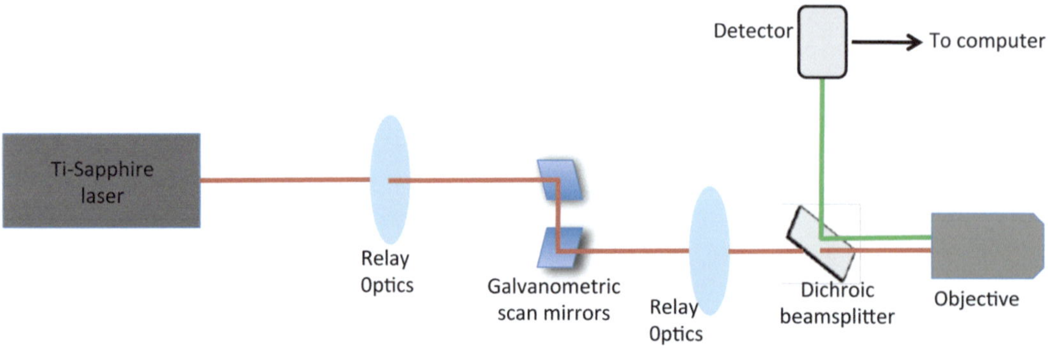

Fig. 22.1 Schematic diagram of the optical pathway in TPLSM. An infrared excitation beam from the Ti-Sapphire laser (*red line*) is focused into *x–y* galvanometric mirrors by a set of relay optics and then expanded and imaged onto the back focal plane of the microscope objective by another set of relay optics. The emitted fluorescence from the sample (*green line*) is collected by the objective and, after separation from the excitation light by a dichroic filter, is directed to a detector

photons with cellular structures results in the scattering of photons in the tissue. This results in reduced penetration of the light and inability to focus light and excite fluorophores deep in the tissue. In addition to the reduced excitation deep in the tissue, emitted photons will also scatter resulting in the blurring of the image and reduced signal to noise ratio.

The development of confocal laser scanning microscopy has improved the ability to view structures deep in the tissue that would be normally severely degraded by scattered light (Pawley 2005). In laser scanning fluorescence microscopy (Fig. 22.1), a laser beam (monochromatic light of a discrete wavelength) is focused through an objective and is scanned across the field of view. The fluorescent light emitted from each point in the scan is transmitted onto a detector (i.e., a photomultiplier tube, PMT). Mapping the intensity on the detector to each point in the two-dimensional scan then yields an image originating from fluorophores. In confocal microscopy optical sectioning is achieved by the placement of an aperture, or a pinhole, in front of a detector (Fig. 22.2b). In this manner, light is focused on a point in the specimen and fluorescence that is emitted from that point passes through the aperture and is therefore being detected. Importantly, fluorescence that is emitted from all out-of-focus points is blocked by the aperture and is therefore not being detected. This results in acquisition of crisp images, with high signal to noise ratio, from each focal plane (Z-plane). Despite its enormous usefulness, confocal microscopy has its limitations especially for live imaging. The main downside is that although focal sectioning is achieved by the rejection of the out-of-focus light, the entire specimen, above and below the focal plane, is being illuminated and therefore fluorophores are excited (Fig. 22.2c). As a result of this excessive excitation, photobleaching (loss of ability to produce fluorescence) and phototoxicity are common problems. In fact, when one steps through the thickness of the preparation (collection of a Z-stack), it is not unusual that the sample towards the end of Z-stack is significantly bleached by the time it is imaged. This happens because all planes of focus are repeatedly illuminated although light is only collected from one focal plane at a time. Such bombardment of a sample by light is problematic for live imaging (in particular with time-lapse imaging) and can cause photodamage and cell death. This limitation is overcome by two-photon laser scanning microscopy (TPLSM) in which excitation is only occurring at the focal plane.

22.1.2 Principles of TPLSM

Fluorescence is based on the ability of some molecules to absorb light at a particular wavelength and to then emit light of lower energy and longer wavelength (Fig. 22.2a). For most useful fluorophores, this happens in the UV/visible range

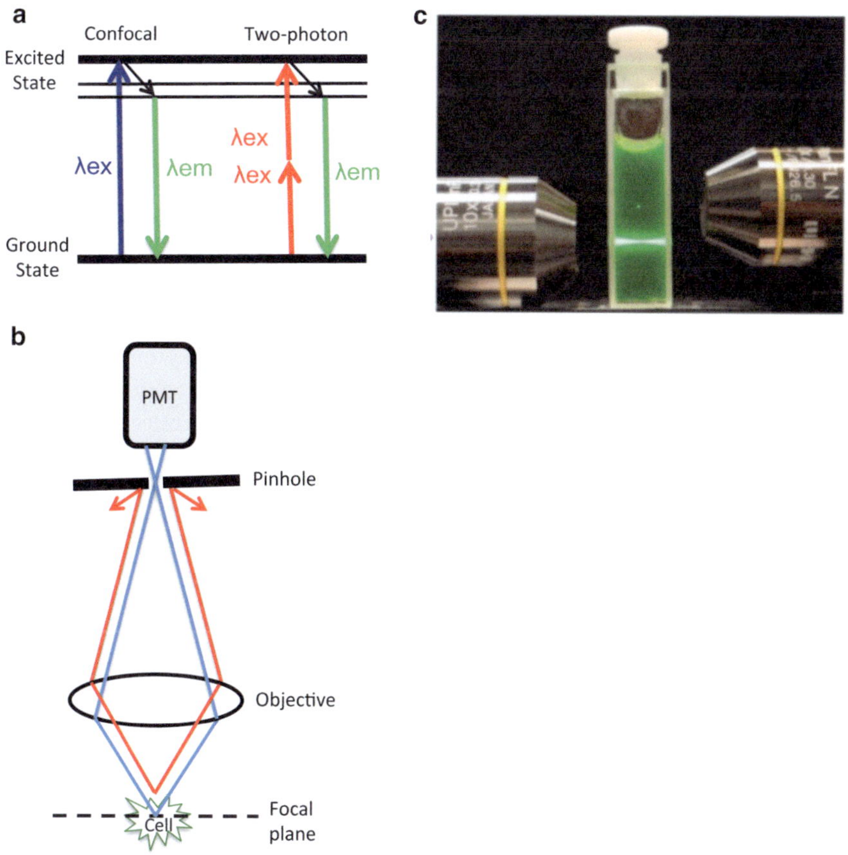

Fig. 22.2 Differences between confocal and two-photon excitation. (**a**) In confocal imaging (one photon), a fluorescent molecule initially at rest (in its ground state) is excited by a high-energy photon (*blue* in the illustration). Then, the molecule reverts back to its ground state while emitting a lower-energy photon (*green* in the illustration) than the original incident photon. In the two-photon case, two lower-energy photons (*red* in the illustration) act nearly simultaneously on the resting fluorescent molecule. A photon with higher energy than the incident photons is emitted. (**b**) Diagram illustrating the confocal principle. Light from a point source is focused on a point in a cell. Fluorescence from the focused point passes through a pinhole in front of a detector (PMT). Fluorescence from other points in the specimen is excluded by the pinhole. (**c**) Experimental illustration of the difference between linear and two-photon excitation. 380 nm (confocal) and 760 nm (two-photon) excitation wavelengths were used to generate fluorescence in a fluorescein solution. Two-photon excitation generates fluorescence only in the focal volume, whereas conventional single-photon excitation generates fluorescence above and below the focal volume (image used by permission, Kevin Belfield, University of Central Florida)

(400–600 nm). At these wavelengths, each photon has sufficient energy to bring a fluorescent molecule to an excited state following its absorption. This linear relationship is the reason that in confocal microscopy fluorophores are being excited all throughout the cone of light above and below the point where the light is focused (Fig. 22.2c). With TPLSM on the other hand, photons of higher wavelength (approximately double the absorption peak of the fluorophore) but lower energy are used. While each photon does not have sufficient energy to excite a fluorophore, when two such photons are almost simultaneously absorbed by a fluorescent molecule, their combined energy brings the fluorophore to an excited state (Fig. 22.2a) (Denk et al. 1990). The probability of two photons being absorbed almost simultaneously is only high enough at the beam focus—where the density of photons is highest. Due to this nonlinear nature of two-photon excitation, fluorophore excitation will effectively only occur at the focal volume, thereby eliminating excitation of out-of-focus fluorophores and achieving optical sectioning (Fig. 22.2c). With two-photon excitation, there is no need for a detection pinhole (Fig. 22.1), because there is no out-of-focus excitation.

Because the probability of two photons interacting nearly simultaneously with a single fluorophore is low, high flux of photons needs to be delivered to the sample. To prevent such high power from destroying the sample, the laser used for TPLSM is pulsed. Pulses of high number of photons travel together for a brief period of time (100 fs) at repetition rate of approximately 100 MHz. While the power during the pulses is high enough to achieve multiphoton excitation, the average power is low. The combination of a pulsed laser with a high numerical aperture objective brings a large number of photons to the region of interest within the sample, thus increasing the probability of multiple photons to simultaneously excite a fluorophore. The Ti-Sapphire lasers most commonly used for TPLSM are tunable to different wavelengths (700–1,200 nm), allowing for excitation of fluorophores with different excitation spectra (Drobizhev et al. 2011).

22.1.3 Advantages of TPLSM

The main advantage of multiphoton imaging is that optical sections can be collected from deeper within a tissue as compared to confocal or wide-field imaging. In general confocal microscopy can image up to 100 μm deep into the tissue while with TPLSM one can image up to 1,000 μm deep in ideal conditions (Helmchen and Denk 2005). This of course depends on the preparation (i.e., amount of fluorophore present) as well as the quality of the imaging system. The reasons for the superior depth imaging with TPLSM are that photons of higher wavelengths are less absorbed by biological samples and are also less scattered compared to the lower wavelengths used for confocal imaging. Because of these properties, the light used for TPLSM can penetrate and excite fluorophores deeper in the tissue.

Emitted photons can also scatter. Although the scattering properties of photons emitted in confocal and multiphoton microscopy are the same (same wavelength, Fig. 22.2a), with the confocal microscope, some useful photons emitted from the focal plane could scatter and not make it through the pinhole, thus reducing the signal. The converse of that is that irrelevant photons emitted from an out-of-focus plane could scatter and find their way through the pinhole, increasing background noise and degrading the image. The absence of a pinhole in TPLSM allows the collection of every emitted photon, even if it was highly scattered since it could only come from the excited focal volume, the region of interest, and is therefore useful signal. Another major advantage of TPLSM is reduced phototoxicity. We have already mentioned that in confocal microscopy optical sectioning is achieved by rejection of emitted light rather than by focal excitation. The result is that much unnecessary light is being delivered to the specimen causing photobleaching. Photons with high energy, as in the UV light used for confocal imaging, can also give rise to highly reactive and destructive free radicals that can cause cell damage when high intensity is used. This could be a problem when the same region is imaged repeatedly over time. The lower-energy infrared light used for TPLSM produces less toxic free radicals throughout the preparation, although phototoxicity at the focal plane is comparable or even could be higher compared with confocal imaging (Ustione and Piston 2011).

22.1.4 Limitations of TPLSM

TPLSM has its limitations. For one, it has slightly lower resolution compared to a confocal microscope. The lateral resolution of both one- and two-photon imaging is diffraction-limited (Pawley 2005). This limit is dictated by the numerical aperture of the objective used and by the wavelength of the excitation light. The axial resolution is approximately twice as long as the lateral resolution or approximately equal to the wavelength of the excitation light. Since TPLSM requires the use of longer wavelengths of light than that of the confocal, single-photon imaging, the theoretical resolution is proportionally worse. The effective resolution of confocal and TPLSM, however, is dependent on a variety of parameters and is not very different (Williams et al. 1994; Ustione and Piston 2011).

Because traditional laser scanning depends on galvanometric mirrors that scan the beam across the specimen (Fig. 22.1), these approaches tend to be slow and cannot capture very fast events (e.g., calcium transients in neurons) at full frame modes. One way around this is to perform line scans, thus increasing temporal information on the expense of spatial information (Yuste and Denk 1995). New technologies such as random access multiphoton imaging that rely on acousto-optical deflectors allow one to acquire images at rates 1,000 times faster than conventional TPLSM (Grewe et al. 2010).

Finally, one major disadvantage of TPLSM is the cost of the lasers necessary to achieve multiphoton excitation. This becomes limiting especially if one would like to simultaneously image more than a couple of fluorophores that have different excitation peaks. One has to remember that confocal microscopy is the best (and also cheaper) choice when imaging thinner specimens. Moreover, if only a single cell layer is to be imaged, a simple wide-field microscope is often an even better solution than confocal microscopy, because there is often no need for optical sectioning in such a thin sample. Nevertheless, TPLSM becomes a must if one needs to perform in vivo imaging of structures in the brain.

22.2 In Vivo Imaging Applications

The applications of in vivo TPLSM imaging have grown significantly over the past decade (Christensen and Nedergaard 2011). By expressing fluorescent proteins under the control of cell type-specific promoters, it is possible to monitor the dynamics of different cell populations in their native environments. In vivo imaging of a variety of brain structures and cells has been performed including neuronal structures, such as dendritic arbors, dendritic spines, and axonal terminals, during brain development and following various sensory stimulation and learning paradigms (Fig. 22.3) (Grutzendler et al. 2002; Trachtenberg et al. 2002; Holtmaat et al. 2005; Xu et al. 2009). In vivo imaging of nonneuronal cells such as astrocytes, microglia, and other immune response cells has also been performed in normal brains, as well as following insult such as infection or stroke (Fig. 22.3) (Davalos et al. 2005; Wang et al. 2006; Brown et al. 2007; McGavern and Kang 2011). Collecting movies of the behavior of cells and interaction between cells in the brain provides a whole new dimension of understanding that is not available by analyzing cells in fixed preparations. In addition to structural imaging, in vivo TPLSM has been used for functional imaging such as measurements of calcium transients in neurons and astrocytes (Fig. 22.3) (Hirase et al. 2004; Wang et al. 2006). Finally another major application is taking advantage of the small/focal excitation afforded by TPLSM to uncage (photoactivate) compounds. For example, glutamate can be uncaged at discrete sites to selectively stimulate neuronal structures of interest. This approach, mainly used in brain slices, has been recently extended to in vivo application (Matsuzaki et al. 2001; Noguchi et al. 2011).

22.3 Methodological Approaches

TPLSM studies require at least some surgical preparation to create a "window" and expose the brain. We will describe the two most common approaches used for brain imaging: the thin skull and the craniotomy preparation. The craniotomy approach, where a portion of the skull is removed, is clearly the more invasive approach and has been shown to cause an inflammatory response (Xu et al. 2007). Nevertheless, multiple researchers that have imaged these windows following a period of recovery and window clearing have successfully used this approach. They were able to image the same structures multiple times over a year (Trachtenberg et al. 2002; Lee et al. 2006). Moreover the craniotomy approach is used when the brain needs to be exposed for dye application (e.g., calcium indicators) or when imaging is combined with electrophysiological recordings from the brain. The thinned skull preparation

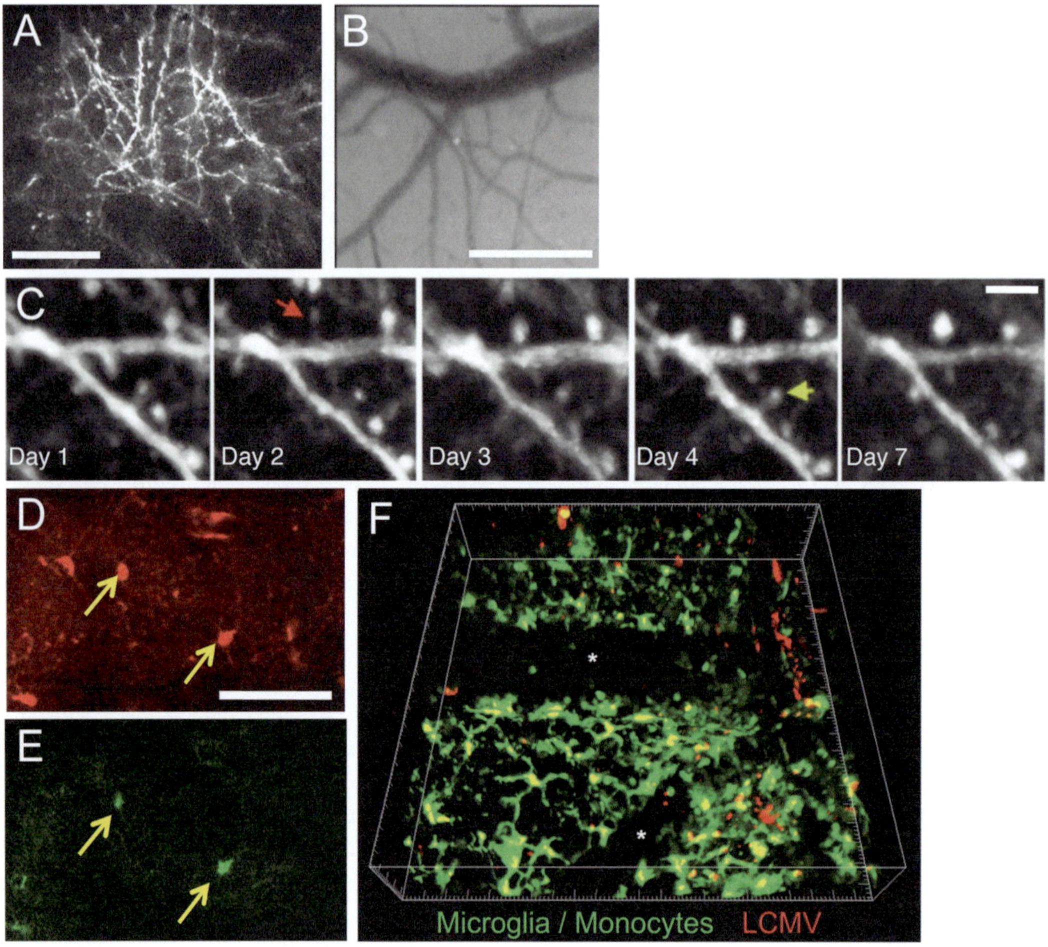

Fig. 22.3 TPLSM in vivo imaging of the brain. (**a**) Low magnification image of apical dendritic tufts of YFP expressing neurons. Bar, 25 μm. (**b**) CCD camera view of the brain vasculature under the thinned skull. The cortical vasculature can be used as a landmark to relocate the imaged region at subsequent time points. Bar, 43 μm. (**c**) The same dendritic spines from 5-week-old mice were imaged over 7 days. A newly formed spine is marked with a *red arrow* and a spine that disappeared after the fourth day is marked with a *yellow arrow*. Bar, 2.5 μm (**d**). Layer I astrocytes loaded with astrocytic marker SR101. (**e**) The calcium indicator, Fluo-4 AM, demonstrates calcium oscillations in a subset of astrocytes shown in (**d**). Bar, 50 μm. (**f**) CX3CR1-GFP mice can be used to visualize innate myeloid sentinels (e.g., microglia and macrophages) in the meninges and brain parenchyma. Circulating monocytes are also visible in these reporter mice. A 3D projection from a representative time lapse shows how brain myeloid cells (*green*) sequester virus (*red*) during a CNS infection. Images (**a–e**) courtesy of Benjamin Reiner and Padmashri Ragunathan from Anna Dunaevsky's lab, University of Nebraska Medical Center. Panel (**f**), courtesy of Dorian McGavern, NINDS, NIH

(skull is thinned to 30 μm) preserves the subcranial milieu intact, avoiding the inflammation induced by surgery (Yang et al. 2011). The downside of this approach is that the bone grows back within a few days and repeated thinning is required if images are to be acquired over weeks and months. In addition the thinned skull window can only be imaged for a limited number of times.

22.3.1 Materials

1. Animals: transgenic mice expressing fluorescent proteins in cells of interest (e.g., Thy1-YFP for neurons and CX3CR1-GFP for immune cells)
2. Chemicals and reagents:
   (a) Ketamine HCl (Hospira, 0409-2051-05)
   (b) Dexdomitor (Pfizer/Orion Pharma, 141–267)

(c) Sterile sodium chloride solution (Sigma, S8776)
(d) Lidocaine HCl 0.5 % and epinephrine 1:200,000 injection USP (Hospira, 0409-3177-01)
(e) Rimadyl (carprofen), Pfizer
(f) Antisedan (Pfizer/Orion Pharma, 141–033)
(g) Artificial cerebrospinal fluid (ACSF) of the following composition (in mM): 126 NaCl, 3 KCl, 1.25 NaH_2PO_4 26 $NaHCO_3$, 2 $CaCl_2$, 1 $MgSO_4.7H_2O$, and 10 glucose (pH 7.4)
(h) Puralube vet ophthalmic ointment (Dechra)

3. Surgical tools and equipment:
   (a) Stereomicroscope
   (b) Mouse stereotaxic instrument (Stoelting, 51730D)
   (c) Dental drill (Aseptico, AEU-10SS)
   (d) Drill burs (SS White, 14717)
   (e) Electrical shaver
   (f) Forceps
   (g) Fine scissors (sharp/curved) (Fine Science Tools, 14061–09)
   (h) Microblade (Miniature Blade Round-Tip Sharp one side, Surgistar, 6400)
   (i) Microblade (Miniature Blade 15° Pointed, Surgistar, 6500)
   (j) Handle for microblades (Fine Science Tools, 10035–00)
4. Materials for window construction:
   (a) Compressed air (Fisher, 23-022-523)
   (b) Alcohol prep pads (PDI, B339)
   (c) Povidine-iodine prep pads (Dynarex, 1108)
   (d) Cotton-tip applicators
   (e) Gelfoam (Patterson Dental, 600–8675)
   (f) Duralay (Reliance Dental Mfg. Co. 2244)
   (g) Super glue
   (h) Round cover glass (#1 thickness, 5 mm) (Warner Instruments, 64–0701)
   (i) Cyanoacrylate epoxy (Loctite, 454)
   (j) Custom head bars

Two-photon laser scanning microscope: Any upright two-photon microscope with a large enough stage to accommodate the mouse can be used.

22.3.2 Procedure

1. Anesthetize mice by intraperitoneal injection of ketamine/Dexdomitor solution (50–75/0.5–1 mg/kg). Wait until animal is not responsive to a pinch of the foot (5–10 min). Monitor the depth of anesthesia by testing the animal's reflexes throughout the surgery and inject more ketamine/Dexdomitor solution when necessary.
2. Inject the mouse with Rimadyl (5 mg/kg) to reduce surgery-induced inflammation. Do not use if the anti-inflammatory drug interferes with the experimental design.
3. Thoroughly shave the hair over most of the scalp and blow any excess hair with compressed air canister. Wipe off the skin with 10 % povidine-iodine solution followed by a sterile alcohol pad. Apply eye ointment to prevent eyes from drying.
4. Place the mouse in a stereotaxic apparatus and immobilize the skull. Make sure not to apply too much pressure with the ear bars.
5. Make a midline incision in the skin exposing the area to be imaged. Apply a few drops of local anesthetic (lidocaine) and after a few minutes remove connective tissue by scraping with a blade the area of the skull to be imaged.

Follow These Steps for Craniotomy

(a) Under a dissecting stereomicroscope, place a drop of ACSF on the skull and use a high-speed drill to make a 3–4 mm diameter circle. The ACSF will dissipate heat produced by drilling. Apply fresh ACSF as needed. Keep drilling until the trough becomes soft and appears to move when pressure is applied on the circle. Do not break through the bone with the drill.
(b) Apply a drop of ACSF on the skull, and using a pointed microblade, carefully cut out the circular piece of skull. If necessary apply ACSF-soaked (20 min) Gelfoam to the craniotomy to stop any bleeding.

At this Point Calcium Indicators or Other Dyes Can Be Applied to the Craniotomy or it Can Be Sealed with Glass

(a) Before sealing the craniotomy, apply a small drop of saline followed by a clean cover glass. Seal the edges of the glass with super glue.

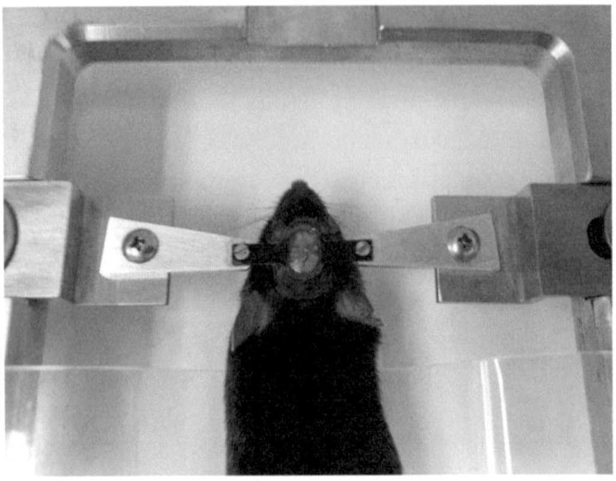

Fig. 22.4 A thinned skull windowed mouse in a custom-built mount. A custom-built plate is secured over the thinned skull window, and the head is immobilized by attaching the bar to a frame that fits on the microscope stage. The region of interest is exposed in the *center*

(b) Attach the custom-made bar over the craniotomy with super glue and dental acrylic (Fig. 22.4).

(c) Inject 0.5 mL of saline subcutaneously and 0.02 mL of Antisedan to reverse the effect of the anesthetics.

Follow These Steps for Thinned Skull Preparation

(a) Apply ACSF to the skull, and under a dissecting stereomicroscope using a high-speed drill, thin a region not larger than 1 × 1 mm². Keep drilling intermittently during the thinning procedure to avoid friction-induced overheating. Apply fresh drops of ACSF as needed. At this point some bleeding from blood vessels in the skull may occur but will stop spontaneously. Keep drilling until most of the spongy layer of the skull is removed.

(b) Switch to thinning the skull with a round-tip microblade. Pass the microblade over the skull with smooth movements at 45° with minimal downward pressure. Do not overthin the skull so that the skull does not collapse.

(c) If imaging will be done for no more than a week, the thinned skull window can be sealed with glass (Marker et al. 2010). Place a small drop of cyanoacrylate glue to the thinned area and cover with a small piece of cover glass. Press on the glass to squeeze out excess glue. Avoid trapping air bubbles or debris.

(d) When thinning is complete, glue the head bar over the thinned area and secure with dental acrylic (Fig. 22.4).

(e) Inject 0.5 mL of saline subcutaneously and 0.02 mL of Antisedan to reverse the effect of the anesthetics.

The head bar can remain on the skull in between imaging sessions. If the same mouse needs to be imaged over months, the head bar can be removed and skull covered by suturing the skin. The same region can be re-thinned weeks to months later.

22.3.3 TPLSM In Vivo Imaging

Tune the laser to the appropriate wavelength (i.e., 920 nm for YFP). Transfer the windowed mouse to the TPLSM microscope and secure the head bar on the adapted microscope stage (Fig. 22.4). Place a drop of water on a glass covered window or ACSF on the thin skulled window. Use a high numerical aperture objective (i.e., ×60 N.A.1.0) to acquire images.

To be able to image the same structures over days, one needs to record the coordinates of the microscope stage at each position. If the microscope and the software do not have this capability, a map of the brain vasculature can be acquired using a CCD camera and together with low magnification fluorescent images can be used on subsequent days to find previously imaged regions (Fig. 22.3a, b).

Measure the laser power at the objective and make sure not to use more than 20–40 mW to avoid photodamage and photobleaching. Identify structures of interest, such a dendritic segment, at a depth of 20–100 μm below the pial surface and acquire high-magnification images (i.e., 47×47 μm; 512×512 pixel; .48 μm step, Fig. 22.3c). The exact imaging parameters, such as magnification, frame size, averaging, and temporal resolution, will vary with the specific application.

Common Problems

- Damage to the brain during surgery is the most common problem. Use clean new drill bits for drilling the skull. The use of ACSF during drilling as well as blowing compressed air can help with cooling the skull. Mechanical damage during thinning with a blade should be avoided by using light strokes and holding the blade at no more than 45°.

- Photodamage and photobleaching is a common problem especially when repeated imaging of the same area is performed. Minimal laser power should be used to acquire images.

22.4 Summary

Two-photon laser scanning microscopy has become the preferred approach for in vivo imaging of brain structures. Due to the low probability of the absorption of two photons simultaneously, TPLSM produces excitation only at the focal plane. This feature results in inherent optical sectioning capability. This use of focal excitation in TPLSM significantly reduces photodamage and photobleaching of the preparation. The use of a high wavelength laser in TPLSM allows one to collect images much deeper from the specimen than possible with other forms of microscopy. The ability to collect both structural and functional images in vivo over time allows researchers to capture real-time events and to relate them to changes in behavior or disease states. The availability of turnkey TPLSM systems (both lasers and imaging systems) has made the approach of in vivo imaging of the brain and other tissues increasingly accessible to scientists.

References

Brown CE, Li P, Boyd JD, Delaney KR, Murphy TH (2007) Extensive turnover of dendritic spines and vascular remodeling in cortical tissues recovering from stroke. J Neurosci 27:4101–4109

Christensen DJ, Nedergaard M (2011) Two-photon in vivo imaging of cells. Pediatr Nephrol 26:1483–1489

Davalos D, Grutzendler J, Yang G, Kim JV, Zuo Y, Jung S, Littman DR, Dustin ML, Gan WB (2005) ATP mediates rapid microglial response to local brain injury in vivo. Nat Neurosci 8:752–758

Denk W, Strickler JH, Webb WW (1990) Two-photon laser scanning fluorescence microscopy. Science 248:73–76

Drobizhev M, Makarov NS, Tillo SE, Hughes TE, Rebane A (2011) Two-photon absorption properties of fluorescent proteins. Nat Methods 8:393–399

Grewe BF, Langer D, Kasper H, Kampa BM, Helmchen F (2010) High-speed in vivo calcium imaging reveals neuronal network activity with near-millisecond precision. Nat Methods 7:399–405

Grutzendler J, Kasthuri N, Gan WB (2002) Long-term dendritic spine stability in the adult cortex. Nature 420:812–816

Helmchen F, Denk W (2005) Deep tissue two-photon microscopy. Nat Methods 2:932–940

Hirase H, Qian L, Bartho P, Buzsaki G (2004) Calcium dynamics of cortical astrocytic networks in vivo. PLoS Biol 2:E96

Holtmaat AJ, Trachtenberg JT, Wilbrecht L, Shepherd GM, Zhang X, Knott GW, Svoboda K (2005) Transient and persistent dendritic spines in the neocortex in vivo. Neuron 45:279–291

Lee WC, Huang H, Feng G, Sanes JR, Brown EN, So PT, Nedivi E (2006) Dynamic remodeling of dendritic arbors in GABAergic interneurons of adult visual cortex. PLoS Biol 4:e29

Marker DF, Tremblay ME, Lu SM, Majewska AK, Gelbard HA (2010) A thin-skull window technique for chronic two-photon in vivo imaging of murine microglia in models of neuroinflammation. J Vis Exp Jove Sep 19;(43).

Matsuzaki M, Ellis-Davies GC, Nemoto T, Miyashita Y, Iino M, Kasai H (2001) Dendritic spine geometry is critical for AMPA receptor expression in hippocampal CA1 pyramidal neurons. Nat Neurosci 4:1086–1092

McGavern DB, Kang SS (2011) Illuminating viral infections in the nervous system. Nat Rev Immunol 11:318–329

Noguchi J, Nagaoka A, Watanabe S, Ellis-Davies GC, Kitamura K, Kano M, Matsuzaki M, Kasai H (2011) In vivo two-photon uncaging of glutamate revealing the structure-function relationships of dendritic spines in the neocortex of adult mice. J Physiol 589:2447–2457

Pawley J (2005) Handbook of biological confocal microscopy, 3rd edn. Springer, Berlin

Trachtenberg JT, Chen BE, Knott GW, Feng G, Sanes JR, Welker E, Svoboda K (2002) Long-term in vivo imaging of experience-dependent synaptic plasticity in adult cortex. Nature 420:788–794

Ustione A, Piston DW (2011) A simple introduction to multiphoton microscopy. J Microsc 243:221–226

Wang X, Lou N, Xu Q, Tian GF, Peng WG, Han X, Kang J, Takano T, Nedergaard M (2006) Astrocytic Ca^{2+} signaling evoked by sensory stimulation in vivo. Nat Neurosci 9:816–823

Williams RM, Piston DW, Webb WW (1994) Two-photon molecular excitation provides intrinsic 3-dimensional resolution for laser-based microscopy and microphotochemistry. FASEB J 8:804–813

Xu HT, Pan F, Yang G, Gan WB (2007) Choice of cranial window type for in vivo imaging affects dendritic spine turnover in the cortex. Nat Neurosci 10:549–551

Xu T, Yu X, Perlik AJ, Tobin WF, Zweig JA, Tennant K, Jones T, Zuo Y (2009) Rapid formation and selective stabilization of synapses for enduring motor memories. Nature 462:915–919

Yang G, Pan F, Parkhurst CN, Grutzendler J, Gan WB (2011) Thinned-skull cranial window technique for long-term imaging of the cortex in live mice. Nat Protoc 5:201–208

Yuste R, Denk W (1995) Dendritic spines as basic functional units of neuronal integration. Nature 375:682–684

Part VII

Neuroelectrophysiology

Techniques for Extracellular Recordings

23

Jingdong Zhang, Jianxun Xia, and Huangui Xiong

Abstract

This chapter provides a general introduction of classifications and applications of the extracellular electrophysiology recording techniques. Among those, recordings of field potentials and compound action potentials on hippocampal brain slices are given as an example to illustrate detailed methods for extracellular recording ex vivo, due to the extensive usage of the rodent hippocampal slices in neuroscience studies. The equipment required for performing the basic experiment is listed in text. The relevant concepts of field excitatory postsynaptic potential (fEPSP), population spikes (PS), and presynaptic volleys (PSV) are described herein. Accordingly, the protocols for recording fEPSP and PS are introduced, along with the analysis and interpretation of the data. The typical results of the recorded fEPSP and PS as waveform figures, as well as the summary of the relevant data as plotting figures, are also illustrated.

In addition, single unit recordings in the cat and rat medullar oblongata are used as examples of an in vivo extracellular recording. The notions and principles of orthodromic- and antidromic- evoked single unit discharge and spontaneous unit discharges are introduced and the protocols for stimulation and recording are provided.

Keywords

Extracellular recording • Field excitatory postsynaptic potential • Population spikes • Unit recording

J. Zhang, M.D., Ph.D. (✉) • J. Xia • H. Xiong
Department of Pharmacology and Experimental Neuroscience, University of Nebraska Medical Center, Durham Research Center, Durham Research Plaza, DRC I 8034, 45th Street at Emile Street, Omaha, NE 68198-5880, USA
e-mail: Jingdong.zhang@unmc.edu; jxia@unmc.edu; hxiong@unmc.edu

H. Xiong and H.E. Gendelman (eds.), *Current Laboratory Methods in Neuroscience Research*, Springer Protocols Handbooks, DOI 10.1007/978-1-4614-8794-4_23,

23.1 Introduction

Extracellular recording is a classical electrophysiological technique that continues to be a widely used and powerful tool in neuroscience research. In general, it can be classified as field potential recording including evoked and behavior-related field potential recordings, population spike (compound action potential) recordings, and single unit recordings. All of these recordings can be carried out either in vivo on anesthetized or awake animals, ex vivo on brain slices or dissected fasciculus of spinal cord or nervous trunk, or in vitro in cultured embryonic neuronal tissue (Legatt et al. 1980; Gray et al. 1995; Juergens et al. 1999; Erickson et al. 2008). The field potential (FP) is dominated by the electrical current flowing from all nearby dendritic synaptic activity within a volume of tissue. A voltage is produced by ion currents moving across the resistance of the extracellular medium during neuronal activity, which can be recorded only when a group of neurons fire in synchronization to generate a "voltage drop" near the tip of electrode. This voltage drop can be measured as a transient fluctuation in voltage between the recording electrode tip and the distant reference wire (Myers et al. 2011). The population spike (PS) is a sum of action potentials near the somata and commonly with larger amplitude than FP if the recorded somata and dendrites are in the same group. The synchronous activities are commonly evoked by stimuli of an afferent pathway or by impulses stemmed from a relevant behavior. Most importantly, the position of the electrode will determine if the recorded signals are PS or FP. For example, the PS is recorded within approximately 50–350 μm between the cells and the tip of the electrode (Legatt et al. 1980; Gray et al. 1995), whereas the FP can be detected within 0.5–3 mm between the dendrites and the electrode tip (Juergens et al. 1999). Meanwhile, the impedance of the electrode is also an element for picking up the signals.

Extracellular unit recording (also termed as single unit recording) allows the researcher to record a single neuronal discharge by positioning a microelectrode near the soma or initial segment. The single unit discharge is an action potential so that it should be all-or-none, but for catching an action potential the impedance of recording electrode should be in a range of 4–10 MΩ. The characteristics of the single unit discharge include fixed latency, fixed amplitude and a large negative peak (Fig. 23.1). When the tip of the electrode is close to the cell membrane especially the initial segment, the positive peak represents recruitment of cations (predominantly sodium ions) towards the recording site. The negative peak will be observed during influx of cations into the soma or the initial segment of the neuron. In order to record an action potential, the distance between the electrode tip and recording soma must be less than 140 μm (Henze et al. 2000). However, the amplitude or latency is subjected to change in cases of administration of reagents that exert effects on ion balance, ion channel permeability, and channel activation or inactivation.

The single unit discharge can be evoked either orthodromically (through physiological excitation of the neuron via extraneous stimuli) or antidromically (through pure biophysical consequence). In the case of orthodromic single unit discharge, the stimulation is applied on an order of presynaptic neurons to the recorded neurons, from which an action potential is elicited through spatial and temporal summation of the excitatory postsynaptic potentials (EPSP). When an electric stimulus is applied onto an axon in a bundle of nerve fibers, the action potential will propagandize in directions towards both the soma and terminal, the former is antidromic and the latter is orthodromic. The stimulation can be applied to either peripheral nerves or central fascicule of nervous fibers and the recording is generally performed within the central nervous system. The stimulation can be electric (using bipolar stimulate electrode), mechanical (stretching, vibrating, pressing, or touching), physical (heat, cold, visual or auditory stimuli), or chemical (such as Formalin, bee venom or capsaicin for pain induction, and Quinine or saccharin for taste stimuli).

The single unit discharge events can be either evoked or spontaneous. Evoked discharges are

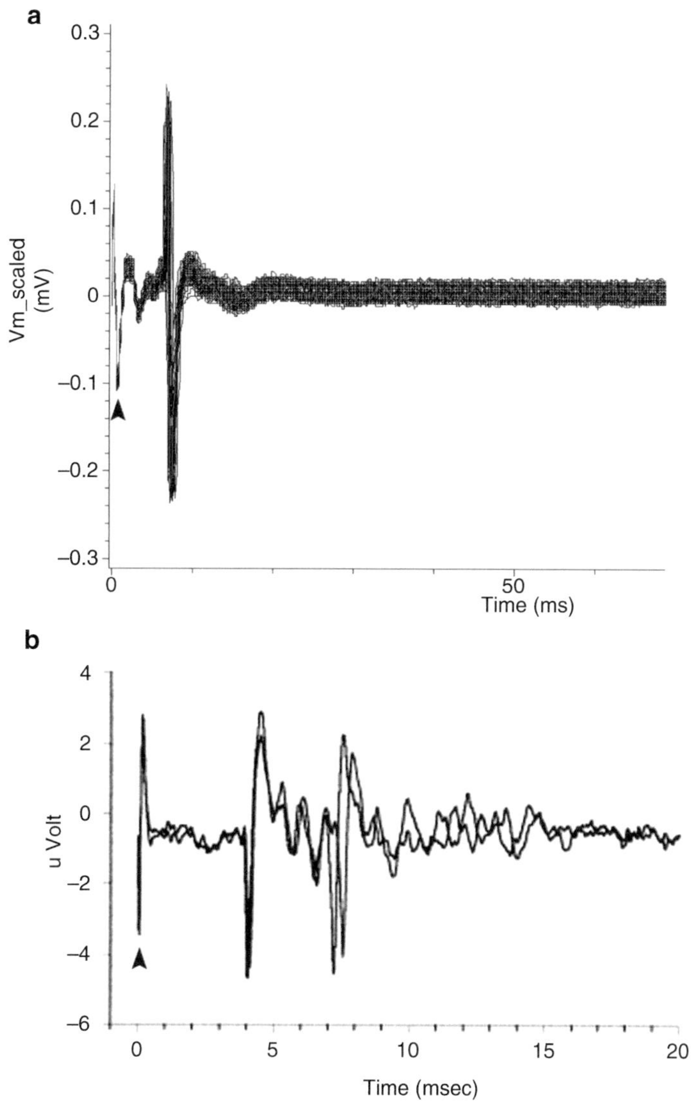

Fig. 23.1 (**a**) An evoked orthodromic single unit discharge recorded from the rat subnucleus oralis of the spinal trigeminal nucleus (Vo). (**b**) Evoked orthodromic single unit activities recorded from the cat interpolaris of the spinal trigeminal nucleus (Vi). *Arrowheads* in (**a**) and (**b**) indicate the stimulating artifact applied onto the masseter nerve (MN)

mostly from the relay neurons without an active role; whereas spontaneous discharges are most likely interneurons or convergent neurons with modulate function. For example, in spinal cord and trigeminal sensory dorsal horns there are wide-dynamic-range neurons with multiple receptive field responses (convergence) that may lead to spontaneous discharge (Mendell 1966; Ro and Capra 1999a). Generally, the evoked unit discharge recording is widely used in studies of neural pathways and the unit discharge is induced by electrical stimulation of a peripheral nerve with primary afferents or a fascicule of connection fibers within the pathway. The single unit discharge can also be evoked by mechanical stimuli as aforementioned, such as stretching a muscle to excite muscle spindles or pinching the skin to elicit pain perception. In these cases, the single unit commonly fires in a pattern of burst discharges. Lastly, recording and analyzing firing rates of spontaneous discharges is widely used in neuro-pharmacological studies combined with administration of receptor antagonists or agonists or neurotoxic substances, through intravenous or intracranial administration.

In this chapter, recordings of FP and PS from rat hippocampal brain slices will be used as examples to show how to conduct extracellular recordings in an in vitro preparation. In vivo single unit recordings in the cat and rat using peripheral and central stimuli will be presented as examples to illustrate the general methods and protocols of the single unit recording. The examples in cat explore the role neurons in the subnucleus interpolaris of the spinal trigeminal nucleus (Vi) and adjacent reticular formation may play in the integration of masticatory muscle proprioception and nociception (Ro and Capra 1999a, b) through recording both evoked orthodromic and antidromic discharges. The examples in rat will describe a recently finding of masticatory—oculomotor pathway and a neural circuit in coordination with the orofacial behaviors by masticator spindle afferents (Luo et al. 2006; Zhang et al. 2011, 2012). Finally, this chapter will introduce an approach of neural tract tracing following electrophysiological recording that is similar to intracellular tracer injection, but more than one neuron around the recording site will be labeled at each time (Luo et al. 2006; Zhang et al. 2012).

Taken together, this chapter provides valuable insights into performing extracellular recordings in a variety of preparations that will allow researchers to design and carry out these studies in their own laboratories.

23.2 Extracellular Recording on Hippocampal Slices

Hippocampal slices prepared from rodent are most widely used in neurobiological research. The evoked and/or spontaneous activities in hippocampal slices can be studied via extracellular recording techniques, in addition to intracellular (sharp electrode) and whole-cell patch recordings. The two most commonly recorded activities in response to electrical stimuli include field excitatory postsynaptic potentials (fEPSP, also termed as FP) evoked from dendritic areas (Fig. 23.2) and population spike (PS) which is an "M" shaped waveform (Figs. 23.2 and 23.3a).

The main objective of this section is to provide readers with basic methods on how to record fEPSP and PS on rodent hippocampal slices.

23.2.1 Essential Equipment (There are many commercial available equipment for extracellular recordings. Those listed hereinbelow are used in the authors' laboratory)

(a) Primary Axopatch 1D amplifier (Molecular Devices)
(b) Secondary quad differential amplifier (EX4-400, Dagan Co.)
(c) Digidata 1322A (Molecular Devices)
(d) Digital storage oscilloscope (VC-6524, Hitachi Denshi)
(e) Isolated stimulator (DS2A, Digitizer Ltd.)
(f) Stimulator (S88, Grass technologies, Astro-Med Inc.)
(g) Motorized micromanipulator (MS-314 controller)
(h) Standard manual control micromanipulator with X, Y and Z axes
(i) Dissecting microscope with >4× magnification capability
(j) Recording chamber (customized)
(k) Electrode puller (P-97, Sutter Instruments)
(l) Borosilicate glass capillaries (1B150F-4, World Precision Instruments)
(m) Microfil nonmetallic syringe needle for filling micropipettes (MF28G67-5, World Precision Instruments)
(n) Bipolar tungsten stimulating electrode (2–4 MΩ, FHC Inc.)
(o) Peristaltic pump (RP-1, Rainin Instrument Co.)
(p) Automatic temperature controller (TC-324B, Warner Instrument Co.)

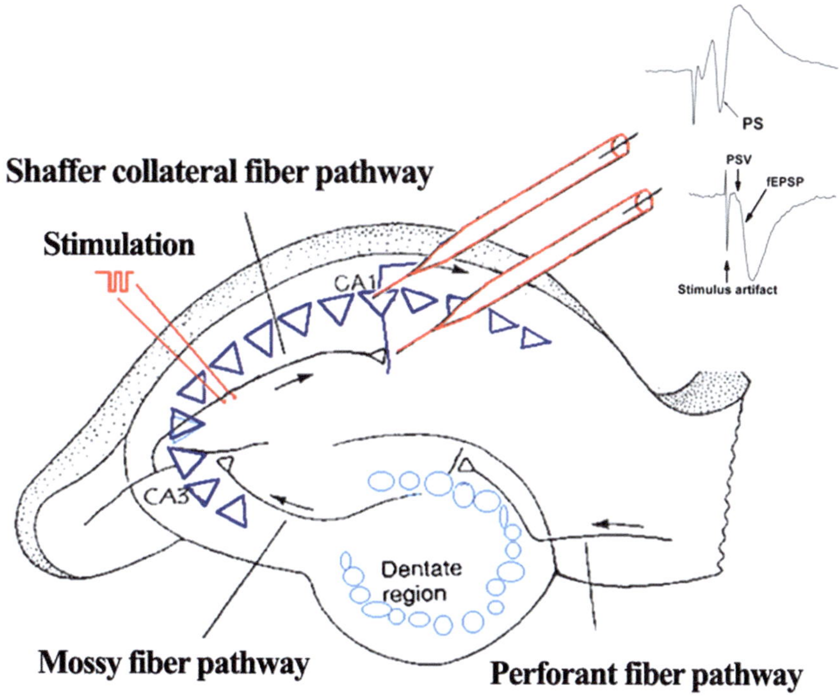

Fig. 23.2 A schematic drawing of a rat hippocampal slice showing tri-synaptic connections and the electrode positions for recording fEPSP and PS

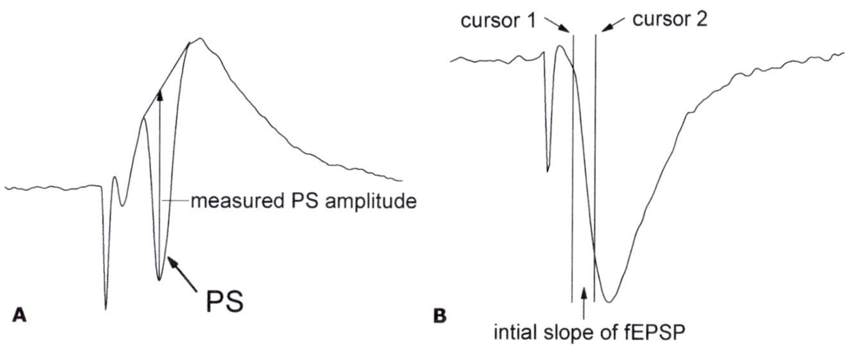

Fig. 23.3 Measuring and analyzing the fEPSP initial slope and PS amplitude. (**a**) The amplitude of PS is measured as the height of a *vertical line* drawn from the negative peak to a *line* connecting the *top* of two positive peaks (*long arrow*). (**b**) The initial slope of a fEPSP is measured between cursors 1 and 2 placed within a range of 30–80 %, or 1–1.5 ms from the beginning of the slope to its peak as indicated by an *upward arrow*

(q) High intensity illuminator (MI-150, Dolan-Jenner Industries)
(r) Active vibration isolation table with Faraday cage (63-541, Technical Manu-facturing Corporation)
(s) Vacuum pump (Pipette-aid, Drummond Scientific Co.)
(t) Data-acquisition and analysis software (Clampex 8.2 and Clampfit 8.2, Molecular Devices; Origin 8.0, Origin Lab Co.)

23.2.2 Methods

For convenience, the recordings of fEPSP and PS in CA1 region of rat hippocampal slices (Gig. 23.2), in response of electrical stimulation of Schaffer collateral (SC) fibers, are used as examples for this section.

1. Transfer brain slice to a recording chamber

   After cutting brain slices, the slices need to be recovered in a incubation container (e.g., a beaker) for at least 1 hr before being used for recordings. To start recording on brain slices, transfer a slice from the incubation container to a recording chamber using a fire-polished wide-mouth Pasteur pipette (Fisher brand, 7mm in diameter) or other wide-mouth pipette). The slice is placed and mechanically secured with an anchor or small weights (preferentially made of platinum) in a "submersion style" recording chamber viewed under a dissecting microscope with an illumination from the bottom of the chamber. The tissue is continuously superfused with oxygenated ACSF heated to 30 °C by an automatic temperature controller (TC-324B, Warner Instrument Co.) and supplied by a peristaltic pump (Rainin Instrument Co.) at the rate of 2 ml/min. The slice is allowed to acclimate and stabilize for 10–15 min before the start of recordings.

2. Positioning of stimulation/recording electrodes and recording of evoked synaptic responses
   (a) Under the dissecting microscope, advance, a bipolar tungsten stimulating electrode (FHC Inc.) mounted on the micromanipulator close to the area containing SC fibers using the rough Y advancer of standard manual control micromanipulator and then further advance the stimulating electrode into slice tissue 50-100 micrometers using the fine Y advancer.
   (b) Set stimulation pulse duration to 20–40μs and stimulation intensity to 20–200μA. Set stimulation frequency as desired (0.05 to 0.2 Hz).
   (c) Using borosilicate glass capillaries and an electrode puller to make recording electrodes. The suitable tip diameter of a recording electrode is about 5 micrometer and has a resistance of 1-2 MΩ when filled with ACSF. Attach the recording electrode to the headstage of an amplifier and advance the recording electrode into the bath solution under microscope. After manual correction of junction potential (set it to zero), further advance the recording electrode to the interested area for recording, such as CA1 dendritic field for recording fEPSP or cell body layer for recording PS).
   (d) The recording electrode is then advanced slowly in small steps (5–10 μm/step) into the slice until the presynaptic volly (PSV) and fEPSP reach maximal amplitudes, which are normally observed at a depth of 100–200 μm beneath the slice surface and can be determined by the distance driven by the micromanipulator. Since the signals recorded extracellularly are much smaller than those recorded intracellularly and are commonly less than 1 mV because of shunting of bath solution, they are amplified with a total of 50–100× by a primary amplifier (Axopatch-1D) combined with a secondary amplifier (EX4-400) and digitized with Digidata 1322A. Signals are sampled at 5 kHz and filtered at 1 Hz for storage on a PC for off-line analysis.

3. Synaptic strength and input-output (I-O) curve
   (a) To generate a I-O curve, a series of stimuli are applied to the SC at an increment of 10 to 50μA and the corresponding voltage responses are recorded in the CA1 area. The levels and number of stimuli may vary, but should be enough to generate a I-O curve (Fig. 23.4).
   (b) The amplitude of a PSV represents the proportion of presynaptic fibers activated, while fEPSP slope provides a measurement of synaptic strength. Plotting a fEPSP slope against a PSV amplitude represents the size of the fEPSP per number of activated afferents and provides a precise

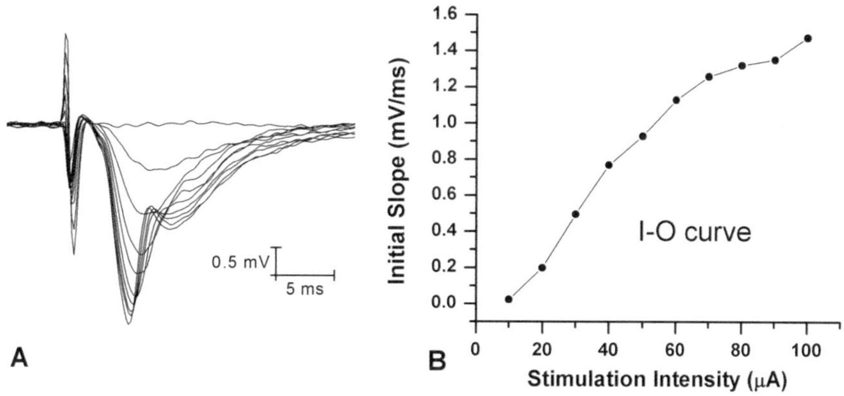

Fig. 23.4 Generation of an input–output (I–O) curve in the CA1 region of a hippocampal slice. (**a**) Superimposed fEPSP traces induced by a series of stimulating pulses with an increment of 10 μA) applied to the SC fiber. (**b**) The I–O curve plotted with the increased fEPSP initial slopes corresponding to the enhanced stimulating intensities shown in panel (**a**)

measurement of synaptic strength (Mathis et al. 2011).

4. Stimulation protocols used to generate synaptic plasticity (LTP and LTD)

   Long-term potentiation (LTP) and long-term depression (LTD) are long-lasting increases and decreases in synaptic transmission efficacy in response to diverse patterns of synaptic activation. Both LTP and LTD are widely considered to be potential mechanisms underlying learning and memory.

   (a) To induce LTP, a test stimulation pulse (20-40μs in duration) at a low frequency (0.05Hz) is applied to to stimulation electrode to evoke a fEPSP and the stimulation intensity is adjusted to the level of generating 40–50 % of the maximal response. Single pulse or paired pulses (50 ms inter-pulse interval) are delivered at 0.05 Hz (once every 20 s) for 30 min to obtain a stable baseline of recording (Fig. 23.6).

   (b) When the baseline recording is stable, a high frequency stimulation (HFS) or a theta burst stimulation (TBS) will be delivered at the time point of 30 min to induce LTP (Fig. 23.6). The train duration of a HFS (100Hz) is either 1000 ms (x2) or 500ms (x4) with a 10 s of inter-train interval. The TBS will be delivered twice at a 10 s interval. Each TBS is composed of 6 bursts with an inter-burst interval of 200 ms and each single theta burst consists of 4 pulses at 100 Hz.

   (c) For LTD induction, a low-frequency stimulation (LFS, 1 Hz, 15 min) will be applied when a stable baseline recording is obtained for 30 min.

23.2.3 Data Analysis

(a) The fEPSP amplitude is computed as the maximal value of the negative-going peak, and the initial slope of fEPSP is measured within the range of 30–80 % or 1–1.5 ms from start of the slope to its peak (Fig. 23.3b).

(b) The PS amplitude is measured by first connecting the two positive-going peaks with a line, followed by measuring the height of a vertical line drawn from the negative peak to the connecting line (Fig. 23.3a). This height represents the PS amplitude (Gholmieh et al. 2004).

(c) LTP or LTD are normally expressed as normalized fEPSP slope (%) or % change in the the initial slope of the fEPSP, or normalized PS amplitude.

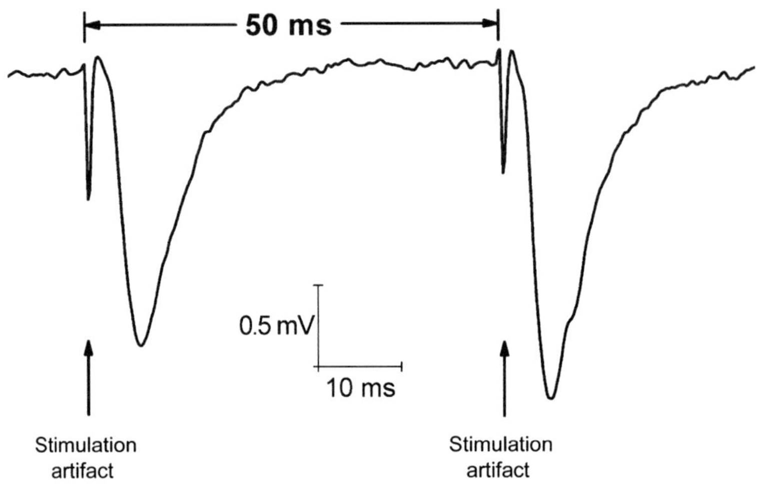

Fig. 23.5 A representative fEPSP recorded in the CA1 dendritic field in response to a paired-pulse stimulation

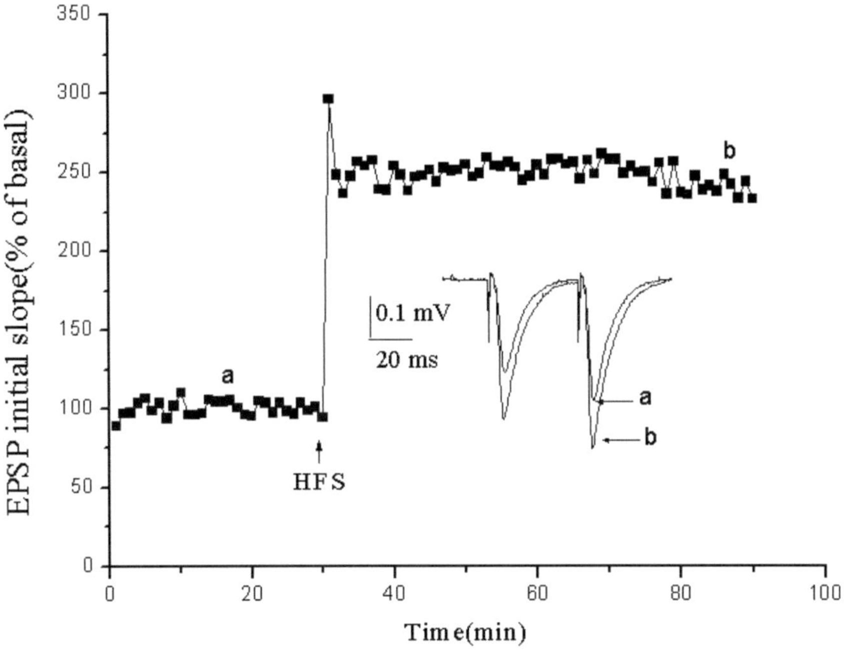

Fig. 23.6 Time course and magnitude of LTP in the Schaffer-collateral to CA1 synapses recorded from rat hippocampal slices. The graph plots the initial slope of falling phase of the evoked fEPSPs recorded from CA1 dendrite field (stratum radium) in response to constant test stimuli, for 30 min before and 60 min following high frequency stimulation (HFS, 100 Hz, 500ms ´ 2). The HFS was delivered at the time indicated by the arrow. Each point in this graph is an average of 3 consecutive sweeps. Representative traces before and after the induction of LTP, taken at the times indicated by letters *a* and *b* respectively, are illustrated as an inset. Note the increase in slope of evoked EPSPs following HFS (HFS induces LTP)

23.2.4 Advantages and Disadvantages of the Hippocampal Extracellular Recording

1. Advantages
   (a) The electrophysiological recording technique described above remains as one of the most popular methods in modern neuroscience research. In contrast to intracellular recording techniques, extracellular recordings have the advantage of stability and flexibility, making the technique applicable to a wide range of research projects.
   (b) Field potential recordings have the best resolution in laminar, densely packed structures, such as the hippocampus, and thus are useful for examining firing behaviors and for determining slice viability (Suter et al. 1999).
   (c) Because they reflect the synchronous activity of many neurons, extracellular recordings also allow information about the coordinated synaptic activity in a population of neurons to be obtained. With detailed anatomical information about the brain areas that are being recorded from, it is helpful to understand the input–output function determining the transformation of information in a neuronal network.
2. Disadvantages
   (a) Cell membrane potential is immeasurable using extracellular recording. Therefore, the entire physiological characteristics of certain group of neurons have to be uncovered by a combination of intracellular or whole-cell patch clamp with extracellular recordings.
   (b) The signals recorded extracellularly are usually small and are difficult to detect, so a secondary amplifier is required to further amplify the signal.

23.3 Extracellular Single Unit Recording in Vivo

23.3.1 Equipment

1. Generally, the recording equipments and the software are almost the same as described in the last section (Sect. 23.2.1). The difference is that instruments for maintaining and monitoring animal's condition are required during in vivo experiment.
2. A combination of respiration, heart rate, blood pressure and end-tidal CO_2 monitors such as PhyioSuite (e.g., Kent Scientific Corporation) or capnography/pulse oximeter (e.g., Harvard Apparatus) is necessary for monitoring the animal's physiological status.
3. An audio monitor is helpful in monitoring unitary firing conditions.
4. For controlling movement of the recording electrode, a single-axis fine-step (2–5 μm/step) hydraulic driving micromanipulator is required for advancing the electrode. Whenever the audio monitor alerts the investigator that the electrode tip is close to a neuron, she or he will get the best position to start recording by adjusting the electrode in a range of micrometers.
5. A thermostatically controlled heating pad with a controlling box is necessary to keep animal warm. This apparatus contains a feedback system with a temperature sensory probe that is either placed along the esophagus or inserted into the rectum to measure body temperature. A homoeothermic circulating water pump with a blanket is also appropriate for the maintenance of body temperature.
6. A respirator is needed for artificial ventilation because a muscle relaxant is sometimes administered to the animal. Small animal ventilator from Harvard Apparatus is good for rat and mouse; while the Kent Scientific respirator for animals >300 g may be used for an adult rat, cat, or rabbit.

23.3.2 Preparation for Recording

1. Prepare recording electrode and stimulation electrode
   (a) The recording electrode is made from a borosilicate glass micropipette with an outside diameter (OD) of 1.5 mm and an inside diameter (ID) of 0.84 mm, or OD of 1.2 mm and ID of 0.7 mm. Pull electrode with a Sutter (horizontal) or Narishige (vertical) puller in tip diameter of 2 or 3 μm.
   (b) Fill electrode with 2 % Pontamine sky blue (PSB) in 0.5 M sodium acetate (pH 7.4), or with 2 % biocytin in saline (pH 7.0–7.4).
   (c) Measure the tip impedance with an Impedance Meter. The electrode resistance should be in a range of 6–10 MΩ. It will be difficult to obtain good unit recordings if the impedance is too high. Dual-amplitude response might be recorded (Fig. 23.7) if the impedance is too low, indicating two units are recorded simultaneously.
   (d) For central stimulation, the use of concentric bipolar stimulation electrodes, such as the stimulating electrode SNE-100/50 (Clark Electromedical Instruments), are recommended.
   (e) For peripheral stimulation of nerve trunk or branch, a hook-like bipolar stimulation electrode is recommended. Such an electrode can be made by two thin silver wires (about 0.2 mm in diameter) embedded in dental cement, then be trimmed with sandpaper-drill and shaped to make the silver-plastic hooks fit the size of the nerve.
2. Prepare animals for surgeries
   (a) Anesthesia. The commonly used anesthetics in lab for cat, rat, and mouse are sodium pentobarbital, chloral hydrate, urenthane, and ketamine plus xylazine. Application of these anesthetics in rats has been compared and documented in detail (Field et al. 1993). In studies on the Vi and adjacent reticular formation 15 mg/kg ketamine plus 1 mg/kg xylazine was administered to cats via intramuscular injection and anesthesia was maintained with 38 mg/kg intravenous (*i.v.*) sodium pentobarbital (Ro and Capra 1999a, b). In our previous studies (Zhang et al. 2011; Zhang et al. 2012), 40 mg/kg sodium pentobarbital was administrated intra-peritoneally (*i.p.*) to the rats for survival surgery and urenthane 1.25 g/kg was given *i.p.* for non-survival surgery.
   (b) In order to reduce tracheal secretion, 0.5 mg/kg atropine sulfate is administrated *i.p.* immediately after injection of anesthetics (Zhang et al. 2012).
   (c) Muscle relaxants are administered *i.v.* during the recording of unit discharge evoked by stimulating muscular nerve. Gallarnine triethiodide (5–10 mg/kg *i.v.*) was administered to the cats in the studies mentioned above, and 30 mg/kg *i.p.* was administered to the rats in our previous works mentioned above.
3. Surgeries
   (a) Tracheotomy: (1) Cut the hair on skin area above larynx and upper trachea. (2) Disinfect the skin with iodine tincture and 70 % ethanol. (3) Cut the skin, dissect the muscles to expose larynx and upper trachea. Expose jugular vein besides trachea and place 2 sutures underneath. (4) Cut a inverted "T" shape opening on the upper trachea. (5) Insert a "Y" shape connective tube into the opening. (6) Close the opening with the tube inside by suture. (7) Suture the muscles and skin after jugular vein cannula had been installed.
   (b) Cannula: (1) Insert a thin silicon tube rostro-caudally into the jugular vein and tighten it with suture. (2) Cut the hair above the femoral artery. (3) Disinfest and cut the skin. (4) Expose the femoral artery. (5) Insert a thin silicon tube rostro-caudally into the artery. (6) Tighten and glue the tube. (7) Suture the connectives and skin.
   (c) Nerve exposure: (1) Cut the hair above the masseter muscle belly. (2) Disinfect and incise the skin. (3) Dissect masseter muscles and expose the masseter nerve.

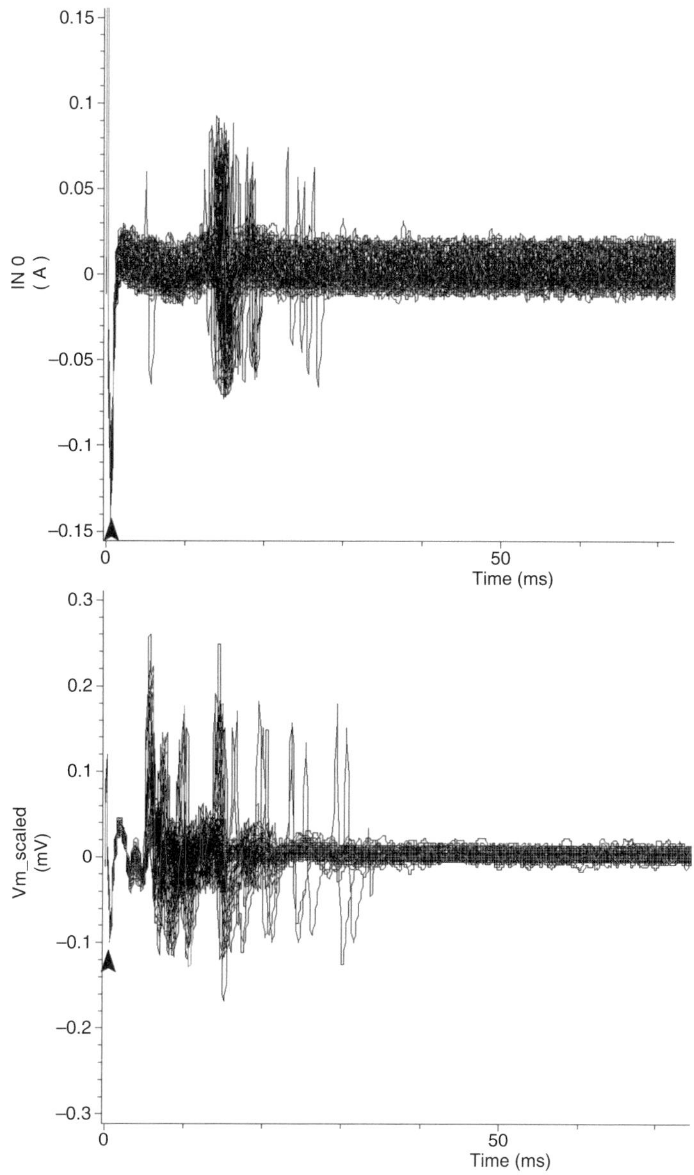

Fig. 23.7 Evoked single unit discharges with two different amplitudes recorded in the same recording site, suggesting two neurons, may be recorded simultaneously due to low impedance of the recording electrode. *Arrowheads* indicate stimulating artifact followed with stimulation of the MN

(4) Hook the silver wires bipolar stimulation electrode onto the nerve. (5) Fix and isolate the electrode from around tissue with Kwik-Cast Silicon sealant glue (World Precision Instruments Inc. Sarasota, FL). (6) Close and suture the muscles and skin.

(d) Craniotomy: (1) Mount the cat or rat onto relevant stereotaxic frame. (2) Clear the hair above the skin of the skull. (3) Disinfect and cut the skin and connectives above and expose the skull. (4) Make craniotomy with a dental drill. (5) For recording of neurons in medulla oblongata, the cerebellum was aspirated away to expose oblongata. (6) Tear away all of arachnoid and pia mater on base of the fourth ventricle. (7) Fill the tissue valley with warm mineral oil (37–39 °C).

23.3.3 Set up Monitoring, Supporting, and Recording Equipment

1. Connecting all cables, wires, and tubes to relevant equipment:
   (a) Connect the tube from the femoral artery to the physiological monitoring equipment. Adjust the equipment to display all necessary information.
   (b) Insert the temperature probe into rectum of the animal, fix the probe onto the tail with tape. Set up the maintaining point at 37 °C and keep checking animal for 15 min until stable.
   (c) Connect the "Y" shape tube inside trachea to the respirator. Clean up the secretion in the trachea with a long and thin cotton swab before connecting "Y" tube to the respirator. For the rat, set up ventilation volume and rate to 2 cm^3/breath at 1.7 Hz. For the cat, set up ventilation parameters (tide volume and frequency) that matches those under natural respiration.
   (d) Connect the tube in the jugular vein with two syringes through a tripass valve. The syringes will contain anesthetic, muscle relaxant, or other reagents used for experimental purposes.
   (e) Connect the wires of the stimulating electrode to the stimulation isolation unit and the stimulator. Apply test pulses to check if the animal's jaw jerks up upon stimulation. If not, check and adjust the stimulation intensity, until the jaw elevation is observed.
   (f) Connect a reference electrode and wire (basically an alligator clamp with a wire) from the animal to the headstage reference electrode receptacle of the Axopatch (as an example) amplifier.
   (g) When all of connections and settings are finished, conduct an inspection for breath and heart rate; end-tidal CO_2, body temperature, and so on. Give a painful stimulus to check the status of anesthesia, adding anesthetic if necessary. Start the electrophysiological recording when the animal's condition remains stable for 10–15 min.
2. Set up selections on each electronic instrument as required by the experiment—for example, for single unit recording in the studies mentioned above:
   (a) Axopatch amplifier: Output mode should be either $I=0$ or at current clamp mode. Set signal filtration to 1–5 kHz. The magnification ratio of input–output should match the range given in the electrode headstage.
   (b) Oscilloscope: Set oscilloscope to normal mode with external triggering.
   (c) The setup of stimulator is dependent upon the experimental design and stimulating parameters may be subject to change throughout the experiment. Routinely, duration of the stimulating square waveform is 0.05–2 ms and amplitude is about 2–5 times of threshold intensity; for evoked single unit recording, the stimulation frequency is about 0.2–0.7 Hz if a "repeated single" mode is selected. Set up the stimulator to train mode if train stimulating is carried out; herein, a train length or duration (e.g., 20–200 ms) with frequency (e.g., square-waveform/s) and train's interval (e.g., 2–50 s) are selected.
3. Configuration of the software. Instruments from different company may use different software. The software needs to be configurated to establish connections between the instrument and computer that the software is installed. Configurate the software according to the instructions provided.

23.3.4 Extracellular Single Unit Recording

1. Positioning the central stimulating electrode and recording electrode.
   (a) In the cat, Ro and Capra (1999a, b) placed the stimulating electrode into ventroposteromedial nucleus of the thalamus

(VPM) contralateral to the recording Vi based on the cat brain atlas, in order to generate antidromic unit discharge in the recording Vi.

(b) In the rat, Zhang et al. (2012) positioned a concentric electrode into the mesencephalic trigeminal nucleus (Vme) based on stereotaxic coordinates of the rat brain (Paxinos and Watson 1998) and on observation of the jaw jerking upward upon stimulus as well.

(c) Positioning the recording electrode: The electrode is positioned according to a brain atlas, and the brain surface anatomic landmark can be used to determine the point of electrode placement. In addition, if the recorded neuron is a motoneuron (such as the oculomotor neuron), contraction of the innervated muscles (e.g., eyeball movement) can be visualized when iontophoresis of the tracer or dye is performed after the recording (Zhang et al. 2011).

(d) Generally, when a new experiment is started, use the coordinates provided by the atlas while making a note of the surface geography. For subsequent recordings, the brain surface anatomic landmark can then be used to correctly position the recording electrode.

2. Finding out the single unit discharge: When the recording electrode is roughly placed in intended area, the moving style needs to be changed to fine-step moving. Stimulating, advancing the electrode in fine-step, listening to the voice of the audio monitor and watching the oscilloscope screen are carried out simultaneously. When a single unit discharge has been evoked, the first sign is voice change in the audio monitor, followed by a spike on the oscilloscope. Decreasing the stimulating intensity until the spike has disappeared and increasing the intensity little by little to determine the threshold (shown as "T" for each unit). Continue to record the discharge for 5–50 sweeps. The latency and amplitude are fixed if it is an evoked single unit discharge.
3. Distinguishing orthodromic single unit discharge from antidromic response: Following frequency and peristimulus time histogram (PSTH) are assumed to identify antidromic discharge in studies by Ro and Capra (Fig. 23.8). However, in some early studies the authors assumed antidromic collision to verify the direction of the impulses and demonstrate the pathway connections (Ro and Capra 1999b). The idea is that antidromic responses can follow higher frequency stimuli than orthodromic discharges, because synaptic transmission is involved in the time course of the orthodromic conduction but antidromic response is a purely biophysical procedure. In this case, PSTH of an antidromic discharge is narrower than that of an orthodromic response. If a pair of stimuli have elicited a pair of antidromically conducted action potentials, the first one will generate an orthodromic spike when it arrives at the soma and the orthodromic one travels back to the same axon. No collision happens when the second antidromic action potential starts to fire after that orthodromic spike has passed. However, if the second antidromic potential left before that orthodromic spike has passed, the two action potentials will have collision on the same axon and both will fade away. In this case the second antidromic potential cannot be recorded at the soma. This technique is termed as "antidromic collision."
4. Examples of burst unit discharge evoked by mechanical stimulation. Stimulation of jaw muscle spindles is taken as examples herein. Lowering down the jaw and pressing the masseter muscle through the skin to elicit a unit response to these stimuli. The recorded neuron will exhibit a burst discharge upon stretching or pressing the jaw muscle (Figs. 23.9 and 23.10). In addition, in some cases a ramp-and-hold apparatus connecting to bottom of the mandible is used to generate this kind of burst firing (Fig. 23.11).
5. Example for effects of train stimuli on spontaneous discharge is displayed by recording in the oculomotor nucleus (III). In addition to evoked single unit discharge recorded in the rat III (Figs. 23.12 and 23.13a–c) upon electric stimulating the masseter nerve, train

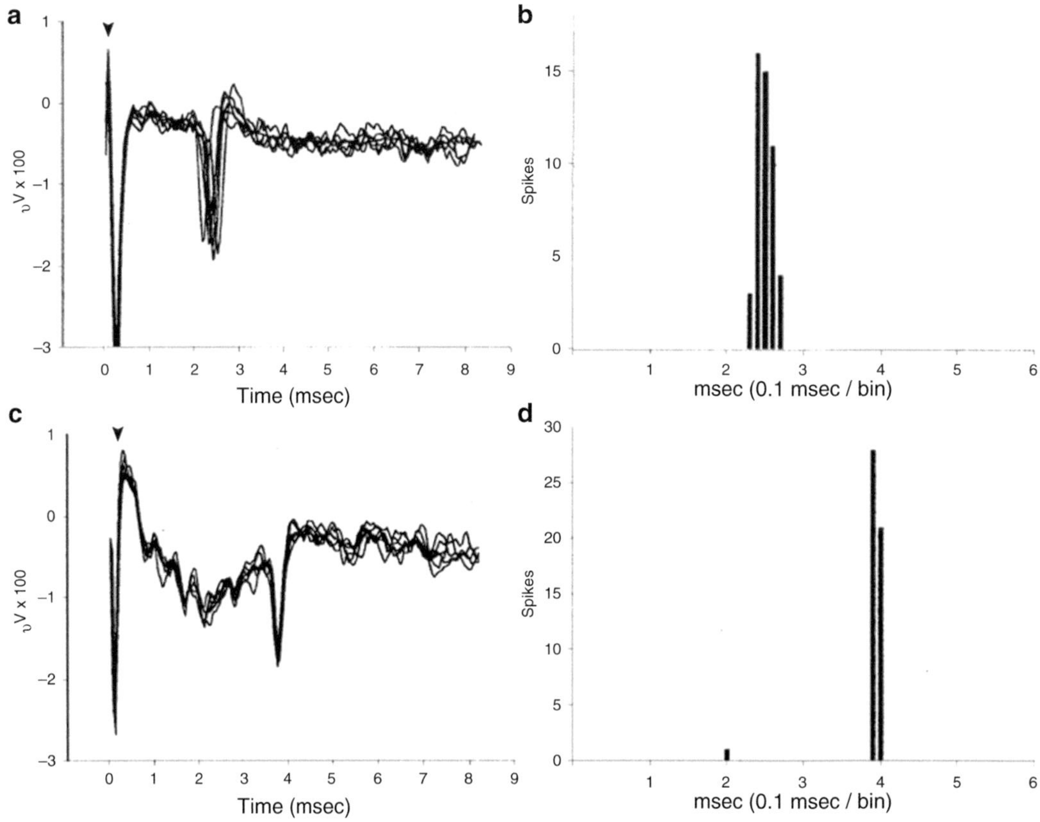

Fig. 23.8 (**a**) An evoked orthodromic single unit discharge recorded from the cat Vi elicited by the MN stimuli. (**b**) Peristimulus time histogram (PSTH) summarized by repeated stimuli and episodic sweep. (**c**) The same single unit responded antidromically to stimulation of the contralateral ventroposteromedial nucleus (VPM) of the thalamus. The ascending projections of the Vi that cross the *midline* and terminate on the VPM is a well-established orofacial somatosensory pathway. (**d**) The PSTH of the antidromic unit discharge is narrower and more consistent than that of orthdromic ones. The following frequency of the antidromic activity is also higher than that of orthodromic one. This figure is cited from Ro and Capra (1999b) with their permission

stimuli is applied to evoke continuous discharges or modulate spontaneous firing rates. Figure 23.13a–c display a set of evoked single unit spikes recorded in the III; Panels d–f exhibit another recording in the III with very slow spontaneous discharge initially but show an excitation inhibition alternative following a couple of train stimulation given manually; Panels g, h show a spontaneous firing inhibited by 6 train s of high frequency stimuli (52 Hz, 200 ms train duration and 3 s intervals). These data can be examples of study interactions or convergence between more than one neural pathway afferents.

23.3.5 Data Analysis

1. Measure the latency and plot PSTH, softwares from different companies are different in processing procedures, but the general processes are as follows
   (a) Place the reference cursor (cursor 1 or 3) at the beginning of the stimulus artifact and the measure cursor (2 or 4) at the start of the spike, respectively. The Δtime shown between cursor 1 and 2 or 3 and 4 is the latency.
   (b) About 50 sweeps are required for plotting a PSTH, measure all of latency in each sweep as stated above.

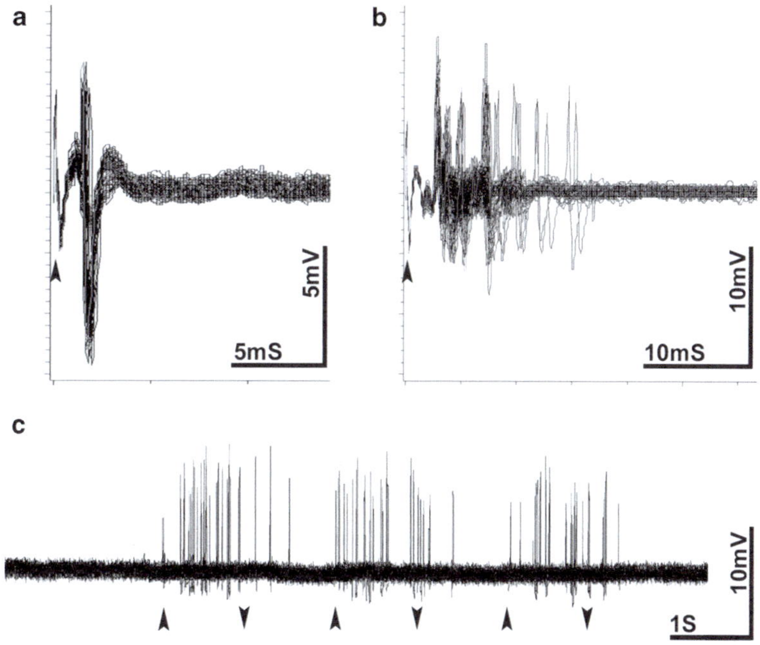

Fig. 23.9 (**a**, **b**) A couple of orthodromic unit firing recorded in the rat Vo following the MN stimuli. (**c**) The unit showing in (**b**) is also responding to mechanical stretching down the jaw. *Arrowheads* in (**a**) and (**b**) point to initiation of the stimuli. *Up-going arrowheads* in (**c**) indicate starting of stretching the jaw; while, *down-going arrowheads* indicate stopping of stretching. This is cited from Zhang et al. (2012)

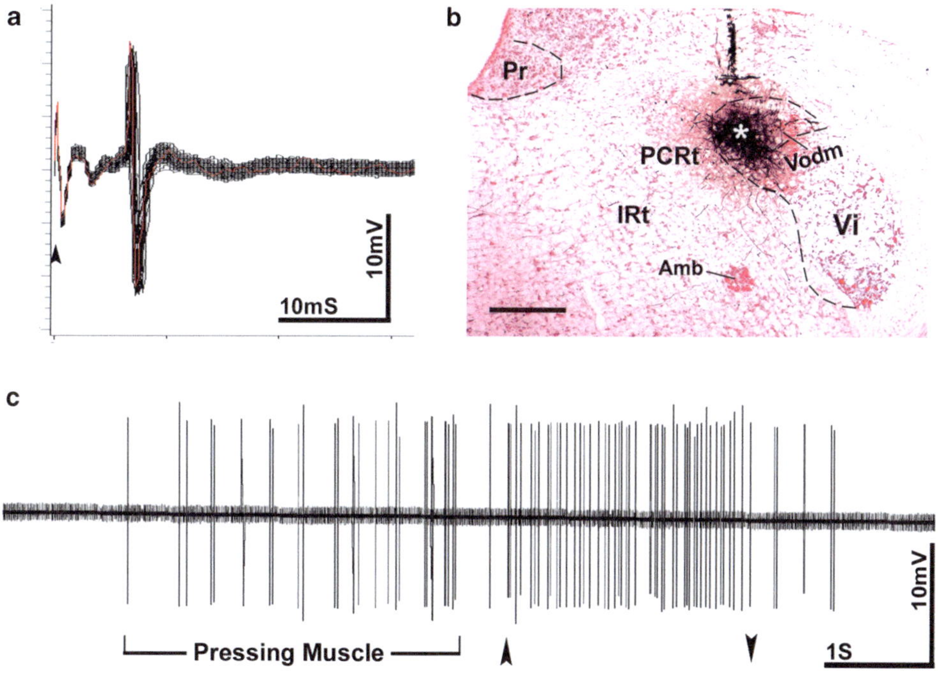

Fig. 23.10 (**a**) A single unit discharge recorded in dorsomedial division of the Vo (Vodm) following stimuli of the MN. (**b**) Biocytin-labeled area of recording is marked by a *white asterisk* in the Vodm. (**c**) This unit is not only excited by the MN stimuli but also activated by pressing the masseter muscle and stretching down the jaw (*up-going and down-going arrowheads* show start and stop stretching) in a pattern of burst responses. Scale bar in b is 150 μm. This is cited from Zhang et al. (2012)

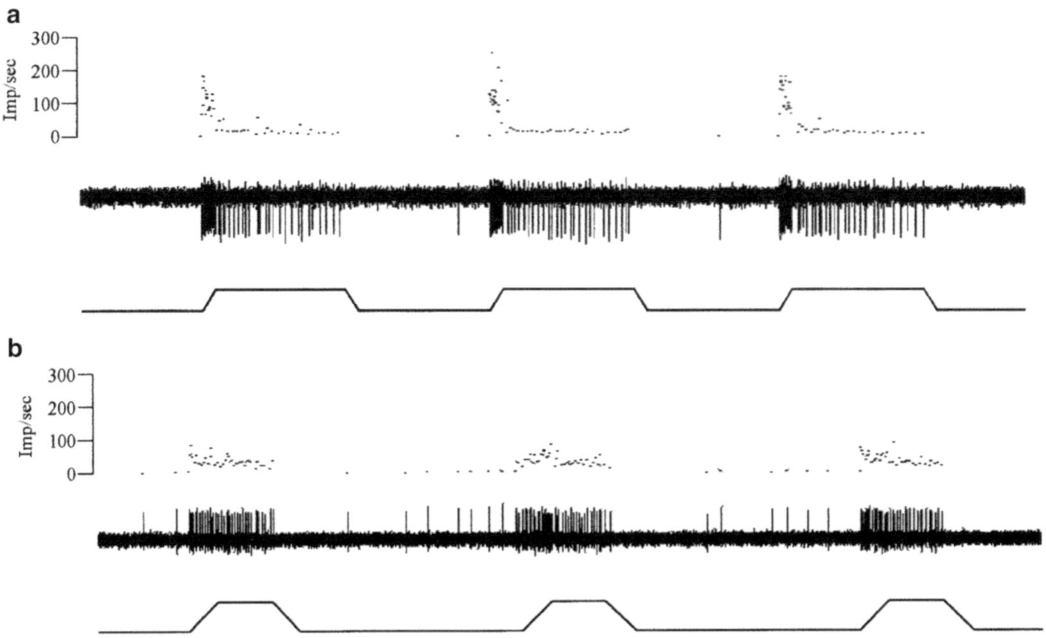

Fig. 23.11 (**a**) Fast adaptive speed sensitive single unit recorded from the cat Vi. *Top line* shows the firing rate corresponding to the ramp activation of the jaw muscle spindles and real spikes are displayed in *middle line*; *bottom line* shows ramp-and-hold curve representing the speed and amplitude of the "ramp" and the time course of the "hold" generated by ramp-and-hold apparatus. (**b**) Slow adaptive, stretching speed irrelevant, and holding strength sensitive single unit recorded from the cat Vi, the *top, middle and bottom lines* exhibit the same information as does in (**a**). This is cited from Ro and Capra (1999b) with their permission

(c) Transfer these data to a statistic software. Set up time course per bin to 0.1 ms to generate a PSTH which can be used to distinguish evoked orthodromic and antidromic discharges.

2. Measure spontaneous discharge frequencies and plot conventional frequency histogram
   (a) Open a gap-free file for spontaneous discharge recording and template it in result-event format.
   (b) Use "threshold search" function under "event detection" to set up the upper and lower threshold values. The upper line should exclude stimuli artifacts and the lower line should exclude noises that are commonly smaller than the signals.
   (c) Select "histogram" function under "analysis" and set up a bin time course, region related to threshold line (either under or above or in between), and type of the histogram. Click "apply" to produce the histogram.
   (d) This function is commonly used to reveal interactions between two or three afferent pathways, or to compare the effects of different reagents (agonists, antagonists, and neurotoxic substances) on a functional pathway.
3. Events count and statistic processing
   (a) Usually, there is a function of "event statistics" in electrophysiology software, the event can be spike of current or voltage, such as action potential in this case.
   (b) To process an event statistics: open a gap-free file that recorded a spontaneous discharge modulated by train stimuli of the afferent impulses, move the cursors to select a period of discharge such as before and after the train stimuli (e.g., use cursors 1 and 2 to bracket control discharges and cursors 3 and 4 to bracket effect events) or before and after *i.v.* delivery of chemical compound.

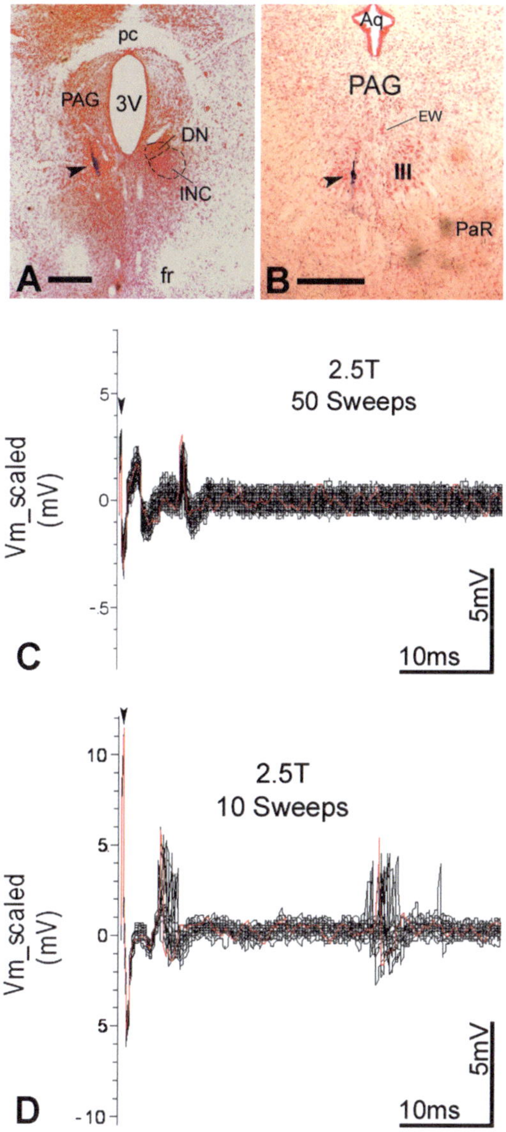

Fig. 23.12 (**a**, **b**) Pontamine sky blue marked recording sites in the INC (**a**) and the III (**b**), indicated by *arrowheads*. (**c**, **d**) Orthodromic single unit responses recorded in the INC (**c**) and the III (**d**), in which *downward arrowheads* indicate the start of electric stimulation on the MN which contains jaw muscle nerve branches. *EW* Edinger-Westphal nucleus, *fr* fasciculus retroflexus, *PaR* pararubral nucleus, *pc* posterior commission. Scale bars in a and b are 200 μm. This is cited from Zhang et al. (2011)

(c) Collect the number and save in Excel, and when several same patterns of the recordings are obtained, process statistics with statistic software.

23.3.6 Visualization of Recording Site, Even Efferent of the Recorded Neurons

1. Before glass micropipettes were widely used, tungsten recording electrodes were the common tool for extracellular recording. In order to view the recording site in these cases, investigators had to increase the stimulating current to induce an electrolysis lesion on the tissues around the tip of the tungsten electrode. Theoretically, the recording site is able to be viewed after counterstaining. Tungsten electrodes are still used in recording extracellular single unit activities in combination with behavior observations of awake animals.
2. Currently, glass micropipettes are more widely used. PSB (Pontamine sky blue) is a commonly used dye for marking recording site. Briefly, fill micropipette electrode with 2 % PSB in 0.5 M sodium acetate (pH 7.2) filled and the PSB will be electrophoresed into tissue after recording (Figs. 23.12a, b and 23.13d, g). Solution pH may influence the polarity of the PSB, therefore negative current may need to be used to result in a visible mark (Zhang et al. 2011).
3. A neural tracer can be injected after the extracellular recording, which will not only mark the recording site but also trace the axon trajectory of the recorded neurons (Figs. 23.14 and 23.15). Development of this method was based on observation of our previous intracellular injections of horseradish peroxidase (HRP) after electrophysiological recording, in which the neurons with the same relay function are always grouped together (Luo et al. 1991; Zhang et al. 1997; Luo et al. 2006). Accordingly, we believe that tracer injection after extracellular recording is more efficacious and feasible, especially when the tracer injection is restricted in an area about 75–150 μm of diameter. This procedure involves filling the electrode with 2 % biocytin in saline (pH 7.0–7.4) and performing iontophoresis after recording (Luo et al. 2006; Zhang et al. 2012). When the recording is accomplished, press the buzz button on remote control several times and start iontophoresis with

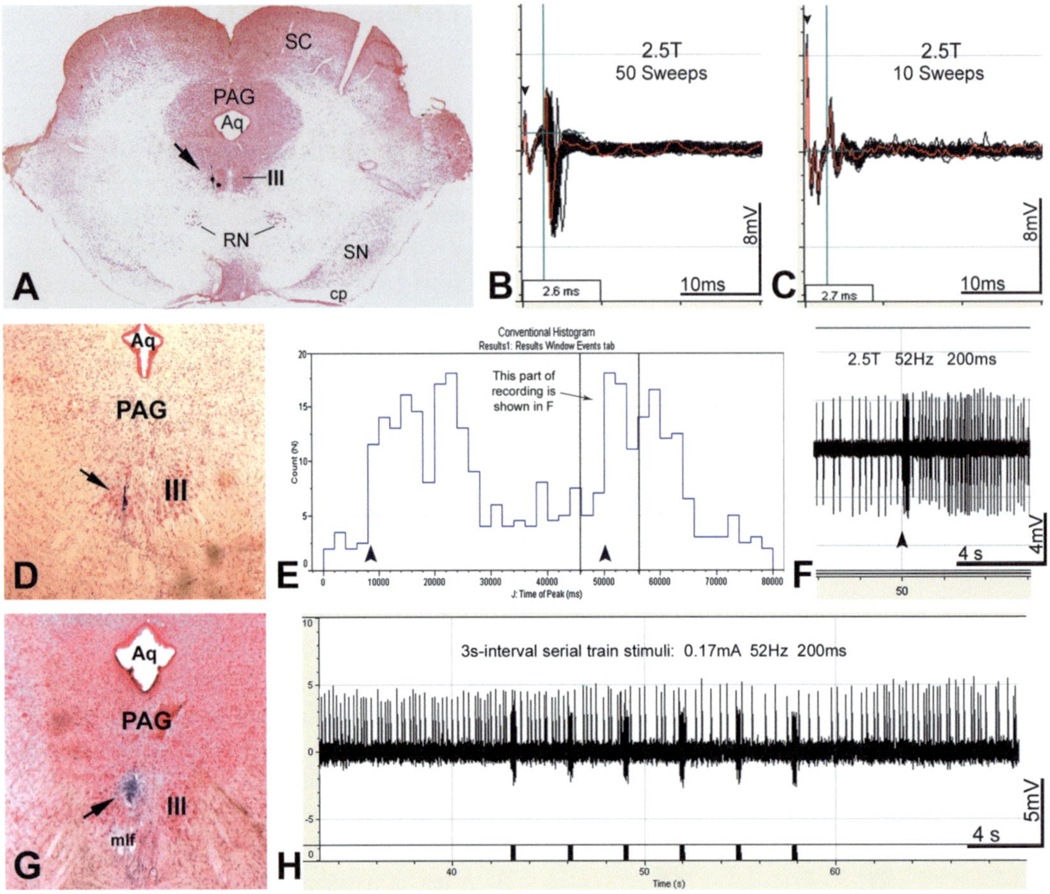

Fig. 23.13 (**a–c**) Orthodromic single unit responses recorded in the III (**a**; *arrowed*). *Downward arrowheads* in (**b** and **c**) indicate stimulus artifact. (**d–f**) A unitary response recorded in the III (**d**; *arrowed*) also shows slow spontaneous discharges before stimulation: (**e**) spikes (number) vs. time-course (ms) histogram exhibits train stimuli enhanced discharging rate (2 s/bin) of this unit but short inhibition is shown in between, *arrowheads* indicate the start of train stimuli; (**f**) real time recording from the frame area (*arrowed*) in (**e**) (from 46,000 to 56,000 ms), start of train stimuli pointed by an *arrowhead*. (**g**) *Sky blue* marked recording site in margin of the III (*arrowed*); (**h**) Real time recording: *upper track* shows serial train stimuli reduced the firing rate of this spontaneous discharges; *bottom track* exhibits artifact of the train stimulation. *cp* cerebral peduncle, *RN* red nucleus, *SC* superior colliculus, *SN* substantia nigra

10–50 μA positive current (250 ms on/off or 200 ms at 2 Hz) for 15–30 min. The protocol for revealing of the labeling is the same as described in Sect. 3.4.2, step 1.

23.3.7 Advantages and Disadvantages of in Vivo Extracellular Single Unit Recording

1. Advantages: Probably the largest benefit of the extracellular single unit recording in vivo is that it can be achieved in an awake, behaving animal. Furthermore, the recording electrode can be mounted in the brain of an animal and set to continuously record the regional activity for a period after the animal has aroused from anesthesia and subjected to behavioral tests (Wilson and McNaughton 1993). In addition to a single electrode, 3-dimensional electrode arrays have been developed to record stereo perspectives of a set of neuronal activity (Du et al. 2009). This makes it possible to simultaneously study a population of neurons interacting with each other through different regional, functional, or sensory inputs in the brains (Bartho et al. 2004).

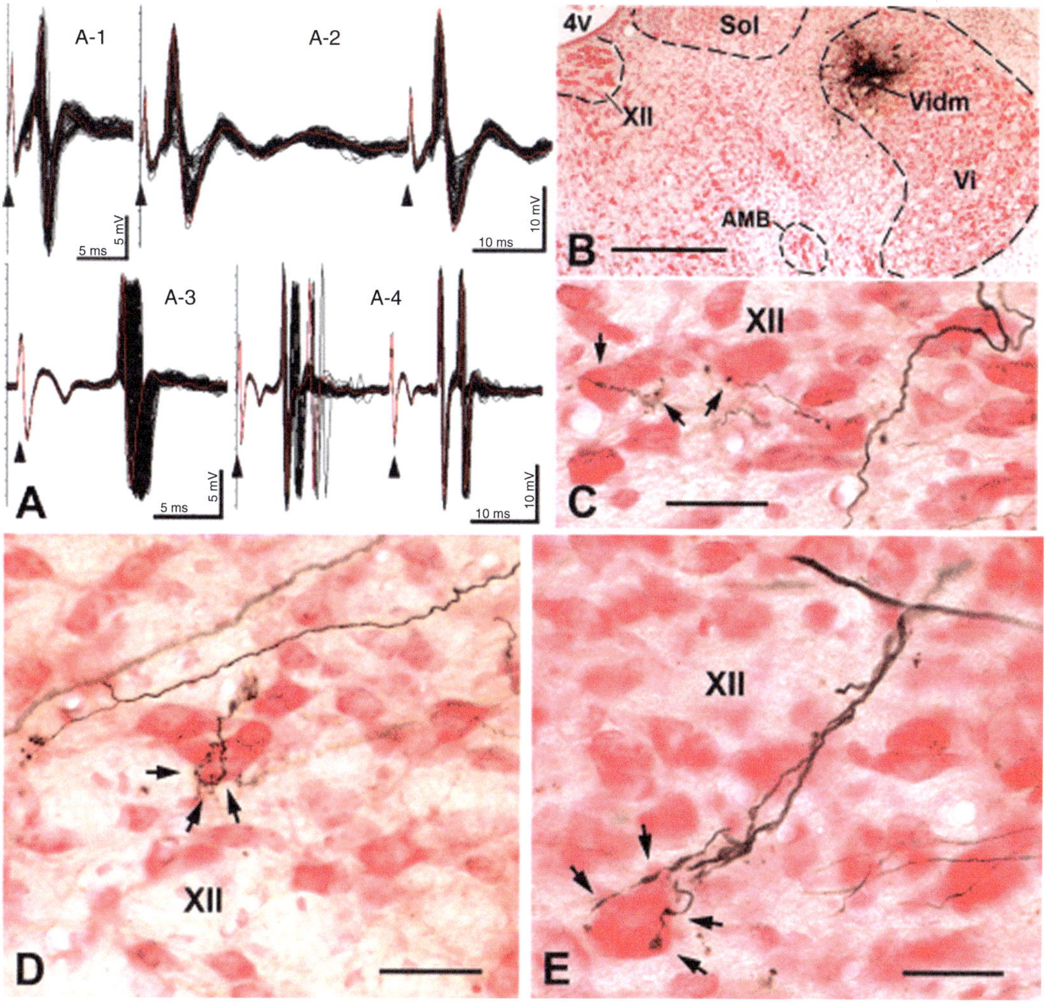

Fig. 23.14 Orthodromic responses of the XII premotor neurons to the masseteric nerve stimulation and connections between the XII premotor neurons and the XII MNs. (**a**) Electric stimulation of the ipsilateral MN induced short-latency orthodromic responses in PCRt (A-1, 2.7 ms), dorsomedial Vi (A-2, 3.5 ms) and Vodm (A-3, 7.0 ms; A-4, 6.5 ms). These orthodromic discharges also responded to jaw displacements and without response to palpation of the teeth, gingiva, and vibrissae. *Note*: in A-4, a single stimulus of the MN evoked dual responses in Vodm. In A-1 to A-3, 100 sweeps were recorded. (**b**) A representative of the recording site labeled by biocytin iontophoresis via the same recording electrode. This recording site is located in Vidm. (**c–e**) Biocytin-labeled axons from XII premotor pools (recording sites) traveled into XII, and they gave rise to axonal collaterals and terminals contacting cell bodies (*arrows*) of XII MNs that were counterstained with neutral *red*. *Sol* solitary tract nucleus, *Amb* ambiguous nucleus, *XII* hypoglossal nucleus. Scale bars are 1 mm in b, 40 μm in C-E. The figure is cited from Luo et al. (2006) with permission of the authors

2. Disadvantages and trouble-shooting: Unlike intracellular recording, extracellular unit recording cannot display EPSP or inhibitory postsynaptic potential (IPSP); it shows only an all-or-none action potential. Therefore, it cannot be applied to show a convergence of both excitatory and inhibitory inputs. Furthermore, membrane potential and post-spike hyper-polarization cannot be observed either. Intracellularly recorded initial segment potential as a definite sign of antidromic activation cannot be revealed, and spike slope analysis is not possible. Large action-potential-like field potentials can be recorded if impedance

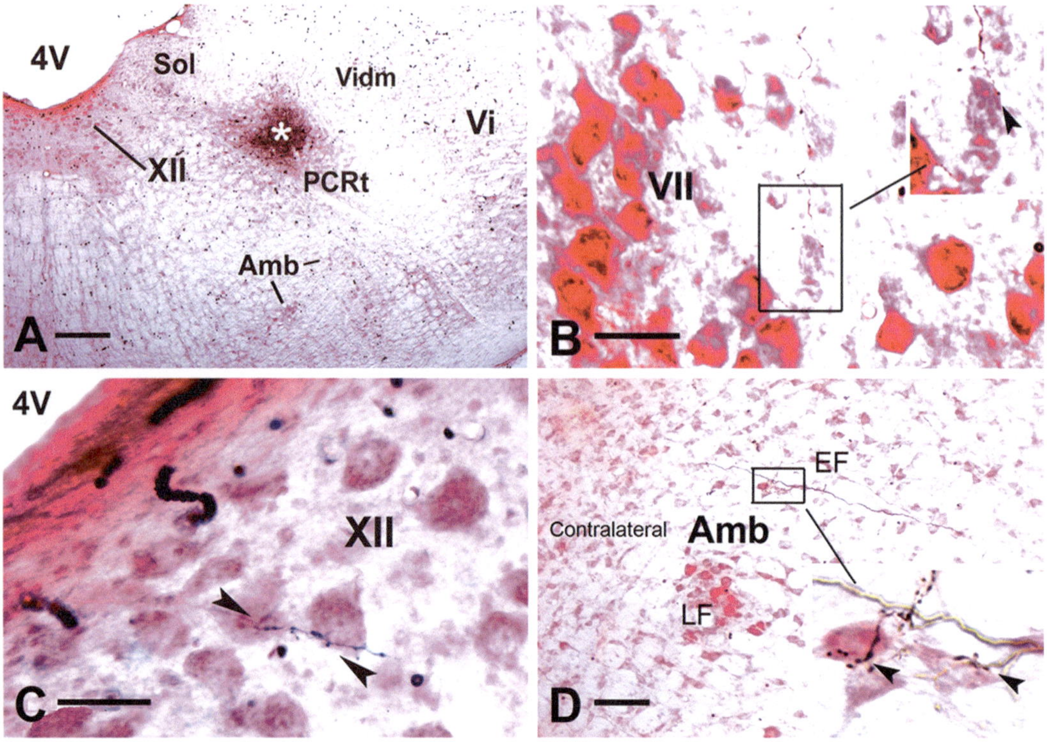

Fig. 23.15 Recording site stained by biocytin (**a**) and distribution of biocytin-labeled fibers and boutons in orofacial motor nuclei (**b–d** and *insets*). (**a**) A *white asterisk* indicates biocytin injection site in the PCRt where typical neuronal responses were recorded. (**b** and *inset*, **c**, **d** and *inset*) Labeled fibers and terminals in close appositions (*arrowheads*) with Nissl-stained motorneurons in the ipsilateralfacial nucleus (**b**), the ipsilateral XII (**c**), and the external formation (EF) of the contralateral Amb (**d**). Scale bars: 225 μm in a, 50 μm in **b–d**. This is cited from Zhang et al. (2012)

of the electrode is relatively low. Hence, determination of the threshold must be conducted each time, in which the unit discharge shows an all-or-none manner, but the threshold of a field potential in stable amplitude is never as consistent as a single unit discharge. Sometimes, a single unitary neuron exhibits a prompt response with a 2–5 ms latency and a delayed response with a 10–20 ms latency (Fig. 23.12; Zhang et al. 2011). Without an audio monitor, some spontaneous discharge can be mistaken as prompt and delayed discharge especially when frequency of the spontaneous firing is relatively consistent. Elongating the time course of the recording window or increasing the number of sweeps in trials may help to distinguish dual responses with spontaneous discharges.

References

Bartho P, Hirase H, Monconduit L, Zugaro M, Harris KD, Buzsaki G (2004) Characterization of neocortical principal cells and interneurons by network interactions and extracellular features. J Neurophysiol 92:600–608

Du J, Riedel-Kruse IH, Nawroth JC, Roukes ML, Laurent G, Masmanidis SC (2009) High-resolution three-dimensional extracellular recording of neuronal activity with microfabricated electrode arrays. J Neurophysiol 101:1671–1678

Erickson J, Tooker A, Tai YC, Pine J (2008) Caged neuron MEA: a system for long-term investigation of cultured neural network connectivity. J Neurosci Methods 175:1–16

Field KJ, White WJ, Lang CM (1993) Anaesthetic effects of chloral hydrate, pentobarbitone and urethane in adult male rats. Lab Anim 27:258–269

Gholmieh G, Courellis S, Dimoka A, Wills JD, LaCoss J, Granacki JJ, Marmarelis V, Berger T (2004) An algo-

rithm for real-time extraction of population EPSP and population spike amplitudes from hippocampal field potential recordings. J Neurosci Methods 136: 111–121
Gray CM, Maldonado PE, Wilson M, McNaughton B (1995) Tetrodes markedly improve the reliability and yield of multiple single-unit isolation from multi-unit recordings in cat striate cortex. J Neurosci Methods 63:43–54
Henze DA, Borhegyi Z, Csicsvari J, Mamiya A, Harris KD, Buzsaki G (2000) Intracellular features predicted by extracellular recordings in the hippocampus in vivo. J Neurophysiol 84:390–400
Juergens E, Guettler A, Eckhorn R (1999) Visual stimulation elicits locked and induced gamma oscillations in monkey intracortical- and EEG-potentials, but not in human EEG. Exp Brain Res 129:247–259
Kauer JA (1999) Blockade of hippocampal long-term potentiation by sustained tetanic stimulation near the recording site. J Neurophysiol 81:940–944
Legatt AD, Arezzo J, Vaughan HG Jr (1980) Averaged multiple unit activity as an estimate of phasic changes in local neuronal activity: effects of volume-conducted potentials. J Neurosci Methods 2:203–217
Li S, Jin M, Koeglsperger T, Shepardson NE, Shankar GM, Selkoe DJ (2011) Soluble Abeta oligomers inhibit long-term potentiation through a mechanism involving excessive activation of extrasynaptic NR2B-containing NMDA receptors. J Neurosci 31:6627–6638
Luo PF, Wang BR, Peng ZZ, Li JS (1991) Morphological characteristics and terminating patterns of masseteric neurons of the mesencephalic trigeminal nucleus in the rat: an intracellular horseradish peroxidase labeling study. J Comp Neurol 303:286–299
Luo P, Zhang J, Yang R, Pendlebury W (2006) Neuronal circuitry and synaptic organization of trigeminal proprioceptive afferents mediating tongue movement and jaw-tongue coordination via hypoglossal premotor neurons. Eur J Neurosci 23:3269–3283
Mathis DM, Furman JL, Norris CM (2011) Preparation of acute hippocampal slices from rats and transgenic mice for the study of synaptic alterations during aging and amyloid pathology. J Vis Exp (49), e2330, doi:10.3791/2330
Mendell LM (1966) Physiological properties of unmyelinated fiber projection to the spinal cord. Exp Neurol 16:316–332
Myers FB, Abilez OJ, Zarins CK, Lee LP (2011) Stimulation and artifact-free extracellular electrophysiological recording of cells in suspension. Conf Proc IEEE Eng Med Biol Soc 2011:4030–4033
Paxinos G, Watson C (eds) (1998) The rat brain in stereotaxic coordinates. Academic, San Diego
Ro JY, Capra NF (1999a) Evidence for subnucleus interpolaris in craniofacial muscle pain mechanisms demonstrated by intramuscular injections with hypertonic saline. Brain Res 842:166–183
Ro JY, Capra NF (1999b) Physiological evidence for caudal brainstem projections of jaw muscle spindle afferents. Exp Brain Res 128:425–434
Suter KJ, Smith BN, Dudek FE (1999) Electrophysiological recording from brain slices. Methods 18:86–90
Wilson MA, McNaughton BL (1993) Dynamics of the hippocampal ensemble code for space. Science 261:1055–1058
Zhang JD, Yoshida A, Shigenaga Y (1997) Ultrastructural analysis of inputs around the soma of an intra-cellularly labeled masseter muscle spindle afferent in cat mesencephalic trigeminal nucleus. J Hirnforsch 38:495–502
Zhang J, Liang H, Luo P, Xiong H (2011) Unraveling a masticatory—oculomotor neural pathway in rat: implications for a pathophysiological neural circuit in human? Int J Physiol Pathophysiol Pharmacol 3:280–287
Zhang J, Luo P, Ro JY, Xiong H (2012) Jaw muscle spindle afferents coordinate multiple orofacial motoneurons via common premotor neurons in rats: an electrophysiological and anatomical study. Brain Res 1489:37–47

"Blind" Patch-Clamp Recordings from Rodent Hippocampal Slices

24

Jianxun Xia and Huangui Xiong

Abstract

Whole-cell patch-clamp recording technique has enabled electrophysiologists to study synaptic transmission and membrane properties of neurons in an isolated preparation as an integral part of synaptic and cellular physiology. Compared with visually guided patch-clamp recording, the "blind" method does not require expensive experimental equipment and is easy to develop. This chapter describes "hands-on" procedures of "blind" patch-clamp recording on rodent hippocampal slices, followed by a discussion of the advantages and potential pitfalls with possible solutions.

Keywords

Brain slices • Hippocampus • Electrophysiology • Whole-cell patch clamp

24.1 Introduction

Tight-seal whole-cell patch-clamp recording techniques (Hamill et al. 1981; Marty and Neher 1983) have been used extensively to study synaptic transmission and membrane properties of neurons in brain slices. The prevalent methods for making whole-cell recordings from brain slices include visually guided and "blind" approaches. While the former requires the assistance of sophisticated infrared microscopy and direct visualization of the cells to be recorded, the "blind" method, the latter, is a low-cost method as it does not require expensive equipment (e.g., infrared microscope) to visualize individual cells. This method was originally developed in experiments on turtle cortex slices (Blanton et al. 1989) and has ever since been used in mammalian brain slices including rodent hippocampal slices. In contrast to a visualized whole-cell recording which can be made by placing an electrode directly onto an individual cell under visual guidance, a "blind" whole-cell recording is usually achieved by slowly advancing the electrode, after entering the slice, until it encounters a cell as detected by an increase of electrode resistance.

The advantages of the "blind" patch method include the following: (1) it does not require

J. Xia • H. Xiong (✉)
Department of Pharmacology and Experimental Neuroscience, University of Nebraska Medical Center, Durham Research Center, Durham Research Plaza, DRC I 8034, 45th Street at Emile Street, Omaha NE 68198-5880, USA
e-mail: jxia@unmc.edu; hxiong@unmc.edu

H. Xiong and H.E. Gendelman (eds.), *Current Laboratory Methods in Neuroscience Research*, Springer Protocols Handbooks, DOI 10.1007/978-1-4614-8794-4_24,

sophisticated microscope. A common dissecting microscope can be used to guide the electrode into the right area of the slice and so it is much cheaper to set up; (2) it is easy to operate and no cleaning of the slice is necessary; (3) cells deep in the slices can be recorded from; and (4) the use of dissecting microscope to guide the electrode location permits relatively wide working space to maneuver the electrode over the recording sites in the recording chamber and makes recordings possible from a large area of the brain slice by freely moving the recording electrode (Okada 2012). The disadvantage is the cells recorded cannot be visually identified.

24.1.1 Objective

The main objective of this chapter is to provide readers with basic methods on how to conduct "blind" whole-cell patch recordings on rodent hippocampal slices, focusing on the procedures utilized in our laboratory.

24.1.2 Equipment

- Axopatch 1D amplifier (Molecular Devices) or equivalent
- Digidata 1322A (Molecular Devices)
- Digital storage oscilloscope (e.g., VC-6524, Hitachi Denshi)
- Isolated stimulator (e.g., DS2A, Digitimer Ltd.)
- Stimulator (S88, Grass technologies, Astro-Med, Inc.)
- Isolated stimulator DS2A (Digitimer Ltd)
- Motorized micromanipulator (e.g., MS-314 control unit, Applied scientific instrumentation)
- Dissecting microscope
- Electrode puller (P-1000, Sutter Instruments)
- Borosilicate glass capillaries (1B150F-4, World Precision Instruments)
- Microfil nonmetallic syringe needle for filling micropipettes (MF28G67-5, World Precision Instruments)
- Tungsten stimulating electrode (2–4 MΩ, UEWMHGSEDNNM, FHC, Inc.)
- Peristaltic pump (RP-1, Rainin Instrument Co.)
- Automatic temperature controller (TC-324B, Warner instrument Co.)
- Active vibration isolation table (63–541, Technical Manufacturing Corporation)
- Data acquisition and analysis software (pClamp 10, Molecular Devices; Origin 8.0, OriginLab Co.)

24.2 Methods

24.2.1 Preparation of Recording Electrodes and Intracellular Solutions

The recording electrodes are pulled on experimental day from a thin-walled borosilicate glass capillaries (1.5 mm o.d., WPI) using a 4–5 loop protocol in Sutter P-1000 puller and are filled through Microfil syringe needle (MF28G67-5, WPI) with intracellular solution containing (in mM) the following: 135 potassium gluconate, 5 KCl, 0.5 $CaCl_2$, 2 $MgCl_2$, 5 EGTA, 5 HEPES, and 5 ATP or 115 cesium gluconate, 15 cesium methanesulfonate, 10 EGTA, 10 HEPES, 5 NaCl, 3 Mg-ATP, 0.3 Tris-GTP, and 5 QX-314, depending on the purpose of experiments. The pH of intracellular solution is adjusted to 7.2–7.3 with KOH or cesium hydroxide and the osmolarity is adjusted to 280–300 mOsm with sucrose if necessary. Ten to 20 mOsm difference between pipette solution and ACSF (normally around 300 mOsm) is beneficial to a tight seal between pipette tip and cell membrane, by virtue of a "sucking" force produced by movement of water across the membrane rupture as the pipette solution dialyzes into the cytoplasm. The cesium salt is included to block certain types of potassium channel to avoid decay of EPSCs by outward potassium current when cell membrane potential is clamped above the reversal potential of potassium. As a derivative of lidocaine, QX-314 blocks voltage-dependent sodium channel, especially in the cases that cells escape the voltage clamping and generate gigantic sodium currents. EGTA or BAPTA is usually included to chelate intracellular calcium ions. GTP is used to conserve G-protein-mediated responses and ATP acts as an energy supplier for other intracellular activities (Walz 2007).

24.2.2 Transfer Hippocampal Slices to the Recording Chamber and Determination of Electrode Resistance

See Chap. 5 for details in preparation of hippocampal slices from rodents.

1. After at least 1 h recovery in an incubation chamber or an incubation beaker, an individual hippocampal slice is transferred, using a fire-polished Pasteur pipette (Fisher brand, 7 mm in diameter), to a recording chamber. The slice is submerged in ACSF, secured with two small weights (preferentially made of platinum), and perfused continuously with oxygenated ACSF preheated to 30 °C with an automatic temperature controller (TC-324B, Warner instrument Co.).
2. To avoid blockade of electrode tip while advance it into the bath and brain tissue, a positive pressure is usually applied to the recording electrode using a 1 mL syringe connected to the electrode holder via polyethylene tubing. The recording electrode is first lowered into the bath when positive pressure is continuously applied.
3. Under current-clamp mode, a small negative current pulse, triggered by a Grass stimulator combined with isolated stimulator (Model DS2A, Digitimer Ltd), is applied through Axopatch 1D amplifier, and the resulted voltage response is monitored on a digital storage oscilloscope. The recording electrode resistance can be determined by dial series resistance compensation knob located on the front panel of Axopatch-1D amplifier to zero (flatten) the voltage response. The recording electrode tip impedance can also be read under the "seal test" function in voltage-clamp mode when small voltage pulses (−1 mV) are applied. After determination of its resistance, the electrode is manually advanced further down to touch tissue surface of the regions of interest, which is indicated by a small increase in the voltage deflection (or small decrease in the current deflection), and further voltage but not current change neutralization is required. Then, the recording electrode could be driven to any desired depth at a pace of 2 μm/step by a hydraulic or motor-driven micromanipulator (Liu et al. 2001).
4. Under the dissecting microscope, the cell body layer in a hippocampal slice can be easily identified. Although individual neurons cannot be seen with a dissecting microscope, the high packing density of neuronal somata within the cell body layer of the hippocampus greatly facilitates successful formation of GΩ seals on neuronal cells.

24.2.3 Obtaining Whole-Cell Recordings

1. A second increase (from 20 to 50 % of initial amplitude) in the voltage response may indicate the electrode tip is approaching a cell. The positive pressure is released and a gentle suction is applied. A progressive increase in voltage responses can be observed if the electrode encounters a neural cell (Fig. 24.1).
2. The successful formation of GΩ seals is validated by the dramatic increase of the voltage deflection (Windhorst and Johansson 1999), which regularly fluctuates more or less than −100 mV as showed on the meter in frontal panel of Axopatch-1D amplifier. Voltage reading less than −60 mV is considered as failure of or loose seal formation. A dramatic decrease of the current deflection (optimally totally cancelled) as monitored in Clampex under voltage-clamp mode also manifests the formation of a GΩ seal.
3. The membrane patch is then ruptured to obtain whole-cell recordings by applying additional brief suction (slight negative pressure, Fig. 24.1). Sometimes cell membrane may be ruptured spontaneously without applying additional suction. A "zap" function can also be used to facilitate rupture in case that cell membrane cannot be easily broken by negative pressure.
4. The resting membrane potential is directly read from the meter on amplifier under

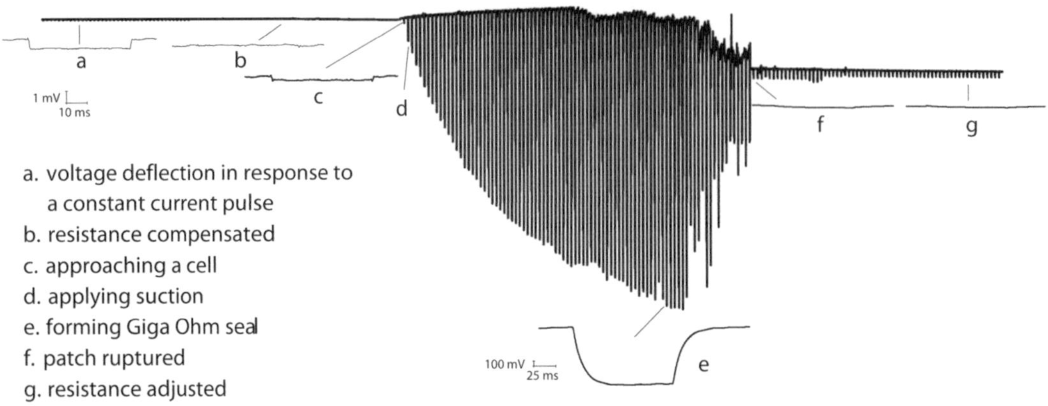

Fig. 24.1 Time course of a "blind" whole-cell recording development in the CA1 region of a rat hippocampal slice. Steps a–g are illustrated and depicted as shown in the figure

current-clamp mode ($I=0$) immediately after formation of whole-cell configuration. Delayed reading may misrepresent the original membrane potential due to dialysis of cell cytoplasm by pipette solution and new ionic equilibrium achieved by exchange of components between pipette solution and cytoplasm (Deng and Xu 2012).

5. To allow the drugs added in the pipette to completely diffuse into the cells and induce their pharmacological actions, a delay of 10–15 min was systematically observed immediately after rupture of cell membrane and prior to recordings. Then the series resistance and capacitance are compensated by adjusting corresponding knobs on the amplifier. In current-clamp mode, typical resting potentials in hippocampal pyramidal cells range from −46 to −70 mV, and series resistance range from 210 to 1,600 MΩ. Cesium-containing intracellular solution normally results in 2–3 times greater series resistances (Blanton et al. 1989).
6. To identify whether the recorded cell is a neuronal cell, a depolarizing current pulse is delivered (injected) to the recorded cell via the recording electrode. If the injection of depolarizing current pulse induces action potential(s) in a recorded cell, it is counted as a neuronal cell. Absence of firing of action potential is normally considered as the marker of nonneuronal cells.
7. To record whole-cell currents, the cell membrane potential can be clamped at different levels, and whole-cell currents can be generated by voltage steps. To record synaptic potentials (under current-clamp mode) or synaptic currents (under voltage-clamp mode), clamp the cell at desired membrane potentials and stimulate necessary nerve fiber pathways (e.g., Schaffer-collateral fiber pathway in the hippocampus) via a stimulating electrode (e.g., tungsten electrodes).

24.3 Potential Problems and Solutions

24.3.1 Successful Stable Recordings

Although it is a convenient method to establish electrophysiological recordings in whole-cell configuration without assistance of optical visualization microscopic equipment and complicated protocols, "blind" patch clamp is relatively difficult to implement compared with recordings in cultured neurons due to the invisibility of neuronal cells and recording electrode tips that may be clogged by brain tissue during manipulation. Thus, to obtain a successful stable recording, great attention should be paid specifically on the following:

1. Young animals (15–30 day old) are usually preferred for experiments, because the neuro-

nal cells from young animals are more resistant to hypoxic and ischemic damage.

2. The pace of recording electrode driven by a micromanipulator should not exceed 2 μm/step in a discontinuous manner; otherwise the rapid-forwarding electrode tip is likely to pierce through and lead to burst of the cells (demonstrated by intensive firing of action potential).
3. The recording electrode tip should not be made very sharp (a 5–10 MΩ resistance is acceptable when filled with potassium- or cesium-containing intracellular solution). A sharp tip usually has a high resistance and is easy to be clogged up when penetrating the brain tissue. The positive pressure applied to a recording electrode also depend on the tip size, with higher to the sharper one and lower to the blunter one to maintain a clean tip.
4. The initial increase of voltage response observed when the recording electrode tip enters into brain tissue often indicates the touch with tissue debris. The fluctuation of voltage can be eliminated by advancing the electrode into deeper tissue of a brain slice; increase in voltage response occurred hereafter represents most likely the electrode tip approaching of a cell. Then a negative pressure is applied to obtain GΩ seal. A new recording electrode should always be replaced if the first attempt to obtain seal fails, because the electrode tip might be contaminated by brain tissue that can prevent the formation of GΩ seal in a subsequent attempt.
5. Cell membrane rupture by negative pressure suction normally leaves membrane debris around the edge of rupture, which can cause resealing or occlusion of established rupture, hence distorting the recording results. An alternative method by applying the "zap" electrical burning function in Axon amplifier can effectively avoid the cell membrane re-fusion.
6. The series resistance should be constantly monitored online during the recording by delivering a transient hyperpolarization current step, because resealing of the rupture commonly occurs and greatly affects the signals recorded. Application of gentle suction or positive pressure is recommended to re-rupture the sealing and reduce the series resistance; however, data with series resistance change >20 % should not be included in analyses.

24.4 Applications

Like visualized whole-cell recording from brain slices, "blind" whole-cell recording technique is a useful technique in the study of the fundamentals of neurophysiology at cellular and simple circuit levels. Taking advantage of this method, the study of synaptic activities and neuronal membrane properties can be performed in laboratories without sophisticated equipment such as infrared microscope and manipulators (e.g., Burleigh manipulators). Thus, the "blind" whole-cell recording technique provides an economic approach in studying neuronal membrane properties and synaptic activities, especially in current challenging fiscal conditions around the world.

References

Blanton MG, Lo Turco JJ, Kriegstein AR (1989) Whole cell recording from neurons in slices of reptilian and mammalian cerebral cortex. J Neurosci Methods 30:203–210

Deng P, Xu ZC (2012) Whole-cell patch-clamp recordings on spinal cord slices. Methods Mol Biol 851:65–72

Hamill OP, Marty A, Neher E, Sakmann B, Sigworth FJ (1981) Improved patch-clamp techniques for high-resolution current recording from cells and cell-free membrane patches. Pflugers Arch 391:85–100

Liu ZW, Li LJ, Liu CG (2001) Blind patch clamp whole-cell recording technique for neurons in hippocampal slices. Sheng Li Xue Bao 53:405–408

Marty A, Neher E (1983) Tight-seal whole-cell recoding. In: Sakmann B, Neher E (eds) Single-channel recording. Plenum, New York, NY, pp 107–122

Okada Y (2012) Patch clamp techniques: from beginning to advanced protocols. Springer, Tokyo

Walz W (2007) Patch-clamp analysis: advanced techniques, 2nd edn. Humana, Totowa

Windhorst U, Johansson H (1999) Modern techniques in neuroscience research. Springer, Berlin

Whole-Cell Patch-Clamp Recording 25

Matthew J. Van Hook and Wallace B. Thoreson

Abstract

Signaling in the nervous system is mediated by changes in voltage across the membranes of individual neurons. Understanding these bioelectrical signals—their ionic basis, modulation, information content, contributions to intercellular communication, and ways in which they encode sensory input and behavioral output—is one of the fundamental tasks of contemporary neuroscience research. A variety of approaches have been developed that enable researchers to measure and manipulate bioelectricity of individual neurons and networks of neurons—a family of experimental approaches collectively termed "electrophysiology." In this chapter, we describe whole-cell patch-clamp recording, an electrophysiological technique that enables researchers to record membrane voltages and, by using voltage-clamp techniques, the ionic currents underlying those voltages. We provide an overview of the background and principles of whole-cell recording as well as a discussion of the basic complement of equipment including air tables, microscopes, Faraday cages, micromanipulators, recording electrodes, amplifiers and other electronics, solutions, and recording pipettes needed for these experiments. We also describe techniques for successful whole-cell recording, including a discussion of perforated patch techniques, and note potential pitfalls and issues of quality control that are key for obtaining high-quality publishable data.

M.J. Van Hook
Department of Ophthalmology and Visual Sciences, University of Nebraska Medical Center, Omaha, NE 68198-5840, USA
e-mail: matt.vanhook@unmc.edu

W.B. Thoreson (✉)
Department of Ophthalmology and Visual Sciences, University of Nebraska Medical Center, Omaha, NE 68198-5840, USA

Department of Pharmacology and Experimental Neuroscience, University of Nebraska Medical Center, Omaha, NE 68198-5840, USA
e-mail: wbthores@unmc.edu

H. Xiong and H.E. Gendelman (eds.), *Current Laboratory Methods in Neuroscience Research*, Springer Protocols Handbooks, DOI 10.1007/978-1-4614-8794-4_25,

Keywords
Electrophysiology • Patch clamp • Single channel • Whole cell • Membrane • Ion channel • Perforated patch

25.1 Introduction and Overview

Since early discoveries that the nervous system uses electrical signals as the basis for transmitting information, neuroscientists have been developing methods to manipulate and record those electrical signals, a family of experimental approaches called electrophysiology. The last 60 years, especially, have seen a blossoming of techniques allowing researchers to make electrical recordings inside individual cells. The most widely used technique today is the whole-cell variety of patch-clamp electrophysiology.

Patch-clamp recording uses glass micropipettes with a relatively large-diameter opening at the tip (~1 μm) and filled with a salt solution to measure the electrical activity of neurons and nonneuronal cells. Bert Sakmann and Erwin Neher originally developed patch clamping as a method for measuring currents flowing through single channels on a small patch of membrane (Neher and Sakmann 1976). They and others later refined and expanded the technique, describing how it was possible to record currents flowing not only through patches of membrane on the cell surface but also from patches of membrane excised from the cell and from channels distributed throughout the whole-cell membrane (Hamill et al. 1981). Since then, patch-clamp recording has become the most widely used method for studying the electrophysiology of individual cells and neuronal networks. Sakmann and Neher were awarded Nobel Prize in Physiology and Medicine in 1991 for developing the patch clamp technique and using it to shed light on the function of ion channels in neuronal signaling.

There are several varieties of patch-clamp recording in common use (Fig. 25.1). These include cell-attached, inside-out and outside-out excised patch recordings, as well as whole-cell patch clamp and its relative, perforated patch.

The first of these, the tight-seal cell-attached configuration, is the starting point for the other types of patch-clamp recording. The cell-attached configuration is formed by pressing the tip of a micropipette up against a cell membrane and applying gentle suction to create a tight seal between the glass and the cell membrane. In cell-attached recordings, one can measure currents through individual channels or populations of channels in that patch of membrane (for more detail, see Chap. 26). Because cell-attached recordings do not rupture a cell's membrane—which washes out the cytoplasm, exchanging it for the contents of the pipette solution—they can also be used as a relatively noninvasive way to assess the spiking behavior of individual neurons. In this application, one measures currents through a small population of voltage-gated sodium and potassium channels as they are activated during action potentials (see Perkins 2006 for principles and applications of this technique).

Excised patch recordings, in which a small patch of membrane is torn away from the rest of the cell, are useful for studying the behavior of individual channels or a small population of channels while manipulating solutions and voltage across the patch. This can be done with either inside-out or outside-out techniques (Fig. 25.1). With inside-out recordings, one tears a patch of membrane away from the cell after establishing a giga-ohm seal, thereby exposing the intracellular side of the patch to the bath. To obtain an outside-out recording, one first ruptures the patch after establishing a giga-ohm seal, entering a whole-cell configuration (see below) and then slowly draws the pipette away so that small pieces of membrane are pulled from the cell. Eventually, this trailing section of membrane breaks away from the rest of the cell and fuses over the end of the pipette so that the extracellular side of a small patch of membrane is exposed to the bath solution.

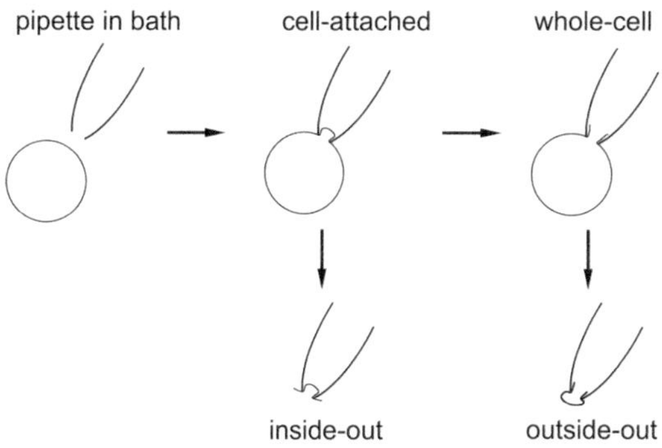

Fig. 25.1 Patch-clamp configurations. To establish cell-attached configuration, a patch pipette is pressed against the cell membrane, and gentle suction is applied to form a tight, giga-ohm seal between the pipette glass and cell membrane. From the cell-attached configuration, pulling back on the pipette tears off a patch of membrane, resulting in the inside-out excised patch configuration. Alternatively, further suction applied in the cell-attached configuration will rupture the patch of membrane, resulting in the whole-cell configuration. The outside-out excised patch configuration is established by pulling back on the pipette while in whole-cell configuration. This allows trailing pieces of membrane to break from the rest of the cell and fuse at the tip of the pipette

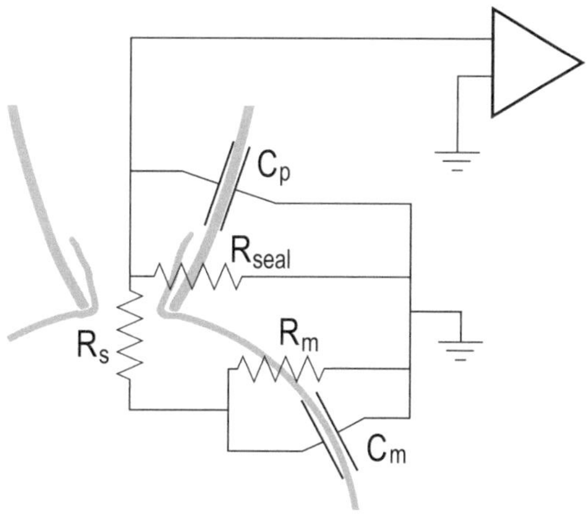

Fig. 25.2 Equivalent circuit of the whole-cell configuration. The thin glass wall of the pipette provides capacitance (C_p) that is nulled out after establishing a giga-ohm seal. This is parallel to the seal resistance (R_{seal}), which results from the tight seal of the pipette glass and the cell membrane. Rupture of the patch results in a low resistance electrical connection with the cell interior. The quality of this connection is indicated by the series resistance (R_s), which is the sum of the access resistance resulting from pieces of debris at the pipette tip and the pipette resistance. The membrane capacitance (C_m) and membrane resistance are indicative of the cell's surface area, and the membrane resistance (R_m) results from open or closed ion channels in the cell's membrane

The focus of this chapter is on whole-cell recording. The whole-cell configuration for voltage- or current-clamp recording is obtained after establishing cell-attached configuration with a tight (>1 GΩ) seal. The patch of membrane is then ruptured, giving electrical access with low enough resistance to allow effective voltage clamp of the cell. In addition to the value of being able to clamp the membrane voltage and measure ionic currents, whole-cell recording allows one to manipulate solutions on both sides of the cell membrane, changing the ionic composition or introducing pharmacological agents. On the extracellular side, this can be done by exchanging the bath solution or by localized rapid application of drugs (e.g., with use of a puffer pipette system). On the intracellular side, the pipette solution typically remains unchanged throughout the course of a recording, although methods have been devised that enable one to exchange the solution in the recording pipette in the middle of an experiment. As considered in detail later, the typical formulation of a pipette solution attempts to mimic the physiological components of the cytoplasm, but it can be modified to enhance currents through certain channels or inhibit currents through others.

The equivalent circuit for the whole-cell configuration is shown in Fig. 25.2. The thin glass wall of the pipette acts as a capacitor, and this capacitance (C_p) should be nulled out after establishing a giga-ohm seal. C_p is parallel to the seal resistance (R_{seal}), which results from the tight seal of the pipette glass to the cell membrane. Rupture of the patch results in a low resistance electrical connection with the cell interior. The quality of this connection is indicated by the series resistance (R_s), which is the sum of the access resistance resulting from pieces of debris at the pipette tip and the pipette resistance. Lipid bilayers have a capacitance of 0.9 μF/cm^2 (Gentet et al. 2000) and so the membrane capacitance (C_m) provides a measure of the cell's surface area. The membrane resistance (R_m) is determined largely by the presence of open or closed ion channels in the cell's membrane.

In later sections of this chapter, we consider ways to assess these different circuit elements.

Voltage clamp is the most common technique for measuring and characterizing ionic currents flowing across a cell's membrane. In voltage-clamp recordings, one "clamps" the voltage across the cell membrane, holding it at a desired set point or varying it as needed to study the properties of various ionic conductances. One can then measure the amount of current needed to maintain the desired transmembrane voltage. Because the current injected by the amplifier is equal and opposite to current flowing across the membrane, voltage-clamp amplifiers provide a direct measure of current flowing through ion channels in the membrane.

Recordings can also be made in the current-clamp configuration in which one measures a cell's membrane potential. This more closely mimics the cell's native state because current clamp allows the membrane potential to vary in response to synaptic inputs or other membrane currents. Under current clamp, one can also inject current through the patch pipette in order to mimic native current input.

In this chapter, we provide a brief summary of patch-clamp recording techniques, highlighting key principles and important technical issues. For a more thorough discussion of the technique and theory of patch-clamp recording, we suggest the reader consult Marty and Neher (2009). Other good resources include "The Axon Guide for Electrophysiology & Biophysics Laboratory Techniques" (freely available from multiple sites online, Molecular Devices 2008) and books by Sakmann and Neher (2009), Molleman (2003), and Kettenman and Grantyn (1992).

25.2 The Patch-Clamp Setup

25.2.1 Microscopes, Air Tables, and Micromanipulators

At the heart of most patch-clamping rigs is a microscope. This enables one to see the preparation, identify cells for recording, and successfully position a recording pipette on the cell. Patch clamping commonly requires one to position a recording pipette on a particular cell or part of a cell, sometimes even on a cell's dendrite. The optics should therefore be of high enough quality to resolve fine cellular details. That said, a simple dissecting microscope may be sufficient for some applications where it is unnecessary to directly visualize a target cell, such as in some brain slice preparations (for a description of this "blind patch" technique, see Castañeda-Castellanos et al. 2006).

However, most patch-clamping experiments require a high-quality microscope for visualizing the target cell and patch pipette. Both upright microscopes (in which the objective lens is positioned above the sample and the condenser is below) and inverted microscopes (in which this arrangement is the opposite) are used for electrophysiology. Inverted microscopes have the advantage of providing a greater working distance above the preparation, allowing more room to position recording electrodes, puffer pipettes, stimulating electrodes, etc. For upright microscopes, water-immersion objective lenses that provide a relatively large working distance have been designed specifically for patch-clamping applications. The microscope body of an upright microscope used for patch-clamp recording is modified so that the objective moves up and down during focusing, but the stage remains fixed in the z-axis. Some laboratories have a camera integrated with the optics of the microscope. This can be used simply as a way to visualize the preparation without using the microscope eyepieces but can also be used to acquire data for imaging experiments (e.g., with calcium-sensitive dyes). Because infrared light penetrates into tissue more deeply than visible light, many investigators use infrared illumination and differential interference contrast (IR-DIC) optics to visualize cells when using slice preparations (Stuart et al. 1993; Sakmann and Stuart 2009). Additionally, when recording from photosensitive cells such as retinal neurons, making use of optogenetic approaches, or using photosensitive reagents, it may be important to use infrared illumination to avoid photoactivation.

The microscope is positioned on a vibration isolation table that uses gas (typically supplied by a tank of nitrogen) to float each leg of a heavy tabletop, thereby minimizing the influence of vibrations transmitted through the floor. Because patch clamping occurs on such a small scale, even small vibrations can easily disrupt a seal, shortening the life of your recording. Therefore, taking every measure to isolate your patch-clamping workspace from vibrations is of crucial importance in making successful recordings. The workspace on the air table is typically surrounded by a Faraday cage, a grounded conductive mesh (usually copper) that shields the setup from picking up electrical noise coming from the environment ("hum" caused by the 60 Hz signal emitted by the power grid in the United States, for instance).

In patch-clamp electrophysiology, the tip of an electrode must be pressed up against the membrane of an individual cell before forming a tight seal and rupturing the membrane patch to obtain a whole-cell recording. To accomplish these fine movements with micron scale precision, the headstage, electrode holder, and electrode are mounted on a micromanipulator. These span a variety of mechanisms and costs, from manual versions such as the Huxley–Wall-type manipulator to more expensive motorized and piezoelectric micromanipulators. We have found that Huxley–Wall manipulators are suitable for the majority of patch-clamp applications.

There are many important considerations in choosing and setting up a micromanipulator (for a good discussion, see Molleman 2003, pp. 52–56). For one, it should be a relatively quick and easy process to change electrodes and return the manipulator to a working position. This enhances productivity and cuts down on frustration when recording pipettes need to be repeatedly changed. Two, to minimize drift and vibration, both of which can hinder successful patching or shorten the life of a recording, it is important to ensure that tubing and wires attached to the electrode holder and headstage have adequate slack so as not to tug on the manipulator or transmit vibrations. Additionally, it helps to firmly attach tubing and wires to the air table or microscope stage to keep the amount of slack from changing mid-recording. Finally, the headstage and pipette holder should be mounted as close to the manipulator as possible so that unwanted small movements at the electrode base are not dramatically amplified at the tip. If possible, it is best to mount the manipulator directly on the microscope stage, rather than on a stand mounted on the table.

25.2.2 Amplifiers and Other Electronics

Patch-clamp recording relies on a specialized amplifier and headstage to record signals and simultaneously provide feedback through the recording pipette (see Sigworth 2009). There are a variety of patch-clamp amplifiers available on the market. The typical amplifier has an assortment of switches and knobs to control recording mode, command potential, filters, gain, capacitance, series resistance compensation, etc. Some newer ones are designed to be operated solely through the computer with software designed by the manufacturer, although several third parties have written substitute programs with additional features (and bugs). Computers are typically used for data acquisition but can also serve to command voltage-clamp and current-clamp stimuli as well as the timing of other events during the recording (light flashes, puffer application of drugs, etc.). This two-way communication by the amplifier and computer is achieved with an AD/DA (analogue-to-digital/digital-to-analogue) converter. Many of these interface with the computer using a card that connects to a PCI slot, although newer versions are beginning to perform this function by way of a USB connection.

The headstage interfaces with the amplifier to transmit electrical signals from the cell and provide feedback and current injection for voltage-clamp and current-clamp recordings. At the end of the headstage is the pipette holder. The pipette holder provides an electrical connection from the pipette to the headstage through a pin and silver wire coated with silver chloride that makes contact with the solution in the pipette. The pipette is held securely in place with rubber gaskets that also provide an airtight chamber, enabling the experimenter to apply suction or pressure to the pipette (see Hamill et al. 1981, Fig. 25.1).

Most components for patch clamping are mounted on a rack or desk next to the air table. Ergonomic considerations should play a substantial role in arranging these components. For instance, commonly used controls such as the mouse, keyboard, and amplifier should be at hand level. It is important to be able to switch between watching the preparation and an oscilloscope (and/or computer monitor) relatively quickly since each gives important and quickly changing information while you attempt to establish a seal and break in.

25.2.3 Superfusion

Most whole-cell patch-clamp experiments require continuous superfusion of the tissue during the recording. This is often accomplished with a simple gravity-fed superfusion system in which a reservoir is placed above the level of the preparation, and the solution is allowed to flow into the recording chamber through tubing. Various pharmacological agents can be applied in this way simply by switching from one reservoir to another. Care should be taken in selecting the tubing of the superfusion lines, especially if they will be used to apply pharmacological agents (see, for instance, Crary et al. 1998). Teflon is often a good choice because it is chemically inert and gas impermeable.

Solution can be drawn off the recording chamber by a variety of methods. The most common approach is to use a Büchner flask attached to a vacuum source in the laboratory. Alternatively, one can use a peristaltic pump in place of a vacuum. Suction can sometimes create electrical noise in a recording, and so, in situations where noise must be minimized, a paper towel or KimWipe can be used as a wick to draw solution from the recording chamber into a waste container. With each of these approaches, care must be taken so that the flow rate is neither too fast, which can introduce drift and vibrations, nor too slow, which limits the flow of fresh superfusate to the preparation. Additionally, it is important to monitor the flow of solution throughout an experiment so that the flow rate from the reservoir does not overwhelm the suction and thus spill superfusate onto the microscope. If the recording chamber does overflow, spilled solution must be promptly cleaned off any microscope components to avoid permanent damage.

25.2.4 Recording Electrodes

Recording electrodes for patch-clamp recording are fabricated in the laboratory using hollow glass pipettes that are heated and pulled to a fine tip using a microelectrode puller. There are many different types of glass but most investigators use borosilicate (see the Axon Guide for a more complete discussion of glass properties). Pipettes range from 1.0 to 2.0 mm outer diameter with varying wall thickness. They can also come with an internal glass filament to assist in filling the tip through capillary action. We typically use borosilicate glass pipettes with 1.2 mm outer diameter, 0.95 mm inner diameter, and an internal glass filament.

To fabricate the pipettes needed for patch-clamp recording, multiple heating stages are usually required. Suitable microelectrode pullers are available from a number of manufacturers. One can assess pipettes by inspection under a microscope and by measuring the electrode resistance. A smooth tip of 1–2 μm diameter is desirable. Resistance values of up to 15 MΩ can be used with small cells that have intrinsically high input resistance (see discussion of series resistance below), but most investigators use pipettes with <10 MΩ resistance. If it is necessary to further smooth the tip, one can fire polish the pipette by heating it in a microforge.

To lower capacitance noise associated with water along the shaft of the electrode, one can coat the shaft with Sylgard #184 (Dow Corning), dental wax, or some other hydrophobic substance. The coating can be applied using a small curved tool and care must be taken to avoid coating the tip. Sylgard is heat-cured, but it is nevertheless a good idea to store Sylgard-coated pipettes with the tips upright to prevent residual uncured Sylgard from flowing onto the tip. It is important to coat electrodes when performing single-channel or capacitance measurements, but this may not be necessary for most whole-cell recording applications.

25.3 Solutions

25.3.1 Extracellular Solution (Superfusate)

Extracellular saline solutions are intended to approximate plasma or cerebrospinal fluid but can be modified to meet specific experimental requirements (e.g., by addition of drugs to isolate certain ion channels or neurotransmitter circuits). Specific solutions used for different experiments in different species and tissues can be found in various publications, but we address a few general points about solution composition below.

To maintain solution pH, one must include pH buffers. The most common pH buffers are HEPES and bicarbonate. When preparing HEPES-buffered solutions, one must adjust the pH to the desired value (typically 7.4) using NaOH and, if desired, oxygenate the solution using 100 % O_2. One reason to use HEPES is that bicarbonate will precipitate with various divalent cations that are often used to study calcium channels (e.g., cobalt, barium). Although amphibian and fish retina tolerate use of HEPES quite well, mammalian neurons often appear to do better with the more physiological buffer, bicarbonate. With bicarbonate as the principal buffer, one does not need to adjust the pH with NaOH since the pH is determined by the amount of CO_2 in the medium. Typically, bicarbonate-buffered solutions are bubbled with 95 % O_2 and 5 % CO_2. With bicarbonate-buffered solutions, one must take care to ensure that the solution is continually well bubbled since the pH will become more alkaline as solution CO_2 levels decline. Supplementing bicarbonate-containing solutions with HEPES can help to clamp the pH and avoid this potential problem.

The PO_2 of a saline solution bubbled with 95–100 % O_2 will rise to very high values, but the oxygen content of saline will nevertheless remain fairly low (~2 vol.%). By contrast, because of the oxygen-storing capacity of hemoglobin, the oxygen content of arterial blood is ~20 vol.%. It is worth recognizing that cells superfused with saline solutions may thus be mildly hypoxic, especially those embedded deep within the tissue.

Mammalian neurons have more stringent metabolic requirements than nonmammalian cells. For studies on mammalian neurons, the medium is often enriched with a number of amino acids, vitamins, and other compounds. One can prepare these solutions in the laboratory, but there are also many commercially available mixtures.

Temperature is another important variable. While it is simpler to conduct experiments at room temperature, it is often helpful to warm or cool the tissue. Tissue from cold-blooded animals will often remain healthier at cool temperatures, and many enzymes in warm-blooded animals do not function optimally at room temperature. The superfusate can be cooled using a Peltier device or by simply placing a small ice pack on the microscope stage. If the solution needs to be warmed, an in-line heater can be placed in the path of the superfusate. Alternatively, the recording chamber itself can be warmed, as can the reservoir of solution. Since gases tend to come out of solution at warmer temperatures, warming the solution in the reservoir rather than using only an in-line heater can avoid the formation of bubbles in the lines.

25.3.2 Intracellular (Pipette) Solutions

Just as extracellular solutions mimic the extracellular environment in vivo, solutions used to fill the glass-recording pipette mimic the intracellular environment. With conventional whole-cell recordings in which one ruptures the membrane, the contents of the pipette diffuse rapidly into the cell. This allows one to manipulate the intracellular environment of the cell to control the levels of different ions (e.g., chloride), alter second messengers (e.g., calcium or nucleotides), and introduce drugs or dyes. The time for compounds to enter the cell depends on the size of the opening (i.e., the tip resistance) and the size of the cell (Marty and Neher 2009). An empirically based formula for the time constant for diffusion was derived by Pusch and Neher (1988). As with extracellular solutions, the specific pipette solutions used for different experiments in different species and tissues can be found in various publications, but we consider a few general issues below.

There is a greater variety in the makeup of pipette solutions than extracellular solutions. One common change is to replace K^+ with Cs^+. Cs^+ does not readily permeate K^+ channels. This change elevates the membrane resistance by reducing K^+ channel conductance and can thus improve one's ability to voltage clamp the cell. K^+ channel blockers like tetraethylammonium (TEA, 10 mM) can also be added to the pipette solution for the same purpose.

Another common change is to manipulate the level of chloride and other anions. By changing chloride levels, one can alter E_{Cl} to either accentuate or diminish inhibitory chloride currents arising from the activation of ionotropic GABA or glycine receptors. Various anions can be used to replace chloride including gluconate, glutamate, methylsulfate, and methanesulfonate. When added to the presynaptic pipette solution, use of glutamate can enhance vesicular loading of glutamate in glutamatergic neurons (Wilson et al. 2005) and thereby enhance the amplitude and stability of postsynaptic currents (Bartoletti and Thoreson 2011). One can prepare 1 M cesium gluconate stock solutions by carefully titrating gluconic acid with CsOH to pH 7.0. One can prepare cesium glutamate solutions by titration of 40 mM glutamic acid with CsOH. For the final pH, pipette solutions are typically buffered to 7.2 using HEPES titrated with either KOH or CsOH.

Pipette solutions typically contain a Ca^{2+} buffer to maintain low intracellular Ca^{2+} levels of 50–100 nM. Most investigators use either EGTA or BAPTA as the principal Ca^{2+} buffer in pipette solutions. EGTA and BAPTA have a similar high affinity for Ca^{2+}, but BAPTA binds Ca^{2+} with 50–100 times faster on-rate (Naraghi 1997). Because of this, BAPTA more effectively limits the spread of Ca^{2+} away from open Ca channels. Solution Ca^{2+} levels can be calculated using freely available programs (e.g., MaxChelator).

ATP, GTP, and phosphocreatine are often added to pipette solutions to provide metabolic fuel for the cell. With ATP, one should also provide magnesium since it is required as a cofactor for Na^+–K^+ ATPase activity.

Ideally, the osmolarity of the pipette solution should match that of the extracellular bathing medium. However, we typically prepare 25 or 50 mL of pipette solution, freezing numerous 0.5 mL aliquots of solution for later use. Thus, a batch of pipette solution lasts much longer than a batch of bathing solution, making it difficult to ensure that both remain precisely matched. It is better to have a pipette solution with slightly lower osmolarity than the bathing solution since a higher osmolarity pipette solution will cause the cell to swell and burst. We therefore adjust the pipette solution osmolarity so that it is ~2 mOsm lower than that of the typical bathing solution. Osmolarity can be measured using a vapor pressure osmometer.

25.4 Procedures for Whole-Cell Recording

25.4.1 Sealing and Rupturing

The process of establishing a whole-cell recording involves carefully maneuvering a recording pipette so that its tip presses gently against the cell membrane, forming a tight seal between the glass of the pipette and the membrane. After seal formation, the patch of membrane inside the pipette is ruptured to provide a physical and electrical connection to the inside of the cell.

25.4.1.1 First Steps: Recording Pipette and Initial Quality Control

The first step is to fill the pipette with enough pipette solution to contact the silver/silver chloride wire. You should try to ensure that there are no bubbles in the pipette, which can serve as resistors and prevent good electrical access to the cell. Pipettes should be filled using a nonmetallic filling needle since EGTA and BAPTA can bind to metal and removal of these compounds from the pipette solution will change its pH. Some investigators filter the pipette solution with a small filter (e.g., 0.22 or 0.45 μm Nalgene or Millex in-line filters) to remove small particles although it is not always necessary. Filling needles are available commercially but one can also

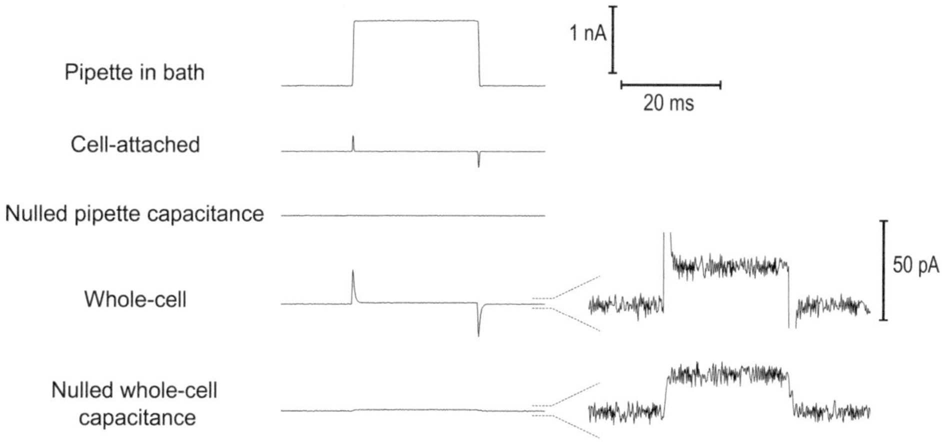

Fig. 25.3 The process of establishing whole-cell recording configuration. The amplifier provides a 10 mV step at 20 Hz. When the 10 MΩ pipette enters the bath, this results in a 1 nA current response due to the pipette resistance. After establishing a giga-ohm seal, the 5 pF pipette capacitance is nulled out. The appearance of larger whole-cell capacitance transients (33 pF) indicates that the patch has been ruptured and that whole-cell configuration has been established. The whole-cell transients are then nulled out with the amplifier circuitry. The current response resulting from the 500 MΩ membrane resistance is apparent in the enlarged traces to the right. This process was modeled with an Axon PATCH-1U model cell

fabricate them by pulling the melted tip of a 1 cm^3 syringe into a fine narrow shaft.

The pipette is mounted in the pipette holder on the micromanipulator and lowered into the bath solution. Because dust and grime may rest on the interface between the air and the bath solution, most patch clampers apply gentle positive pressure to the pipette before lowering it into the bath. In the absence of any experimenter-applied positive pressure, capillary action of the pipette provides a net negative pressure, which can draw debris into the pipette, clogging or coating the tip and hindering your ability to establish a seal. Positive pressure can be applied from your mouth or by attaching a 1 mL syringe to the pressure line. This positive pressure provides a small continuous flow of pipette solution out of the pipette tip, gently blowing away potential contaminants. Because debris on the pipette tip will almost certainly prevent successful sealing, one must use a fresh recording pipette every time you attempt to patch a cell.

As you lower the pipette into the bath, set the amplifier to provide a 10 mV step (at 20 Hz or so) and set the voltage-clamp command potential to 0 mV. To make a 10 mV difference, the amplifier will put out as much current as necessary (within its operating limits) in inverse proportion to the resistors between the pipette electrode and the reference electrode in the bath (according to Ohm's law, $V = I \times R$). Thus, when the electrode is out of the bath and separated from the reference electrode by a practically infinite resistor (air), the amplifier will not need any current to generate a 10 mV potential. When the pipette enters the bath solution, the opening of the pipette tip becomes the only major resistor separating the pipette electrode from the reference electrode. Thus, the amplifier needs to produce considerably more current in order to make a 10 mV potential between the pipette and reference electrode, and this will be reflected as a square current response on the oscilloscope (Fig. 25.3). The measure of resistance provided by the current response to this 10 mV step is the primary readout in the process of establishing a whole-cell recording, giving you an indication to your progress and the quality of your recording.

As the pipette tip enters the bath, you have an opportunity to assess the quality of your recording pipette. If the resistance is a little low, the tip may be too big to establish a stable seal on the cell membrane. If it is very low (<1 MΩ), it is likely that you bumped the pipette tip and broke it or that it was incorrectly pulled (see above, Sect. 2.4). A tip resistance that is too high makes it more difficult to maintain low resistance electrical access throughout your recording and can also make it hard to rupture into the cell. A large tip resistance can also be an indication that an otherwise suitable pipette has become clogged, either by a bubble or by a piece of dust.

25.4.1.2 Junction Potentials

Once you have passed this first stage of quality control, you must adjust the baseline of the current response so that it rests at 0 pA. This nulls out the liquid–liquid and liquid–metal junction potentials. The liquid–metal junction potential is a voltage difference that develops as a result of having a metal electrode in a salt solution and is accomplished using the DC (direct current) offset controller on the amplifier. This offset potential is minimized by proper coating of the pipette and reference electrodes with silver chloride. If this offset changes dramatically from one recording to the next or drifts excessively after being nulled out, this is an indication that the AgCl coat on your pipette electrode or reference electrode is beginning to wear off and one or the other needs to be re-coated. This is best done by lightly sanding the silver wire and attaching it to the positive terminal of a battery. Another piece of silver is attached to the negative terminal and both wires are placed in a solution of 0.9 % NaCl and allowed to sit for 30+ min or until a dull gray coat of AgCl has accumulated on the wire. It should then be washed and dried before use (see Alvarez-Leefmans 1992).

Another source of offset potential is the liquid–liquid junction potential. This junction potential develops at the interface between two different salt solutions (the pipette and extracellular solutions). The liquid–liquid junction potential can be removed along with the liquid–metal junction potential by offset adjustments in the amplifier. However, after you rupture into the cell, the contents of the pipette replace the cytosol and erase any liquid–liquid junction potential. Thus, the actual holding potential of the cell is the potential specified by the amplifier plus the liquid junction potential that existed prior to sealing onto the cell. The liquid junction potential varies with the composition of both solutions and therefore needs to be calculated (e.g., using the liquid junction potential calculator in pCLAMP software) or measured for each combination of pipette and bath solutions (Neher 1992; Kenyon 2002). To measure the liquid junction potential, one measures the potential of a pipette placed in the bath and then determines the change in that potential when the bath solution is replaced with the pipette solution, thereby erasing the liquid junction potential. Most researchers typically account for this offset (which often ranges up to 15 mV) in the data-analysis stage, although it should also be taken into account when designing voltage-clamp protocols.

25.4.1.3 Establishing a Giga-Ohm Seal

The next step in the process of making a whole-cell recording is to approach the target cell with the patch pipette. If you are targeting cells deep within a slice of tissue, you may need to increase the positive pressure to keep the pipette tip clear of debris. With isolated cells, strong positive pressure is unnecessary. As the pipette gets closer to the cell, it may be necessary to reduce the positive pressure to avoid blowing the cell away.

Once the pipette tip nears the cell, you must bring it into contact with the cell so that it "dimples" the membrane. This is evident as a crescent-shaped deformation of the membrane as the pipette presses into it and the slight positive pressure pushes the membrane away from the tip.

Before dimpling the pipette onto the cell membrane, you need to consider where on the cell to place the pipette. Ideally, this should be the region of the cell's membrane at which the pipette is perpendicular to a line tangent to the cell membrane. Contacting the cell at this angle helps maintain a stable seal by ensuring that the pipette is not tugging on the membrane.

When the pipette dimples onto the cell membrane, you should see a small decrease in the seal test current indicating an increase in resistance at the pipette tip. This is because the cell membrane is providing additional resistance between the pipette and reference electrodes and the amplifier needs to put out less current to make a 10 mV potential. Once you see this slight resistance increase (the desired resistance increase will vary depending on your cell type and recording conditions, but a few MΩ is fairly typical), you can release the positive pressure. When you do this, you should see a further increase in resistance as the membrane begins to seal with the pipette tip. You want this resistance to climb to >1,000 MΩ (a GΩ seal), so that aside from any remaining pipette capacitance transients, the current response is nearly a flat line (Fig. 25.3). Again, this is because a very high resistance means that the amplifier needs little current to generate a 10 mV potential. Sometimes, releasing the positive pressure will be sufficient for the cell membrane to quickly and spontaneously seal, bringing the resistance above 1 GΩ. In many cases, however, you will need to provide a very gentle negative pressure (suction) to draw some of the membrane into the pipette and make a tight seal. Do this very slowly and gently as you watch the seal resistance to monitor your progress. You can apply suction with your mouth or by using a syringe (typically 1 mL). Too little suction will do little to establish a tight seal while too much runs the risk of rupturing the patch of membrane before it has had a chance to seal tightly onto the pipette. After the resistance has climbed by 100 MΩ or more, hyperpolarizing the patch to the command potential you will use in the recording (−70 mV, for instance) may help in forming a giga-ohm seal. After a seal has formed, release the negative pressure and let the seal stabilize for minute or two. You now have a "cell-attached" patch. If you are truly attached to a cell, you may be able to see some single-channel activity from channels captured in this patch.

After establishing a giga-ohm seal, you should null out transients associated with the pipette capacitance (Fig. 25.3). This can be done with controls on the amplifier that subtract a signal equal in amplitude and kinetics and opposite in sign from the current traces. While some amplifiers controlled through a computer interface automatically measure and compensate for pipette transients with the click of a button, other amplifiers require use of separate controls for amplitude and time constant to perform this subtraction. When uncompensated, capacitance transients will appear as small, fast current responses at the beginning and end of the seal test voltage step (Fig. 25.3). During your experiments, uncompensated capacitance discharge at the beginning and end of voltage commands runs the risk of saturating the amplifier and leading to the appearance of transients in the current records. Moreover, uncompensated capacitance will take time to charge at the beginning of a voltage command, drawing current that would otherwise be used to charge the cell's membrane and thereby distorting the kinetics of the membrane response.

Once you have nulled out the pipette capacitance transients (Fig. 25.3) and before establishing a whole-cell configuration by breaking in, you should switch to a hyperpolarized command potential close to the cell's presumed resting potential (if you had not already done so to help with seal formation). In most cases, this will be the command potential you use for your experiment.

25.4.1.4 Rupturing the Patch

Rupturing the patch of membrane in the pipette, or "breaking in," makes the pipette continuous with the intracellular space of the cell. You will know you have broken in by the sudden appearance of transient capacitance events in response to the 10 mV pulses superimposed on a small square current response corresponding to the cell's membrane resistance (Fig. 25.3). The capacitance transients reflect charging and discharging of the cell's membrane capacitance and resistance through the series resistance at the tip of the pipette. The square current response is the result of the cell's membrane resistance. For cells with a very high membrane resistance (i.e., >1 GΩ), the square response may not be readily apparent in the current recording.

The most common method for breaking in involves applying suction to the patch. When applying suction by mouth, this seems to work best by using quick pulses of increasing intensity. Rupture often occurs with the release of suction. You can also apply suction with a 1–2 mL syringe, drawing back on the plunger and then abruptly releasing pressure by detaching the syringe from the line. Try increasingly large suction steps. Alternatively, many patch-clamp amplifiers have a control that applies a large, fast voltage pulse that causes a dielectric breakdown of the membrane patch ("zap" on some amplifiers). In some cases, simultaneous suction and the voltage pulses work well. If there is no zap circuit in the amplifier, you can try using the electrophysiology software to apply a large voltage step (1–2 V, 0.1–10 ms) or briefly toggle between voltage and current clamp. It is essential when breaking in that you watch the seal test current response and stop as soon as you see the whole-cell capacitance transients, lest you damage your cell. For similar reasons as nulling out the pipette capacitance transients, it can also be helpful to null out the whole-cell capacitance transients. This process is the same as the one used for pipette capacitance except that it uses a different set of controls on the amplifier.

Sometimes during the course of breaking in, you will see a large square current response to the voltage step, indicating that the resistance has declined dramatically and you have lost the seal. In this case, you should start over with a new recording pipette. Although you might have luck successfully recording from the same target cell, you may be better off finding a new cell. On other occasions, you will break in but will lose the seal as well. This will be apparent as whole-cell capacitance transients superimposed on a large square current response. For most experiments, these "leaky" recordings are of little use.

25.4.2 Quality Control

Accurate measurements of membrane currents require accurate voltage clamp. A common cause of poor voltage clamp is excessive resistance at the tip of the recording pipette (series resistance or R_s). R_s is the sum of the pipette resistance and the access resistance resulting from pieces of membrane and other debris near the pipette tip. Because R_s is in series with the membrane resistance (R_m; Fig. 25.2), R_s introduces a steady-state voltage error determined by the ratio of R_s:R_m. For example, R_s=40 MΩ will introduce a steady-state error of 10 % when clamping a cell with R_m=400 MΩ. In addition to steady-state errors, high R_s limits the amount of current that can be injected into the cell and thus limits the speed at which the change in membrane potential follows the command potential. It is therefore important to monitor the quality of the recording throughout your experiment, keeping careful watch on the R_s. R_s can be calculated from the height of the initial capacitative transient (I_c) evoked by a voltage step using Ohm's law ($V=I_c \times R_s$). Increasing R_s will therefore be apparent as a decrease in amplitude and broadening of whole-cell transients.

If R_s rises to an unacceptable level, meaning that the ruptured patch is beginning to reseal or becoming clogged with other debris, you may be able to break back in by reapplying suction or a zap. In some cases, this saves your recording, allowing you to continue acquiring quality data. In other cases, attempting to re-rupture results in loss of the seal. Despite this risk, in cases where R_s has climbed above acceptable levels, it may salvage an otherwise unusable recording.

You can also compensate for R_s using circuits in the amplifier. Because R_s compensation involves positive feedback circuits, they can become unstable and oscillate as you approach 100 % compensation. A circuit developed by Sherman et al. (1999) that provides more stable R_s compensation is incorporated into the Alembic amplifier design.

An additional source of errors in whole-cell voltage-clamp recording arises from the difficulty of maintaining adequate voltage clamp at the tips of long cellular processes (Williams and Mitchell 2008; Bar-Yehuda and Korngreen 2008). The inability to clamp distal processes is called poor space clamp. Blocking membrane conductances, especially K^+ channels, can improve space clamp by increasing membrane resistance. In a well-clamped cell, activation of voltage-gated Na^+ channels will evoke an inward current that inactivates rapidly. Because Na^+ channels will remain inactivated for as long as the membrane remains depolarized, the activation of repetitive spikes by maintained depolarization suggests that regions of the cell are not being effectively clamped at a constant depolarized potential and is therefore evidence of poor space clamp. Good space clamp requires a single electrotonic compartment. This can be shown by the ability to fit the charging curves accompanying small (<10 mV) voltage changes in either voltage or current clamp with a single exponential curve. Two or more exponentials needed to fit the charging curve suggests that there are multiple electrotonic compartments indicative of poor space clamp.

25.4.3 Perforated Patch Recording

While exchanging the intracellular milieu with the contents of the pipette solution gives a certain degree of control over the intracellular environment, it will inevitably wash out components important for cellular function. This commonly leads to a gradual degradation of the quality of the recording or rundown of cellular processes under study. To get around this pitfall, researchers sometimes turn to perforated patch recordings. In perforated patch, a channel-forming compound is given time to insert into the membrane patch after cell-attached configuration is established (Marty and Neher 2009; Molleman 2003). These pores in the membrane patch are sufficient to give electrical access to the cell membrane but allow for the diffusion of only small ions, preventing changes in endogenous signaling components needed for regulating cellular function. The antibiotics nystatin and amphotericin are often used as perforating agents (Rae et al. 1991). These nonspecific ion channels allow free movement of monovalent ions but maintain endogenous second messengers and calcium levels. Amphotericin has a higher conductance than nystatin but is also more expensive. Another antibiotic that is sometimes used is gramicidin (Kyrozis and Reichling 1995). Because gramicidin is permeable only to monovalent cations, it has the potential benefit of maintaining endogenous chloride levels. The saponin beta-escin has also been used for perforated patch recording although the pores are much larger than those created by pore-forming antibiotics (Fan and Palade 1998).

Nystatin and amphotericin can be dissolved in DMSO (50 mg/mL) and then added to the pipette solution (final concentrations: 500 μg/mL nystatin, 250 μg/mL amphotericin). Gramicidin can be dissolved in ethanol (5 mg/mL) and then added to the pipette solution (5 μg/mL). Fresh antibiotic should be purchased every 6 months and fresh solutions prepared very 3 h.

The presence of these pore-forming agents interferes with seal formation and so some investigators fill the tip of the pipette with antibiotic-free solution and then backfill with antibiotic-containing solution. Our impression is that it is a bit easier to seal with gramicidin than nystatin, amphotericin, or escin.

For perforated patch recordings, one does not need to rupture the membrane patch. Successful perforation and access into the cell can be observed by the development of whole-cell capacitance transients. Full perforation can often be achieved within 5–10 min.

25.4.4 Voltage-Clamp Protocols

Once you have successfully obtained a whole-cell recording, it is time to begin your experiment. Whole-cell voltage-clamp recordings can be used to measure ion currents flowing through voltage-dependent and ligand-gated ion channels. A large array of different protocols has been devised to allow use of patch-clamp techniques for a multitude of different measurements. The simplest protocols involve measurements of currents evoked by activation of ligand-gated channels in the presence of a constant membrane voltage or measurement of the activation of voltage-dependent channels by application of voltage steps. In addition to measurements of membrane currents and voltage, approaches have been developed to allow rapid and continuous measurement of membrane capacitance changes (Neher and Marty 1982; Gillis 2009). These techniques allow measurements of synaptic vesicle exocytosis, even down to the level of single-vesicle fusion events (Klyachko et al. 2008; Gillis 2009; Neher and

Marty 1982). For a more complete discussion of the design of protocols for voltage-clamp experiments, see Molleman (2003). Additional details can be found by consulting published studies.

25.5 Conclusions

The precise measurements of electrophysiological activity provided by patch-clamp recording techniques allow powerful insights into cellular function. As presented in this chapter, whole-cell patch-clamp recording can be used to examine the behavior of individual cells. As considered in the next chapter, patch-clamp recording can also be used to measure the activity of single proteins via single-channel recordings. The power of patch-clamp recording techniques has been extended by various innovations and combination with other techniques. For example, by recording simultaneously from multiple cells, one can study the mechanisms of synaptic communication and properties of neuronal networks (Van Hook and Thoreson 2013). Patch pipettes can be used to introduce dyes for anatomical tracing or optical measurements with ion-sensitive or activity-dependent dyes (Babai and Thoreson 2009; Thoreson et al. 2003; Luo et al. 2012). Patch clamping can be useful for molecular biology studies by allowing the extraction of mRNA from single cells for RT-PCR (Moyer and Jonas 2009). Excised patches containing ion channels (e.g., glutamate receptor channels) at the tip of a patch pipette can be used as "sniffer" pipettes to measure cell secretion (Copenhagen and Jahr 1989), and excised patches with cyclic nucleotide-gated channels can be inserted inside cells to measure intracellular cyclic nucleotide levels (Trivedi and Kramer 1998). Patch-clamp recording techniques can also be used to study ion channels in the membranes of cell organelles such as mitochondria (Kinnally et al. 1987; Sorgato et al. 1987). The exquisitely detailed insights into cell function provided by patch clamp techniques ensure that it will remain a key approach in neuroscience research for many years to come.

References

Alvarez-Leefmans FJ (1992) Extracellular reference electrodes. In: Kettenman H, Grantyn R (eds) Practical electrophysiological methods. Wiley, New York, pp 171–182

Babai N, Thoreson WB (2009) Horizontal cell feedback regulates calcium currents and intracellular calcium levels in rods photoreceptors of salamander and mouse rod photoreceptors. J Physiol 587:2353–2364

Bartoletti TM, Thoreson WB (2011) Quantal amplitude at the cone ribbon synapse can be adjusted by changes in cytosolic glutamate. Mol Vis 17:920–931

Bar-Yehuda D, Korngreen A (2008) Space-clamp problems when voltage clamping neurons expressing voltage-gated conductances. J Neurophysiol 99:1127–1136

Castañeda-Castellanos DR, Flint AC, Kriegstein AR (2006) Blind patch clamp recordings in embryonic and adult mammalian brain slices. Nat Protoc 1:532–542

Copenhagen DR, Jahr CE (1989) Release of endogenous excitatory amino acids from turtle photoreceptors. Nature 341:536–539

Crary JI, Gordon SE, Zimmerman AL (1998) Perfusion system components release agents that distort functional properties of rod cyclic nucleotide-gated ion channels. Vis Neurosci 15:1189–1193

Fan JS, Palade P (1998) Perforated patch recording with beta-escin. Pflugers Arch 436:1021–1023

Gentet LJ, Stuart GJ, Clements JD (2000) Direct measurement of specific membrane capacitance in neurons. Biophys J 79:314–320

Gillis KD (2009) Techniques for membrane capacitance measurements. In: Recording S-C (ed) Second Edition (Sakmann B and Neher E, eds). Springer, New York, pp Pp155–Pp197

Hamill OP, Marty A, Neher E, Sakmann B, Sigworth FJ (1981) Improved patch-clamp techniques for high-resolution current recording from cells and cell-free membrane patches. Pflugers Arch 391:85–100

Kenyon, JL (2002) Primer on Junction Potentials for the Patchologist. http://www.medicine.nevada.edu/physio/docs/revised_primer_on_junction_potentials_3e.pdf

Kettenman H, Grantyn R (1992) Practical electrophysiological methods. Wiley, New York, NY

Kinnally KW, Tedeschi H, Mannella CA (1987) Evidence for a novel voltage-activated channel in the outer mitochondrial membrane. FEBS Lett 226:83–87

Klyachko V, Zhang Z, Jackson M (2008) Low-noise recording of single-vesicle capacitance steps in cell-attached patches. Methods Mol Biol 440:283–295

Kyrozis A, Reichling DB (1995) Perforated-patch recording with gramicidin avoids artifactual changes in intracellular chloride concentration. J Neurosci Methods 57:27–35

Luo J, Boosalis BJ, Thoreson WB, Margalit E (2012) A comparison of optical and electrophysiological

methods for recording retinal ganglion cells during electrical stimulation. Curr Eye Res 37:218–227
Marty A, Neher E (2009) Tight-seal whole-cell recording. In: Sakmann B, Neher E (eds) Single-channel recording, 2nd edn. Springer, New York, pp 31–51
Molecular Devices (2008) The Axon guide: a guide for electrophysiology & biophysics laboratory techniques.
Molleman A (2003) Patch clamping: an introductory guide to patch clamp electrophysiology. Wiley, West Sussex, England
Moyer J, Jonas P (2009) Polymerase chain reaction analysis of ion channel expression in single neurons of brain slices. In: Sakmann B, Neher E (eds) Single-channel recording, 2nd edn. Springer, New York, pp 357–373
Naraghi M (1997) T-jump study of calcium binding kinetics of calcium chelators. Cell Calcium 22:255–268
Neher E (1992) Correction for liquid junction potentials in patch clamp experiments. Methods Enzymol 207: 123–131
Neher E, Marty A (1982) Discrete changes of cell membrane capacitance observed under conditions of enhanced secretion in bovine adrenal chromaffin cells. Proc Natl Acad Sci U S A 79:6712–6716
Neher E, Sakmann B (1976) Single-channel currents recorded from membrane of denervated frog muscle fibres. Nature 260:799–802
Perkins KL (2006) Cell-attached voltage-clamp and current-clamp recording and stimulation techniques in brain slices. J Neurosci Methods 154:1–18
Pusch M, Neher E (1988) Rates of diffusional exchange between small cells and a measuring patch pipette. Pflugers Arch 411:204–211
Rae J, Cooper K, Gates P, Watsky M (1991) Low access resistance perforated patch recordings using amphotericin B. J Neurosci Methods 37:15–26
Sakmann B, Neher E (2009) Single-channel recording, 2nd edn. Springer, New York
Sakmann B, Stuart G (2009) Patch-pipette recordings from the soma, dendrites, and axon of neurons in brain slices. In: Sakmann B, Neher E (eds) Single-channel recording, 2nd edn. Springer, New York, pp 199–211
Sherman AJ, Shrier A, Cooper E (1999) Series resistance compensation for whole-cell patch-clamp studies using a membrane state estimator. Biophys J 77: 2590–2601
Sigworth FJ (2009) Electronic design of the patch clamp. In: Sakmann B, Neher E (eds) Single-channel recording, 2nd edn. Springer, New York, pp 95–126
Sorgato MC, Keller BU, Stühmer W (1987) Patch-clamping of the inner mitochondrial membrane reveals a voltage-dependent ion channel. Nature 330: 498–500
Stuart GJ, Dodt HU, Sakmann B (1993) Patch-clamp recordings from the soma and dendrites of neurons in brain slices using infrared video microscopy. Pflugers Arch 423:511–518
Thoreson WB, Bryson EJ, Rabl K (2003) Reciprocal interactions between calcium and chloride in rod photoreceptors. J Neurophysiol 90:1747–1753
Trivedi B, Kramer RH (1998) Real-time patch-cram detection of intracellular cGMP reveals long-term suppression of responses to NO and muscarinic agonists. Neuron 21:895–906
Van Hook MJ, Thoreson WB (2013) Simultaneous whole cell recordings from photoreceptors and second-order neurons in an amphibian retinal slice preparation. J Vis Exp e50007
Williams SR, Mitchell SJ (2008) Direct measurement of somatic voltage clamp errors in central neurons. Nat Neurosci 11:790–798
Wilson NR, Kang J, Hueske EV, Leung T, Varoqui H, Murnick JG, Erickson JD, Liu G (2005) Presynaptic regulation of quantal size by the vesicular glutamate transporter VGLUT1. J Neurosci 25:6221–6234

Part VIII

Immunohistochemistry and Autoradiography

Analysis of Receptor Binding and Quantitative Autoradiography

26

L. Charles Murrin

Abstract

Localizing and quantifying receptors in discrete, small tissue regions requires a method of visualizing these receptors at the light microscopic level combined with a means for measuring receptor number over a wide range. Quantitative autoradiography provides both. This technique requires radioligands with high specific activity that are also very selective for the receptor or receptors in question. A hallmark of such radioligands is a high affinity for the receptor and, usually, a low affinity for other similar receptors. It is also necessary to develop methods to allow selective and measureable labeling of the receptor, similar to any radioligand binding approach. This includes establishing the binding kinetics, both association rate and dissociation rate, under a specific set of conditions and demonstrating the radioligand binds with an appropriate pharmacology. Conditions must also maintain tissue integrity to a sufficient degree to allow identification of particular tissue regions, cellular groupings or nuclei.

It is also possible to demonstrate the functional linkage of receptors to other proteins using the quantitative autoradiographic approach. Receptors linked to G proteins, particularly Gi and/or Go, are especially amenable to this approach. Again, it is necessary to establish optimum conditions for each receptor subtype.

Through the use of these techniques, it is possible to examine receptor localization, density, and functional linkage as well as changes in these parameters during development or due to disease processes or drug treatments. This chapter describes the necessary steps to use these techniques correctly.

L.C. Murrin, Ph.D. (✉)
Department of Pharmacology and Experimental Neuroscience, University of Nebraska College of Medicine, 985800 Nebraska Medical Center, Omaha, NE 68198-5800, USA
e-mail: cmurrin@unmc.edu

H. Xiong and H.E. Gendelman (eds.), *Current Laboratory Methods in Neuroscience Research*, Springer Protocols Handbooks, DOI 10.1007/978-1-4614-8794-4_26,

Keywords
Receptor analysis • Radioligand binding • Quantitative autoradiography • Receptor pharmacology • Receptor characterization • Methods • Receptor anatomy

26.1 Introduction

Receptors are of major interest in studying biological systems because they are key components in cell-to-cell signaling. They are acted upon by signaling moieties and play a vital role in transducing signals from one segment of a biological system to other segments. While receptors are usually acted upon by small soluble molecules, such as neurotransmitters or hormones, some receptors interact with other peptides or proteins on other cells. While the term *receptor* is most commonly applied to proteins or protein complexes such as these, the methods described in this chapter can also be applied to other molecules that interact with diffusible substances, including transporters and enzymes. Throughout the rest of this chapter, the term *receptor* will include this broader array of binding sites.

Because receptors play a vital role in regulating the function of biological systems, it is important to understand them as thoroughly as possible. This includes determining where receptors are located and how many receptors exist in a particular region of an organ or on particular cells. Knowledge of the number of receptors in a particular tissue provides important clues to their functional role in that tissue.

It is also important to characterize receptors in terms of their affinity for a variety of drugs, i.e., their pharmacology. Determination of naturally occurring or exogenous molecules that interact with a receptor allows further study of functional aspects of the receptors. This may include their response to continued interaction with agonists (e.g., desensitization) or to lack of stimulation by agonists (e.g., supersensitivity), which can be studied with the techniques described in this chapter. Intracellular responses produced through receptors can also be examined, although these assays are not discussed here (see Sim et al. 1995, 1997; Happe et al. 2001; Laitinen 2003 for further information). In addition, knowledge of which compounds interact with a receptor sets the stage for the design of new and potentially better compounds both for experimental purposes and for clinical use, such as treatment of diseases. All of this information can be obtained from receptor binding assays.

This chapter presents a set of criteria for establishing the binding to or labeling of one particular binding site or receptor by a specific radioligand, and methods that can be used to demonstrate these criteria. The goal is to define conditions that allow one to analyze a specific receptor in terms of quantity, pharmacology, localization, and adaptation to disease states or drug treatment. As the term *specific* implies, the conditions allow identification of one receptor or one group of receptors to the exclusion of all others. For example, this may mean establishing conditions necessary to label the D2-family of dopamine receptors to the exclusion of the D-1 family, serotonin receptors, and any other catecholamine receptors or labeling D2 receptors to the exclusion of D3, D4, and any other receptors. It is important that these conditions provide consistent data that are repeatable. This chapter will describe the basic methods for analyzing the properties of receptors in tissues. Greater detail can be obtained from texts focusing on this procedure (Yamamura et al. 1985; Haylett 2003).

26.2 Components of a Radioligand Binding Assay

26.2.1 Radioligands

To analyze a receptor of interest using radioligand binding, the radioligand must have high specificity for the receptor in question with a good specific to nonspecific binding ratio and a high specific activity.[1] The quantity of radioligand bound to a tissue preparation under a particular set of conditions (buffer, temperature, time, etc.) is defined as total binding. The quantity of radioligand bound under these same conditions but in the presence of a high concentration of another nonradioactive drug that also binds to the receptor in question is defined as nonspecific binding. Total binding minus nonspecific binding defines *specific* binding. Specificity of a ligand for a receptor will be discussed later under saturation and competition studies. Factors involved in choosing an appropriate radioisotope will be covered next.

While it is possible to use radioligands with specific activities (the amount or radioactivity incorporated into drug molecules) as low as 10 Ci/mmol, practically speaking specific activity should be 30 Ci/mmol or higher. In general, the higher the specific activity, the greater the difference between specific and nonspecific binding and, so, the easier it is to conduct the assay. The ratio of specific to nonspecific binding should be minimally 1:1 and preferably 3:1 or greater, with the best ligands having ratios approaching 10:1. Again, the higher the ratio, the easier it is to conduct the assay. Low specific to nonspecific binding ratios usually require more tissue and more radioligand to attain acceptable results. These can be limiting factors due to scarcity of tissue or due to increased cost of tissue or ligand.

The isotope used in a radioligand has several consequences. The most commonly used isotopes are ^{3}H (tritium) and ^{125}I. Other isotopes that have been used are ^{14}C, ^{32}P, ^{33}P, and ^{35}S. For radioligand binding experiments, tritium is the most commonly used isotope. Synthesis of tritium-containing molecules is relatively easy given modern synthetic chemistry techniques. Substitution of tritium atoms for hydrogen atoms (^{1}H) rarely alters the properties of a molecule such that the binding characteristics of the ligand to the receptor are changed. Tritium has major advantages in its half-life, 12.3 years, and in the fact that tritium decay is rarely destructive to the remaining molecules of radioligand. As a result, tritiated ligands can be purchased and used over a period of many months to years, depending on the stability of the parent molecule. If experiments will take place over an extended period of time, this can provide significant cost savings. Finally, almost any molecule can be tritiated. The upper limit of specific activity for commercially available tritiated molecules is about 120 Ci/mmol, although most radioligands fall in the 50–80 Ci/mmol range.

The iodine isotope used routinely to synthesize radioligands is ^{125}I. The major advantage of radioiodinated ligands is the ability to achieve specific activities up to 2,200 Ci/mmol. This provides greatly enhanced sensitivity compared to other isotopes. It also allows some studies, particularly autoradiographic studies, to be completed far more quickly than studies using tritiated isotopes. Radioiodinated ligands also have significant potential disadvantages. First, since iodine is very large, addition of this bulky atom may alter the binding characteristics of the ligand. For instance, enkephalin molecules, which can be readily iodinated on their tyrosine, lose their ability to bind to opioid receptors when ^{125}I is added. This issue must be carefully assessed when using this isotope. Another disadvantage is the half-life, 59.4 days. In addition, the energy of ^{125}I decay is sufficient to destroy nearby molecules in solution. As a result of these problems that lead to relatively rapid loss of radioligand, radioiodinated ligands

[1] Ci/mmol (Curies per millimole) is the standard measure of the level of radioactivity incorporated into a molecule, how "hot" it is. A Curie is defined as 3.7×10^{10} disintegrations (radioactive decay events) per second or 2.22×10^{12} disintegrations per minute. While the International System of Units supports defining radioactivity levels as Becquerels (Bq) or one disintegration per second, the Curie remains the most common unit of measure in biological materials. A millimole is 10^{-3} mol of a molecule.

must be used quickly after synthesis, usually within one half-life. It is a common finding that radioiodinated ligands become useless for ligand binding studies after 45–60 days, probably due to a combination of the above factors. Due to the energy of the radio emissions, greater safety precautions must be taken when using radioiodinated ligands. A final shortcoming is that a limited group of molecules can be radioiodinated. Usually they must contain another atom for which iodine can be exchanged, such as chlorine or fluorine, or they must contain an amino acid that can be iodinated, i.e., tyrosine or serine.

^{35}S has seen limited use in radioligand binding assays, primarily because relatively few ligands contain sulfur (see Gonsiorek et al. 2006, for example). Incorporation of a radioactive sulfur atom into a molecule requires a more difficult synthetic process compared to incorporating tritium or ^{125}I. The relatively short half-life of ^{35}S, just over 87 days, means each preparation has a limited life span. The major use of ^{35}S is incorporation into nucleotides in place of a phosphorus, e.g., [^{35}S] GTPγS. The latter has been used in functional assays for some G-protein coupled receptors (Sim et al. 1997; Happe et al. 2001).

The phosphorous isotopes, ^{32}P and ^{33}P, have shortcomings similar to ^{35}S. Relatively few ligands contain phosphorous, incorporation into the ligand requires more difficult syntheses, and their half-lives are short. Because of the high energy of its radio emissions, ^{32}P can expose film a considerable distance from the point source and so provides poor resolution at the anatomic level. It also requires greater safety precautions than the other isotopes discussed. The use of carbon isotopes, specifically ^{14}C, is uncommon because of synthetic issues for incorporating ^{14}C into a drug and because it is difficult and/or very expensive to incorporate enough ^{14}C to obtain a useful specific activity.

26.2.2 Tissue

It is, of course, necessary to have tissue expressing the receptor of interest, preferably in high concentrations. This is usually not an issue because receptors generally become of interest based on tissue expression, especially high levels in specific tissues or regions. Cultured cells that express the receptor either naturally or through transfection are frequently available or can be developed. Using tissue culture approaches allows the binding characteristics, including pharmacology, of the receptor to be established in detail with relative ease. It should be kept in mind, however, that artificial expression of a receptor in a cell may produce a receptor with somewhat different properties than is found naturally.

A variety of tissue preparations can be used depending on the receptor and the goal of the experiments. The most commonly used preparation is a crude membrane preparation. Tissue is homogenized in a hypoosmotic buffer, centrifuged at $6{,}000 \times g$, and the supernatant further centrifuged at $25{,}000 \times g$. The resulting pellet is a crude membrane preparation. Usually such preparations can be stored at −20 or −80 °C for at least a month without loss of receptor binding activity, although there are instances in which receptor binding is lost rapidly within a few days (Sanders et al. 2005). This provides a preparation free of nuclear and cytoplasmic receptors and has proven very useful in characterization of membrane-bound receptors. A further refined preparation can be made using sucrose density centrifugation to provide more purified preparations of cytoplasmic membranes free of mitochondria, but these preparations usually require more starting tissue to provide enough material for an assay. Tissue can also be prepared to provide synaptosomal preparations, nuclear fractions or cytoplasmic fractions for the study of soluble receptors. For autoradiographic studies examining the anatomical localization of receptors, fresh frozen tissue sections are routinely used.

26.2.3 Incubation Conditions

Incubation conditions for an assay include the tissue preparation, discussed above, the buffer, temperature, and duration of the assay. The latter will be discussed below under kinetic studies. There are an almost infinite number of possibilities for conditions. However, most studies employ one of the common approaches. If incubation conditions have been established in previous studies, these are an obvious starting point and they allow studies to be compared across laboratories.

The buffer can be one of the most variable aspects of incubation conditions, but in practice the list of those used is relatively short. Many early studies, when radioligand binding was a new technique, used a simple 50 mM Tris buffer. This has the advantages of being simple to make and having good buffering capacity, and this buffer is still commonly used. The disadvantage is that a simple Tris buffer is not physiological because the buffering agent is not a physiological compound and this buffer does not reflect the normal ionic environment. Nevertheless, Tris buffers do provide a useful and easy way to characterize a receptor in initial studies.

As the technique developed, more studies employed buffers containing a physiological complement of ions. This allows binding conditions to reflect more closely the normal environment of the tissue. The move to more physiological buffers (e.g., Krebs-Ringer Bicarbonate; Sigma-Aldrich, St. Louis, MO) became prevalent when it was realized that the ionic composition of the buffer can alter the binding properties of the receptor and/or the ligands being used (Pert and Snyder 1974; Deupree et al. 1996). This is particularly important in studies of transporter molecules, many of which are sodium dependent (e.g., see Happe and Murrin 1993).

Physiological temperature is 37 °C and the earliest radioligand binding studies used this temperature. This requires the use of a shaking water bath to maintain and equalize the temperature throughout the sample. It was subsequently discovered that incubation at room temperature produces essentially the same results in most cases and frequently provides better sample-to-blank ratios. Radioligand and tissue degradation are slower at lower temperatures, making it easier to use longer incubation times. This proved to be particularly important when using peptides as ligands. In some cases, such as the study of benzodiazepine receptors, it was necessary to do experiments at temperatures as low as 4 °C to produce useful specific to nonspecific binding ratios. Again, the low temperature reduced or eliminated degradation of ligands, particularly those susceptible to cleavage by enzymes found naturally in the tissue.

26.2.4 Separation of Bound Ligand from Free Ligand

To measure radioligand binding to receptors, it is necessary to separate bound ligand from free ligand. There are multiple approaches possible including dialysis, centrifugation, filtration, and simple washing of the tissue.

Dialysis is rarely used today, partially because it is a slow process. Also, because dialysis is based upon diffusion of free ligand into a very large volume outside the dialysis bag, it requires a ligand that is bound to its receptor essentially irreversibly. Otherwise, all ligand will eventually dissociate from the receptor and specific binding will be lost.

Centrifugation is used primarily in situations involving a ligand with a relatively low affinity for its receptor. Centrifugation has the advantage of maintaining the tissue in contact with the ligand in solution throughout the separation process, minimizing loss of specific binding due to diffusion of the ligand away from its receptor since equilibrium between bound and free ligand is maintained until the tissue is thoroughly pelleted. The major disadvantage of centrifugation is that some free ligand (not specifically bound to tissue) will be trapped in the pellet. As a result, the sample-to-blank ratio is often lower in the centrifugation method than it is in the filtration method. Centrifugation is also slower than filtration.

Filtration is the most common method used to separate bound ligand from free ligand. This approach is fast, usually produces the best sample-to-blank ratios, and allows processing of a large number of samples quickly with available filter machines (e.g., Brandel, Gaithersburg, MD). While many types of filter have been used in the past, optimum results are usually obtained with glass fiber filters (Whatman, Piscataway, NJ). There are a variety of glass fiber filters available differing in thickness, pore size, and overall size. The most commonly used are GF/B and GF/C (Whatman) or their equivalents. Filters are placed on a filter machine, a vacuum is applied, and samples are poured or aspirated onto the filters. In most cases the complete separation of membranes from incubation medium occurs in less than 2 s. The samples can then be washed with ice-cold buffer to further remove nonspecifically bound ligand, and the filters are then air-dried for 15–30 s via the vacuum. A common protocol is to wash each filter three times with 5 mL of ice-cold buffer. Use of buffer at 4 °C greatly reduces or eliminates diffusion of the radioligand from its receptor during the wash step. Following removal from the filter machine, filters are usually allowed to air dry for a few hours or overnight before being analyzed for radioactivity by liquid scintillation counter or gamma counter.

Precautions must be taken with the filtration method. It is important that the tissue/membranes are captured by the filter paper and cannot pass through to any appreciable degree. Surprisingly, radioligands will sometimes bind to filters, increasing background. This nonspecific filter binding can, paradoxically, be displaced by the competing ligands used in the assay, producing a false increase in specific binding values. This possibility must be assessed carefully. If this problem arises, it is usually possible to eliminate this filter binding by soaking the filters in a low concentration of polyethyleneimine in the wash buffer (0.1–0.2 %) for 5 or more minutes prior to the assay. The buffer used for the wash step in a filtration assay is usually the same buffer as used in the receptor binding phase, but it is possible to use a very simple buffer such as 50 mM Tris to equal effect.

The previous material frequently mentioned *saturable and non-saturable binding*, *sample-to-blank ratios,* and *specific to nonspecific binding* comparisons. It is important to define these terms, although their meaning is context dependent.

Radioligands can bind to more than one site and, in fact, most radioligands bind to multiple sites, although with differing affinities. *Saturable* binding represents binding of a radioligand to a tissue preparation that has an upper limit, indicating there are a finite number of binding sites within the tissue preparation. This is a necessary property of all receptors, indicating they are not present in infinite quantity, and it supports a physiological relevance for the binding site. Saturable binding includes binding to receptors of interest in a particular study but also may include binding to other receptors or proteins. Saturable binding of a radioligand can be displaced by the same unlabeled ligand. *Non-saturable* binding includes binding to molecules that are present in what appear to be infinite numbers, such as membrane lipids, or sequestration of a ligand in a lipid membrane. Non-saturable binding cannot be displaced by competing ligands and is always part of the background noise in an assay.

Specific binding sites are always saturable. In addition, they represent the particular binding site(s) or receptor(s) one wishes to analyze. Most radioligands bind to several receptors. For instance, many drugs that bind to dopamine D2 receptors also bind to serotonin 5HT2, alpha1 or alpha2 adrenergic receptors, cholinergic muscarinic receptors, or a combination of these. In a situation in which one wishes to study only D2 receptors, binding to all of these sites is saturable but only binding to D2 receptors is specific. Binding to all other receptors as well as non-saturable binding is *nonspecific*. Binding to specific sites can be blocked by an unlabeled ligand that binds very selectively to the specific site in question. This defines nonspecific binding for the assay. Nonspecific binding is binding in the presence of a competing ligand and represents non-saturable binding or the combination of non-saturable binding and nonspecific binding.

Assays and radioligands are designed to maximize specific binding and minimize nonspecific binding, leading to the optimum *sample-to-blank ratio*. To establish a blank binding value or background, a competing ligand is used at a high concentration. If its affinity (K_d) for the receptor is known, it should be used at a concentration 100-fold or more greater than its K_d. This will block 99 %+ of the specific receptor sites. It is best in choosing a ligand to establish specific binding (determine background binding) to use a competing ligand with a different chemical structure than the radioligand. This reduces the likelihood that the unlabeled ligand will bind to the same secondary, saturable, nonspecific sites as the radioligand, sites that are not of interest for the particular assay.

It is important to optimize binding conditions in order to analyze only one particular receptor or binding site. While the assay for a particular receptor is being established, this process may require several iterations. As each parameter in the assay is defined, it may alter parameters determined previously. While this usually is not a major factor, it is important to verify the criteria for specific receptor binding when the final assay conditions have been established and modify the assay if these criteria are no longer met.

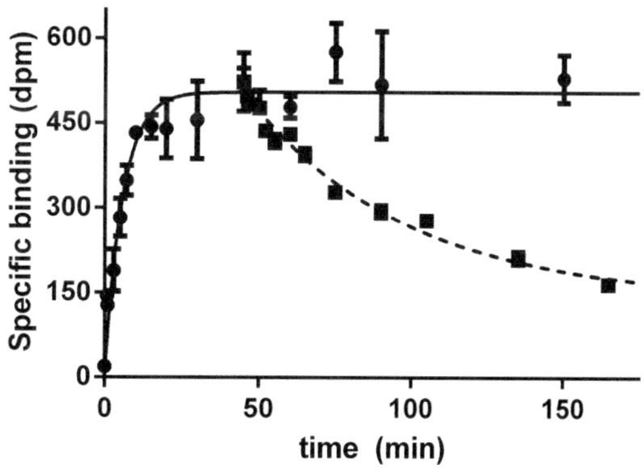

Fig. 26.1 Kinetic data for [^{3}H]cytisine binding to α4*β2* nicotinic receptors. Data are adapted from Happe et al. (1994) and show cytisine associates (*solid curve, filled circles*) with its receptor rapidly and dissociates (*dashed line, filled squares*) slowly, indicating a high affinity. The kinetic data result in K_d=0.45 nM, consistent with saturation studies and with previous studies using membranes. Background binding levels have been subtracted, resulting in specific binding

26.3 Criteria for Specific Receptor Binding

Establishing specific radioligand binding under a particular set of conditions requires that several criteria be met. These are saturability, appropriate pharmacology, and appropriate localization. The specific set of experiments needed to clearly establish conditions for specific binding are well defined.

26.3.1 Kinetic Studies

The first experiments to be carried out usually are studies of the kinetics of radioligand binding to a receptor. This includes both association rate and dissociation rate studies:

1. Association rate studies provide several pieces of information. Radioligand binding should reach a steady state, as predicted by the law of mass action. The time to reach steady state is the minimum incubation time necessary for the rest of the studies. If association rate studies are extended for a long period of time, they can indicate whether there is degradation of the receptor or the ligand with extensive incubation, as demonstrated by a decrease in binding from the maximum level. If this occurs, it places an outer limit on the amount of time allowed for the assay.

Association rate studies demonstrate the changing levels of radioligand binding to a receptor with time (solid line, Fig. 26.1). A fixed concentration of radioligand, preferably near the K_d value of the ligand if it is known, is incubated with tissue at a chosen temperature, and the assay is terminated at multiple time points. Initially times between 0 and 30 min can be used and duration adjusted if necessary. Studies should be carried out for a sufficient time to clearly demonstrate binding has reached equilibrium or a steady-state level. Establishing the time to reach steady state is critical because the equations used to analyze saturation and competition studies assume steady-state conditions apply.

Analysis of association rate studies provides the association rate for the radioligand, k_1. The association rate for a ligand is concentration dependent and this must be taken into account in subsequent studies using multiple concentrations (saturation studies) or a different concentration of ligand. In some instances it may be necessary to adjust the time of incubation to assure steady state is reached at all concentrations, although this usually is not a major factor.

2. Dissociation rate studies provide two useful parameters. First, from a practical point they indicate how rapidly a radioligand dissociates from its receptor. This indicates whether standard techniques of terminating an assay, such as filtration, are useful or whether, in the case of ligands that dissociate extremely rapidly (seconds), other measures must be taken to minimize radioligand dissociation. These other measures include centrifugation, which keeps receptor and radioligand in contact throughout the termination step, and handling each sample individually and very quickly. The latter approach is frequently needed to measure glutamate binding to its receptors, for example. Dissociation rate studies can also demonstrate when a ligand is irreversibly bound to a receptor.

   Dissociation rate studies begin with incubation of tissue and radioligand to steady-state conditions, established in association rate studies. Dissociation is usually initiated by addition to each sample of a very large concentration of unlabeled ligand, more than 100-fold greater than the concentration of radioligand, maintaining all other conditions constant. The binding study is then terminated at multiple time points after addition of the unlabeled ligand and the level of radioligand binding is measured. With increasing time, there should be a decrease in the amount of radioligand bound (dashed line, Fig. 26.1). An alternative approach is to dilute each sample in a very large ("infinite") volume of incubation buffer so that the concentration of the radioligand is reduced 100-fold or more. This approach is usually difficult from a physical perspective and is not often used.

   Analysis of dissociation rate data provides the dissociation rate for a radioligand, k_{-1} or k_2. Dissociation rate is concentration independent. The two kinetic rates, k_1 and k_{-1}, provide a measure of radioligand affinity for its receptor: $k_{-1}/k_1 = K_d$, the dissociation constant. This value should be consistent with the radioligand affinity (K_d) determined with saturation studies.

In summary, kinetic studies determine the time necessary for radioligand binding to reach steady state under a particular set of conditions, and so the minimum incubation time for an assay. They indicate how rapidly a ligand dissociates from its receptor, information necessary to determine the optimum approach to terminating the assay. Finally, the two parameters determined from the assay, k_1 (association rate) and k_{-1} (dissociation rate; also k_2) provide a measure of radioligand affinity for the receptor ($k_{-1}/k_1 = K_d$). See Yamamura et al. (1985) for detailed methods of determining K_d from k_1 and k_{-1}.

26.3.2 Saturation Studies

Saturation studies establish three important factors for specific radioligand binding. First, they provide proof that specific binding of the radioligand to the site of interest is, in fact, saturable and therefore involves a site that is finite in quantity, a characteristic of all biological molecules. Second, saturation studies provide an additional method of determining the affinity, K_d, of a radioligand for its binding site or receptor. Third, saturation studies can help us determine if a radioligand, under a particular set of conditions, is binding to one or more receptors.

A saturation study consists of incubating tissue in the presence of increasing concentrations of radioligand until steady-state conditions are reached. For each sample a corresponding blank should be included in the assay. The final protocol for a specific saturation study should include radioligand concentrations ranging from at least tenfold below to tenfold above the K_d value of the radioligand. The concentration range needed for a study is determined by preliminary studies. A rule of thumb is it requires at least 6, preferably 8, radioligand concentrations to accurately determine the K_d value of a radioligand for a single site, and 12–18 concentrations to determine K_d values for two sites. These concentrations should be equally distributed above and below the K_d value(s) as much as possible.

In situations in which it is not practical to use a radioligand at tenfold greater than its K_d value, alternate approaches can be used. This is frequently the case when using radioiodinated ligands because using radioligand at a concentration tenfold greater than its K_d value may be exorbitantly

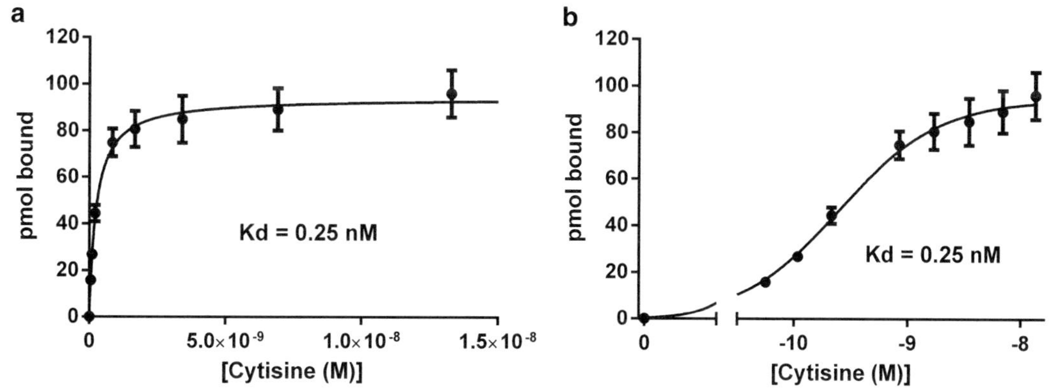

Fig. 26.2 Saturation study data [^{3}H]cytisine. Data are adapted from Happe et al. (1994). Both figures present the same data using a linear *x* axis (**a**) and a logarithmic *x* axis (**b**). Data were analyzed by nonlinear regression analysis

expensive and/or involve the use of unacceptably high levels of radioactivity. In these cases the radioligand can be diluted with unlabeled ligand, reducing specific activity. This allows use of higher concentrations of radioligand in the assay at reduced, but still useful, levels. Another option is to use a fixed concentration of radioligand and use unlabeled ligand at varying concentrations so the final result is a range of ligand concentrations at varying specific activities. This is, in effect, a competition study. These two approaches assume the labeled and unlabeled ligands have identical properties at the receptor. Finally, it is possible to combine these two approaches, as described elsewhere (Bylund and Murrin 2000).

Graphical approaches have been developed to analyze saturation data (Scatchard 1949; Rosenthal 1967). However, nonlinear regression analysis of the data is the most accurate way to determine the affinity (K_d) of a ligand for its receptor and the number of binding sites (B_{max}) for a radioligand. This is most readily accomplished through the use of programs developed for this purpose, as detailed below.

Presentation of saturation data usually employs a sigmoid dose–response curve with a logarithmic *x* axis. As shown in Fig. 26.2a, data can be plotted on a linear scale. This approach provides a clear visual indication that the binding is saturable and gives a good measure of the B_{max}, but it is difficult to visually interpolate the K_d value (50 % of B_{max}) with this graph. Use of a sigmoid dose–response curve (Fig. 26.2b) makes interpolation of the K_d value relatively easy, and it also provides a good indication of the B_{max}. Both of these graphs plot curves generated via nonlinear regression analysis.

26.3.3 Pharmacology

The final major parameter of radioligand binding to be examined is its pharmacology. These studies rely upon the availability of a number of drugs different from the radioligand and with known specificity for the receptor in question. Such an array of drugs may not be available when first studying a new receptor.

Pharmacological analyses, or competition/inhibition assays, provide several types of information. They confirm that the specific binding measured represents binding to a particular receptor. The rank order of the competing ligands in inhibiting the radioligand binding should correspond to their known affinities for the receptor in question. This also provides further support for the specificity of the radioligand binding. Once it is established that a radioligand binds to a particular receptor under a specific set of conditions, inhibition or competition assays can be used to screen an array of drugs for their ability to bind to that receptor. This provides the opposite type of information, the affinity of a

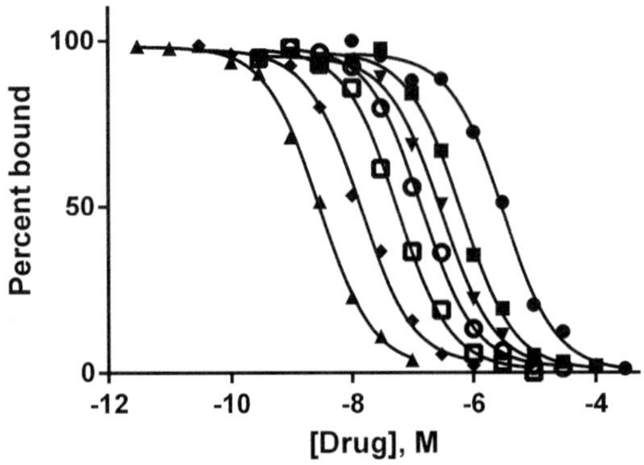

Fig. 26.3 Example of competition curves for a series of drugs competing against a radioligand at one binding site. The *curves* represent drugs with affinities (K_i values) ranging from 3 nM (farthest *left, filled upright triangles*) to 3 μM (farthest *right, filled circles*)

Table 26.1 The affinity (K_d) and the concentration (B_{max}) of dopamine transporter in rat striatum using [^{125}I]RTI-55 as ligand

Tissue preparation	K_d (nM)	B_{max} (pmol/mg tissue)
Membranes	0.79 ± 0.05	1.16 ± 0.19
Sections	14.9 ± 0.9	0.96 ± 0.11

There is a significant decrease in measured affinity (increase in K_d) in tissue sections compared to homogenates. The B_{max} values are similar, indicating the same site was labeled in both preparations. All assay conditions (buffer, ligand concentration, temperature) were maintained the same in both assays

Table 26.2 Differences in K_i for drugs inhibiting the binding of [^{125}I]RTI-55 in tissue sections and homogenates

Drug	Sections	Membranes
	K_i (nM)	K_i (nM)
GBR12909	35	1
Mazindol	237	10
Win 35,428	320	45
Cocaine	3070	311

There were significant differences in K_i values when comparing these two tissue preparations. In each set of studies, the same buffers and drug concentrations were used for sections and membranes

panel of drugs for a particular receptor, and is the basis for screening drugs for activity at a receptor.

Pharmacological analysis uses competition assays in which a fixed concentration of radioligand is present in each sample and varying concentrations of unlabeled drug are used to compete with the radioligand for its binding to a receptor. With increasing concentrations of the unlabeled competing drug, less of the radioligand will bind, defining a competition curve. The rules of thumb about the number of ligand concentrations and the method of analysis parallel those for saturation studies. Figure 26.3 is an example of competition curves.

These studies determine the IC_{50} values for the drugs, i.e., the concentration of drug inhibiting 50 % of radioligand binding at a fixed radioligand concentration. The IC_{50} values depend on the concentration of the radioligand used in a particular assay and so will change if radioligand concentration varies. The IC_{50} values can be converted to K_i values using the Cheng-Prusoff equation (Cheng and Prusoff 1973). The K_i value defines the actual affinity of the unlabeled drug for the receptor in question and should be consistent between differing labs.

It is worth emphasizing again the need to reestablish specific binding criteria when assay conditions are changed. In the examples used above, the affinity and pharmacology of [^{3}H] cytisine binding to α4*β2* nicotinic receptors in membranes and in tissue sections are essentially identical (see also Happe et al. 1994). On the other hand, binding of [^{125}I]RTI-55 to the dopamine transporter in homogenates and tissue sections demonstrated important differences in affinity of the radioligand and competing unlabeled ligands (Tables 26.1, and 26.2). If binding parameters have changed and this is not detected, interpretation of the data could be misleading.

26.3.4 Appropriate Localization

If a radioligand is thought to bind to a particular receptor, it follows that specific binding should be present in tissues expressing that receptor in proportion to the level of receptor expression.

Also, there should be no specific binding in tissues that do not express the receptor. Appropriate localization is another parameter to confirm the specificity and identity of the radioligand binding being studied. This is relatively easy to demonstrate in cultured cells or large segments of tissue, e.g., heart versus lung or cerebral cortex versus striatum. In studies of the CNS, the general rule is that binding to a particular receptor should parallel the presence of the neurotransmitter for that receptor. This is generally the case, although there are numerous instances of very weak parallels (Herkenham and McLean 1986).

26.4 Analysis of Receptor Binding Data

The equations derived for the analysis of radioligand binding data are based on the law of mass action. The mathematical basis for analysis of receptor binding data has been explained in detail elsewhere (Kenakin 1997; e.g., see Haylett 2003) and is beyond the scope of this chapter. The interaction of a ligand with its receptor is not linear with respect to time or ligand (and receptor) concentration, and nonlinear regression analysis is the most accurate method for binding analyses. Because of the difficulty of using these approaches by hand, methods were derived to transform the data so that a linear graph is produced, allowing easier analysis. Transformations have the inherent problem of introducing distortions into the data and it is sometimes difficult to compensate for these distortions through experimental design. Nevertheless, these graphic approaches do provide useful information in certain cases and they are easy to comprehend visually. Fortunately, programs have been developed to specifically analyze receptor binding data, beginning with LIGAND (Munson and Rodbard 1980) and developing into very sophisticated and easy to use graphic programs, such as Prism (GraphPad Software, San Diego, CA) and SigmaPlot (Systat Software, San Jose, CA). These programs provide detailed analysis of radioligand binding kinetics, saturation studies, and detailed pharmacological/competition studies. They include analyses to provide a statistical basis for conclusions on whether the radioligand binding involves one or more binding sites. For more detailed information about nonlinear regression and statistical analysis, the reader is referred to the sources mentioned above or to detailed online information (http://www.graphpad.com/Support/support.cfm).

26.5 Quantitative Autoradiography and Radioligand Binding

Radiation due to radioactive materials will produce an exposure pattern on photographic film when closely apposed, producing a radiograph. When a tissue or other specimen is labeled with a radioactive substance and is apposed to film, it produces an image on the film that reflects the distribution of the radioactive material within the tissue, an autoradiograph.

Quantitative autoradiography applies the techniques and advantages of radioligand binding to anatomy, allowing detailed quantitative analysis of receptors and other radioligand binding sites at the light microscopic and electron microscopic levels. The range over which receptors can be quantified in tissue sections is much greater with quantitative autoradiography than with other visualization techniques, a major advantage. This has proven to be extremely useful in studying neurotransmitter and hormonal systems and their receptors in the central nervous system. It has also been used in the study of other organs with a well-defined heterogeneous structure. Extensive discussions of this technique are available (Rogers 1979; Boast et al. 1986). Electron microscopic studies have significant technical differences compared to light microscopic studies, and the reader is referred to the review by Rogers for further details.

Quantitative autoradiography has a number of significant advantages compared to homogenate or cell binding assays. It is spatially accurate and

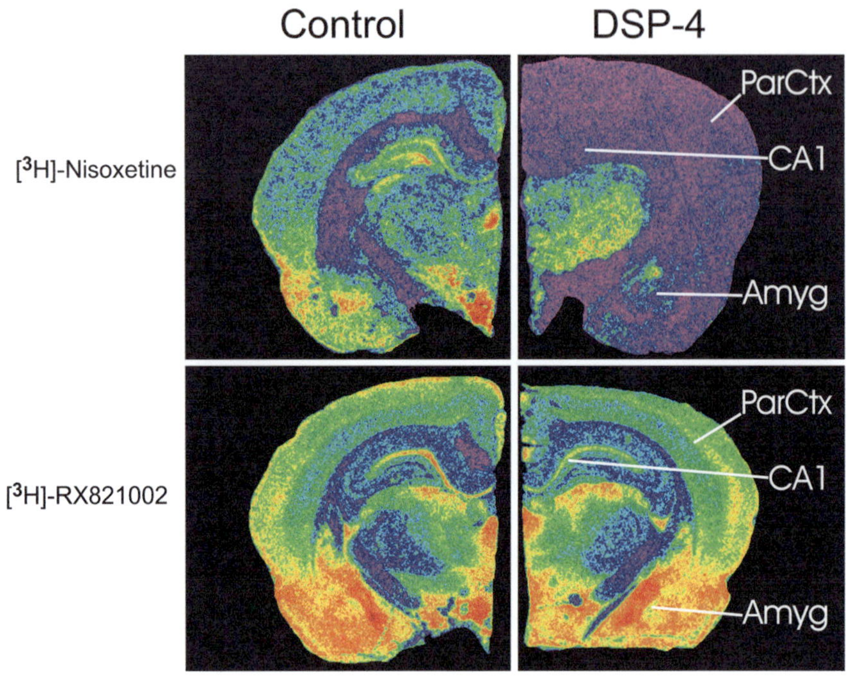

Fig. 26.4 Example of an autoradiograph, which illustrates the detail of information available through this procedure. The left side presents data from control rats while the right side presents data from rats administered DSP-4, a toxin that specifically lesions noradrenergic neurons originating in the locus coeruleus. The top figures present autoradiographs of the norepinephrine transporter (NET—labeled with [^{3}H]nisoxetine), and the bottom figures present autoradiographs of alpha-2 adrenergic receptors (A2AR—labeled with [^{3}H]RX821002). DSP-4 lesions noradrenergic innervation in many brain regions, hence eliminating NET, while producing a small increase in postsynaptic A2AR (unpublished image; see Sanders et al. 2011). Note the detailed information available in very small brain regions. *Amyg* amygdala, *CA1* CA1 region of hippocampus, *ParCtx* parietal cortex

allows quantitative analysis of receptors in very small tissue regions, which would be impractical or impossible to isolate with other approaches. Because tissue sections are very thin, multiple sections can be obtained from the same region, even in small brains such as mouse brain. This, in turn, allows more detailed analysis of a particular receptor or analysis of several receptors in the same region of the same animal (Fig. 26.4). Related to this, studies can be carried out on very small samples, such as those that may be available in human studies. This is of particular value when the tissue in question is limited in availability. The use of thin sections also reduces the number of samples or animals necessary to repeat a study several times for statistical validity. Finally, there is little disruption and loss of receptors in tissue sections, as may be found in homogenates (Norman et al. 1989).

The disadvantages of quantitative autoradiography compared to membrane or cell culture approaches include requiring more time from the beginning to the end of an experiment, i.e., days to weeks instead of hours to days. Specific to nonspecific binding ratios sometimes are lower than in homogenate assays, although there are cases where the opposite is true. Also, it is not possible to increase the quantity of tissue in an assay to facilitate the detection of receptors present in very low concentrations. Finally, when using tritiated ligands and comparing receptor density from region to region, it is necessary to consider quench correction (Geary and Wooten 1985; Happe and Murrin 1990).

The fundamentals of radioligand binding are the same when used for quantitative autoradiography as for studies using membranes or cells. The factors to consider and the criteria for specific binding are the same with a few additional caveats.

The ligand used in autoradiography is nearly always first identified and characterized using membrane binding studies, primarily because the latter are much quicker experiments. The isotopes with which the ligand is radiolabeled are also similar, i.e., tritium and ^{125}I account for the vast majority. In autoradiographic studies, however, there is a much greater difference between these two. While the time required to quantify samples using these two isotopes is similar for membrane binding studies, a great deal more time is required to produce an image on film with tritium compared to ^{125}I. Whereas an iodinated ligand may produce useful film exposure in 3–5 days, a tritiated ligand may require 2–4 weeks, and in some cases longer. The time required depends, of course, on the density of the receptors in the tissue. If brain regions with widely varying receptor densities are to be examined, autoradiographic studies may require two or more exposures of different lengths because the film used (e.g., Kodak BioMax MS) has a linear response to radioactivity over a limited range.

Quantitative autoradiography uses fresh frozen tissue sections mounted on microscope slides (Young and Kuhar 1979). Tissue is sectioned in a cryostat, thaw-mounted onto slides, and maintained in the frozen state (–20 °C or lower) until use. Receptors will begin to deteriorate if subjected to multiple freeze-thaw cycles. While tissue sections can be maintained in the frozen state for weeks or months without noticeable changes in binding for most receptors, there are instances in which the receptor begins to deteriorate relatively quickly. This was found to be the case for the norepinephrine transporter (Sanders et al. 2005). Fixation of tissue should be avoided because normal fixatives destroy the ability of receptors to bind ligands. While very light fixation may have no significant effect on binding in a few cases, this varies from receptor to receptor and is best avoided. Section thickness varies from lab to lab but generally falls in the range of 6–20 μm. Incubation conditions must be established for the specific thickness in use, and tissue thickness should be kept constant throughout a related series of studies.

Incubation conditions must be optimized for autoradiographic studies. While these conditions are frequently similar to conditions for membrane binding studies, this is not always the case. In particular, since a tissue section is a solid, albeit thin, piece of tissue, it is physically quite different from a preparation that is dispersed throughout the incubation medium. Frequently more time is needed to reach steady-state binding in tissue sections compared to homogenates. Dissociation rates may also be slower. Finally, while saturation and competition assays used to measure drug affinity for receptors usually produce results similar to homogenate binding assays, there can be unexpected differences (Tables 26.1 and 26.2). These possibilities again emphasize the need to establish optimum binding conditions when a new preparation is used.

Labeling of tissue sections follows the same procedures as membrane and cell binding assays, i.e., incubation to steady state followed by washing with buffer at 4 °C. Sections on slides are usually washed by dipping in a large volume of ice-cold buffer (50 mL+) at least three times. An important addition is a final brief rinse in ice-cold H_2O to remove buffer salts. Otherwise the salts will collect on and, in some cases, interact with the photographic film, making analysis difficult or impossible. Sections are then allowed to dry, usually under a stream of cool air or overnight.

It is necessary to include well-calibrated standards with each film because the process of film developing is not exact and may introduce differences in the signal (density of grains) from film to film. Commercial standards are available (American Radiolabeled Chemicals, St. Louis, MO) or they can be made by hand (Miller and Zahniser 1987). Standards for tritium or ^{14}C can be calibrated for use with short-lived isotopes (^{35}S, ^{125}I), avoiding the need to purchase or make new standards every few months (Miller and Zahniser 1987).

Analysis of autoradiographs requires use of an image analysis system (e.g., MCID, Linton, Cambridge, England or ImageJ, http://rsbweb.nih.gov/ij/download.html). These are computer based and require a well-regulated camera system to capture and analyze the image. See Chapters 1 and 2 in Boast et al. (1985). A useful site with further resources can be found at www.nitrc.org.

Quantitative autoradiography can also be used to measure receptor function for G-protein-coupled receptors. The techniques for this approach are similar in many ways to those described above. Discussion of these approaches is beyond the scope of this chapter, but further information is readily available (Sim et al. 1995, 1997; Happe et al. 2001).

26.6 Summary

The use of radioligand binding studies to analyze receptors (and other targets) is a very useful experimental technique. It allows quantification and characterization of receptors over a wide range of receptor densities and with greater precision than is possible with other approaches. The use of the autoradiographic approach allows the anatomy of a system to be included in a study, enabling receptor analysis in very small tissue regions. This has proven to be of particular value in the study of the central nervous system, allowing correlation of the function of particular brain regions with specific neurotransmitter systems and providing new insights into how known or future drugs may alter these functions. Radioligand binding is a powerful technique enabling detailed analyses of receptors and receptor changes due to drug treatments, diseases, or other differences in biological conditions.

References

Boast CA, Snowhill EW, Altar CA (1985) Quantitative receptor autoradiography. In: Neurology and neurobiology, vol. 19. Alan R. Liss, New York

Bylund DB, Murrin LC (2000) Radioligand saturation binding experiments over large concentration ranges. Life Sci 67:2897–2911

Cheng YC, Prusoff WH (1973) Relationship between the inhibition constant (K_i) and the concentration of inhibitor which causes 50 percent inhibition (I_{50}) of an enzymatic reaction. Biochem Pharmacol 22: 3099–3108

Deupree JD, Hinton KA, Cerutis DR, Bylund DB (1996) Buffers differentially alter the binding of [^{3}H]rauwolscine and [^{3}H]RX821002 to *alpha-2* adrenergic receptor subtypes. J Pharmacol Exp Ther 278:1215–1227

Geary WA, Wooten GF (1985) Regional tritium quenching in quantitative autoradiography of the central nervous system. Brain Res 336:334–336

Gonsiorek W, Hesk D, Chen SC, Kinsley D, Fine JS, Jackson JV, Bober LA, Deno G, Bian H, Fossetta J, Lunn CA, Kozlowski JA, Lavey B, Piwinski J, Narula SK, Lundell DJ, Hipkin RW (2006) Characterization of peripheral human cannabinoid receptor (hCB2) expression and pharmacology using a novel radioligand, [^{35}S]Sch225336. J Biol Chem 281:28143–28151

Happe HK, Bylund DB, Murrin LC (2001) Agonist-stimulated [^{35}S]GTPγS autoradiography: optimization for high sensitivity. Eur J Pharmacol 422:1–13

Happe HK, Murrin LC (1990) Tritium quench in autoradiography during postnatal development of rat forebrain. Brain Res 525:28–35

Happe HK, Murrin LC (1993) High-affinity choline transport sites: use of [^{3}H]hemicholinium-3 as a quantitative marker. J Neurochem 60:1191–1201

Happe HK, Peters JL, Bergman DA, Murrin LC (1994) Localization of nicotinic cholinergic receptors in rat brain: autoradiographic studies with [^{3}H]cytisine. Neuroscience 62:929–944

Haylett DG (2003) Direct measurement of drug binding to receptors. In: Foreman JC, Johansen T (eds) Textbook of receptor pharmacology. CRC, Boca Raton, pp 153–180

Herkenham M, McLean S (1986) Mismatches between receptor and transmitter localizations in the brain. In: Boast CA (ed) Quantitative receptor autoradiography. Alan R. Liss, New York, pp 137–171

Kenakin T (1997) Molecular pharmacology. A short course. Blackwell Science, Cambridge

Laitinen JT (2003) [35S]GTPγS autoradiography: a powerful functional approach with expanding potential for neuropharmacological studies on receptors coupled to Gi family of G proteins. Curr Neuropharmacol 2:191–206

Miller JA, Zahniser NR (1987) The use of ^{14}C-labeled tissue paste standards for the calibration of ^{125}I-labeled ligands in quantitative autoradiography. Neurosci Lett 81:345–350

Munson PJ, Rodbard D (1980) LIGAND: a versatile computerized approach for characterization of ligand-binding systems. Anal Biochem 107:220–239

Norman AB, Borchers MT, Wachendorf TJ, Price AL, Sanberg PR (1989) Loss of D_1 and D_2 dopamine receptors and muscarinic cholinergic receptors in rat brain following in vitro polytron homogenization. Brain Res Bull 22:633–636

Pert CB, Snyder SH (1974) Opiate receptor binding of agonists and antagonists affected differentially by sodium. Mol Pharmacol 10:868–879

Rogers AW (1979) Techniques of autoradiography. Elsevier, Amsterdam

Rosenthal HE (1967) A graphic method for the determination and presentation of binding parameters in a complex system. Anal Biochem 20:525–532

Sanders JD, Happe HK, Bylund DB, Murrin LC (2005) Development of the norepinephrine transporter in the rat CNS. Neuroscience 130:107–117

Sanders JD, Happe HK, Bylund DB, Murrin LC (2011) Changes in postnatal norepinephrine alter alpha-2 adrenergic receptor development. Neuroscience 192: 761–772

Scatchard G (1949) The attractions of proteins for small molecules and ions. Ann N Y Acad Sci 51:660–672

Sim LJ, Selley DE, Childers SR (1995) *In vitro* autoradiography of receptor-activated G proteins in rat brain by agonist-stimulated guanylyl 5'-[γau^{35}S] thio]-triphosphate binding. Proc Natl Acad Sci U S A 92:7242–7246

Sim LJ, Selley DE, Childers SR (1997) Autoradiographic visualization in brain of receptor-G protein coupling using [^{35}S]GTPγS binding. Methods Mol Biol 83:117–132

Yamamura HI, Enna SJ, Kuhar MJ (1985) Neurotransmitter receptor binding. Raven, New York

Young WS, Kuhar MJ (1979) A new method for receptor autoradiography: [^{3}H]opioid receptors in rat brain. Brain Res 179:255–270

Techniques in Immunohistochemistry and Immunocytochemistry

27

Yang Yuan and Jyothi Arikkath

Abstract

Immunohistochemistry and immunocytochemistry are emerging as powerful tools to supplement the knowledge obtained by routine histological techniques. These have been significantly aided by technical advances in microscopy. The combination of immunolabeling techniques and microscopy now allows detailed insights into tissue and cellular architecture and is proving to be extremely valuable for understanding the micro architecture of the brain in health and disease. In this chapter, we discuss some of the considerations and techniques for immunohistochemistry of the brain and dissociated, cultured neurons from the hippocampus and cortex.

Keywords

Immunolabeling • Immunocytochemistry • Immunohistochemistry • Antibody • Fluorescence microscopy

27.1 Introduction

A key step forward in understanding the function of the brain is our ability to visualize its tissue structure and molecular architecture with a high level of precision. It is clear that the three-dimensional organization of tissue and molecular components of tissue is vital to normal physiology and function and numerous alterations are observed in neurodegenerative, neuropsychiatric, and neurodevelopmental disorders (Penzes et al. 2011; Emoto 2011). Immunohistochemistry techniques, combined with the recent advances in microscopy, offer the ability to precisely localize molecules within the larger framework of the tissue. Moreover, recent technical advances enable the precise quantitation of these molecules and offer unprecedented views of the molecular and cellular architecture of the brain (Taraska and Zagotta 2010; Rust et al. 2006; Ding et al. 2009; Micheva and Smith 2007). In this chapter, we discuss some of the basic techniques for immunofluorescent histochemistry and cytochemistry.

Y. Yuan • J. Arikkath (✉)
Developmental Neuroscience, Munroe-Meyer Institute, University of Nebraska Medical Center, Omaha, NE 68198-5960, USA
e-mail: yang.yuan@unmc.edu; Jyothi.arikkath@unmc.edu

H. Xiong and H.E. Gendelman (eds.), *Current Laboratory Methods in Neuroscience Research*, Springer Protocols Handbooks, DOI 10.1007/978-1-4614-8794-4_27,

27.2 Antibody Generation and Specificity

27.2.1 Antibodies

A key requirement for immunohistochemistry or immunocytochemistry is the availability of antibodies that specifically react with the antigen of interest. Antibodies (known as immunoglobulin or Ig) have a "Y"-shaped basic structure including four polypeptides—two heavy chains (50 kD) and two light chains (23 kD) (Fig. 27.1). The heavy and light chains are held together by disulfide bonds and noncovalent interactions. The tips of "Y" (Fab) are antigen binding sites which give antibody specificity for each antigen. The Fc region allows the antibody to bind to cells or complement. The structure and function of the heavy chain determine the category into which the antibody is classified. Five major isotypes of antibodies are known, namely, IgG, IgE, IgM, IgA, and IgD.

IgG is the predominant immunoglobulin in the serum and extracellular fluid (about 75–80 %). It has a half-life of 20–23 days (human) and is produced by the secondary immune response. It is the only isotype that can pass through the human placental barriers and is key in immune protection in the neonatal anti-infection immune system. IgM represents about 5–10 % of the total Ig. It is the largest antibody which mainly exists in the blood. IgM is more likely to activate complement than IgG. Its synthesis and secretion is the earliest in the development of an individual. IgA mainly exists in the gastrointestinal tract and bronchial secretions, colostrum, saliva, and tears. It participates in local immune mucosa. IgD represents the lowest content of the total Ig (about 1 %). Its functional roles are less clear. IgE is secreted by the plasma cells of submucous lymphatic organization. It can cause type I hypersensitivity reaction.

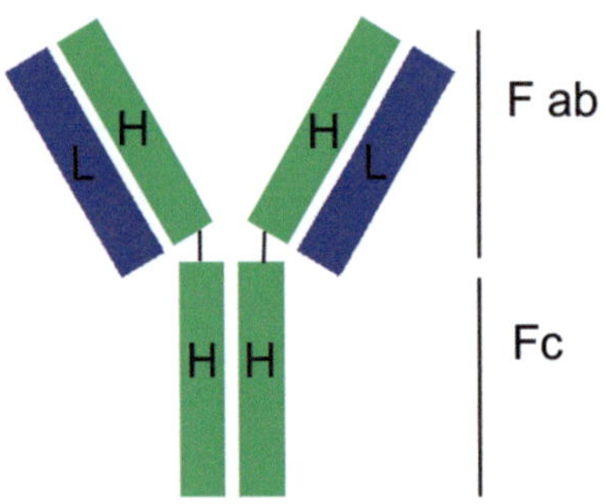

Fig. 27.1 Structure of an antibody. The antibody has a Y-shaped structure and includes two heavy (H) and 2 light (L) chains. The tips of the Y are the Fab regions, which allow antibody binding to antigen. The Fc regions allow the antibody to bind to receptors or complement molecules

27.2.2 Polyclonal and Monoclonal Antibodies

Polyclonal antibodies are antibodies that are directed against a mixture of epitopes of the antigen, while monoclonal antibodies are specific to a single epitope. When an antigen is injected into a host mammal such as a human, mouse, rat, rabbit, or horse, the antigen is broken down into a series of short peptides. Then, these peptides attach to major histocompatibility complex class II (MHC II) molecules on the surface of B cells. T cell can stimulate these B cell to turn into plasma cells and produce antibody. Each B cell only produces antibodies specific to a single epitope, and different B cells can produce antibodies directed to different epitopes on the same protein. Hence, the resulting antibody pool is polyclonal, i.e., it has antibodies to multiple epitopes of the same protein (Fig. 27.2).

The generation of monoclonal antibodies takes advantage of the ability of a single B cell to produce antibodies to a single epitope. To generate monoclonal antibodies specific for a single epitope of an antigen, single B cells are fused with a cancer cell line (e.g., myeloma) to generate hybridomas that are then cultured individually. Unfused B cells do not survive in the culture because they have a definite life span. Single hybridoma cells are isolated by the incorporation of hypoxanthine-aminopterin-thymidine (HAT) in the media that kill unfused cancer cell (e.g., myeloma cell). The hybridoma clones are then cultured individually (Fig. 27.3). When cell clones that all produce the same antibody are

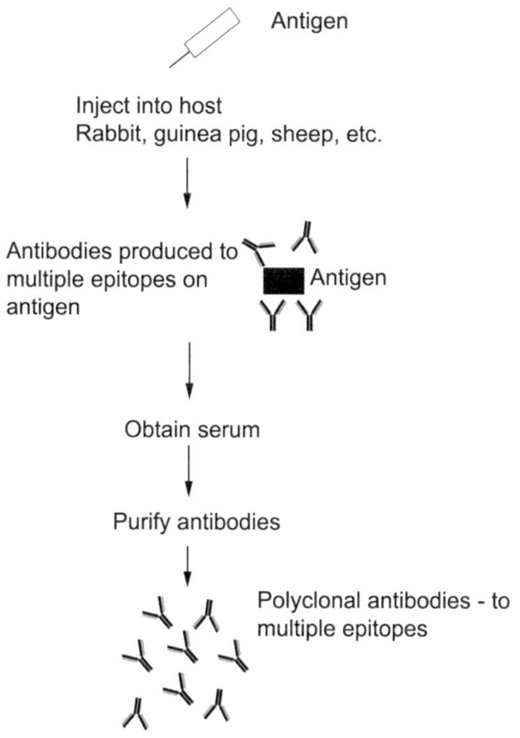

Fig. 27.2 Generation of polyclonal antibodies. The antigen of interest is injected into a host animal. The immune response generated by the host results in the generation of antibodies that recognize multiple epitopes on the antigen

generated, these antibodies are called monoclonal antibodies (Taraska and Zagotta 2010). Table 27.1 summarizes some characteristics of monoclonal and polyclonal antibodies.

27.2.3 Specificity of Antibodies

Specificity is the ability of an individual antibody or a population of antibody molecules to react with only one epitope or a group of epitopes within a single protein or antigen. Antibody specificity is an extremely important consideration, since it can drastically change the interpretation of the experiment. The best way to assess the specificity of an antibody is to test it on tissue from a knockout animal or cells expressing an siRNA or shRNA to the protein of interest.

27.3 Immunohistochemistry

Immunohistochemistry is a technique that takes advantage of the availability of antibodies specific to certain proteins and allows the detection of these proteins in native tissue by fluorescent or other labeling and microscopy. In the following sections, we describe the general protocol for immunohistochemistry using fluorescent probes.

27.3.1 Procedures for Immunohistochemistry

Immunofluorescent histochemistry can be achieved by the following sequence of steps (Fig. 27.4). The tissue is fixed and embedded. Next, the tissue is sectioned (typically 10–50 μm thick) using a cryostat, vibratome, or sliding vibratome. The resulting sections are stained with primary antibodies of interest followed by washing and incubation with fluorescently conjugated secondary antibodies. The primary antibody is the antibody that binds to the epitope of interest. The secondary antibody is an antibody conjugated to a fluorophore or enzyme, which can bind to the primary antibody and hence allows detection of the primary antibody. After further washing, the sections are ready to be mounted for microscopy (Fig. 27.4). The general methodology for staining adult mouse brain sections follows.

27.3.2 Immunohistochemistry on Cryostat Cut Sections

1. The mouse is deeply anesthetized with anesthetic (e.g., isoflurane) and transcardially perfused with 40 mL of prewarmed 1× PBS.
2. The brain is removed from the skull and put into 5 mL 4 % paraformaldehyde and incubated at 4 °C overnight. When the brain is removed at this point, it needs to look white. Any indication of blood suggests that the perfusion was not adequate.
3. The brain is put into 5 mL 30 % sucrose in 1× PBS at 4 °C overnight.

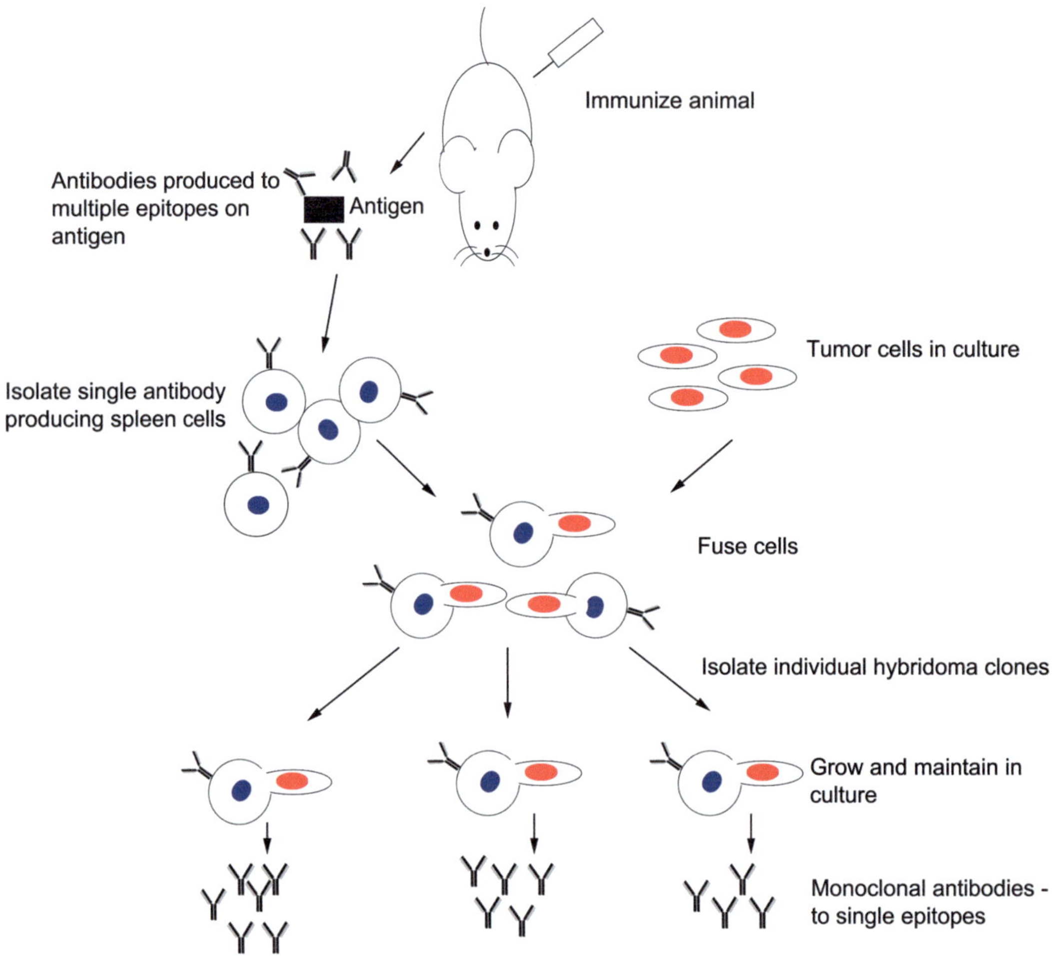

Fig. 27.3 Generation of monoclonal antibodies. The host cell is injected with the antigen. The animal generates antibodies to multiple epitopes on the antigen. Antibody-producing B cells from the animal are isolated and fused with tumor cells in culture. Individual monoclonal antibody-producing hybridoma clones are isolated and maintained in culture

4. The brain is placed in Tissue-Tek® OCT™ compound and rapidly frozen in liquid nitrogen. The frozen tissue can be wrapped in foil and stored in −80 °C until ready for sectioning.
5. On the day of experiment, the frozen brain is transferred to −20 °C for 2 h to allow the temperature of the frozen tissue block to equilibrate to the temperature of the cryostat, then cut in desired thickness (typically 10–50 μm) sections, and mounted on glass slides suitable for immunohistochemistry (e.g., Superfrost plus microscope slide or regular slides coated with VECTABOND™ Reagent which can enhance the adherence of tissue sections to glass slides).
6. Slides with mounted sections are placed in 37 °C heater for 2 h or overnight at room temperature.
7. Dried slides are placed in 1× PBS for 10 min at room temperature.
8. Antigen retrieval techniques can be performed at this point if necessary (see antigen retrieval protocols below).
9. The sections are incubated with blocking buffer (e.g., 5 % goat serum in 1× PBS) with 0.1 % triton X-100 for 1 h at room temperature.

Table 27.1 Some characteristics of monoclonal and polyclonal antibodies

	Monoclonal antibody	Polyclonal antibody
Homogeneity	High	Low
Specificity	High	Variable
Antibody production	Unlimited	Limited
Amounts of antibody generated	Unlimited quantity	Limited quantity
Affinity	High	Variable
Cross-reactivity to nonspecific epitopes	Low	Can be high
Time required to generate antibody	Long	Short
Cost to generate antibody	Very expensive	Less expensive

The choice of serum is dependent on the species of animal that the secondary antibody was raised in.

10. Sections are incubated in a humidified chamber with appropriately diluted primary antibody in blocking buffer (e.g., 5 % goat serum in 1× PBS) at 4 °C overnight in a humidified chamber. It is critical to ensure that the sections are well covered in antibody and do not dry out during the incubation.
11. Sections are washed three times at 5 min for each wash with 1× PBS at room temperature.
12. Sections are incubated in dark with 100 μL of an appropriately diluted fluorophore-conjugated secondary antibody in blocking buffer (e.g., 5 % goat serum in 1× PBS) for 1–2 h at room temperature. Secondary antibodies conjugated to various fluorophores are commercially available. At this point onwards, it is important to note that the staining is light sensitive and all manipulations should be performed in the dark.
13. Sections are washed three times at 5 min each in the dark with 1× PBS at room temperature.
14. After 10 s incubation in 100 μL 4,6-Diamidino-2-phenylindole dihydrochloride hydrate (DAPI) solution (1:1,000), sections are washed 30 s with 1× PBS. DAPI labels nuclei.
15. Slides are mounted in the dark with glass coverslips using mounting media (e.g., antifade). For prolonged storage, the coverslips are sealed to the slide with nail polish.
16. Stained slides are ready for microscopy. For long-term storage, the slides can be stored at −80 for several months.

27.3.3 Immunohistochemistry on Paraffin-Embedded Sections

Frozen sections, as described above, retain their antigens well. However, they have two limitations. One is that any ice crystals formed in the tissue during the freezing procedure can destroy tissue architecture. Two, technical limitations prevent the cutting of sections that are less than 10 μm thick. Embedding the tissue in paraffin overcomes both of these limitations. However, in a number of cases, the antigenicity of the epitope is compromised and antigen retrieval techniques may be required for visualizing the antigen of interest.

Protocol

1. The mouse is anesthetized and perfused as described above for frozen tissue.
2. Tissue is incubated in 4 % paraformaldehyde or methanol (100 % methanol to acetic acid = 3:1) at 4 °C overnight.
3. The brain is transferred to 70 % ethanol for several hours and then placed in Tissue-Tek processing system overnight.
4. The brain is embedded with paraffin and then stored at room temperature.
5. Sections are cut in desired thickness (typically 5–10 μm) using microtome and mounted on glass slides suitable for immunohistochemistry (e.g., Superfrost plus microscope slide). Cut sections can be stored and covered, for years at room temperature.
6. Slides with mounted sections are placed in 37 °C heater for 2 h or overnight at room temperature.
7. Slides are deparaffinized by xylene twice for 5 min each.
8. The slides are hydrated with 100 % ethanol twice for 5 min each, 95 % ethanol twice for 5 min each, and 70 % ethanol twice for 5 min each and then rinsed in distilled water.
9. The steps 8–15 described for frozen tissue are followed.

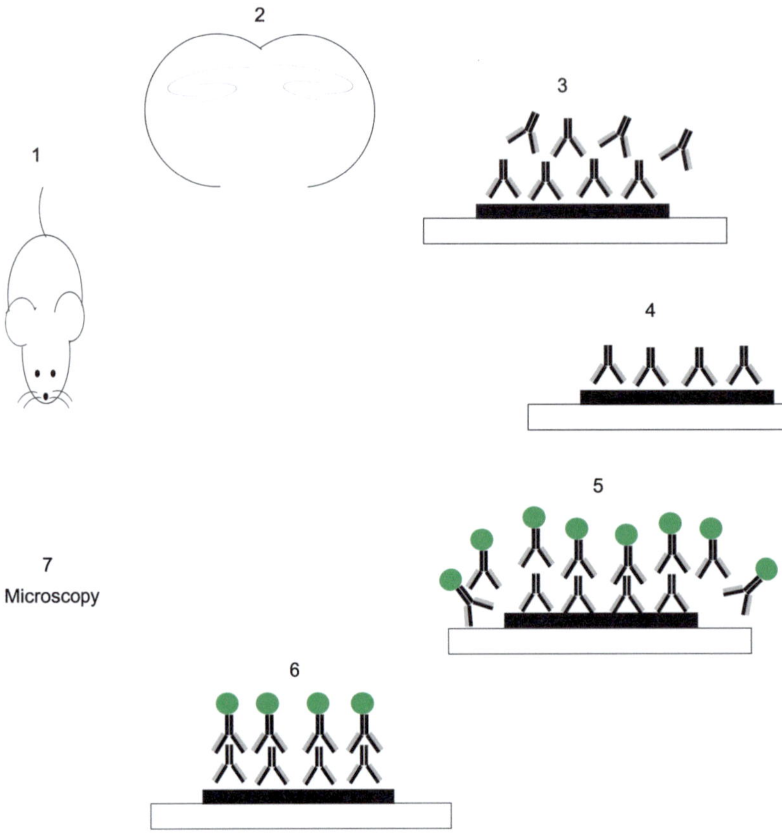

Fig. 27.4 Overview of immunohistochemistry protocol. (1) Perfusion of mouse. (2) Cut brain sections. (3) Incubate sections with primary antibody to protein of interest. (4) Wash to remove nonspecifically bound material. (5) Incubate with fluorescently conjugated secondary antibody. (6) Wash to remove nonspecifically bound material and mount. (7) Sections are ready for microscopy

27.3.4 Immunohistochemistry for Floating Sections

Floating sections can be obtained by cutting frozen tissue on a sliding microtome. These sections have the advantages of frozen sections and retain their antigenicity well. Moreover, because of the nature of the staining, thicker sections (40–50 μm) can be well stained. However, these sections are fragile and cumbersome to handle.

1. The mouse is anesthetized and perfused as described above.
2. The brain is incubated in 4 % paraformaldehyde overnight followed by overnight incubations in 15 % sucrose and 30 % sucrose. At this point, the brains can be stored in an antifreeze solution for several months at −20 °C if necessary. If stored in antifreeze, the brain must be reincubated in 30 % sucrose overnight before the next steps.
3. The brain is cut on a sliding microtome and free-floating sections are collected.
4. Steps 8–15 described for the frozen sections above are followed.
5. Using a fine brush, the sections are mounted on slides with mounting media and coverslips placed carefully. Once the slide is dry enough, the edges of the coverslip can be sealed with nail polish. The slides can be stored at −80 for several months.

The advantages and disadvantages of each type of sectioning technique are described in Table 27.2.

Table 27.2 Advantages and disadvantages of the three common sectioning techniques for immunohistochemistry

Tissue sectioning technique	Advantages	Disadvantages
Cryostat sections	Good quality and easy to handle sections	1. Can introduce freezing artifacts
		2. Inadequately perfused tissue can be difficult to cut
		3. Thick sections do not allow good penetration of antibody
Paraffin sections	Excellent morphology	1. Long and labor-intensive procedure
		2. Some antibodies that work well on cryostat and sliding microtome sections may require antigen retrieval
Sliding microtome	Allows good penetration of antibody	1. Sections difficult to handle
	OCT mounting not necessary	2. Washing steps are prone to damage tissue

27.3.5 Techniques for Signal Amplification

In some cases, the staining of the tissue may be quite weak. This may either be due to the properties of the primary antibody or the low concentration of the protein in native tissue. In either case, amplification of the signal may be advantageous. The reader is referred to several excellent reviews for these techniques (Wang et al. 1999; Ness et al. 2003).

27.3.6 Antigen Retrieval Techniques

When tissue is fixed by PFA, many amino acid residues form aldehyde bonds within or between molecules. This may result in the reduction or loss of antigenic determinants, leading to poor immunostaining. Antigen retrieval allows the recovery of antigenic determinants in the tissue.

One of most common techniques of antigen retrieval includes warming the sections in a sodium citrate buffer, followed by washing in PBS. This is followed by immunostaining as described above. Alternatively, sections can be treated with a proteolytic enzyme, such as proteinase K or trypsin, to reveal the antigen.

27.4 Immunocytochemistry on Cultured Hippocampal Neurons

This protocol is described for primary rat hippocampal and cortical neurons cultured according to standard protocols. Briefly, hippocampi or cortices are isolated from E18 rats, dissociated and plated on polylysine-coated coverslips. These can be maintained in culture for up to a few weeks. The reader is referred to excellent protocols for culturing these neurons (Kaech and Banker 2006).

27.4.1 Procedures

1. The coverslips with live neurons are placed in a prewarmed (37 °C) solution of 4 % paraformaldehyde and 4 % sucrose in PBS for 10 min.
2. The coverslips are washed twice in PBS for 5 min each.
3. The coverslips are treated with a solution of 0.1 % triton X-100 in 1× PBS for 10 min. This allows cell permeabilization and subsequent antibody uptake.
4. The coverslips are incubated in blocking solution of 5 % BSA in PBS for 1 h at room temperature.
5. Following this, the coverslips are incubated in primary antibody, washed, incubated in secondary antibody, and washed as described above for steps 8–14 for the frozen tissue immunohistochemistry.
6. Coverslips are mounted facedown on slides in antifade. Once they are slightly dry, the edges are sealed with nail polish. The mounted coverslips can be stored at –80 for several months.

27.4.2 Double Immunolabeling

Simultaneous labeling of two separate antigens on a single section to differentially detect both

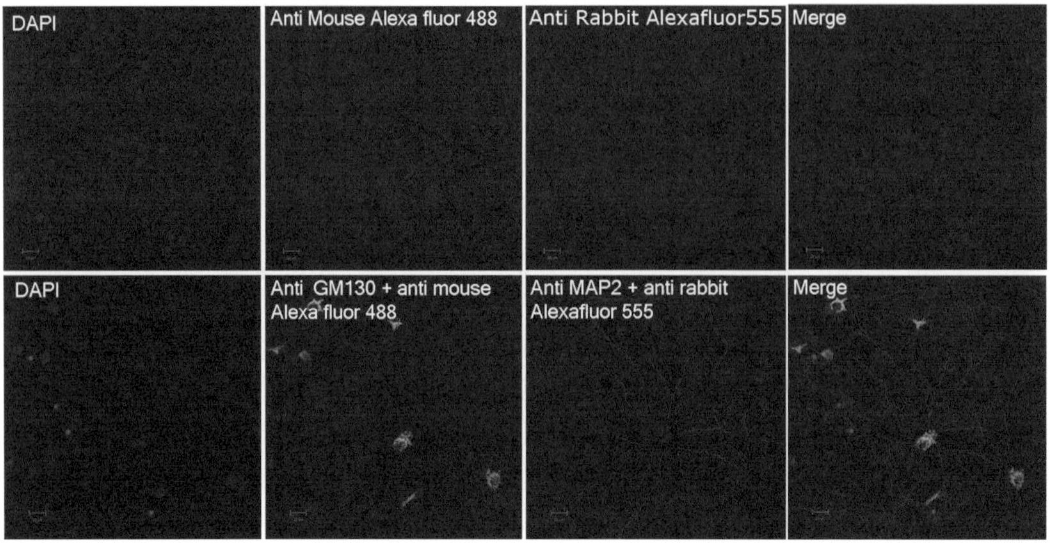

Fig. 27.5 Rat hippocampal neurons in culture stained with rabbit primary antibodies to MAP2, a dendritic marker, and anti-rabbit Alexa Fluor 555-conjugated secondary antibodies (red); mouse primary antibodies to GM130, a Golgi marker, and anti-mouse Alexa Fluor 488-conjugated secondary antibodies (*green*); and DAPI (*blue*) to visualize nuclei. Note that cells incubated with only fluorophore-conjugated secondary antibody (top) yield no signal, indicating specificity of staining

antigens can be successfully performed using the techniques outlined above. The requirements for the success of this technique are the availability of specific primary antibodies that do not cross-react with each other and the availability of fluorescently conjugated secondary antibodies tagged with different fluorophores whose emission spectra can be clearly distinguished.

27.4.3 Microscopy

Once the sections or coverslips with neurons are stained, they are amenable to imaging by fluorescence microscopy or confocal microscopy. Figure 27.5 shows an example of cultured rat hippocampal neurons (DIV 28) labeled with (1) rabbit primary antibodies to MAP2, a dendritic marker, and secondary antibodies conjugated to Alexa Fluor 555 (red); (2) mouse primary antibodies to GM130, a marker for the Golgi apparatus, and secondary antibodies conjugated to Alexa Fluor 488 (green); and (3) DAPI to visualize nuclei (blue) imaged with a point-scanning confocal microscope. Note that the staining of the tissue or cells with only fluorescently conjugated secondary antibodies (top panel) yields no signal, indicating that the observed staining is specific to the protein and antibody of interest. Figure 27.6 shows an example of mouse brain section stained with antibodies to MAP2 and GM130 and DAPI with secondary conjugated fluorophores as described for the cultured neurons. Figure 27.6a is a low magnification (10×) triple-stained image of the hippocampus. Figure 27.6b is a higher magnification of the hippocampus (40×). Figure 27.6c is a Z-stack composite of five images taken 0.96 μm apart.

27.5 Troubleshooting

The protocols for immunohistochemistry and immunocytochemistry are well standardized and have a high probability of success. It is important to titrate the primary and secondary antibodies to

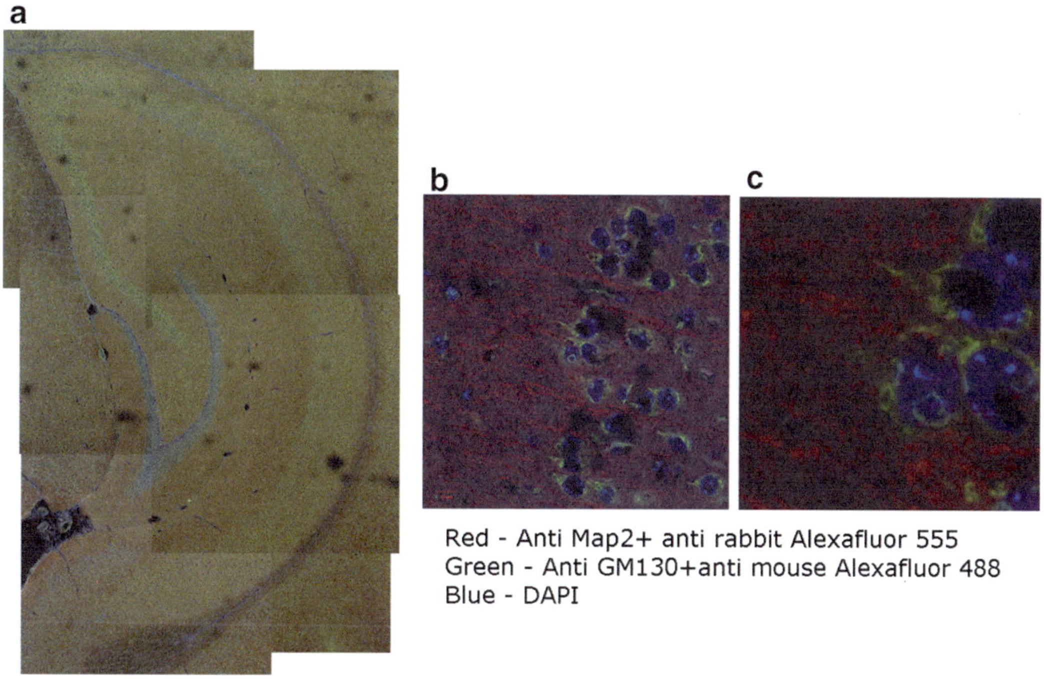

Fig. 27.6 Example of cryostat cut mouse brain sections stained with antibodies to MAP2 (*red*), a dendritic marker; GM130 (*green*), a Golgi marker; and DAPI (blue) to visualize nuclei stained as described for Fig. 27.5. (**a**) Low magnification image of the hippocampus (10×). (**b**) High magnification of a region of the hippocampus. (**c**) Z-stacked composite of five images taken 0.96 μm apart

ensure optimal staining. It is also important to stain tissues in parallel with secondary antibody only and no primary antibody to ensure that the observed staining is specific and not an artifact due to nonspecific binding of the secondary antibody. If successful staining is not observed, we recommend troubleshooting as described below.

Problem: No/Weak Staining

Possible reasons: (1) Primary and secondary antibodies do not match. (2) Primary or secondary antibody is not active. (3) Concentration of antibody (either primary or secondary) is too low. (4) Time of incubation is inadequate. (5) Section has inadequate deparaffinization (for paraffin cut sections). (6) Tissue overfixation.

Solutions: (1) Use appropriate secondary antibody. For example, if primary antibody is raised in mouse, the appropriate secondary antibody would be anti-mouse. (2) Use new batch of antibody. (3) Increase the concentration of antibody. (4) Increase the time of incubation. (5) Deparaffinize section longer or use fresh xylene. (6) Decrease time of postfixation. Use antigenic retrieval if tissue has been overfixed already.

Problem: Overstaining

Possible reasons: (1) The concentration of antibody (primary and/or secondary) is too high. (2) Time of incubation is too long. (3) Adequate washing steps not followed.

Solutions: (1) Decrease concentration of the antibody. (2) Incubate for shorter period of time. (3) Wash sections well with several volumes of buffer for each wash.

Problem: High Background

Possible reasons: (1) Antibody is not specific. (2) Inadequate fixation. (3) Inadequate washing.

Solutions: (1) Use specific antibody or higher concentration. (2) Increase the time of fixation. (3) Wash sections several times with adequate volumes of buffer.

27.6 Materials

Chemicals

1. Anesthetic (e.g., isoflurane)
2. 1× phosphate-buffered saline (PBS) at pH 7.4
3. 100 % xylene
4. 100 % ethanol, 95 % ethanol, 70 % ethanol
5. 4 % paraformaldehyde (PFA)
6. 100 % methanol
7. Acetic acid
8. 30 % sucrose
9. 5 and 1 % bovine serum albumin (BSA)
10. 5 % goat serum
11. Triton X-100
12. Tissue-Tek® OCT™
13. DAPI (4,6-Diamidino-2-phenylindole dihydrochloride hydrate)
14. Mounting media—ProLong® Gold antifade reagent
15. Primary antibodies
16. Secondary antibodies conjugated to appropriate fluorophore

Equipment

1. Superfrost plus microscope slides
2. Cryostat/sliding microtome/rotary microtome
3. Glass coverslips
4. Forceps
5. Scissor
6. 60 cm^3 syringe
7. Extension tubing with lure lock ends
8. Stopcocks
9. Butterfly needles
10. Fluorescence microscope
11. Tissue-Tek® VIP® 5 Vacuum Infiltration Processor
12. Tissue-Tek® TEC™ 5 Tissue Embedding Console System
13. Microwave
14. Incubator

References

Ding JB, Takasaki KT, Sabatini BL (2009) Supraresolution imaging in brain slices using stimulated-emission depletion two-photon laser scanning microscopy. Neuron 63(4):429–437

Emoto K (2011) Dendrite remodeling in development and disease. Dev Growth Differ 53(3):277–286

Kaech S, Banker G (2006) Culturing hippocampal neurons. Nat Protoc 1(5):2406–2415

Micheva KD, Smith SJ (2007) Array tomography: a new tool for imaging the molecular architecture and ultrastructure of neural circuits. Neuron 55(1):25–36

Ness JM, Akhtar RS, Latham CB, Roth KA (2003) Combined tyramide signal amplification and quantum dots for sensitive and photostable immunofluorescence detection. J Histochem Cytochem 51(8): 981–987

Penzes P, Cahill ME, Jones KA, VanLeeuwen JE, Woolfrey KM (2011) Dendritic spine pathology in neuropsychiatric disorders. Nat Neurosci 14(3): 285–293

Rust MJ, Bates M, Zhuang X (2006) Sub-diffraction-limit imaging by stochastic optical reconstruction microscopy (STORM). Nat Methods 3(10):793–795

Taraska JW, Zagotta WN (2010) Fluorescence applications in molecular neurobiology. Neuron 66(2): 170–189

Wang G, Achim CL, Hamilton RL, Wiley CA, Soontornniyomkij V (1999) Tyramide signal amplification method in multiple-label immunofluorescence confocal microscopy. Methods 18(4):459–464

Part IX

Analysis of Gene Expression

Protein Identification by Mass Spectrometry: Proteomics

28

Melinda Wojtkiewicz, Kelley Barnett, and Pawel Ciborowski

Abstract

Identification, characterization, and quantitation of changes of proteins induced by various types of external and/or internal stimuli are the major objectives of proteomics. Because proteins consist of 20 amino acids, their sequencing is more complicated than nucleic acids, which are a combination of four nucleotides. Proteins and peptides cannot be amplified and/or hybridized *in vitro* as nucleic acids. Moreover, proteins may have multiple posttranslationally modified sites and multiple types of modifications, such as phosphorylation, nitration, acetylation, and methylation, just to mention a few. These facts have a decisive impact on analytical methods used for protein sequencing and characterization. Before the development of soft ionization (the formation of ions without breaking any chemical bonds) techniques, proteins and peptides were sequenced using the Edman degradation method, based on the sequential removal of N-terminal amino acids (Edman 1970). This method, although still used (Mari et al. 2010), has many limitations such as an inability to sequence peptides that have the N-terminus blocked (Chen et al. 2008). Fragmentation of relatively short peptides (usually 6–30 amino acids in length) in gas phase in mass spectrometers exponentially increased the confidence with which we are able to determine peptide sequence and reveal other structural features, including posttranslational modifications (Biemann 1988; Strachunskii et al. 1992; Appella et al. 2000; Meng et al. 2005; Paizs and Suhai 2005). A number of specific proteolytic enzymes are available to fragment larger proteins into shorter peptides more suitable for high-quality analysis in tandem mass spectrometry (MS/MS) (see Fig. 28.3). The most commonly used enzyme, trypsin, specifically cleaves the peptide bond C-terminal to

M. Wojtkiewicz • K. Barnett • P. Ciborowski (✉)
Department of Pharmacology and Experimental Neuroscience, 985800 Nebraska Medical Center, University of Nebraska College of Medicine, Omaha, NE, 68198-5800, USA
e-mail: pciborowski@unmc.edu

H. Xiong and H.E. Gendelman (eds.), *Current Laboratory Methods in Neuroscience Research*, Springer Protocols Handbooks, DOI 10.1007/978-1-4614-8794-4_28,

arginine and lysine residues with any amino acid excluding proline (Thiede et al. 2005). The Arg/Lys–Pro peptide bond is cleaved at a very low rate; however, if such a peptide is well ionized, it still can be detected and sequenced by mass spectrometry (Pottiez et al. 2010)

Keywords
Neural stem cell • Neural progenitor cells • Neurosphere • Differentiation

28.1 Introduction

Identification, characterization, and quantitation of changes of proteins induced by various types of external and/or internal stimuli are the major objectives of proteomics. Because proteins consist of 20 amino acids, their sequencing is more complicated than nucleic acids, which are a combination of four nucleotides. Proteins and peptides cannot be amplified and/or hybridized in vitro as nucleic acids. Moreover, proteins may have multiple posttranslationally modified sites and multiple types of modifications, such as phosphorylation, nitration, acetylation, and methylation, just to mention a few. These facts have a decisive impact on analytical methods used for protein sequencing and characterization. Before the development of soft ionization (the formation of ions without breaking any chemical bonds) techniques, proteins and peptides were sequenced using the Edman degradation method, based on the sequential removal of N-terminal amino acids (Edman 1970). This method, although still used (Mari et al. 2010), has many limitations such as an inability to sequence peptides that have the N-terminus blocked (Chen et al. 2008). Fragmentation of relatively short peptides (usually 6–30 amino acids in length) in gas phase in mass spectrometers exponentially increased the confidence with which we are able to determine peptide sequence and reveal other structural features, including posttranslational modifications (Biemann 1988; Strachunskii et al. 1992; Appella et al. 2000; Meng et al. 2005; Paizs and Suhai 2005). A number of specific proteolytic enzymes are available to fragment larger proteins into shorter peptides more suitable for high-quality analysis in tandem mass spectrometry (MS/MS) (see Fig. 28.3). The most commonly used enzyme, trypsin, specifically cleaves the peptide bond C-terminal to arginine and lysine residues with any amino acid excluding proline (Thiede et al. 2005). The Arg/Lys–Pro peptide bond is cleaved at a very low rate; however, if such a peptide is well ionized, it still can be detected and sequenced by mass spectrometry (Pottiez et al. 2010).

28.2 Objective

The objective of this chapter is to provide a very concise overview of the measurement of proteins and peptides by mass spectrometry. The chapter begins with an introduction to two principle techniques of protein and peptide ionization, a summary of mass analyzers, and the principles of sample preparation. It is followed by a step-by-step description of a laboratory exercise to help students in understanding the basic concept of using mass analyzers for the measurement of proteins and peptides. We expect that after performing this basic mass spectrometry laboratory exercise, students will have a general overview of the methodology and, if interested, will broaden their knowledge. Literature citations provided in this chapter are just a beginning for more in-depth reading.

28.3 Ionization Techniques: ESI and MALDI

Peptide analysis by mass spectrometry requires that peptides are first ionized into the gas phase. The two most common methods of ionization

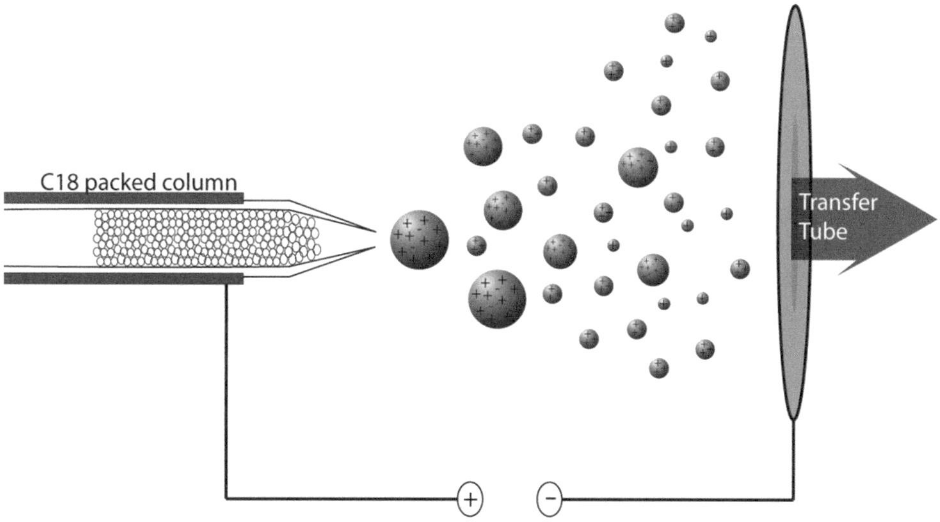

Fig. 28.1 Principals of ESI. Analytes (peptides) are eluted by gradient of ACN from C_{18} microcapillary columns in the form of positively charged droplets. As droplets are travelling through transfer tube (Thermo Scientific ion traps) at high temperature and in vacuum, solvents are quickly evaporated leaving analytes charged and ready to be drawn by entrance lenses

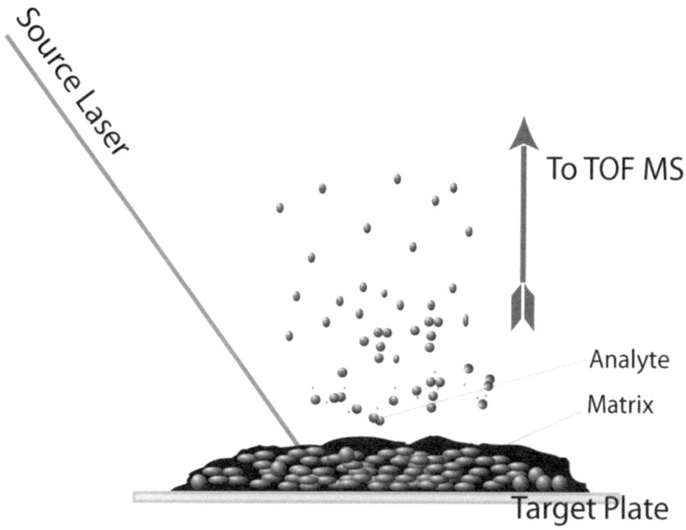

Fig. 28.2 Ionization by MALDI. Analytes (peptides) are mixed with matrix on the target plate. A source laser provides the energy for evaporation and transfer of positive charge from matrix to analyte. Positively charged ions will fly toward the negatively charged electrodes of the mass spectrometer

used in mass spectrometry of proteins and peptides (Ahmed 2008) are electrospray ionization (ESI) (Yates 1998; Wilm 2011) and matrix-assisted laser desorption ionization (MALDI) (Roepstorff 2000; Seidler et al. 2010). ESI uses an electrostatic charge placed on a liquid sample to create a cloud of highly charged droplets (Fig. 28.1). MALDI uses a UV laser beam to transfer a charge from the matrix to a solid-phase sample (Fig. 28.2). These methods complement each other in that they ionize partially overlapping subsets of the peptides of interest. Examples of ESI and MALDI generated spectra are shown in Fig. 28.3.

The main advantages of using ESI are the high sensitivity (in the order of attomoles to femtomoles) and the soft ionization of peptides that allows for the formation of gas-phase ions without

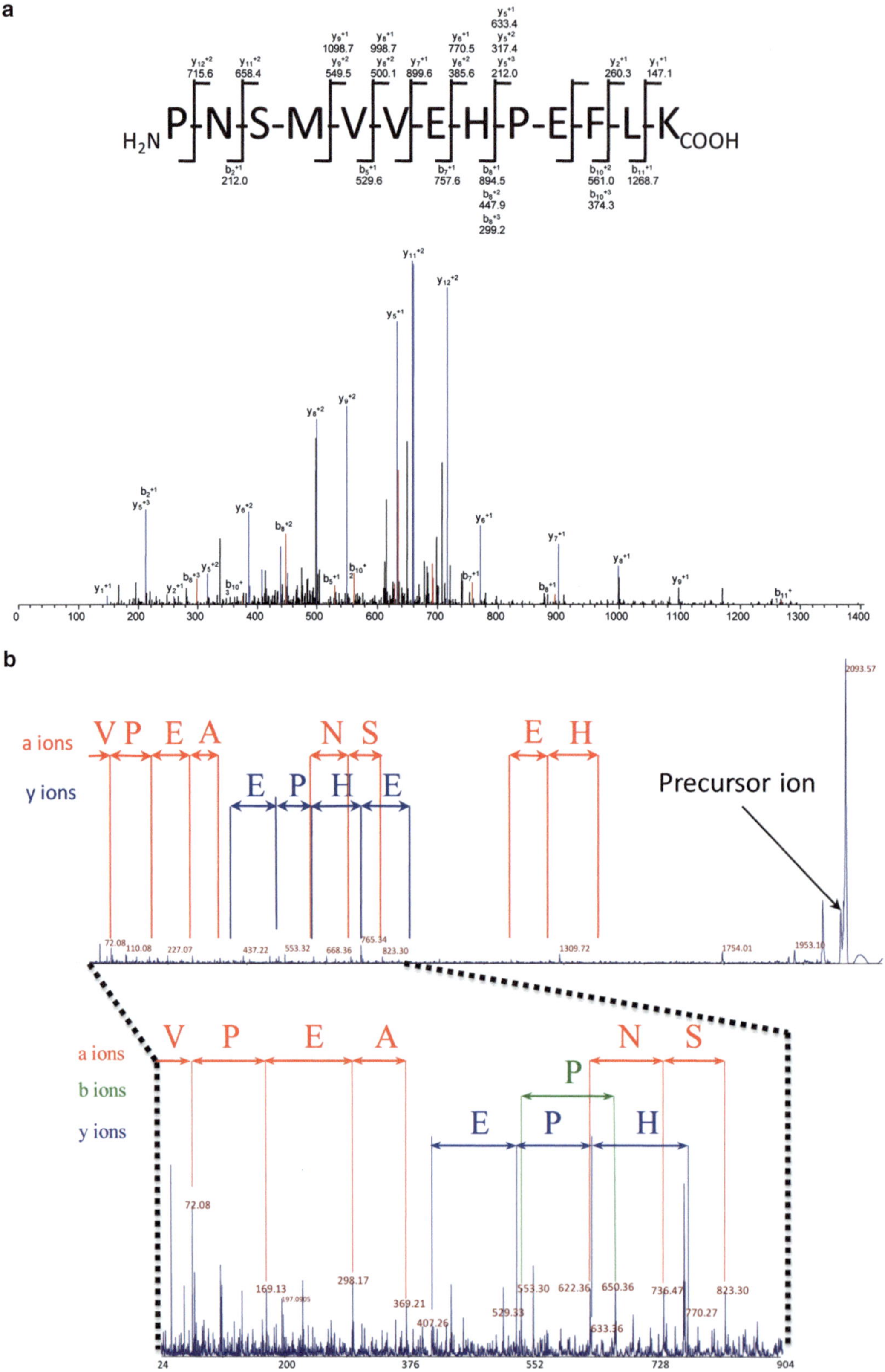

Fig. 28.3 Examples of spectra generated by mass spectrometers. (**a**) A typical MS/MS spectrum produced by fragmentation in ion trap. Peptide ions were generated by ESI. Peptide sequence with fragment ion assignment is shown. (**b**) A typical MS/MS spectrum produced by fragmentation in collision cell by collision-induced dissociation (CID). Peptide ions were generated by MALDI. Peptide sequence with fragment ion assignment is shown

in-source fragmentation. If a protein or peptide is well ionized, ESI has a practical mass range up to approximately 70,000 Da; however, complex mixtures can reduce sensitivity. ESI begins with a liquid sample passing through a charged capillary to create a cloud of charged droplets. To achieve flow, high-pressure liquid chromatography (HPLC) systems are used for a nano-flow rate of 250 nL/min, which is sprayed into the orifice of mass spectrometer. These HPLC systems are usually used to separate the components of a solution prior to mass analysis. The column used in our example is a PicoFrit C18 column-emitter from New Objectives, Inc. (Woburn, MA); however, home-packed columns are routinely used in other labs. The type of mass spectrometer used determines the placement of the column in the system. In this example, the column is placed after the ionization source. This allows sample to leave the column and become ionized directly into the orifice without running through extra lines (which may cause mixing in dead volumes and peak tailing). The nanospray system allows for the use of less solvent, concentrating the sample and decreasing the amount of time needed for the solvent to evaporate before entering the mass spectrometer. As the solvent evaporates, the density of the charged ions increases until the repulsive forces of the positively charged ions exceed the force of the droplet surface tension (Rayleigh limit) creating tiny charged droplets (Grimm and Beauchamp 2002; Li et al. 2005), which are attracted to the oppositely charged entrance of the mass spectrometer.

When using the MALDI platform, sample ionization is supported by either applying sample in solution onto the MALDI plate and layering with matrix solution over the sample or mixing the sample with a matrix solution prior to spotting. For MALDI analysis, there are many matrices available; but for short peptides, alpha-cyano-4-hydroxycinnamic acid (CHCA) is usually used, and for proteins, sinapinic acid (SPA) appears to be better. Upon drying, the mixture of sample and matrix forms crystals, and a laser beam provides the energy needed to transfer charge from the matrix to the sample peptide. Due to high temperature, analytes are evaporated and form a cloud of ions in the gas phase that can be further transferred to a drift tube where their mass is determined by their time of flight.

28.4 Mass Analyzers

There are many types of mass spectrometers (mass analyzers) available for research purposes, and users of this textbook are encouraged to read additional literature. Table 28.1 summarizes characteristics and utilities of mass spectrometers used in proteomic experiments, protein/peptide identification, and characterization of posttranslational modifications.

MALDI-TOF (time of flight) mass spectrometry can provide two complementary alternatives

Table 28.1 Characteristics and performance of mass spectrometers commonly used in proteomic experiments. Adapted from Domon and Aebersold (2006)

	IT–LIT	TOF–TOF	QQ–LIT	Q–Q–TOF	Q–Q–Q	FT-ICR
Mass accuracy	Low	Good	Medium	Good	Medium	Excellent
Resolving power	Low	High	Low	Good	Low	Very high
Sensitivity	Good	High	High		High	Medium
Dynamic range	Low	Medium	High	Medium	High	Medium
ESI	Yes		Yes	Yes	Yes	Yes
MALDI	Optional	Yes		Optional		
MS/MS	Yes	Yes	Yes	Yes	Yes	Yes
Throughput	Moderate	Very high	High	High	High	High

IT–LIT ion trap–linear ion trap, *TOF–TOF* time of flight–time of flight, *QQ–LIT* quadrupole–quadrupole–linear ion trap, *Q–Q–TOF* quadrupole–quadrupole–time of flight, *Q–Q–Q* triple quadrupole, *FT-ICR* Fourier-transformed ion cyclotron resonance for exploring the structural features of your protein or peptide. One is to use the peptide fingerprinting method based on a set of peaks determined from a tryptic digest of the peptide of interest. This approach is best when dealing with a mixture of one to two proteins and is frequently used when identifying proteins in spots excised from two-dimensional electrophoresis. In this case, fragmenting the peptides is not necessary; however, mass accuracy is critical within 10 ppm (parts per million); and the number of experimental peaks needs to match greater than 30 % of the number of theoretical peaks. The second approach when using the MALDI is to use the TOF–TOF (MS/MS) mode to fragment the peptides and derive their amino acid sequences. It is very similar to ESI-nano-LC-MS/MS in that it uses a full scan followed by TOF–TOF scans of the most abundant precursor ions. With this approach, low-complexity samples can be analyzed because high-quality spectra for only two unique peptides are needed. These samples, which do not need fractionation, are deposited manually on the MALDI target plate, while the high-complexity samples can be fractionated using a nano-LC system and then automatically spotted on the target plate using a plate spotter such as the Tempo LC-MALDI spotter from AB Sciex (Foster City, CA).

28.5 Sample Preparation for Mass Spectrometry

Sample preparation is extremely important when using mass spectrometry to analyze peptides and proteins (Granvogl et al. 2007; Ahmed 2009; Gundry et al. 2009; Silberring and Ciborowski 2010). Workspaces should be kept as clean as possible and gloves worn at all times. Impurities from samples (such as lipids or glycans) or from processing (such as detergent, dust, or even keratin from your hair) can lead to signal suppression.

Pre-analysis fractionation of samples, also called off-line fractionation, is often completed before mass spectrometry analysis. In most profiling experiments, samples are very complex and fractionated into one or more dimensions (e.g., immunodepletion of most abundant proteins from serum/CSF) (Rozek et al. 2007, 2008). It is very common to separate peptides by isoelectric focusing, such as on the OFFGEL (Agilent, Inc., Santa Clara, CA) system. During purification, one needs to be aware that every step of sample preparation leads to a loss of material. This must be taken under consideration when calculating the initial amount of protein. Gel electrophoresis is also commonly used to purify peptide/protein samples. When extracting proteins from a gel, staining with Coomassie blue is suggested. If dealing with very small amounts of proteins/peptides, silver or Sypro Ruby or other fluorescent-based gel stains may be used for visualization. However, mass spectrometry-friendly silver staining of gels is least recommended due to poor recovery of digested peptides. A simple in-gel tryptic digestion and reverse-phase ZipTip® (Millipore, Billerica, MA) sample cleanup allow desalting and concentration of the sample. If possible, it is best to resuspend samples in 0.1 % formic acid in water, 50 % acetonitrile (ACN), or 50 % methanol. For ESI, samples are resuspended into autosampler vials with approximately 6–10 μL of sample, depending on concentration.

28.6 Reagents: General Considerations

Because of high and constantly increasing sensitivity of mass spectrometers, quality of reagents is important. Only use glass pipettes to prepare solutions. Acidic solvents can cause plastics and even the volume markings from glass pipettes to leave behind contaminants that will damage the results. For ESI analyses, use only MS grade or Optima (Thermo Fisher Scientific, Waltham, MA) solvents because HPLC grade is not sufficient for quality mass spectrometry results. For MALDI, Milli-Q (Millipore, Billerica, MA) water should be used for spotting solutions because HPLC grade water can vary in salt concentration and cause adduct ions to appear in the spectra.

28.7 Protocol: Laboratory Exercise

There are many protocols available describing enzymatic and chemical fragmentation of proteins for mass spectrometry analysis. Because this is not a focus of this chapter, we direct readers to literature resources (McGregor and De Souza 2006; Bodzon-Kulakowska et al. 2007; Canas et al. 2007; Luque-Garcia and Neubert 2007).

28.7.1 Sample Cleanup Using ZipTip®

ZipTips® are micropipette tips with a layer of chemically active resin embedded into the tip in order to facilitate the concentration and purification of peptide and protein samples. This is routinely done for any type of mass spectrometry analysis, either ESI or MALDI. For peptide purification, a reverse-phase C_{18} resin is routinely used. For proteins, less hydrophobic resins, such as C_4 is usually employed. Tips with strong cationic exchange (SCX) resins are used less frequently. Below is a detailed protocol for peptide cleanup:

- Prepare reagents:
  - Wetting solution: 100 % Optima LC/MS ACN (Fisher) in water.
  - Resuspension buffer: 0.5 % trifluoroacetic acid (TFA) in water.
  - Equilibration/bash solutions: 0.1 % TFA in water.
  - Elution solution: 50 % ACN and 0.1 % TFA in water.
- Resuspend sample in 15 μL of the resuspension buffer.
- Pipette 10 μL of wetting solution into ZipTip®. Dispense to waste and repeat.
- Pipette 10 μL of equilibration solution into ZipTip®. Dispense to waste and repeat twice.
- Bind the peptide/protein to ZipTip® by fully aspirating and dispensing sample ten times.
- Aspirate 10 μL wash solution into ZipTip®. Dispense to waste and repeat.
- Carefully aspirate and dispense 10 μL elution solution through ZipTip® at least six times without introducing air into the sample. Transfer to autosampler vial and speedvac to dryness.
- At this stage, sample may be stored at −20 °C until ready to run on mass spectrometer. Immediately before running samples, resuspend in 6 μL 0.1 % formic acid in Optima® LC/MS water (Fisher) and mix well by pipetting up and down. Be sure there are no air bubbles in the autosampler vial when you load the sample on the mass spectrometer so that the sample needle is able to withdraw a consistent injection volume.

28.7.2 Mass Analysis Using ESI and Ion Trap

There are many types of mass spectrometers appropriate for proteomics work along with multiple types of parameter settings on each instrument. In our example, a LTQ Orbitrap XL (Fisher), a hybrid mass spectrometer, is used, which couples a slower but high-resolution Orbitrap with a faster ion trap. Attached to the mass spectrometer is the Eksigent (Dublin, CA) nano-LC system equipped with two alternating peptide traps and a PicoFrit C18 column-emitter from New Objective (Woburn, MA). This setup allows one sample to be analyzed on the column while another is loaded onto a trap and washed with a solution of 98 % water and 2 % ACN further supplemented with 0.1 % formic acid for ion pairing. The peptide traps also protect the column from contaminant buildup. Peptides are eluted using a 60-min linear gradient of 0–60 % ACN with 0.1 % formic acid.

For all applications, calibration is performed using standards provided by the manufacturer to ensure each precursor is at an accurate mass. For specific proteomics applications, the instrument is tuned using a direct infusion of angiotensin in order to optimize the ion path of the peptides into the mass spectrometer.

The acquisition method is in a data-dependant mode with one precursor scan in the Orbitrap, followed by fragmentation of the five most abundant peaks in a part of the instrument called collision-induced dissociation (CID), detected in the LTQ. Resolution of the precursor scan is set to 60,000, scanning from 300 to 2,000 charge to mass ratio (*m/z*). Precursor peaks with a minimum signal count of 50,000 are dynamically excluded after two MS/MS selections for 60 s within a range ±10 ppm. Monoisotopic precursor selection (MIPS) should be enabled. No charge state rejection is necessary but previously found background peaks should be included in a mass rejection list. Collision energy can be set to 35 using an isolation width of 2 *m/z* and an activation *Q* of 0.250. To begin data acquisition on the LTQ Orbitrap XL:

- Make sure the tops of all autosampler vials are clean. Place samples in cooled autosampler tray.
- Set up the sequence of samples' analysis in Xcalibur software. From the Roadmap homepage, select "Sequence Setup." On the main screen will be a spreadsheet to enter information about the samples and the sequence they will be run in.
  - Under file name: enter a sample name.
  - Under path: double click the cell to select where the .raw file will be saved.
  - Inst Meth (Instrument Method): double click the cell to select method to use with the sample.
    - Important: when rotating between two peptide traps, two different methods must be programmed and alternated in the sequence so that the 10-port valve switches between loading and eluting the two traps.
  - Position: for row 1, enter the position of the sample named in row 2. In row 2, enter the position of the sample named in row 3, etc. This is also due to the alternating peptide traps.
    - Use letters and numbers indicated on autosampler tray to describe positions, for example, A04.
  - To be sure the instrument is ready, click the "Status" tab on the left navigation bar. Both LC channels, the LTQ Orbitrap XL MS and the NanoLC-AS1 Autosampler, should say "ready to download."
  - To return to sequence setup, select the "Acquisition Queue" tab. Check that your sequence is correct then go to "Actions" and "Run Sequence." When window opens, be sure all of your samples are included in the "Run rows" box. To begin the sequence, select "OK."
  - Your loaded sequence will appear in the left navigation bar. Once "sequence row #1" turns green, your sample acquisition has begun.
  - After completion of your MS run, the system will provide a .raw file that can be analyzed with proteomic software packages, such as Proteome Discoverer (Fisher).

28.8 Mass Analysis Using MALDI-TOF–TOF

Similarly to ESI mass spectrometers in which peptides or other analytes are eluted from a reverse-phase column directly into mass spectrometer, plate spotters perform the same chromatographic task, mixing fractions with matrix of choice and depositing mixtures on target plates for batch acquisition in MALDI-TOF–TOF instrument. In this laboratory exercise we describe both techniques: manual spotting with single spot analysis and automatic spotting using Tempo LC spotter from AB Sciex with automated analysis. In both cases for mass analysis, we used AB Sciex 4800 MALDI-TOF–TOF instrument.

- Mix and spot several calibration spots containing Calibration Mix 1 and 2 (AB Sciex) in CHCA matrix solution. Manually spot sample peptide onto plate by pipetting 0.5 μL of sample and letting it dry, and then apply another 0.5 μL of sample. Let air-dry completely between each application. Then repeat for a total of 1 μL of matrix, also allowing it to dry completely between additions.
- Open the Reflector positive MS acquisition method and modify the method for a range of 400–3,000 *m/z*. Open both an MS internal and MS default calibration processing method, using a peak list for the calibration mix spotted. Use the internal processing file to update the MS operating mode default calibration. Perform a plate calibration as well.

- Open the MS/MS 1 kV method and both a MS/MS internal and default calibration processing methods with a peak list for the fragmentation of angiotensin (1,296 *m/z*) in the internal method. Perform an MS/MS internal calibration to update the MS/MS operating mode default calibration.
- Open an interpretation method and modify it to exclude common background peaks observed in calibration spots or in blank matrix spots. This method will use a single precursor scan to select the five most abundant peaks, which will then be analyzed by MS/MS. Be sure to include the MS/MS acquisition method and MS/MS default processing method in the interpretation method used.

If a MALDI plate spotter is used, the following preparation methods are used:

- Inject samples prepared for analysis onto a C18 peptide trap followed by a microcapillary C18 column (New Objective). Load the peptides with millipore water with 0.1 % TFA and elute using a 90-min linear gradient of 0–80 % ACN with 0.1 % TFA. (Collected fragments can either be spotted manually or with the automated spotter.)
- Prepare a fresh matrix solution of a 1:1 mixture of 10 μg/mL each of CHCA and 2,5-dihydroxybenzoic acid (DHB) in 60 % ACN with 0.1 % TFA. Samples can be automatically mixed with matrix on the spotter or layered with equal parts of matrix solution and spotted on the plate in 0.5 μL increments as with manual spotting.
- Manually spot additional blank and 100 fmol BSA spots for calibration, method verification, and a peak rejection list. Each sample should be preceded by a calibration in both MS and MS/MS modes. Also, check the plate calibration for shot accuracy. The BSA sample should be run in both interactive and automatic mode to verify the interpretation method and sensitivity of acquisition methods.
- Create an automatic job run using the reflector positive acquisition method to determine precursor masses and then use the interpretation method for peak selection. Fragmentation in MS/MS will run automatically.

28.9 Data Interpretation

The output of results from sample analysis by mass spectrometry is called a spectrum. Signals generated by measured ions are displayed on a two-dimensional graph in which the *x*-axis represents the value of mass-to-charge ratio (*m/z*) and the *y*-axis represents the signal intensity. It is important to note that signal intensity scale is always 100 %, which represents the most abundant ion. Each spectrum contains information concerning how many ions constitute 100 % and is expressed as total ion current (TIC). Because ionized proteins and peptides might carry more than one charge (proton), their size is expressed as mass-to-charge ratio. Therefore, a peptide with molecular weight of 1,000 Da will be measured at *m/z* 1,000 as singly charged molecular species ($1{,}000^{+1}$) but *m/z* 500 as doubly charged molecular species (500^{+2}).

Spectra may be interrogated (examined) manually to reveal composition of a sample or the structural features of a single molecule, if the molecule was subjected to fragmentation in collision cell of mass spectrometer. Manual interrogation is time-consuming, making this approach suitable only for in-depth examination of a limited number of spectra. If many spectra need to be analyzed, which is the case in proteomic profiling experiments, algorithms and computers are needed.

In recent years, there has been a rapid development of existing and new software packages assisting with database searches of spectra generated by mass spectrometry analyses. Currently, output files provide detailed information about MS/MS fragmentation, mass error, sequence coverage, probability of false-positive identifications, and many other pieces of information. More importantly, this type of software provides us with an ability to set both our search parameters and filter thresholds based on necessary criteria. Taking this under consideration, it is critical that search parameters are always noted along with results so that data is comparable to other experiments in other labs. In this section, we will perform a database search using Proteome Discoverer

1.2, which is based on the SEQUEST algorithm and offered by Thermo Fisher Scientific, along with their ion trap mass spectrometers. There are many databases available to search data. In this example, we use the NCBI database.

- Download database from ftp://ftp.ncbi.nih.gov/blast/db/FASTA/nr.gz. Once this file has been downloaded and unzipped, sub-databases can be created in Proteome Discoverer to meet the specific needs of your search. This sub-database is created using keywords to include or exclude in results. For example, we normally exclude keratins to speed up the search. To be fully functional, these sub-databases will also need to be indexed in the system. Monoisotopic mass type should be selected for LTQ Orbitrap data and average mass type for ion traps such as the LTQ system.
- The parameters are an important component to reduce false-positive hits (criteria is too unrestricting) and to prevent losing data (criteria too tight). Individual investigators determine search parameters. Typically for ion trap, we use a peptide tolerance of 2.0 amu and fragment ion tolerance of 1 amu, using only the b- and y-ion series in the calculations. Search the spectra obtained using Proteome Discoverer against a NCBI database indexed for human proteins. Set up a static modification for oxidation (M) and a dynamic modification for carboxymethyl (C) before beginning your search.
- Filtering output is also important to avoid excessive data. The following are suggested filters for the above example:

Peptide	Delta CN	0.100
Peptide	Sp—preliminary score	500.0
Peptide	Xcorr vs. charge state	1.50, 2.00, 2.50, 3.00
Protein	Protein probability	0.001

28.10 Most Common Problems/Pitfalls

There are many problem areas in proteins and peptides mass spectrometry analyses of. One is inherent, such as inability of any given protein or peptide to be ionized and fragmented to reveal the amino acid sequence or the presence and/or location of posttranslational modification. Another inherent problem originates from the limited amount of initial material. This is particularly important in clinical proteomics when diagnostic laboratories and quite often multiple researchers share material obtained from a patient. A second problem area is contamination introduced during sample preparation that causes signal suppression. For example, polymer contamination is common because various reagents that are polymeric in nature are used to solubilize samples. Another important issue is a carryover of samples from one analysis to another. Carryover requires extensive washing of the LC-MS/MS system, which costs a lot of time considering that it is under nano-flow. Usually, if mass spectrometer is properly calibrated, tuned, and kept clean with regular preventive maintenance services, sample carryover and contamination do not create much of a problem. The most frequent are problems associated with function of nano-flow. Small deviation from proper flow of mobile phase may have quite profound impact on sensitivity of protein and peptide detection.

28.11 Discussion/Summary

The intension of this chapter is to guide students to perform a simple experiment showing potential of mass spectrometry-based protein and peptide analysis. It provides students with minimal information required to understand the underlying laboratory principles of the exercise and to perform such exercise in a step-by-step manner. Important steps that may have a big impact on the final outcome of a mass spectrometry experiment were discussed. This chapter also describes the use of two specific instruments: Thermo Fisher Scientific LTQ Orbitrap XL and AB Sciex 4800 MALDI-TOF–TOF; however, it must emphasize that there are many other high-quality instruments available that can be used. The reader should be aware that field of mass spectrometry-based protein and peptide analysis is very dynamic and excellent papers are published every day. This includes review papers and monographic books that include more recent findings than those cited here. Students are encouraged to explore new information using electronic resources.

References

Ahmed FE (2008) Utility of mass spectrometry for proteome analysis: part I. Conceptual and experimental approaches. Expert Rev Proteomics 5:841–864

Ahmed FE (2009) Sample preparation and fractionation for proteome analysis and cancer biomarker discovery by mass spectrometry. J Sep Sci 32:771–798

Appella E, Arnott D, Sakaguchi K, Wirth PJ (2000) Proteome mapping by two-dimensional polyacrylamide gel electrophoresis in combination with mass spectrometric protein sequence analysis. EXS 88:1–27

Biemann K (1988) Contributions of mass spectrometry to peptide and protein structure. Biomed Environ Mass Spectrom 16:99–111

Bodzon-Kulakowska A, Bierczynska-Krzysik A, Dylag T, Drabik A, Suder P, Noga M, Jarzebinska J, Silberring J (2007) Methods for samples preparation in proteomic research. J Chromatogr B Analyt Technol Biomed Life Sci 849:1–31

Canas B, Pineiro C, Calvo E, Lopez-Ferrer D, Gallardo JM (2007) Trends in sample preparation for classical and second generation proteomics. J Chromatogr A 1153:235–258

Chen W, Yin X, Yin Y (2008) Rapid and reliable peptide de novo sequencing facilitated by microfluidic chip-based Edman degradation. J Proteome Res 7:766–770

Domon B, Aebersold R (2006) Mass spectrometry and protein analysis. Science 312:212–217

Edman P (1970) Sequence determination. Mol Biol Biochem Biophys 8:211–255

Granvogl B, Ploscher M, Eichacker LA (2007) Sample preparation by in-gel digestion for mass spectrometry-based proteomics. Anal Bioanal Chem 389:991–1002

Grimm RL, Beauchamp JL (2002) Evaporation and discharge dynamics of highly charged droplets of heptane, octane, and p-xylene generated by electrospray ionization. Anal Chem 74:6291–6297

Gundry RL, White MY, Murray CI, Kane LA, Fu Q, Stanley BA, Van Eyk JE (2009) Preparation of proteins and peptides for mass spectrometry analysis in a bottom-up proteomics workflow. Curr Protoc Mol Biol Chapter 10:Unit10.25

Li KY, Tu H, Ray AK (2005) Charge limits on droplets during evaporation. Langmuir 21:3786–3794

Luque-Garcia JL, Neubert TA (2007) Sample preparation for serum/plasma profiling and biomarker identification by mass spectrometry. J Chromatogr A 1153:259–276

Mari A, Ciardiello MA, Tamburrini M, Rasi C, Palazzo P (2010) Proteomic analysis in the identification of allergenic molecules. Expert Rev Proteomics 7:723–734

McGregor E, De Souza A (2006) Proteomics and laser microdissection. Methods Mol Biol 333:291–304

Meng F, Forbes AJ, Miller LM, Kelleher NL (2005) Detection and localization of protein modifications by high resolution tandem mass spectrometry. Mass Spectrom Rev 24:126–134

Paizs B, Suhai S (2005) Fragmentation pathways of protonated peptides. Mass Spectrom Rev 24:508–548

Pottiez G, Haverland N, Ciborowski P (2010) Mass spectrometric characterization of gelsolin isoforms. Rapid Commun Mass Spectrom 24:2620–2624

Roepstorff P (2000) MALDI-TOF mass spectrometry in protein chemistry. EXS 88:81–97

Rozek W, Ricardo-Dukelow M, Holloway S, Gendelman HE, Wojna V, Melendez L, Ciborowski P (2007) Cerebrospinal fluid proteomic profiling of HIV-1-infected patients with cognitive impairment. J Proteome Res 6:4189–4199

Rozek W, Horning J, Anderson J, Ciborowski P (2008) Sera proteomic biomarker profiling in HIV-1 infected subjects with cognitive impairment. Proteomics Clin Appl 2:1498–1507

Seidler J, Zinn N, Boehm ME, Lehmann WD (2010) De novo sequencing of peptides by MS/MS. Proteomics 10:634–649

Silberring J, Ciborowski P (2010) Biomarker discovery and clinical proteomics. Trends Analyt Chem 29:128

Strachunskii LS, Prokhorenkov PI, Novikova Iu A, Krechikova OI, Iakusheva LV (1992) [Use of cefoperazone in clinical practice]. Antibiot Khimioter 37:19–21

Thiede B, Hohenwarter W, Krah A, Mattow J, Schmid M, Schmidt F, Jungblut PR (2005) Peptide mass fingerprinting. Methods 35:237–247

Wilm M (2011) Principles of electrospray ionization. Mol Cell Proteomics 10(M111):009407

Yates JR 3rd (1998) Mass spectrometry and the age of the proteome. J Mass Spectrom 33:1–19

DNA Microarrays as a Tool for Neurosciences Research

29

James D. Eudy and Lynette Smith

Abstract

DNA microarray technology is a useful tool to further our understanding of underlying biological pathways governing the behavior of cells and tissues. In fact, microarrays have evolved over the last 10 years or so to become one of the major technologies utilized in biomedical research. This chapter will provide a brief history of the development of DNA microarray technology, a brief survey of widely accepted microarray platforms, and then expand upon critical components of DNA microarray experimentation necessary to ensure optimal outcomes in a given microarray experiment. Topics covered will include the necessity of a good initial experimental design, quality control issues with respect to input RNA or DNA, appropriate use of biological replicates relevant to a given biological question, and analysis strategies employed to extract meaningful results from a microarray experiment.

Keywords

DNA microarray • Gene expression • RNA • DNA • Normalization • Hybridization • Analysis

J.D. Eudy, Ph.D. (✉)
Department of Genetics, Cell Biology, and Anatomy, University of Nebraska Medical Center, 1027 Durham Research Center II, 985915 Nebraska Medical Center, Omaha, NE, 68198-5915, USA
e-mail: jdeudy@unmc.edu

L. Smith, M.S.
Department of Biostatistics, College of Public Health, University of Nebraska Medical Center, Omaha, NE, 68198-5915, USA
e-mail: lmsmith@unmc.edu

29.1 Introduction and Background

DNA microarray technology is a powerful technology utilized for a variety of molecular biological applications including gene expression measurements, genomic copy number ascertainment, and identification of DNA–protein interactions relevant to transcriptional control mechanisms (Leung and Cavalieri 2003; Bulyk

H. Xiong and H.E. Gendelman (eds.), *Current Laboratory Methods in Neuroscience Research*, Springer Protocols Handbooks, DOI 10.1007/978-1-4614-8794-4_29,

2006; Edelmann and Hirschhorn 2009). It is a useful tool to further our understanding of underlying biological pathways governing the behavior of cells and tissues. In fact, microarrays have evolved over the last 10 years or so to become one of the major technologies utilized in biomedical research. The first manuscripts describing the proof of principle experiments were published in the mid-1990s by the members of a laboratory at Stanford University headed by Pat Brown (Schena et al. 1995; DeRisi et al. 1996, 1997). These manuscripts described the profiling of the relative expression levels of genes in the plant *Arabidopsis, a melanoma cell line*, and the yeast organism *Saccharomyces cerevisiae*. During the approximate 16 years since those initial manuscripts, DNA microarrays have become a mainstream technology used routinely in both basic research and clinical laboratories worldwide. In fact, a PubMed database search at the time of this writing with the keyword "DNA microarray" yielded a result of over 56,500 articles identified relevant to DNA microarray technology, an observation that serves to substantiate the widespread adoption of the technology in the scientific community.

All DNA microarrays, regardless of the platform, exploit the complementary base pairing properties of nucleic acids to achieve their particular goal, whether the goal is measuring transcriptional activity in RNA samples, comparing the genomic DNA content between two DNA samples, or determination of DNA–protein binding interactions. In the case of glass DNA microarrays designed to monitor gene expression levels, a population of fluorescently labeled nucleic acids derived from the RNA extracted from the biological samples under investigation is hybridized to oligonucleotides representing the entire expressed genetic content of a given organism located upon the glass array. The procedure facilitates the genome-wide measurement of the relative levels of gene transcription in the samples under study providing a "snapshot" of the overall biological transcriptional state of the cells, which in turn is reflective of the biological state of the cells at that particular moment in time.

Gene transcription profiling can yield important insights relevant to a myriad of different biological questions. If changes in transcriptional activity of certain genes may be contributing to the phenotype being investigated, then gene transcriptional profiling may be employed as part of an overall scientific study to gain insight into a particular biological question. Though the procedures of DNA microarray technology have become relatively standardized in recent years, care must still be taken to ensure that experiments are conducted appropriately. All of the steps in the overall procedure of the microarray experiment must be performed carefully and with appropriate forethought. The above being stated, microarrays are a valuable tool and often provide exciting insights to further our understanding of biological processes.

29.2 Objectives

This chapter will provide an overview of the major steps comprising a DNA microarray experiment designed to measure gene transcription. Topics covered include the necessity of a good initial experimental design, quality control issues with respect to input RNA, and appropriate use of biological replicates relevant to a given biological question. An outline of labeling, hybridization, scanning, and analysis strategies employed to extract meaningful results from an experiment will be provided. With regard to microarray data analysis, readers will learn about the need for normalization and filtering of data and subsequent statistical tests used to define differentially expressed genes. The reader will also learn about the clustering of genes to identify patterns of coordinately expressed genes. This outline will be provided in the context of both the spotted glass array and Affymetrix array technologies (Affymetrix Inc., Santa Clara, CA).

29.3 Procedures/Methodological Approaches

There are a variety of important considerations during a microarray experiment that merit close attention to ensure a positive result. First and foremost, it is important to recognize that the gene expression profile one obtains is really a "snapshot" of the transcriptional state of the cells or tissue at the precise time that the RNA was extracted. Other things the researcher needs to be cognizant of include the need to ascertain the integrity and purity of the RNA harvested. It is also essential to utilize good molecular biological technique during processing of samples throughout all of the steps of the procedure including the labeling, hybridization, scanning, and analysis steps. It is beyond the scope and space limitations to exactly detail each of the steps required to perform a microarray experiment; however, the goal is to provide relevant information for critical parameters.

To perform microarray experiments from start to finish requires substantial DNA microarray-specific instrumentation and infrastructure in addition to standard molecular biological instrumentation common in many individual researcher laboratories such as centrifuges, water baths, and UV spectrometers. Typically most researchers interface with their local microarray core facility to have the labeling and hybridizations performed and often consult with statisticians and bioinformaticists to assist with data analysis. The relationship is best if it is an interactive one between all parties.

29.3.1 Experimental Design: Biological Replicates and Appropriate Controls

Before a gene array experiment is begun, the researcher should determine what information they hope to gain. For example, "what happens at the transcriptional level when one adds growth factor A to these neurons in cell culture?" Important parameters might include the concentration of the growth factor to be added and the time point(s) the RNA is to be harvested. As the microarray will be measuring the relative levels of genes at exactly the time point that the cells were harvested, ideally the investigator will have preliminary information regarding their own particular system to guide these choices. Another consideration is that most microarrays available today contain oligonucleotides representing all the genes within that particular organism's genome. These arrays will be assaying the transcriptional activity within the cells reflective of all the processes taking place at that given time point. A very carefully designed experiment is essential in order to generate the data to answer the cell process under investigation. As an experienced scientist knows, it is imperative to include all the appropriate controls one needs for a given experiment designed to answer a specific question. Naturally this fundamental tenet also applies to microarray experiments.

Perhaps one of the most important considerations is the one of biological replication within the experiment. Simply stated, microarray experiments should include at least three or four biological replicates of both the control sample and the experiment sample(s). To determine the sample size necessary for an Affymetrix experiment comparing expression levels between two or more groups, the coefficient of variation (CV), defined as the standard deviation (SD) divided by the mean, can be used in combination with the fold change the researcher wishes to detect to determine the number of biological replicates per group (Han et al. 2004). For the RMA normalization method (discussed later in the chapter) with an animal model, Han et al. found the 95th percentile of the within group CV of all genes to be 0.15. This can be incorporated into a sample size equation given by Han et al. as

$$n = 2 * \frac{\ln\left(\mathrm{CV}^2 + 1\right)\left(z_{1-\frac{\alpha}{2}} + z_{1-\beta}\right)^2}{\ln\left(f\right)^2} + 2,$$

where CV is the coefficient of variation, f is the fold change one would like to detect, and $z_{1-\gamma}$ is the $100(1-\gamma)$th percentile of the standard normal distribution. With CV of 0.15, fold change of 2, a significance (α) level of 0.001 ($z_{1-\frac{a}{2}} = 3.29$), and a power ($1-\beta$) of 0.95 ($z_{1-\beta}=1.64$), we get a per group sample size $n=4.25$ which can be rounded to 5 (4 or 5) biological replicates per group. McShane et al. have sample size tables for animal model microarray studies with spotted arrays; to detect a fold change of 2, they determine that a sample size of 4–5 biological replicates per group is needed to obtain adequate power (McShane et al. 2003). Sample size determination for microarray prediction or clustering problems is beyond the scope of this book, and consultation with a biostatistician is recommended (Simon et al. 2002; Dobbin et al. 2008).

For purposes of illustration, let us consider a simple experiment, an experiment containing two factors. Let the first factor be "growth factor A" and the time point of "3 hours" be the second factor. In this case, an appropriate experimental design might begin with the careful distribution of equal number of cells into six cell culture plates to initiate the experiment, three control plates and three experimental plates. The plates should all be treated exactly the same throughout the experiment with the exception that the three experimental plates receive the growth factor. This might be interpreted that the only difference is the addition of the growth factor. The growth factor is likely suspended in a liquid solution (vehicle) of some sort. The control plates should receive the same volume of vehicle minus the growth factor. The vehicle alone might elicit a transcriptional change in the cells, and due to the nature of a genome-wide microarray, these transcriptional changes will be measured as well. If the experiment is set up appropriately, then the genes identified between the samples will correspond to those specific to addition of growth factor.

In closing, a carefully designed experiment is absolutely critical to the successful outcome of a microarray experiment. The experiment should be designed to control for the specific biological phenomenon the researcher is wishing to measure. While this consideration is obvious to an experienced researcher, it is particularly relevant to the situation where one is querying the entire genomic content of the organism.

29.3.2 RNA Preparation and Quality Assessment

RNA is a labile molecule and good molecular biological technique is crucial to avoid the degradation of the RNA during extraction and storage. Cells or tissues comprising the experiment should be harvested and either immediately go into the extraction procedure or be stored in liquid nitrogen or another medium that protects the RNA from degradation, such as RNA*later* (Ambion Inc., Austin, TX). RNase-free gloves and other instruments used in the extraction as well as the tubes and counters should be RNase-free. There are a variety of commercial kits available for RNA extraction, and each will contain detailed instructions which should be adhered to carefully. The resultant RNA should be resuspended in the solution provided by the commercial kit as they typically consist of RNase-free molecular grade H_2O containing Tris–EDTA to chelate the Mg^{2+} ions present, which in turn helps to keep any existing Mg^{2+}-dependent RNases from functioning. RNA should be stored at a concentration of 50 ng–1 μg/μL at −80 °C until use.

It is imperative that the RNA be evaluated with respect to integrity and purity prior to use in the microarray experiment. These parameters include (1) the potential degradation of the RNA, (2) the potential presence of genomic DNA contamination, and (3) the potential presence of organic contaminants. RNA degradation in most cases can be avoided if proper techniques are adhered to during isolation. There are, however, differences in endogenous levels of RNase between different tissue types that can make the isolation of intact RNA from certain tissues more challenging than others. For example, human and other mammalian pancreatic tissue contains relatively high levels of the RNase enzyme, and it is prudent to be cognizant of this property. The potential presence of chemical contaminants introduced during the RNA isolation procedure also deserves mention. Many isolation procedures utilize phenolic or other organic compounds which are designed to denature proteins in the cell and tissue preparations during RNA isolation. If these compounds remain in the final RNA preparation, they can certainly inhibit critical enzymatic reactions during the labeling procedure performed prior to hybridization.

Two of the most widely utilized methods to assess the integrity of the RNA sample and to ascertain if there is any genomic DNA contamination include the traditional agarose gel electrophoresis stained with ethidium bromide and, alternatively, an instrument specifically designed for the analysis of RNA. One particular instrument that has been widely adopted for use in molecular biology laboratories is the Agilent 2100 Bioanalyzer (Agilent Technologies, Santa Clara, CA). Both of these techniques exploit the evaluation of the relative ratios of the 28S and 18S RNA subunits of the ribosome molecule. Details regarding how to prepare, process, and interpret the data from agarose gels are widely available and should be consulted if one uses this method of integrity assessment. With respect to the RNA analysis-specific instrumentation, consult the instrument instruction manual.

The potential presence of chemical contaminants in the RNA sample is most often evaluated using UV spectroscopy. The relative ratios of absorbance at 260 nm versus 280 nm (A260/280) and 230 nm (A260/230) should be 1.8–2.0. Phenol has a maximum absorbance in the 280 and 230 range, and A260/280 ratios below 1.8 are indicative of RNA that may still contain phenol and these samples should not be used. It is best to re-isolate RNA from new samples prior to performing the microarray experiment. If the RNA would be difficult to obtain again from hard-to-obtain samples, an additional purification of the existing sample might sufficiently "clean up" the sample for subsequent use.

29.3.3 DNA Microarray Platforms: Glass Oligonucleotide and Affymetrix Arrays

Two platforms that are widely used today include the glass oligonucleotide containing microarray and the Affymetrix platforms (Affymetrix Inc., Santa Clara, CA). This chapter will discuss both platforms, as they are used extensively in academic and commercial DNA microarray core facilities.

When glass microarray technology was first developed, individual core laboratories often printed their own microarrays containing gene-specific cDNAs or oligonucleotides. At the time of this writing, many core facilities purchase commercially available slides from a variety of vendors and process these slides for the researchers accessing the core lab. While glass microarrays vary with respect to genetic content, some typical generalities include long gene-specific DNA oligonucleotides in the range of 50–70 bp and a glass surface coated with a resin that facilitates binding of the oligonucleotides to the slide surface, for example, an amino silane or epoxy. Oligonucleotides might be spotted on the arrays using contact printing, but some microarrays are generated using an inkjet printing process. Glass microarrays are usually stable for up to 6 months if stored in a desiccator at room temperature. Care should be taken when handling the slides to avoid touching the surface of the array and disrupting the individual spots comprising the oligonucleotides.

Affymetrix arrays are a very popular type of microarray and are used extensively in many academic and commercial service core laboratories. These arrays differ from glass microarrays in several key ways: (1) the oligonucleotides are short oligonucleotides; 25 bp in length, (2) each gene is represented by a probe set consisting of multiple unique oligonucleotides targeting different sequences within the gene; and (3) the oligonucleotides are synthesized directly upon the array by one nucleotide at a time by photolithography. The first generation of Affymetrix arrays utilized probe sets located in the last 600 bp of the 3′ region of the gene and were comprised of 11 pairs of oligonucleotides. The individual pairs contained "perfect match" and "mismatch" probes which differ from the perfect match probe by one nucleotide located in the middle of the oligonucleotide. This design was intended to control for potential cross hybridization between genes of similar sequence. The more recent version of Affymetrix array does not utilize the paired oligonucleotide strategy but uses a probe set consisting of only perfect match probes distributed across the coding region of the gene.

29.3.4 Sample Processing: Labeling of Target, Hybridization to Array, Scanning, and Image Capture

Glass microarrays are often referred to as two-color arrays as two samples are labeled with cyanine (Cy3 or Cy5) dyes and competitively hybridized simultaneously to the microarray (Fig. 29.1). Following hybridization the slides are washed to remove nonspecific binding and scanned with a dual-laser scanner to generate 16-bit .tiff images. The labeling procedures today are by and large completed using an "indirect labeling strategy" and kitted reagents from commercial vendors. A typical flow might involve beginning with 100–500 ng of total RNA from both the test and control samples which are reverse transcribed using oligo-dT and random hexamer primers. The resultant cDNA is subjected to in vitro transcription in the presence of aminoallyl-dUTP to generate antisense aminoallyl-containing RNA (aRNA). The aRNA containing the aminoallyl groups is chemically coupled to the appropriate fluorophore (Cy3 or Cy5) which contain *N*-hydroxysuccinimide (NHS) esters (hence the term "indirect labeling") and are purified by column chromatography. Probes are mixed and hybridized to the arrays in hybridization buffer (often containing 3–5× saline-sodium citrate (SSC) and formamide in the range of 25–40 %) overnight at approximately 42–48 °C depending upon the length of the oligonucleotide on the array and specific hybridization buffer used. Hybridization specifics to include temperature and hybridization buffer composition vary from laboratory to laboratory and are provided by the vendor of the microarray. Following hybridization, the slides are washed in a series of steps that increase in the degree of stringency until nonspecific binding is removed and specific binding is retained. The slides are then scanned using a dual-laser scanner that excites at the 532 (Cy3) and 635 nm (Cy5) wavelengths. Resolution varies from scanner to scanner but a very popular scanner employed in many core facilities known as a GenePix 4000b scanner scans at a resolution of 5 μm (Axon 4000b, Molecular Devices Inc, Sunnyvale, CA). Separate images for Cy3 and Cy5 are generated, overlaid, and subsequently stored as 16-bit image .tiff files.

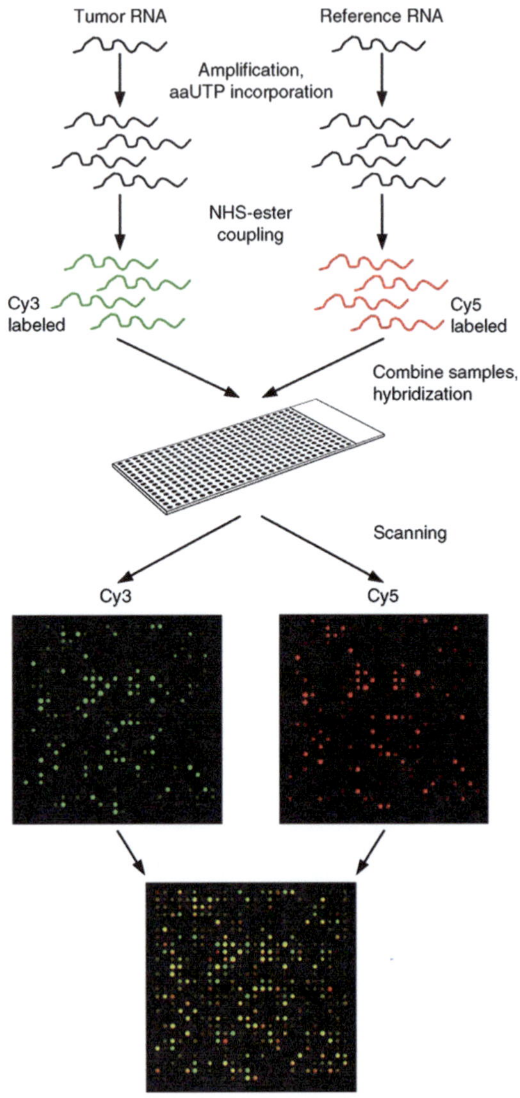

Fig. 29.1 Principle of a two-color microarray experiment. RNA samples from tumor and control tissues are individually amplified, labeled with different fluorescent dyes, and hybridized to a single DNA microarray. The fluorescence intensity is measured for each probe in both samples, and relative gene expression levels can be calculated. By Ghadimi, B. Michael; Grade, MarianBook: Cancer Gene Profiling Book Published: 2009-01-01 DOI: 10.1007/978-1-59745-545-9 Chapter: 16

Affymetrix arrays are processed using a complete instrumentation system from Affymetrix consisting of a hybridization oven, a staining and washing station, and a GC 3000 scanner. When RNA samples are prepared for hybridization to Affymetrix arrays, the procedure employed depends upon the type of expression array to which the samples are being hybridized (for more information visit the Affymetrix website). Affymetrix arrays are often referred to as "single-color arrays" due to the fact that each individual RNA sample is hybridized to its own array. The labeling procedure employs reverse transcription and depending upon the type of array either in vitro transcription in the presence of biotin-labeled ribonucleotide analog to generate biotinylated aRNA or in vitro transcription followed by cDNA conversion, fragmentation, and subsequent end labeling of the cDNA with biotin. Following hybridization to the array, the arrays are stained with streptavidin (which binds the biotinylated molecules) coupled to the fluorophore phycoerythrin and scanned using the Affymetrix 3000 scanner at a resolution of up to 0.51 μm to generate 16-bit image files.

29.3.5 Image Analysis: Extraction of Raw Intensity Values

When working with glass arrays, the dual-color .tiff images are overlaid with a file containing location information for each of the genes. In the case of Axon 4000b scanner, the file utilized for this is known as a gene array list (.gal) file. The .gal file is loaded and overlaid onto the image for extraction of the raw intensity values located within each of the 5 μm pixels located within and collectively comprising the "spot" corresponding to the gene. Cy5 and Cy3 values are measured for each of the pixels located within the spot, and values corresponding to a variety of metrics are calculated including mean and median values of pixels, standard deviation, and total signal. For downstream analysis, most investigators use the background corrected median Cy5/median Cy3 ratio values.

Affymetrix image analysis and intensity value extraction is automatically performed during the scanning procedure. After scanning is performed, a set of files is generated including .dat and .cel files. The files to be utilized in the downstream analysis are the .cel files.

29.3.6 Normalization

Glass oligo arrays and Affymetrix arrays both require normalization prior to further analysis in order to remove systematic variation other than differential expression. Variation can be present in different Cy3 and Cy5 labeling efficiency, scanning, and other experimental parameters. Since the platforms are very different, the normalization methods required differ depending on the platform used (Simon et al. 2003).

Prior to normalization for glass oligo arrays, the researcher can do spot filtering which filters out "bad" spots on the array; this does not filter out an entire gene for each array, but instead replaces a particular gene with a missing value on an individual array. Types of optional spot filters that can be applied are intensity filters (removing low-intensity spots that are near background levels), spot size filters (removing extremely small spots), and spot flag filters (bad spots, control spots, etc. that were flagged in the image analysis).

Some normalization methods available for glass oligo arrays are global median normalization and lowess normalization. With global median normalization, the median $\log_2$ ratio $[\log_2(Cy5/Cy3)]$ is calculated for each array and that amount is subtracted from each $\log_2$ ratio on the array so that the overall median $\log_2$ expression for each array is zero. With lowess normalization, a smooth nonlinear function (lowess) is fit to the M-A plot The M-A plot is a plot of $\log_2$ ratio (Cy5/Cy3) on the y-axis versus average log intensity $([\log_2(Cy5)+\log_2(Cy3)]/2)$ on the x-axis. The lowess function is then subtracted from the un-normalized $\log_2$ ratios for the array in order to obtain the normalized $\log_2$ ratios. Pre- and post-normalization, M-A plots of each array should be examined to determine if the normalization is adequate. One hopes to see random scatter around a $\log_2$ ratio of zero across the range of average $\log_2$ intensity values.

There are several different types of normalization methods available for Affymetrix arrays and a variety of software packages available to implement them (MAS5, PLIER, GC-RMA, dChip (Li and Hung Wong 2001)). One of the most commonly used normalization methods is RMA (robust multichip average) which converts probe level data to a gene-level expression value (Irizarry et al. 2003). The default RMA background correction method corrects the perfect match (PM) probe intensities by using a model based on the assumption that the observed intensities are the sum of signal and noise. Quantile normalization (default method) is used to normalize the PM probes, and the calculation of summary expression measures is done using the median polish method, which fits a multichip linear model to the data, and gives the expression on the $\log_2$ scale (Irizarry et al. 2003). RMA can be implemented using the *rma* function of the *affy* package of the Bioconductor project in the R programming language but is also available in other software packages including BRB-ArrayTools and Expression Console from Affymetrix (Bioconductor, BRB-ArrayTools; R Development Core Team 2008).

29.3.7 Gene Filtering

Prior to analysis for differential expression, prediction, or other analyses, it is common to filter genes that are not changing from array to array. In general, the idea of gene expression analysis is to identify genes that are changing between conditions (class comparison), good predictors of a characteristic (class prediction), or useful in identifying new groups (class discovery). The genes that are most useful in these types of analysis are genes that are most variable; so by filtering out genes that are not changing much between arrays, we are focusing our attention on the most informative genes. Types of filters commonly used are fold change filters, variability filters, and percent missing filters (Simon et al. 2003).

29.3.8 Differential Expression

A common type of microarray analysis is class comparison analysis where the goal is to determine a set of genes that are differentially expressed between two or more groups (classes). There are a variety of methods available to do this. In order to compare continuous variables between two classes, the most common statistical test is the *t*-test. Alternatives to the *t*-test include the nonparametric Wilcoxon rank sum test and permutation tests, which can be used if the data does not follow a normal distribution. In the context of microarray data analysis, for each gene being measured, the researcher would like to compare mean expression between classes using a *t*-test (or another statistical test). In this case, statistical tests are typically modified so that each gene is not considered on its own but in the context of the other genes measured in the microarray experiment. Some of these tests incorporate borrowed information from all the other genes measured in the microarray experiment resulting in a more powerful comparison. Most microarray software packages include one or more tools (R Development Core Team 2008) for comparing differential expression between classes, including limma in Bioconductor, SAM software, dChip, and random-variance *t*-tests (Smyth et al. 2003; Smyth 2005; Tusher et al. 2001; Li and Hung Wong 2001; Wright and Simon 2003). Many of these methods can be extended to more than two classes, for example, ANOVA models and time series models.

In conjunction with conducting the statistical tests comes the challenge of determining which of the genes are differentially expressed. If the researcher is looking at the expression of one gene between two classes, a statistical test is performed and the *p*-value is computed, and if it is less than 0.05 significance level, then the conclusion of the test is that the gene is differentially expressed between the two classes. In a microarray experiment, thousands of genes are measured at one time, and under the null hypothesis that is being tested (no genes are differentially expressed), on average 5 % of the genes will have *p*-values less than 0.05 by chance. For example, if 10,000 genes are measured, on average 500 genes will have *p*-values less than 0.05 by chance and are false positives. This is a multiple comparison problem. Typically in microarray experiments, researchers use methods to control the false discovery rate (FDR), the proportion of genes that are false positives out of the number of genes that are differentially expressed (Simon et al. 2003). One way to control the FDR is using the Benjamini–Hochberg method (Benjamini and Hochberg 1995). This method orders the *p-values* from the statistical tests and then applies an adjustment to the *p*-values based on their ordering. The SAM method implements a permutation method to adjust for multiple comparisons and control the FDR (Tusher et al. 2001).

29.3.9 Clustering

Clustering methods are useful in class discovery microarray experiment. Most often researchers have predefined classes, and they wish to determine genes that are differentially expressed between the classes. In a class discovery experiment, the researchers no longer have predefined classes, and they are exploring the data in order to see if the gene expression profiles of the samples can cluster the samples together into meaningful groups. Since this analysis does not consider prespecified groups or other external information, it is called an "unsupervised method." Types of clustering that are used in microarray analysis are hierarchical clusters, self-organizing maps (SOMs), and principal component analysis; here we will discuss hierarchical clustering and SOMs (Eisen et al. 1998; Tamayo et al. 1999). As in class comparison analysis, the researcher should filter out genes that are uninformative before further analysis, when trying to find genes that are similar and different, one doesn't want to include genes that are the same across all arrays, so variability filters and/or fold change filters are advised. Prior to clustering, another data processing step is often preformed, median centering of genes. This involves calculating the median value for each gene across arrays and subtracting that value from each of the expression values for that gene, so that after subtracting the median is now zero across the arrays.

In hierarchical clustering, a measure of distance is used to describe how close or far apart the arrays/genes are. A popular distance measurement used in microarray experiments is 1-correlation. In hierarchical clustering, one usually performs an agglomerative method, where each array begins as its own cluster and arrays that are most similar based on the distance measure are put in a cluster together. Next, a way to combine clusters is needed; this is called a linkage method. In average linkage, the average distance (correlation) between all pairs of arrays, one from each cluster, is used. Complete linkage uses the maximum distance between an array of one cluster and an array of another cluster. Both of these linkage methods are performed in microarray hierarchical clustering. The clusters are then displayed as a dendrogram, a tree structure, and can be associated with a heatmap as well. The length of the vertical lines connecting the clusters tells how similar the clusters are and is a useful part of interpreting the cluster.

SOMs can be used to cluster together genes that behave in the same way across the arrays (Tamayo et al. 1999). They are often used to cluster genes in time series experiments and can classify genes that behave similarly over time, for example, genes that increase, decrease, or show another pattern over time. The number of clusters must be prespecified by the analyst and the clusters are laid out on a grid whose size must also be specified. If the experiment has three time points and the researcher would like to cluster the genes that behave similarly over time, there are several patterns that can be specified, increases over time, decreases over time, increase then decrease, decrease then increase, increase to the second time point then stay the same, decrease to the second time point then stay the same, or stay the same then increase or decrease. From this, there are about eight patterns to be identified, so eight clusters are specified, with possible grid formations of 2×4 or 4×2. Several clusters numbers and grid sizes should be examined before settling on a final product.

Sometimes clustering methods and heatmaps are used in a supervised analysis, where the differentially expressed genes are selected to use in the clustering programs instead of just the most variable genes. Because genes that are different between classes are preselected for inclusion in the clustering program, this is considered supervised analysis. The clusters that result are primarily a way of displaying the results of the differential expression analysis that has already been done. This can also be useful however. In a SOMs analysis, the researcher can select the genes that are differentially expressed and then use SOMs to display the patterns.

29.4 Typical Results

The typical results of a microarray experiment will generate a list of genes that are differentially expressed between the control and experimental group(s). Criteria for differential expression should include both a quantitative change in relative expression also known as "fold change" and a determination of the degree of significance expressed as a *p*-value. If more than two conditions exist, the researcher might also cluster data to identify groups of coordinately expressed genes. Extrapolating biological meaning from this list of genes or of these clusters can be challenging or, in some instances, quite obvious. For example, if the experiment is designed to see what cytokine genes are being upregulated after addition of a specific growth factor, one can simply look at the gene list generated and the corresponding value of the change. If the goal is not as straightforward as the last example, then determining biological meaning from a list of genes might be more challenging.

To illustrate the typical results of a spotted microarray experiment, we provide the following simple microarray study. Total brain RNA profiled against total skeletal muscle RNA.

29.4.1 Materials and Methods

Three biological replicates of human brain total RNA were profiled against three biological replicates of human skeletal muscle RNA. For each tissue type and biological replicate, 1 μg of total RNA was subjected to reverse transcription and subsequent in vitro transcription to generate aminoallyl aRNA using an Amino Allyl MessageAmp II kit (Ambion Inc, Austin, TX). Five micrograms of each of the aminoallyl aRNA samples was coupled to either Cy3 or Cy5 following recommendations by the manufacture. A dye-swap strategy was employed to control for any potential dye-bias effects in the following manner. Two of the skeletal muscle aRNA samples were labeled with Cy5 and one with Cy3. Two of the brain aRNA samples were labeled with Cy3 and one with Cy5. Brain versus skeletal muscle samples were mixed together and fragmented to 100 bp using 10× fragmentation buffer (Ambion Inc., Austin, TX). Fragmented samples were resuspended with 50 μL of hybridization buffer #2 (Ambion Inc., Austin, TX) and hybridized to three microarrays containing 39,200 60-mer oligonucleotides overnight at 42 °C overnight. Following hybridization the slides were washed one time with a solution 2× SSPE, 0.1 % SDS at 42 °C for 2 min followed by a wash with 0.1× SSPE, 0.1 % SDS at 42 °C for 2 min. Slides were scanned at a resolution of 5 μm with a 4000b Axon Scanner (Molecular Devices, Sunnyvale, CA) to generate multi-image .tiff files. Images were analyzed with GenePix 6.1 software to generate background corrected Cy5 median/Cy3 median values for all spots.

Normalization and analysis of gene expression data was done using BRB-ArrayTools developed by Dr. Richard Simon and Amy Peng (BRB-ArrayTools). Prior to analysis, median background was subtracted and an intensity filter was applied: if at least one of the Cy5 and Cy3 channels was below an intensity of 10, the spot was set to missing. Normalization was then done to "center" each array using lowess smoother and a $\log_2$ transformation was applied to all ratios. M-A plots were examined to check if the normalization method was adequate. Gene filters were then applied; a gene was excluded if the percent of spots missing or filtered out exceeded 50 % and empty spots and control spots were excluded. This array contains duplicate spots of the same genes; for this analysis, duplicate spots have not been averaged and are treated as separate genes.

Random-variance paired *t*-tests were used to determine which genes are differentially expressed between brain and skeletal muscle. Paired *t*-tests were chosen because both the brain and skeletal muscle tissue were hybridized to the

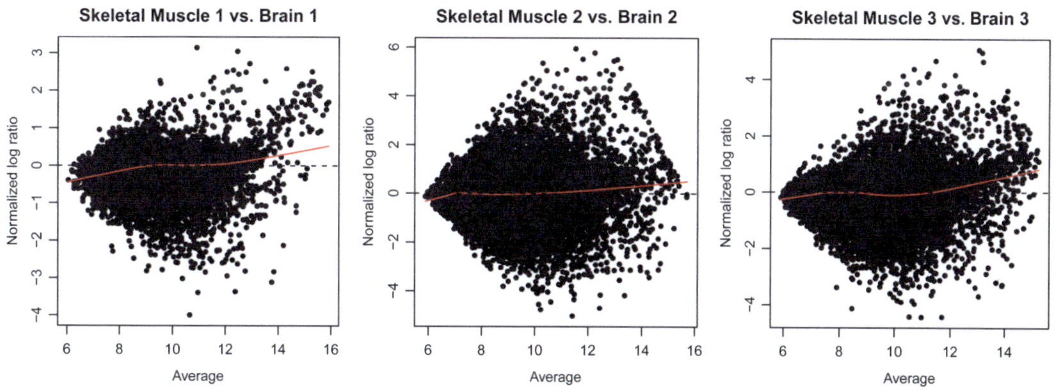

Fig. 29.2 M-A plot of skeletal muscle versus brain. The *x*-axis is the average intensity of the *red* and *green* channels; the *y*-axis is the normalized log2 ratio of the *red*/*green* channels. The first panel corresponds to the first array, the second panel corresponds to the second array, and the third panel corresponds to the third

same arrays, so the samples are naturally paired. (If the brain and skeletal muscle samples each were hybridized with a reference sample, for a total of six arrays, we would have chosen an unpaired *t*-test.) The random-variance paired *t*-test allows sharing information among genes about variation without assuming that all genes have the same variance, which gives a more accurate estimate of the variability when sample sizes are small (Wright and Simon 2003). A FDR of 10 % was used to indicate differential expression (Benjamini and Hochberg 1995).

A total of 39,200 genes were included on the array. After spot and gene filtering, 24,993 genes were left for analysis. M-A plots were examined to check if the normalization method was adequate (Fig. 29.2). We expect random scatter around zero with no obvious trends. 2,683 genes were found to be significantly different between the brain and skeletal muscle groups with FDR less than 10 %. The genes were sorted by fold change to separate the over- and underexpressed genes. For purposes of illustration and due to space limitations, only the top ten and low ten genes differentially expressed between the two tissues are listed here, we sorted by *p*-value within the overexpressed gene set to select the top ten overexpressed genes, and we did likewise in the underexpressed group. The top ten overexpressed and underexpressed genes in the brain samples compared to skeletal muscles can be found in Table 29.1. The table contains the original ranking by FDR, the unique ID, the gene symbol, the estimate of the FDR, the geometric mean intensity of brain and skeletal muscle groups, and the fold change of the brain/skeletal muscle.

29.5 Most Common Problems/Pitfalls and Solutions

Microarray experiments, if conducted properly, often lead to very interesting findings and further the knowledge of the biological process under investigation. However, many investigators have experienced a low-quality outcome. Some of the most common problems encountered can be avoided. When characterizing why an experiment was not performed at the desired quality, begin by examining the initial experimental design. Was the experiment really designed with the appropriate forethought? Did the investigator have sufficient prior knowledge of critical parameters regarding his or her biological system under study? For example, consider the example of a time course study. If the investigator wanted to determine the transcriptional outcome of adding a certain chemical compound to cells in culture and harvested the RNA after 24 h and saw no change in gene expression, perhaps it is due to having harvested the cells at the wrong time point. Perhaps 12 or 48 h might have been more appropriate. As mentioned earlier

Table 29.1 Contains the top ten overexpressed (a) and underexpressed (b) genes in the brain compared to skeletal muscle

Number	Overall rank	Unique ID	Gene symbol	FDR for class effect	Geometric mean expression in class brain	Geometric mean expression in class skeletal	Fold change (brain/skeletal)
(a)							
1	1	PH_hs_0000559	GPM6B	0.0001	55.0091	4.0955	13.432
2	7	PH_hs_0018034	ANKS1B	0.0001	9.5931	0.723	13.268
3	12	PH_hs_0016696	KCNC2	0.0003	2.6153	0.2454	10.657
4	16	PH_hs_0002327	SYNPR	0.0003	14.3458	1.7238	8.322
5	21	PH_hs_0000096	SOX2	0.0003	5.3317	0.446	11.954
6	33	PH_hs_0007357	HDGFRP3	0.0005	8.1107	0.9601	8.448
7	34	PH_hs_0000309	VSNL1	0.0005	86.9215	13.5529	6.413
8	35	PH_hs_0011122	SULT4A1	0.0006	9.4349	1.5225	6.197
9	36	PH_hs_0028961	GJA1	0.0006	64.4063	10.8602	5.930
10	42	PH_hs_0030957	FAM107A	0.0006	58.0892	9.2769	6.262
(b)							
1	2	PH_hs_0046015	PDLIM3	0.0001	5.1546	150.7505	0.034
2	3	PH_hs_0004701	KLF10	0.0001	0.7249	10.1164	0.072
3	4	PH_hs_0026721	MKNK2	0.0001	4.1261	65.7656	0.063
4	5	PH_hs_0048894	MYH1	0.0001	5.3176	152.0331	0.035
5	6	PH_hs_0008881	CACNG1	0.0001	0.5791	7.1479	0.081
6	8	PH_hs_0026211	ANKRD1	0.0001	1.5833	26.9888	0.059
7	9	PH_hs_0017066	ASB5	0.0001	1.1981	35.7125	0.034
8	10	PH_hs_0012655	APOBEC2	0.0002	0.37	6.3861	0.058
9	11	PH_hs_0022034	NEXN	0.0002	0.8525	11.9223	0.072
10	13	PH_hs_0003524	CAMP	0.0003	0.2906	3.9705	0.073

The genes are ordered by original rank by adjusted *p*-value. Included in the table is the original rank out of all 2,683 significant genes; gene identifiers; the estimated FDR for each gene; the geometric mean intensity in the brain group and in the skeletal muscle group; and the fold change, brain/skeletal muscle

in the chapter, a good experimental design is critical to a positive outcome. Other problems that commonly arise relate to the quality of the input RNA. It is critical to assess the potential degradation and chemical contamination issues prior to utilizing the RNA in the experiment. For an investigator in which the isolation of RNA is a new procedure, trial runs should be performed to gain proficiency in the procedure. Some of the problems might arise from the processing of the arrays in the laboratory. The procedure of labeling, hybridization, washing, and scanning has to be adhered to closely by the technologists performing the procedure. In addition, technologists should also be appropriately trained to conduct the procedures. Proficiency testing for the technologist new to the microarray procedure and who is performing the procedure should also be utilized.

Inadequate sample size is also a pitfall that researchers run into in conducting microarray experiments. Often the number of biological replicates for cell line studies would not be adequate for animal studies and what is adequate for animal studies is not adequate in human studies. The researcher needs to think carefully about what sort of fold changes they would like to detect and how much variability is present in his or her study. After the microarray experiment has been performed, time should be invested in the normalization step and examining the quality assurance measures. If one normalization method does not appear to be working for your particular microarray experiment, you can try others and may have more success. In the data analysis step, the researcher needs to take into account the experimental design in the analysis. Are the samples paired or independent? For spotted arrays, is there a common reference group or dye swap that needs to be considered? How many groups are present and which type of statistical test is most appropriate? These are common issues in the analysis phase of the experiment that can be overcome by careful thought and consultation with experts if necessary.

29.6 Discussion/Summary

DNA microarray technology has contributed tremendous amounts of information and has helped to elucidate critical molecular underpinnings of many diseases as well as normal biological processes. When DNA microarray technology was first developed, it was recognized immediately by the scientific community that it provided an unprecedented opportunity to ask and subsequently answer biological questions previously unattainable due to current technological limitations. It is taken for granted that the ability to profile the transcriptional activity of the entire genome simultaneously was only "wishful thinking" less than 20 years ago. While microarrays are now routinely used in biological research, one of the current challenges is the extraction of biological meaning from large datasets. This is where pathway analysis and other computational strategies may help. As more information is gained regarding the functions of the proteins encoded by genes being queried, then biologists will be better prepared to extract meaning from the experiment. In addition, more computational methods to identify gene networks reliably are needed. In closing, when conducted appropriately, DNA microarray technology can be an extremely useful tool for the biological researcher wishing to gain insight into his or her biological system under study.

References

Benjamini Y, Hochberg Y (1995) Controlling the false discovery rate: a practical and powerful approach to multiple testing. J R Stat Soc Ser B 57:289–300

BRB-Array Tools version 4.1.0 stable release developed by Dr. Richard Simon and Amy Peng Lam

Bulyk ML (2006) DNA microarray technologies for measuring protein-DNA interactions. Curr Opin Biotechnol 17(4):422–430

DeRisi J, Penland L, Brown P, Bittner M, Meltzer P, Ray M, Chen Y, Su Y, Trent J (1996) Use of a cDNA microarray to analyze gene expression patterns in human cancer. Nat Genet 14:457–460

DeRisi J, Iyer V, Brown P (1997) Exploring the metabolic and genetic control of gene expression on a genomic scale. Science 278:680–686

R Development Core Team (2008) R: A language and environment for statistical computing. R Foundation for Statistical Computing, Vienna. ISBN 3-900051-07-0. URL http://www.R-project.org

Dobbin KK, Zhao Y, Simon RM (2008) How large a training set is needed to develop a classifier for microarray data? Clin Cancer Res 14(1):108–114

Edelmann L, Hirschhorn I (2009) Clinical utility of array cGH for the detection of chromosomal imbalances associated with mental retardation multiple congenital anomalies. Ann N Y Acad Sci 1151: 157–166

Eisen MB, Spellman PT, Brown PO, Botstein D (1998) Cluster analysis and display of genome-wide expression patterns. Proc Natl Acad Sci U S A 95(25): 14863–14868

Han E, Wu Y, McCarter R, Nelson JF, Richardson A, Hilsenbeck SG (2004) Reproducibility, sources of variability, pooling, and sample size: important considerations for the design of high-density oligonucleotide array experiments. J Gerontol A Biol Sci Med Sci 59(4):306–315

Irizarry RA, Bolstad BM, Collin F, Cope LM, Hobbs B, Speed TP (2003) Summaries of Affymetrix GeneChip probe level data. Nucleic Acids Res 1(4):e15

Leung YF, Cavalieri D (2003) Fundamentals of cDNA microarray data analysis. Trends Genet 19(11): 649–659

Li C, Hung Wong W (2001) Model-based analysis of oligonucleotide arrays: expression index computation and outlier detection. Proc Natl Acad Sci U S A 98:31–36

McShane LM, Shih JH, Michalowska AM (2003) Statistical issues in the design and analysis of gene expression microarray studies of animal models. J Mammary Gland Biol Neoplasia 8(3):359–374

Schena M, Shalon D, Davis R, Brown P (1995) Quantitative monitoring of gene expression patterns with a complimentary DNA microarray. Science 270:467–470

Simon R, Radmacher MD, Dobbin K (2002) Desing of studies using DNA microarrays. Genet Epidemiol 23:21–36

Simon RM, Korn EL, McShane LM, Radmacher MD, Wright GW, Zhao Y (2003) Design and analysis of DNA microarray investigations. Springer, New York, NY

Smyth GK (2005) Limma: linear models for microarray data. In: Gentleman R, Carey V, Dudoit S, Irizarry R,

Huber W (eds) Bioinformatics and computational biology solutions using R and bioconductor. Springer, New York
Smyth GK, Yang Y-H, Speed TP (2003) Statistical issues in microarray data analysis. Methods Mol Biol 224:111–136
Tamayo P, Slonim D, Mesirov J, Zhu Q, Kitareewan S, Dmitrovsky E, Lander ES, Golub TR (1999) Interpreting patterns of gene expression with self-organizing maps: methods and application to hematopoietic differentiation. Proc Natl Acad Sci U S A 96(6):2907–2912
Tusher V, Tibshirani R, Chu G (2001) Significance analysis of microarrays applied to transcriptional responses to ionizing radiation. Proc Natl Acad Sci U S A 98:5116–5121
Wright GW, Simon R (2003) A random variance model for detection of differential gene expression in small microarray experiments. Bioinformatics 19: 2448–2455

Metabolomics

30

Yazen Alnouti

Abstract

The interaction between biological systems and their environments is complex and often involves mechanisms beyond changes in the genome and/or the proteome at the metabolome level. Metabolomics is a field of science, which involves the comprehensive quantitative and qualitative profiling of multiple metabolites and their interactions with environmental variables of interest such as diet, disease, environment, or exposure to chemicals. Monitoring these metabolites represents the closest end-point to understand the cell dynamics at the functional level as compared to genomics or proteomics. Metabolomics is heavily applied in the identification of biomarkers for disease diagnosis and prognosis, drug toxicity and efficacy, genetic polymorphisms, drug metabolism, and fluxomics. These applications are discussed in detail in this chapter.

There are at least 3,000 metabolites in the human body and up to 200,000 metabolites in the plant kingdom, which are very different in their physical and chemical properties such as pK_a, polarity, solubility, and size. The comprehensive quantitative analyses of the metabolome and handling the amount of data generated from such analyses pose substantial challenges. Therefore, metabolomics is a technique-driven enterprise that heavily relies on the analytical and bioinformatics support of data generation and analysis.

There is no single global analytical technique that combines all the characteristics required for comprehensive quantitative analyses of the metabolome. However, mass spectrometry (MS) and nuclear magmatic resonance (NMR) represent the analytical techniques of choice in metabolomics. The theory, application, and experimental considerations of these techniques, which are relevant to metabolomics, are presented in this chapter.

Y. Alnouti, Ph.D. (✉)
Department of Pharmaceutical Sciences,
University of Nebraska Medical Center,
Omaha, NE, 68198-6025, USA
e-mail: yalnouti@unmc.edu

H. Xiong and H.E. Gendelman (eds.), *Current Laboratory Methods in Neuroscience Research*,
Springer Protocols Handbooks, DOI 10.1007/978-1-4614-8794-4_30,

Finally, the raw data generated from the quantitative analysis of the metabolome requires several stages of processing to improve its quality, reduce its size, and convert it to a format that can be subject to statistical analysis. Different strategies of data processing and statistical analysis are discussed in this chapter.

Keywords
Metabolomics • LC–MS • UPLC • HPLC • NMR • Structure elucidation • PCA

30.1 Introduction

30.1.1 History of Metabolomics

The determination of the human genome in the 1990s has opened wide doors to the scientific community to study the relationship between genetics and disease diagnosis, causes, progression, pathogenesis, and treatment. The principal emphasis in biomedical studies in that era focused on the simultaneous determination of gene expression changes between subjects and the correlation of these changes to end-points of clinical significance, and this type of study has been given the name of transcriptomics or genomics. Hopes were very high that genetic differences might be able to account for all disease processes, which may lead to new diagnostic approaches and better personalized therapies. The revolution in molecular biology through the fields of genomics/transcriptomics was facilitated by the development of high-throughput and automated analytical methods for the array analysis of DNA/ RNA.

The subsequent advances in the analytical techniques for protein detection and quantification have led to the evolution of the field of proteomics. The proteome is far more complex and dynamic than the genome and the transcriptome. Therefore, the analytical support for proteomics is associated with more challenges and demands than those associated with genomics. However, since qualitative and/or quantitative changes at the protein level are more likely to lead to corresponding changes in the function of a biological system than changes at the transcriptome level, there was considerable motivation to surmount the technical difficulties associated with proteomics.

During the past 10 years, the complexity of the interaction between biological systems and their environments has gained increasing recognition through advances in the fields of molecular biology and epigenetics. It is now realized that biological systems have very advanced mechanisms to interact with their environments beyond changes in gene expression or even protein translation. These mechanisms include chemical modification of DNA, posttranslational modification of proteins, and the translocation and stability of translated proteins. Therefore, it became accepted that we need to work at a higher level than the genome, transcriptome, or the proteome to better understand the interaction between biological systems and their environments.

Cells operate via a complex network of small molecules/metabolites that drive their biochemical processes. Metabolomics is a field of science, which involves the comprehensive quantitative and qualitative profiling of multiple metabolites and their interactions with environmental variables of interest such as diet, disease, environment, or exposure to chemicals. Monitoring these metabolites represents the closest end-point to understand the cell dynamics at the functional level as compared to genomics or proteomics. Therefore, metabolic biomarkers are closely identifiable with real biological end-points and provide a global systems interpretation of biological effects. In addition, metabolomics' cost per sample and per analyte is lower than those associated with other "omics." Furthermore, metabolomics has the advantage of requiring minimally invasive or noninvasive biofluids such as blood and urine as opposed to tissue biopsies. These advantages, especially the concept of proximity to functional end-points, led to research efforts which parallel those associated with genomics and proteomics but at the metabolome level.

Even though the term is relatively new, the concept of metabolomics in the sense of monitoring metabolite's response to certain treatments both in vitro and in vivo models is not a new field. Such efforts date back to the 1960s, when Horning et al. investigated the changes in the urinary profiles of a wide range of endogenous compounds (Dalgliesh et al. 1966), including aromatic acids and steroids (Horning et al. 1969), under different pathological conditions. These efforts were then extended to study metabolite profiles in human saliva (Ward et al. 1976). Several studies on "quantitative metabolic profiling" nuclear magmatic resonance (NMR) and mass spectrometry (MS) were then applied to cardiac diseases (Kopp et al. 1980), inborn errors of metabolism (Ward et al. 1976; Millington et al. 1990), and exposure to hepatotoxic drugs (Sequeira et al. 1990).

The term "metabonomics," however, was not established until it was formally defined in 1999 by Jeremy Nicholson and his colleagues as "the quantitative measurement of the dynamic multiparametric metabolic response of living systems to pathophysiological stimuli or genetic modification" (Nicholson et al. 1999). In their pioneering studies, Nicholson and colleagues were focused on the application of metabolite profiling in assessing drug toxicity. In 2001, Oliver Fiehn, who was focused on the application of metabolite profiling in plant biology, established the term "metabolomics" and stated that, "A comprehensive and quantitative analysis of all metabolites could help researchers understand such systems. Since such an analysis reveals the metabolome of the biological system under study, this approach should be called metabolomics." (Fiehn 2001). These days, most scientists use the two terms interchangeably to describe the same field. However, literature search reveals that the number of publications using the term "metabolomics" is dramatically more than those using the term "metabonomics." Therefore, we will be using the term "metabolomics" throughout this chapter.

30.1.2 Global and Targeted Approaches of Metabolomics

In a simplistic view, a metabolomics experiment is a comparative analysis of samples to identify if these samples can be distinguished on the basis of quantitative and/or qualitative differences in their metabolic profiles. In general, two strategies can be followed in metabolite profiling analysis to perform such experiment. The first strategy targets a limited number of metabolites or a subset of known metabolites that either have similar structures or involved in a specific metabolic pathway. The metabolites of interest have been already selected and their absolute concentrations are measured usually to prove a hypothesis based on a priori information about the targeted metabolites (Want et al. 2010). This approach is referred to as "targeted" metabolomics.

The second strategy relies on a holistic approach towards the profiling of as many metabolites as possible without a prior knowledge of the identity of these metabolites. The decision of what metabolites may be of interest is taken after the generation of unbiased data for all metabolites. This "non-targeted" approach produces a comprehensive list of metabolites with absolute or relative quantification. These metabolic changes can then be mapped to specific pathways and provide biomarkers and/or mechanistic information.

The two approaches have different requirements. In targeted analysis, sample preparation and analytical conditions are optimized to maximize recovery and sensitivity for the particular metabolites of interest. In addition, targeted analyses are usually thoroughly validated to ensure high accuracy, precision, and robustness of the method. Rigorous validation is a pre-requirement because the method itself is usually used to validate the findings of non-targeted studies. In contrast, non-targeted analyses rely on "generic" experimental conditions and "high resolution" instruments that will minimize loss of metabolites and result in the detection of the largest portion of the metabolome given the wide variation in the physicochemical properties of the various metabolites. In addition, non-targeted analyses rely heavily on bioinformatics support to process and interpret the huge amount of data generated.

30.2 Applications of Metabolomics

Metabolomics aims to characterize the metabolic profiles in biological systems as a part of the whole systems biochemistry and thus promises to provide crucial information related to understanding not only the mechanisms of diseases and their treatment but also the efficacy and toxicity of these treatments. Therefore, applications of metabolomics are diverse and include various models spanning from cell lines to genetically manipulated animal models, and take place at various stages including both preclinical to clinical. These will be described in detail in the following sections.

30.2.1 Preclinical Discovery and Development

The advent of combinatorial chemistry in pharmaceutical industry resulted in a tremendous increase in the rate of synthesis of new chemical entities (NCEs). This new pipeline puts tremendous strain on downstream high-throughput methods to screen the array of compounds for efficacy and toxicity to find the best candidates to move forward along the drug development process. Consequently, moving attrition earlier into the process became a primary goal of the pharmaceutical industry. Therefore, biomarkers of both efficacy and safety are in high demand as a means to shorten and simplify clinical development.

Most adverse drug reactions (ADRs) are related to liver or kidney toxicities. Serum transaminases for hepatic toxicity or serum urea nitrogen or creatinine for renal toxicity are well-established peripheral biomarkers. To be accepted as alternatives to these gold standards, metabolomics approaches should provide more sensitive and/or specific biomarkers. In addition, metabolomics approaches have a great advantage of requiring minimal sample volumes rather than terminal large samples, which can be critical in small animal models such as mice. Furthermore, a metabolomics biomarker could be identified in a more accessible biological fluid such as saliva or urine.

The Consortium of Metabonomic Toxicology (COMET) has comprehensively assessed the utility of metabolomics in preclinical drug toxicity (Lindon et al. 2003). The major outcome of COMET was the generation of standardized ^{1}H NMR spectra from urine obtained from rats treated with 80 known toxins of various target organs. Organ-specific targets were identified and they were used with relatively high selectivity and sensitivity in the prediction of liver and kidney toxicity (Ebbels et al. 2007).

Significant advances in the application of metabolomics have certainly been made in the area of efficacy and toxicity biomarkers. However, it still has a long way to go before it can be considered standard practice. The fact that many toxins produce the same changes in metabolic profiles regardless of their mechanism has dampened the enthusiasm of toxicologists to use these metabolomics-based markers in toxicity studies. These commonly changing metabolites, including 2-oxoglutarate, acetate, citrate, creatine, creatinine, glucose, hippurate, lactate, and succinate, have been often referred to as the "usual suspects" to metabolomics investigations (Robertson 2005). These compounds are indeed general biomarkers of change because they are altered in many toxicities and disease states. However, they are not specific enough to be useful as diagnostic biomarkers.

One of the most important applications of metabolomics to preclinical pharmaceutical discovery is identifying *markers of efficacy*. Once efficacy markers are identified, they can be used to set high-throughput and cost-effective assays to screen potential drug candidates for efficacy against relevant diseases at early stages of drug discovery. The combination of metabolomics approaches with use of knock-out or knock-down mice has become a valuable tool which provides a deeper understanding of the mechanisms of drug efficacy. For example, using a metabolomics approach to quantify the lipid metabolome in obese (NZO × NON)F1 male mice revealed rosiglitazone-induced hypolipidemia by disrupting the mobilization of liver lipids, which exerted tissue-specific effects on lipid metabolism. It was concluded that the phenotypes induced by rosiglitazone are mediated by multiple tissue-specific metabolic variables. The discovery of this multifactorial mechanism of action was facilitated by the quantification of the plasma lipid metabolome, which reflected more general metabolome changes in tissues (Boros et al. 2003).

30.2.2 Disease Biomarkers

In 2001, an NIH working group standardized the definition of a biomarker as "a characteristic that is objectively measured and evaluated as an indicator of normal biological processes, pathogenic processes, or pharmacologic responses to a therapeutic intervention." Measurements of specific endogenous metabolites, such as glucose, urea, and cholesterol, to help in the diagnosis of diseases have long been a part of the practice of medicine. Metabolomics has a great potential to improve diagnostic capabilities, including repertoires of early warning indicators of disease risk and pathogenesis, as well as to enhance capabilities for monitoring and surveillance of disease control programs. Furthermore, metabolomics can be performed on bodily fluids readily accessible to the physician, including urine, blood, saliva, breath condensate, sweat, ascites, and amniotic and cerebrospinal fluid. One of the most well-established areas of metabolomics-based biomarker studies of disease is the diagnosis of inborn errors of metabolism in children (Shlomi et al. 2009). In addition, several publications and review articles have addressed the potential of disease diagnosis using metabolomics approaches in the areas of coronary cardiovascular diseases (Brindle et al. 2002; Waterman et al. 2010), neuroscience (Quinones and Kaddurah-Daouk 2009), general oncology (Claudino et al. 2007), meningitis (Coen et al. 2005), autoimmune diseases (Seeger 2009), parasitic infections (Wang et al. 2004), renal cell carcinoma (Moka et al. 1998), ovarian (Odunsi et al. 2005), colorectal (Wang et al. 2010), and breast cancer (Oakman et al. 2011), tuberculosis (Parida and Kaufmann 2010), cystic fibrosis (Wetmore et al. 2010), hepatobiliary (Gowda 2010), metabolic syndrome (Oresic 2010), and liver diseases and transplant (Duarte et al. 2005), rheumatoid arthritis (Roy et al. 2004), infertility (Bender et al. 2010), and eye diseases (Young and Wallace 2009).

30.2.3 Pharmacometabolomics

Pharmacogenomics is a well-established field that aims to understand and predict the effect of genetic make-up and genetic polymorphisms of individuals on pharmacological, toxicological, and pharmacokinetic properties of drugs. Similarly, the metabolism status of individuals may affect their response to drugs. Therefore, metabolomics has a great potential to guide individualized therapy. Unlike pharmacogenomics, metabolomics approaches are sensitive to both genetic and environmental influences that determine the metabolic fingerprint of an individual. It has been demonstrated that the urinary metabolic profiles of individuals can provide signatures, or metabolic phenotypes, which allow for the early detection of liver toxicity and may even predict individuals' risk for liver toxicity before exposure to acetaminophen, a typical hepatotoxin, based on bacterially generated metabolites (Clayton et al. 2006, 2009). This application of metabolomics is called "pharmacometabolomics" (Winnike et al. 2010).

30.2.4 Drug Metabolism

Parallel to the need for high-throughput efficacy and toxicity screens, similar needs exist for metabolite identification (ID) of NCEs. Targeted metabolite ID is based on the LC–MS/MS analysis of ions that belong to predicted metabolic modifications of the parent compound. For example, the predicted parent masses of +16, +80, and +176 Da would suggest hydroxylation, sulfation, or glucuronidation, respectively. However, this approach is ineffective for identification of unusual/unexpected metabolites, metabolites resulting from a combination of metabolic pathways, or for metabolites that have different fragmentation patterns than the parent compound. This problem can be eliminated by the use of radiolabeled compounds; however, these are not always available or may not be available in sufficient amounts. Alternatively, metabolomics analyses will identify any components responsible for distinguishing the treatment group from untreated groups without any previous knowledge of the compound being investigated. The differences between vehicle-treated and xenobiotic-treated mice are primarily defined by the presence of the xenobiotic and its metabolites. Through the application of multivariate analysis (MVA), separation of vehicle-treated and xenobiotic-treated sample groups can be achieved and the structure of responsible components (metabolites) identified. This is the same approach used for the analysis of endogenous metabolites, but applied to identify the metabolites of xenobiotics instead. Therefore, both predictable and unpredictable metabolic transformations can be detected without the need for radiolabels (Plumb et al. 2003).

30.2.5 Global System Biology: The Integration of "Omics"

In global system biology, data and results from different "omics" studies at different levels of molecular biology, such as transcriptomics, proteomics, and metabolomics, are integrated. For example, both transcriptomic and metabolic changes following the administration of bromobenzene, a hepatotoxin, to rats were studied. It was found that changes in levels of genes and metabolites were related to the degree of necrosis, providing putative novel markers of hepatotoxicity. The integrated analysis of hepatic transcriptomics and plasma metabolomics was able to more sensitively detect changes related to hepatotoxicity and discover biomarkers (Heijne et al. 2005).

A combination of metabolomics and proteomics approaches was employed in a study of acute hepatotoxicity induced by valproic acid (VA) in rats. NMR metabolomics analysis of urine as well as liver tissue showed that VA caused an increase in glucose concentrations. Subsequent proteomics studies identified two proteins that are involved in the conversion of glycogen to glucose, glycogen phosphorylase and amylo-1,6-glucosidase, which were upregulated by VA administration. It was also demonstrated that glucose reabsorption via kidney tubules flux from glucose to glycogen in the liver is reduced. Therefore, combining metabolomics, proteomics, and metabolic flux platforms led to a more precise mechanistic understanding of changes in glucose homeostasis induced by VA (Beger et al. 2009).

30.2.6 Functional Genomics

Functional genomics aim to reveal the roles of genes discovered by the complete genome sequence of several organisms by moving from DNA sequence to biological functions. Many of these genes are called "silent" genes because they do not show any overt phenotype, when they are deleted from the genome. This, however, does indicate that these genes do not have a function but rather indicates they may have an invisible phenotype. Comparing the intracellular concentrations of metabolites between wild type and knock-out cells was shown to reveal phenotypes for proteins active in metabolic regulation by identifying the site of action, in the metabolic network, of a silent gene (Raamsdonk et al. 2001). Metabolomics also has been used to demonstrate biochemical sequelae of mutations in animal models of diabetes/metabolic syndrome (Plumb et al. 2005; Connor et al. 2010), apoptosis (Lee and Britz-McKibbin 2010), idiosyncratic liver injury (Maddox et al. 2006), metabolic syndrome/diabetes: diabetic db/db, nonalcoholic fatty liver diseases (Dumas et al. 2006), apolipoprotein E-null (atherosclerosis model) mice (Leo and Darrow 2009), and farnesoid-X-receptor-null (FXR) mice (Cho et al. 2010).

30.2.7 Fluxomics

Metabolic network and pathway models can be based on metabolic profiling over time using metabolomics approaches (i.e., fluxomics) (Gottschalk et al. 2008; Meadows et al. 2008). Methods for quantifying intracellular fluxes are important for understanding the interactions of the pathways in metabolic networks. Flux estimation methods may be based on the monitoring of intracellular metabolites over time, labeling experiments, or a combination of metabolite profiling and labeling experiments. The labeling experiments are typically performed by incubation with labeled glucose and/or amino acids. The metabolomics platforms, NMR (Sauer et al. 1999) and MS (Marx et al. 1996), combined the capabilities of both the simultaneous measurements of all metabolite levels and determination of isotopic pattern.

30.3 Analytical Techniques in Metabolomics

DNA/RNA are derived from a combination of four different nucleosides and proteins are derived from the combination of 20 amino acids. In contrast, there are at least 3,000 metabolites in the human body (Duarte et al. 2007) and up to 200,000 metabolites in the plant kingdom (Fiehn 2001). These metabolites include salts, sugars, acids, bases, lipids, steroids, and other compounds, and therefore are very different in their physical and chemical properties such as pK_a, polarity, solubility, and size. In addition, the dynamic range of these metabolites is more than 9 orders of magnitude (picomolar–millimolar range). Therefore, comprehensive analyses of the metabolome pose substantial challenges to the analytical support of metabolomics studies (Dunn et al. 2005). An ideal analytical technique for the support of metabolomics would provide as comprehensive and sensitive a metabolic profile as possible and would not be biased towards particular classes of metabolites. In addition, the technique should have a sufficiently broad dynamic range and provide both quantitative and qualitative results for the structural elucidation of the analytes of interest. Finally, the untargeted nature of metabolomics experiments typically requires a large number of samples to produce statistically significant results. Therefore, high-throughput analytical support with minimum or no sample preparation is required in metabolomics. There is no single analytical technique that combines all such characteristics and, thus, a combination of multiple and complementary analytical techniques is necessary.

Several such analytical techniques are employed and include infrared (IR) (Kaderbhai et al. 2003), ultraviolet (UV) (Zhang et al. 2011), coulometric, fluorescence, and electrochemical (Gamache et al. 2004) detection. However, mass spectrometry (MS) and NMR represent the analytical techniques of choice in metabolomics. Excellent reviews are available on the technical aspects and metabolomics applications of MS (Villas-Boas et al. 2005; Lenz and Wilson 2007; Monton and Soga 2007; Kiefer et al. 2008; Lu et al. 2008; Han et al. 2009; Wu et al. 2009; Junot et al. 2010; Drexler et al. 2011; Lei et al. 2011) and NMR (Kaddurah-Daouk et al. 2008; Powers 2009; Kim et al. 2010; Schripsema 2010; Zhang et al. 2010). In addition, excellent books are available in this area (Robertson and Lindon 2005; Tomita and Nishioka 2005; Lindon et al. 2007; Nielsen and Jewett 2007; Weckwerth 2007; Griffiths 2008; Knapp and Cabrera 2009).

Both NMR and MS techniques have advantages and limitations (Lenz and Wilson 2007). Therefore, it is unrealistic to think that one single analytical technique can characterize the metabolome. Instead, the two techniques combined provide complementary data that cover a larger portion of the metabolome than data provided by each technique separately.

30.3.1 Nuclear Magmatic Resonance

Some nuclei, such as ^{1}H, ^{13}C, ^{15}N, ^{19}F, ^{29}Si, and ^{31}P, have angular momentum properties because the spinning positively charged nuclei generate a magnetic field, which is characterized by a magnetic momentum. The magnetic properties of these nuclei result in three unique features for every molecule in NMR spectra, which can then be used to identify unknown analytes. These features include chemical shift, multiplicity of the splitting pattern, and the magnitudes of the splittings.

The NMR instrument is simply composed of a magnet and a probehead. The strength of the magnetic field determines the resolution, sensitivity, as well as the price of the NMR instrument. Most metabolite profiling studies reported in the literature are carried out at magnetic field strengths higher than 600 MHz. The sample is loaded into the probehead and placed in the center of the magnet. The probehead contains a coil that both sends radiofrequency (RF) pulses to the sample and detects the NMR signals returning from the excited nuclei. Probeheads are available with different coil sizes that can support sample volumes down to 2 mL (Schlotterbeck et al. 2002). High-throughput and automated sample handling systems are available that support flow injections from 96-well format (Lindon et al. 2003).

NMR has several advantages that make it an attractive analytical technique in metabolomics including: (1) Minimum matrix effect: NMR is less sensitive to matrix constituents and solvents compared to MS. Therefore, preparation of biofluid samples for NMR studies requires minimum sample preparation and typically only requires dilution of the sample with simple buffers containing some deuterium oxide. (2) Unbiased detection: In contrast to MS, NMR has the same sensitivity to all detected analytes and therefore relative concentrations of metabolites can be obtained by simple integration and absolute concentrations by comparison with an internal standard (IS). (3) Reproducibility: NMR data has relatively high reproducibility between experiments as well as between various laboratories. (4) Dynamic range: The dynamic range of NMR is only limited by sensitivity and can achieve 4–6 orders of magnitude. (5) Nondestructive detection, which allows conservation of samples for multiple usage.

Even though NMR is the most or one of the most used analytical techniques in metabolomics, NMR has some disadvantages including: (1) Sensitivity: NMR is generally 10- to 100-fold less sensitive than MS and requires at least micrograms of an analyte to be detected. (2) Speed: Acquisition rate of NMR is slower than MS, which limits its compatibility with online hyphenation of separation frontends. (3) Cost: NMR instrumentation is generally more expensive than MS. (4) Diversity: The commercially available NMR platforms are not as diverse as those offered by MS instrumentation. (5) NMR has difficulty identifying large molecules such as proteins.

30.3.2 Mass Spectrometry (MS)

Mass spectrometry is a group of analytical techniques that separate analytes in their gaseous state according to their mass to charge ratio (m/z). Properties of MS that make it an attractive technique in metabolomics include: (1) Versatility: A diversity of ionization sources and mass analyzers are available, which allows the analysis of a wide variety of molecules including polar and nonpolar as well as small and large molecular weights. (2) Selectivity: Multiple levels of selectivity based on mass determination of analytes as well as their fragmentation pattern. (3) Compatibility with liquid and gas separation frontends: A variety of ionization sources allows efficient online hyphenation with gas chromatography (GC), liquid chromatography (LC), and capillary electrophoresis (CE), which adds another level of selectivity to MS techniques. (4) Cost: The acquisition and maintenance of MS instrumentation is relatively inexpensive when compared to other analytical techniques such as NMR. (5) Technical demand: Operating MS instrumentation and software is relatively not technically demanding. (6) Sensitivity: MS is much more sensitive than NMR.

Despite its widespread use in metabolomics, MS has some disadvantages including: (1) Matrix effect: Other endogenous molecules, especially salts, compete over the charge in the MS source, which may suppress/enhance the signal intensity of the analytes of interest. Therefore, in contrast to NMR, elaborate sample preparation and chromatographic separations are required in MS-based metabolomics. (2) MS data are highly dependent on the type of MS used, operation conditions, and chromatography and sample preparation conditions, which raises issues with data reproducibility and makes comparison between different labs or different studies more difficult than with the use of NMR (Gika et al. 2010). (3) Separation of isobaric compounds: Isobaric compounds that share the same mass cannot generally be separated by MS analyzers, except the ion mobility MS. However, isobaric compounds can be separated by MS/MS if they have different fragmentation patterns. (4) Dynamic range: MS dynamic range is typically restricted to 3 orders of magnitude. (5) Biased representation of analytes: Analytes ionize in different efficiencies in the MS source. Therefore, various analytes of the same concentration may produce different instrument responses depending on their ionization efficiency, which means quantification of analytes is not possible without authentic standards.

30.3.2.1 The "Matrix Effect" Obstacle

LC–MS studies are complicated by the phenomenon known as "matrix effect" whereby the presence of co-eluting substances suppresses or enhances the ionization of a particular analyte. Therefore, the same amount of analyte results in different MS responses based on the matrix in which it exists (Taylor 2005). Several strategies can be used to decrease/prevent the matrix effect including: (1) The application of more selective sample preparation techniques can prevent the co-extraction of as many endogenous matrix components as possible. (2) The improvement of LC separation can enhance the resolution of analytes of interest from other charge-competing matrix components which may decrease/prevent the matrix effect on the ionization of the analytes of interest. However, meticulous sample preparation and chromatography conditions may markedly lower the throughput of the metabolomics analyses. (3) The use of capillary LC reduces the exposure of MS to amount of biological matrix, thereby reducing the matrix effect. (4) Isotope-labeled internal standards (ISs) have the same retention time (RT) and behave the same as their corresponding unlabeled analytes. Therefore, these ISs are expected to encounter the same matrix effect as there are corresponding unlabeled analytes, which compensates for the variation in injection, sample preparation, instrumental parameters, and matrix effects (Matuszewski et al. 2003). This approach is not practical in metabolomics experiments because a large number of analytes, which are not known yet, are analyzed. (5) In vivo stable isotope labeling in cell cultures by incubation with ^{2}H, ^{13}C, and ^{15}N, which leads to complete substitution of the naturally occurring isotopes of these elements with these stable isotopes, can be used. These labeled cells can provide ISs for all endogenous components by mixing the labeled cells with cells of interest in known proportions and determining the ratio of the peak area of any metabolite in cells to that of the corresponding peak area of stable isotope from the labeled cells (Kim et al. 2005).

30.3.2.2 Instrumentation

LC–MS Ionization Sources

All MS instrumentation requires an ionization source responsible for converting analyte molecules into ions in the gas phase. Both cations and anions can be detected in MS by selecting positive or negative ionization modes.

Ionization in atmospheric pressure ionization (API) sources allows the direct coupling of LC systems to MS, where LC effluents directly flow into the ionization source. The most widely used API techniques include electrospray ionization (ESI), atmospheric pressure chemical ionization (APCI), and photon pressure ionization (PPI). All API techniques are considered soft ionization techniques that result in minimum in-source fragmentation. In addition API techniques support the formation of multiply charged ions, which expands their mass dynamic range and allows the analysis of both small and large molecules. APCI, ESI, and PPI are complementary ionization techniques and it was shown that using a dual ESI/APCI source resulted in a 34 % increase in the coverage (number of features detected) of the metabolome of human erythrocytes (Syage et al. 2004; Nordstrom et al. 2008).

API especially ESI is the most commonly used ionization source in metabolomics. ESI utilizes a high-electric field and high temperature to evaporate and produce charged droplets from sprayed solutions. The major drawback of API is the "matrix effect" phenomenon, which results from the competition of co-eluting molecules during the ionization process. This leads to the signal suppression of analytes by endogenous sample components, especially salts and lipids. Therefore, sample preparation steps are required before LC–MS analysis of biological samples to control the matrix effect on ESI signal.

The throughput LC–MS approach can be a limiting factor in large-scale metabolomics studies because LC separation and MS detection are performed sequentially. In addition, the requirement of sample preparation for LC–MS to reduce the matrix effect can also be time-consuming. Therefore, special emphasis was placed on the development of direct ionization techniques that support direct sample analysis without prior sample preparation or chromatography separation. Due to these efforts, a family of API techniques has emerged recently and is collectively known as "ambient ionization" techniques. The biggest application of ambient ionization is in field-portable MS as well as imaging and non-proximate analyte detection in biology. Desorption electrospray ionization (DESI) and direct analysis in real time (DART) are considered the prototype ambient ionization techniques. Both techniques are successfully utilized in metabolomics (Jackson et al. 2008; Zhou et al. 2010). A large number of variations of ambient ionization techniques have appeared such as extractive ESI (EESI), atmosphere solid analysis probe (ASAP), desorption atmospheric pressure chemical ionization (DAPCI), electrospray laser desorption/ionization (ELDI), flowing atmospheric pressure afterglow (FAPA), dielectric barrier discharge ionization (DBDI), and plasma-assisted desorption/ionization (PADI).

MS Analyzers

The mass analyzer is the part responsible for the separation of ions after their generation by the MS source. Mass analyzers operate in very different ways, but the net result is the separation of ions in space or in time according to their mass to charge (*m/z*) ratio. Resolution, accuracy, and speed are the most important features that characterize mass analyzers. Mass accuracy is always associated with resolution. Super-high resolution (>100,000) and accuracy (<1 ppm) analyzers include orbitrap and Fourier transform-ion cyclotron (FT-ICR). However, because these are trapping/scanning analyzers, the high resolution is obtained at the expense of the duration of the scan speed or duty cycle. In contrast, because it performs array rather than scanning separation, the time of flight (TOF) analyzer provides relatively high resolution (up to 20,000) and accuracy (<5 ppm) at shorter duty cycles. Low-resolution MS analyzers include quadrupole and 3D-(Paul) and linear-ion traps. Clearly, higher resolution/accuracy leads to more confidence in structural elucidation. In addition, more efficient duty cycle is required to keep up with chromatography frontends. For example, an MS duty cycle of <500 ms is required to generate ten data points for the construction of a 5-s UPLC peak.

Additional level of selectivity in MS is achieved by tandem MS (MS/MS). MS/MS can be achieved in time, where precursor masses are selected, fragmented, and resulting product ions are detected sequentially by one mass analyzer, or achieved in space, where two MS analyzers are separated in space and precursor ions are separated by the first analyzer, while product ions are separated by the second analyzer. Several types of scans can be performed during an MS/MS experiment: (1) product ion scans provide spectra of all fragments (product ions) resulting from the fragmentation of a specific precursor ion; (2) precursor ion scans provide spectra of all precursor ions that result in the production of a specific product ion; (3) neutral loss scans provide spectra of all molecules that fragment in a way which results in a specific loss of mass; and (4) product ions can be trapped and further fragmented, and the process can be repeated for an "*n*" number of cycles (MS^n).

30.3.3 Separation Frontends

The simplest way of introducing a sample into an MS instrument is direct injection or infusion of liquid samples. Therefore, mass represents the only basis for metabolite separation, which puts more demand on the resolution of the MS. In addition, the matrix effect resulting from ionization competition may misrepresent the actual abundance of analytes in the sample as well as decrease the results' reproducibility. Several examples are available in the literature of direct infusion MS applications (Enot et al. 2008; Han et al. 2008; Overy et al. 2008; Koulman et al. 2009). However, the matrix effect of samples generated in metabolomics studies limits the application of this approach. In addition, isobaric metabolites, which can be separated by LC, cannot be differentiated in direct infusion MS. Therefore, in metabolomics, MS is mostly operated after hyphenation with a separation frontend rather than as a standalone instrument. Separation frontends typically hyphenated to MS include gas chromatography, liquid chromatography, and capillary electrophoresis (Issaq et al. 2008).

30.3.3.1 Gas Chromatography

Gas chromatography (GC) provides tremendous resolving power and high sensitivity. Electron impact ionization (EI) and chemical ionization (CI) are the most common ionization sources in GC–MS analysis. EI is considered a hard ionization technique, which results in the complete in-source fragmentation of analytes, whereas CI is a softer ionization technique that may enable the detection of the intact molecular ions. In addition, the GC–MS interphase is well configured and is associated with minimal practical problems. Furthermore, GC–MS databases and reference libraries are available to aid in structural elucidation of metabolites of interest. However, only volatile analytes can be analyzed by GC and therefore extensive sample preparation and derivatization is needed to increase the volatility of nonvolatile analytes such as salts, polar metabolites, and large molecules (Halket et al. 2005). Typically, derivatization procedures include methoxymation to cover the carbonyl groups followed by silylation to mask active hydrogen atoms such as OH and NH_2 groups (Halket et al. 2005; Kanani et al. 2008; Pasikanti et al. 2008). Therefore, derivatization, extensive sample preparation, and long chromatographic analysis time make GC–MS a rather low-throughput technique for metabolomics.

30.3.3.2 Liquid Chromatography

Liquid chromatography (LC) has lower resolution than GC, but has the advantages of minimal sample preparation requirement and high throughput. In addition, LC columns are commercially available in a wide variety of selectivities, mechanisms of separations, and sizes, which support the analysis of all types of analytes. Mechanisms of LC separations include reverse phase, normal phase, ion exchange, ion pairing, chirality, size exclusion, hydrophilic interaction chromatography (HILIC), as well as a mixture of various modes. This wide range of available mobile and stationary phases makes it difficult to standardize conditions on which to build useful reference libraries as in GC–MS.

HPLC systems have traditionally been operated with pumps that are limited to 6,000 psi, and RP columns packed with 3–5 μm particles. Reducing the particle size and inter-particle spaces markedly reduce longitudinal diffusion, which increase the number of theoretical plates; therefore, chromatographic resolution. However, this increase in resolution is achieved at the expense of increasing back pressure (Plumb et al. 2007). Fittings that connect HPLC lines typically start leaking at pressures exceeding 5,000 psi. More recently, the UPLC system was introduced, which has special fittings made of gold that can handle pressure up to 15,000 psi resulting from the use of columns packed with <2 mm. This decrease in particle size results in a marked increase in theoretical plates, extremely sharp peaks, high sensitivity, and high resolution at short run times (Lenz and Wilson 2007). Figure 30.1 demonstrates representative chromatograms and principal component analyses (PCAs) of one sample generated by a metabolomics study and analyzed by HPLC and UPLC systems. The higher sensitivity and resolution of the UPLC system over HPLC resulted in the detection of more features in mouse urine.

Reversed phase (RP) columns are the most widely used columns in metabolomics. However, many metabolites and other endogenous molecules

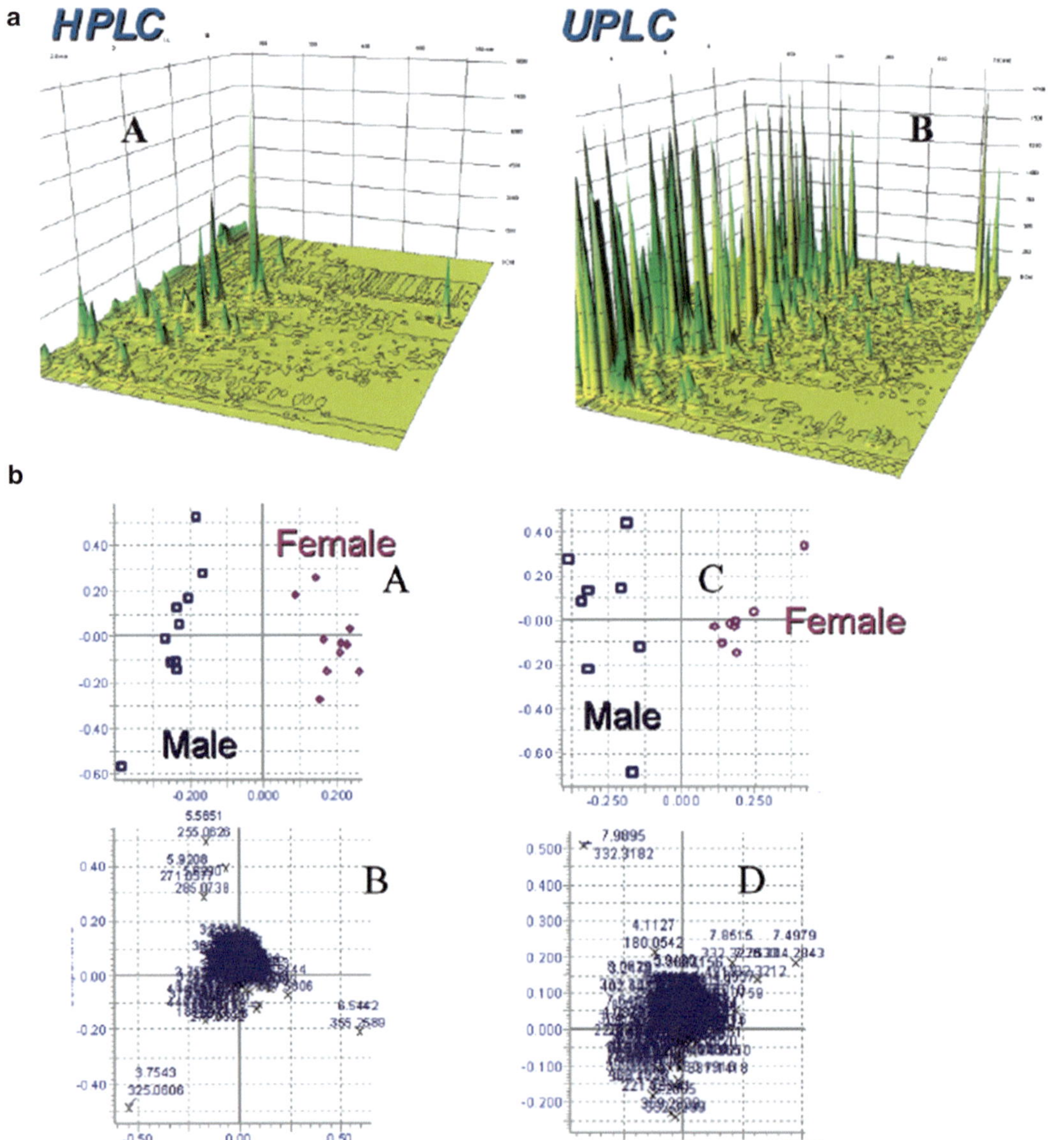

Fig. 30.1 (**a**) Three-dimensional mass chromatograms for (*A*) HPLC–MS and (*B*) UPLC–MS of white male mouse urine showing retention time, *m*/*z*, and intensity. (**b**) Comparison of the PCA for HPLC–MS and UPLC–MS data obtained from male and female white mouse urines. (*A*) PCA score plot for HPLC analysis, (*B*) the corresponding loadings plot for HPLC, (*C*) the scores plot by UPLC, and (*D*) the corresponding loadings plot for UPLC. Reproduced with permission from Wilson et al. (2005)

are polar molecules that are not retained well on RP columns. Therefore, alternative retention mechanisms were introduced. HILIC utilizes NP-like columns using RP-like mobile phases, which are compatible with MS. HILIC uses polar NP stationary phases, such as silica- or polymer-based packings that have been modified with amide, diol, amino, or zwitterionic groups. In HILIC, polar analytes have high retention and are eluted with a gradient increase in the aqueous proportion of the mobile phase (Alpert 1990). Therefore, HILIC is an ideal choice for the LC–MS analysis of polar analytes and has been widely used in metabolomics (Tolstikov et al. 2007; Cubbon et al. 2010;

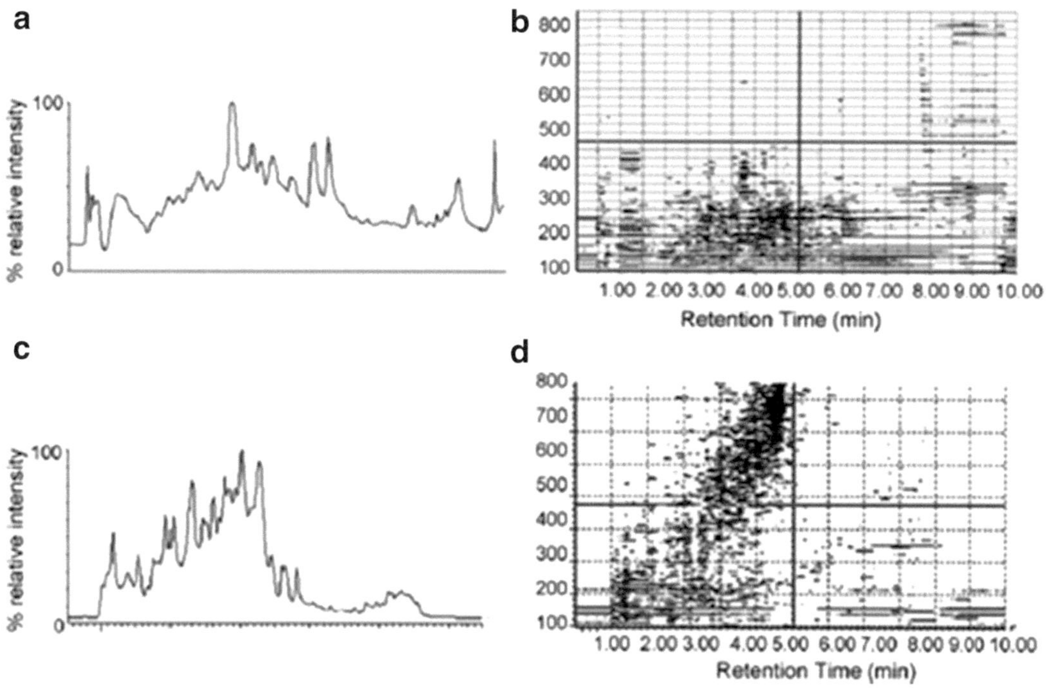

Fig. 30.2 Typical mass chromatogram of Zucker rat urine analyzed by (**a**) HPLC and (**c**) capillary LC–MS with the corresponding two-dimensional ion maps, (**b**) HPLC, and (**d**) capillary LC. Reproduced with permission from Granger et al. (2005)

Spagou et al. 2010). In addition, multidimensional separation approaches can be used, such as 2D chromatography, where two columns with different selectivity and mechanism of retention are connected in tandem (Wang et al. 2008).

Capillary LC using capillary columns packed with smaller particles and low-flow rates (nL/min) provides high-chromatographic resolution, high-peak capacity, minimal sample requirement, and increased signal to noise due to more concentrated peaks and reduced ion suppression. In a comparison study, both HPLC– and capillary LC–MS metabolomics analyses were performed on the same samples (Granger et al. 2005). The capillary method reveals the presence of higher number of analytes in rat urine compared to conventional HPLC separation (Fig. 30.2).

30.3.3.3 Capillary Electrophoresis

Capillary electrophoresis (CE) has very high resolving capabilities and provides a wide variety of selectivities that are different than LC. In addition, CE has minimal sample and solvent requirements. Therefore, CE is an environmentally friendly technique, which can be complimentary to HPLC. However, CE is still not widely used in metabolomics because of issues related to the system stability. The matrix components of biological samples continuously adsorb to the internal walls of the separation capillary over time, which causes serious reproducibility concerns in retention time and peak area of analytes (Ullsten et al. 2006). In addition, errors related to the high voltage power source in CE prevent its use for large batches of samples. Regardless, several CE-based metabolomics applications are available (Ramautar et al. 2008; Ramautar et al. 2011).

30.4 Experimental Protocols

30.4.1 Sample Handling

Sample collection, preparation, and storage can introduce large variability to NMR data. Some provisions are recommended during these steps to prevent the alteration of the endogenous metabolite profile, maintain the quality of biological samples, and avoid the introduction of new sources of variation and errors to the data.

1. Sample collection: In general, vials with stabilizers and unknown additives should be avoided. More importantly, the same type of vials should be used throughout the study. Additionally, different sample types require different considerations. For urine samples, special efforts should be made to prevent or decrease contamination with feces, hair, food, and dander, which is usually of concern when using metabolic cages. During collection, urine samples should be kept cold in vials pretreated with bacteriostatic agents such as sodium azide (Holland et al. 2005). In addition, amber-colored sterilized tubes are preferred for urine collection to protect light-sensitive analytes.

   Plasma samples are generated from blood samples after treatment with anticoagulants such as heparin or small molecules such as EDTA (ethylene-diamine-tetraacetic acid) and citrate. These anticoagulants should be avoided because they produce strong signals in the NMR spectrum, which interfere with many endogenous metabolites. Protease inhibitor cocktails, routinely used in proteomics studies, should also be avoided for the same reason. Plasma samples should be also kept ice-cold during sample collection and aliquoting of samples should be done prior to freezing and storage. Thawing should be done on ice, in a cold water bath or at room temperature, not by heating. Standard operating procedures (SOPs) for the collection of high-quality serum and plasma samples are available (Tuck et al. 2009).
2. Sample storage: It is generally recommended to store all biological samples in metabolomics studies in −80 °C freezers for long-term storage, minimize freeze–thaw cycles, and keep samples cold during preparation and analysis. For short-term storage, metabolomics samples are typically stored in −20 °C freezers. Deprez et al. monitored the short- and long-term stability for a large number of analytes in plasma samples and did not observe any changes in NMR spectra up to 6 months storage at −80 °C (Deprez et al. 2002).
3. Sample preparation: Sample preparation procedures should ideally be reproducible, robust, easy to perform, have high recovery of analytes, and do not introduce artifacts to the sample. Sample preparation strategies are vastly different between targeted and non-targeted approaches. For non-targeted metabolomics, the goal is to extract as many metabolites as possible and reduce potential loss of metabolites by using minimum amount of sample extraction. Protein precipitation (PP) and liquid–liquid extraction (LLE) with organic solvents such as methanol, acetonitrile, and chloroform as well as solid phase extraction (SPE) under "generic" conditions are capable of extracting a wide variety of analytes, and therefore are largely used in sample preparation of non-targeted metabolomics. Targeted approaches usually utilize more elaborate sample preparation techniques that aim to maximize the recovery of specific analytes of interest and avoid the co-extraction of as many endogenous metabolites as possible.

In contrast to other analytical techniques including MS, metabolite profiling of biofluids by NMR requires minimal sample preparation. However, the inorganic metal ion composition and pH of biological fluids, especially of urine, show a great variation between samples obtained from different individuals as well as samples obtained from the same individual at different time points. This may have a marked effect on the signal intensity and chemical shift of metabolites. Therefore, desirable sample preparation procedures will help normalize for some of these sources of variation. Chelating agents such as EDTA may help normalizing bivalent ion composition, while the use of buffers such as phosphate helps normalize samples' pH. A buffer containing 100–200 mM phosphate is typically added to urine samples to normalize their pH (Lindon et al. 2003). In most cases, sample preparation of NMR samples includes pH adjustment with a buffer, dilution with water, and the addition of D_2O, trimethylsilyl propionate (TSP), and sodium azide (Keun et al. 2002).

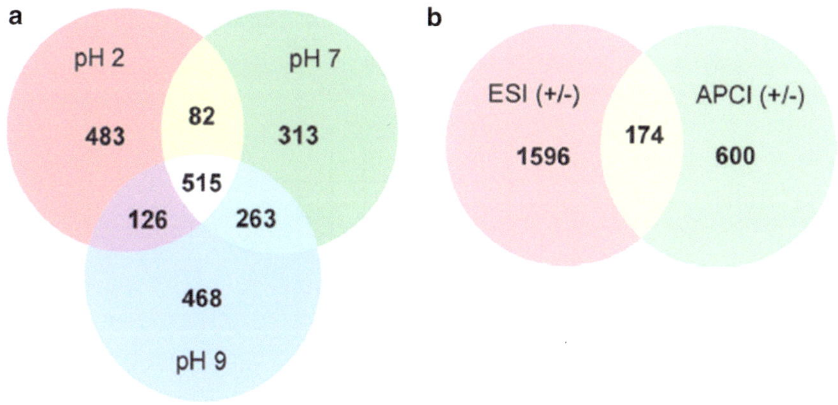

Fig. 30.3 Effect of the pH used during sample extraction (**a**) and the ionization source (**b**) on the features detected by LC–MS. Reproduced with permission from Sana et al. (2008)

30.4.2 MS protocols

30.4.2.1 Sample Preparation

Because of the "matrix effect" phenomenon, more elaborate, technically demanding, and time-consuming sample preparation is required for MS-based analysis. Therefore, more sophisticated extraction conditions are required to ensure the co-extraction of the metabolome and yet remove the undesired endogenous compounds in the matrix such as proteins and small salts.

Protein precipitation (PP) with organic solvents such as methanol or acetonitrile is used in most metabolomics applications. In addition, LLE with two immiscible solvents such as chloroform/methanol has been applied for global extraction of the metabolome (Jiye et al. 2005; Want et al. 2006). Variations of these methods with pH adjustment using acid or base have also been applied (Rabinowitz and Kimball 2007). SPE using various packings has also been used for sample preparation in metabolomics (Dixon et al. 2011). Several studies have compared various extraction techniques for metabolomics applications (Lin et al. 2007; Wu et al. 2008; El Rammouz et al. 2010; Shin et al. 2010). These studies demonstrated that rather than one sample preparation method being better than the other, the various methods resulted in the detection of different metabolite features. Therefore, a comprehensive extraction of all components of the metabolome would require multiple extraction conditions, separation procedures, and ionization source and modes (Want et al. 2006). Figure 30.3 demonstrates the effect of the extraction conditions and the ionization source on the features detected by LC–MS.

30.4.2.2 Data Reproducibility

There are more sources for the variability of data produced by MS compared to NMR, which has been shown to be a relatively stable and reproducible analytical platform. (1) Chromatographic behavior of analytes including RT and peak shape varies with time due to changes in flow rates, column performance, mobile phase composition and pH, and column and ambient temperature. Therefore, LC changes should be continuously monitored using exogenous or endogenous ISs. (2) Mass shifts occur over time and, therefore, lock mass standards are used to ensure high mass accuracy is achieved throughout the experiment. (3) MS response is known to decrease after multiple injections of biological samples due to source contamination. Therefore, there are three features of the analytical system that are required to be stable to achieve this: retention time (RT), signal intensity, and mass accuracy. These limitations should be recognized and efforts should be made to control them.

Given the many factors causing data variation in LC–MS experiments, it is clearly necessary to

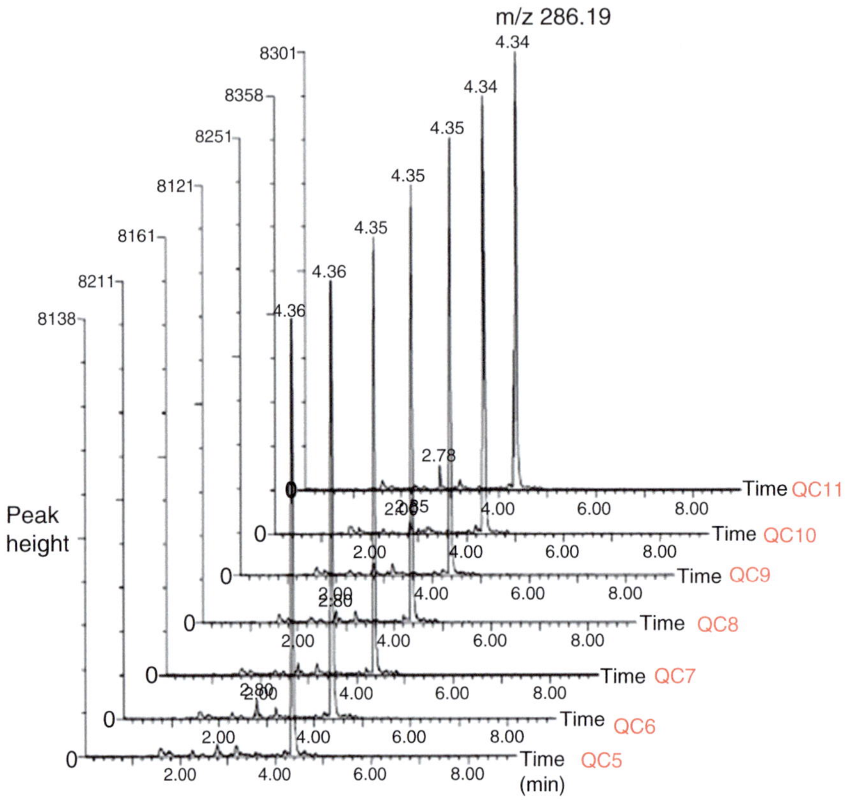

Fig. 30.4 Extracted ion chromatograms of an ion (*m/z* 286.19, retention time 4.3 min) in seven sequential QC samples run in between of 67 human urine samples demonstrating the repeatability in peak height and retention time. Reproduced with permission from Gika et al. (2008)

use elaborate quality control (QC) procedures to ensure the validity of the final result. Typically, exogenous ISs are added to all samples in fixed quantities to normalize for variation in sample preparation and matrix effect. However, because hundreds/thousands of analytes are analyzed, which are mostly unknown a priori, a different approach is followed. A single biological QC sample is prepared by mixing equal portions from the actual samples to be analyzed. This "pooled" sample is used to provide a representative "mean" sample containing all the analytes. Aliquots of this QC samples are run before study samples to ensure system conditioning and then run at regular intervals throughout the samples batch. A set of endogenous peaks is selected as ISs to monitor for consistency in RT and instrument response throughout the experiment (Zelena et al. 2009). In addition, CV can be calculated for the intensities of all the ions found in the QC samples rather than a selected set of ions. Figure 30.4 shows an example of a stable ion in human urine with highly reproducible retention time and MS response. In contrast, highly variable ions may not be good candidates as biomarkers. Typically these unstable peaks have low intensity, which is sometimes close to the cut-off level of the peak picking algorithm, and/or have low retention time close to the system dead-volume.

30.4.2.3 Structural Elucidation of Unknown Metabolites

The ultimate goal of metabolomics study is to identify metabolites responsible for the differentiation between treatment groups. Therefore, once metabolites of interest have been determined from the data analysis, they need to be identified. A panel of complementary spectroscopic methods including NMR and MS is required to reach this goal. One approach is to split the HPLC flow between NMR and MS instruments, which provide complimentary spectra simultaneously from both platforms (Tang et al. 2009). 2D proton–proton and proton–carbon NMR experiments could be used to aid in the structural elucidation of unknown metabolites. However, 2D experiments are time-consuming and, more importantly, large quantity of pure analytes is generally needed for the use of NMR in structural elucidation. Therefore, structural elucidation in metabolomics heavily relies on MS and MS/MS data.

When identity is still not reached using these strategies, preparative LC or TLC may be performed to obtain analytes in a more pure form, which can be subjected for further NMR and MS analysis. Obviously, this process is time-consuming and should be undertaken only for the identification of worthy biomarkers. Finally, standards should be synthesized for comparison before final confirmation of identity, and until this is done any identification must be considered provisional.

Structural Elucidation Using MS

The first step in structural elucidation is the proposal of a molecular formula (elemental composition) of the unknown metabolite. The rule of thumb is that the number of possible formulas increases exponentially with increasing the ion mass and decreases with the increase in MS resolution. Accurate mass measurements of 5-ppm or better are needed instead of nominal mass measurements. However, even with ultra-high resolution instruments, 15 different molecular formulas for mass are possible for a >600 Da, within a 1-ppm window. In such a case the nitrogen rule, charge state, and isotopic patterns can also be used to further narrow down the list of possible molecular formulas (Werner et al. 2008).

All single-stage MS-related strategies may not be enough to identify the structure of a molecule. In this regard, MS/MS experiments provide valuable points for structural elucidation. MS/MS adds another level for MS selectivity based on the fragmentation pattern of molecules. Several MS/MS experiments can be performed, including product ion, precursor ion, neutral loss, and MS^n, which provide rich and complimentary information that can lead to identifying the structure. Algorithms are available to provide candidate structures for unknown analytes using these criteria (Loftus et al. 2008).

Metabolomics Databases

One of the biggest advantages of GC–MS analysis is the availability of well-established structural databases. The largest GC–MS spectra libraries are available from Wiley (~400,000 spectra) and the National Institute of Standards and Technology (NIST) (~200,000 spectra). In contrast, several LC–MS platforms are used in metabolomics and MS/MS spectra of metabolites are highly dependent on type of these LC–MS systems and the conditions, under which they are operated. Therefore, LC–MS databases are still under development and are also often not interchangeable between different instruments or laboratories. Consequently, metabolomics databases are not as powerful as protein or gene databases in terms of unknown molecular identification.

Despite all challenges in compiling LC–MS libraries, several MS-based metabolite databases are available. Some of these databases link metabolites to biochemical pathways, and therefore can aid in the mechanistic interpretation of identified biomarkers such as the Kyoto Encyclopedia of Genes and Genomes (KEGG) and HumanCyc databases. The other category of LC–MS metabolomics library includes those that compile MS, MS–MS, and retention time of metabolites without providing any biological sense to the identified structure such as ARM, HMDB, MELTIN, and MassBank databases. Table 30.1 lists some metabolomics-related databases.

Other Approaches

Determination of the number of exchangeable hydrogen atoms (i.e., attached to O, N, and S atoms) by determining the shift in isotopic pattern of unknown analytes facilitates their structural elucidation. Furthermore, changes in the isotopic pattern of product ions during H/D exchange

Table 30.1 A list of widely used metabolomics-related databases

The Human Metabolome Data Base (HMDB)	http://bigg.ucsd.edu/
The BIGG Database	http://lipidbank.jp/
LipidBank	http://www.lipidmaps.org/
LipidMaps	http://www.lipidmaps.org/
MassBank	http://www.massbank.jp/?lang=en
METLIN	http://metlin.scripps.edu/
Fiehn GC–MS Database	http://fiehnlab.ucdavis.edu/Metabolite-Library-2007/
Madison Metabolomics Consortium Database (MMCD)	http://mmcd.nmrfam.wisc.edu/
NMR Metabolomics Database of Linkoping, Sweden (MDL)	http://www.liu.se/hu/mdl/main/
The Platform for RIKEN Metabolomics (PRIMe)	http://prime.psc.riken.jp/
SUGABASE (a carbohydrate-NMR database)	http://www.boc.chem.uu.nl/sugabase/sugabase.html
National Institute of Standards and Technology (NIST)-physical reference data	http://www.nist.gov/pml/data/index.cfm
NMR database of lignin and cell wall model compounds	http://ars.usda.gov/Services/docs.htm?docid=10491
NMR Shift DB	http://nmrshiftdb.nmr.uni-koeln.de/
AraCyc	http://www.arabidopsis.org/biocyc/index.jsp
METACYC	http://metacyc.org/
BIOCYC	http://biocyc.org/
IUBMB-Nicholson Metabolic Maps, Minimaps & Animaps	http://www.iubmb-nicholson.org/
Kyoto Encyclopedia of Genes and Genomes (KEGG)	http://www.genome.jp/kegg/pathway.html

experiments help identify the position of particular functional groups with exchangeable protons. H/D exchange can be performed in the liquid phase (Kamel et al. 2003) using deuterated mobile phases or in the gas phase (Campbell et al. 1995) using deuterated gases in the source or the collision cell. The extent and rate of H/D exchanges increase with the basicity of the used gas. Therefore, it is possible to selectively label functional groups of the molecule in the gas phase, which may further help in structural elucidation. Another approach for structural elucidation of unknown metabolites relies on derivitization approaches. Derivatization is a traditionally known strategy to enhance analyte volatility for GC analysis. In addition, derivatization can be used to improve the ionization efficiency, identify the position of a functional group, trap reactive metabolites, and improve chromatographic behavior of compounds (Liu and Hop 2005).

30.5 Data Processing

Raw data generated by NMR and MS requires several stages of processing to improve its quality and convert it to a format that can be subject to statistical analysis. Data processing transforms the data in a way that subsequent analyses and modeling are made easier, more robust, and more accurate. MS raw data are generally reported in a table format consisting of retention time, *m/z* value, and ion abundance. The overall task of data processing is to detect all peaks and collapse the three-dimensional data into a two-dimensional data matrix (a peak-table). The final results and conclusions reached from metabolomics experiments are dependent not only on how they were generated but more importantly on the steps taken during data processing. Actually, the same data can be processed differently using different approaches to reach completely different conclusions. Various tools are available for data processing, which may be divided into the following steps:

30.5.1 Binning/Bucketing/Rounding

When comparing several samples using chemometric methods, it is important for a particular metabolite to have the same *m/z* value in all samples; otherwise it will be treated as a different ion in each sample. Therefore, rather than reporting data separately at every reported *m/z* value, data can be pooled from a window of *m/z* called "bin" or "bucket." Ion intensity for a rounded *m/z* value can be defined as sum of all intensities binned together or alternatively computed with interpolation from a continuous spectrum (Anderle et al. 2004; Radulovic et al. 2004). NMR data are also divided into 4-ppm-wide bins, which will generate 200–250 "bins" from a typical 10 ppm NMR spectrum.

30.5.2 Noise Reduction

All spectral signals contain some level of noise related to many sources such as the buffers and solvents used, or the random noise associated with detectors. Noise reduction methods aim at removing random noise from the measurement signal. Two major approaches are typically followed for noise reduction: baseline subtraction and filtering or smoothing (Fredriksson et al. 2007).

In baseline subtraction, baseline removal starts with modeling the baseline shape followed by subtracting the noise shape from the raw signal. Filtering aims to increase the signal-to-noise ratio (S/N) by enhancing the informative parts of a signal and by reducing the noise. Smoothing results from bunching a number of data points and replacing them by one value, such as their mean, which represents all bunched points, thereby replacing the need for individual values. Various filters for data smoothing are available including simple moving average, median, or geometric mean, and the Savitzky-Golay filter, which depends on fitting a higher-order polynomial to the data over a local window (Wang et al. 2003; Radulovic et al. 2004). Excessive filtering may oversmooth the peak or increase its width or skewness, which may merge peaks that were otherwise well resolved.

30.5.3 Peak Detection

Peak detection aims to differentiate "true" signal from noise. Traditionally, peak detection algorithms search local maxima for signals that exceed a predefined threshold signal-to-noise ratio (SNR). Threshold SNR is typically determined from the noise level in a signal-free region of the chromatogram and can either be consistent throughout the spectrum (Patterson et al. 2010) or be range-specific (Wang et al. 2003). A peak is sensed when a local maximum increases in amplitude over a limited range in *m/z* or time (Hastings et al. 2002). However, some peaks have complex patterns that make peak identification using these traditional approaches impractical. Therefore, newer strategies have been developed based on transformation into wavelet domain, which can enhance the effective SNR by using additional information provided by the wavelet coefficients (Du et al. 2006).

30.5.4 Peak Deconvolution

In LC– and GC–MS, one analyte may produce multiple ions with different *m/z* values. This is due to the detection of isotopic peaks, the formation of in-source fragments, ion adducts, and/or multiple-charge state ions. Therefore, deconvolution methods assign different ions to the same analyte, which reduces the complexity of chromatograms and avoid the detection of redundant features (Hermansson et al. 2005). Deconvolution algorithms utilize the fact that different fragments from the same molecule have the same retention time (Sinha et al. 2004).

30.5.5 Peak Alignment

Alignment aims to match corresponding peaks across multiple samples within a specified *m/z* and retention time error/shift window. Alignment of MS data is more problematic than NMR due to the low reproducibility of RT associated with LC or GC needed in MS. In certain cases, drift may occur along the *m/z* axis as well, although this is far less of a problem than variations in time. To compare a particular metabolite between several samples, it has to be defined by the exact same *m/z* and RT features; otherwise it will be recognized as different peaks in each sample.

Most alignment methods work in pair-wise fashion by aligning samples against either a

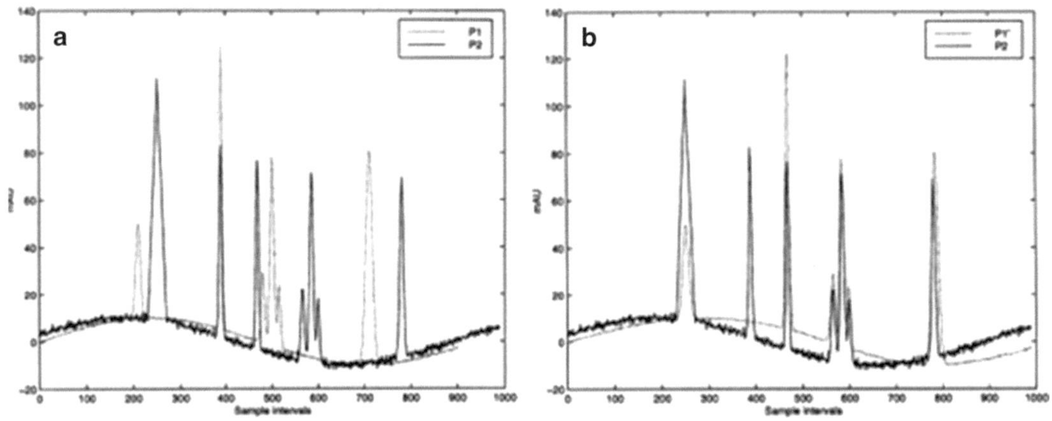

Fig. 30.5 Two chromatograms (**a**) prior to aligning and (**b**) after aligning using COW. Reproduced with permission from Smedsgaard et al. (1998)

selected reference sample or a template. The first step in time alignment is usually dividing chromatograms into segments of equal or variable lengths. These segments then undergo multiple steps of shifting, shrinking, and stretching (wrapping) to achieve an optimum overlap that maximizes the correlation coefficient between the segments of a spectrum and a target (Nielsen et al. 1998; Forshed et al. 2005). This approach is applied by the correlation optimized warping (COW) algorithm (Fig. 30.5).

30.5.6 Normalization

LC– and GC–MS signal is inherently associated with high variability. As discussed earlier, this is due to many factors related to extraction recovery, chromatography, and more importantly matrix effect. Several normalization approaches can be followed to reduce this variability. Global normalization approaches rely on using the same normalization factor for all analytes, whereas local normalization approaches use different normalization schemes for different analytes. The local normalization schemes may vary according to *m/z*, RT, or abundance (Bijlsma et al. 2006). For example, a chromatogram can be divided into zones, with peaks in every zone normalized to ISs that elute in the same zone (Katajamaa and Oresic 2005).

To reduce this variability, the peak area of analytes is typically normalized to the peak area of an exogenous IS, which is spiked in equal amounts to all samples. However, because hundreds or even thousands of analytes, which are mostly unknown a priori, are analyzed, different approaches can be followed. First, endogenous peaks belonging to metabolites believed not to vary between samples are used as endogenous ISs (Baggerly et al. 2003). This is similar to using a "housekeeping" gene such as GAPDH to normalize RNA data. Second, peak intensities can be normalized to the total ion current (TIC) of every sample (the summed intensities over all time points). The assumption here is that the total concentration of analytes is constant under all conditions so the concentration increase in a specific group of metabolites is balanced by a decrease of another group (Wang et al. 2003). Third, analyte peak areas can be normalized to the local TIC of different regions of the chromatogram (Forshed et al. 2003).

30.5.7 Software

Several commercial and open-source software programs are available for automatic sample processing including smoothing, alignment, denoising, deconvolution and extraction of peaks, as well as statistical analysis such as PCA (Lange et al. 2008). All NMR and MS manufacturers produce metabolomics software to process data generated by their own instrumentation. These instrument-specific software include MarkerLynx (Waters), MassHunter (Agilent Technologies), MarkerView (Applied Biosystems), SIEVE (ThermoFisher Scientific), and ProfileAnalysis (Bruker Daltonics). These software can perform data processing as well as statistical analysis.

On the other hand, open-source software packages, such as MSFACTS, MET-IDEA, metaXCMS, MZmine, MathDAMP, and MetAlign, are designed to process metabolomics data, irrespective of their source. To use these software packages, raw data files must be first converted to universal data formats such as netCDF, mzXML, mzML, mzData, and ASCII (Katajamaa and Oresic 2007). After data processing, the output of these software packages can be exported into software such as SIMCA or MATLAB for MVA such as PCA. An updated list of software and tools for processing metabolomics data can be found at the metabolomics society website (http://www.metabolomicssociety.org/software.html).

30.6 Statistical Analysis

Metabolomic experiments generate enormous amounts of quantitative and qualitative data that can be overwhelming, time-consuming, and not easy to process or interpret. Therefore, sophisticated statistical tools are required to reduce the data and extract useful information from into a workable and understandable format. Without bioinformatics and statistical support and data reduction tools, the generated data can be impossible to interpret and trends very difficult to find.

The major purpose of statistical data analysis in metabolomics is pattern recognition, including identification of similarities and differences in data. Statisticians apply different multivariate statistical analyses and pattern-recognition tools to distill the large amounts of data in an effort to find discriminative features that differentiate treatment groups. Multiple features of the metabolomics profile may discriminate treatment groups and MVA may indicate which of these metabolites are more important to identify its structure. Several statistical methods are applied in metabolomics. Specific reviews comparing these methods are available (Lindon et al. 2001; Trygg et al. 2007; Mahadevan et al. 2008).

MVA methods can be divided into unsupervised and supervised MVA. In unsupervised MVA, sample classification is unknown to the analytical software, while in supervised MVA this information is provided to the software for the purpose of model construction. Among all MVA approaches, the unsupervised method of PCA is the most commonly used in metabolomics.

PCA is a variable reduction procedure which attempts to simplify complex and diverse relationships that exist among a set of observed variables by uncovering common latent factors that link together the seemingly unrelated variables. These artificial variables (called principal components) are constructed from linear combinations of optimally weighted observable variables in such a way that the resulting components are uncorrelated and account for a maximal amount of variance in the data set. The first component should account for most of the variation in the data set, followed by the second component and so on. The weight of each observable variable is called "loading" and it determines the variable contribution to the total component variance. Variables with higher loadings have greater influence on a component and are therefore responsible for sample clustering or separation.

The results of PCA can be summarized by two plots: (1) Scores plot: Scores plot is a scatter plot of the scores of the components that account for most data variation, typically, the first two components. Scores plot reveals any clusters, trends, outliers, groups, etc. in the samples. (2) Loading plot: Scatter plot shows the contribution of the individual variables to the trends observed in the scores plot, which can be used to identify the metabolites primarily responsible for the samples group separation. Each point in the loading plot represents a variable/feature including its intensity and its corresponding chemical shift for NMR or retention time and *m/z* for MS. Therefore, PCA is always recommended as starting point for

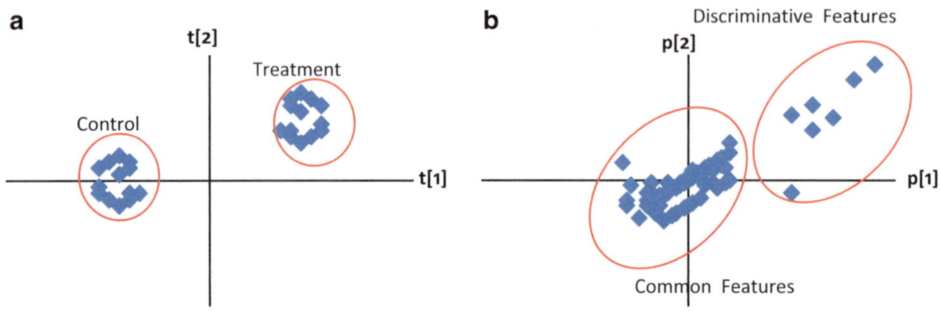

Fig. 30.6 (**a**) Score plots of the first two components and (**b**) loading plot of all features, features with highest contribution to the clustering trend in the scores plot (discriminative features), as well as features with low contribution to the clustering trend (common features) are *circled*

analyzing multivariate data and will rapidly provide sample clusters and features responsible for these clusters. Figure 30.6 depicts the scores and loading plots resulting from PCA of a hypothetical set of samples. The scores plot shows the samples can be clustered into control and treatment groups based on the first two components (t[1] and t[2]). The loading plot shows discriminative features that demonstrate higher x and/or y values in the loading plot than common features.

References

Alpert AJ (1990) Hydrophilic-interaction chromatography for the separation of peptides, nucleic acids and other polar compounds. J Chromatogr 499:177–196

Anderle M, Roy S, Lin H, Becker C, Joho K (2004) Quantifying reproducibility for differential proteomics: noise analysis for protein liquid chromatography-mass spectrometry of human serum. Bioinformatics 20:3575–3582

Baggerly KA, Morris JS, Wang J, Gold D, Xiao LC, Coombes KR (2003) A comprehensive approach to the analysis of matrix-assisted laser desorption/ionization-time of flight proteomics spectra from serum samples. Proteomics 3:1667–1672

Beger RD, Hansen DK, Schnackenberg LK, Cross BM, Fatollahi JJ, Lagunero FT, Sarnyai Z, Boros LG (2009) Single valproic acid treatment inhibits glycogen and RNA ribose turnover while disrupting glucose-derived cholesterol synthesis in liver as revealed by the [U-C-13(6)]-d-glucose tracer in mice. Metabolomics 5:336–345

Bender K, Walsh S, Evans ACO, Fair T, Brennan L (2010) Metabolite concentrations in follicular fluid may explain differences in fertility between heifers and lactating cows. Reproduction 139:1047–1055

Bijlsma S, Bobeldijk I, Verheij ER, Ramaker R, Kochhar S, Macdonald IA, van Ommen B, Smilde AK (2006) Large-scale human metabolomics studies: a strategy for data (pre-) processing and validation. Anal Chem 78:567–574

Boros LG, Brackett DJ, Harrigan GG (2003) Metabolic biomarker and kinase drug target discovery in cancer using stable isotope-based dynamic metabolic profiling (SIDMAP). Curr Cancer Drug Targets 3:445–453

Brindle JT, Antti H, Holmes E, Tranter G, Nicholson JK, Bethell HWL, Clarke S, Schofield PM, McKilligin E, Mosedale DE, Grainger DJ (2002) Rapid and noninvasive diagnosis of the presence and severity of coronary heart disease using H-1-NMR-based metabonomics. Nat Med 8:1439–1444

Campbell S, Rodgers MT, Marzluff EM, Beauchamp JL (1995) Deuterium exchange reactions as a probe of biomolecule structure. Fundamental studies of cas phase H/D exchange reactions of protonated glycine oligomers with D2O, CD3OD, CD3CO2D, and ND3. J Am Chem Soc 117:12840–12854

Cho JY, Matsubara T, Kang DW, Ahn SH, Krausz KW, Idle JR, Luecke H, Gonzalez FJ (2010) Urinary metabolomics in Fxr-null mice reveals activated adaptive metabolic pathways upon bile acid challenge. J Lipid Res 51:1063–1074

Claudino WM, Quattrone A, Biganzoli L, Pestrin M, Bertini I, Di Leo A (2007) Metabolomics: available results, current research projects in breast cancer, and future applications. J Clin Oncol 25:2840–2846

Clayton TA, Lindon JC, Cloarec O, Antti H, Charuel C, Hanton G, Provost JP, Le Net JL, Baker D, Walley RJ, Everett JR, Nicholson JK (2006) Pharmaco-metabonomic phenotyping and personalized drug treatment. Nature 440:1073–1077

Clayton TA, Baker D, Lindon JC, Everett JR, Nicholson JK (2009) Pharmacometabonomic identification of a significant host-microbiome metabolic interaction affecting human drug metabolism. Proc Natl Acad Sci U S A 106:14728–14733

Coen M, O'Sullivan M, Bubb WA, Kuchel PW, Sorrell T (2005) Proton nuclear magnetic resonance-based

metabonomics for rapid diagnosis of meningitis and ventriculitis. Clin Infect Dis 41:1582–1590
Connor SC, Hansen MK, Corner A, Smith RF, Ryan TE (2010) Integration of metabolomics and transcriptomics data to aid biomarker discovery in type 2 diabetes. Mol Biosyst 6:909–921
Cubbon S, Antonio C, Wilson J, Thomas-Oates J (2010) Metabolomic applications of HILIC-LC-MS. Mass Spectrom Rev 29:671–684
Dalgliesh CE, Horning EC, Horning MG, Knox KL, Yarger K (1966) A gas–liquid-chromatographic procedure for separating a wide range of metabolites occuring in urine or tissue extracts. Biochem J 101:792–810
Deprez S, Sweatman BC, Connor SC, Haselden JN, Waterfield CJ (2002) Optimisation of collection, storage and preparation of rat plasma for 1H NMR spectroscopic analysis in toxicology studies to determine inherent variation in biochemical profiles. J Pharm Biomed Anal 30:1297–1310
Dixon E, Clubb C, Pittman S, Ammann L, Rasheed Z, Kazmi N, Keshavarzian A, Gillevet P, Rangwala H, Couch RD (2011) Solid-phase microextraction and the human fecal VOC metabolome. PLoS One 6:e18471
Drexler DM, Reily MD, Shipkova PA (2011) Advances in mass spectrometry applied to pharmaceutical metabolomics. Anal Bioanal Chem 399:2645–2653
Du P, Kibbe WA, Lin SM (2006) Improved peak detection in mass spectrum by incorporating continuous wavelet transform-based pattern matching. Bioinformatics 22:2059–2065
Duarte IF, Stanley EG, Holmes E, Lindon JC, Gil AM, Tang HR, Ferdinand R, McKee CG, Nicholson JK, Vilca-Melendez H, Heaton N, Murphy GM (2005) Metabolic assessment of human liver transplants from biopsy samples at the donor and recipient stages using high-resolution magic angle spinning H-1 NMR spectroscopy. Anal Chem 77:5570–5578
Duarte NC, Becker SA, Jamshidi N, Thiele I, Mo ML, Vo TD, Srivas R, Palsson BO (2007) Global reconstruction of the human metabolic network based on genomic and bibliomic data. Proc Natl Acad Sci U S A 104:1777–1782
Dumas ME, Barton RH, Toye A, Cloarec O, Blancher C, Rothwell A, Fearnside J, Tatoud R, Blanc V, Lindon JC, Mitchell SC, Holmes E, McCarthy MI, Scott J, Gauguier D, Nicholson JK (2006) Metabolic profiling reveals a contribution of gut microbiota to fatty liver phenotype in insulin-resistant mice. Proc Natl Acad Sci U S A 103:12511–12516
Dunn WB, Bailey NJ, Johnson HE (2005) Measuring the metabolome: current analytical technologies. Analyst 130:606–625
Ebbels TMD, Keun HC, Beckonert OP, Bollard ME, Lindon JC, Holmes E, Nicholson JK (2007) Prediction and classification of drug toxicity using probabilistic modeling of temporal metabolic data: the consortium on metabonomic toxicology screening approach. J Proteome Res 6:4407–4422
El Rammouz R, Letisse F, Durand S, Portais JC, Moussa ZW, Fernandez X (2010) Analysis of skeletal muscle metabolome: evaluation of extraction methods for targeted metabolite quantification using liquid chromatography tandem mass spectrometry. Anal Biochem 398:169–177
Enot DP, Lin W, Beckmann M, Parker D, Overy DP, Draper J (2008) Preprocessing, classification modeling and feature selection using flow injection electrospray mass spectrometry metabolite fingerprint data. Nat Protoc 3:446–470
Fiehn O (2001) Combining genomics, metabolome analysis, and biochemical modelling to understand metabolic networks. Comp Funct Genomics 2:155–168
Forshed J, Schuppe-Koistinen I, Jacobsson SP (2003) Peak alignment of NMR signals by means of a genetic algorithm. Anal Chim Acta 487:189–199
Forshed J, Torgrip RJ, Aberg KM, Karlberg B, Lindberg J, Jacobsson SP (2005) A comparison of methods for alignment of NMR peaks in the context of cluster analysis. J Pharm Biomed Anal 38:824–832
Fredriksson M, Petersson P, Jornten-Karlsson M, Axelsson BO, Bylund D (2007) An objective comparison of pre-processing methods for enhancement of liquid chromatography-mass spectrometry data. J Chromatogr A 1172:135–150
Gamache PH, Meyer DF, Granger MC, Acworth IN (2004) Metabolomic applications of electrochemistry/mass spectrometry. J Am Soc Mass Spectrom 15:1717–1726
Gika HG, Macpherson E, Theodoridis GA, Wilson ID (2008) Evaluation of the repeatability of ultra-performance liquid chromatography-TOF-MS for global metabolic profiling of human urine samples. J Chromatogr B Analyt Technol Biomed Life Sci 871:299–305
Gika HG, Theodoridis GA, Earll M, Snyder RW, Sumner SJ, Wilson ID (2010) Does the mass spectrometer define the marker? A comparison of global metabolite profiling data generated simultaneously via UPLC-MS on two different mass spectrometers. Anal Chem 82:8226–8234
Gottschalk M, Ivanova G, Collins DM, Eustace A, O'Connor R, Brougham DF (2008) Metabolomic studies of human lung carcinoma cell lines using in vitro H-1 NMR of whole cells and cellular extracts. NMR Biomed 21:809–819
Gowda GAN (2010) Human bile as a rich source of biomarkers for hepatopancreatobiliary cancers. Biomark Med 4:299–314
Granger J, Plumb R, Castro-Perez J, Wilson ID (2005) Metabonomic studies comparing capillary and conventional HPLC-oa-TOF MS for the analysis of urine from Zucker obese rats. Chromatographia 61:375–380
Griffiths WJ (2008) Metabolomics, metabonomics and metabolite profiling. RSC, Cambridge
Halket JM, Waterman D, Przyborowska AM, Patel RK, Fraser PD, Bramley PM (2005) Chemical derivatization and mass spectral libraries in metabolic profiling by GC/MS and LC/MS/MS. J Exp Bot 56:219–243
Han J, Danell RM, Patel JR, Gumerov DR, Scarlett CO, Speir JP, Parker CE, Rusyn I, Zeisel S, Borchers CH (2008) Towards high-throughput metabolomics using

ultrahigh-field Fourier transform ion cyclotron resonance mass spectrometry. Metabolomics 4:128–140

Han J, Datla R, Chan S, Borchers CH (2009) Mass spectrometry-based technologies for high-throughput metabolomics. Bioanalysis 1:1665–1684

Hastings CA, Norton SM, Roy S (2002) New algorithms for processing and peak detection in liquid chromatography/mass spectrometry data. Rapid Commun Mass Spectrom 16:462–467

Heijne WHM, Lamers RJAN, van Bladeren PJ, Groten JP, van Nesselrooij JHJ, van Ommen B (2005) Profiles of metabolites and gene expression in rats with chemically induced hepatic necrosis. Toxicol Pathol 33:425–433

Hermansson M, Uphoff A, Kakela R, Somerharju P (2005) Automated quantitative analysis of complex lipidomes by liquid chromatography/mass spectrometry. Anal Chem 77:2166–2175

Holland NT, Pfleger L, Berger E, Ho A, Bastaki M (2005) Molecular epidemiology biomarkers—sample collection and processing considerations. Toxicol Appl Pharmacol 206:261–268

Horning MG, Chambaz EM, Brooks CJ, Moss AM, Boucher EA, Horning EC, Hill RM (1969) Characterization and estimation of urinary steroids of the newborn human by gas-phase analytical methods. Anal Biochem 31:512–531

Issaq HJ, Abbott E, Veenstra TD (2008) Utility of separation science in metabolomic studies. J Sep Sci 31:1936–1947

Jackson AU, Werner SR, Talaty N, Song Y, Campbell K, Cooks RG, Morgan JA (2008) Targeted metabolomic analysis of Escherichia coli by desorption electrospray ionization and extractive electrospray ionization mass spectrometry. Anal Biochem 375:272–281

Jiye A, Trygg J, Gullberg J, Johansson AI, Jonsson P, Antti H, Marklund SL, Moritz T (2005) Extraction and GC/MS analysis of the human blood plasma metabolome. Anal Chem 77:8086–8094

Junot C, Madalinski G, Tabet JC, Ezan E (2010) Fourier transform mass spectrometry for metabolome analysis. Analyst 135:2203–2219

Kaddurah-Daouk R, Kristal BS, Weinshilboum RM (2008) Metabolomics: a global biochemical approach to drug response and disease. Annu Rev Pharmacol 48:653–683

Kaderbhai NN, Broadhurst DI, Ellis DI, Goodacre R, Kell DB (2003) Functional genomics via metabolic footprinting: monitoring metabolite secretion by Escherichia coli tryptophan metabolism mutants using FT-IR and direct injection electrospray mass spectrometry. Comp Funct Genomics 4:376–391

Kamel AM, Zandi KS, Massefski WW (2003) Identification of the degradation product of ezlopitant, a non-peptidic substance P antagonist receptor, by hydrogen deuterium exchange, electrospray ionization tandem mass spectrometry (ESI/MS/MS) and nuclear magnetic resonance (NMR) spectroscopy. J Pharm Biomed Anal 31:1211–1222

Kanani H, Chrysanthopoulos PK, Klapa MI (2008) Standardizing GC-MS metabolomics. J Chromatogr B Analyt Technol Biomed Life Sci 871:191–201

Katajamaa M, Oresic M (2005) Processing methods for differential analysis of LC/MS profile data. BMC Bioinformatics 6:179

Katajamaa M, Oresic M (2007) Data processing for mass spectrometry-based metabolomics. J Chromatogr A 1158:318–328

Keun HC, Ebbels TM, Antti H, Bollard ME, Beckonert O, Schlotterbeck G, Senn H, Niederhauser U, Holmes E, Lindon JC, Nicholson JK (2002) Analytical reproducibility in (1)H NMR-based metabonomic urinalysis. Chem Res Toxicol 15:1380–1386

Kiefer P, Portais JC, Vorholt JA (2008) Quantitative metabolome analysis using liquid chromatography-high-resolution mass spectrometry. Anal Biochem 382:94–100

Kim JK, Harada K, Bamba T, Fukusaki E, Kobayashi A (2005) Stable isotope dilution-based accurate comparative quantification of nitrogen-containing metabolites in Arabidopsis thaliana T87 cells using in vivo N-15-isotope enrichment. Biosci Biotechnol Biochem 69:1331–1340

Kim HK, Choi YH, Verpoorte R (2010) NMR-based metabolomic analysis of plants. Nat Protoc 5:536–549

Knapp JS, Cabrera WL (2009) Metabolomics: metabolites, metabonomics, and analytical technologies. Nova, Hauppauge, NY

Kopp SJ, Glonek T, Erlanger M, Perry EF, Barany M, Perry HM Jr (1980) Altered metabolism and function of rat heart following chronic low level cadmium/lead feeding. J Mol Cell Cardiol 12:1407–1425

Koulman A, Cao M, Faville M, Lane G, Mace W, Rasmussen S (2009) Semi-quantitative and structural metabolic phenotyping by direct infusion ion trap mass spectrometry and its application in genetical metabolomics. Rapid Commun Mass Spectrom 23: 2253–2263

Lange E, Tautenhahn R, Neumann S, Gropl C (2008) Critical assessment of alignment procedures for LC-MS proteomics and metabolomics measurements. BMC Bioinformatics 9:375

Lee R, Britz-McKibbin P (2010) Metabolomic studies of radiation-induced apoptosis of human leukocytes by capillary electrophoresis-mass spectrometry and flow cytometry: adaptive cellular responses to ionizing radiation. Electrophoresis 31:2328–2337

Lei Z, Huhman D, Sumner LW (2011) Mass spectrometry strategies in metabolomics. J Biol Chem 286: 25435–25442

Lenz EM, Wilson ID (2007) Analytical strategies in metabonomics. J Proteome Res 6:443–458

Leo GC, Darrow AL (2009) NMR-based metabolomics of urine for the atherosclerotic mouse model using apolipoprotein-E deficient mice. Magn Reson Chem 47:S20–S25

Lin CY, Wu HF, Tjeerdema RS, Viant MR (2007) Evaluation of metabolite extraction strategies from

tissue samples using NMR metabolomics. Metabolomics 3:55–67
Lindon JC, Holmes E, Nicholson JK (2001) Pattern recognition methods and applications in biomedical magnetic resonance. Prog Nucl Mag Res Spectrosc 39:1–40
Lindon JC et al (2003) Contemporary issues in toxicology the role of metabonomics in toxicology and its evaluation by the COMET project. Toxicol Appl Pharmacol 187:137–146
Lindon JC, Nicholson JK, Holmes E (2007) The handbook of metabonomics and metabolomics, 1st edn. Elsevier, Amsterdam
Liu DQ, Hop CECA (2005) Strategies for characterization of drug metabolites using liquid chromatography-tandem mass spectrometry in conjunction with chemical derivatization and on-line H/D exchange approaches. J Pharm Biomed Anal 37:1–18
Loftus N, Miseki K, Iida J, Gika HG, Theodoridis G, Wilson ID (2008) Profiling and biomarker identification in plasma from different Zucker rat strains via high mass accuracy multistage mass spectrometric analysis using liquid chromatography/mass spectrometry with a quadrupole ion trap-time of flight mass spectrometer. Rapid Commun Mass Spectrom 22: 2547–2554
Lu X, Zhao X, Bai C, Zhao C, Lu G, Xu G (2008) LC-MS-based metabonomics analysis. J Chromatogr B Analyt Technol Biomed Life Sci 866:64–76
Maddox JF, Luyendyk JP, Cosma GN, Breau AP, Bible RH Jr, Harrigan GG, Goodacre R, Ganey PE, Cantor GH, Cockerell GL, Roth RA (2006) Metabonomic evaluation of idiosyncrasy-like liver injury in rats cotreated with ranitidine and lipopolysaccharide. Toxicol Appl Pharmacol 212:35–44
Mahadevan S, Shah SL, Marrie TJ, Slupsky CM (2008) Analysis of metabolomic data using support vector machines. Anal Chem 80:7562–7570
Marx A, deGraaf AA, Wiechert W, Eggeling L, Sahm H (1996) Determination of the fluxes in the central metabolism of Corynebacterium glutamicum by nuclear magnetic resonance spectroscopy combined with metabolite balancing. Biotechnol Bioeng 49: 111–129
Matuszewski BK, Constanzer ML, Chavez-Eng CM (2003) Strategies for the assessment of matrix effect in quantitative bioanalytical methods based on HPLC-MS/MS. Anal Chem 75:3019–3030
Meadows AL, Kong B, Berdichevsky M, Roy S, Rosiva R, Blanch HW, Clark DS (2008) Metabolic and morphological differences between rapidly proliferating cancerous and normal breast epithelial cells. Biotechnol Prog 24:334–341
Millington DS, Kodo N, Norwood DL, Roe CR (1990) Tandem mass spectrometry: a new method for acylcarnitine profiling with potential for neonatal screening for inborn errors of metabolism. J Inherit Metab Dis 13:321–324
Moka D, Vorreuther R, Schicha H, Spraul M, Humpfer E, Lipinski M, Foxall PJD, Nicholson JK, Lindon JC (1998) Biochemical classification of kidney carcinoma biopsy samples using magic-angle-spinning H-1 nuclear magnetic resonance spectroscopy. J Pharm Biomed Anal 17:125–132
Monton MR, Soga T (2007) Metabolome analysis by capillary electrophoresis-mass spectrometry. J Chromatogr A 1168:237–246, discussion 236
Nicholson JK, Lindon JC, Holmes E (1999) “Metabonomics”: understanding the metabolic responses of living systems to pathophysiological stimuli via multivariate statistical analysis of biological NMR spectroscopic data. Xenobiotica 29: 1181–1189
Nielsen JH, Jewett MC (2007) Metabolomics: a powerful tool in systems biology. Springer, Berlin
Nielsen NPV, Carstensen JM, Smedsgaard J (1998) Aligning of single and multiple wavelength chromatographic profiles for chemometric data analysis using correlation optimised warping. J Chromatogr A 805:17–35
Nordstrom A, Want E, Northen T, Lehtio J, Siuzdak G (2008) Multiple ionization mass spectrometry strategy used to reveal the complexity of metabolomics. Anal Chem 80:421–429
Oakman C, Tenori L, Biganzoli L, Santarpia L, Cappadona S, Luchinat C, Di Leo A (2011) Uncovering the metabolomic fingerprint of breast cancer. Int J Biochem Cell Biol 43(7):1010–1020
Odunsi K, Wollman RM, Ambrosone CB, Hutson A, McCann SE, Tammela J, Geisler JP, Miller G, Sellers T, Cliby W, Qian F, Keitz B, Intengan M, Lele S, Alderfer JL (2005) Detection of epithelial ovarian cancer using H-1-NMR-based metabonomics. Int J Cancer 113:782–788
Oresic M (2010) Systems biology strategy to study lipotoxicity and the metabolic syndrome. Biochim Biophys Acta 1801:235–239
Overy DP, Enot DP, Tailliart K, Jenkins H, Parker D, Beckmann M, Draper J (2008) Explanatory signal interpretation and metabolite identification strategies for nominal mass FIE-MS metabolite fingerprints. Nat Protoc 3:471–485
Parida SK, Kaufmann SH (2010) The quest for biomarkers in tuberculosis. Drug Discov Today 15:148–157
Pasikanti KK, Ho PC, Chan EC (2008) Gas chromatography/mass spectrometry in metabolic profiling of biological fluids. J Chromatogr B Analyt Technol Biomed Life Sci 871:202–211
Patterson AD, Gonzalez FJ, Idle JR (2010) Xenobiotic metabolism: a view through the metabolometer. Chem Res Toxicol 23:851–860
Plumb RS, Stumpf CL, Granger JH, Castro-Perez J, Haselden JN, Dear GJ (2003) Use of liquid chromatography/time-of-flight mass spectrometry and multivariate statistical analysis shows promise for the detection of drug metabolites in biological fluids. Rapid Commun Mass Spectrom 17:2632–2638
Plumb RS, Granger JH, Stumpf CL, Johnson KA, Smith BW, Gaulitz S, Wilson ID, Castro-Perez J (2005) A rapid screening approach to metabonomics using

UPLC and oa-TOF mass spectrometry: application to age, gender and diurnal variation in normal/Zucker obese rats and black, white and nude mice. Analyst 130:844–849

Plumb R, Mazzeo JR, Grumbach ES, Rainville P, Jones M, Wheat T, Neue UD, Smith B, Johnson KA (2007) The application of small porous particles, high temperatures, and high pressures to generate very high resolution LC and LC/MS separations. J Sep Sci 30:1158–1166

Powers R (2009) NMR metabolomics and drug discovery. Magn Reson Chem 47(suppl 1):S2–S11

Quinones MP, Kaddurah-Daouk R (2009) Metabolomics tools for identifying biomarkers for neuropsychiatric diseases. Neurobiol Dis 35:165–176

Raamsdonk LM, Teusink B, Broadhurst D, Zhang NS, Hayes A, Walsh MC, Berden JA, Brindle KM, Kell DB, Rowland JJ, Westerhoff HV, van Dam K, Oliver SG (2001) A functional genomics strategy that uses metabolome data to reveal the phenotype of silent mutations. Nat Biotechnol 19:45–50

Rabinowitz JD, Kimball E (2007) Acidic acetonitrile for cellular metabolome extraction from Escherichia coli. Anal Chem 79:6167–6173

Radulovic D, Jelveh S, Ryu S, Hamilton TG, Foss E, Mao YY, Emili A (2004) Informatics platform for global proteomic profiling and biomarker discovery using liquid chromatography-tandem mass spectrometry. Mol Cell Proteomics 3:984–997

Ramautar R, Mayboroda OA, Deelder AM, Somsen GW, de Jong GJ (2008) Metabolic analysis of body fluids by capillary electrophoresis using noncovalently coated capillaries. J Chromatogr B 871:370–374

Ramautar R, Mayboroda OA, Somsen GW, de Jong GJ (2011) CE-MS for metabolomics: developments and applications in the period 2008–2010. Electrophoresis 32:52–65

Robertson DG (2005) Metabonomics in toxicology: a review. Toxicol Sci 85:809–822

Robertson DG, Lindon JC (2005) Metabonomics in toxicity assessment. Taylor & Francis, Boca Raton, FL

Roy SM, Anderle M, Lin H, Becker CH (2004) Differential expression profiling of serum proteins and metabolites for biomarker discovery. Int J Mass Spectrom 238:163–171

Sana TR, Waddell K, Fischer SM (2008) A sample extraction and chromatographic strategy for increasing LC/MS detection coverage of the erythrocyte metabolome. J Chromatogr B 871:314–321

Sauer U, Lasko DR, Fiaux J, Hochuli M, Glaser R, Szyperski T, Wuthrich K, Bailey JE (1999) Metabolic flux ratio analysis of genetic and environmental modulations of Escherichia coli central carbon metabolism. J Bacteriol 181:6679–6688

Schlotterbeck G, Ross A, Hochstrasser R, Senn H, Kuhn T, Marek D, Schett O (2002) High-resolution capillary tube NMR. A miniaturized 5-microL high-sensitivity TXI probe for mass-limited samples, off-line LC NMR, and HT NMR. Anal Chem 74:4464–4471

Schripsema J (2010) Application of NMR in plant metabolomics: techniques, problems and prospects. Phytochem Anal 21:14–21

Seeger K (2009) Metabolic changes in autoimmune diseases. Curr Drug Discov Technol 6:256–261

Sequeira S, So PW, Everett JR, Elcombe CR, Kelvin AS, Nicholson JK (1990) 1H-NMR spectroscopy of biofluids and the investigation of xenobiotic-induced changes in liver biochemistry. J Pharm Biomed Anal 8:945–949

Shin MH, Lee do Y, Liu KH, Fiehn O, Kim KH (2010) Evaluation of sampling and extraction methodologies for the global metabolic profiling of Saccharophagus degradans. Anal Chem 82:6660–6666

Shlomi T, Cabili MN, Ruppin E (2009) Predicting metabolic biomarkers of human inborn errors of metabolism. Mol Syst Biol 5:263

Sinha AE, Hope JL, Prazen BJ, Nilsson EJ, Jack RM, Synovec RE (2004) Algorithm for locating analytes of interest based on mass spectral similarity in GC x GC-TOF-MS data: analysis of metabolites in human infant urine. J Chromatogr A 1058:209–215

Smedsgaard J, Nielsen NPV, Carstensen JM (1998) Aligning of single and multiple wavelength chromatographic profiles for chemometric data analysis using correlation optimised warping. J Chromatogr A 805:17–35

Spagou K, Tsoukali H, Raikos N, Gika H, Wilson ID, Theodoridis G (2010) Hydrophilic interaction chromatography coupled to MS for metabonomic/metabolomic studies. J Sep Sci 33:716–727

Syage JA, Hanold KA, Lynn TC, Horner JA, Thakur RA (2004) Atmospheric pressure photoionization. II. Dual source ionization. J Chromatogr A 1050:137–149

Tang HR, Xiao CN, Wang YL (2009) Important roles of the hyphenated HPLC-DAD-MS-SPE-NMR technique in metabonomics. Magn Reson Chem 47:S157–S162

Taylor PJ (2005) Matrix effects: the Achilles heel of quantitative high-performance liquid chromatography-electrospray-tandem mass spectrometry. Clin Biochem 38:328–334

Tolstikov VV, Fiehn O, Tanaka N (2007) Application of liquid chromatography-mass spectrometry analysis in metabolomics: reversed-phase monolithic capillary chromatography and hydrophilic chromatography coupled to electrospray ionization-mass spectrometry. Methods Mol Biol 358:141–155

Tomita M, Nishioka T (2005) Metabolomics: the frontier of systems biology. Springer, Tokyo

Trygg J, Holmes E, Lundstedt T (2007) Chemometrics in metabonomics. J Proteome Res 6:469–479

Tuck MK, Chan DW, Chia D, Godwin AK, Grizzle WE, Krueger KE, Rom W, Sanda M, Sorbara L, Stass S, Wang W, Brenner DE (2009) Standard operating procedures for serum and plasma collection: early detection research network consensus statement standard operating procedure integration working group. J Proteome Res 8:113–117

Ullsten S, Danielsson R, Backstrom D, Sjoberg P, Bergquist J (2006) Urine profiling using capillary electrophoresis-mass spectrometry and multivariate data analysis. J Chromatogr A 1117:87–93

Villas-Boas SG, Mas S, Akesson M, Smedsgaard J, Nielsen J (2005) Mass spectrometry in metabolome analysis. Mass Spectrom Rev 24:613–646

Wang WX, Zhou HH, Lin H, Roy S, Shaler TA, Hill LR, Norton S, Kumar P, Anderle M, Becker CH (2003) Quantification of proteins and metabolites by mass spectrometry without isotopic labeling or spiked standards. Anal Chem 75:4818–4826

Wang YL, Holmes E, Nicholson JK, Cloarec O, Chollet J, Tanner M, Singer BH, Utzinger J (2004) Metabonomic investigations in mice infected with Schistosoma mansoni: an approach for biomarker identification. Proc Natl Acad Sci U S A 101:12676–12681

Wang Y, Wang J, Yao M, Zhao X, Fritsche J, Schmitt-Kopplin P, Cai Z, Wan D, Lu X, Yang S, Gu J, Haring HU, Schleicher ED, Lehmann R, Xu G (2008) Metabonomics study on the effects of the ginsenoside Rg3 in a beta-cyclodextrin-based formulation on tumor-bearing rats by a fully automatic hydrophilic interaction/reversed-phase column-switching HPLC-ESI-MS approach. Anal Chem 80: 4680–4688

Wang HL, Tso VK, Slupsky CM, Fedorak RN (2010) Metabolomics and detection of colorectal cancer in humans: a systematic review. Future Oncol 6:1395–1406

Want EJ, O'Maille G, Smith CA, Brandon TR, Uritboonthai W, Qin C, Trauger SA, Siuzdak G (2006) Solvent-dependent metabolite distribution, clustering, and protein extraction for serum profiling with mass spectrometry. Anal Chem 78:743–752

Want EJ, Coen M, Masson P, Keun HC, Pearce JT, Reily MD, Robertson DG, Rohde CM, Holmes E, Lindon JC, Plumb RS, Nicholson JK (2010) Ultra performance liquid chromatography-mass spectrometry profiling of bile acid metabolites in biofluids: application to experimental toxicology studies. Anal Chem 82:5282–5289

Ward ME, Politzer IR, Laseter JL, Alam SQ (1976) Gas chromatographic mass spectrometric evaluation of free organic acids in human saliva. Biomed Mass Spectrom 3:77–80

Waterman CL, Kian-Kai C, Griffin JL (2010) Metabolomic strategies to study lipotoxicity in cardiovascular disease. Biochim Biophys Acta 1801:230–234

Weckwerth W (2007) Metabolomics: methods and protocols. Humana Press, Totowa, NJ

Werner E, Heilier JF, Ducruix C, Ezan E, Junot C, Tabet JC (2008) Mass spectrometry for the identification of the discriminating signals from metabolomics: current status and future trends. J Chromatogr B Analyt Technol Biomed Life Sci 871:143–163

Wetmore DR, Joseloff E, Pilewski J, Lee DP, Lawton KA, Mitchell MW, Milburn MV, Ryals JA, Guo L (2010) Metabolomic profiling reveals biochemical pathways and biomarkers associated with pathogenesis in cystic fibrosis cells. J Biol Chem 285:30516–30522

Wilson ID, Nicholson JK, Castro-Perez J, Granger JH, Johnson KA, Smith BW, Plumb RS (2005) High resolution "ultra performance" liquid chromatography coupled to oa-TOF mass spectrometry as a tool for differential metabolic pathway profiling in functional genomic studies. J Proteome Res 4:591–598

Winnike JH, Li Z, Wright FA, Macdonald JM, O'Connell TM, Watkins PB (2010) Use of pharmaco-metabonomics for early prediction of acetaminophen-induced hepatotoxicity in humans. Clin Pharmacol Ther 88:45–51

Wu H, Southam AD, Hines A, Viant MR (2008) High-throughput tissue extraction protocol for NMR- and MS-based metabolomics. Anal Biochem 372:204–212

Wu ZM, Huang ZQ, Lehmann R, Zhao CX, Xu GW (2009) The application of chromatography-mass spectrometry: methods to metabonomics. Chromatographia 69:S23–S32

Young SP, Wallace GR (2009) Metabolomic analysis of human disease and its application to the eye. J Ocul Biol Dis Infor 2:235–242

Zelena E, Dunn WB, Broadhurst D, Francis-McIntyre S, Carroll KM, Begley P, O'Hagan S, Knowles JD, Halsall A; HUSERMET Consortium, Wilson ID, Kell DB (2009) Development of a robust and repeatable UPLC-MS method for the long-term metabolomic study of human serum. Anal Chem 15;81(4):1357–64. doi: 10.1021/ac8019366. PMID: 19170513

Zhang S, Nagana Gowda GA, Ye T, Raftery D (2010) Advances in NMR-based biofluid analysis and metabolite profiling. Analyst 135:1490–1498

Zhang HP, Wang Y, Gu X, Zhou JY, Yan C (2011) Metabolomic profiling of human plasma in pancreatic cancer using pressurized capillary electrochromatography. Electrophoresis 32:340–347

Zhou M, McDonald JF, Fernandez FM (2010) Optimization of a direct analysis in real time/time-of-flight mass spectrometry method for rapid serum metabolomic fingerprinting. J Am Soc Mass Spectrom 21:68–75

Bioinformatic Methods and Resources for Neuroscience Research

31

Chittibabu Guda

Abstract

Students majoring in life sciences are seldom exposed to the use of bioinformatic tools in their regular coursework. Exponential growth of data in the biomedical research disciplines, including neuroscience, warrants the need for training undergraduate and graduate students in the use of bioinformatic and neuroinformatic tools. Two main objectives of this chapter are to provide an overview of important bioinformatic and neuroinformatic resources and to explain the usage of the commonly used bioinformatic data analysis tools. For each major tool, the theory behind the methodology is briefly described to enable the user to understand how the program works. The bioinformatic resources described in this article refer to the most commonly used tools and databases for sequence homology search, multiple sequence alignment, protein domain analysis, gene set enrichment analysis, pathway analysis, and interaction network analysis. In addition to the general-purpose tools and databases, a list of neuroinformatic specific resources is provided.

Keywords

Neuroinformatics • Bioinformatics • Computational neuroscience

C. Guda, Ph.D (✉)
Department of Genetics, Cell Biology and Anatomy, University of Nebraska Medical Center, Omaha, NE, USA

Bioinformatics and Systems Biology Core Facility, University of Nebraska Medical Center, Omaha, NE, USA

Member, UNMC Fred and Pamela Buffet Cancer Center & Eppley Institute, University of Nebraska Medical Center, Omaha, NE, USA

Courtesy Faculty, Department of Biochemistry and Molecular Biology, College of Medicine, University of Nebraska Medical Center, 6015 Durham Research Center, Omaha, NE, USA
e-mail: babu.guda@unmc.edu

31.1 Introduction and Background

The use of bioinformatic and neuroinformatic tools has become indispensable in the neuroscience research owing to the exponential growth of data from high-throughput studies. The emergence of systems biology in the past decade warrants elaborate computational processing of experimental results that includes integration of different data types and meta-analyses against

H. Xiong and H.E. Gendelman (eds.), *Current Laboratory Methods in Neuroscience Research*, Springer Protocols Handbooks, DOI 10.1007/978-1-4614-8794-4_31,

curated (reviewed by experts) and federated (multiple, related databases united) data sources to interpret the data. Hence, bioinformatics/neuroinformatics has become an integral part of neuroscience. While computational biologists strive to develop tools for data analyses, user training is very important to avail the full utility of such tools. Different tools are being developed on different computer platforms such as UNIX, Windows, or Mac OS, and basic understanding of these systems is required for efficient and wide usage of such tools. On the other hand, web-based bioinformatic tools are the most popular among the research community for their ease of use and accessibility. Typically, simple searches using the default parameters provided by a web-based program require no special skills, but simple searches are not sufficient to get the most out of the tools and databases. Many programs require manipulation of search parameters that require special training and thorough understanding of how each program works. In this article, we provide an overview of the common bioinformatic and neuroinformatic tools and explain the theory behind these tools in an easy-to-understand format.

31.1.1 Objectives

Two main objectives of this chapter are to provide an overview of important bioinformatic and neuroinformatic resources and to explain to the reader the usage of the most commonly used bioinformatic data analysis tools. For each major tool, the theory behind the methodology will be briefly described to enable the user to understand how the program works. Tools covered in this chapter include the Blast suite of programs to search for sequence homology, multiple sequence alignment programs (ClustalW), protein domain analysis (Pfam), gene set enrichment analysis (GSEA), pathway analysis (KEGG), and interaction network analysis (Cytoscape). A selective list of primary and curated databases and the data analysis tools is provided in Table 31.1. A list of neuroinformatic resources that are specifically used in neuroscience research is provided in Table 31.2.

31.2 Data Resources and Analysis Tools

A myriad of data resources and analysis tools are available in the public domain for commonly used tasks in bioinformatic research. Primary or archival databases are predominantly maintained by publicly funded institutions such as the National Center for Biotechnology Information (NCBI) and European Bioinformatics Institute (EBI). GenBank and UniProt from NCBI and EBI, respectively, cover most of the primary data on nucleic acids and proteins, while Protein Data Bank (PDB) covers information on macromolecular structures. These archival databases offer tools and services for direct deposit of experimental data by researchers. A variety of specialized or derived databases are created by extracting data from the primary databases and applying analysis tools or additional annotations to create value-added secondary databases. NCBI, EBI, and PDB also maintain in-house research divisions, whose mission is to process, archive, analyze, annotate, and disseminate data to educate and benefit the research community. As such, NCBI, EBI, and PDB support and maintain "one-stop-shop" federated databases and web portals to access a variety of specialized databases as well as the data analysis tools (Table 31.1).

31.2.1 Popular Bioinformatic Resources

A list of popular general-purpose bioinformatic tools and resources is provided in Table 31.1. These include major web portals such as NCBI and EBI; curated databases such as GenBank, UniProt, KEGG (Kyoto Encyclopedia of Genes and Genomes); and popular utilities such as Blast (basic local alignment search tool) server, Entrez (the life sciences search engine), and SRS (Sequence Retrieval System). Note that the list provided in Table 31.1 is not exclusive, but represents a sample of the most commonly used tools. In Sect. 31.2.3, the theory behind some of the commonly used methods is provided.

Table 31.1 A list of commonly used resources for bioinformatic/neuroinformatic research

Database name	Description	URL
Web portals		
NCBI	National Center for Biotechnology Information	http://www.ncbi.nlm.nih.gov
EBI	European Bioinformatics Institute	http://www.ebi.ac.uk
DDJB	DNA Data Bank of Japan	http://www.ddbj.nig.ac.jp
RCSB	Research Collaborative for Structural Bioinformatics	http://www.rcsb.org
Curated databases		
GenBank	Genetic sequence database	http://www.ncbi.nlm.nih.gov/genbank
UniProt	A knowledgebase of the universe of proteins	http://www.uniprot.org
Ensembl	A vertebrate genome database	http://uswest.ensembl.org/index.html
OMIM	A compendium of human genes and phenotypes	http://www.ncbi.nlm.nih.gov/omim
RefSeq	Reference Sequence Database	http://www.ncbi.nlm.nih.gov/RefSeq
dbSNP	Database of Short Genetic Variations	http://www.ncbi.nlm.nih.gov/snp
GEO	A public repository of gene expression data	http://www.ncbi.nlm.nih.gov/geo
IntEnz	Integrated Relational Enzyme Database	http://www.ebi.ac.uk/intenz
InterPro	An integrated database of protein signatures	http://www.ebi.ac.uk/interpro
IntAct	An interacting protein database	http://www.ebi.ac.uk/intact
Reactome	A curated pathway database	http://www.reactome.org
BioModels	A repository of computational biological models	http://www.ebi.ac.uk/biomodels-main
GO	Gene Ontology database	http://www.geneontology.org
PRIDE	Proteomics Identifications database	http://www.ebi.ac.uk/pride
KEGG	Kyoto Encyclopedia of Genes and Genomes	http://www.genome.jp/kegg
Utilities/tools		
PubMed	Biomedical literature search tool	http://www.ncbi.nlm.nih.gov/pubmed
Entrez	The life sciences search engine	http://www.ncbi.nlm.nih.gov/sites/gquery
Blast	Sequence homology search tool	http://blast.ncbi.nlm.nih.gov/Blast.cgi
NCBI ToolBox	Data model and programming resources	www.ncbi.nlm.nih.gov/IEB/ToolBox
NCBI FTP site	Web portal to download data from NCBI	http://www.ncbi.nlm.nih.gov/Ftp
ClustalW2	Multiple sequence alignment program	http://www.ebi.ac.uk/Tools/msa/clustalw2
SRS	Sequence Retrieval System	http://srs.ebi.ac.uk
Broad Institute Software	Open-source bioinformatic software	http://www.broadinstitute.org/scientific-community/software
UCSC Genome Browser	A web-based portal for browsing genomes	http://genome.ucsc.edu
BIRN	Biomedical Informatics Research Network	http://www.birncommunity.org
GSEA	Gene Set Enrichment Analysis	http://www.broadinstitute.org/gsea
Pfam	Protein family database	http://pfam.janelia.org
Cytoscape	Network visualization and analysis tool	http://www.cytoscape.org

Table 31.2 A list of resources dedicated for neuroinformatic research

Database name	Description	URL
NIF	Neuroscience Information Framework	http://www.neuinfo.org
INCF	International Neuroinformatics Coordinating Facility	http://www.incf.org
Digital Brain Atlas	Scalable brain atlas	http://scalablebrainatlas.incf.org
Allen Brain Atlas	A collection of gene expression/neuroanatomical data	http://www.brain-map.org
Whole Brain Catalog	An open-source virtual catalog of the mouse brain	http://wholebraincatalog.org
Human Brain Atlas	Brain diversity bank with stained and MRI sections	https://www.msu.edu/~brains/brains/human
NeuronDB	A searchable database of neuronal properties	http://senselab.med.yale.edu/NeuronDB
NeuroMorpho	A curated inventory of digitally reconstructed neurons	http://neuromorpho.org/neuroMorpho
NITRC	A source for neuroimaging tools and resources	http://www.nitrc.org
BrainSpan	Atlas of the developing human brain	http://www.brainspan.org
Neuro Diseases	A catalog of identified neurological disorders	http://www.ninds.nih.gov/disorders
GENSAT	Gene expression atlas of adult CNS in the mouse	http://www.gensat.org
Human Brain Project	Simulation of the human brain	http://www.humanbrainproject.eu
CoCoMac	Collations of Connectivity Data on the Macaque Brain	http://cocomac.org
BrainInfo	A knowledgebase on the structures in the brain	http://braininfo.rprc.washington.edu
BMI-PF	Brain Machine Interface Platform	http://bmi.neuroinf.jp
Aging Genes DB	Genes affecting age-related neurological diseases	http://www.uwaging.org/genesdb

31.2.2 Neuroinformatic Resources

Of the many definitions used to describe neuroinformatics, the simplest one states that it is a field that includes building databases and tools for understanding the nervous system (Huerta et al. 1993). Advances made over the past 2 decades have resulted in the development of sophisticated neuroinformatic tools and databases as enlisted in Table 31.2. These include databases containing real and virtual brain maps, scalable brain atlases, digitally reconstructed neurons, catalogs of the neurological disorders, etc., and sophisticated tools to interact with these databases. The Neuroscience Information Framework (NIF), an initiative of the NIH Blueprint for Neuroscience Research, maintains a web portal that offers a dynamic inventory of web-based neuroscience resources that include data, materials, and utilities (Bandrowski et al. 2012). Similarly, the International Neuroinformatics Coordinating Facility (INCF) offers a collaborative platform to the international neuroinformatic research community for sharing data and computing resources (Bjaalie and Grillner 2007). Another popular resource, the Allen Brain Atlas contains an extensive collection of gene expression and neuroanatomical data and a novel suite of searching and visualization tools (Jones et al. 2009). In addition to human brain, other neuroinformatic resources covering the brains of mouse and macaque are also available (Geschwind 2004; Kötter 2004).

31.2.3 Data Analysis Tools in Bioinformatics/ Neuroinformatics

This section explains the theory behind popular bioinformatic tools and describes their usage. These tools cover searching for sequence homology, identification of domains/motifs in protein sequences, aligning multiple sequences, gene enrichment analysis of microarray datasets, and pathway and network analysis.

31.2.3.1 Sequence Homology Search

Looking for sequence similarity of a query sequence (search for homology) against a set of sequences in a database is the most commonly performed task in bioinformatics. Knowledge of the underlying components of a biological database and the search engine will help perform advanced power searches to get the most out of a database. To understand how homology searching of a nucleotide or protein database works, below we introduce the reader to the basic theoretical concepts of homology searching.

1. Basic Concepts of Sequence Homology Searching

   The basic elements of a sequence homology search include a query sequence, a database, the search algorithm, and a set of measures to evaluate the outcome of the search. The query is typically a nucleic acid or amino acid sequence and the database contains a collection of sequences that are used to search against the query sequence. There are two major comprehensive databases in the public domain; one is the "nr" database (stands for nonredundant) maintained by the NCBI (for both proteins and nucleotides) and the second one is the "UniProt" database (for proteins) maintained by the EBI. These two databases overlap significantly and cross-reference to each other, but are not identical. Nevertheless, either of these databases is sufficient for general searches. A number of specialized or curated databases such as RefSeq or Swiss-Prot are also available in the public domain, and the user can also create a customized database that can be formatted for searching.

   The core of an efficient search algorithm is its scoring function, which quantifies the extent of similarity between a pair of sequences in a comparison. Scoring function in a search algorithm (such as Blast) is designed to reward similarity and penalize dissimilarity using numeric scores so that similar sequence pairs will score higher and vice versa. A scoring function mainly keeps track of three parameters that include matches, mismatches, and gaps between two sequences. Matches are always rewarded; mismatches can be rewarded, penalized, or considered neutral depending upon the evolutionarily accepted mutation rate between a pair of nucleic or amino acids. For this purpose, several types of nucleic acid and amino acid substitution matrices were developed such as PAM, BLOSUM, and Gonnet series (Dayhoff et al. 1978; Henikoff and Henikoff 1992; Gonnet et al. 1992). A substitution matrix will have substitution scores for all possible pairs of nucleic or amino acids including the identical pairs. Hence, a nucleic acid substitution matrix contains 10 scores and an amino acid substitution matrix contains 210 scores that range from negative to positive. A scoring function uses scores from one of the substitution matrices during sequence comparison to determine the raw score, which is also known as the "similarity score." The raw similarity score is used to determine the metrics to evaluate if two sequences are homologous or not.

Two of the most widely used metrics of sequence similarity are expectation value (*E*-value) and Bit score. These measures are determined from the raw score after normalizing to the length of the match, size of the database, etc. The higher the raw score, the lesser the *E*-value and the higher the Bit score. Hence, two identical or near identical sequences can get an *E*-value of "zero" (the least possible *E*-value). *E*-values are calculated for all sequence pairs compared in a database search; these values can range from zero to higher. An *E*-value of one means that the same sequence pair can also be matched at random with the same score at least once in the current database. Hence, matches with higher *E*-values are not reliable because the chances of finding the same match at random are also higher. *E*-values are not normalized to the size of the database; hence, for the same query and the same database sequence match, searches against differently sized databases will generate different *E*-values. Generally, the chances of finding random matches in smaller databases are smaller; hence, when searching smaller databases, matches with much smaller *E*-values are more reliable compared to bigger databases. On the other hand, Bit scores are normalized to the size of the database and hence will have consistent values across databases. Usually, a Bit score of 40 or higher for a matched pair of sequences is considered reliable irrespective of the database size.

31.2.3.2 Different Types of Blast Programs

Basic local alignment search tool (Blast) suite contains a variety of software utilities to search nucleic acid or amino acid sequences (Altschul et al. 1990). It offers several flavors of the original Blast program that vary by the type of query sequence used and the type of database searched. The web-based Blast program can be accessed at http://blast.ncbi.nlm.nih.gov. Different types of Blast programs are described below:

1. *Blastn*: A nucleotide query sequence (DNA/RNA) can be searched against a nucleotide database.
   (a) *Megablast*: This is a nucleotide Blast specifically optimized to work much faster with highly similar sequences such as the genomic sequences. Megablast was created to handle the large volume of similar sequences from high-throughput sequencing (Zhang et al. 2000).
2. *Blastp*: An amino acid query sequence can be searched against a protein database. This is a basic method that works well to identify closely related sequences. Nevertheless, more powerful variations of Blastp that include Psi-Blast and Phi-Blast can perform highly sensitive and highly specific searches, respectively (Altschul et al. 1997).
   (a) *Psi-Blast*: This is Position-Specific Iterated Blast designed to identify remote homologs of protein sequences from a database. Psi-Blast is very sensitive compared to Blastp, because it uses a profile-based method to build position-specific scoring matrix (PSSM) based on the conservation pattern of amino acids in the alignment (see Sect. 31.4.1 for more details on profiles). The first step of Psi-Blast is the same as Blastp; matching sequences identified in the first step are aligned to build a PSSM, which will be used as the query to identify more matches in the second step. In the subsequent steps, a new PSSM is generated at each iteration and searched against the database until the program converges, i.e., no more new sequences are identified by the program. The efficiency of Psi-Blast in finding

remote homologs is attributed to its use of PSSMs (as against generic substitution matrices used by Blastp) and the multiple iterations in the searching process.

(b) *Phi-Blast*: Phi-Blast stands for Pattern-Hit Iterated Blast that uses a sequence pattern (in a "regular expression" format used by PROSITE database) as input along with the query sequence (http://www.ncbi.nlm.nih.gov/blast/html/PHIsyntax.html). This program searches a database like a regular "Blastp" with additional constraints imposed to match the given pattern in the target sequences. The addition of pattern match requirement in the search process helps to identify family-specific proteins with higher specificity. However, this method is useful only when family-specific patterns are already known.

(c) *RPS-Blast*: Reverse Position-Specific Blast is used to search a query sequence against a database of pre-calculated PSSMs (Marchler-Bauer et al. 2002). The direction of the search is reversed in this case (a sequence against a database of PSSMs) compared to Psi-Blast, where a PSSM is searched against a database of sequences and hence the name "reverse position-specific." For annotating an entire proteome, RPS-Blast is very useful because each PSSM represents a protein family and the corresponding functions of all the sequences in a proteome can be identified in a single search.

3. *Blastx*: A nucleotide query sequence is searched against a protein database using the "blastx" program. The input sequence is translated into amino acid sequence in 6 frames (three frames each way of $5' \rightarrow 3'$ and $3' \rightarrow 5'$) and each translated product is searched against a protein sequence database. Blastx program is especially useful to search EST (Expressed Sequence Tag) sequences against protein sequence databases to identify potential coding regions.
4. *Tblastn*: This program works in the opposite direction of "Blastx," where each nucleotide sequence in the target database is translated in 6 frames to search against the protein query sequence. Using this program, peptides identified from proteomic studies can be searched against nucleotide databases to identify corresponding open reading frames.
5. *Tblastx*: This program performs a two-way translation of both nucleotide query and the nucleotide database sequences and compares the 6×6 frame translation products at the protein level. This is the most time-consuming of all the programs in the Blast suite due to the permutations of searches involved.

31.2.4 Local Installation of Blast Software

The NCBI web server hosts the Blast search service for protein or nucleic acid sequences. Web-based programs typically offer only a limited number of default program options for the user, thus limiting their full potential use. Moreover, the web-based Blast offers only a limited selection of target databases for searching. For advanced users, NCBI also offers a downloadable version of the Blast program that is compatible with many operating systems including Windows, Mac, and Linux. A downloadable package can be locally installed with minimal configuration settings to unleash the full power of these programs. Major advantages of using the local stand-alone version include the ability to run large-scale batch searches on local servers, ability to better manipulate the search parameters and create customized target databases, and the ability to read and process Blast output data in a tab-delimited text format that can be opened with an Excel spreadsheet or a text editor or uploaded into a database for further processing. Some useful links to access Blast programs and databases are given below.

NCBI Toolkit: http://www.ncbi.nlm.nih.gov/books/NBK7160/

Blast program download: ftp://ftp.ncbi.nlm.nih.gov/blast/executables/blast+/LATEST/

Blast DB download: ftp://ftp.ncbi.nlm.nih.gov/blast/db/

Blast Manual: http://blast.ncbi.nlm.nih.gov/BLAST_guide.pdf

31.3 Multiple Sequence Alignment

Multiple sequence alignments (MSA) are used to identify local and global similarities among more than two sequences by looking at the conserved patterns of nucleotide bases or amino acids along the columns of an alignment. A true multiple alignment program should make an all-against-all comparison of sequences in a set to find the most optimal alignment. But, this approach is not practical to align large number of sequences due to the exponential time complexity involved. Hence, the popular MSA programs use pairwise alignment strategies to progressively align sequences and build MSAs using heuristic guide trees. The best scoring alignment pair is used as the seed for initial alignment and subsequent sequences are progressively added in such a way that the better alignments are added first and the weaker alignments are added towards the end to avoid creation of too many gaps by the latter. The cost of computation required for the progressive approaches is relatively small and hence multiple alignments for large families can be run very fast. Due to this practicality, progressive alignments such as ClustalW and MUSCLE have become very popular (Thompson et al. 1994; Edgar 2004). ClustalW is a hugely popular MSA program that is available freely on many different operating systems. Web-based versions of several MSA programs can be accessible from the EBI's tools website (http://ebi.ac.uk/tools/sequence.html). An example of input and output sequences using ClustalW program is illustrated in Fig. 31.1. Most of these programs can also be downloaded from the EBI's FTP site (ftp://ftp.ebi.ac.uk/pub/software) and installed on local servers for integration with other programs.

31.4 Family/Domain Homology Search

Blast searches are typically used to detect pairwise similarity between two sequences; but to identify the common features/patterns of a set of related sequences (known as families), statistical

a

```
>P89927_9CAUD/239-284
FLGYEARINFTGLGDGLVSIETSHQVGAELDKLTAWLDERGWAYYY
>Q81CE8_BACCR/228-273
KLGYTSKIISRGDNQGLVYFETDYRQGNELDKATAWLDTKGIKYFY
>Q736Q2_BACC1/228-273
ALGYESRIISYGDKQGLVRFETAYRQGNELDRATAWLDAKGLKYFY
>A7GPG5_BACCN/224-268
ERNIKASIIFEGKNGNPYVLTEKMSNPEMDKFTAWLDERGWYYEY
>A9VS47_BACWK/222-266
EHGTKGKVVSDPLTGLAYLQTEILPNGELDKITAWMDERNWWYEY
>A9VPQ0_BACWK/206-253
ERGTKGKVVVNPLTGLAYIQTEVLPNSELDKITWWMDTRPGGKWWYEY
```

b

```
CLUSTAL 2.1 multiple sequence alignment


Q81CE8_BACCR/228-273      KLGYTSKIISRGDNQGLVYFETDYRQGNELDKATAWLDTK---GIKYFY 46
Q736Q2_BACC1/228-273      ALGYESRIISYGDKQGLVRFETAYRQGNELDRATAWLDAK---GLKYFY 46
P89927_9CAUD/239-284      FLGYEARINFTGLGDGLVSIETSHQVGAELDKLTAWLDER---GWAYYY 46
A9VS47_BACWK/222-266      EHGTKGKVVSDPL-TGLAYLQTEILPNGELDKITAWMDER---NWWYEY 45
A9VPQ0_BACWK/206-253      ERGTKGKVVVNPL-TGLAYIQTEVLPNSELDKITWWMDTRPGGKWWYEY 48
A7GPG5_BACCN/224-268      ERNIKASIIFEGK-NGNPYVLTEKMSNPEMDKFTAWLDER---GWYYEY 45
                            .  . :       *   . *    . *:*: * *:* :      * *
```

Fig. 31.1 An example showing the input and output of ClustalW program. (**a**) Unaligned input sequences in FASTA format. (**b**) Colored output of the multiple sequence alignment. Columns with *stars* represent identical conserved residues; columns with *double dots* indicate similar but not identical conserved residues; columns with *single dots* represent partially conserved residues of the same kind, while other columns indicate no significant conservation of residues. Search was performed using ClustalW2 program accessible at http://www.ebi.ac.uk/Tools/msa/clustalw2

modeling is required. Profiles and hidden Markov models (HMMs) are commonly used methods that convert the distinct sequence features of a protein or nucleic acid family into a numeric model. Each model is distinct and represents only the family of sequences that was used for building that model. The advantages of building such models are that they are family specific and more robust than using single sequence searches, and they offer a superior computational advantage to annotate large-scale datasets.

31.4.1 Profile-Based Methods

To build a profile, a set of closely related sequences is aligned using an MSA program such as ClustalW. In an MSA, conserved columns generally correspond to a distinct functional or structural feature of a family. In other words, certain nucleotide bases or amino acids are more important at certain positions than others and this confers the family-specific importance of these amino acid positions. Profile-based methods try to capture this position-specific information for each column position in the alignment (Gribskov et al. 1987). In a nutshell, frequency counts of each nucleotide base or amino acid in each column are used to assign column-based weights for all pairs of substitutions; thus, the base or amino acid substitution scores vary for the same pair by their column position depending on their conservation pattern in a given column. This is in contrast to using substitution scores from a BLOSUM or PAM matrix, which are constant for a pair of bases or amino acids irrespective of their column position in the alignment. Hence, profile-based methods are more sensitive in finding remote homologs compared to regular Blast-based searches. However, to build a statistically significant model to represent a family or domain, at least five or more sequences are needed. The Psi-Blast method in the Blast suite uses profile-based approach to build position-specific scoring matrices (PSSMs) during its iterative searches and hence Psi-Blast searches are more sensitive than regular Blast searches.

31.4.2 Hidden Markov Models

In a profile, each column in an MSA is treated independently of another column in the model building and scoring process. HMMs are similar to profiles except that an additional parameter known as "transition probability" is also determined to link the transition from one column to the other. In simple terms, the independent columns only reflect the conservation of amino acids or nucleotide bases within a column (known as emission probabilities), while a transition probability also keeps track of what comes next given what exists before. Conceptually, such models are more relevant for biological sequences because the order of sequence elements is very important for the structure and function of both protein and nucleic acids. HMMs have been extensively used for a variety of problems in molecular biology (reviewed by Yoon 2009) that include gene finding (Borodovsky and Lomsadze 2011), prediction of posttranslational modification sites (Pierleoni et al. 2008), and building databases of protein family profiles (Punta et al. 2012). One of the most popular databases built using HMMs is the Pfam database (Punta et al. 2012). Pfam is a protein family database that contains Pfam-A covering all the annotated protein sequence families in the public domain and Pfam-B for the unannotated families. Pfam-A contains about 15,000 HMMs, each representing a protein superfamily (clan), a family, or a domain. Pfam-A domains have been extensively used in bioinformatic research for annotating the functions of proteins based on their constituent domains. In Blast searches, a query sequence is searched against a database of sequences, while in a Pfam search, a query sequence is searched against a database of profile HMMs. A protein can have multiple functions based on its domain composition. Pfam searches are more sensitive, precise, and informative because all the functions of a protein can be known in a single search. The Pfam web server can be accessed at http://pfam.janelia.org, and the underlying software, HMMER, can be downloaded at http://hmmer.janelia.org. Currently, Pfam-A database can annotate the functions of over 70–80 % of the proteins in the proteomes of the sequenced genomes. An example of a Pfam-A

a

```
>sp|Q9UBS5|GABR1_HUMAN Gamma-aminobutyric acid type B receptor subunit 1
MLLLLLLAPLFLRPPGAGGAQTPNATSEGCQIIHPPWEGGIRYRGLTRDQVKAINFLPVD
YEIEYVCRGEREVVGPKVRKCLANGSWTDMDTPSRCVRICSKSYLTLENGKVFLTGGDLP
ALDGARVDFRCDPDFHLVGSSRSICSQGQWSTPKPHCQVNRTPHSERRAVYIGALFPMSG
GWPGGQACQPAVEMALEDVNSRRDILPDYELKLIHHDSKCDPGQATKYLYELLYNDPIKI
ILMPGCSSVSTLVAEAARMWNLIVLSYGSSSPALSNRQRFPTFFRTHPSATLHNPTRVKL
FEKWGWKKIATIQQTTEVFTSTLDDLEERVKEAGIEITFRQSFFSDPAVPVKNLKRQDAR
IIVGLFYETEARKVFCEVYKERLFGKKYVWFLIGWYADNWFKIYDPSINCTVDEMTEAVE
GHITTEIVMLNPANTRSISNMTSQEFVEKLTKRLKRHPEETGGFQEAPLAYDAIWALALA
LNKTSGGGGRSGVRLEDFNYNNQTITDQIYRAMNSSSFEGVSGHVVFDASGSRMAWTLIE
QLQGGSYKKIGYYDSTKDDLSWSKTDKWIGGSPPADQTLVIKTFRFLSQKLFISVSVLSS
LGIVLAVVCLSFNIYNSHVRYIQNSQPNLNNLTAVGCSLALAAVFPLGLDGYHIGRNQFP
FVCQARLWLLGLGFSLGYGSMFTKIWWVHTVFTKKEEKKEWRKTLEPWKLYATVGLLVGM
DVLTLAIWQIVDPLHRTIETFAKEEPKEDIDVSILPQLEHCSSRKMNTWLGIFYGYKGLL
LLLGIFLAYETKSVSTEKINDHRAVGMAIYNVAVLCLITAPVTMILSSQQDAAFAFASLA
IVFSSYITLVVLFVPKMRRLITRGEWQSEAQDTMKTGSSTNNNEEEKSRLLEKENRELEK
IIAEKEERVSELRHQLQSRQQLRSRRHPPTPPEPSGGLPRGPPEPPDRLSCDGSRVHLLY
K
```

b

c

Family	Description	Entry type	Clan	Envelope		Alignment		HMM		Bit score	E-value
				Start	End	Start	End	From	To		
Sushi	Sushi domain (SCR repeat)	Domain	n/a	34	96	55	96	**17**	56	18.9	0.0012
Sushi	Sushi domain (SCR repeat)	Domain	n/a	109	157	123	157	**21**	56	30.5	3e-07
ANF_receptor	Receptor family ligand binding region	Family	CL0144	187	543	188	536	**2**	**340**	221.3	1.4e-65
7tm_3	7 transmembrane sweet-taste receptor of 3 GCPR	Family	n/a	601	860	601	858	1	**236**	163.9	3.4e-48

Fig. 31.2 An example showing the input and output of a Pfam search. (**a**) Protein sequence of GABA type B receptor subunit 1. (**b**) Graphic output of Pfam search representing the domains of GABA receptor protein. (**c**) Tabular output of Pfam search showing the start and end positions of the sequence matching with Pfam models and corresponding scores. This protein contains four domain regions that include two repeats of Sushi domain that have cell adhesion function, ANF_receptor domain that binds to a wide range of receptors, and the 7tm_3 domain which functions as a class C G-protein-coupled receptor. As illustrated, the functions of individual domains of a protein can be readily identified by Pfam database search, which was performed at http://pfam.janelia.org

database search for a GABA receptor protein is shown in Fig. 31.2.

31.5 Gene Set Enrichment Analysis

GSEA interprets important sets of genes in an experiment rather than individual genes from the differentially expressed gene list in two datasets (Subramanian et al. 2005). Ranked genes from a differentially expressed gene list "L" are compared against a predefined gene set "S" (gene sets are defined by a common biological function, chromosomal location, or regulation) to calculate their relevance to the genes in the predefined sets. GSEA software also comes with an accompanied database called Molecular Signature Database (MSigDB) that contains over 6,700 predefined and annotated gene sets to use with the GSEA software. These gene sets are created based on the chromosome positional information (C1), curated pathway information (C2), cis-regulatory motifs (C3), cancer expression neighborhood information (C4), or GO (Gene Ontology) term similarity (C5). An enrichment score (ES) is computed by walking down the ranked gene list L and increasing a running-sum statistic until ES reaches the maximum. The magnitude of increment (to ES) when a gene in L is present in a gene set S depends upon the extent of correlation of the gene with a phenotype in a set. The higher the ES score, the bigger or the stronger the gene set that is enriched. ES reflects the degree to which a gene set is overrepresented at the top or bottom of a ranked list of genes. In simple terms, the genes that contribute to most part of the ES are considered the core set of genes and these genes are included in the leading-edge subset. Leading-edge subset analysis can be carried out by a stand-alone version of GSEA-P software. GSEA and the accompanying MSigDB database that contains the gene sets can be downloaded from the Broad Institute's web page (http://www.broadinstitute.org/gsea).

31.6 Pathway Analysis

KEGG is one of the first and the most popular resource of its kind for pathway analysis (Kanehisa et al. 2012). It offers a knowledgebase containing a number of federated databases covering pathways, diseases, drugs, orthology groups, genomes, genes, and ligands and also data analysis tools for querying and mapping databases and pathway prediction. Currently, KEGG offers hundreds of reference pathway maps covering different areas such as metabolism, organismal systems, human diseases, and drug development. KEGG provides generic pathway reference maps that can be customized for viewing by species or by orthology groups, enzymes, or reactions in that pathway. All the components of the pathways, i.e., substrates, enzymes, products, genes, and proteins, are internally and externally mapped to allow easy extraction of information. MetaCyc is another popular web-based resource that contains a huge collection of curated metabolic pathways (Caspi et al. 2012). Some commercial software tools such as Ingenuity Pathway Analysis (IPA, http://www.ingenuity.com) and Pathway Studio from Ariadne Genomics (http://www.ariadnegenomics.com) are also available with integrated public and proprietary tools for comprehensive pathway analysis.

31.7 Network Analysis

Cytoscape is one of the most versatile open-source software tools available for network visualization and data integration and analysis (Smoot et al. 2011). Cytoscape program is built based on network graph theory with biological entities such as cells, genes, or proteins represented as nodes and the interactions between them represented as edges. The node and edge attributes represent various functional properties or expression values of nodes and edges. Attribute values can be used to filter data, to control the visual aspects of nodes and edges, and to perform complex network computations and other analyses. A key feature of Cytoscape is that it allows diverse types of attribute data to be visualized over the nodes and edges of a network. Since Cytoscape is an open-source software, a myriad of plug-ins have been developed to work with this software, which facilitate integration of a variety of data attributes generated by other bioinformatic tools. Cytoscape can be accessed from http://www.cytoscape.org, and the program can be downloaded under GNU's lesser general public license (LGPL) at http://www.cytoscape.org/download.html.

31.8 Conclusion

The bioinformatic resources described in this article refer to the most commonly used tools and databases for sequence homology search, MSA, protein domain analysis, GSEA, pathway analysis, and interaction network analysis. These basic tools are repeatedly used to process data in various biomedical research disciplines that include neuroinformatics. Students are encouraged to visit the stated websites and practice data analysis using real data examples. Resources listed in Table 31.2 cover the major neuroinformatic tools and databases that are rapidly evolving to serve the neuroinformatic research community.

References

Altschul SF, Gish W, Miller W, Myers EW, Lipman DJ (1990) Basic local alignment search tool. J Mol Biol 215:403–410

Altschul SF, Madden TL, Schäffer AA, Zhang J, Zhang Z, Miller W, Lipman DJ (1997) Gapped BLAST and PSI-BLAST: a new generation of protein database search programs. Nucleic Acids Res 25:3389–3402

Bandrowski AE, Cachat J, Li Y, Müller HM, Sternberg PW, Ciccarese P, Clark T, Marenco L, Wang R, Astakhov V, Grethe JS, Martone ME (2012) A hybrid human and machine resource curation pipeline for the Neuroscience Information Framework. Database 2012:bas005

Bjaalie JG, Grillner S (2007) Global neuroinformatics: the International Neuroinformatics Coordinating Facility. J Neurosci 27:3613–3615

Borodovsky M, Lomsadze A (2011) Eukaryotic gene prediction using GeneMark.hmm-E and GeneMark-ES. Curr Protoc Bioinformatics Chapter 4:Unit 4.6.1–10

Caspi R, Altman T, Dreher K, Fulcher CA, Subhraveti P, Keseler IM, Kothari A, Krummenacker M, Latendresse M, Mueller LA, Ong Q, Paley S, Pujar A, Shearer AG, Travers M, Weerasinghe D, Zhang P, Karp PD (2012) The MetaCyc database of metabolic pathways and enzymes and the BioCyc collection of pathway/genome databases. Nucleic Acids Res 40:D742–D753

Dayhoff MO, Schwartz RM, Orcutt BC (1978) A model of evolutionary change in protein. Atlas Protein Seq Struct 5:345–352

Edgar RC (2004) MUSCLE: multiple sequence alignment with high accuracy and high throughput. Nucleic Acids Res 32:1792–1797

Geschwind D (2004) GENSAT: a genomic resource for neuroscience research. Lancet Neurol 3:82

Gonnet GH, Cohen MA, Benner SA (1992) Exhaustive matching of the entire protein sequence database. Science 256:1443–1445

Gribskov M, McLachlan AD, Eisenberg D (1987) Profile analysis: detection of distantly related proteins. Proc Natl Acad Sci U S A 84:4355–4358

Henikoff S, Henikoff JG (1992) Amino acid substitution matrices from protein blocks. Proc Natl Acad Sci U S A 89:10915–10919

Huerta MF, Koslow SH, Leshner AI (1993) The Human Brain Project: an international resource. Trends Neurosci 16:436–438

Jones AR, Overly CC, Sunkin SM (2009) The Allen brain atlas: 5 years and beyond. Nat Rev Neurosci 10:821–828

Kanehisa M, Goto S, Sato Y, Furumichi M, Tanabe M (2012) KEGG for integration and interpretation of large-scale molecular datasets. Nucleic Acids Res 40:D109–D114

Kötter R (2004) Online retrieval, processing, and visualization of primate connectivity data from the CoCoMac database. Neuroinformatics 2:127–144

Marchler-Bauer A, Panchenko AR, Shoemaker BA, Thiessen PA, Geer LY, Bryant SH (2002) CDD: a database of conserved domain alignments with links to domain three-dimensional structure. Nucleic Acids Res 30:281–283

Pierleoni A, Martelli PL, Casadio R (2008) PredGPI: a GPI-anchor predictor. BMC Bioinformatics 23:392

Punta M, Coggill PC, Eberhardt RY, Mistry J, Tate J, Boursnell C, Pang N, Forslund K, Ceric G, Clements J, Heger A, Holm L, Sonnhammer EL, Eddy SR, Bateman A, Finn RD (2012) The Pfam protein families database. Nucleic Acids Res 40:D290–D301

Smoot ME, Ono K, Ruscheinski J, Wang PL, Ideker T (2011) Cytoscape 2.8: new features for data integration and network visualization. Bioinformatics 27:431–432

Subramanian A, Tamayo P, Mootha VK, Mukherjee S, Ebert BL, Gillette MA, Paulovich A, Pomeroy SL, Golub TR, Lander ES, Mesirov JP (2005) Gene set enrichment analysis: a knowledge-based approach for interpreting genome-wide expression profiles. Proc Natl Acad Sci U S A 102:15545–15550

Thompson JD, Higgins DG, Gibson TJ (1994) CLUSTAL W: improving the sensitivity of progressive multiple sequence alignment through sequence weighting, position specific gap penalties and weight matrix choice. Nucleic Acids Res 22:4673–4680

Yoon BJ (2009) Hidden Markov models and their applications in biological sequence analysis. Curr Genomics 10:402–415

Zhang Z, Schwartz S, Wagner L, Miller W (2000) A greedy algorithm for aligning DNA sequences. J Comput Biol 7:203–214

Part X

Animal Models: Behavior and Pathology

Motor Function in Rodent Models of Neurodegenerative Disorders

32

Jessica A. Hutter Saunders, Max V. Kuenstling, Robert A. Weir, R. Lee Mosley, and Howard E. Gendelman

Abstract

Motor dysfunction and behavior abnormalities are often present in neurological diseases and disorders. Rodent models of neurological diseases may display motor dysfunctions or behavior and cognitive deficits similar to human disease, but these are often difficult to measure and interpret. Nevertheless, tests of motor function and behavior are important tools for characterizing mouse models of diseases and for measuring motor function and cognitive deficits after treatment. In this chapter we will describe some of the most commonly-used tests of motor dysfunction, behavior, and cognition.

Keywords

Behavior • Motor dysfunction • Rotarod • Open field • Morris water maze • Balance beam • Paw prints • Novel object recognition

32.1 Introduction and Objectives

Motor function and cognitive abnormalities are common manifestations of neurodegenerative disorders that include, but are not limited to: Parkinson's and Alzheimer's diseases (PD and AD), Huntington's disease, essential tremor, amyotrophic lateral sclerosis (ALS), and transmissible spongiform encephalopathies (prion diseases). Rodent models of these diseases may recapitulate motor and cognitive deficits seen in human disease. Although motor and cognitive deficits are often difficult to measure and interpret, they are necessary to uncover disease mechanisms and to develop new therapeutics. Herein we describe commonly used tests of rodent motor dysfunction, behavior, and cognition including methods for measuring movement, coordinated motor functions, muscle strength, and memory and learning. We present protocols and expected results for motor function tests such as open field, stride length, rotarod, balance beam, grip strength, Morris water maze (MWM), and novel object recognition (NOR) tests. For each, interpretation

J.A.H. Saunders • M.V. Kuenstling • R.A. Weir
R.L. Mosley • H.E. Gendelman, M.D. (✉)
Department of Pharmacology and Experimental Neuroscience, Center for Neurodegenerative Disorders, University of Nebraska Medical Center, Omaha, NE 68198-5880, USA
e-mail: hegendel@unmc.edu

H. Xiong and H.E. Gendelman (eds.), *Current Laboratory Methods in Neuroscience Research*, Springer Protocols Handbooks, DOI 10.1007/978-1-4614-8794-4_32,

of results and test pitfalls that commonly arise during experimentation are discussed.

32.2 Measurement of Gross Activity and Behavior

32.2.1 Open Field Activity Test

Rodents' natural tendency to explore novel environments may be affected by disease. Thus, measures of gross activity and behavior are commonly used to monitor disease onset and progression. One example, is the open field activity test (OFAT) that was first conducted using a box with a grid painted on the floor where an observer simply recorded the number of times the rodent crossed a line (Hall 1934). Currently available automated systems provide reliable and objective measures of multiple activities such as total distance traveled, average speed of movement, time spent in specific areas, rearing, and small repetitive movements. Most systems consist of a round or square arena that is connected to a computer with data-collection software. Activity may be recorded by video, or the arena is equipped with photo beams, which determine and record the movements of the mouse across *X*, *Y*, and *Z* coordinates over time. Here, we discuss the use of automated systems that employ photo beams on two levels.

1. Materials
   - Tru Scan Activity System by Coulbourn Instruments or similar automated device
   - A computer with the appropriate software package (Tru Scan) to record the data
   - Water and 70 % (v/v) ethanol to clean the arena
2. Procedure
   (a) Choose and select behaviors to be recorded within the computer program before conducting the experiment.
   (b) Before disease onset or treatment, a pretest is conducted. Place the mouse in the clean arena, place the lid on the arena, and start recoding.
   (c) Clean the arena between tests with 70 % ethanol followed by water to decrease olfactory cues. Twenty-four hours after pretest data has been collected, administer treatments.

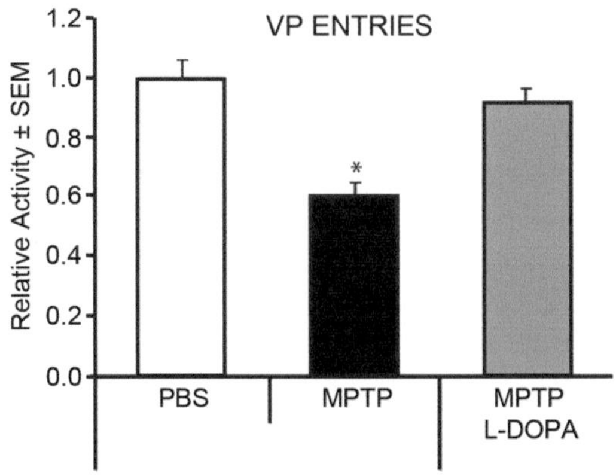

Fig. 32.1 OFAT results. Mice were treated with PBS alone, the neurotoxin (MPTP) alone, or MPTP and then L-DOPA for 1 week before the open field test. Open field was used to measure the number of times the mouse entered the vertical plane (reared). Before treatments, mice were tested for basal level of activity (pretest). One week after treatment, mice were tested again (posttest). Data from the posttest were normalized to data from the pretreatment test and are expressed as the relative activity to the PBS group. The relative number of vertical plane entries (rears) was decreased in mice treated with MPTP compared to those treated with PBS and those treated with MPTP and then L-DOPA

   (d) Measure open field activity after disease onset in the same manner as the pretests were performed.
3. Typical results and data analysis

   Before treatment, a pretest is conducted to measure baseline of activity for each mouse. One day later, mice are randomly assigned to three treatment groups. Controls are administered given phosphate buffered saline (PBS), while the two experimental groups are treated with 1-methyl-4-phenyl-1,2,3,6-tetrahydropyridine (MPTP). MPTP leads to the selective death of dopaminergic neurons and parallels the histopathologic abnormalities seen in PD. One of the MPTP-treated groups is administered L-3,4-dihydroxyphenylalanine (L-DOPA), the most commonly used PD therapeutic, daily for 1 week. Posttreatment data are collected 1 week following treatments. The total number of vertical plane moves is shown in Fig. 32.1. MPTP-treated mice stand on their hind limbs (rear) less than PBS-injected controls. MPTP-treated mice given L-DOPA for 1 week show an increase in the number of rears. Data from the posttreatment test are normalized to data from the pretreatment results for each mouse and the mean of each of the neurotoxin-treated groups are compared to the mean of the PBS-treated controls.

4. Summary

OFAT is a simple method for collecting large amounts of data without being overly labor-intensive. However, as with other tests, the outcome of the OFAT is dependent on many factors and anxiety can affect results. The exploratory behavior, and thus the level of activity in the arena, can be significantly decreased by stress (Brooks and Dunnett 2009). Therefore, care should be taken to eliminate noise and decrease light and olfactory cues. The researcher should always clean the floor and sides of the arena with 70 % ethanol and then water between each mouse trial. To further decrease anxiety, the researcher should handle and habituate the animals to the arena before collecting data. If administering a drug or treatment, such as a neurotoxin, the researcher can conduct the experiment in a paired manner as described above, where data is collected for each mouse before the treatment and then again after the treatment. Be cognizant of when the experiment is performed in relation to the treatment; some may cause acute deficits (i.e., post-surgery recovery), which are not related to death or dysfunction of neurons *per se*. For instance, MPTP is a neurotoxin that causes destruction of the nigrostriatal axis. It is also a derivative of meperidine, which induces a narcoleptic state that is evident hours and even days after administration. Thus, lack of motor activity measured more proximal to the time of administration probably is more indicative of a suppressive effect of drug on the CNS rather than loss of dopaminergic neurons.

32.3 Tests of Coordinated Motor Function

Coordinated motor function can be measured in many ways, making it important to choose tests that suit the specific needs of the experimenter. One must consider the validity, reliability, utility, and sensitivity of the test (Brooks and Dunnett 2009). Most of the following tests are easy to conduct and are inexpensive; with the rotarod being the only test described that requires purchasing special equipment.

32.3.1 Paw Print Test

The ink paw print test is used to assess gait patterns of mice. The simplest form of this test requires no more than waterproof ink, a walkway, and a sheet of white paper. Waterproof ink (such as India ink) is applied to the feet of the individual mouse, which is placed on a paper-lined gangway and allowed to walk to the end and into their home cage or a dark box (Crawley and Paylor 1997; Carter et al. 2001; Crawley 2008; Brooks and Dunnett 2009). A bright light can be placed at the beginning of the gangway if the mouse does not readily walk towards the box. The total length of the gangway should be long enough to accurately sample the gait. Depending on the mouse model and the goal of the experiment, the researcher may wish to use different colored inks for the front and back paws to independently measure forepaw and hindpaw movement. A gangway of 50 cm in length and 10 cm wide with sides at least 10 cm high is recommended. The average stride length and width can be determined by measuring the distance from the center of one paw print to the next on the same or opposite side. The recorded stride lengths of a mouse can be averaged to attain a mean stride length for each mouse.

1. Materials
   - India ink
   - Paint brush
   - White paper
   - Flat, raised surface or gangway
   - The home cage of the mouse or a box to which they have been habituated
   - A light (if needed)
2. Procedure
   (a) Place the white paper on the gangway and tape the edges to immobilize the paper.
   (b) Pick up and immobilize the mouse, and paint the bottom of the feet with ink.
   (c) Gently place the mouse on the gangway and let it walk to its home cage at the other end.
   (d) Collect the paper and mark it with the subject's identification number; let ink dry.

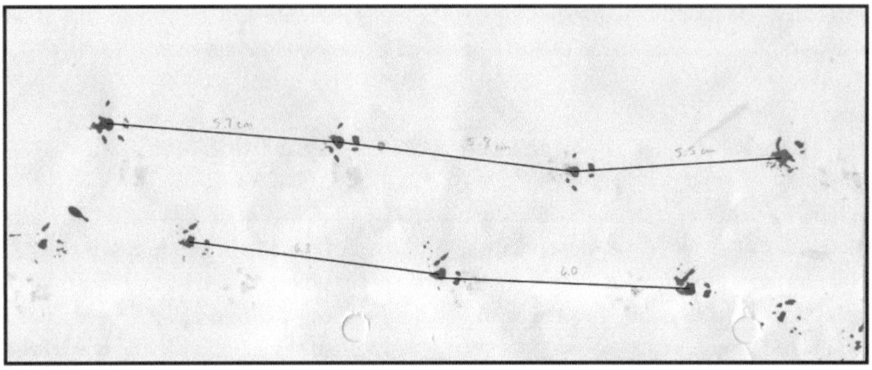

Fig. 32.2 Paw print tests. One simple way to determine gait disturbances is to measure the distance between steps with a ruler

(e) Place a new sheet of paper on the gangway for the next subject.

(f) Repeat this procedure for each mouse, alternating between treatment groups.

3. Typical results and data analysis

The steps at the beginning and end of the paper are discounted in analysis as they are often affected by the researcher putting the mouse on the paper or the mouse slowing to maneuver into the cage or box at the end of the gangway. After prints are collected for all of the subjects, the average stride length per subject is determined (Fig. 32.2). In this test, a ruler is used to measure the distance between the centers of each paw print. The researcher calculates the mean stride length and width for each subject, and the mean stride length of each experimental group of mice is compared against the control group (wild type or untreated). If the mean stride length of the experimental group differs from the control, this is indicative of a motor deficit.

32.3.2 Balance Beam Test

Originally developed in 1980 as a method to assess motor deficits in older rats (Wallace et al. 1980), the balance beam is an excellent tool for evaluating motor coordination and balance of rodents under a range of conditions. In addition to assessing the effects of aging, the test can gauge the influence of CNS lesions on motor function and as a tool to characterize the motor phenotypes of mice that include transgenic models (Carter et al. 1999; Kiernan et al. 1999). The beam-walking test (also referred to as the raised-bridge or raised-beam test) measures the ability of a rodent to maintain balance while it traverses an elevated, narrow beam to reach an enclosed safe platform. Measurements can include latency in crossing the beam as well as the number of footslips made. Some studies record failure to successfully cross the beam (frequency of falls) and this is used as another dependent variable.

Early versions of the test use a series of beams with decreasing diameters (consisting of both square and round cross-sections) to progressively vary the difficulty of the task (Perry et al. 1995; Carter et al. 1999). Newer versions of the test utilize a flat beam that becomes progressively narrower as it reaches the safe enclosure, as well as introduce modifications that increase a mouse's propensity to successfully cross the beam rather than clinging on to the beam (Schallert et al. 2002). These modifications have included the simple change of raising the goal platform for a slight incline of the beam (as the tendency to run upwards to escape is normal for mice), or the inclusion of a small ledge on either side of the beam such that when a paw slip occurs, grip of the beam is retained.

A major advantage of the test is that most apparatuses used in this test can be made in-house easily and with little expense; although as a consequence, the materials are not available com-

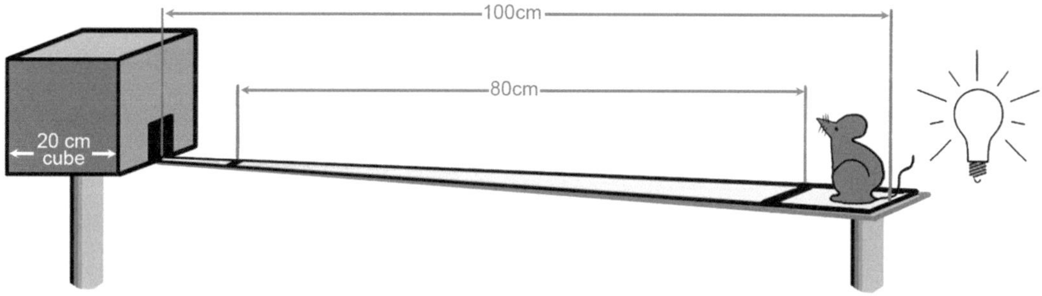

Fig. 32.3 Setup of balance beam apparatus. A simple and inexpensive setup for the beam-walking test utilizing a narrowing 100 cm beam with a ledge, leading to a dark safe box. A slight incline as well as placement of a bright lamp near the start position encourages movement across the beam. *Dark markings*, denoting the middle 80 cm of the beam, are drawn to insure that any measurements can easily be made within these marks

mercially and slight variations between apparatus setup and design may lead to discrepancies between independent studies. While a single investigator with the help of two video cameras can carry out the protocol, we suggest two investigators work in tandem in order to facilitate data collection, as foot slips must be recorded from both sides of the beam in addition to recording the overall time to cross. The following protocol (modified from Carter et al. 2001) for mice acts as an example of how to setup the apparatus, train the rodents, and finally test different experimental groups.

1. Materials
   - 70 % (v/v) ethanol (to clean apparatus between subjects)
   - Beams: 1-m long strips of smooth wood, in any of the following configurations:
     - Flat beam that becomes progressively narrower with additional ledge
     - Square cross-sections with varying sizes (i.e., 5, 12, and 28 mm wide)
     - Round cross-sections with varying sizes (i.e., 11, 17, and 28 mm in diameter)
   - Support stand to hold up start section of beam
   - Safe box/platform (dark box, i.e., 20 × 20 × 20 cm with 4 × 4 cm entrance) on support stand
   - Bright lamps (i.e., 60-W)
   - Video camera, tripod, and timer
2. Procedure

   Setup of test apparatus (Fig. 32.3)

   (a) Draw two dark lines; one 10 cm from the beginning of the start section, and one 10 cm from the end of the beam. This defines an 80 cm region in the middle of the beam from which latency and footslip measurements should be taken (to eliminate any adverse measurements due to hesitation associated with initiation of the test or entry into the safe box).

   (b) Place the "start" of the beam on the support stand, and place the safe box at the other end.

   (c) Position lamps above and to the side of the beam start in order to create an aversive stimulus (*i.e.*, mice will escape bright light to reach dark box).

   (d) Position video camera on tripod 1–2 m away so full length of beam is in viewfinder.
3. Training phase

   (a) Transfer mice to experimental room to habituate for 60 min, keeping temperature and lighting constant to the end of testing.

   (b) Turn on lamps and video camera. Prepare for start of trials:
   - With two investigators present, one should operate the camera and record number of footslips on the camera side of the beam. The other investigator places and removes the mice, records traversal latency, and notes the number of footslips on their side of the beam. A video camera is still recommended

as many of these movements occur so quickly that they are readily missed.

  - With one investigator, each trial can be repeated allowing viewing of footslips from the left side and then the right, or record each trial with a camera on each side of the beam and note footslips from the recording.

(c) Place mouse at start of beam, record latency to traverse the beam (up to 60 s).

  - Run four consecutive trials before returning mouse to its home cage.
  - Use widest (least difficult) beam for training.
  - For the first trials, mice may need a slight nudge with a finger in order to fully cross the beam. For later trials, mice will learn to rapidly cross to the safety box.

(d) Using 70 % ethanol, clean apparatus between animals, allowing alcohol to evaporate completely. If a mouse defecates or urinates during trials, the beam must be cleaned.

(e) Repeat for all remaining mice. Return mice to holding room until next day for training. Thoroughly clean apparatus with 70 % ethanol at the end of the day.

(f) Perform a total of 16 trials per mouse, 4 trials per day for 4 consecutive days.

4. Testing phase

(a) Transfer mice to experimental room to habituate for 60 min, keeping temperature and lighting constant to the end of testing.

(b) Run two consecutive trials per mouse per beam to be tested.

  - If testing small and medium beams with square and round cross-sections, run two trials each for a total of 8 trials.
  - Start with widest beam and move to smaller beams.
  - Record latency to cross the beam and the number of times the hind feet slip off.

(c) Using 70 % ethanol, clean apparatus between animals, allowing alcohol to evaporate completely. If a mouse defecates or urinates during trials, the beam must be cleaned.

(d) Repeat two trials for each beam on all remaining mice.

(e) Return mice to holding room. Clean apparatus.

(f) Analyze data.

5. Typical results and data analysis

For each mouse, calculate the mean of each measurement (latency, right and left hind footslips) taken in the middle 80 cm for the two trials per beam. Any mouse that does not cross within 60 s is given a maximum score of 60 s for analysis. Falls, while rare in normal mice, may occur in experimental or transgenic mice and can be recorded as a failure. Certain situations may occur during testing that require specific considerations for data analysis. Some mice will not traverse the beam; however, creating a slight incline as well as placing bright lamps at the start of the test should encourage the rodents to traverse. Some rodents, especially those with severe phenotypes, will have difficulty in completing the test. If a mouse has great difficulty with the task, use the widest beam available; as narrower ones are necessary only for fine-tuned classification. If a subject takes longer than 60 s to traverse the beam, remove it from the beam and place it in the safety box for a few seconds. The subject can then be retested immediately or returned to their cage for later testing. If you measure an extreme number of footslips on a mouse, or if the subject lies flat while clasping its legs around the beam, record the maximum number of footslips (such as 55) for analysis. Finally, if a subject falls from the beam more than once, do not continue to test the animal on other, more difficult beams. Instead, record the maximum latency (60 s) and footslips (e.g., 55) for any subsequent trials.

The beam-walking test can be used to assess changes in motor coordination following the implementation of a variety of experimental parameters, such as traumatic brain injury or treatment with pharmaceutical agents. Those subjects that have undergone treatment or injury that would disrupt motor function should show significant changes in traversal time and number of footslips (both should be increased) as compared to control trials. To control for inter subject variability, each subject can be their own control by taking measurements before and after treatment. The mean of each subject's difference in latency/footslips from before and after treatment can then be collected from each treatment group and analyzed as compared to other groups. To compare two dependant sample means, a repeated measures *t*-test ("paired *t*-test") is sufficient. In order to compare multiple independent groups, it would be best to use two- or three-way analysis of variance (ANOVA) which would depend on the number of variables such groups, age, and/or task difficulties; or should the study be designed for analyzing performance changes over time, Dunnett's test may prove more useful (Winer et al. 1991).

32.3.3 Rotarod

One of the most popular and easy-to-use tests for coordinated motor function is the rotarod. This test assesses the ability of mice to walk on a suspended rotating rod. A decreased latency to fall from the rod is indicative of motor dysfunction. This test is relatively simple to conduct, yet it can be administered and analyzed in several different ways. In its most common form, rodents are trained and tested at several different fixed speeds (measured in rpm) (Rozas et al. 1997, 1998; Carter et al. 2001; LeDoux 2005; Crawley 2008). Alternatively, the rod can be set to accelerate to a defined speed over a defined period of time using the accelerating rotarod test (Jones and Roberts 1968; Sedelis et al. 2000; Keshet et al. 2007). Most rotarods are equipped with infrared beams or weight sensors that detect when the mouse has fallen from the rod. The latency to fall is either logged automatically by a computer or manually by an observer.

1. Materials
   - Rotarod (e.g., Accuroto Rotarod by AccuScan Instruments Inc., Rotamex by Columbus Instruments, or ROTOR-ROD by San Diego Instruments)
   - 70 % (v/v) ethanol to clean the rod and floor of the equipment
2. Procedures
   (a) Habituate mice to handling and walking on the rotarod before the onset of motor dysfunction or administration of treatments by first placing them on the still rod and then by allowing them to walk on the rotating rod a several speeds for at least 90 s per speed. No negative reinforcement (e.g., electric shock) is needed when training of mice; when mice fall off the rod during training, simply place them back on the rod until the training session is complete. You can train the mice to walk on the rod at more speeds than they will be tested. The actual rpm used will depend on the diameter of the rod used (i.e., the smaller the rod, the higher the rpm is needed to produce the same walking speed as with a larger rod).
   (b) After habituation, an optional baseline of rotarod performance can be obtained. Test the mice at speeds that are not too easy and not too difficult; most mice should be able to stay on the rod for the duration of the test.
   (c) To test the accelerating rotarod performance of a mouse, place the mouse on the rod and turn on the rotarod so that it accelerates from a low rpm to a medium rpm. Zero rpm to 40 rpm for a mouse-sized rod is common. For the fixed-speed rotarod test, turn on the rotarod and place the mouse on the rotating rod at the lowest speed to start.
   (d) Log the time it takes for the mouse to fall off the rotating rod.
   (e) For the accelerating rotarod test, conduct three trials per mouse with at least 5 min rest between trials. For the fixed-speed rotarod test, conduct three trials per speed (usually 3–4 speeds) with at least 5 min rest between each trial.
3. Typical results and data analysis

   Mice with a motor coordination deficit will not stay on the rod for as long as mice without

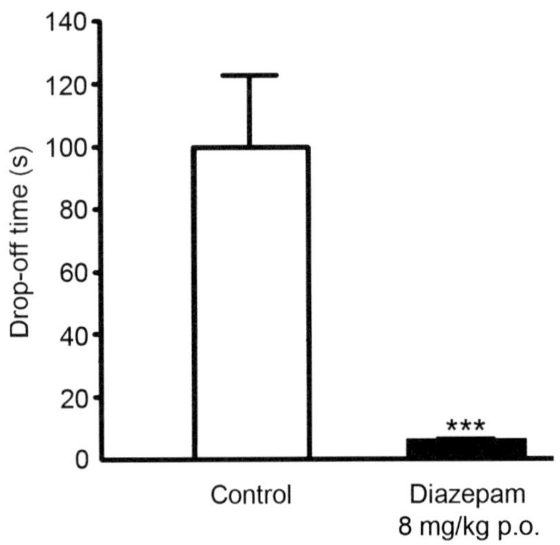

Fig. 32.4 Rotarod data. Effects of diazepam (p.o.) on rotarod performance in the rat (Porsolt et al. 2006)

a deficit. Each mouse is scored in seconds for the length of time on the rod, and the final performance score is an average of two or three trials per speed (Jones and Roberts 1968; Carter et al. 2001; Keshet et al. 2007). See Fig. 32.4 for an example of decreased latency to fall after diazepam treatment as compared to control mice (Porsolt et al. 2006). For data analysis where n is greater than 30, a paired *t*-test can be used if data is collected prior to disease onset or treatments (i.e., the means are dependent). For independent means where n is less than 30 but the groups have equal variances and normal distribution, an independent sample's *t*-test should be conducted. Because some mice have the tendency to simply not participate, this may skew the data, and thus data that is not normally distributed should be analyzed using the Mann–Whitney test. To preclude this problem some researchers choose to train more mice than necessary and then exclude mice that do not participate (Brooks and Dunnett 2009).

4. Summary of tests of motor coordination

The paw print test and beam-walking tests are inexpensive, yet reliable tools that can be utilized to assess motor function, coordination, and balance in rodents under a variety of experimental schemes. For both tests, almost all equipment used can be built in-house, and the use of various choices in gangway or beam design allow for multiple ways in which these tests could be implemented to get valuable results. Data collection and analysis is relatively simple, and can be performed by one or two investigators within a short time span. While standardization of testing apparatuses is minimal (as none are available commercially due to the ease at building the setup independently), both the paw print and beam-walking tests are used widely in neuroscience research. Rotarod is also commonly used in neuroscience research to determine deficits in coordinated motor function. While requiring specialized equipment, the widespread use of this test makes it relatively easy to locate proper equipment. Gait analysis can also be performed using more sophisticated equipment such as a treadmill with a video capture system (Hampton and Amende 2010). For example, Columbus Instruments makes several models of treadmills, which consist of a clear running surface for a single rodent, and Noldus Information Technology offers a popular model called the CatWalk. These systems allow for efficient analysis of foot placement, stride length and width, toe spread, print length, and other parameters.

All of these tests are capable of revealing abnormalities in motor function, but the nature of these tests do not allow the researcher to determine directly which neuronal population is affected by disease. Moreover, the results of these tests may be greatly affected by comorbidities and external factors. Thus, it is important for the investigator to consider all factors that may influence results. Lack of motivation, muscle weakness, poor health, deficits in cognition, or impaired motor learning may lead to poor motor coordination performance as assessed by the abovementioned tests and may not be indicative of a motor neuron deficit *per se*. Furthermore, be aware of the environment in which motor coordination tests are being performed. External stressors such as bright lights and loud noises should be minimized since they may increase anxiety and affect the outcome of the tests. Overall, when conducted with care, the paw print, beam-walking, and rotarod tests are reliable tests of motor coordination and are applicable to many models of neurological disease.

32.4 Measurements of Muscle Strength

Decreased muscle strength and endurance occurs in many neurological diseases including ALS, muscular dystrophy, stroke, and prion disease. Thus, mouse models of these diseases often display decreased muscle strength or tone. For example, the ALS transgenic mouse model, G93A superoxide dismutase 1 (SOD1), demonstrates hind limb weakness by 5 months of age and paralysis in one or more limbs within the following 2 weeks (Gurney et al. 1994). Loss of muscle strength in these mice can be used in combination with other tests to determine disease onset and treatment efficacy. Here, we describe two commonly used tests for measuring muscle strength and endurance.

32.4.1 Paw Grip Endurance Test

The paw grip endurance (PaGE) test is designed to assess muscle strength and endurance of mice (Weydt et al. 2003). It is generally conducted using the lid of a conventional mouse-housing cage, although special equipment can be purchased. Mice are placed on the lid and the lid is turned upside down so that the mouse is hanging and the latency to fall is recorded (Crawley 2007).

1. Materials
   - A wire lid from a conventional rodent housing cage
   - Timer
2. Procedure:
   (a) Place the mouse in the center of the wire lid.
   (b) Gently jar or shake the lid to induce the mouse to grip the lid.
   (c) Turn the lid upside down (180°).
   (d) Record the latency until the mouse releases either the hind limbs or falls off completely. A maximum cut off should be established before testing (i.e., 90 s).
   (e) Test each mouse three times.

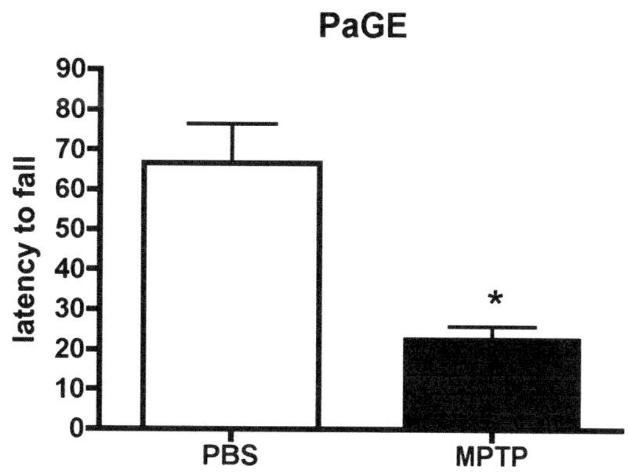

Fig. 32.5 Paw grip endurance test. At 1 week post-intoxication, mice intoxicated with MPTP (24 mg/kg × 4) display decreased latency to fall from the cage lid when compared to control mice injected with PBS

3. Typical results and data analysis

   Mice with decreased neuromuscular strength should have a shorter latency to fall than normal mice, such as mice following treatment with MPTP (Fig. 32.5). The longest latency to fall should be used in data analysis (Weydt et al. 2003). Use Student's *t*-test to compare two groups and an ANOVA to compare three or more groups.

32.4.2 Grip Force Recordings

The grip force-recording test measures forelimb muscle strength with the use of a dynamometer (Cabe et al. 1978), which can be set to measure peak tension. Peak tension is the force applied by the mouse to the grip immediately before release of the grip. This test takes advantage of the tendency for mice to reach out for an object when suspended by the tail, and thus it does not require training before the test is conducted and can be used for studies with a repeated measures design (Cabe et al. 1978; LeDoux 2005).

1. Materials
   - Digital force gauge by Chatillon mounted on a grip strength platform with a "front grip" attached to the dynamometer (San Diego Instruments, San Diego, CA 92126)
2. Procedure
   (a) Position the force scale horizontally and lower the mouse by the tail towards the

apparatus until it grasps the front grip with both forelimbs.
(b) Steadily pull the mouse backward in a horizontal plane.
(c) Record the peak tension.
(d) Repeat the test three consecutive times per mouse, with 30–90 s rest between trials. Exclude and repeat trials in which the rodent grasps with only one paw.
(e) The measurement of strength (peak tension) for each mouse is the mean of the three trials.

3. Typical results and data analysis

Reduced muscle strength should be evident as lower peak tension. Compare the mean peak tension of each group using Student's *t*-test to compare two groups and an ANOVA for three or more groups.

32.5 Tests of Cognitive Function

Several neurological disorders, including AD, traumatic brain injury, stroke, Down syndrome, and PD, affect cognition or mental acuity. This can be measured in rodents using the MWM, which measures spatial memory acquisition, and NOR, which tests memory.

32.5.1 Novel Object Recognition

The NOR test was first introduced in 1988 as a way to assess a rat's ability to discriminate a novel object from a familiar one (Ennaceur and Delacour 1988), eventually becoming a common method of testing object memory in rodents under a variety of experimental conditions. In this test, consisting of three phases, the difference in the time spent exploring a novel as compared to a familiar object is measured and compared between experimental and control groups. Following a familiarization phase, where the subjects become accustomed to the open field-testing chamber, the rodents begin the sample phase where they are placed in the same chamber and exposed to two identical samples. Finally in the test (or choice) phase, the subject is presented with two different objects: a familiar object used in the sample phase and a novel object.

Exploration of an object is most commonly defined as having the subject's nose oriented towards, and within two centimeters of, the object. A rodent's exploratory preference towards a novel object is thought to indicate that the familiar object exists in memory, and therefore repeated exploration of the familiar object is not needed (Dere et al. 2006). This rationalization has prompted the use of NOR as an effective way to analyze memory function in animals. Certain considerations should be made in designing the experiment, as findings have shown that results can be influenced by the properties of the objects used. Rodents will often show a preference for objects on which they can climb compared to those they cannot. Furthermore, the shape, texture, color/brightness, and especially odor of the objects can influence the memory formation of an object as well as the preference for exploring them. Much attention should be afforded to object selection, with all efforts to choose objects "different" enough without imparting any preference bias. Objects must be cleaned thoroughly between trials, as well as placed in the same location and orientation in both phases of the test.

Some studies have demonstrated that younger animals show a preference for familiarity over novelty (Roder et al. 2000). This often occurs when the original object is only briefly presented, or if multiple novel objects are repeatedly introduced next to the familiar object. This phenomena is often seen in early trials but is gradually replaced by a consistent preference for the novel object following repeated exposures to the same environment. Finally, much discussion has occurred on whether this test involves working or episodic memory. The general consensus is that episodic memory is likely the form being tested (Ennaceur 2010), as the animals are exposed to a specific object in a specific place at a particular time (what, where, and when, respectively) rather than keeping the object in memory for a brief period of time to insure will be available for use later on, as it is likely the subjects do not expect to see the object again.

The following can be used as a basic protocol for designing a NOR experiment, although the exact details of the test would greatly depend on the hypothesis.

1. Materials
   - 70 % (v/v) ethanol (to clean apparatus between subjects)
   - Rectangular chamber (high, dark walls)
   - Objects (special considerations, see data analysis)
     - Two "identical" objects
     - One separate object, somewhat similar in size/shape, yet distinctly different
   - Video camera, tripod, timer
   - Optional: Computer tracking software (i.e., EthoVisionXT)
2. Procedure

   Setup of test apparatus
   (a) Place testing apparatus (rectangular chamber) in room
      - High walls, made of dark/smooth material, will encourage greater object exploration
        - *Optional*: Draw line down middle of chamber, separating objects into two "areas" in order to assess motor function (see Data Analysis)
      - Chamber should not be too large, and no bedding should be placed in the chamber
        - For rats, floor space should be 40×40 cm ($l \times w$)
        - For mice, floor space should be 25×25 cm ($l \times w$)

   (b) Mount video camera above the chamber on a tripod so entire arena is captured.

   (c) *For sample and test phases*: Objects should be "attached" to chamber in some way (to prevent displacement of the object), placed in separate corners of the chamber (back left and back right).
3. Familiarization phase

   (a) Transfer animals from holding room into experimental room to habituate for 60 min, keeping temperature and lighting constant to the end of testing.

   (b) Familiarize animals by individually placing them in the *empty* chamber, allowing them to explore the box for 5 min a day, for 2 days (start next phase within 24 h to insure effectiveness).
4. Sample phase

   (a) Transfer mice from holding room into experimental room to habituate for 60 min, keeping temperature and lighting constant to the end of testing.

   (b) Place (attach) the two "identical" (sample) objects in the back left and right corners of the chamber.

   (c) Turn on video camera and begin recording. Place a card with animal's identification number within view of the camera.

   (d) Place the animal in the cage specifically as follows (to prevent placement bias):
      - Set them at the mid-point of the front wall, opposite the sample objects
      - Face the animal towards the front wall, away from the objects, parallel to the side walls

   (e) The investigator should immediately step back from the chamber, so that their presence is not seen and remain in that similar location throughout all testing procedures.

   (f) After the sample exposure time (e.g., 10 min), remove the animal from the chamber and return to home cage (keep cages in experimental room throughout).

   (g) Clean apparatus thoroughly with 70 % ethanol to remove scent trails.

   (h) Repeat the procedure until all subjects have been exposed to the sample objects.

   (i) Begin the test phase following predetermined training-to-testing interval:
      - 1 h is commonly used for most robust object recognition
      - Other times such as 3, 6, 12, or 24 h can be used; however, with time there is a loss of object recognition.
5. Test phase

   (a) Replace *one* of the familiar objects with the novel (test) object.

   (b) Turn on video camera and begin recording. Place a card with animal's identification number within view of the camera.

   (c) Place the animal in the chamber as described in step 4 of Sample Phase (above)

(d) Step away from the chamber. Test for a short duration (e.g., 3–5 min). Longer durations decrease the reliability of the data, as the novel object becomes increasingly familiar through the duration of the test session.
(e) Remove the animal from the chamber, clean apparatus thoroughly with 70 % ethanol.
(f) Repeat the procedure for all remaining subjects.
(g) Analyze data.

6. Data analysis and typical results

The scoring of the trials can be made from the video either manually with a timer or through specialized tracking software. Manually, an observer (who needs to be blinded to the experimental conditions of the animals) will assign a timer to each object. An animal is determined to be interacting with an object when its nose is in contact with or directed to the object within a defined distance (e.g., 2 cm). The investigator will record the amount of time the subject is "interacting" with each object by manually observing the video. There are a number of ways to analyze the results; however, the most common measures include dividing the novel object interaction time by the total interaction time with both objects (discrimination ratio) or by subtracting the familiar object interaction time from the novel object interaction time (difference score). In this way, a discrimination ratio well above 0.5 or a positive difference score reflects recognition of the familiar object, as there was more exploration of the novel object. Inversely, a lower discrimination ratio or a difference score around zero or below may represent no preference for the novel object, therefore reflecting a lack of memory for the familiar object. Additional behaviors can be measured as a way to assess motor behavior in addition to memory, these include latency to first contact with object, number of times animal crosses line separating the areas, number of rearing events (standing on rear paws), or the number of times animal comes in contact with an object. If the researcher uses computer software for animal tracking, the velocity of movement and the total distance traveled by the animal may also be determined. It is important to note that cognitive function may influence motor testing, and differences in motor function may have an effect memory testing and the animal's apparent ability to discriminate between the objects. This test is often used to study effects of aging on memory, positive/negative cognitive effects of pharmaceutical compounds (i.e., amnesic drugs), or to compare and describe newly discovered mutant mouse strains. When running these experiments, it may become necessary to rule out other explanations that may simulate memory deficits. For instance, a novel mutant strain may not be able to discriminate a novel object the same as wild type mice; and while this may be due to a memory deficit, it also could be due to differences in stress reactivity, exploration efficiency, or the ability or of the animal to perceive the environment, or possibly to more overt differences in eyesight or motor function.

7. Summary

NOR has become a popular technique to study learning and memory. It is relatively easy to set up and perform with the results either obtained manually or with computer software assistance. The results can be easily compared between experimental and control groups, although discussion of their meaning should include whether or not the treatment causes changes in other factors that may indirectly affect the animal's ability to discriminate between objects. Also, the exact processes that are involved in this "memory" are not completely understood, with much debate concerning whether it involves working or episodic memory. Nevertheless, this test can be used under a variety of experimental schemes, and has therefore become a common way to use a rodent's tendency to interact with novel objects as a way to study memory formation and retention.

32.5.2 Morris Water Maze

The MWM is an experimental paradigm designed to test rodent spatial memory acquisition (Morris 1984). Although engineered with rats in mind, the MWM has also become a staple of behavioral neuroscience testing for mice. The MWM has been shown through countless studies to be very reliable and accurate as a method for testing and assessing spatial cognition (Vorhees and Williams 2006). By its very essence as a water-based test, the MWM eliminates the potential pitfalls created by the rodents' natural olfactory abilities, while simultaneously creating a more level playing field for motor performance by utilizing swimming. Although many other biasing variables exist that may alter the outcome of the test, a prudent experimenter can minimize these through consistency and proper planning, and thus provide reliable data.

1. Materials
   - Circular water tank (ideally 1.2–1.8 m in diameter, and at least 1.2 m deep)
   - Water
   - Platform (clear or opaque, usually 10–12 cm in diameter)
   - Ruler (for measuring water depth relative to platform)
   - Visual cues for spatial orientation (e.g., page-sized symbols)
   - High-resolution video recording apparatus with a bird's-eye view of the tank
   - Hardware and software for video tracking with adequate storage space
   - *Optional*:
   - Nontoxic tempera paint or powdered nonfat milk (for creating an opaque liquid)
   - Thermometer
2. Procedure
   (a) Preparing the water maze for the hidden platform test

   As these trials will be recorded and stored, it is important to assure that the tank is in the same place consistently day-to-day, and that applicable parameters are defined in the software (such as quadrant mapping as shown in Fig. 32.6). Visual cues should be affixed in static positions.

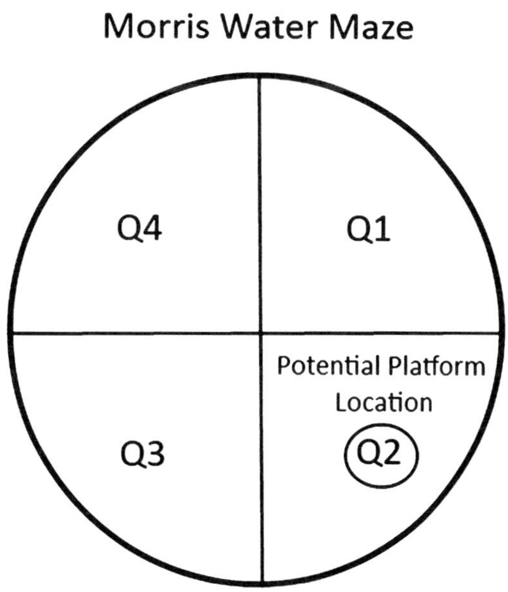

Fig. 32.6 Morris Water Maze (MWM) layout. The MWM may be divided into quadrants for experimental purposes, as shown. The platform should be placed several inches from the tank wall, equidistant from the *X* and *Y* axes. With Q2 as the location of the platform, Q4 should be used as the drop point for the probe test

   If the experimenter will be sitting near the tank, it is important to stay out of view and not to be overly fragrant, as this can provide an olfactory cue.
   - Place the platform in the tank (premark the bottom of the tank to guarantee consistent platform placement).
   - Fill the tank with water (20–26 °C) until approximately 1.25–1.9 cm above the top of the platform to allow the animal to stand on the platform comfortably (this depth may be adjusted depending on the size of the animal, but it is important to maintain consistent conditions).
   - If using an opaque platform, the nontoxic tempera paint or powdered nonfat milk should be uniformly mixed into the water to create opacity.

   (b) Performing the MWM test (10-day trial)

   The experimenter should be blind to treatment conditions of the mice. Four trials and a probe test will be performed each day. Starting quadrants should be rotated (e.g., Day 1—Q1, Day 2—Q3, etc., excluding the quadrant containing the platform).

Trials 1–4:

- Initialize your video-tracking software to record each trial.
- Lower the mouse gently into the starting quadrant, facing the wall, without letting the head of the animal go under the water.
- The trial ends after either 60 s have elapsed or the mouse finds and remains on the platform for 20 s. Record the time it took for the mouse to find the platform.
- Regardless of the outcome, the mouse should spend at least 10–20 s on the platform before retrieval to train it to reach the platform.
- Repeat the trial three more times, and then dry the mouse and return it to its cage.

Probe test:

- Before undergoing the probe test, make sure that at least 30 min have passed since the last trial.
- Remove the platform, but be aware of its former location.
- Lower the mouse into the tank in Q4, and record the number of annulus crossings in 60 s.
- Allow mice to rest 24 h between each trial day. For the final day of the trial, do not perform the probe test until 24 h later. Alternatively, some experimenters may prefer to only do a probe test following day 10.

(c) Common pitfalls and obstacles

- Not all mice will stay on the platform during training. Early in training, gently stabilize them on the platform for several seconds until they learn the paradigm.
- During the test interval, some mice will perform abnormal circling patterns, essentially circumscribing the perimeter of the tank. Mice may also find the platform, but immediately jump off and continue swimming. Note any of these abnormal behaviors on your tracking sheet.
- The water's surface level above the apex of the platform can play a very important role. If mice are not able to breathe comfortably when standing on the platform, they will not likely stay. If they struggle too greatly to get on the platform, they may likewise be deterred. Two possible remedies include using a more graspable material for the platform, or adjusting the water level at the beginning of day one of the trials, and noting this depth for use in future trials.
- Water temperature may also play an important role in the outcome of this experiment. As noted previously, 20–26 °C is optimal. Water that is too warm does not provide an incentive for the mouse to find the platform. Water that is too cold may cause panic or otherwise disrupt the experiment.

3. Summary

The MWM provides a highly reliable form of cognitive assessment for rodent models and is used commonly in neuroscience research. Although some potential obstacles exist in administering the test, these may generally be overcome with forethought. Because the efficacy of this memory and learning test depends on normal motor function, it is important to use rodents that have no motor impairments. If due preparation and consistency are employed, this design should provide pertinent and useful results (as shown in Fig. 32.7) for better understanding different treatments and transgenic conditions.

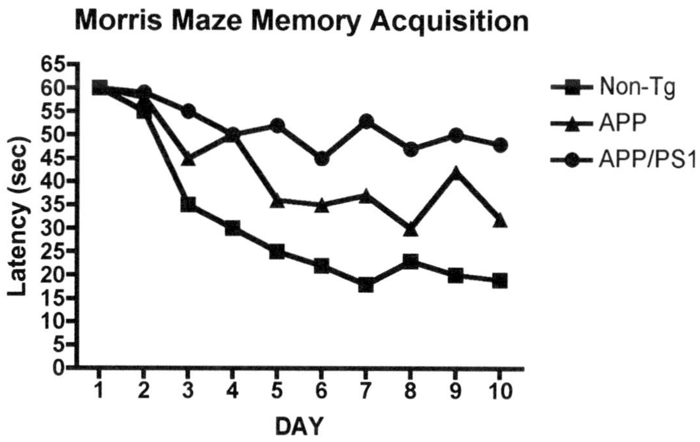

Fig. 32.7 MWM results. One of the most common ways of presenting MWM data is through a line graph with clearly defined points. Typically, the *y*-axis will represent latency in seconds, while the *x*-axis will correspond with the day the test results were recorded. As can be inferred, a higher latency corresponds with impaired memory acquisition/cognitive impairment such as those that are found in APP and APP/PS1 models of AD

32.6 Conclusions

While the execution and analysis of most tests for behavior and motor function are straightforward, many factors must be considered as confounders. Numerous factors to consider when conducting motor function and behavior analyses on mice with neurological disease have been described in detail (Schellinck et al. 2010). First, the environment in which the animals are housed and tested may affect results. Lighting, temperature, noise level, type of bedding and enrichment, frequency of cage changes, type of food, location of housing in relationship to the testing room should be vigorously controlled. Stressful encounters such as the use of ear tags, ear punching, and toe clipping should be avoided. All mice must be housed under the same conditions and provided with the same enrichment, which should be described in as much detail as possible in the methods section of publications (Schellinck et al. 2010). More difficult to control is the social environment of the home cage. Some strains of mice must be housed singly after adolescence due to aggressive behavior towards other mice, while most strains become stressed if housed individually (D'Arbe et al. 2002; Bartolomucci et al. 2003; Moragrega et al. 2003). Additional factors to consider for behavior and motor function testing include body weight of the subject, age, strain of rodent, and motivation (McFadyen et al. 2003; Schellinck et al. 2010). Mouse strain can affect learning and motor function as well. For example, visual and hearing deficits occur in some transgenic mouse models of neurological disease that can affect outcomes of specific tests (i.e., a blind mouse cannot properly respond to visual cues of the MWM) (Buhot et al. 2001; Garcia et al. 2004; Schellinck et al. 2010). Thus, it is important to choose tests that are compatible with the animal model and to be aware of the general health, sensory functions, and motor capabilities of the rodents (Karl et al. 2003).

References

Bartolomucci A, Palanza P, Sacerdote P, Ceresini G, Chirieleison A, Panerai AE, Parmigiani S (2003) Individual housing induces altered immuno-endocrine responses to psychological stress in male mice. Psychoneuroendocrinology 28:540–558

Brooks SP, Dunnett SB (2009) Tests to assess motor phenotype in mice: a user's guide. Nat Rev Neurosci 10:519–529

Buhot MC, Dubayle D, Malleret G, Javerzat S, Segu L (2001) Exploration, anxiety, and spatial memory in transgenic anophthalmic mice. Behav Neurosci 115:455–467

Cabe PA, Tilson HA, Mitchell CL, Dennis R (1978) A simple recording grip strength device. Pharmacol Biochem Behav 8:101–102

Carter RJ, Lione LA, Humby T, Mangiarini L, Mahal A, Bates GP, Dunnett SB, Morton AJ (1999) Characterization of progressive motor deficits in mice transgenic for the human Huntington's disease mutation. J Neurosci 19:3248–3257
Carter RJ, Morton J, Dunnett SB (2001) Motor coordination and balance in rodents. Curr Protoc Neurosci Chapter 8:Unit 8.12–Unit 18.12
Crawley JN (2007) What's wrong with my mouse?: Behavioral phenotyping of transgenic and knockout mice. Wiley, Hoboken, NJ
Crawley JN (2008) Behavioral phenotyping strategies for mutant mice. Neuron 57:809–818
Crawley JN, Paylor R (1997) A proposed test battery and constellations of specific behavioral paradigms to investigate the behavioral phenotypes of transgenic and knockout mice. Horm Behav 31:197–211
D'Arbe M, Einstein R, Lavidis NA (2002) Stressful animal housing conditions and their potential effect on sympathetic neurotransmission in mice. Am J Physiol Regul Integr Comp Physiol 282:R1422–R1428
Dere E, Kart-Teke E, Huston JP, De Souza Silva MA (2006) The case for episodic memory in animals. Neurosci Biobehav Rev 30:1206–1224
Ennaceur A (2010) One-trial object recognition in rats and mice: methodological and theoretical issues. Behav Brain Res 215:244–254
Ennaceur A, Delacour J (1988) A new one-trial test for neurobiological studies of memory in rats. 1: behavioral data. Behav Brain Res 31:47–59
Garcia MF, Gordon MN, Hutton M, Lewis J, McGowan E, Dickey CA, Morgan D, Arendash GW (2004) The retinal degeneration (rd) gene seriously impairs spatial cognitive performance in normal and Alzheimer's transgenic mice. Neuroreport 15:73–77
Gurney ME, Pu H, Chiu AY, Dal Canto MC, Polchow CY, Alexander DD, Caliendo J, Hentati A, Kwon YW, Deng HX et al (1994) Motor neuron degeneration in mice that express a human Cu, Zn superoxide dismutase mutation. Science 264:1772–1775
Hall CS (1934) Emotional behavior in the rat. 1. Defecation and urination as measures of individual differences in emotionality. J Comp Physiol 18:385–403
Hampton TG, Amende I (2010) Treadmill gait analysis characterizes gait alterations in Parkinson's disease and amyotrophic lateral sclerosis mouse models. J Mot Behav 42:1–4
Jones BJ, Roberts DJ (1968) The quantitative measurement of motor inco-ordination in naive mice using an accelerating rotarod. J Pharm Pharmacol 20: 302–304
Karl T, Pabst R, von Horsten S (2003) Behavioral phenotyping of mice in pharmacological and toxicological research. Exp Toxicol Pathol 55:69–83
Keshet GI, Tolwani RJ, Trejo A, Kraft P, Doyonnas R, Clayberger C, Weimann JM, Blau HM (2007) Increased host neuronal survival and motor function in BMT Parkinsonian mice: involvement of immunosuppression. J Comp Neurol 504:690–701
Kiernan BW, Garcion E, Ferguson J, Frost EE, Torres EM, Dunnett SB, Saga Y, Aizawa S, Faissner A, Kaur R, Franklin RJ, Ffrench-Constant C (1999) Myelination and behaviour of tenascin-C null transgenic mice. Eur J Neurosci 11:3082–3092
LeDoux MS (2005) Animal models of movement disorders. Elsevier Academic, Burlington, MA
McFadyen MP, Kusek G, Bolivar VJ, Flaherty L (2003) Differences among eight inbred strains of mice in motor ability and motor learning on a rotorod. Genes Brain Behav 2:214–219
Moragrega I, Carrasco MC, Vicens P, Redolat R (2003) Spatial learning in male mice with different levels of aggressiveness: effects of housing conditions and nicotine administration. Behav Brain Res 147:1–8
Morris R (1984) Developments of a water-maze procedure for studying spatial learning in the rat. J Neurosci Methods 11:47–60
Perry TA et al (1995) Cognitive and motor function in transgenic mice carrying excess copies of the 695 and 751 amino acid isoforms of the amyloid precursor protein gene. Alzheimers Res 1:5–14
Porsolt R, Castagné V, Dürmüller N, Lemaire M, Moser P, Roux S, France C (2006) Central nervous system (CNS) safety pharmacology studies. In: Vogel HG, Hock F, Maas J, Mayer D (eds) Drug discovery and evaluation. Springer, Berlin, pp 15–60
Roder BJ, Bushneil EW, Sasseville AM (2000) Infants' preferences for familiarity and novelty during the course of visual processing. Infancy 1:491–507
Rozas G, Guerra MJ, Labandeira-Garcia JL (1997) An automated rotarod method for quantitative drug-free evaluation of overall motor deficits in rat models of parkinsonism. Brain Res Brain Res Protoc 2:75–84
Rozas G, Lopez-Martin E, Guerra MJ, Labandeira-Garcia JL (1998) The overall rod performance test in the MPTP-treated-mouse model of Parkinsonism. J Neurosci Methods 83:165–175
Schallert T, Woodlee MT, Fleming SM (2002) Disentangling multiple types of recovery from brain injury. In: Krieglstein J (ed) Pharmacology of cerebral ischemia. Medpharm Scientific Publishers, Stuttgart, pp 201–216
Schellinck HM, Cyr DP, Brown RE (2010) Chapter 7—How many ways can mouse behavioral experiments go wrong? Confounding variables in mouse models of neurodegenerative diseases and how to control them. In: Brockmann HJ, Ropre TJ, Naguib M, Wynne-Edwards KE, Mitani JC, Simmons LW (eds) Advances in the study of behavior. Academic, New York, pp 255–366
Sedelis M, Hofele K, Auburger GW, Morgan S, Huston JP, Schwarting RK (2000) MPTP susceptibility in the mouse: behavioral, neurochemical, and histological analysis of gender and strain differences. Behav Genet 30:171–182
Vorhees CV, Williams MT (2006) Morris water maze: procedures for assessing spatial and related forms of learning and memory. Nat Protoc 1:848–858
Wallace JE, Krauter EE, Campbell BA (1980) Motor and reflexive behavior in the aging rat. J Gerontol 35:364–370
Weydt P, Hong SY, Kliot M, Moller T (2003) Assessing disease onset and progression in the SOD1 mouse model of ALS. Neuroreport 14:1051–1054
Winer B, Brown R, Michels K (1991) Statistical principles in experimental design. McGraw-Hill, New York

Humanized Mice

33

Larisa Y. Poluektova and Edward Makarov

Abstract

Human-specific viral infections of CNS require the presence of human host cells. For modeling of HIV-1-associated neurocognitive diseases, the presence of human cells of macrophage lineage that support viral replication is necessary. This process is recognized as HIV-1 encephalitis and mirrored in immunodeficient mice by transplantation of HIV-1-infected human macrophages in deep brain nuclei. However, the development of minor cognitive/motor decline can be observed in the absence of encephalitis. These types of pathology could be imitated in chronically infected animals. The stable engraftment of human hematopoietic stem cells and development of human immune system facilitated establishment of chronic HIV-1 infection. The chapter describes both approaches to mimic human HIV-1-associated pathologies, requirements for mouse strain background, and morphologic evaluation of brain pathology.

Keywords

Immunocompromised mice • *scid* mutation • Macrophages • Hematopoietic stem cells • HIV-1 • Encephalitis • Neuroinflammation • Microglia • Peripheral blood lymphocytes

33.1 Introduction

33.1.1 Major Approaches for "Humanization" of Immunocompromised Mice to Study Human-Specific Viruses Affecting CNS

"Humanized" mice are a promising translational model for studying human hematopoiesis and immunity. Humanization of mice started from the

L.Y. Poluektova (✉) • E. Makarov
Department of Pharmacology and Experimental Neuroscience, University of Nebraska Medical Center, 985880 Nebraska Medical Center, Omaha, NE 68198-5880, USA
e-mail: lpoluekt@unmc.edu; makarove@unmc.edu

H. Xiong and H.E. Gendelman (eds.), *Current Laboratory Methods in Neuroscience Research*, Springer Protocols Handbooks, DOI 10.1007/978-1-4614-8794-4_33,

severe combined immunodeficiency (*scid*) mutation in C.B-17 mice (Bosma et al. 1983) through the use of transplanted peripheral blood lymphocytes (hu-PBL) (Mosier et al. 1988) and by the transplantation of a human fetal thymus/liver sandwich under the kidney capsule (SCID-hu Thy/Liv) leading to local human thymocytes production and differentiation (McCune et al. 1988; Namikawa et al. 1988). The transfer of the *scid* mutation to a nonobese diabetic (NOD/LtSz) NOD/LtSz-*scid* mouse background in the United States (Greiner et al. 1995; Hesselton et al. 1995; Serreze et al. 1995; Shultz et al. 1995) and NOD/Shi-*scid* in Japan (Koyanagi et al. 1997) created a "gold standard" for assaying human hematopoietic stem cell (HSC) biology in the xenotransplantation model (McDermott et al. 2010). Their usefulness has been enhanced by the development of new stocks of immunodeficient hosts on NOD/Shi-*scid* and NOD/LtSz-*scid* mice that lack the IL-2 receptor common gamma chain $IL2R\gamma_c^{null}$. These stocks of mice were named NOG and NSG, respectively, for distinguishing purposes. Additional immunodeficient strains were generated on NSG and Balb/c backgrounds by introduction of two lymphoid-specific recombination-activating genes (RAG1 and RAG2) (Oettinger 1996). A genetic cross of $\gamma_c^{-/-}$ and (RAG2) deficiency in a Balb/c strain background produced mice doubly homozygous for the γ_c and RAG2 alleles ($\gamma_c^{-/-}$/RAG2$^{-/-}$, BRG, or double knockout, DKO) (Goldman et al. 1998; Traggiai et al. 2004). The lack of adaptive immune function, multiple defects in innate immunity, and heightened levels of human hematolymphoid engraftment made these multiple strains of immunodeficient mice attractive for creation of human–mouse chimeric "humanized" animals. They support studies in many areas of immunology including autoimmunity, transplantation, infectious diseases, and cancer (Brehm et al. 2010a, b, c; Libby et al. 2010; Pino et al. 2010). These models are particularly valuable in experimentation where there is no appropriate small animal model of the human disease, as in the case of human-specific viral infections and specifically for HIV-1 (Van Duyne et al. 2009; Brehm et al. 2010b; de Jong et al. 2010; Zhang et al. 2010; Berges and Rowan 2011; Robinet and Baumert 2011; Sato and Koyanagi 2011; Washburn et al. 2011).

33.2 Macrophages Infected by HIV-1 in Mouse Brain as a Model of HIV-1 Encephalitis

The model is based on the injection of human monocyte-derived macrophages (MDM) into the basal ganglia of mouse brains. Infected human MDM are able to induce pathology similar to the human HIV-1 encephalitis (HIVE) (Persidsky et al. 1996; Persidsky and Gendelman 2003). MDM can be infected with different strains of macrophage-tropic viruses. Infected cells can be used 24 h after infection to study the efficacy of antiretroviral drugs. MDM exposed to the virus for 4–18 h, washed and transplanted into the mouse brain, will survive and support viral replication. At this time only a few cells will be productively infected, and the actual spread of viral infection between MDM will occur in vivo. The area around the injection line will be affected by viral toxins and inflammatory products secreted by human MDM. This area is reminiscent to the human HIV-1 encephalitic damage. If peripherally applied compounds effectively penetrate into the brain and sufficiently inhibit viral replication, we will observe a reduction in the number of infected human cells and also the inflammatory reaction surrounding them. MDM can be used after 7 days of in vitro infection and, when they are 100 % infected, could be transplanted into the mouse brain. In such a setting HIV-1-infected MDM are a strong source of virus, viral toxins, and inflammatory factors. It is impossible to stop HIV-1 replication, but this setting is useful for studies of anti-inflammatory and adjunctive therapeutics. Human macrophages will survive up to 28 days as multinucleated infected cells. The uninfected macrophages will migrate from the site of injection and will be distributed along the blood vessels.

Table 33.1 Comparison of C.B-17-scid and Balb/c-Rag2$^{-/-}\gamma_c^{-/-}$ mice for studies of neuroinflammation and neurodegeneration

	C.B-17/SCID	Balb/c-Rag-2$^{-/-}\gamma_c^{-/-}$
Genotype	DNA repair enzyme DNA-dependent protein kinase is shorter at C terminus and the molecule lost 95 % of activity	Double knockout, recombination-activating gene RAG2, and IL-2 γ chain cytokine receptor (CD132) were deleted
	Transcript of DNA-PK present in all cell types (Collis et al. 2005)	
Phenotype of CNS	Cultured hippocampal neurons are hypersensitive to apoptosis induced by exposure to topoisomerase inhibitors, amyloid beta peptide (Ab), and glutamate (Culmsee et al. 2001)	Rag2 not present in neurons and not detected in CNS (Chun et al. 1991)
	Increased vulnerability of hippocampal CA1 and CA3 neurons was observed in adult scid mice after kainate-induced seizures (Neema et al. 2005)	
	More resistant to apoptosis of neurons due to interruption of DNA-PK-mediated activation of p53 (Chechlacz et al. 2001; Vemuri et al. 2001)	
Immune system phenotype	Very low number of mature T and B cells. Normal amount of natural killer cells (Blunt et al. 1995; Tournoy et al. 2000)	Alymphatic mostly due to the absence of common cytokine receptor gamma chain (CD132)
	Macrophage-mediated elimination of human cells (Shibata et al. 1998)	No T, B, and NK cells
	Monocytes are vulnerable to oxidative stress (Bauer et al. 2011)	Signaling from IL-2, IL-4, IL-7, IL-9, IL-15, and IL-21 is abolished
Role in murine HIVE pathogenesis?	Increased sensitivity of neurons to glutamate toxicity and reduced sensitivity to oxidative stress-mediated apoptosis?	Reduced ability to eliminate human cells?
		Reduced inflammatory responses?

33.2.1 Material and Methods

1. Applicable strains of mice.

   The levels of innate immunity impairment will influence survival of human macrophages and force mouse glial cell activation (inflammation) in opposite directions. C.B-17-scid mice retain low numbers of T cells, B cells, NK cells, and Ig-producing cells in the spleen and lymph nodes and have low levels of IgM/IgG in plasma. These animals develop a stronger inflammatory reaction in response to the injection of human macrophages. BRG or NSG/NOG animals do not have T cells, B cells, and NK cells, and IgG/IgM levels are undetectable. These animals have a less pronounced inflammation in the experimental setting when injected with human HIV-1-infected macrophages. The age of mice should not exceed 4–5 weeks to allow direct injection through the bones without significant traumatic injury. Four-week-old males C.B-17/scid (Charles River Laboratories (Wilmington, MA)), BRG (Central Institute of Experimental Animals (Dr. Mamoru Ito, Kawasaki, Japan)), NSG (stock # 05557) (Jackson Laboratories (Bar Harbor, ME)), and NOG (Taconic/CIEA) should be considered as the optimal strains. When testing blood–brain barrier penetrance and suppression of viral replication by antiretrovirals, as well as testing of adjunctive neuroprotective therapeutics and planning for the long survival of human macrophages, NSG/NOG mice are preferable. C.B-17/scid might be more informative for the development of inflammatory reactions in the brain after injection of human macrophages. The comparison of these two genetic backgrounds and phenotypes of the CNS and immune systems are shown in Table 33.1.

2. Human monocyte-derived macrophages cell preparation
   (a) Monocytes are obtained from HIV-1, HIV-2, and hepatitis B seronegative donor leukopaks and are purified by countercurrent centrifugal elutriation (reviewed in Chap. 12).
   (b) Monocytes are cultivated in DMEM (Sigma-Aldrich, St. Louis, MO, USA) supplemented with 10 % heat-inactivated pooled human serum, 1 % glutamine, 50 mg/mL gentamicin, 10 mg/mL ciprofloxacin (Sigma-Aldrich), and 1,000 U/mL highly purified recombinant human MCSF (generous gift from Genetics Institute, Cambridge, MA, USA) in Teflon flasks at a concentration of 2×10^6 cells/mL for 7 days. Half media exchanges are performed every other day during the 7-day period.
3. Cell infection and injection
   (a) After 7 days in culture, MDM are infected with HIV-1_{ADA} at a multiplicity of infection of 0.01.
   (b) A 1:10 dilution of HIV-1_{ADA} is required for a 4-h incubation of virus with MDM, and a 1:20 dilution of HIV-1_{ADA} is the standard for an overnight (18–20 h) infection. The viral titers were assayed on human MDM and determined to be 10^{50} tissue culture-infectious dose$_{50}$ (TCID$_{50}$) per milliliter.
   (c) A cytospin procedure should be performed after the incubation period to ensure proper infection of cells. In order to confirm productive infection of MDM, cytospins are stained for HIV-1p24 protein. Cytospin preparations of HIV-1-infected or uninfected cells were fixed in cold acetone/methanol (1:1) for 10 min at −20 °C, and then anti-p24 MAb (1:10; Dako, Carpinteria, CA, USA) was applied for 45 min at room temperature, followed by the Dako EnVision HRP/DAB+ polymer-based system for development (Chap. 26). Cells used 24 h after infection for the intracranial (i.c.) injection ($1–2 \times 10^6$) should be cultured for additional 4–5 days. Cells used after 7 days of infection can be stained before injection to ensure productive infection.

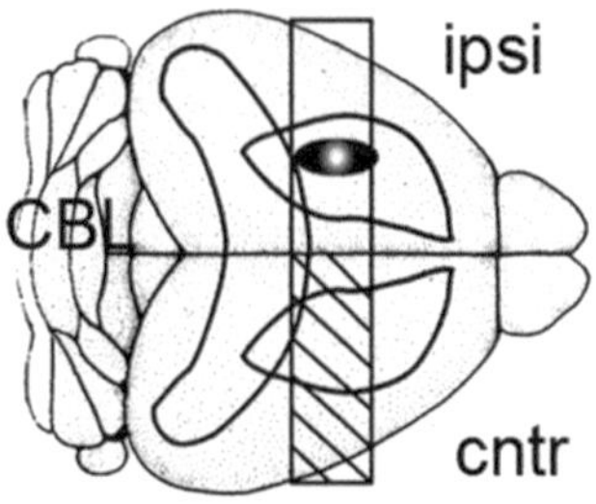

Fig. 33.1 Location of injection point and areas for comparative analysis by immunohistochemistry, PCR, and FACS. The injection point is located on the ipsilateral (ipsi) side of the brain. *CBL* cerebellum

   (d) HIV-1-infected MDM (1.5×10^5 cells in 5 μL) are stereotactically injected i.c. into caudate–putamen (Fig. 33.1). The stereotactic injection technique is described in Chap. 2 of this book.
   (e) The coordinates for inoculation are near the basal ganglia (nucleus caudate–putamen, striatum) located 3.5 mm behind the bregma, 3.5 mm lateral from the sagittal midline at a depth and angle of 4.0 mm and 35° from the vertical line.
4. Tissue processing.

   The brains of HIVE animals can be collected for neurohistologic evaluation (Chaps. 1 and 4), flow cytometry (Chap. 13), and nucleic acids and proteins detection (Chaps. 14 and 15). Injection of HIV-1-infected MDM into the putamen/caudate nuclei induces inflammation in surrounding areas and affects other brain regions involved in the neuronal connectivity map. Significant changes could be found in the hippocampus, thalamus, and corpus callosum by day 7 after injection. From 14 to 28 days after inoculation, brain tissue damage will heal and morphological changes will be presented as a line of scar tissue at the place of injection. These animals can be used for magnetic resonance imaging (MRI, Chap. 19), single-photon emission computed tomography [(SPECT)-CT, Chap. 20], as well as behavior studies (Chap. 32).

33.3 Human Peripheral Blood Lymphocyte Reconstitution as a Model of Neuroinflammation Associated with Graft-Versus-Host Disease

As mentioned in the previous chapter, the mouse genetic background is important for human cell reconstitution. The ability to engraft hu-PBL is optimally accomplished on mice that do not have their own lymphocytes, like animals with a common cytokine receptor chain knockout Balb/c-Rag2$^{-/-}\gamma_c^{-/-}$ and NOG/NSG (Lepus et al. 2009). They are not able to reject human cells. However, human cells can proliferate in the mouse environment and induce graft-versus-host disease (GVHD) with brain involvement. The classical models of normal mouse brain involvement are characterized by infiltration of meninges, choroid plexus, and perivascular and "angiitis"-like infiltration (Hickey and Kimura 1987; Kajiwara et al. 1991; Padovan et al. 2001; Sostak et al. 2004, 2010). In the rat model the changes in gene expression patterns were observed as soon as 3 days after GVH induction (Furukawa et al. 2004).

NSG mice reconstituted with human mature peripheral blood lymphocytes (hu-PBL) showed strong brain involvement. Brain pathology is characterized by infiltration of meninges, choroid plexus, and perivascular and "angiitis"-like infiltration. The examples of brain pathology are shown on Fig. 33.2. Hu-PBL animals can be infected with HIV-1. However, the migration of human CD3+ T cells to the brain occurred irrespective of HIV-1 infection. In the brain, human

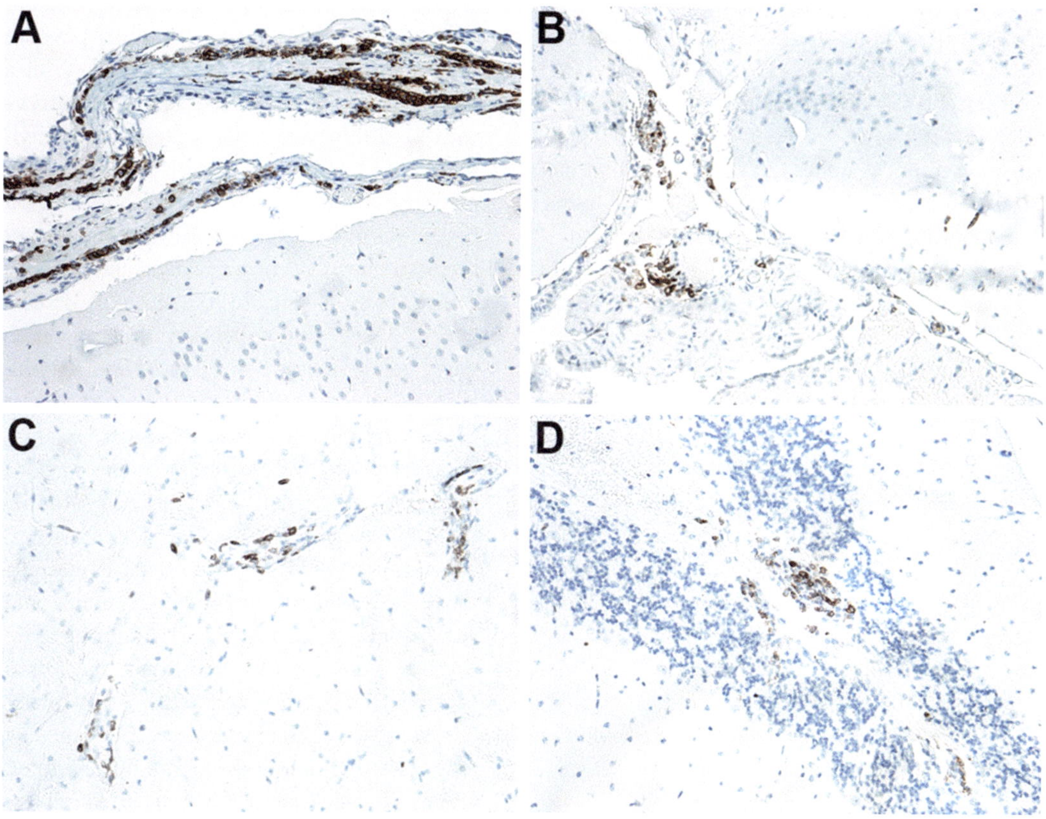

Fig. 33.2 Human cell distribution in brain of hu-PBL mice. (**a**) Leptomeninges infiltrated with human-activated lymphocytes. (**b**) Choroid plexus with human-activated cells. (**c**, **d**) Brain parenchyma with perivascular invasion of human-activated cells. All sections were stained with antibodies to human HLA-DR and counterstained with hematoxylin. Original magnification ×200

CD3+ T cells were found in the leptomeninges, perivascular areas, and the parenchyma (Koyanagi et al. 1997; Poluektova et al. 2004). Occasionally mice can develop "multiple sclerosis"-like lesions of white matter tracks (Poluektova et al. 2004).

Some aspects of human vs. mouse GVH reaction in the brain recapitulate known observations for retroviral infections of the CNS. The lesion pattern, at least in the initial stages for lentiviruses, seems to be similar; it consists of meningitis; perivascular infiltrations, especially of the deep white matter; and inflammation of the choroid plexus.

Under experimental conditions, human PBL in the mouse environment become highly activated. Their phenotype undergoes profound changes from naïve CD45RA+ cells to activated CD45R0+ effectors with different patterns of HIV-1 co-receptor expression (Fais et al. 1999). The selection of HIV-1 strain and time of administration depend on the goal of the experiment (Gorantla et al. 2005a; 2007; 2010a).

The xeno-GVHD is fatal for immunodeficient mice. The longevity of mouse survival depends on strain of mice, amount of transplanted human cells, compatibility of mouse and human major and minor MHC antigens, and sex. In the case of a severe reaction, the mice will not survive more than 2–3 weeks after reconstitution. With a mild reaction they can survive over 4–5 weeks (Schroeder et al. 2011).

33.3.1 Material and Methods

1. Human lymphocyte preparation

   Lymphocytes are obtained from HIV-1, HIV-2, and hepatitis B seronegative donor leukopaks and are purified by countercurrent centrifugal elutriation (reviewed in Chap. 12). 20×10^6 cells/mL in PBS solution, not exceeding 300 μL total volume, are injected into the mouse peritoneal cavity.
2. HIV-1 infection of hu-PBL mice

   Intraperitoneal injections of HIV-1 stock are performed at selected concentrations of infectious particles (tissue culture infective dose 50 %, $TCID_{50}$) in the volume of 100 μL/mouse. The macrophage-tropic CCR5-utilizing HIV-1_{ADA} at a $TCID_{50}$ of 10^4 provides sustained 100 % infection of the hu-PBL mice.
3. FACS analysis of hu-PBL mouse peripheral blood and single splenocyte cell suspension

   The repopulation of the mouse body by human PBL is monitored by FACS analysis of mouse peripheral blood. Fifty to two hundred microliters are collected by facial vein bleeding in EDTA-containing tubes (BD Bioscience, San Jose, CA, USA).
4. Facial vein bleeding procedure

   The mouse is not anesthetized during the procedure. Using two fingers, grab the mouse, drawing the skin firmly at the nape of the neck very close to the head, making sure that the mouse's head is in straight alignment with its body. The head cannot be angled down or to the side. Slightly compress the blood vessels of the neck on the contralateral side (opposite to the puncture site). Locate the small, circular hairless area ventrally on the lower jaw and medial to the angle of the mandible. With a 5-mm lancet, puncture the skin slightly dorsally (1–2 mm) to a shallow depth of approximately 1–2 mm. Blood will drip from this puncture site and can be collected in a blood collection tube containing EDTA. Blood will seep at a rate of 1 drop per second and between 5 and 10 drops (up to 200 μL) will be collected. Once the collection is completed, apply gentle pressure to the puncture site with a gauze sponge to aid in blood clotting.
5. Single-cell splenocyte suspension preparation

   The spleens of normal 4–8-week-old NSG mice are very small. In mice reconstituted with human PBL, the size of spleens varies, depending on the levels of human cell engraftment. It could be enlarged 5–50 times and occasionally reach the size of 2 cm in length. Spleens, bone marrow, and thymuses should be collected in ice-cold PBS. Tissues are homogenized in a sterile petri dish, and the suspension is filtered through a 40-μM filter. Cells are pelleted by centrifugation at 1,200 rpm for 10 min, supernatant removed, and pellets resuspended in 1 mL of phenol-free PBS supplemented with 2 % fetal bovine serum. Cell concentrations are adjusted to

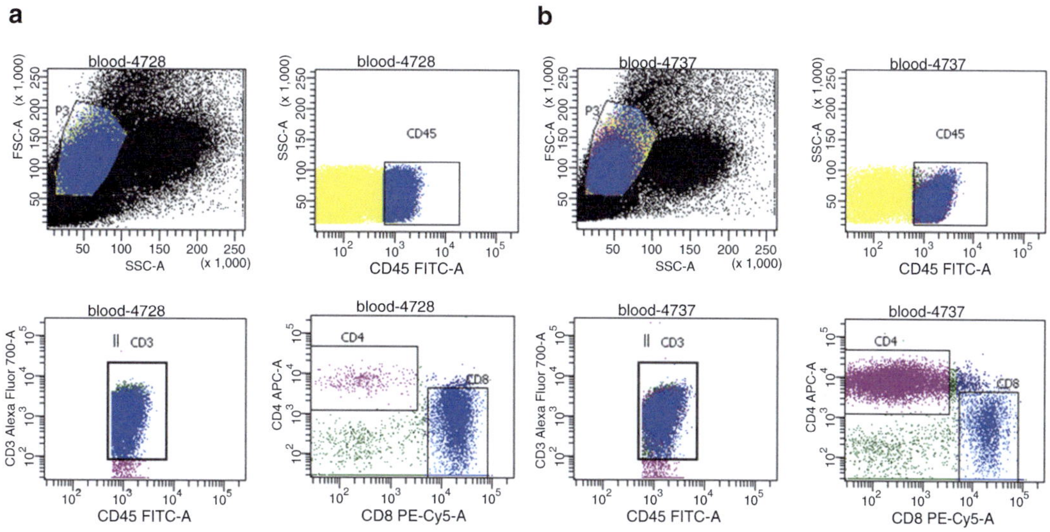

Well: 4728 **HIV-1 infected mouse**

Population	#Events	%Parent	%Total
All Events	977,741	####	100.0
P3	84,812	8.7	8.7
CD45	25,154	29.7	2.6
CD3	22,990	91.4	2.4
CD8	21,041	91.5	2.2
CD4	405	1.8	0.0

Well: 4737 **uninfected control mouse**

Population	#Events	%Parent	%Total
All Events	239,230	####	100.0
P3	45,348	19.0	19.0
CD45	30,875	68.1	12.9
CD3	28,821	93.3	12.0
CD8	11,142	38.7	4.7
CD4	14,802	51.4	6.2

Fig. 33.3 FACS analysis of the blood of hu-PBL mice. (**a**) Gating strategy and analysis of human CD4 and CD8 cells in HIV-1-infected hu-PBL mouse 2 weeks postinfection and 3 weeks after reconstitution. (**b**) The same analysis of uninfected mouse blood collected at the same time

$1–2 \times 10^6$ cells/mL, and 50–100 μL of suspension are used for the FACS analysis.

6. Staining for FACS analysis

   Fifty microliters of mouse blood collected in EDTA-containing tubes or 50 μL of single-cell splenocyte or bone marrow suspensions is mixed with the specific panel of antibodies. These panels should include antihuman CD45 pan leucocyte antigen to separate mouse and human cell populations. The HIV-1 infection panel should include antibodies to human CD3, CD4, and CD8 antigens. An example of the analysis of peripheral blood of uninfected and HIV-1-infected animals is shown in Fig. 33.3. The final dilution of antibodies will be determined by preliminary screening with hu-PBL.

   Due to fatality, animals should be evaluated by the GVHD assessment scale, which considers weight loss, general appearance of fur, and mobility. GVHD diagnosis is given to a mouse with 10 % weight loss, appearance of ruffled fur, and limited mobility. As requested by the IACUC committee, mice are euthanized when weight loss >20 % or when limited mobility or disruption of general appearance is evident in the mice.

33.4 Severe HIVE Model Based on Combination of Two Approaches: Hu-PBL Reconstitution and Intracranial Injection of HIV-1-Infected Macrophages

The combination of peripheral reconstitution with hu-PBL and the intracranial MDM transplantation of cells derived from the same donor are utilized to study human adaptive immune responses (Poluektova et al. 2002; Gorantla et al. 2005b). In this model animals developed significant inflammation in the brain in addition to a severe peripheral GVH reaction. Human PBL should be transplanted first. By 7 days after transplantation, lymphocytes will populate mouse tissues and will be present in blood circulation. At that time HIV-1-infected MDM can be injected into the brain. Human lymphocytes will enter the site of injection during surgery due to the traumatic injury-induced increased permeability of blood–brain barrier. Significant accumulation of lymphocytes around blood vessels and their interaction with human macrophages can be detected. Due to severe brain damage by infiltrating human lymphocytes, this model had no significant application.

33.5 Chimeric Mice Transplanted with Human Hematopoietic Stem Cells

NSG/NOG mice provide a unique opportunity to study HIV-1-associated neuropathology at the natural setting of disease progression (Poluektova 2012). These animals can survive for a long time, carry a human immune system, and be chronically infected with HIV-1. Transplanted human HSC will find their niche in the mouse bone marrow. HSC are able to retain self-renewal properties and generate all types of human blood cells. Generated human monocytes will populate the mouse brain, meninges, and perivascular spaces. HIV-1 infection will induce activation of mouse glial cells and lead to neurologic impairment. The reconstitution of mice with human HSC can be performed by two ways. The first way is to transplant HSC intrahepatically into newborn pups, which are irradiated with 1.0 Gy. There is no mortality after irradiation. The female mother can be separated to another cage, and the irradiation of the litter can be performed in only a few minutes, if a 60-Co source is available. However, if the source of irradiation is not available, chemical partial depletion of mouse HSCs can be performed. Another way to reconstitute mice is to transplant HSC into adult 4–5-week-old mice by intravenous infusion of HSC. In order to make a niche for HSC, mice first need to be treated by treosulfan (analog of busulfan, widely clinically utilized chemotherapeutic agent for myeloablation)(Sjoo et al. 2006; Sauer et al. 2007). The dose of treosulfan can be at a non-myeloablative range from 1 to 3 g/kg. For example, animals can receive three daily injections of 1.5 g/kg i.p., and HSC injected 24 h after the last dose of treosulfan. Irradiation at birth leads to diminished body weight of animals and development of cataracts. NSG/NOG mice develop deafness by 3–4 weeks of age. Vision is not affected. All these factors should be taken in consideration for analysis of brain pathology and testing of mouse behavior.

33.5.1 Material and Methods

1. Human CD34+ cell isolation from cord blood
   (a) Human umbilical cord blood is collected with anticoagulant heparin (15–20 mL), diluted with sterile Dulbecco's PBS, without calcium and magnesium in a 1:2 ratio, and is loaded over lymphocyte separation medium (LSM) and spun at room temperature at 400 g using the lowest deceleration setting for 35 min.
   (b) The white cell layer is collected in sterile centrifuge tubes and diluted up to 50 mL in sterile PBS to be spun at 1,400 rpm for 8 min in order to wash off all LSM.
   (c) CD34+ cells are isolated from mononuclear cells using the MACS separation system with anti-CD34 microbeads (Miltenyi Biotec, Auburn, CA, USA), MACS magnetic separation apparatus, and MACS LS separation columns (Miltenyi Biotec, Auburn, CA, USA).
   (d) Count and check the viability of the CD34+ cells using trypan blue and a hemacytometer, resuspend them in sterile PBS, and keep them on ice until transplantation.
   (e) Stain a small proportion of the isolated CD34+ cells with the mouse antihuman monoclonal antibodies to CD34 and CD3, and analyze by flow cytometry for CD34 expression and residual CD3+ T cells to assess the efficacy of magnetic bead isolation. This protocol yields greater than

90–95 % pure CD34+ cells and less than 1.0 % CD3+ T cells. If necessary, depletion of CD3 T cells can be performed using anti-CD3 MACS beads.

(f) Unused cells can be frozen for future use in freezing media with special isopropanol freezing containers, which freeze the cells at 1 °C/min overnight, and then transferred to a liquid nitrogen tank.

2. NSG mice transplantation with HSC
   (a) Monitor pregnant females for the birth of new litters. Ensure that the pups are nurtured and fed by their mother by checking newborn stomachs for signs of milk and confirm normal motor function. Engraftment procedures are performed within 24–48 h on postnatal newborn pups.
   (b) Remove mother to another cage and place cage with nest under irradiation device. Irradiate pups with 100 cGy whole-body irradiation. Return mother into the cage.
   (c) Transplantation of HSC can be done 4–24 h after irradiation. To anesthetize, place irradiated pups with nest material on a sterile petri dish over ice for 5–10 min, until gross movement ceases.
   (d) Load 30 μL of the CD34+ HSC cell suspension into a 0.5-mL syringe fitted with a 28-G ½ needle.
   (e) Intrahepatic injection: ~10^5 CD34+ cells are injected directly into the visibly dark liver taking care not to pierce the liver all the way through. The number of HSC will influence the quality and speed of reconstitution.
   (f) Efficient reconstitution can be accomplished with ~10 true stem cells, and the usual ratio of high potency cells in a CD34+ population is 1:1,000 to 1:10,000.
   (g) Place the pups on a warming pad or under an incandescent lamp for 1–2 min and return into the cage.
   (h) Wean pups between 3 and 4 weeks of age and verify engraftment at 12–16 weeks of age by FACS.
   (i) Human tissues and mice engrafted with human tissues should be considered potential biological hazards and handled with proper personal protective equipment in an animal biosafety level 2 (ABSL2) room, in accordance with governmental and institutional biosafety guidelines.

3. Human CD34+ cell isolation from fetal liver
   (a) Remove the liquid in which the liver tissue was shipped, transfer tissue to a gentleMACS C tube (Miltenyi Biotec), and wash the tissue with X-VIVO 10 media (Lonza) at least three times or until small tissue particles are not visible.
   (b) Add the three enzymes, hyaluronidase, collagenase, and deoxyribonuclease (Sigma-Aldrich® Corporation), to help break up the tissue, and dissociate for 1 min using the gentleMACS Dissociator (Miltenyi Biotec).
   (c) Place the suspension into a 37° incubator for 15 min so that the enzymes can completely break down the tissue.
   (d) Filter the suspension using a 40-μm cell strainer (BD Falcon, Bedford, MA, USA), and continue from step 1 in the cord blood isolation protocol where PBS is added and layered over LSM.

4. Dynamics of reconstitution and FACS analysis

   Transplantation of HSC into the newborn mouse pup liver leads to a brief period of expansion of CD34+ cells inside the liver hematopoietic environment, which will end by 7–8 days of mouse age. Human HSC will occupy the available niches in bone marrow where human hematopoiesis will persist. At the same time, T cell progenitors/precursors will seed the underdeveloped mouse thymus, allowing human thymocytes to be developed in the mouse thymic environment. The dynamics of human immune system establishment will be monitored by FACS analysis of mouse peripheral blood for the presence of human lymphocytes and monocytes. The staining should include antihuman CD45 pan leucocyte antibodies to separate human cells from mouse. Human CD45+ cells can be further analyzed for the presence of CD3+ T cells

(CD4+ and CD8+), B cells, and CD14+ monocytes. The frequency of blood collection depends on the goal of experiments. For the sustained survival of animals, collections should be done by facial vein bleeding, not more often than every 2–4 weeks and at a volume collected not exceeding 100 μL.

The pattern of human cell populations in mouse blood is as follows: human monocytes (CD14+ cells) and B cells (CD19+) will appear in circulation by first 2–3 months. Human CD3−CD4+ lymphoid tissue inducers also could be found during first 2 months post HSC transplantation. The development of thymopoiesis will manifest by the human CD3+ cells in circulation, and a proportion of B cells will gradually decline. By 4–5 months post-transplantation, animals will have stable populations of human cells in bone marrow and spleen. Splenic periarterial spaces will be reminiscent of mouse white pulp follicular structures containing B cells and plasma cells, which produce human immunoglobulins. The deletion (complete or partial) of common cytokine receptor gamma chain in NSG/NOG mice precludes signaling of all lymphopoietic cytokines (IL-2, IL-7, IL-15, and others) and the development of secondary lymphoid structures and groups of lymph nodes. Chimeric NSG/NOG mice will have cervical, thoracic, and mesenterial lymph nodes, but not Peyer's patches, nasal associated, nor subcutaneous lymph nodes (Denton et al. 2012). The human immune system in the mouse environment is not completely functional, and several interesting pathologic conditions could be further observed in aging chimeric animals. Some of them include the development of chronic autoimmune reactions, hair loss, eventual weight loss, and skin damage. Such conditions were found after 5–6 months of age and were associated with the decline of regulatory T cells (Onoe et al. 2011), augmentation of autoimmune antibodies (Chang et al. 2012), and expansion of mast cells driven by activation of human CD4+ lymphocytes (Kambe et al. 2004; Tanaka et al. 2012).

Humanized mice have ongoing bone marrow human hematopoiesis and human monocytes are present in circulation (Gorantla et al. 2007; 2010c; Tanaka et al. 2012). They become detectable at 2–4 weeks after HSC transplantation as a first and independent lineage of human cells. Example of peripheral blood FACS analysis of a reconstituted mouse is shown on Fig. 33.3. Tissues from humanized mice can be collected and processed as described above.

5. Brain repopulation by human cells

   Monocytes are able to populate mouse brain meninges and enter perivascular spaces. By 6–7 months of age, the repopulation of meninges could reach 5–30 %. The list of available antibodies to detect human cells in mouse brain includes the following: CD68, HLA-DR, CD45 (all from Dako), and CD163 (Vector Laboratories, Burlingame, CA, USA). The description of brain pathology in humanized mice was published (Gorantla et al. 2010b; 2012a, b; Dash et al. 2011).

33.6 Conclusion

"Humanized" mice offer a window into aspects of human physiology that are otherwise inaccessible. The best available methods for mouse brain humanization rely on cell transplantation into immunodeficient mice. The initial step was the direct transplantation of macrophages, which resembled the function and effects of HIV-1 infection in human brains. In light of the significant progress that has been achieved in developing new mouse strains for humanization, mouse models for human neuron, astrocyte, and real microglial reconstitution are on the horizon. Undeniably, mice transplanted with human neuronal and glial cells will have significant and practical applications as the function of human cells will be persistent for weeks/months. These advances may enable manufacturing of reproducible in vivo models for diverse human-specific viral infections, drug development, and other essential research applications.

References

Bauer M, Goldstein M, Christmann M, Becker H, Heylmann D, Kaina B (2011) Human monocytes are severely impaired in base and DNA double-strand break repair that renders them vulnerable to oxidative stress. Proc Natl Acad Sci USA 108: 21105–21110

Berges BK, Rowan MR (2011) The utility of the new generation of humanized mice to study HIV-1 infection: transmission, prevention, pathogenesis, and treatment. Retrovirology 8:65

Blunt T, Finnie NJ, Taccioli GE, Smith GC, Demengeot J, Gottlieb TM, Mizuta R, Varghese AJ, Alt FW, Jeggo PA et al (1995) Defective DNA-dependent protein kinase activity is linked to V(D)J recombination and DNA repair defects associated with the murine scid mutation. Cell 80:813–823

Bosma GC, Custer RP, Bosma MJ (1983) A severe combined immunodeficiency mutation in the mouse. Nature 301:527–530

Brehm MA, Shultz LD, Greiner DL (2010a) Humanized mouse models to study human diseases. Curr Opin Endocrinol Diabetes Obes 17:120–125

Brehm MA, Bortell R, Diiorio P, Leif J, Laning J, Cuthbert A, Yang C, Herlihy M, Burzenski L, Gott B, Foreman O, Powers AC, Greiner DL, Shultz LD (2010b) Human immune system development and rejection of human islet allografts in spontaneously diabetic NOD-Rag1null IL2rgammanull Ins2Akita mice. Diabetes 59:2265–2270

Brehm MA, Cuthbert A, Yang C, Miller DM, DiIorio P, Laning J, Burzenski L, Gott B, Foreman O, Kavirayani A, Herlihy M, Rossini AA, Shultz LD, Greiner DL (2010c) Parameters for establishing humanized mouse models to study human immunity: analysis of human hematopoietic stem cell engraftment in three immunodeficient strains of mice bearing the IL2r gc null mutation. Clin Immunol 135:84–98

Chang H, Biswas S, Tallarico AS, Sarkis PT, Geng S, Panditrao MM, Zhu Q, Marasco WA (2012) Human B-cell ontogeny in humanized NOD/SCID gammac(null) mice generates a diverse yet auto/poly- and HIV-1-reactive antibody repertoire. Genes Immun 13:399–410

Chechlacz M, Vemuri MC, Naegele JR (2001) Role of DNA-dependent protein kinase in neuronal survival. J Neurochem 78:141–154

Chun JJ, Schatz DG, Oettinger MA, Jaenisch R, Baltimore D (1991) The recombination activating gene-1 (RAG-1) transcript is present in the murine central nervous system. Cell 64:189–200

Collis SJ, DeWeese TL, Jeggo PA, Parker AR (2005) The life and death of DNA-PK. Oncogene 24:949–961

Culmsee C, Bondada S, Mattson MP (2001) Hippocampal neurons of mice deficient in DNA-dependent protein kinase exhibit increased vulnerability to DNA damage, oxidative stress and excitotoxicity. Mol Brain Res 87:257–262

Dash PK, Gorantla S, Gendelman HE, Knibbe J, Casale GP, Makarov E, Epstein AA, Gelbard HA, Boska MD, Poluektova LY (2011) Loss of neuronal integrity during progressive HIV-1 infection of humanized mice. J Neurosci 31:3148–3157

de Jong YP, Rice CM, Ploss A (2010) New horizons for studying human hepatotropic infections. J Clin Invest 120:650–653

Denton PW, Olesen R, Choudhary SK, Archin NM, Wahl A, Swanson MD, Chateau M, Nochi T, Krisko JF, Spagnuolo RA, Margolis DM, Garcia JV (2012) Generation of HIV latency in humanized BLT mice. J Virol 86:630–634

Fais S, Lapenta C, Santini SM, Spada M, Parlato S, Logozzi M, Rizza P, Belardelli F (1999) Human immunodeficiency virus type 1 strains R5 and X4 induce different pathogenic effects in hu-PBL-SCID mice, depending on the state of activation/differentiation of human target cells at the time of primary infection. J Virol 73:6453–6459

Furukawa H, Yamashita A, del Rey A, Besedovsky H (2004) c-Fos expression in the rat cerebral cortex during systemic GvH reaction. Neuroimmunomodulation 11:425–433

Goldman JP, Blundell MP, Lopes L, Kinnon C, Di Santo JP, Thrasher AJ (1998) Enhanced human cell engraftment in mice deficient in RAG2 and the common cytokine receptor gamma chain. Br J Haematol 103:335–342

Gorantla S, Che M, Gendelman HE (2005a) Isolation, propagation, and HIV-1 infection of monocyte-derived macrophages and recovery of virus from brain and cerebrospinal fluid. Methods Mol Biol 304:35–48

Gorantla S, Santos K, Meyer V, Dewhurst S, Bowers WJ, Federoff HJ, Gendelman HE, Poluektova L (2005b) Human dendritic cells transduced with herpes simplex virus amplicons encoding human immunodeficiency virus type 1 (HIV-1) gp120 elicit adaptive immune responses from human cells engrafted into NOD/SCID mice and confer partial protection against HIV-1 challenge. J Virol 79:2124–2132

Gorantla S, Sneller H, Walters L, Sharp JG, Pirruccello SJ, West JT, Wood C, Dewhurst S, Gendelman HE, Poluektova L (2007) Human immunodeficiency virus type 1 pathobiology studied in humanized BALB/c-Rag2-/-gammac-/- mice. J Virol 81:2700–2712

Gorantla S, Makarov E, Roy D, Finke-Dwyer J, Murrin LC, Gendelman HE, Poluektova L (2010a) Immunoregulation of a CB2 receptor agonist in a murine model of neuroAIDS. J Neuroimmune Pharmacol 5:456–468

Gorantla S, Makarov E, Finke-Dwyer J, Castanedo A, Holguin A, Gebhart CL, Gendelman HE, Poluektova L (2010b) Links between progressive HIV-1 infection of humanized mice and viral neuropathogenesis. Am J Pathol 177:2938–2949

Gorantla S, Makarov E, Finke-Dwyer J, Gebhart CL, Domm W, Dewhurst S, Gendelman HE, Poluektova LY (2010c) CD8+ cell depletion accelerates HIV-1 immunopathology in humanized mice. J Immunol 184:7082–7091

Gorantla S, Poluektova L, Gendelman HE (2012a) Rodent models for HIV-associated neurocognitive disorders. Trends Neurosci 35:197–208

Gorantla S, Gendelman HE, Poluektova LY (2012b) Can humanized mice reflect the complex pathobiology of HIV-associated neurocognitive disorders? J Neuroimmune Pharmacol 7:352–362

Greiner DL, Shultz LD, Yates J, Appel MC, Perdrizet G, Hesselton RM, Schweitzer I, Beamer WG, Shultz KL, Pelsue SC et al (1995) Improved engraftment of human spleen cells in NOD/LtSz-scid/scid mice as compared with C.B-17-scid/scid mice. Am J Pathol 146:888–902

Hesselton RM, Greiner DL, Mordes JP, Rajan TV, Sullivan JL, Shultz LD (1995) High levels of human peripheral blood mononuclear cell engraftment and enhanced susceptibility to human immunodeficiency virus type 1 infection in NOD/LtSz-scid/scid mice. J Infect Dis 172:974–982

Hickey WF, Kimura H (1987) Graft-vs.-host disease elicits expression of class I and class II histocompatibility antigens and the presence of scattered T lymphocytes in rat central nervous system. Proc Natl Acad Sci USA 84:2082–2086

Kajiwara K, Hirozane A, Fukumoto T, Orita T, Nishizaki T, Kamiryo T, Ito H (1991) Major histocompatibility complex expression in brain of rats with graft-versus-host disease. J Neuroimmunol 32:191–198

Kambe N, Hiramatsu H, Shimonaka M, Fujino H, Nishikomori R, Heike T, Ito M, Kobayashi K, Ueyama Y, Matsuyoshi N, Miyachi Y, Nakahata T (2004) Development of both human connective tissue-type and mucosal-type mast cells in mice from hematopoietic stem cells with identical distribution pattern to human body. Blood 103:860–867

Koyanagi Y, Tanaka Y, Tanaka R, Misawa N, Kawano Y, Tanaka T, Miyasaka M, Ito M, Ueyama Y, Yamamoto N (1997) High levels of viremia in hu-PBL-NOD-scid mice with HIV-1 infection. Leukemia 11(suppl 3): 109–112

Lepus CM, Gibson TF, Gerber SA, Kawikova I, Szczepanik M, Hossain J, Ablamunits V, Kirkiles-Smith N, Herold KC, Donis RO, Bothwell AL, Pober JS, Harding MJ (2009) Comparison of human fetal liver, umbilical cord blood, and adult blood hematopoietic stem cell engraftment in NOD-scid/gammac-/-, Balb/c-Rag1-/-gammac-/-, and C.B-17-scid/bg immunodeficient mice. Hum Immunol 70:790–802

Libby SJ, Brehm MA, Greiner DL, Shultz LD, McClelland M, Smith KD, Cookson BT, Karlinsey JE, Kinkel TL, Porwollik S, Canals R, Cummings LA, Fang FC (2010) Humanized nonobese diabetic-scid IL2rgammanull mice are susceptible to lethal Salmonella Typhi infection. Proc Natl Acad Sci USA 107:15589–15594

McCune JM, Namikawa R, Kaneshima H, Shultz LD, Lieberman M, Weissman IL (1988) The SCID-hu mouse: murine model for the analysis of human hematolymphoid differentiation and function. Science 241: 1632–1639

McDermott SP, Eppert K, Lechman ER, Doedens M, Dick JE (2010) Comparison of human cord blood engraftment between immunocompromised mouse strains. Blood 116:193–200

Mosier DE, Gulizia RJ, Baird SM, Wilson DB (1988) Transfer of a functional human immune system to mice with severe combined immunodeficiency. Nature 335:256–259

Namikawa R, Kaneshima H, Lieberman M, Weissman IL, McCune JM (1988) Infection of the SCID-hu mouse by HIV-1. Science 242:1684–1686

Neema M, Navarro-Quiroga I, Chechlacz M, Gilliams-Francis K, Liu J, Lamonica K, Lin SL, Naegele JR (2005) DNA damage and nonhomologous end joining in excitotoxicity: neuroprotective role of DNA-PKcs in kainic acid-induced seizures. Hippocampus 15:1057–1071

Oettinger MA (1996) Cutting apart V(D)J recombination. Curr Opin Genet Dev 6:141–145

Onoe T, Kalscheuer H, Danzl N, Chittenden M, Zhao G, Yang YG, Sykes M (2011) Human natural regulatory T cell development, suppressive function, and postthymic maturation in a humanized mouse model. J Immunol 187:3895–3903

Padovan CS, Gerbitz A, Sostak P, Holler E, Ferrara JL, Bise K, Straube A (2001) Cerebral involvement in graft-versus-host disease after murine bone marrow transplantation. Neurology 56:1106–1108

Persidsky Y, Gendelman HE (2003) Mononuclear phagocyte immunity and the neuropathogenesis of HIV-1 infection. J Leukoc Biol 74:691–701

Persidsky Y, Limoges J, McComb R, Bock P, Baldwin T, Tyor W, Patil A, Nottet HSLM, Epstein L, Gelbard H, Flanagan E, Reinhard J, Pirruccello SJ, Gendelman HE (1996) Human immunodeficiency virus encephalitis in SCID mice. Am J Pathol 149:1027–1053

Pino S, Brehm MA, Covassin-Barberis L, King M, Gott B, Chase TH, Wagner J, Burzenski L, Foreman O, Greiner DL, Shultz LD (2010) Development of novel major histocompatibility complex class I and class II-deficient NOD-SCID IL2R gamma chain knockout mice for modeling human xenogeneic graft-versus-host disease. Methods Mol Biol 602:105–117

Poluektova LY (2012) Murine models for neuroAIDS. In: Gendelman HE, Grant I, Everall IP, Fox HS, Gelbard HA, Lipton SA, Swindells S (eds) The neurology of AIDS, 3rd edn. Oxford University Press, New York, pp 414–431

Poluektova LY, Munn DH, Persidsky Y, Gendelman HE (2002) Generation of cytotoxic T cells against virus-infected human brain macrophages in a murine model of HIV-1 encephalitis. J Immunol 168:3941–3949

Poluektova LY, Gorantla S, Gendelman HE (2004) Studies of adaptive immunity in a murine model of HIV-1 encephalitis. In: Gendelman HG, Grant I, Lipton SA, Swindells S (eds) Neurology of AIDS, 2nd edn. Oxford University Press, New York, pp 297–309

Robinet E, Baumert TF (2011) A first step towards a mouse model for hepatitis C virus infection containing a human immune system. J Hepatol 55:718–720

Sato K, Koyanagi Y (2011) The mouse is out of the bag: insights and perspectives on HIV-1-infected humanized mouse models. Exp Biol Med 236:977–985

Sauer M, Zeidler C, Meissner B, Rehe K, Hanke A, Welte K, Lohse P, Sykora KW (2007) Substitution of cyclophosphamide and busulfan by fludarabine, treosulfan and melphalan in a preparative regimen for children and adolescents with Shwachman-Diamond syndrome. Bone Marrow Transplant 39:143–147

Schroeder K et al (2011) Report from the EPAA workshop: in vitro ADME in safety testing used by EPAA industry sectors. Toxicol In Vitro 25:589–604

Serreze DV, Leiter EH, Hanson MS, Christianson SW, Shultz LD, Hesselton RM, Greiner DL (1995) Emv30null NOD-scid mice. An improved host for adoptive transfer of autoimmune diabetes and growth of human lymphohematopoietic cells. Diabetes 44:1392–1398

Shibata S, Asano T, Noguchi A, Naito M, Ogura A, Doi K (1998) Peritoneal macrophages play an important role in eliminating human cells from severe combined immunodeficient mice transplanted with human peripheral blood lymphocytes. Immunology 93:524–532

Shultz LD, Schweitzer PA, Christianson SW, Gott B, Schweitzer IB, Tennent B, McKenna S, Mobraaten L, Rajan TV, Greiner DL et al (1995) Multiple defects in innate and adaptive immunologic function in NOD/LtSz-scid mice. J Immunol 154:180–191

Sjoo F, Hassan Z, Abedi-Valugerdi M, Griskevicius L, Nilsson C, Remberger M, Aschan J, Concha H, Gaughan U, Hassan M (2006) Myeloablative and immunosuppressive properties of treosulfan in mice. Exp Hematol 34:115–121

Sostak P, Reich P, Padovan CS, Gerbitz A, Holler E, Straube A (2004) Cerebral endothelial expression of adhesion molecules in mice with chronic graft-versus-host disease. Stroke 35:1158–1163

Sostak P, Padovan CS, Eigenbrod S, Roeber S, Segerer S, Schankin C, Siegert S, Saam T, Theil D, Kolb HJ, Kretzschmar H, Straube A (2010) Cerebral angiitis in four patients with chronic GVHD. Bone Marrow Transplant 45:1181–1188

Tanaka S, Saito Y, Kunisawa J, Kurashima Y, Wake T, Suzuki N, Shultz LD, Kiyono H, Ishikawa F (2012) Development of mature and functional human myeloid subsets in hematopoietic stem cell-engrafted NOD/SCID/IL2rgammaKO mice. J Immunol 188: 6145–6155

Tournoy KG, Depraetere S, Pauwels RA, Leroux-Roels GG (2000) Mouse strain and conditioning regimen determine survival and function of human leucocytes in immunodeficient mice. Clin Exp Immunol 119:231–239

Traggiai E, Chicha L, Mazzucchelli L, Bronz L, Piffaretti JC, Lanzavecchia A, Manz MG (2004) Development of a human adaptive immune system in cord blood cell-transplanted mice. Science 304:104–107

Van Duyne R, Pedati C, Guendel I, Carpio L, Kehn-Hall K, Saifuddin M, Kashanchi F (2009) The utilization of humanized mouse models for the study of human retroviral infections. Retrovirology 6:76

Vemuri MC, Schiller E, Naegele JR (2001) Elevated DNA double strand breaks and apoptosis in the CNS of scid mutant mice. Cell Death Differ 8:245–255

Washburn ML, Bility MT, Zhang L, Kovalev GI, Buntzman A, Frelinger JA, Barry W, Ploss A, Rice CM, Su L (2011) A humanized mouse model to study hepatitis C virus infection, immune response, and liver disease. Gastroenterology 140:1334–1344

Zhang L, Meissner E, Chen J, Su L (2010) Current humanized mouse models for studying human immunology and HIV-1 immuno-pathogenesis. Sci China Life Sci 53:195–203

Animal Models for PD and ALS

34

Max V. Kuenstling, Adam M. Szlachetka, and R. Lee Mosley

Abstract

Current research into neurodegenerative disorders relies heavily on animal models that recapitulate human disease. This chapter focuses on genetic and toxin-based animal models of Parkinson's disease (PD) and amyotrophic lateral sclerosis (ALS). Early models failed to duplicate many of the hallmarks of these diseases, such as the loss of dopaminergic neurons in PD and motor neuron loss in ALS, and were often dismissed as newer models were developed. Nevertheless, even current models may not fully mimic all facets of disease etiology, progression, or pathology, necessitating continued research into developing models that more closely replicate the clinical manifestations of the disease. For both PD and ALS, genetic models have been generated based on the clinical reports of familial forms of the disease. For PD, models have been created using genes coding for α-synuclein, parkin, LRRK2, DJ-1, and PINK1, while mutations in genes coding for SOD1, FUS/TLS, and TDP-43 have been used for ALS models. For either disease, fewer than 10 % of all cases are linked to genetic causes, whereas the majority of cases are termed sporadic and are likely caused by environmental factors or a combination of environmental factors and genetically imposed susceptibilities. Therefore, a large number of animal models are toxin-based, utilizing chemicals or metals that have been correlated with human disease. As a whole, current animal models serve as adequate substitutes to their human counterparts, providing valuable tools for translational and basic science research in PD and ALS, with the caveat that most models typically do not recapitulate the totality of the human disease. Therefore, this chapter intends to describe the more commonly used animal models for these diseases and relate their limitations.

M.V. Kuenstling • A.M. Szlachetka • R.L. Mosley, Ph.D. (✉)
Department of Pharmacology and Experimental Neuroscience, Center for Neurodegenerative Disorders, University of Nebraska Medical Center, Omaha, NE 68198-5930, USA
e-mail: rlmosley@unmc.edu

H. Xiong and H.E. Gendelman (eds.), *Current Laboratory Methods in Neuroscience Research*, Springer Protocols Handbooks, DOI 10.1007/978-1-4614-8794-4_34,

Keywords
Parkinson's disease • Amyotrophic lateral sclerosis • Animal models • α-Synuclein • LRRK2 • Parkin • 6-OHDA • MPTP • SOD1 • FUS/TLS • TDP-43

34.1 Introduction

Parkinson's disease (PD) is a progressive neurodegenerative disorder that is characterized by degeneration of dopaminergic (DAergic) neurons that originate in the substantia nigra (SN) and innervate the caudate nucleus/putamen. Additionally, a hallmark of PD is the accumulation of proteinaceous inclusions called Lewy bodies (LBs) within some DAergic neurons. Although the etiology of PD is unknown, environmental and genetic factors in combination are thought to play critical roles. Molecular genetics has identified several genes causing familial PD, including those coding for proteins that are found in LBs such as α-synuclein and parkin, as well as those implicated in mitochondrial function, such as leucine-rich repeat kinase 2 (*LRRK2*), *DJ-1*, and phosphatase and tensin homolog (*PTEN*)-induced putative kinase 1 (*PINK1*). However, familial cases of PD represent only approximately 10 % of the total cases. The mechanism(s) by which non-familial PD (also termed sporadic or idiopathic PD) develops has not been delineated. Animal models provide tools by which mechanisms of disease progression and symptom onset may be uncovered and utilized in approaches to therapeutic strategies. For PD, these animal models either incorporate genetic manipulation of genes associated with familial PD or utilize neurotoxins that target the mitochondrial or proteosomal function of DAergic neurons.

Amyotrophic lateral sclerosis (ALS) is a neurodegenerative disease characterized by the selective death of motor neurons in the motor cortex, brainstem, and spinal cord. Approximately 5 % of the cases are termed familial ALS (fALS), and are typically autosomal-dominant, while 90 % of the cases present no family history and are considered sporadic ALS (sALS). Both are clinically indistinguishable as upper and lower motor neurons are ultimately lost. Earliest models used toxic levels or deficiencies of metals such as lead or mercury as a way to study the disease. Genetic mutations in superoxide dismutase 1 (*SOD1*), fused in sarcoma/translocated in liposarcoma (*FUS/TLS*), and transactivation response (*TAR*) DNA-binding protein with molecular weight 43 kDa (*TDP-43*) genes have been identified in 25–35 % of the fALS cases. For many years, the gold standard for modeling fALS has been murine models of human *SOD1* mutations; however, more recent models have been developed which may better recapitulate different aspects of ALS disease progression.

This chapter focuses on animal models that have been genetically manipulated to simulate familial disease genotypes; treated with neurotoxins to target DAergic neuron loss in PD or motor neuron loss in ALS; or used to study therapeutic modulation of those model disease outcomes. Thus the objective of this chapter is to acquaint the student with the more commonly used animal models for PD and ALS.

34.2 Animal Models of PD

PD is an ancient disorder, for which symptoms have been described as far as 5,000 B.C. in the Indian medical doctrine of *Ayurveda* and in medical texts of Chinese, Romans, Persians, and Greeks. Indeed the ancient Greek physician Claudius Galen (129–200 A.D.) in 174 A.D. described the tremor at rest and is credited by some for the term "shaking palsy." In 1817, James Parkinson more accurately described the clinical phenotypic features of this disease in his "Essay on the Shaking Palsy." Over 100 years later, the pathological features of severe loss of DAergic neuronal soma in the substantia nigra pars compacta (SNpc) and termini in the caudate nucleus/putamen of the dorsal striatum; loss of dopamine neurotransmitter; development of proteinaceous inclusions, called Lewy bodies (LBs); and neuroinflammatory processes were linked with this disorder. Taking his name, Parkinson's disease is currently the second most common age-related neurodegenerative disease after Alzheimer's disease and the most common neurodegenerative motor disorder, but is characterized by a number of motor and non-motor symptoms. While the disease can be accurately diagnosed, and symptoms can be alleviated using levo-(L)-3,4-dihydroxyphenylalanine (L-DOPA), current therapeutics can neither cure the disorder nor completely stop disease progression. A majority of cases are sporadic, with age and rural residence having the highest risk factors; the latter thought likely due in part to exposure to pesticides, herbicides, or heavy metals; however, familial disorders comprise 10 % of total cases. Therefore, animal models have been created that assess pathogenesis resulting from genetic modifications, as well as neurotoxic insult (Table 34.1).

34.2.1 Genetic Models of PD

While most patients present with the sporadic form of PD, a number of cases have been observed in families with high incidences of the disease. While the genetic forms of this commonly sporadic disease are rare, shared phenotypic similarities between the two forms offer the possibility that disease pathogenesis may be shared as well. This eventually led to the identification of numerous mutations in genes causing familial PD (Fleming et al. 2005). Moreover, polymorphisms in certain genes have been shown to increase the risk of sporadic PD (Farrer et al. 2001), and mutations have been found in early-onset PD patients with no family history of the disease (Lucking et al. 2000). From genetic discoveries, countless animal models have been created, utilizing a variety of null mutations, point mutations, or extra copies in one or more of these PD-related genes.

The genes manipulated in models for autosomal-dominantly inherited PD include gain-of-function mutations in α-synuclein (*SNCA*) and *LRRK2*. These models utilize transgenic mice with extra copies of the gene or delivery of the genetic material through lenti- or adeno-associated virus (AAV). Loss-of-function mutations in autosomal-recessive inherited PD genes such as *PARK2*, *PINK1*, and *DJ-1* cause early-onset PD in models created from null mutation or genetic knock-out (KO) of these genes in mice. In addition, region- or neuron-specific gene modifications have been developed to disrupt the dopamine (DA) system by targeting genes involved in cellular function, such as the gene mitochondrial transcription factor A (*TFAM*) in the MitoPARK model. While few of the current genetic models entirely recapitulate the major symptoms of PD, in particular LB inclusions or the loss of DAergic neurons, slight effects on the DA system have been identified as well as larger disturbances in DA levels in the striatum, DA transporter (DAT) efficiency, or motor function (Fleming et al. 2005). The protein products of *SNCA*, *PARK2*, and *DJ-1* seem to play a role in the ubiquitin-proteasome system (UPS) pathway (Takahashi et al. 2001); a convincing finding considering that the composition of LB inclusions is primarily α-synuclein and ubiquitin. Studies involving these genetic models have already begun to increase our understanding of the numerous molecular events underlying PD.

Given the relatively low incidence of familial PD, genetic studies in afflicted humans are somewhat challenging and provide a greater impetus for the development of genetic animal models. While most molecular studies have been performed in rodent models, in particular mice, invertebrate animals such as *Drosophila melanogaster* and *Caenorhabditis elegans* (*C. elegans*) have proven quite useful for genetic study (Gasser 2009). Although they possess fewer neurons and their cellular structures are smaller than rodents and primates, *Drosophila* and humans have similar neuron–glial systems, many genetic homologs, and conserved molecular pathways. Much of the latest research has addressed mechanisms by which pathogenic mutations disrupt the normal function of proteins associated with PD risk or disease progression.

Table 34.1 Rodent models of PD

Model	PD pathology	PD symptoms	Advantages	Disadvantages
Genetic				
α-Synuclein	Reduced striatal DA levels 20–70 % loss of DA neurons in SNpc Inclusions	Motor impairment	Transgenic or viral-vector delivery Good construct validity	Viral-vector model is stereotactically injected, increasing variability
LRRK2	Slight DA pathology No loss of DA neurons	Motor impairment	Progressive age-dependent motor impairment	Very recent model Only slight DA pathology
Parkin	Slight reduction in striatal DA levels Slight to major DA pathology	Slight motor impairment	Direct relation to ubiquitin-proteasome system Good construct validity	Only slight DA pathology
PINK1	Minimal DA pathology	Motor impairment	Interaction of PINK1 and Parkin	Minimal DA pathology
DJ-1	Minimal DA pathology	Slight motor impairment	Increased sensitivity to MPTP	Minimal DA pathology
KO of α-synuclein, Parkin, PINK1, DJ-1	Little or no DA pathology	No motor impairment	Show slight increased sensitivity to MPTP in some models	Lacks most PD features
MitoPARK (DAT-cre, Tfam lox/lox)	Reduced striatal DA levels >70 % loss of DA neurons in SNpc Inclusions with little LB resemblance	Motor impairment	Age-dependent onset Chronic progressive symptoms Good construct validity	Complex/high-cost breeding scheme Neurodegeneration in areas other than SNpc (VTA)
Neurotoxic				
6-OHDA	Reduced striatal DA levels >70 % loss of DA neurons in SNpc No inclusions	Motor impairments with bilateral lesion Quantifiable turning behavior with unilateral lesion	Works in mice, rats, monkeys Well characterized Used in dyskinesia models	Requires stereotactic injection, increasing variability Acute neuron degeneration Poor construct validity
MPTP	Reduced striatal DA levels >70 % loss of DA neurons in SNpc Inclusions only with chronic administration	Motor impairments	Lipophilic, allowing systemic administration Works in mice Well characterized Good construct validity	Extremely toxic to humans; therefore, dangerous to administer Reduced reliability
Paraquat	Reduced striatal DA levels 20–70 % loss of DA neurons in SNpc No inclusions	Motor impairments	Systemic administration	Systemic toxicity Not well characterized Low construct validity
Rotenone	Reduced striatal DA levels >70 % loss of DA neurons in SNpc No inclusions	Motor impairments	Systemic administration	Systemic toxicity Low construct validity
Epoxomicin	Reduced striatal DA levels 20–70 % loss of DA neurons in SNpc Inclusions	Motor impairments	Systemic administration	Inconsistent findings in rats Low construct validity
LPS	20–70 % loss of DA neurons in SNpc Reduced striatal termini No inclusions	Unclear on motor impairments	Driven only by inflammation Susceptibility to other toxins is increased Chronic model	Few studies in mice Minimal results yet on motor function

Source: Adapted from Terzioglu and Galter (2008)

34.2.1.1 α-Synuclein

The first genetic links to PD were found in families that had multiple copies of or missense mutations in the gene coding for α-synuclein (α-syn) expression (*SNCA*), leading to early-onset PD. These mutations in *SNCA* such as A53T (i.e., threonine substituted for alanine at position 53), A30P, or E46K cause familial PD in an autosomal-dominant fashion (Polymeropoulos et al. 1997; Kruger et al. 1998; Gelman et al. 2003; Lees et al. 2009). Similar pathologies are seen in brains from these patients as compared to those with sporadic PD. Although mutations in *SNCA* have not been found in sporadic PD (Lynch et al. 1997; Chan et al. 1998; Mizuno et al. 2001), the basis for the idea that *SNCA*-based genetic models may characterize the sporadic form of the disease partially lies in the fact that LBs comprise α-syn aggregates (Spillantini et al. 1997, 1998). Most of the current mouse models either exploit overexpression of the human wild-type (WT) *SNCA* under a variety of promoters or express the disease-associated A53T or A30P mutations (Richfield et al. 2002).

Human α-syn is a 140-amino acid protein, widely expressed throughout the nervous system, found primarily in presynaptic nerve terminals, closely associated with synaptic vesicles (Maroteaux et al. 1988; George et al. 1995), and is believed to be involved in vesicle or chaperone functions, as well as neurotransmitter release (Kahle et al. 2002). It has been suggested that α-syn binds reversibly to components of the vesicular trafficking machinery (Jensen et al. 1998, 1999, 2000), and may act to regulate the readily releasable neurotransmitter pool (Abeliovich et al. 2000). The native protein was thought to have no secondary or tertiary structure, but has a tendency to misfold following a variety of modifications or mutations. More recently, α-syn was shown to exist physiologically as an α-helically folded tetramer that is more resistant to forces of aggregation compared to recombinant molecules (Bartels et al. 2011). Modifications and mutations of native proteins are likely deleterious since certain misfolded forms of α-syn have been shown to be toxic (especially fibrils and protofibrils) and lead to cellular death in in vitro studies (Cookson 2005). These findings led to the fabrication of a number of *SNCA*-based animal models.

Early studies in *Drosophila* and *C. elegans* were used to connect genetic modifications with downstream effects. While these organisms have less complex nervous systems and do not endogenously express *SNCA*, they have proven useful due to their ease in manipulation and relative low-cost in larger studies. In the nematode, *C. elegans*, overexpression of the protein causes rapid DAergic neuronal loss as well as reduced motor function (Lakso et al. 2003; Kuwahara et al. 2006); however, intraneuronal inclusion bodies are typically not found.

Studies using overexpression of both mutated and WT human *SNCA* in *Drosophila* show loss of DAergic neurons and motor impairment that follow an age-dependent pattern as well as the presence of LB-like α-syn-containing inclusions (Feany and Bender 2000). The behavioral abnormalities seen in these models can be corrected using L-DOPA or DA agonists (Pendleton et al. 2002), suggesting their utility for genetic screens for new PD-linked genes or as an early test for therapeutic efficacy.

Since the initial discovery of the link between *SNCA* and PD, a number of mouse lines have been created that overexpress α-syn under an assortment of promoter schemes. Transgenic mice, overexpressing WT human *SNCA* under the neuron-specific platelet-derived growth factor β (PDGFβ), display a reduction in tyrosine hydroxylase (TH) and DA in the striatum as well as a reduction in motor function (Masliah et al. 2000). Mice expressing α-syn from *SNCA* containing the A30P or A53T mutations under the TH promoter exhibit loss in striatal DA and neurons in the SNpc as well as impaired motor performance (Richfield et al. 2002; Thiruchelvam et al. 2004). Under the mouse prion promoter (mPrP), expression of the A53T mutant *SNCA* in mice showed significant protein modifications such as ubiquitination and phosphorylation with aggregated and truncated forms of the protein, and a progressive, age-dependent degeneration of DAergic neurons that was analogous to the human disease; however, neurodegeneration was confined to neither only DAergic neurons nor the nigrostriatal axis (Martin et al. 2006). In KO mice of *Snca*, no major neuronal abnormalities are found, presynaptic terminal function is normal, and the mice survive to adulthood, but mice exhibited slower locomotion than normal controls (Abeliovich et al. 2000). Therefore, loss of α-syn function most probably does not lead to neurodegeneration. Thus, mutations or protein oxidation leading to the modification, misfolding, or aggregation of α-syn are more likely a leading cause of PD neuropathology and disease progression.

In addition to the use of transgenic animals to model PD, extrinsic α-syn can also be delivered using viral vectors, such as lentiviruses and AAVs. The delivery/overexpression of A53T mutant or WT SNCA using AAVs in the SNpc of rats leads to a progressive age-dependent reduction of DAergic neurons and motor skills, as well as the development of α-syn inclusions (Kirik et al. 2002). These vectors must be stereotactically injected; therefore, rats have often been the subjects for these studies due to their larger size, although other rodents have served as recipients (Kirik et al. 2002; Klein et al. 2002; Lo Bianco et al. 2002; Lauwers et al. 2003; Theodore et al. 2008). While the viral vector models may be useful for specific studies, substantial limitations exist, one of which is that each animal requires stereotactic injection, so production of large numbers of subjects expressing identical amounts of protein in a similar anatomical location with little variability is extremely difficult.

34.2.1.2 *LRRK2*

Recently, *LRRK2* was found mutated in certain families with an autosomal-dominant late-onset PD (Zimprich et al. 2004). No less than nine pathogenic mutations have been found within *LRRK2* thus far (Paisan-Ruiz et al. 2004; Khan et al. 2005), with the most common being the G2019S substitution, which accounts for ~5 % of familial PD and is even detected in 1.5 % of sporadic cases (Gilks et al. 2005; Healy et al. 2008). *LRRK2* encodes for a rather large protein (dardarin) that contains a serine/threonine kinase as well as a GTPase domain, and is often localized to membranous structures (Biskup et al. 2006). The function and expression patterns remain unclear, and early studies detected *LRRK2* mRNA throughout most of the nigrostriatal DAergic pathway, but not in the SNpc (Galter et al. 2006). However, quantitative-PCR confirmed *LRRK2* expression within the human SNpc (Melrose et al. 2006). Furthermore, LRRK2 protein has been found in LBs of sporadic PD patients (Zhu et al. 2006). LRRK2 protein exhibits kinase activity in vitro, and both the I2020T and G2019S mutations enhance the kinase activity (West et al. 2005; Gloeckner et al. 2006). Although these data demonstrate that *LRRK2* mutations may lead to a toxic gain of function via increased kinase activity, this has yet to be revealed in any in vivo model.

The *Drosophila* orthologue of *LRRK2* (dLRRK2) showed increased expression in DAergic neurons in fly brains (Imai et al. 2008). When constructs containing mutations similar to the human form of the gene (i.e., G2019S) are over-expressed in these flies, a late-onset loss of DAergic neurons occurs along with reduced motor function and reduced life-span (Liu et al. 2008). Motor impairment can be alleviated with treatment; however, loss of DAergic neurons remains unchanged. Similarly, in *C. elegans* models expressing the G2019S mutation, DA loss is greater than from expression of WT *LRRK2* (Saha et al. 2009). Transgenic mice expressing WT or mutant forms of LRRK2 show slight pathology in nigrostriatal projections; however, neither show acute degeneration of DAergic neurons, but do demonstrate a progressive age-dependent loss of motor function that can be improved with DA replacement (Lin et al. 2009). *Lrrk2* KO mice, as in *Snca* KO mice, live to adulthood and show no major abnormalities (Andres-Mateos et al. 2009). Current studies have yet to show conclusive *LRRK2* involvement in PD pathogenesis, suggesting that *LRRK2* may increase susceptibility to PD in the presence of other genetic or environmental factors.

34.2.1.3 Parkin

Parkin has been found to be the predominant causative gene in autosomal-recessive juvenile parkinsonism (ARJP) as well as early-onset recessive PD (Kitada et al. 1998). Mutations within the gene, in addition to being a cause for familial PD, are also present in 77 % of sporadic cases in which onset of disease occurs before age 20 (Lucking et al. 2000). Forming a 465-amino acid protein, parkin is an E3 ubiquitin ligase that acts as a substrate-recognition molecule that targets proteins for degradation by the UPS (Shimura et al. 2000). The localization patterns of parkin are not fully understood; however, it has been proposed that the protein associates with lipid rafts as well as postsynaptic densities within neurons. When mutated, loss of function is believed to lead to accumulation of parkin's substrates that cannot be degraded by the UPS, leading to cellular toxicity (Shimura et al. 2000; Sriram et al. 2005). This suggests that without parkin's ubiquitin-ligase activity, substrates are no longer ubiquitinated leading to an accumulation of these proteins in neurons. The selectivity for DA neurons could be based on the accumulation of certain substrates such as synphilin-1 and a glycosylated form of α-syn (von Coelln et al. 2004), or through the action of DA itself, as it can covalently modify parkin to form adducts that functionally inactivate parkin (LaVoie et al. 2005).

This hypothesis is complicated by findings that KO mice lacking parkin show minimal changes in the number of DAergic neurons or their survival (Goldberg et al. 2003; Itier et al. 2003), indicating that other factors may also play a role in pathogenesis. Still, parkin-null mutations in both *Drosophila* and mice lead to motor deficiencies (Greene et al. 2003), and mice show slight nigrostriatal deficits such as elevated DA levels and changes in synaptic excitability (Goldberg et al. 2003). Furthermore, overexpression of human mutant parkin in flies causes DAergic neurodegeneration that is age-dependent in conjunction with locomotor defects (Sang et al. 2007; Wang et al. 2007). Similarly, mice expressing the Q311X parkin mutation show late-onset loss of motor function as well as reduction in striatal DA levels due to a significant loss of DAergic neurons (Lu et al. 2009). While these models may not entirely recapitulate human parkinsonism, they prove useful in addressing mechanisms of UPS disruption and possible PD onset and progression.

34.2.1.4 *PINK1*

Mutations in *PINK1* lead to a variety of phenotypes, including both early- and late-onset PD that are indistinguishable from sporadic cases (Hatano et al. 2004; Valente et al. 2004), and represent the second most frequent cause of autosomal-recessive early-onset PD after parkin. PINK1 is a 581-amino acid protein that contains a serine/threonine kinase domain, is localized to mitochondrial membranes by a targeting motif located at the N-terminus (Silvestri et al. 2005), and is expressed throughout the human brain (Gandhi et al. 2006). It is hypothesized that PD-causative mutations in PINK1 might lead to loss of protein function.

The earliest PINK1 studies used loss-of-function mutations in *Drosophila* in order to elucidate its role in mitochondrial function. These mutants induce degeneration of both DAergic neurons and myocytes (Clark et al. 2006), suggesting severe mitochondrial dysfunction due to enlargement of cristae. Interestingly, new data suggest that PINK1 and parkin interact to protect mitochondrial function, with PINK1 acting upstream of parkin. PINK1 mutant phenotypes share marked similarities with parkin mutants, and parkin overexpression is able to rescue mitochondrial defects found in PINK1 deficiency, while PINK1 overexpression does not rescue the phenotypes of parkin mutants (Park et al. 2006). Mouse models deficient in PINK1 display mitochondrial dysfunction, but numbers of DAergic neurons and striatal levels of DA are similar to WT animals and parkin KO mice (Gispert et al. 2009). While this model may not successfully replicate PD features, it has increased awareness that complex interactions of PD-linked genes are likely involved in PD onset and progression.

34.2.1.5 *DJ-1*

While the prevalence is more rare than *PINK1* or *PARK2*, mutations in *DJ-1* can also lead to autosomal-recessive early-onset PD (Bonifati et al. 2004). This gene was originally found in two consanguineous PD pedigrees; one family carrying a deletion that prevented protein function, while the other contained a missense mutation leading to a proline-induced kink in an α-helix and accumulation in the mitochondria (Bonifati et al. 2003). DJ-1 has been shown to be expressed largely in the brain, localized in the cytosol, mitochondrial matrix, and intermembranous space (Zhang et al. 2005), and is suggested to act as an antioxidant and redox-dependent chaperone (Zhou et al. 2006). Extensive oxidation of DJ-1, as found with increased oxidative stress of the nigrostriatal axis and observed in the brains of PD patients, leads to its inactivation and eventual aggregation (Choi et al. 2006). Loss-of-function and double-deletion mutations in *Drosophila* yield no change in number of DAergic neurons compared to controls (Menzies et al. 2005; Meulener et al. 2005). As seen in *Pink1* and parkin KO mice, *DJ-1* KO mice show neither motor abnormalities nor the loss of DAergic neurons in the SNpc (Goldberg et al. 2005; Andres-Mateos et al. 2007); however, DJ-1 mutant mice show increased sensitivity to 1-methyl-4-phenyl-1,2,3,6 tetrahydropyridine (MPTP) toxicity (Kim et al. 2005). This suggests that a loss of DJ-1 function does not directly cause PD, rather susceptibility to parkinsonism increases due to mitochondrial dysfunction.

34.2.1.6 MitoPARK

Recently created, an innovative rodent animal model for PD uses conditional knockouts to fine-tune gene expression in a region-specific manner. In order to accomplish this, mice express cre recombinase (cre) under the control of a specific promoter, such as the TH promoter in order to target all catecholamine neurons (Gelman et al. 2003; Lindeberg et al. 2004) or the DAT promoter allowing DA neurons to be targeted (Zhuang et al. 2005; Backman et al. 2006; Turiault et al. 2007). The next step is to induce the deletion of a particular gene by breeding one of the cre-expressing mice with mice homozygous for a floxed gene (both chromosomal copies are flanked by LoxP sites). A variety of floxed genes could be used; however, in MitoPark mice the mitochondrial transcription factor A (*Tfam*) gene is deleted (Ekstrand et al. 2007). These mice have respiratory chain-deficient DAergic neurons due to the cell-specific eradication of TFAM, leading to degeneration of DAergic nerve terminals in the striatum and a progressive DAergic neurodegeneration in the SNpc; however, this is also observed in the ventral tegmental area (VTA). Nevertheless, this results in marked reduction in DA levels in the nigrostriatal system and an increase of DA turnover as observed in PD. In addition, MitoPark mice show motor impairments that are resolved following L-DOPA treatment; however, mice respond differently to the treatment depending on the progression of the symptoms, reminiscent of PD patients (Ekstrand et al. 2007). Finally, MitoPark mice exhibit inclusions within degenerating neurons similar to hallmark LBs of PD; however, these inclusions do not show SNCA immunoreactivity and form in MitoPark *Snca*$^{-/-}$ mice.

This inventive new model has a number of advantages over many other models. First, only DAergic neurons are targeted, similar to toxin models such as 6-hydroxydopamine (6-OHDA) and MPTP. Secondly, this model is reproducible and reliable with complete penetrance and negligible variability. Finally, the time course of the disease progression is slow, allowing more progressive and chronic neurodegeneration that parallels the disease course in human PD. Still, there exist some disadvantages. Neurodegeneration is not restricted to DAergic neurons of only the SNpc, but also within the VTA. The cost of animal care is extremely high, due to the slow development of the phenotypes (about 5 months), and the model requires a complex breeding scheme in which only 25 % of the offspring have the required genotype (Ekstrand and Galter 2009). Additionally, not yet determined in this model is the extent of induced inflammation or the effects of anti-inflammatory therapeutics. Nevertheless, this genetic model has pushed the boundary of the capability of animal models for PD. Hopefully, newer and more novel model constructs will gradually lead to better recapitulation of PD onset and progression.

34.2.2 Neurotoxic Models of PD

Some of the earliest attempts at creating animal models of PD utilized a variety of neurotoxins, rather than genetics, to reproduce the pathology, behavioral abnormalities, and molecular changes that characterize PD. The earliest animal models used neurotoxins that had actions specific to the DA system, such as 6-OHDA and MPTP (Schober 2004). Following associative studies linking pesticides with the incidence of PD (Ferraz et al. 1988; Ascherio et al. 2006), more recent toxin-induced models have used agents such as paraquat, rotenone, and maneb, as well as proteosomal inhibitors such as epoxomicin, which elicit a more general toxicity following systemic exposure. Through an exclusively inflammation-driven model, intoxication with lipopolysaccharide (LPS) can lead to SNpc-specific DAergic neurodegeneration in a chronic and progressive manner. While no model recapitulates every feature of the disease seen in humans, models utilizing environmental risk factors (neurotoxins) allow researchers to get a better understanding of the molecular events leading to the more prevalent form of parkinsonism, sporadic PD.

34.2.2.1 6-OHDA

The hydroxylated derivative of the neurotransmitter DA was used in the first animal model of PD over 40 years ago (Ungerstedt 1968). Because 6-OHDA is unable to cross the blood–brain barrier (BBB), systemic administration is ineffective necessitating the drug be administered by stereotactic injection at the median forebrain bundle (MFB), the striatum, or the SN in order to properly target the nigrostriatal DAergic pathway (Javoy et al. 1976; Jonsson 1983). Following injection, 6-OHDA irreversibly causes DA depletion following neuronal death through a combination of reactive oxygen species (ROS) and toxic quinines (Bove et al. 2005). Injection into the MFB or SN causes DA neurodegeneration within 24 h (Jeon et al. 1995), while intrastriatal injection leads to a progressive retrograde degeneration of nigrostriatal neurons (Berger et al. 1991; Sauer and Oertel 1994; Przedborski et al. 1995). The specificity of 6-OHDA towards catecholamine neurons is due to its preferential uptake by DA and noradrenalin transporters (Luthman et al. 1989). Intraneuronally, 6-OHDA remains in the cytosol, leading to the inactivation of a variety of macromolecules through the production of ROS (Cohen and Werner 1994). Similar to PD, the 6-OHDA model leads to the killing of DAergic neurons, while non-DAergic neurons remain unaffected; however, LB inclusions are not found in brain tissues of these animals (Bove et al. 2005).

Although the pathology induced by 6-OHDA does not accurately portray that of PD, the toxin is still widely used as a hemiparkinson rat model to produce changes in motor behavior. This model is created by a unilateral injection of the agent into the SN, leaving the contralateral side as a control (Ungerstedt 1971). In this model, systemic administration of a DAergic receptor agonist, such as apomorphine or amphetamine, yields a quantifiable circling behavior which is dependent on the extent of the lesion (Ungerstedt and Arbuthnott 1970). This quantifiable motor behavior has become a major asset for screening new therapeutic drugs (Jiang et al. 1993), as well as an assessment of gene therapy (Bjorklund et al. 2002) and regeneration of damaged DAergic pathways.

34.2.2.2 MPTP

In 1982, drug addicts developed irreversible and rapid parkinsonian symptoms that were indistinguishable from sporadic PD following the self-administration of what was thought to be synthetic heroin (Langston et al. 1983). MPTP was found to be the neurotoxic contaminant (Fig. 34.1), a by-product produced by accident during the illicit synthesis of 1-methyl-4-phenyl-4-propionoxypiperidine (MPPP) (meperidine, Demerol) in a basement laboratory (Davis et al. 1979). Years later, postmortem studies on these individuals showed severe loss of neurons in the SN, but an absence of LBs, which may have been due to the young age of the individuals as well as the acute exposure to the neurotoxin (Langston et al. 1999). This incident, while tragic, led to the utilization of MPTP to induce experimental DAergic neurodegeneration in rodent and

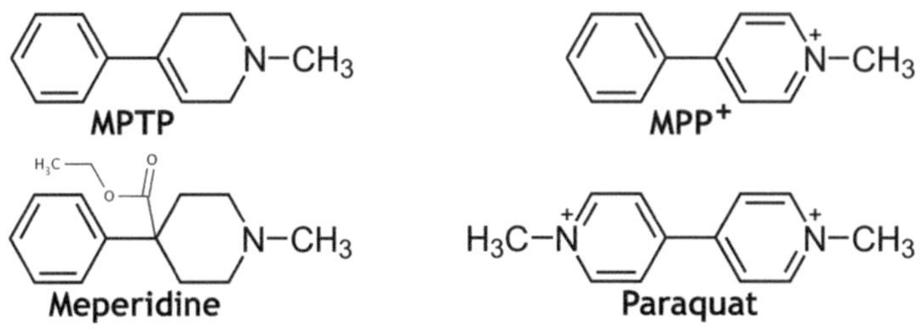

Fig. 34.1 Structural similarities between MPTP, meperidine, MPP^+, and paraquat. Only slight differences exist between meperidine and the contaminating substance MPTP, which has proved to be useful as a model of PD; while a second *N*-methyl-pyridinium group on paraquat is all that differs from the chemical structure for MPP^+

nonhuman primate animal models (Langston et al. 1984; Markey et al. 1984; Langston and Irwin 1986; Kopin and Markey 1988).

MPTP is a pro-toxin and is highly lipophilic, allowing it to easily cross the BBB after systemic administration (Markey et al. 1984). MPTP is then oxidized by monoamine oxidase B (MAO-B) in astroglia and serotonergic neurons forming the active toxic metabolite 1-methyl-4-phenyl-2,3-dihydropyridium ion (MPP^+). Following release into the extracellular space, MPP^+ is selectively transported into DA neurons by the DAT (Javitch et al. 1985; Mayer et al. 1986), where it may meet one of three fates. First, it can exist in the cytosol interacting with negatively charged cytosolic molecules (Klaidman et al. 1993). Second, it can be transported by the vesicular monoamine transporter-2 (VMAT2) and sequestered inside the synaptosomal vesicles (Liu et al. 1992). Finally, MPP^+ can become concentrated within the mitochondria of DA neurons due to the mitochondrial transmembrane potential (Ramsay and Singer 1986). While vesicular sequestration of MPP^+ protects cells from MPTP-induced death (Liu et al. 1992), transportation into the mitochondria allows MPP^+ to inhibit complex I of the electron transport chain (Nicklas et al. 1985). This inhibition quickly leads to decreased ATP production in the striatum and ventral midbrain (Chan et al. 1991; Fabre et al. 1999), increased neuronal injury, and neurodegeneration. In addition, complex I inhibition may cause widespread oxidative stress by disrupting the flow of electrons leading to the production of ROS, especially superoxide (Hasegawa et al. 1990, 1997). MPTP administration also induces a significant micro-glia reaction in the SNpc, immediately after intoxication whereby the microglial neuroinflammatory responses act to hasten dopaminergic neurodegenerative events and worsen lesion development (Liberatore et al. 1999). Postmortem studies showed the presence of this microglial-mediated inflammatory response many years after the initial exposure, suggesting that a primary insult may trigger a progressive process of neurodegeneration that is facilitated by focused neuroinflammation. Microglia not only become activated but also show increased expression of major histocompatibility complex (MHC) and inducible nitric oxide synthase (iNOS), and appear to multiply in number following MPTP insult, suggesting selective recruitment or proliferation (Kurkowska-Jastrzebska et al. 1999b). In addition, the substantial influx of $CD4^+$ and $CD8^+$ T lymphocytes into the SNpc and striatum in PD patients and MPTP-treated mice (Kurkowska-Jastrzebska et al. 1999a; Benner et al. 2004, 2008; Brochard et al. 2009; Reynolds et al. 2010) as well as the loss of susceptibility to MPTP in immune- and T cell-deficient mice, and the restoration of MPTP susceptibility by the reconstitution of total lymphocyte and $CD4^+$ lymphocyte (Benner et al. 2008; Brochard et al. 2009) suggest a significant role for the adaptive immune system and T cells in dopaminergic neurodegeneration. A substantive amount of data clearly demonstrate that induced and natural $CD4^+$ regulatory T cells (Tregs) have the capacity to ameliorate MPTP-induced microglial inflammatory responses and provide protection to virtually the entire nigrostriatal dopaminergic axis; including dopaminergic nigral neuron bodies, as well as striatal termini

(Benner et al. 2004, 2008; Boska et al. 2005, 2007; Laurie et al. 2007; Reynolds et al. 2007, 2008, 2009a, b, 2010; Kosloski et al. 2010; Hutter-Saunders et al. 2011). MPTP-induced inflammation is also responsible for modifying encephalic proteins, such as α-synuclein and ubiquitin that drain along with inflammatory mediators to the lymphatic tissues and activate antigen presenting cells, which present modified self-antigens to T cells, lead to evasion of self-tolerance, and the production of antigen-specific effector T cells (Teffs) (Benner et al. 2008; Reynolds et al. 2009a, 2010). For instance, nitrated-α-syn (*N*-α-syn)-specific Th1 and Th17 proinflammatory Teffs exacerbate MPTP-induced inflammatory responses and worsen the lesion. Moreover, *N*-α-syn-specific Teffs diminish the suppressive capabilities of natural Tregs; however, in the MPTP model, these Teffs in the presence of exogenously induced Tregs potentiate Treg function and neuronal protection. Thus, these data demonstrate that the innate and adaptive immune systems and their effects on each other may well influence neurodegeneration and neuroprotection in the MPTP model.

The MPTP animal model replicates most of the pathological features in PD with most comparative data derived from studies in monkeys (Forno et al. 1993), as very few human MPTP cases have been autopsied. As in PD, the monkey model provides preferential degeneration of DAergic nerve terminals in the putamen over those in the caudate nucleus (Moratalla et al. 1992). Studies in both MPTP-intoxicated monkeys and mice show a preponderance of DAergic neurodegeneration more extensive in the SNpc than the VTA (Herrero et al. 1993). MPTP-intoxicated monkeys respond well to L-DOPA and DA receptor agonist therapies (Langston and Irwin 1986); however, other hallmarks of PD such as intraneuronal LB inclusions are mostly absent. Chronic MPTP infusion in mice using minipumps has been shown to generate intracellular inclusion bodies, although these are different than the LBs found in human brain tissue (Fornai et al. 2005). Nevertheless, the use of MPTP in mice shows pathological dysfunction in the enteric nervous system (ENS) that is also observed in PD patients. More recent studies have shown that MPTP intoxication decreases the number of TH-positive enteric neurons in these animals (Anderson et al. 2007; Natale et al. 2010), suggesting that the MPTP mouse model may be valuable for exploration of extranigral pathology in PD.

Studies addressing therapeutic assessment with the MPTP model have utilized nonhuman primates. A major cause of PD motor dysfunction was found to be due to hyperactivity in the subthalamic nucleus using electrophysiology to study MPTP-treated monkeys (Bergman et al. 1990). This original finding led to the creation of the chronic high-frequency stimulation procedure (deep brain stimulation) that is currently used to improve motor functions in PD patients who no longer respond to DA replacement therapy (Limousin et al. 1998). The use of MPTP in rats has proven to be of decreased value, as injections of MPTP at comparable doses used in mice and primates show little to no neurodegeneration in rats (Giovanni et al. 1994), possibly due to differences in levels of VMAT2. While mice are susceptible to MPTP, they are less sensitive than primates, thus require much higher equivalent doses per weight (Przedborski et al. 2001). While the MPTP monkey model is valuable for the assessment of novel therapies, a variety of practical considerations makes the mouse MPTP model more widely used in studies that assess the molecular mechanisms of DAergic neurodegeneration.

34.2.2.3 Paraquat and Maneb

The herbicide 1,1′-dimethyl-4,4′-bipyridinium (paraquat) has a structure comparable to MPP^+ (Fig. 34.1), suggesting this chemical should be added to the list of possible risk factors for PD (Di Monte et al. 1986). While being structurally similar, paraquat does not localize in DA neurons following systemic administration (Widdowson et al. 1996a, b), due in part to its lack of selectivity for DAT and difficulty in crossing the BBB (Shimizu et al. 2001). Regardless, systemic administration in mice triggers DAergic neurodegeneration in the SNpc as well as induces SNCA-containing inclusion formation (Manning-Bog et al. 2002; McCormack et al. 2002), which is thought to be due to the creation of superoxide radicals (Day et al. 1999). In addition, reduction in striatal DA nerve terminals and motor behavior changes can be observed after paraquat administration (Brooks et al. 1999). Manganese ethylenebis-dithiocarbamate (maneb) is a fungicide that is a suspected neurotoxic agent for DAergic neurons. The use of maneb and paraquat in combination amplifies neurotoxic effects on the DAergic system in mice (Thiruchelvam et al. 2000b). Furthermore, epidemiological studies have observed that the use of maneb and paraquat together in similar geographical areas is widespread, suggesting that large numbers of people may be exposed to the combined action of these neurotoxins (Thiruchelvam et al. 2000a). While paraquat may not have DAergic-specific action, it does replicate a number of PD features following relatively simple systemic administration and should prove to be a useful model to study the molecular underpinnings of human PD.

34.2.2.4 Rotenone

A natural cytotoxic chemical extracted from the root of *Lonchocarpus* spp., a genus of tropical plants, rotenone has widespread use as both a fish poison and insecticide. It is highly lipophilic and can therefore rapidly cross cellular membranes including the BBB (Talpade et al. 2000). Acting in a manner similar to MPP^+, rotenone binds and inhibits complex I in the mitochondria preventing the movement of electrons from complex I to ubiquinone in the electron transfer chain. Unlike MPP^+, rotenone is only mildly toxic towards humans, and exhibits a short half-life in the environment due to its instability as a compound. Chronic low-dose intravenous administration of rotenone to rats creates features of PD, including slowly progressing DAergic neurodegeneration (Betarbet et al. 2000) as well as intracellular inclusions similar to LBs, both being immunoreactive for SNCA and ubiquitin (Betarbet et al. 2000; Sherer et al. 2003). Since rotenone can enter most cells, preferential susceptibility of DAergic neurons is thought due to the predisposition of those cells to oxidative stress.

In contrast, acute rotenone intoxication spares DAergic neuron loss (Ferrante et al. 1997), while causing reductions in other CNS neuronal populations (Hoglinger et al. 2003). These findings question the concept of a preferential sensitivity of DAergic neurons and hint that rotenone may exert more pervasive neurotoxicity. Indeed, observations of abnormal motor behaviors in rotenone-injected rats that show either no loss or insignificant loss of nigral DAergic neurons support the effects of rotenone beyond the nigrostriatal axis (Lapointe et al. 2004) and suggest that rotenone-induced damage to non-nigral neurons may result in motor dysfunction. While the rodent rotenone model recapitulates a number of pathophysiological features of human PD, including the loss of gastrointestinal myenteric neurons (Drolet et al. 2009), the use of rotenone results in relatively low levels of DAergic neurodegeneration, is technically difficult leading to low reproducibility (Betarbet et al. 2000), and results in the death of many animals from acute toxicity that is distinct from CNS dysfunction.

34.2.2.5 Epoxomicin

A more recent rodent model targets the proteasome system to induce neuronal neurodegeneration. Following systemic administration of the proteosomal inhibitor, epoxomicin, rats displayed a number of key features of PD such as DAergic neurodegeneration within the SN, intraneuronal aggregates that were immunoreactive for α-syn and ubiquitin, as well as reduced DAergic termini in the striatum (McNaught et al. 2004). Conversely, a separate independent study found that systemic administration of epoxomicin was ineffective in both rats and monkeys (Kordower et al. 2006). Further studies have found that stereotactic injection of proteosomal inhibitors blocks DAergic neuronal death in animals that had received MPP^+ or rotenone, but generated α-syn inclusions in those neurons (Sawada et al. 2004). These inconsistent findings necessitate further study into the use of proteosomal inhibitors as a model of PD in animals.

34.2.2.6 LPS

Postmortem analysis of PD brains, as well as a number of epidemiological studies, suggests that neuroinflammation plays a large role in the progression of PD (McGeer et al. 1988; Liu 2006). Endotoxins, such as LPS, have been shown to drive a microglial immune response, leading to widespread inflammation. As neurons lack expression of TLR4 (Lehnardt et al. 2003), the major source of LPS-induced signal transduction, LPS has no apparent direct effect on DAergic neurodegeneration; therefore, resulting PD pathogenesis in an animal model given a dose of LPS is thought to be solely inflammation-driven. Multiple forms of the LPS-induced model have been created to directly study the effects of inflammation on the induction of nigrostriatal dopaminergic neurodegeneration. The earliest in vivo LPS animal models used a single injection of microgram quantities of LPS into the SN of rats, leading to pronounced loss of DAergic neurons in the SNpc (Lu et al. 2000). Injection of this endotoxin causes an acute, but irreversible, loss of neurons, while levels of non-DA neurons in the SNpc as well as DAergic neurons in the VTA remain unaffected (Castano et al. 1998; Liu et al. 2000).

In order to produce a more chronic period of inflammation, in hopes of more progressive neurodegeneration, nanogram amounts of LPS were delivered to rats using a mini-pump (Gao et al. 2002). No loss of DAergic neurons was observed during the 2-week dosing period and only a slight loss was detected by 4 weeks; however, by 6 weeks, significant losses of nigral DAergic neurons were detectable (McCoy et al. 2006). This suggests that short-lived neuroinflammation occurring early in life may lead to a delayed, but progressive, DAergic neurodegeneration in the SNpc.

Finally, another LPS animal model aims to understand if the presence of endotoxin has any effect on the fetal development of the nigrostriatal DA system. This concept stems from the findings that some women suffer from cervical bacterial infections during pregnancy, possibly leading to LPS and other toxins having an effect on the development of the fetus (Romero et al. 1989; Dammann and Leviton 1997). In this model, LPS is given to pregnant rats through in utero injection at gestation day 10.5 (Ling et al. 2002). Following birth, these pups show significant decreases in numbers of nigral DAergic neurons and DAergic striatal termini. Furthermore, in utero exposure of LPS increases susceptibility of the pups to other neurotoxins, such as 6-OHDA (Ling et al. 2004). Overall, the LPS PD model may yet become a vital system to understand the underlying mechanisms leading to neuroinflammation and ultimately neurodegeneration. Since this model is purely inflammation-driven, it is unique among the numerous existing animal PD models.

34.3 Animal Models of ALS

ALS is an adult-onset, quickly progressing neurodegenerative disease with selective loss of motor neurons in the motor cortex, brainstem, and spinal cord that leads to paralysis and eventual death. The majority of cases are sporadic (sALS), without any clear family history. About 5–10 % of ALS patients inherit the familial form of the disease (fALS) in an autosomal-dominant fashion. Both fALS and sALS patients present with muscle weakness and atrophy, as well as spasticity due to the loss of both upper and lower motor neurons. Following paralysis and the loss of respiratory muscle function, death typically ensues within 3–5 years after initial clinical signs. Due to numerous possible etiologies of sALS, most of the current research models focus on genetic etiologies. Still, the majority of early research was performed using observational or spontaneous cases of ALS in both humans and animals, until technology and knowledge allowed more complex models with genetic manipulation (Table 34.2).

34.3.1 Early Environmental Models

Early ALS models were based on environmental factors, some of which were already known to cause the disease in humans. ALS-like features can be induced in animals following intoxication with heavy metals such as lead, mercury, and aluminum, or through a deficiency of other metals such as calcium and magnesium. Lead poisoning has long been known to cause ALS-like syndromes in humans (Conradi et al. 1982). These patients show a variety of ALS features that can

Table 34.2 Rodent models of ALS

Model	ALS pathology	ALS symptoms	Advantages	Disadvantages
Environmental				
Metals (Pb, Hg, Al, Ca/Mg)	Neurofilaments loss of anterior horn motor cells	Motor impairment	Easy to administer through a variety of routes	Large variability within results
Toxins (IDPN)	Neurofilaments spheroids	Slight motor impairment	Duplicates spheroid formation in ALS	Only slight motor impairment
Genetic				
Wobbler	Degeneration of motor neurons	"Wobbly" gait	Progressive motor weakness	Widespread degeneration
		Progressive motor weakness	Motor neuron loss	Non-ALS features
Nmd/Pmn	Progressive loss of motor neurons	Progressive paralysis	Displays ALS-like features in mice	Human mutation causes different disease, not ALS
	Muscle atrophy	Short life-span		
SOD1	Loss of motor neurons	Hindlimb weakness	Well characterized	Therapeutic results do not translate to human disease
	Neurofibrils formed	Paralysis and death	Good construct validity	
FUS/TLS	No loss of motor neurons	Early death	Promising results in *Drosophila*	Very early model
	Progressive neurodegeneration?			No rodent models available yet
TDP-43	Loss of motor neurons	"Swimming" gait	Displays ALS-like features in mice	Results from overexpression, not mutation
	Inclusions formed	Weight loss		
VAPB	No motor loss	No motor phenotypes	Possible interactions with TDP-43	No features of ALS
	TDP-43 protein accumulations	Normal survival		
Dynactin	Loss of motor neurons	Motor phenotypes	Displays ALS-like features in mice	Early model
	Inclusions formed			Homozygous model embryonic lethal

Source: Adapted from Terzioglu and Galter (2008)

be improved by chelating agents. Inversely, classical ALS patients, without lead exposure, fail to benefit from these agents (Mitchell 1987). Combault, who performed some of the first studies on lead intoxication of animals, found widespread loss of myelin in the peripheral nerves of guinea pigs (Combault 1880). Twelve years later Nissl described atrophy of anterior horn motor cells in lead-intoxicated rabbits (Nissl 1892). More recent studies in rats and rhesus monkeys showed demyelination of peripheral nerves, along with intranuclear lead inclusion bodies in the anterior horn cells as well as in the SN.

A syndrome has been described that mimics ALS in patients exposed to organic or inorganic mercury (Barber 1978; Adams et al. 1983). This led to various animal models assessing the effects of mercury on motor neurons (Arvidson 1992; Pamphlett et al. 1998; Pamphlett and Kum-Jew 2001). These animals showed vacuole formation in dorsal ganglia, as well as motor dysfunction such as weakening of the hind legs and paresis at later stages, yet showed no lesions in anterior horn motor neurons. Aluminum has also been used as a model of neurofibrillary degeneration. Administration of this metal leads to large numbers of neurofibrillary tangles, ALS-like symptoms, and the development of neurofilament-containing spheroids in ventral horn axons (Troncoso et al. 1982). Epidemiological studies of metals and trace elements in Western Pacific populations, wherein a high incidence of motor neuron disease exists, showed not only excess levels of certain metals but also environmental deficiencies of calcium and magnesium (Garruto et al. 1984; Yase 1987). Rats fed a low Ca/Mg diet showed degeneration of anterior horn cells as well as muscle atrophy reminiscent of ALS (Nakagawa et al. 1977), suggesting that environmental deficiencies cause mobilization of these metals from the bone and eventually lead to motor neuron disease.

In addition to using metals to induce ALS-like pathologies in animal models, the administration of toxins such as β,β1-iminodipropionitrile (IDPN) induces similar pathologies by disrupting normal motor axonal transport. Rats that received IDPN show accumulations of neurofilament in proximal axons, leading to the creation of spheroids (axonal swelling) and atrophy of the distal portions of the axon (Clark et al. 1980). Spheroid formation in this model resembles that in the tissues of ALS patients (Griffin and Price 1980). While these early environmental factor-based models replicated some features of ALS, they have been largely supplanted in favor of newer, more relevant genetic models that more similarly recapitulate the disease.

34.3.2 Genetic Models

Just as researchers found correlations of ALS-like symptoms with metal toxicity, much of the early genetic models were based on observations of disease progression and pathologies similar to ALS. The Wobbler mouse resulted from a spontaneous mutation at the Institute of Animal Genetics in Edinburgh in the late 1960s. These mice show progressive motor weakness along with a "wobbly" gait and degeneration of motor neurons of the brainstem and spinal cord that varies with genetic background (Duchen and Strich 1968). While initially considered a good ALS model, primary neurodegeneration was found to be more widespread than the motor neurons (Rathke-Hartlieb et al. 1999) and demonstrated disrupted spermatogenesis (Heimann et al. 1991); features not presented in human ALS.

Other early genetic models include the "Neuromuscular degeneration" (Nmd) and "Progressive motor neuronopathy" (Pmn) mouse models, which show ALS-like symptoms due to spontaneous mutations. Nmd mice possess an autosomal-recessive mutation causing loss of motor neurons that eventually leads to muscle atrophy (Cook et al. 1995). These mice become progressively paralyzed, and survive no longer than 4 weeks following birth. Pmn mice show progressive neurodegeneration in motor axons, leaving cell bodies intact and also succumb early, rarely surviving beyond 6–7 weeks (Schmalbruch et al. 1991). Nonetheless, genetic mutations of the same causative genes in humans cause diseases and symptoms much different than those presented in Nmd and Pmn mice, suggesting these models do not serve as entirely accurate models of ALS, but may prove useful in other diseases.

Following major breakthroughs in genetics, a number of newer genetic models were created that exploited mutations discovered in the genomes of patients with typical Mendelian fALS. The fact that these mutations were already known to cause ALS in humans pushed these models ahead of earlier models. As more mutations and associative genes in ALS were discovered, researchers induced similar models or genetically transferred human mutations to animals. Genes associated with risk of ALS that have been utilized in model systems include *SOD1*, fused in sarcoma/translocated in liposarcoma (*FUS/TLS*), transactivation response (*TAR*) DNA-binding protein with molecular weight 43 kDa *TDP-43*, vesicle-associated membrane protein (*VAMP*)-associated protein B (*VAPB*), and Dynactin (*DCNT1*); most of which, in part, recapitulate the pathology and phenotypes of ALS.

34.3.2.1 Superoxide Dismutase 1

One of the first mutations found in fALS was a G93A substitution in the *SOD1* gene. Researchers created the first transgenic ALS model by overexpressing this mutated form of human *SOD1* in mice (Gurney et al. 1994). The resulting phenotype consists of progressive hind limb weakness that leads to paralysis and eventually death, almost flawlessly paralleling the human disease. In contrast, no phenotype is observed in mice that overexpress WT *SOD1*. Following the original G93A mutant, a number of other mouse models were created utilizing other mutations including G27R, G85R, G86R, and D90A (Joyce et al. 2011); all of which show similar phenotypes of the G93A mutant suggesting a toxic "gain of function."

The normal WT *SOD1* enzyme uses copper to convert superoxide into oxygen and hydrogen peroxide, while the loading of copper within the enzyme is performed by a specific copper chaperone (CCS). In order to determine if disruption of oxidative function is the cause of *SOD1* toxicity, researchers crossbred G93A mutants with KO mice lacking CCS (Subramaniam et al. 2002). They found no change in life-span, signifying the initiating factor of toxicity is something other than oxidative stress in *SOD1* mice. Moreover, site-specific substitution of copper-binding His residues to inhibit Cu^{2+} complex formation also failed to slow disease progression, indicating that copper-binding ions do not play a significant role in the toxic "gain of function." In addition, overexpression of mutant forms of *SOD1* in either motor neurons or glial cells alone does not lead to motor phenotypes (Gong et al. 2000; Pramatarova et al. 2001); suggesting interplay between the cell types is required. In order to demonstrate the necessity of nonneuronal involvement, a study used a transgenic mouse containing a floxed G37R *SOD1* gene under the action of Cre recombinase driven by either Islet-1 transcription factor (*Isl1*), a homeodomain transcription factor expressed in motor neurons and dorsal root and cranial sensory ganglia, or CDllb expression by microglia (Boillee et al. 2006). Expression of *SOD1* by motor neurons leads to early phase disease onset and disease progression, whereas expression of *SOD1* by microglia has little effect on early phase disease, but significantly affected late disease progression. These data serve to illustrate the distinct phases that can be defined by the action of *SOD1* expression by different cell types, and that the latter supports a non-cell-autonomous mechanism of motor neurodegeneration.

The original G93A mutant mouse has been the mainstay in ALS research; however, it has recently come under scrutiny. Some researchers have suggested that the phenotype seen with these mice results from the overexpression of SOD1 protein (24 copies of the *SOD1* gene) rather than the effect of the mutation (Shibata 2001). One model derived from the G93A strain, known as the G93Adl or *G1del* line, surfaced following spontaneous loss of G93A *SOD1* copy number while transferring the transgene to a congenic strain at the Jackson Laboratory (Teuling et al. 2008). These mice show progressive loss of motor neurons, astrogliosis, and degeneration of fiber tracks, but carry only 8–10 copies of the human G93A mutant gene, and show delayed progression of phenotype beginning at approximately 30 weeks of age (Alexander et al. 2004); a phenotype that is more similar to human ALS in comparison to the original G93A mutant that displays early onset at around 90 days. Thus, these mice may more accurately model ALS, or at the very least prove advantageous in therapeutic studies aimed at early-stage or presymptomatic familial cases.

Assessment of new ALS therapeutic strategies has almost exclusively used the mutant *SOD1* model. Transgenic G93A and H46R *SOD1* mutations in rats, with their larger size, have allowed the evaluation of complex procedures involving the spinal cord, such as implantation of intrathecal catheters for chronic therapy. However, *SOD1* mutation models may not be appropriate for evaluating therapeutic modalities. Minocycline, a third generation tetracycline, inhibits microglial activation and showed diminished levels of activated microglia and increased survival in three separate studies using minocycline to treat different mutant *SOD1* models (Kriz et al. 2002; Van Den Bosch et al. 2002; Zhu et al. 2002). Nevertheless, a subsequent clinical study showed no efficacy and was actually deemed harmful in human subjects of sALS (Gordon et al. 2007). This questions the validity of the *SOD1* models; however, a number of possibilities may be responsible for failed clinical trials. As the immune and neurological systems of mice and man differ largely, the effects in mice may have been small enough that similar effects could be missed in a sizable and genetically heterogeneous human cohort, or may only be efficacious in familial cases. Additionally, minocycline treatment was initiated before disease onset in mice, whereas clinical trials used patients who already had ALS. All together, the *SOD1* mouse will likely remain the mainstay of ALS research for some time, until the discovery of a model that can better recapitulate both familial and sporadic forms of ALS and show a successful translation of therapeutics from rodent to human.

34.3.2.2 Fused in Sarcoma/ Translocated in Liposarcoma

In 2009, two independent studies reported missense mutations (R521C and H517Q) in the *FUS/TLS* gene that was associated with 4–5 % of the patients with fALS (Kwiatkowski et al. 2009; Vance et al. 2009). No *FUS/TLS* mutations have been found in 293 sporadic cases surveyed. The FUS/TLS protein is thought to regulate transcription and control RNA metabolism; as a result KO mice die shortly after birth and show abnormal neuronal spine morphology, but no loss of motor neurons (Hicks et al. 2000). At the time of this writing, no transgenic rodent models are available that overexpress WT or mutant *FUS/TLS*, although there have been initial reports of mutants in yeast and findings of mutations in *Drosophila* that lead to progressive neurodegeneration (Chen et al. 2011; Ju et al. 2011). However, whether this model fully reproduces the human disease in animals has yet to be determined.

34.3.2.3 Transactivation Response DNA-Binding Protein with Molecular Weight 43 kDa (TDP-43)

Initially, TDP-43 was found to be a major component of intraneuronal inclusions in tissues from sALS patients (Arai et al. 2006; Neumann et al. 2006). Eventually, 33 different dominant mutations of the gene were found in 22 unrelated families as well as 29 sporadic cases (Banks et al. 2008). These missense mutations occur in the C-termini of the protein, and are present in less than 5 % of familial cases (Lagier-Tourenne and Cleveland 2009). The precise role of TDP-43 is not fully understood, although it is thought to act with FUS/TLS in a variety of RNA processes (Buratti and Baralle 2008). Under the control of the mPrP, overexpression of A315T mutant TDP-43 in mice leads to a "swimming" gait and weight loss after 3 months (Wegorzewska et al. 2009), which are consequences of aggregate protein formation and subsequent degeneration of axons in the dorsal corticospinal tract and lateral columns.

Whether these deleterious results were due to overexpression of TDP-43 or expression of the mutation is contentious, since mice expressing the WT TDP-43 under control of the mPrP have not been produced, yet overexpression of the WT gene under the Thy-1 promoter results in loss of motor neurons that leads to gait dysfunctions as well as nuclear and cytoplasmic inclusions (Wils et al. 2010). This suggests that overexpression of TDP-43, and not the mutation, may be sufficient for neurotoxicity. While the current state of this model may be of limited use for fALS studies, future development of a more sophisticated methodology (such as a knock-out/knock-in model) may solve this issue and create a superior rodent model that better recapitulates ALS disease features.

34.3.2.4 Vesicle-Associated Membrane Protein-Associated Protein B

A slow progressing, late-onset atypical form of ALS was observed in a Brazilian family, which was discovered to be associated with a missense mutation in the *VAPB* gene (Nishimura et al. 2004). Transgenic mice overexpressing the WT or the P56S mutant form of *VAPB* in the brain show neither motor deficit phenotypes nor changes in survival; however, they do show accumulations of TDP-43 in the cytoplasm of motor neurons (Tudor et al. 2010). In *Drosophila*, transgenic expression of a mutant homologue of the human VAPB gene (DVAP-33A) leads to a number of the hallmarks seen in human disease, including neuronal death and aggregate formation, as well as motor dysfunction (Chai et al. 2008). This suggests a putative interplay between these gene products, and continued investigation into these effects, and the effects of other gene combinations, may prove beneficial to ALS research.

34.3.2.5 Dynactin

Members of a European family were afflicted with an autosomal-dominant, late-onset motor neuron disease, which was determined to be associated with a G59S mutation in a binding domain of *DCNT1* (Puls et al. 2003). This "loss-of-function" mutation leads to the functional disruption and ultimate aggregation of the protein, dynactin, in cell culture studies (Levy et al. 2006). Hemizygous G59S Dynactin mice present ALS-like phenotypes after 10 months of age with loss of spinal motor neurons following a massive accumulation of proteins at neuromuscular junctions (Lai et al. 2007). However, further examination of this system is necessary to assess its utility as a model of ALS.

34.4 Sample Protocol: Acute MPTP Intoxication

It would be impossible to outline the steps involved in creating each and every animal model discussed above in a single chapter. Therefore, this sample protocol is provided for one of the more commonly used animal models of PD, the acute MPTP mouse model. The experiment involves two groups, PBS- and MPTP-treated. The effect MPTP intoxication has on DAergic neurons is assessed 7 days postinjection. Although briefly discussed here, excellent reviews describing the MPTP protocol and safety procedures in greater detail should be consulted prior to initiating studies with this neurotoxic agent (Przedborski et al. 2001; Jackson-Lewis and Przedborski 2007).

WARNING: Special care should be taken when handling MPTP as it is highly neurotoxic. The period for high-risk contamination from MPTP is from the initial injection until the time that MPTP and its metabolites are no longer found in the excreta of treated animals, which is typically after 3 days. Nevertheless, in the interest of precaution this period of high-risk should be increased to 5 days post-MPTP injection. Potential risks of exposure to MPTP are through direct contact with the animal, inner surfaces of the animal cage, bedding material, and animal dander. The experiment should be performed in a procedure room designated only for MPTP use, containing a chemical fume hood and enclosed glove box. During the experiment, two trained investigators should be present in the room at all times. Personal protective equipment required for the experiment includes the following: Tyvek® coverall, chemical goggles, double gloves, double shoe covers, and a chemical respirator fitted with HEPA filter cartridges. MPTP can be purchased from Sigma Chemical Company (St. Louis, MO) as the nonvolatile, hydrochloride salt conjugate. MPTP should be dissolved in solvent when used and only in quantities necessary for the daily procedure, as MPTP oxidizes at room temperature and must be discarded after 1 day.

MPTP can be neutralized by using a 1 % bleach (sodium hypochlorite) solution in water.

34.5 General Procedure

1. Move mice into the designated MPTP procedure room at least 2 days before injection to allow the animals to acclimate and stabilize dopaminergic processes that may be altered due to changing to a new environment. Animal health checks will need to be performed daily through the end of the experiment.
2. On the day of MPTP injection, record the weight of each mouse. The identity of each mouse can be denoted by making small marks on their tail with a permanent marker (e.g., three marks for mouse #3). These weights will be used to calculate the amount of MPTP solution needed and to determine the volume for injection in each mouse. Table 34.3 provides a sample of the weights table used for this example protocol.
3. Using the recorded mouse weights, calculate the total amount of MPTP-HCl to weigh and the amount of PBS necessary in which to dissolve the salt. Table 34.4 provides a sample calculation spreadsheet for a dose of 18 mg of MPTP/kg mouse body weight (18 mg/kg) of the mice weighed in step 2. Note that MPTP dosages are based on the amount of MPTP free base from the MPTP-HCl salt; thus, more MPTP-HCl is required to obtain the correct dose of MPTP as free base. For instance, to deliver a dose of 18 mg/kg of MPTP free base, 21.12 mg/kg of MPTP-HCl should be utilized.
4. Weigh out the appropriate amount of MPTP-HCl (note calculations involving free base vs. salt conjugate) and dissolve in an appropriate volume of PBS. The volume of PBS utilized should provide the dose of MPTP-HCl to be delivered in 10 mL/kg (10 μL/g of body weight). For instance, to deliver an 18 mg/kg dose of MPTP (21.12 mg/kg dose of MPTP-HCl), a 25 g (0.025 kg) mouse should receive 250 μL of a 2.112 mg/mL solution of MPTP-HCl in PBS. The solution should be sterile or passed through a 0.22 μm filter prior to injection. All manipulations involving MPTP powder and filter sterilization should be performed within a glove box and in the presence of at least one other trained person. After the MPTP solution is prepared, decontaminate all materials that came in contact with the powder using 1 % bleach.
5. For acute MPTP intoxication, a total of four injections are given over the course of 8 h. Injections are spaced by 2 h and given intraperitoneally (i.p.) or subcutaneously (s.c.), while alternating sides (i.e., first injection in the right abdominal area, second in the left) to minimize injection-site irritation. Control mice are injected with PBS using a volume of 10 mL/kg body weight (10 μL/g) (see Injection Volume, Table 34.3). Mice receiving MPTP are injected in a chemical fume hood with the prepared MPTP solution in a volume to deliver 10 mL/kg body weight (10 μL/g). To inject, hold the mouse over the cage or absorbent bench paper to absorb excreta or any dropped or spilled MPTP solution. After MPTP injections are

Table 34.3 MPTP protocol: Mouse Weights

Treatment	Cage	Mouse ID	Weight (g)	Injection volume (μL)
PBS	1	1	24.5	250
		2	19.8	200
		3	22.4	220
MPTP	2	1	21.8	220
		2	23.1	230
		3	20.6	210

Table 34.4 MPTP protocol: calculation worksheet

Total mice (MPTP+PBS)		MPTP calculations	
Number	6	MPTP free base (mg/kg)	18.0
Total weight (g)	132	MPTP-HCl (mg/kg)	21.12
Average weight (g)	22.0	Theoretical (mg)	5.53
		Overage 10 % (mg)	6.09
Total MPTP -treated mice		Concentration of MPTP/PBS (mg/mL)	2.11
Number	3	Volume PBS for solution (mL)	2.9
Total weight (g)	66	Actual amount weighed (mg)	6.5
Average weight (g)	21.8	Actual volume of PBS (mL)	3.1

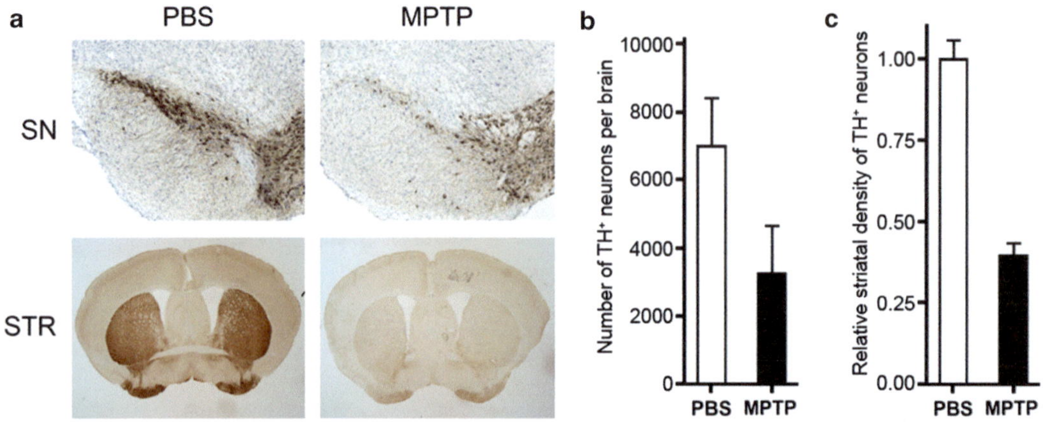

Fig. 34.2 Nigral and striatal degeneration following MPTP intoxication. (**a**) Photomicrographs of TH-immunostained sections of the substantia nigra (SN) and of the striatum (STR) from mice treated with PBS or MPTP, obtained on day 7 posttreatment. (**b**) Stereological quantification of surviving TH^+ neurons in the SNpc of mice treated with PBS or MPTP. Values represent mean number of TH^+ neurons ± SEM. (**c**) Densitometry analysis of TH^+ termini within the striatum. Values represent mean relative percentage to PBS striatal density ± SEM. (LM Kosloski and HE Gendelman, University of Nebraska Medical Center, unpublished data, 2013)

complete, decontaminate all materials that came in contact with the MPTP solution using 1 % bleach and dispose as biohazardous material.

6. MPTP-injected animals should remain in the same cage for at least 5 days postinjection unless otherwise stated in an approved IACUC protocol. After 5 days, mice can be transferred to new cages, and the old cages must be decontaminated. To decontaminate the cages, spray the inside and outside of cages with 1 % bleach, and soak soiled bedding for 24 h in a volume of 1 % bleach to cover the bedding before disposing of the bedding as biohazardous material and returning the cages for washing.

For this example, MPTP- and PBS-treated mice were terminally anesthetized with 200 mg/kg of sodium pentobarbital i.p. on day 7 posttreatment and transcardially perfused with PBS followed by 4 % paraformaldehyde in PBS. Brains were removed and post-fixed for 24 h in 4 % paraformaldehyde/PBS. Fixed brains were cryopreserved in 30 % sucrose/PBS for 48 h, embedded in optimal cutting temperature (OCT) medium, and maintained at −80 °C prior to sectioning. OCT-embedded brains were cut in a "cryostat" to yield 30 μm thick sections. Sections were collected as free-floating tissues and immunostained with rabbit IgG anti-TH antibody followed by biotinylated goat IgG anti-rabbit IgG and peroxidase-conjugated biotin–avidin complex (ABC Kit #PK4000, Vector Laboratories). Tyrosine hydroxylase-expressing nigral neurons or striatal termini were visualized with 3,3′-diaminobenzidine (DAB). Immunostained sections were counterstained with 0.1 % thionin/acetate buffer for Nissl bodies, dehydrated, cleared, dried, and covered with a glass coverslip. Numbers of TH^+ $Nissl^+$ neurons in the substantia nigra were evaluated by stereological analysis (see Chap. 4) and the densities of TH^+ neuronal termini in the striatum were assessed by digital image analysis. Immunohistochemistry revealed that by 7 days post-MPTP intoxication, numbers of TH^+ neurons within the SNpc and TH^+ termini in the striatum were noticeably depleted in the MPTP-treated mice compared to the PBS controls (Fig. 34.2a). Stereological analysis confirmed observations in the SN by demonstrating that the numbers of TH^+ $Nissl^+$ neurons in the SN from MPTP-treated mice were over 60 % depleted compared to those from PBS control mice (Fig. 34.2b). Similarly, densitometric analysis revealed that the density of the TH^+ termini in the striatum was lost after MPTP treatment compared to PBS controls (Fig. 34.2c).

34.6 Summary

A number of animal models exist for PD and ALS that exploit either genetic or environmental factors, allowing the study of the processes underlying disease pathology or the assessment of novel therapeutic strategies. Initially, animal models were created from correlative findings linking environmental toxins with disease pathogenesis. For PD, early toxin-based models (6-OHDA and MPTP) targeted the DA system directly, while more recent environmental models result in more general toxicity (paraquat, rotenone, or epoxomicin) or recapitulate DAergic neurodegeneration driven by inflammation (LPS). These have proven quite useful for studying the nigrostriatal disruption in idiopathic PD. Less helpful have been the ALS toxin-based models, which attempted to use intoxication with heavy metals (lead, mercury, or aluminum) or deficiencies of other metals (calcium and magnesium), and have been abandoned in favor of newer, more relevant genetic models that more accurately simulate features of the disease.

For both PD and ALS, mutations discovered in the genomes of patients with a familial form of the disease have driven the creation of genetic models that exploit the overexpression, knockdown, or mutation of a variety of disease-linked genes. Gain-of-function mutations and overexpression of autosomal-dominantly inherited genes (*SNCA* or *LRRK2*) or knock-out and loss-of-function mutations of autosomal-recessive genes (parkin, *PINK1*, or *DJ-1*) have resulted in DAergic neuronal loss, LB inclusions, and minor motor disruption; consistent with the pathological hallmarks of PD. Additionally, region- or neuron-specific gene modification using the MitoPARK model allows direct targeting of the DA system in specific areas such as the SNpc or striatum. The earliest genetic animal model created to recapitulate the features of ALS involved the exploitation of *SOD1*, yet remains widely used today due to its use in both basic science and translational ALS research. More recent models have surfaced that manipulate the latest gene targets found in fALS cases (*FUS/TLS*, *TDP-43*, *VAPB*, or *DCNT1*). Whether these models fully reproduce the human disease in animals has yet to be determined.

The generation of many animal models over the last 40 years has had a profound effect on PD and ALS research. While many have failed to recapitulate the human disease at all, several recent models are capable of reproducing some of the motor and pathological disruptions. Nevertheless, none of the current animal models fully mimic the onset or etiology, progression, and pathology of PD or ALS; therefore, only partial insight may be gleaned from the results of studies using these animal models. Further work in developing models that more closely replicate most, if not all, of the clinical features of the disease is still needed. A model that is relatively inexpensive and simple to maintain, not toxic to researchers, shows good construct validity, and translates well for human therapeutics would be ideal; especially if following systemic administration the animal develops a chronic progressive disease phenotype. Nevertheless, the current animal models serve as more than adequate substitutes to their human counterparts, providing valuable tools for translational and basic science research in PD and ALS.

References

Abeliovich A, Schmitz Y, Farinas I, Choi-Lundberg D, Ho WH, Castillo PE, Shinsky N, Verdugo JM, Armanini M, Ryan A, Hynes M, Phillips H, Sulzer D, Rosenthal A (2000) Mice lacking alpha-synuclein display functional deficits in the nigrostriatal dopamine system. Neuron 25:239–252

Adams CR, Ziegler DK, Lin JT (1983) Mercury intoxication simulating amyotrophic lateral sclerosis. JAMA 250:642–643

Alexander GM, Erwin KL, Byers N, Deitch JS, Augelli BJ, Blankenhorn EP, Heiman-Patterson TD (2004) Effect of transgene copy number on survival in the G93A SOD1 transgenic mouse model of ALS. Brain Res Mol Brain Res 130:7–15

Anderson G, Noorian AR, Taylor G, Anitha M, Bernhard D, Srinivasan S, Greene JG (2007) Loss of enteric dopaminergic neurons and associated changes in colon motility in an MPTP mouse model of Parkinson's disease. Exp Neurol 207:4–12

Andres-Mateos E, Perier C, Zhang L, Blanchard-Fillion B, Greco TM, Thomas B, Ko HS, Sasaki M, Ischiropoulos H, Przedborski S, Dawson TM, Dawson VL (2007) DJ-1 gene deletion reveals that DJ-1 is an atypical peroxiredoxin-like peroxidase. Proc Natl Acad Sci U S A 104:14807–14812

Andres-Mateos E, Mejias R, Sasaki M, Li X, Lin BM, Biskup S, Zhang L, Banerjee R, Thomas B, Yang L, Liu G, Beal MF, Huso DL, Dawson TM, Dawson VL (2009) Unexpected lack of hypersensitivity in LRRK2 knock-out mice to MPTP (1-methyl-4-phenyl-1,2,3,6-tetrahydropyridine). J Neurosci 29:15846–15850

Arai T, Hasegawa M, Akiyama H, Ikeda K, Nonaka T, Mori H, Mann D, Tsuchiya K, Yoshida M, Hashizume Y, Oda T (2006) TDP-43 is a component of ubiquitin-positive tau-negative inclusions in frontotemporal lobar degeneration and amyotrophic lateral sclerosis. Biochem Biophys Res Commun 351:602–611

Arvidson B (1992) Accumulation of inorganic mercury in lower motoneurons of mice. Neurotoxicology 13:277–280

Ascherio A, Chen H, Weisskopf MG, O'Reilly E, McCullough ML, Calle EE, Schwarzschild MA, Thun MJ (2006) Pesticide exposure and risk for Parkinson's disease. Ann Neurol 60:197–203

Backman CM, Malik N, Zhang Y, Shan L, Grinberg A, Hoffer BJ, Westphal H, Tomac AC (2006) Characterization of a mouse strain expressing Cre recombinase from the 3′ untranslated region of the dopamine transporter locus. Genesis 44:383–390

Banks GT, Kuta A, Isaacs AM, Fisher EM (2008) TDP-43 is a culprit in human neurodegeneration, and not just an innocent bystander. Mamm Genome 19:299–305

Barber TE (1978) Inorganic mercury intoxication reminiscent of amyotrophic lateral sclerosis. J Occup Med 20:667–669

Bartels T, Choi JG, Selkoe DJ (2011) Alpha-synuclein occurs physiologically as a helically folded tetramer that resists aggregation. Nature 477:107–110

Benner EJ, Mosley RL, Destache CJ, Lewis TB, Jackson-Lewis V, Gorantla S, Nemachek C, Green SR, Przedborski S, Gendelman HE (2004) Therapeutic immunization protects dopaminergic neurons in a mouse model of Parkinson's disease. Proc Natl Acad Sci U S A 101:9435–9440

Benner EJ, Banerjee R, Reynolds AD, Sherman S, Pisarev VM, Tsiperson V, Nemachek C, Ciborowski P, Przedborski S, Mosley RL, Gendelman HE (2008) Nitrated alpha-synuclein immunity accelerates degeneration of nigral dopaminergic neurons. PLoS One 3:e1376

Berger K, Przedborski S, Cadet JL (1991) Retrograde degeneration of nigrostriatal neurons induced by intrastriatal 6-hydroxydopamine injection in rats. Brain Res Bull 26:301–307

Bergman H, Wichmann T, DeLong MR (1990) Reversal of experimental parkinsonism by lesions of the subthalamic nucleus. Science 249:1436–1438

Betarbet R, Sherer TB, MacKenzie G, Garcia-Osuna M, Panov AV, Greenamyre JT (2000) Chronic systemic pesticide exposure reproduces features of Parkinson's disease. Nat Neurosci 3:1301–1306

Biskup S, Moore DJ, Celsi F, Higashi S, West AB, Andrabi SA, Kurkinen K, Yu SW, Savitt JM, Waldvogel HJ, Faull RL, Emson PC, Torp R, Ottersen OP, Dawson TM, Dawson VL (2006) Localization of LRRK2 to membranous and vesicular structures in mammalian brain. Ann Neurol 60:557–569

Bjorklund LM, Sanchez-Pernaute R, Chung S, Andersson T, Chen IY, McNaught KS, Brownell AL, Jenkins BG, Wahlestedt C, Kim KS, Isacson O (2002) Embryonic stem cells develop into functional dopaminergic neurons after transplantation in a Parkinson rat model. Proc Natl Acad Sci U S A 99:2344–2349

Boillee S, Yamanaka K, Lobsiger CS, Copeland NG, Jenkins NA, Kassiotis G, Kollias G, Cleveland DW (2006) Onset and progression in inherited ALS determined by motor neurons and microglia. Science 312:1389–1392

Bonifati V, Rizzu P, van Baren MJ, Schaap O, Breedveld GJ, Krieger E, Dekker MC, Squitieri F, Ibanez P, Joosse M, van Dongen JW, Vanacore N, van Swieten JC, Brice A, Meco G, van Duijn CM, Oostra BA, Heutink P (2003) Mutations in the DJ-1 gene associated with autosomal recessive early-onset parkinsonism. Science 299:256–259

Bonifati V, Oostra BA, Heutink P (2004) Linking DJ-1 to neurodegeneration offers novel insights for understanding the pathogenesis of Parkinson's disease. J Mol Med (Berl) 82:163–174

Boska MD, Lewis TB, Destache CJ, Benner EJ, Nelson JA, Uberti M, Mosley RL, Gendelman HE (2005) Quantitative 1H magnetic resonance spectroscopic imaging determines therapeutic immunization efficacy in an animal model of Parkinson's disease. J Neurosci 25:1691–1700

Boska MD, Hasan KM, Kibuule D, Banerjee R, McIntyre E, Nelson JA, Hahn T, Gendelman HE, Mosley RL (2007) Quantitative diffusion tensor imaging detects dopaminergic neuronal degeneration in a murine model of Parkinson's disease. Neurobiol Dis 26: 590–596

Bove J, Prou D, Perier C, Przedborski S (2005) Toxin-induced models of Parkinson's disease. NeuroRx 2:484–494

Brochard V, Combadiere B, Prigent A, Laouar Y, Perrin A, Beray-Berthat V, Bonduelle O, Alvarez-Fischer D, Callebert J, Launay JM, Duyckaerts C, Flavell RA, Hirsch EC, Hunot S (2009) Infiltration of CD4+ lymphocytes into the brain contributes to neurodegeneration in a mouse model of Parkinson disease. J Clin Invest 119:182–192

Brooks AI, Chadwick CA, Gelbard HA, Cory-Slechta DA, Federoff HJ (1999) Paraquat elicited neurobehavioral syndrome caused by dopaminergic neuron loss. Brain Res 823:1–10

Buratti E, Baralle FE (2008) Multiple roles of TDP-43 in gene expression, splicing regulation, and human disease. Front Biosci 13:867–878

Castano A, Herrera AJ, Cano J, Machado A (1998) Lipopolysaccharide intranigral injection induces inflammatory reaction and damage in nigrostriatal dopaminergic system. J Neurochem 70:1584–1592

Chai A, Withers J, Koh YH, Parry K, Bao H, Zhang B, Budnik V, Pennetta G (2008) HVAPB, the causative gene of a heterogeneous group of motor neuron diseases in humans, is functionally interchangeable with its Drosophila homologue DVAP-33A at the neuromuscular junction. Hum Mol Genet 17:266–280

Chan P, DeLanney LE, Irwin I, Langston JW, Di Monte D (1991) Rapid ATP loss caused by 1-methyl-4-phenyl-1,2,3,6-tetrahydropyridine in mouse brain. J Neurochem 57:348–351

Chan P, Tanner CM, Jiang X, Langston JW (1998) Failure to find the alpha-synuclein gene missense mutation (G209A) in 100 patients with younger onset Parkinson's disease. Neurology 50:513–514

Chen Y, Yang M, Deng J, Chen X, Ye Y, Zhu L, Liu J, Ye H, Shen Y, Li Y, Rao EJ, Fushimi K, Zhou X, Bigio EH, Mesulam M, Xu Q, Wu JY (2011) Expression of human FUS protein in Drosophila leads to progressive neurodegeneration. Protein Cell 2:477–486

Choi J, Sullards MC, Olzmann JA, Rees HD, Weintraub ST, Bostwick DE, Gearing M, Levey AI, Chin LS, Li L (2006) Oxidative damage of DJ-1 is linked to sporadic Parkinson and Alzheimer diseases. J Biol Chem 281:10816–10824

Clark AW, Griffin JW, Price DL (1980) The axonal pathology in chronic IDPN intoxication. J Neuropathol Exp Neurol 39:42–55

Clark IE, Dodson MW, Jiang C, Cao JH, Huh JR, Seol JH, Yoo SJ, Hay BA, Guo M (2006) Drosophila pink1 is required for mitochondrial function and interacts genetically with parkin. Nature 441:1162–1166

Cohen G, Werner P (1994) Free radicals, oxidative stress, and neurodegeneration. In: Calne DB (ed) Neurodegenerative diseases. W.B. Saunders, Philadelphia, pp 139–161

Combault M (1880) Contribution a l'etude anatomique de la n'evite parenchymateuse subaigue et chronique: n'evite segmentaire p'eri-axile. Arch Neurol (Paris) 1:11–38

Conradi S, Ronnevi LO, Narris FH (1982) Motor neuron disease and toxic metals. In: Rowland LP (ed) Human motor neuron disease. Raven, New York, pp 201–231

Cook SA, Johnson KR, Bronson RT, Davisson MT (1995) Neuromuscular degeneration (nmd): a mutation on mouse chromosome 19 that causes motor neuron degeneration. Mamm Genome 6:187–191

Cookson MR (2005) The biochemistry of Parkinson's disease. Annu Rev Biochem 74:29–52

Dammann O, Leviton A (1997) Does prepregnancy bacterial vaginosis increase a mother's risk of having a preterm infant with cerebral palsy? Dev Med Child Neurol 39:836–840

Davis GC, Williams AC, Markey SP, Ebert MH, Caine ED, Reichert CM, Kopin IJ (1979) Chronic parkinsonism secondary to intravenous injection of meperidine analogues. Psychiatry Res 1:249–254

Day BJ, Patel M, Calavetta L, Chang LY, Stamler JS (1999) A mechanism of paraquat toxicity involving nitric oxide synthase. Proc Natl Acad Sci U S A 96:12760–12765

Di Monte D, Sandy MS, Ekstrom G, Smith MT (1986) Comparative studies on the mechanisms of paraquat and 1-methyl-4-phenylpyridine (MPP+) cytotoxicity. Biochem Biophys Res Commun 137:303–309

Drolet RE, Cannon JR, Montero L, Greenamyre JT (2009) Chronic rotenone exposure reproduces Parkinson's disease gastrointestinal neuropathology. Neurobiol Dis 36:96–102

Duchen LW, Strich SJ (1968) An hereditary motor neurone disease with progressive denervation of muscle in the mouse: the mutant 'wobbler'. J Neurol Neurosurg Psychiatry 31:535–542

Ekstrand MI, Galter D (2009) The MitoPark mouse—an animal model of Parkinson's disease with impaired respiratory chain function in dopamine neurons. Parkinsonism Relat Disord 15(suppl 3):S185–S188

Ekstrand MI, Terzioglu M, Galter D, Zhu S, Hofstetter C, Lindqvist E, Thams S, Bergstrand A, Hansson FS, Trifunovic A, Hoffer B, Cullheim S, Mohammed AH, Olson L, Larsson NG (2007) Progressive parkinsonism in mice with respiratory-chain-deficient dopamine neurons. Proc Natl Acad Sci U S A 104:1325–1330

Fabre E, Monserrat J, Herrero A, Barja G, Leret ML (1999) Effect of MPTP on brain mitochondrial H2O2 and ATP production and on dopamine and DOPAC in the striatum. J Physiol Biochem 55:325–331

Farrer M, Maraganore DM, Lockhart P, Singleton A, Lesnick TG, de Andrade M, West A, de Silva R, Hardy J, Hernandez D (2001) Alpha-synuclein gene haplotypes are associated with Parkinson's disease. Hum Mol Genet 10:1847–1851

Feany MB, Bender WW (2000) A Drosophila model of Parkinson's disease. Nature 404:394–398

Ferrante RJ, Schulz JB, Kowall NW, Beal MF (1997) Systemic administration of rotenone produces selective damage in the striatum and globus pallidus, but not in the substantia nigra. Brain Res 753:157–162

Ferraz HB, Bertolucci PH, Pereira JS, Lima JG, Andrade LA (1988) Chronic exposure to the fungicide maneb may produce symptoms and signs of CNS manganese intoxication. Neurology 38:550–553

Fleming SM, Fernagut PO, Chesselet MF (2005) Genetic mouse models of parkinsonism: strengths and limitations. NeuroRx 2:495–503

Fornai F, Schluter OM, Lenzi P, Gesi M, Ruffoli R, Ferrucci M, Lazzeri G, Busceti CL, Pontarelli F, Battaglia G, Pellegrini A, Nicoletti F, Ruggieri S, Paparelli A, Sudhof TC (2005) Parkinson-like syndrome induced by continuous MPTP infusion: convergent roles of the ubiquitin-proteasome system and alpha-synuclein. Proc Natl Acad Sci U S A 102: 3413–3418

Forno LS, DeLanney LE, Irwin I, Langston JW (1993) Similarities and differences between MPTP-induced parkinsonism and Parkinson's disease. Neuropathologic considerations. Adv Neurol 60:600–608

Galter D, Westerlund M, Carmine A, Lindqvist E, Sydow O, Olson L (2006) LRRK2 expression linked to dopamine-innervated areas. Ann Neurol 59:714–719

Gandhi S, Muqit MM, Stanyer L, Healy DG, Abou-Sleiman PM, Hargreaves I, Heales S, Ganguly M, Parsons L, Lees AJ, Latchman DS, Holton JL, Wood NW, Revesz T (2006) PINK1 protein in normal human brain and Parkinson's disease. Brain 129:1720–1731

Gao HM, Jiang J, Wilson B, Zhang W, Hong JS, Liu B (2002) Microglial activation-mediated delayed and progressive degeneration of rat nigral dopaminergic neurons: relevance to Parkinson's disease. J Neurochem 81:1285–1297

Garruto R, Yanagihara R, Gajdusek D, Arion D (1984) Concentrations of trace and essential elements in garden soil and drinking water in the Western Pacific. National Taiwan University Press, Taipei

Gasser T (2009) Molecular pathogenesis of Parkinson disease: insights from genetic studies. Expert Rev Mol Med 11:e22

Gelman DM, Noain D, Avale ME, Otero V, Low MJ, Rubinstein M (2003) Transgenic mice engineered to target Cre/loxP-mediated DNA recombination into catecholaminergic neurons. Genesis 36:196–202

George JM, Jin H, Woods WS, Clayton DF (1995) Characterization of a novel protein regulated during the critical period for song learning in the zebra finch. Neuron 15:361–372

Gilks WP, Abou-Sleiman PM, Gandhi S, Jain S, Singleton A, Lees AJ, Shaw K, Bhatia KP, Bonifati V, Quinn NP, Lynch J, Healy DG, Holton JL, Revesz T, Wood NW (2005) A common LRRK2 mutation in idiopathic Parkinson's disease. Lancet 365:415–416

Giovanni A, Sonsalla PK, Heikkila RE (1994) Studies on species sensitivity to the dopaminergic neurotoxin 1-methyl-4-phenyl-1,2,3,6-tetrahydropyridine. Part 2: central administration of 1-methyl-4-phenylpyridinium. J Pharmacol Exp Ther 270:1008–1014

Gispert S et al (2009) Parkinson phenotype in aged PINK1-deficient mice is accompanied by progressive mitochondrial dysfunction in absence of neurodegeneration. PLoS One 4:e5777

Gloeckner CJ, Kinkl N, Schumacher A, Braun RJ, O'Neill E, Meitinger T, Kolch W, Prokisch H, Ueffing M (2006) The Parkinson disease causing LRRK2 mutation I2020T is associated with increased kinase activity. Hum Mol Genet 15:223–232

Goldberg MS, Fleming SM, Palacino JJ, Cepeda C, Lam HA, Bhatnagar A, Meloni EG, Wu N, Ackerson LC, Klapstein GJ, Gajendiran M, Roth BL, Chesselet MF, Maidment NT, Levine MS, Shen J (2003) Parkin-deficient mice exhibit nigrostriatal deficits but not loss of dopaminergic neurons. J Biol Chem 278:43628–43635

Goldberg MS, Pisani A, Haburcak M, Vortherms TA, Kitada T, Costa C, Tong Y, Martella G, Tscherter A, Martins A, Bernardi G, Roth BL, Pothos EN, Calabresi P, Shen J (2005) Nigrostriatal dopaminergic deficits and hypokinesia caused by inactivation of the familial parkinsonism-linked gene DJ-1. Neuron 45:489–496

Gong YH, Parsadanian AS, Andreeva A, Snider WD, Elliott JL (2000) Restricted expression of G86R Cu/Zn superoxide dismutase in astrocytes results in astrocytosis but does not cause motoneuron degeneration. J Neurosci 20:660–665

Gordon PH, Moore DH, Miller RG, Florence JM, Verheijde JL, Doorish C, Hilton JF, Spitalny GM, MacArthur RB, Mitsumoto H, Neville HE, Boylan K, Mozaffar T, Belsh JM, Ravits J, Bedlack RS, Graves MC, McCluskey LF, Barohn RJ, Tandan R (2007) Efficacy of minocycline in patients with amyotrophic lateral sclerosis: a phase III randomised trial. Lancet Neurol 6:1045–1053

Greene JC, Whitworth AJ, Kuo I, Andrews LA, Feany MB, Pallanck LJ (2003) Mitochondrial pathology and apoptotic muscle degeneration in Drosophila parkin mutants. Proc Natl Acad Sci U S A 100:4078–4083

Griffin J, Price D (1980) Proximal axonopathies induced by toxic chemicals. In: Spencer P, Schaumburg H (eds) Experimental and clinical neurotoxicology. Williams and Wilkins, Baltimore, pp 161–178

Gurney ME, Pu H, Chiu AY, Dal Canto MC, Polchow CY, Alexander DD, Caliendo J, Hentati A, Kwon YW, Deng HX et al (1994) Motor neuron degeneration in mice that express a human Cu, Zn superoxide dismutase mutation. Science 264:1772–1775

Hasegawa E, Takeshige K, Oishi T, Murai Y, Minakami S (1990) 1-Methyl-4-phenylpyridinium (MPP+) induces NADH-dependent superoxide formation and enhances NADH-dependent lipid peroxidation in bovine heart submitochondrial particles. Biochem Biophys Res Commun 170:1049–1055

Hasegawa E, Kang D, Sakamoto K, Mitsumoto A, Nagano T, Minakami S, Takeshige K (1997) A dual effect of 1-methyl-4-phenylpyridinium (MPP+)-analogs on the respiratory chain of bovine heart mitochondria. Arch Biochem Biophys 337:69–74

Hatano Y, Li Y, Sato K, Asakawa S, Yamamura Y, Tomiyama H, Yoshino H, Asahina M, Kobayashi S, Hassin-Baer S, Lu CS, Ng AR, Rosales RL, Shimizu N, Toda T, Mizuno Y, Hattori N (2004) Novel PINK1 mutations in early-onset parkinsonism. Ann Neurol 56:424–427

Healy DG et al (2008) Phenotype, genotype, and worldwide genetic penetrance of LRRK2-associated Parkinson's disease: a case-control study. Lancet Neurol 7:583–590

Heimann P, Laage S, Jockusch H (1991) Defect of sperm assembly in a neurological mutant of the mouse, wobbler (WR). Differentiation 47:77–83

Herrero MT, Hirsch EC, Kastner A, Ruberg M, Luquin MR, Laguna J, Javoy-Agid F, Obeso JA, Agid Y (1993) Does neuromelanin contribute to the vulnerability of catecholaminergic neurons in monkeys intoxicated with MPTP? Neuroscience 56:499–511

Hicks GG, Singh N, Nashabi A, Mai S, Bozek G, Klewes L, Arapovic D, White EK, Koury MJ, Oltz EM, Van Kaer L, Ruley HE (2000) Fus deficiency in mice results in defective B-lymphocyte development and activation, high levels of chromosomal instability and perinatal death. Nat Genet 24:175–179

Hoglinger GU, Feger J, Prigent A, Michel PP, Parain K, Champy P, Ruberg M, Oertel WH, Hirsch EC (2003) Chronic systemic complex I inhibition induces a hypokinetic multisystem degeneration in rats. J Neurochem 84:491–502

Hutter-Saunders JA, Mosley RL, Gendelman HE (2011) Pathways towards an effective immunotherapy for Parkinson's disease. Expert Rev Neurother 11:1703–1715

Imai Y, Gehrke S, Wang HQ, Takahashi R, Hasegawa K, Oota E, Lu B (2008) Phosphorylation of 4E-BP by LRRK2 affects the maintenance of dopaminergic neurons in Drosophila. EMBO J 27:2432–2443

Itier JM et al (2003) Parkin gene inactivation alters behaviour and dopamine neurotransmission in the mouse. Hum Mol Genet 12:2277–2291

Jackson-Lewis V, Przedborski S (2007) Protocol for the MPTP mouse model of Parkinson's disease. Nat Protoc 2:141–151

Javitch JA, D'Amato RJ, Strittmatter SM, Snyder SH (1985) Parkinsonism-inducing neurotoxin, N-methyl-4-phenyl-1,2,3,6-tetrahydropyridine: uptake of the metabolite N-methyl-4-phenylpyridine by dopamine neurons explains selective toxicity. Proc Natl Acad Sci U S A 82:2173–2177

Javoy F, Sotelo C, Herbet A, Agid Y (1976) Specificity of dopaminergic neuronal degeneration induced by intracerebral injection of 6-hydroxydopamine in the nigrostriatal dopamine system. Brain Res 102:201–215

Jensen PH, Nielsen MS, Jakes R, Dotti CG, Goedert M (1998) Binding of alpha-synuclein to brain vesicles is abolished by familial Parkinson's disease mutation. J Biol Chem 273:26292–26294

Jensen PH, Hager H, Nielsen MS, Hojrup P, Gliemann J, Jakes R (1999) Alpha-synuclein binds to Tau and stimulates the protein kinase A-catalyzed tau phosphorylation of serine residues 262 and 356. J Biol Chem 274:25481–25489

Jensen PH, Islam K, Kenney J, Nielsen MS, Power J, Gai WP (2000) Microtubule-associated protein 1B is a component of cortical Lewy bodies and binds alpha-synuclein filaments. J Biol Chem 275:21500–21507

Jeon BS, Jackson-Lewis V, Burke RE (1995) 6-Hydroxydopamine lesion of the rat substantia nigra: time course and morphology of cell death. Neurodegeneration 4:131–137

Jiang H, Jackson-Lewis V, Muthane U, Dollison A, Ferreira M, Espinosa A, Parsons B, Przedborski S (1993) Adenosine receptor antagonists potentiate dopamine receptor agonist-induced rotational behavior in 6-hydroxydopamine-lesioned rats. Brain Res 613:347–351

Jonsson G (1983) Chemical lesioning techniques: monoamine neurotoxins. In: Bjorklund A, Hokfelt T (eds) Handbook of chemical neuroanatomy. Elsevier, Amsterdam, pp 463–507

Joyce PI, Fratta P, Fisher EM, Acevedo-Arozena A (2011) SOD1 and TDP-43 animal models of amyotrophic lateral sclerosis: recent advances in understanding disease toward the development of clinical treatments. Mamm Genome 22:420–448

Ju S, Tardiff DF, Han H, Divya K, Zhong Q, Maquat LE, Bosco DA, Hayward LJ, Brown RH Jr, Lindquist S, Ringe D, Petsko GA (2011) A yeast model of FUS/TLS-dependent cytotoxicity. PLoS Biol 9:e1001052

Kahle PJ, Haass C, Kretzschmar HA, Neumann M (2002) Structure/function of alpha-synuclein in health and disease: rational development of animal models for Parkinson's and related diseases. J Neurochem 82:449–457

Khan NL et al (2005) Mutations in the gene LRRK2 encoding dardarin (PARK8) cause familial Parkinson's disease: clinical, pathological, olfactory and functional imaging and genetic data. Brain 128:2786–2796

Kim RH, Smith PD, Aleyasin H, Hayley S, Mount MP, Pownall S, Wakeham A, You-Ten AJ, Kalia SK, Horne P, Westaway D, Lozano AM, Anisman H, Park DS, Mak TW (2005) Hypersensitivity of DJ-1-deficient mice to 1-methyl-4-phenyl-1,2,3,6-tetrahydropyrindine (MPTP) and oxidative stress. Proc Natl Acad Sci U S A 102:5215–5220

Kirik D, Rosenblad C, Burger C, Lundberg C, Johansen TE, Muzyczka N, Mandel RJ, Bjorklund A (2002) Parkinson-like neurodegeneration induced by targeted overexpression of alpha-synuclein in the nigrostriatal system. J Neurosci 22:2780–2791

Kitada T, Asakawa S, Hattori N, Matsumine H, Yamamura Y, Minoshima S, Yokochi M, Mizuno Y, Shimizu N (1998) Mutations in the parkin gene cause autosomal recessive juvenile parkinsonism. Nature 392:605–608

Klaidman LK, Adams JD Jr, Leung AC, Kim SS, Cadenas E (1993) Redox cycling of MPP+: evidence for a new mechanism involving hydride transfer with xanthine oxidase, aldehyde dehydrogenase, and lipoamide dehydrogenase. Free Radic Biol Med 15:169–179

Klein RL, King MA, Hamby ME, Meyer EM (2002) Dopaminergic cell loss induced by human A30P alpha-synuclein gene transfer to the rat substantia nigra. Hum Gene Ther 13:605–612

Kopin IJ, Markey SP (1988) MPTP toxicity: implications for research in Parkinson's disease. Annu Rev Neurosci 11:81–96

Kordower JH, Kanaan NM, Chu Y, Suresh Babu R, Stansell J III, Terpstra BT, Sortwell CE, Steece-Collier K, Collier TJ (2006) Failure of proteasome inhibitor administration to provide a model of Parkinson's disease in rats and monkeys. Ann Neurol 60:264–268

Kosloski LM, Ha DM, Hutter JA, Stone DK, Pichler MR, Reynolds AD, Gendelman HE, Mosley RL (2010) Adaptive immune regulation of glial homeostasis as an immunization strategy for neurodegenerative diseases. J Neurochem 114:1261–1276

Kriz J, Nguyen MD, Julien JP (2002) Minocycline slows disease progression in a mouse model of amyotrophic lateral sclerosis. Neurobiol Dis 10:268–278

Kruger R, Kuhn W, Muller T, Woitalla D, Graeber M, Kosel S, Przuntek H, Epplen JT, Schols L, Riess O (1998) Ala30Pro mutation in the gene encoding alpha-synuclein in Parkinson's disease. Nat Genet 18:106–108

Kurkowska-Jastrzebska I, Wronska A, Kohutnicka M, Czlonkowski A, Czlonkowska A (1999a) The inflammatory reaction following 1-methyl-4-phenyl-1,2,3,6-tetrahydropyridine intoxication in mouse. Exp Neurol 156:50–61

Kurkowska-Jastrzebska I, Wronska A, Kohutnicka M, Czlonkowski A, Czlonkowska A (1999b) MHC class II positive microglia and lymphocytic infiltration are present in the substantia nigra and striatum in mouse model of Parkinson's disease. Acta Neurobiol Exp (Wars) 59:1–8

Kuwahara T, Koyama A, Gengyo-Ando K, Masuda M, Kowa H, Tsunoda M, Mitani S, Iwatsubo T (2006) Familial Parkinson mutant alpha-synuclein causes dopamine neuron dysfunction in transgenic Caenorhabditis elegans. J Biol Chem 281:334–340

Kwiatkowski TJ Jr et al (2009) Mutations in the FUS/TLS gene on chromosome 16 cause familial amyotrophic lateral sclerosis. Science 323:1205–1208

Lagier-Tourenne C, Cleveland DW (2009) Rethinking ALS: the FUS about TDP-43. Cell 136:1001–1004

Lai C, Lin X, Chandran J, Shim H, Yang WJ, Cai H (2007) The G59S mutation in p150(glued) causes dysfunction of dynactin in mice. J Neurosci 27:13982–13990

Lakso M, Vartiainen S, Moilanen AM, Sirvio J, Thomas JH, Nass R, Blakely RD, Wong G (2003) Dopaminergic neuronal loss and motor deficits in Caenorhabditis elegans overexpressing human alpha-synuclein. J Neurochem 86:165–172

Langston JW, Irwin I (1986) MPTP: current concepts and controversies. Clin Neuropharmacol 9:485–507

Langston JW, Ballard P, Tetrud JW, Irwin I (1983) Chronic parkinsonism in humans due to a product of meperidine-analog synthesis. Science 219:979–980

Langston JW, Forno LS, Rebert CS, Irwin I (1984) Selective nigral toxicity after systemic administration of 1-methyl-4-phenyl-1,2,5,6-tetrahydropyrine (MPTP) in the squirrel monkey. Brain Res 292:390–394

Langston JW, Forno LS, Tetrud J, Reeves AG, Kaplan JA, Karluk D (1999) Evidence of active nerve cell degeneration in the substantia nigra of humans years after 1-methyl-4-phenyl-1,2,3,6-tetrahydropyridine exposure. Ann Neurol 46:598–605

Lapointe N, St-Hilaire M, Martinoli MG, Blanchet J, Gould P, Rouillard C, Cicchetti F (2004) Rotenone induces non-specific central nervous system and systemic toxicity. FASEB J 18:717–719

Laurie C, Reynolds A, Coskun O, Bowman E, Gendelman HE, Mosley RL (2007) CD4+ T cells from copolymer-1 immunized mice protect dopaminergic neurons in the 1-methyl-4-phenyl-1,2,3,6-tetrahydropyridine model of Parkinson's disease. J Neuroimmunol 183:60–68

Lauwers E, Debyser Z, Van Dorpe J, De Strooper B, Nuttin B, Baekelandt V (2003) Neuropathology and neurodegeneration in rodent brain induced by lentiviral vector-mediated overexpression of alpha-synuclein. Brain Pathol 13:364–372

LaVoie MJ, Ostaszewski BL, Weihofen A, Schlossmacher MG, Selkoe DJ (2005) Dopamine covalently modifies and functionally inactivates parkin. Nat Med 11:1214–1221

Lees AJ, Hardy J, Revesz T (2009) Parkinson's disease. Lancet 373:2055–2066

Lehnardt S, Massillon L, Follett P, Jensen FE, Ratan R, Rosenberg PA, Volpe JJ, Vartanian T (2003) Activation of innate immunity in the CNS triggers neurodegeneration through a Toll-like receptor 4-dependent pathway. Proc Natl Acad Sci U S A 100:8514–8519

Levy JR, Sumner CJ, Caviston JP, Tokito MK, Ranganathan S, Ligon LA, Wallace KE, LaMonte BH, Harmison GG, Puls I, Fischbeck KH, Holzbaur EL (2006) A motor neuron disease-associated mutation in p150Glued perturbs dynactin function and induces protein aggregation. J Cell Biol 172:733–745

Liberatore GT, Jackson-Lewis V, Vukosavic S, Mandir AS, Vila M, McAuliffe WG, Dawson VL, Dawson TM, Przedborski S (1999) Inducible nitric oxide synthase stimulates dopaminergic neurodegeneration in the MPTP model of Parkinson disease. Nat Med 5:1403–1409

Limousin P, Krack P, Pollak P, Benazzouz A, Ardouin C, Hoffmann D, Benabid AL (1998) Electrical stimulation of the subthalamic nucleus in advanced Parkinson's disease. N Engl J Med 339:1105–1111

Lin X, Parisiadou L, Gu XL, Wang L, Shim H, Sun L, Xie C, Long CX, Yang WJ, Ding J, Chen ZZ, Gallant PE, Tao-Cheng JH, Rudow G, Troncoso JC, Liu Z, Li Z, Cai H (2009) Leucine-rich repeat kinase 2 regulates the progression of neuropathology induced by Parkinson's-disease-related mutant alpha-synuclein. Neuron 64:807–827

Lindeberg J, Usoskin D, Bengtsson H, Gustafsson A, Kylberg A, Soderstrom S, Ebendal T (2004) Transgenic expression of Cre recombinase from the tyrosine hydroxylase locus. Genesis 40:67–73

Ling Z, Gayle DA, Ma SY, Lipton JW, Tong CW, Hong JS, Carvey PM (2002) In utero bacterial endotoxin exposure causes loss of tyrosine hydroxylase neurons in the postnatal rat midbrain. Mov Disord 17:116–124

Ling ZD, Chang Q, Lipton JW, Tong CW, Landers TM, Carvey PM (2004) Combined toxicity of prenatal bacterial endotoxin exposure and postnatal 6-hydroxydopamine in the adult rat midbrain. Neuroscience 124:619–628

Liu B (2006) Modulation of microglial pro-inflammatory and neurotoxic activity for the treatment of Parkinson's disease. AAPS J 8:E606–E621

Liu Y, Roghani A, Edwards RH (1992) Gene transfer of a reserpine-sensitive mechanism of resistance to N-methyl-4-phenylpyridinium. Proc Natl Acad Sci U S A 89:9074–9078

Liu B, Jiang JW, Wilson BC, Du L, Yang SN, Wang JY, Wu GC, Cao XD, Hong JS (2000) Systemic infusion of naloxone reduces degeneration of rat substantia nigral dopaminergic neurons induced by intranigral injection of lipopolysaccharide. J Pharmacol Exp Ther 295:125–132

Liu Z, Wang X, Yu Y, Li X, Wang T, Jiang H, Ren Q, Jiao Y, Sawa A, Moran T, Ross CA, Montell C, Smith WW (2008) A Drosophila model for LRRK2-linked parkinsonism. Proc Natl Acad Sci U S A 105:2693–2698

Lo Bianco C, Ridet JL, Schneider BL, Deglon N, Aebischer P (2002) Alpha-synucleinopathy and selective dopaminergic neuron loss in a rat lentiviral-based model of Parkinson's disease. Proc Natl Acad Sci U S A 99:10813–10818

Lu X, Bing G, Hagg T (2000) Naloxone prevents microglia-induced degeneration of dopaminergic substantia nigra neurons in adult rats. Neuroscience 97:285–291

Lu XH, Fleming SM, Meurers B, Ackerson LC, Mortazavi F, Lo V, Hernandez D, Sulzer D, Jackson GR, Maidment NT, Chesselet MF, Yang XW (2009)

Bacterial artificial chromosome transgenic mice expressing a truncated mutant parkin exhibit age-dependent hypokinetic motor deficits, dopaminergic neuron degeneration, and accumulation of proteinase K-resistant alpha-synuclein. J Neurosci 29:1962–1976
Lucking CB, Durr A, Bonifati V, Vaughan J, De Michele G, Gasser T, Harhangi BS, Meco G, Denefle P, Wood NW, Agid Y, Brice A (2000) Association between early-onset Parkinson's disease and mutations in the parkin gene. N Engl J Med 342:1560–1567
Luthman J, Fredriksson A, Sundstrom E, Jonsson G, Archer T (1989) Selective lesion of central dopamine or noradrenaline neuron systems in the neonatal rat: motor behavior and monoamine alterations at adult stage. Behav Brain Res 33:267–277
Lynch T, Farrer M, Hutton M, Hardy J (1997) Genetics of Parkinson's disease. Science 278:1212–1213
Manning-Bog AB, McCormack AL, Li J, Uversky VN, Fink AL, Di Monte DA (2002) The herbicide paraquat causes up-regulation and aggregation of alpha-synuclein in mice: paraquat and alpha-synuclein. J Biol Chem 277:1641–1644
Markey SP, Johannessen JN, Chiueh CC, Burns RS, Herkenham MA (1984) Intraneuronal generation of a pyridinium metabolite may cause drug-induced parkinsonism. Nature 311:464–467
Maroteaux L, Campanelli JT, Scheller RH (1988) Synuclein: a neuron-specific protein localized to the nucleus and presynaptic nerve terminal. J Neurosci 8:2804–2815
Martin LJ, Pan Y, Price AC, Sterling W, Copeland NG, Jenkins NA, Price DL, Lee MK (2006) Parkinson's disease alpha-synuclein transgenic mice develop neuronal mitochondrial degeneration and cell death. J Neurosci 26:41–50
Masliah E, Rockenstein E, Veinbergs I, Mallory M, Hashimoto M, Takeda A, Sagara Y, Sisk A, Mucke L (2000) Dopaminergic loss and inclusion body formation in alpha-synuclein mice: implications for neurodegenerative disorders. Science 287:1265–1269
Mayer RA, Kindt MV, Heikkila RE (1986) Prevention of the nigrostriatal toxicity of 1-methyl-4-phenyl-1,2,3,6-tetrahydropyridine by inhibitors of 3,4-dihydroxyphenylethylamine transport. J Neurochem 47:1073–1079
McCormack AL, Thiruchelvam M, Manning-Bog AB, Thiffault C, Langston JW, Cory-Slechta DA, Di Monte DA (2002) Environmental risk factors and Parkinson's disease: selective degeneration of nigral dopaminergic neurons caused by the herbicide paraquat. Neurobiol Dis 10:119–127
McCoy MK, Martinez TN, Ruhn KA, Szymkowski DE, Smith CG, Botterman BR, Tansey KE, Tansey MG (2006) Blocking soluble tumor necrosis factor signaling with dominant-negative tumor necrosis factor inhibitor attenuates loss of dopaminergic neurons in models of Parkinson's disease. J Neurosci 26: 9365–9375
McGeer PL, Itagaki S, Boyes BE, McGeer EG (1988) Reactive microglia are positive for HLA-DR in the substantia nigra of Parkinson's and Alzheimer's disease brains. Neurology 38:1285–1291
McNaught KS, Perl DP, Brownell AL, Olanow CW (2004) Systemic exposure to proteasome inhibitors causes a progressive model of Parkinson's disease. Ann Neurol 56:149–162
Melrose H, Lincoln S, Tyndall G, Dickson D, Farrer M (2006) Anatomical localization of leucine-rich repeat kinase 2 in mouse brain. Neuroscience 139:791–794
Menzies FM, Yenisetti SC, Min KT (2005) Roles of Drosophila DJ-1 in survival of dopaminergic neurons and oxidative stress. Curr Biol 15:1578–1582
Meulener M, Whitworth AJ, Armstrong-Gold CE, Rizzu P, Heutink P, Wes PD, Pallanck LJ, Bonini NM (2005) Drosophila DJ-1 mutants are selectively sensitive to environmental toxins associated with Parkinson's disease. Curr Biol 15:1572–1577
Mitchell JD (1987) Heavy metals and trace elements in amyotrophic lateral sclerosis. Neurol Clin 5:43–60
Mizuno Y, Hattori N, Mori H, Suzuki T, Tanaka K (2001) Parkin and Parkinson's disease. Curr Opin Neurol 14:477–482
Moratalla R, Quinn B, DeLanney LE, Irwin I, Langston JW, Graybiel AM (1992) Differential vulnerability of primate caudate-putamen and striosome-matrix dopamine systems to the neurotoxic effects of 1-methyl-4-phenyl-1,2,3,6-tetrahydropyridine. Proc Natl Acad Sci U S A 89:3859–3863
Nakagawa S, Yoshida S, Suematsu C, Shimizu E, Hirohata T, Kumamoto T, Yase Y, Kawai K, Iwata S (1977) The calcium-magnesium-deficient rat: a study on the distribution of calcium in the spinal cord using the electron probe microanalyser. Experientia 33: 1225–1226
Natale G, Kastsiushenka O, Fulceri F, Ruggieri S, Paparelli A, Fornai F (2010) MPTP-induced parkinsonism extends to a subclass of TH-positive neurons in the gut. Brain Res 1355:195–206
Neumann M, Sampathu DM, Kwong LK, Truax AC, Micsenyi MC, Chou TT, Bruce J, Schuck T, Grossman M, Clark CM, McCluskey LF, Miller BL, Masliah E, Mackenzie IR, Feldman H, Feiden W, Kretzschmar HA, Trojanowski JQ, Lee VM (2006) Ubiquitinated TDP-43 in frontotemporal lobar degeneration and amyotrophic lateral sclerosis. Science 314:130–133
Nicklas WJ, Vyas I, Heikkila RE (1985) Inhibition of NADH-linked oxidation in brain mitochondria by 1-methyl-4-phenyl-pyridine, a metabolite of the neurotoxin, 1-methyl-4-phenyl-1,2,5,6-tetrahydropyridine. Life Sci 36:2503–2508
Nishimura AL, Mitne-Neto M, Silva HC, Richieri-Costa A, Middleton S, Cascio D, Kok F, Oliveira JR, Gillingwater T, Webb J, Skehel P, Zatz M (2004) A mutation in the vesicle-trafficking protein VAPB causes late-onset spinal muscular atrophy and amyotrophic lateral sclerosis. Am J Hum Genet 75:822–831

Nissl F (1892) Uber die veraenderungen der ganglienzellen am facialiskern des kaninchens nach ausreissung der nerve. Allg Z Psychiatr 48:197–198

Paisan-Ruiz C et al (2004) Cloning of the gene containing mutations that cause PARK8-linked Parkinson's disease. Neuron 44:595–600

Pamphlett R, Kum-Jew S (2001) Mercury vapor uptake into the nervous system of developing mice. Neurotoxicol Teratol 23:191–196

Pamphlett R, Slater M, Thomas S (1998) Oxidative damage to nucleic acids in motor neurons containing mercury. J Neurol Sci 159:121–126

Park J, Lee SB, Lee S, Kim Y, Song S, Kim S, Bae E, Kim J, Shong M, Kim JM, Chung J (2006) Mitochondrial dysfunction in Drosophila PINK1 mutants is complemented by parkin. Nature 441:1157–1161

Pendleton RG, Parvez F, Sayed M, Hillman R (2002) Effects of pharmacological agents upon a transgenic model of Parkinson's disease in Drosophila melanogaster. J Pharmacol Exp Ther 300:91–96

Polymeropoulos MH, Lavedan C, Leroy E, Ide SE, Dehejia A, Dutra A, Pike B, Root H, Rubenstein J, Boyer R, Stenroos ES, Chandrasekharappa S, Athanassiadou A, Papapetropoulos T, Johnson WG, Lazzarini AM, Duvoisin RC, Di Iorio G, Golbe LI, Nussbaum RL (1997) Mutation in the alpha-synuclein gene identified in families with Parkinson's disease. Science 276:2045–2047

Pramatarova A, Laganiere J, Roussel J, Brisebois K, Rouleau GA (2001) Neuron-specific expression of mutant superoxide dismutase 1 in transgenic mice does not lead to motor impairment. J Neurosci 21: 3369–3374

Przedborski S, Levivier M, Jiang H, Ferreira M, Jackson-Lewis V, Donaldson D, Togasaki DM (1995) Dose-dependent lesions of the dopaminergic nigrostriatal pathway induced by intrastriatal injection of 6-hydroxydopamine. Neuroscience 67:631–647

Przedborski S, Jackson-Lewis V, Naini AB, Jakowec M, Petzinger G, Miller R, Akram M (2001) The parkinsoniantoxin1-methyl-4-phenyl-1,2,3,6-tetrahydropyridine (MPTP): a technical review of its utility and safety. J Neurochem 76:1265–1274

Puls I, Jonnakuty C, LaMonte BH, Holzbaur EL, Tokito M, Mann E, Floeter MK, Bidus K, Drayna D, Oh SJ, Brown RH Jr, Ludlow CL, Fischbeck KH (2003) Mutant dynactin in motor neuron disease. Nat Genet 33:455–456

Ramsay RR, Singer TP (1986) Energy-dependent uptake of N-methyl-4-phenylpyridinium, the neurotoxic metabolite of 1-methyl-4-phenyl-1,2,3,6-tetrahydropyridine, by mitochondria. J Biol Chem 261:7585–7587

Rathke-Hartlieb S, Schmidt VC, Jockusch H, Schmitt-John T, Bartsch JW (1999) Spatiotemporal progression of neurodegeneration and glia activation in the wobbler neuropathy of the mouse. Neuroreport 10:3411–3416

Reynolds AD, Banerjee R, Liu J, Gendelman HE, Mosley RL (2007) Neuroprotective activities of CD4+CD25+ regulatory T cells in an animal model of Parkinson's disease. J Leukoc Biol 82:1083–1094

Reynolds AD, Glanzer JG, Kadiu I, Ricardo-Dukelow M, Chaudhuri A, Ciborowski P, Cerny R, Gelman B, Thomas MP, Mosley RL, Gendelman HE (2008) Nitrated alpha-synuclein-activated microglial profiling for Parkinson's disease. J Neurochem 104:1504–1525

Reynolds AD, Stone DK, Mosley RL, Gendelman HE (2009a) Nitrated {alpha}-synuclein-induced alterations in microglial immunity are regulated by CD4+ T cell subsets. J Immunol 182:4137–4149

Reynolds AD, Stone DK, Mosley RL, Gendelman HE (2009b) Proteomic studies of nitrated alpha-synuclein microglia regulation by CD4+CD25+ T cells. J Proteome Res 8:3497–3511

Reynolds AD, Stone DK, Hutter JA, Benner EJ, Mosley RL, Gendelman HE (2010) Regulatory T cells attenuate Th17 cell-mediated nigrostriatal dopaminergic neurodegeneration in a model of Parkinson's disease. J Immunol 184:2261–2271

Richfield EK, Thiruchelvam MJ, Cory-Slechta DA, Wuertzer C, Gainetdinov RR, Caron MG, Di Monte DA, Federoff HJ (2002) Behavioral and neurochemical effects of wild-type and mutated human alpha-synuclein in transgenic mice. Exp Neurol 175: 35–48

Romero R, Manogue KR, Mitchell MD, Wu YK, Oyarzun E, Hobbins JC, Cerami A (1989) Infection and labor. IV. Cachectin-tumor necrosis factor in the amniotic fluid of women with intraamniotic infection and preterm labor. Am J Obstet Gynecol 161:336–341

Saha S, Guillily MD, Ferree A, Lanceta J, Chan D, Ghosh J, Hsu CH, Segal L, Raghavan K, Matsumoto K, Hisamoto N, Kuwahara T, Iwatsubo T, Moore L, Goldstein L, Cookson M, Wolozin B (2009) LRRK2 modulates vulnerability to mitochondrial dysfunction in Caenorhabditis elegans. J Neurosci 29:9210–9218

Sang TK, Chang HY, Lawless GM, Ratnaparkhi A, Mee L, Ackerson LC, Maidment NT, Krantz DE, Jackson GR (2007) A Drosophila model of mutant human parkin-induced toxicity demonstrates selective loss of dopaminergic neurons and dependence on cellular dopamine. J Neurosci 27:981–992

Sauer H, Oertel WH (1994) Progressive degeneration of nigrostriatal dopamine neurons following intrastriatal terminal lesions with 6-hydroxydopamine: a combined retrograde tracing and immunocytochemical study in the rat. Neuroscience 59:401–415

Sawada H, Kohno R, Kihara T, Izumi Y, Sakka N, Ibi M, Nakanishi M, Nakamizo T, Yamakawa K, Shibasaki H, Yamamoto N, Akaike A, Inden M, Kitamura Y, Taniguchi T, Shimohama S (2004) Proteasome mediates dopaminergic neuronal degeneration, and its inhibition causes alpha-synuclein inclusions. J Biol Chem 279:10710–10719

Schmalbruch H, Jensen HJ, Bjaerg M, Kamieniecka Z, Kurland L (1991) A new mouse mutant with progressive motor neuronopathy. J Neuropathol Exp Neurol 50:192–204

Schober A (2004) Classic toxin-induced animal models of Parkinson's disease: 6-OHDA and MPTP. Cell Tissue Res 318:215–224

Sherer TB, Kim JH, Betarbet R, Greenamyre JT (2003) Subcutaneous rotenone exposure causes highly selective dopaminergic degeneration and alpha-synuclein aggregation. Exp Neurol 179:9–16

Shibata N (2001) Transgenic mouse model for familial amyotrophic lateral sclerosis with superoxide dismutase-1 mutation. Neuropathology 21:82–92

Shimizu K, Ohtaki K, Matsubara K, Aoyama K, Uezono T, Saito O, Suno M, Ogawa K, Hayase N, Kimura K, Shiono H (2001) Carrier-mediated processes in blood–brain barrier penetration and neural uptake of paraquat. Brain Res 906:135–142

Shimura H, Hattori N, Kubo S, Mizuno Y, Asakawa S, Minoshima S, Shimizu N, Iwai K, Chiba T, Tanaka K, Suzuki T (2000) Familial Parkinson disease gene product, parkin, is a ubiquitin-protein ligase. Nat Genet 25:302–305

Silvestri L, Caputo V, Bellacchio E, Atorino L, Dallapiccola B, Valente EM, Casari G (2005) Mitochondrial import and enzymatic activity of PINK1 mutants associated to recessive parkinsonism. Hum Mol Genet 14:3477–3492

Spillantini MG, Schmidt ML, Lee VM, Trojanowski JQ, Jakes R, Goedert M (1997) Alpha-synuclein in Lewy bodies. Nature 388:839–840

Spillantini MG, Crowther RA, Jakes R, Hasegawa M, Goedert M (1998) Alpha-synuclein in filamentous inclusions of Lewy bodies from Parkinson's disease and dementia with Lewy bodies. Proc Natl Acad Sci U S A 95:6469–6473

Sriram SR, Li X, Ko HS, Chung KK, Wong E, Lim KL, Dawson VL, Dawson TM (2005) Familial-associated mutations differentially disrupt the solubility, localization, binding and ubiquitination properties of parkin. Hum Mol Genet 14:2571–2586

Subramaniam JR, Lyons WE, Liu J, Bartnikas TB, Rothstein J, Price DL, Cleveland DW, Gitlin JD, Wong PC (2002) Mutant SOD1 causes motor neuron disease independent of copper chaperone-mediated copper loading. Nat Neurosci 5:301–307

Takahashi K, Taira T, Niki T, Seino C, Iguchi-Ariga SM, Ariga H (2001) DJ-1 positively regulates the androgen receptor by impairing the binding of PIASx alpha to the receptor. J Biol Chem 276:37556–37563

Talpade DJ, Greene JG, Higgins DS Jr, Greenamyre JT (2000) In vivo labeling of mitochondrial complex I (NADH:ubiquinone oxidoreductase) in rat brain using [(3)H]dihydrorotenone. J Neurochem 75:2611–2621

Terzioglu M, Galter D (2008) Parkinson's disease: genetic versus toxin-induced rodent models. FEBS J 275: 1384–1391

Teuling E, van Dis V, Wulf PS, Haasdijk ED, Akhmanova A, Hoogenraad CC, Jaarsma D (2008) A novel mouse model with impaired dynein/dynactin function develops amyotrophic lateral sclerosis (ALS)-like features in motor neurons and improves lifespan in SOD1-ALS mice. Hum Mol Genet 17:2849–2862

Theodore S, Cao S, McLean PJ, Standaert DG (2008) Targeted overexpression of human alpha-synuclein triggers microglial activation and an adaptive immune response in a mouse model of Parkinson disease. J Neuropathol Exp Neurol 67:1149–1158

Thiruchelvam M, Richfield EK, Baggs RB, Tank AW, Cory-Slechta DA (2000a) The nigrostriatal dopaminergic system as a preferential target of repeated exposures to combined paraquat and maneb: implications for Parkinson's disease. J Neurosci 20:9207–9214

Thiruchelvam M, Brockel BJ, Richfield EK, Baggs RB, Cory-Slechta DA (2000b) Potentiated and preferential effects of combined paraquat and maneb on nigrostriatal dopamine systems: environmental risk factors for Parkinson's disease? Brain Res 873:225–234

Thiruchelvam MJ, Powers JM, Cory-Slechta DA, Richfield EK (2004) Risk factors for dopaminergic neuron loss in human alpha-synuclein transgenic mice. Eur J Neurosci 19:845–854

Troncoso JC, Price DL, Griffin JW, Parhad IM (1982) Neurofibrillary axonal pathology in aluminum intoxication. Ann Neurol 12:278–283

Tudor EL, Galtrey CM, Perkinton MS, Lau KF, De Vos KJ, Mitchell JC, Ackerley S, Hortobagyi T, Vamos E, Leigh PN, Klasen C, McLoughlin DM, Shaw CE, Miller CC (2010) Amyotrophic lateral sclerosis mutant vesicle-associated membrane protein-associated protein-B transgenic mice develop TAR-DNA-binding protein-43 pathology. Neuroscience 167:774–785

Turiault M, Parnaudeau S, Milet A, Parlato R, Rouzeau JD, Lazar M, Tronche F (2007) Analysis of dopamine transporter gene expression pattern—generation of DAT-iCre transgenic mice. FEBS J 274:3568–3577

Ungerstedt U (1968) 6-Hydroxy-dopamine induced degeneration of central monoamine neurons. Eur J Pharmacol 5:107–110

Ungerstedt U (1971) Postsynaptic supersensitivity after 6-hydroxy-dopamine induced degeneration of the nigro-striatal dopamine system. Acta Physiol Scand Suppl 367:69–93

Ungerstedt U, Arbuthnott GW (1970) Quantitative recording of rotational behavior in rats after 6-hydroxy-dopamine lesions of the nigrostriatal dopamine system. Brain Res 24:485–493

Valente EM et al (2004) Hereditary early-onset Parkinson's disease caused by mutations in PINK1. Science 304:1158–1160

Van Den Bosch L, Tilkin P, Lemmens G, Robberecht W (2002) Minocycline delays disease onset and mortality in a transgenic model of ALS. Neuroreport 13: 1067–1070

Vance C et al (2009) Mutations in FUS, an RNA processing protein, cause familial amyotrophic lateral sclerosis type 6. Science 323:1208–1211

von Coelln R, Dawson VL, Dawson TM (2004) Parkin-associated Parkinson's disease. Cell Tissue Res 318:175–184

Wang C, Lu R, Ouyang X, Ho MW, Chia W, Yu F, Lim KL (2007) Drosophila overexpressing parkin R275W mutant exhibits dopaminergic neuron degeneration and mitochondrial abnormalities. J Neurosci 27:8563–8570

Wegorzewska I, Bell S, Cairns NJ, Miller TM, Baloh RH (2009) TDP-43 mutant transgenic mice develop features of ALS and frontotemporal lobar degeneration. Proc Natl Acad Sci U S A 106:18809–18814

West AB, Moore DJ, Biskup S, Bugayenko A, Smith WW, Ross CA, Dawson VL, Dawson TM (2005) Parkinson's disease-associated mutations in leucine-rich repeat kinase 2 augment kinase activity. Proc Natl Acad Sci U S A 102:16842–16847

Widdowson PS, Farnworth MJ, Simpson MG, Lock EA (1996a) Influence of age on the passage of paraquat through the blood-brain barrier in rats: a distribution and pathological examination. Hum Exp Toxicol 15: 231–236

Widdowson PS, Farnworth MJ, Upton R, Simpson MG (1996b) No changes in behaviour, nigro-striatal system neurochemistry or neuronal cell death following toxic multiple oral paraquat administration to rats. Hum Exp Toxicol 15:583–591

Wils H, Kleinberger G, Janssens J, Pereson S, Joris G, Cuijt I, Smits V, Ceuterick-de Groote C, Van Broeckhoven C, Kumar-Singh S (2010) TDP-43 transgenic mice develop spastic paralysis and neuronal inclusions characteristic of ALS and frontotemporal lobar degeneration. Proc Natl Acad Sci U S A 107: 3858–3863

Yase Y (1987) The pathogenetic role of metals in motor neuron disease—the participation of aluminum. In: Cosi V, Kato A, Parlette W, Pinelli P, Poloni M (eds) Amyotrophic lateral sclerosis: therapeutic, psychological and research aspects. Plenum, New York, pp 89–96

Zhang L, Shimoji M, Thomas B, Moore DJ, Yu SW, Marupudi NI, Torp R, Torgner IA, Ottersen OP, Dawson TM, Dawson VL (2005) Mitochondrial localization of the Parkinson's disease related protein DJ-1: implications for pathogenesis. Hum Mol Genet 14: 2063–2073

Zhou W, Zhu M, Wilson MA, Petsko GA, Fink AL (2006) The oxidation state of DJ-1 regulates its chaperone activity toward alpha-synuclein. J Mol Biol 356: 1036–1048

Zhu S, Stavrovskaya IG, Drozda M, Kim BY, Ona V, Li M, Sarang S, Liu AS, Hartley DM, Wu DC, Gullans S, Ferrante RJ, Przedborski S, Kristal BS, Friedlander RM (2002) Minocycline inhibits cytochrome c release and delays progression of amyotrophic lateral sclerosis in mice. Nature 417:74–78

Zhu X, Siedlak SL, Smith MA, Perry G, Chen SG (2006) LRRK2 protein is a component of Lewy bodies. Ann Neurol 60:617–618, author reply 618–619

Zhuang X, Masson J, Gingrich JA, Rayport S, Hen R (2005) Targeted gene expression in dopamine and serotonin neurons of the mouse brain. J Neurosci Methods 143:27–32

Zimprich A et al (2004) Mutations in LRRK2 cause autosomal-dominant parkinsonism with pleomorphic pathology. Neuron 44:601–607

Animal Models of Alzheimer's Disease

35

Tomomi Kiyota

Abstract

Alzheimer's disease (AD) ranks as a leading cause of death among the progressive neurodegenerative diseases. It is the most common form of dementia found in the elderly population and is characterized by deterioration in mental function and abnormal behavior including wandering, irritability, and aggression. The anatomical features of AD are brain shrinkage caused by massive neuronal loss and dilatation of lateral ventricles. The neuropathological hallmarks include senile plaques containing deposits of amyloid-β peptide (Aβ) and neurofibrillary tangles (NFTs) consisting of hyperphosphorylated forms of the microtubule-associated protein tau. Both Aβ and tau can be misfolded to generate their toxic forms resulting in neuroinflammation and neurodegeneration. Normally Aβ is a soluble peptide but can be misfolded to form toxic oligomeric Aβ and fibrillar Aβ, causing amyloid deposits and cerebral amyloid angiopathy (CAA). Abnormally phosphorylated tau also forms soluble toxic oligomers and insoluble NFTs intraneuronally. Since the first murine model of AD was reported one and a half decades ago, many other models have been generated by modification of transgenes that cause familial AD. These models have contributed to significant basic research and numerous therapeutic innovations to treat AD and halt disease progression. Although most of the models mirror featured aspects of AD pathogenesis such as β-amyloidosis, NFT formation, cognitive dysfunction, and/or synaptic loss, no perfect model matching human AD is currently available. In this chapter, representative animal models for AD research and their pathological and behavioral features are described.

T. Kiyota, Ph.D. (✉)
Department of Pharmacology and Experimental Neuroscience, 985930 Nebraska Medical Center, University of Nebraska Medical Center, Omaha, NE 68198-5930, USA
e-mail: tkiyota@unmc.edu

H. Xiong and H.E. Gendelman (eds.), *Current Laboratory Methods in Neuroscience Research*, Springer Protocols Handbooks, DOI 10.1007/978-1-4614-8794-4_35,

Keywords
Alzheimer's disease • Transgenic mouse models • Amyloid-β precursor protein • Presenilin • Amyloid-β peptides • Tau pathology

35.1 Overview of Alzheimer's Disease Pathogenesis and Animal Models

Alzheimer's disease (AD) is the most prevalent neurodegenerative disorder in the world (Selkoe 1991). The diagnostic neuropathological hallmarks of AD include senile plaques, which are the accumulation of misfolded amyloid-β peptides (Aβ) in the extracellular space, and perivascular deposits of Aβ in vascular endothelial cells as cerebral amyloid angiopathy (CAA). Aβ is processed by sequential cleavage of amyloid-β precursor protein (APP), which is a single-pass transmembrane protein with the C-terminus of the Aβ region located within the intramembranous domain; by β-site APP cleaving enzymes (BACE-1 and BACE-2) at the N-terminus of the Aβ; and by a γ-secretase complex consisting of presenilin (PS) 1 or 2, nicastrin, aph-1, and pen-2 at the C-terminus of the Aβ (Haass 2004). The process wherein Aβ is abnormally deposited is called β-amyloidosis. Another neuropathological hallmark is neurofibrillary tangles (NFTs), which are intraneuronal aggregation of misfolded tau protein. Tau is a microtubule-associated protein that has numerous functions, including stabilizing microtubules, regulating neurite outgrowth, and axonal transport (Drubin and Kirschner 1986; Drechsel et al. 1992). NFTs are formed as a result of intraneuronal polymerization of tau abnormally hyperphosphorylated by kinases such as glycogen synthase kinase-3 (GSK-3) and cyclin-dependent kinase 5 (CDK5) (Mandelkow et al. 1993; Drewes 2004).

Currently, the most prominent theory for the causation of AD is the amyloid hypothesis (Hardy and Selkoe 2002; Tanzi and Bertram 2005). In the healthy brain, the non-amyloidogenic pathway is predominant in releasing a secreted amino-terminal fragment of APP (APPα) from the cell surface and leaving a short carboxy-terminal APP fragment (C83). The pathway is mediated by α-secretase such as a disintegrin and metalloprotease (ADAM) 9, 10, or 17 and tumor necrosis factor-α-converting enzyme (Mattson 2004). In the AD brain, the amyloidogenic pathway is predominant in producing Aβ. Accumulation of Aβ—especially in its most pathogenic form of oligomeric Aβ such as dimers, dodecamers, and others—leads to reactive astrogliosis (Hou et al. 2011), microglial activation (Maezawa et al. 2011), tau hyperphosphorylation causing NFT formation (De Felice et al. 2008; Jin et al. 2011), neuronal toxicity, impaired hippocampal long-term potentiation (LTP), and eventually synaptic and neuronal loss (Walsh et al. 2002; Walsh and Selkoe 2004b; Lesne et al. 2006; Shankar et al. 2008). These losses and axonal pathology are most of the key neuropathological features leading to dementia (DeKosky and Scheff 1990; Masliah 2001; Scheff and Price 2001; Raff et al. 2002). Furthermore, impaired hippocampal neurogenesis in the adult brain also contributes to deficits in learning and memory (Haughey et al. 2002b; Tatebayashi et al. 2003; Verret et al. 2007; Lazarov and Marr 2010).

Although the most common form (>95 %) of AD patients is sporadic and late onset in the elderly population, approximately less than 5 % is familial AD (FAD), in which mutations in APP, PS-1, or PS-2 gene can lead to early-onset FAD (Rocchi et al. 2003). Numerous evidence supports the amyloid hypothesis in FAD patients, since inherited forms of FAD linked with APP or PS gene mutations are associated with an increased production of Aβ1-42 (Aβ42), which is more aggregation-prone form of Aβ than a shorter form Aβ1-40 (Aβ40) (Yoshiike et al. 2003; Jan et al. 2008). In mutant APP overexpressing transgenic (Tg) mouse models of AD, abundant Aβ deposits develop in the brain, along with memory deficits and impaired hippocampal neurogenesis (Haughey et al. 2002a, b; Morgan 2007). Active or passive immunization of Aβ, or other therapeutic prevention to clear Aβ, is beneficial and improves cognitive function in experimental mice (Wilcock and Colton 2008; Town 2009; Morgan 2011). Concerning tau pathology in AD, tau is hyperphosphorylated by several kinases such as GSK-3, whose activity is increased by interference of insulin or of the Wnt pathway (Townsend et al. 2007; Magdesian et al. 2008). In Tg mouse models overexpressing both mutant APP and tau, oligomeric Aβ formation precedes and accentuates tau-related pathology, resulting in both β-amyloidosis and tau pathology (Frautschy et al. 1991; Gotz et al. 2001; Lewis et al. 2001; Oddo et al. 2003b). Removal of accumulated Aβ ameliorates cognitive deficits as well as tau pathology (Oddo et al. 2006; McKee et al. 2008). These findings suggest that NFT formation is downstream of Aβ aggregation; thus, the amyloid hypothesis is strongly supported as an explanation for AD pathogenesis.

Massive neuronal loss and brain atrophy are major diagnostic signs and pathogenic features of AD. The mechanism of how these primary pathogenic events take place is not fully understood yet, but toxic forms of oligomeric Aβ are thought to trigger synaptic loss, nerve damage, and neuronal cell death (Volles et al. 2001; Volles and Lansbury 2002; Walsh and Selkoe 2004a). In a set of in vitro experimental models, Aβ dimers, trimers, and larger oligomers produced from cultured neurons inhibit LTP and damage spines (Klein et al. 2001; Walsh and Selkoe 2004a; Townsend et al. 2006; Selkoe 2008). Naturally formed Aβ dodecamers in the brain of APP Tg mice promote memory deficits in an age-dependent manner (Lesne et al. 2006). Aβ trimers and dodecamers produced from cultured cells appear to share common synaptotoxic properties (Selkoe 2008). Moreover, Aβ dimers obtained from human cerebrospinal fluid disrupt synaptic plasticity and inhibit hippocampal LTP in vivo (Klyubin et al. 2008). These studies suggest that misfolded Aβ oligomers (2–12 mers) seem to be the most predominant contributors in AD pathogenesis including tau pathology (Lacor et al. 2007).

35.2 Experimental Animal Models of Alzheimer's Disease

In order to properly examine the disease, ideal animal models of AD would need to exhibit these important attributes: (1) developing Aβ plaques and CAA; (2) predominantly producing high-dose Aβ42; (3) promoting Aβ oligomer formation; (4) producing intraneuronal accumulation of Aβ; (5) inducing hyperphosphorylation of tau and developing intraneuronal tangles; (6) enhancing neurodegeneration with synaptic loss, axonal damage, and defects in neurogenesis; and (7) showing behavioral (learning and memory) and electrophysiological (LTP, synaptic plasticity) impairments that correspond to the pattern of neurodegeneration. Although numerous animal models of AD—mostly Tg mouse models with featured aspects of AD pathogenesis—have been developed, there is currently no single model able to mimic all the aspects observed in AD. Possibly the best model is an aged monkey (Price et al. 1997) but is accompanied with high cost and requires significant time for studies. Therefore, it is necessary to choose an animal model of AD that is appropriate and best fit for the study-specific aims, symptom components, and outcomes.

The most popular animal models of AD are Tg mice overexpressing APP, PS-1, and/or PS-2 with single or multiple FAD-linked mutations (Hutton and Hardy 1997; Cruts and Van Broeckhoven 1998; Rocchi et al. 2003; Bertoli-Avella et al. 2004; Pastor and Goate 2004). APP Tg and APP/PS double-Tg mice exhibit key features of AD such as high levels of Aβ and formation of Aβ deposition, but NFTs are not generated. Therefore, mouse models overexpressing tau or a mutant form of tau with NFT formation are utilized to investigate the role of tau dysfunction and NFTs. Although severe neuronal loss is not found in any current mouse model, impaired neurogenesis is observed in some of the models.

35.3 Mouse Models of Amyloid-β Plaque Deposition

Most cases of AD are late-onset sporadic AD (>95 %) and not associated with any FAD mutation. However, the neuropathology observed in sporadic cases is essentially identical to that observed in early-onset FAD cases. Human Aβ aggregates to generate Aβ deposition in the brain, but murine Aβ cannot, even if overexpressed (Jankowsky et al. 2007). Thus, Tg mouse models overexpressing human APP (hAPP) transgenes with FAD mutations to produce human Aβ are used very often. Alternative splicing in the control of hAPP gene expression produces three major isoforms containing 770, 751, or 695 amino acids, but the 695 isoform is predominantly produced in the brain as shown in their mRNA levels: APP 770:751:695 mRNA ratio is 1:10:20 in cortex (Tanaka et al. 1989). These Tg models show Aβ overproduction, especially increased Aβ42 causing a higher tendency to aggregate than Aβ40 (Walsh and Selkoe 2004b). Representative Tg mouse models of AD are summarized in Table 35.1.

The first successful Tg mouse model of AD was reported in 1995. This mouse overexpresses a hAPP minigene containing the Indiana mutation (V717F), driven by a platelet-derived growth factor-β (PDGF) promoter, thus named PDAPP mouse (line 109) (Games et al. 1995). This model exhibits tenfold increased expression of APP transgene compared to endogenous mouse APP. Since the Indiana mutation increases γ-secretase processing of Aβ42 compared to Aβ40, there is robust Aβ accumulation and deposition along with thioflavin-S-positive (TS+) plaques with dense cores. These processes begin between 6 and 9 months of age and are associated with dystrophic neuritis, astrocytosis, and microgliosis (Games et al. 1995). Synaptic loss, impaired synaptic plasticity, and age-related learning deficits are observed in both young and aged mice (Larson et al. 1999; Chen et al. 2000; Dodart et al. 2000). Hyperphosphorylated tau, but not NFT, is also found in the brain (Masliah et al. 2001).

The Tg2576 strain is the most widely used Tg model and well characterized in many aspects of AD. This mouse strain overexpresses an APP transgene with the Swedish mutation (APP 695, K670N/M671L) fivefold higher than endogenous mouse APP under the control of the hamster prion protein (PrP) promoter, which drives expression broadly in the nervous system (Hsiao et al. 1996). The Swedish mutation contributes to accelerated production of both Aβ40 and Aβ42, since the mutation enhances BACE-1 cleavage of APP (Citron et al. 1992). Aβ levels in the brain start to increase at 5 and 6 months, Aβ deposits and TS + plaques with dense cores start to develop between 9 and 12 months (Hsiao et al. 1996; Lesne et al. 2006), and abundant CAAs are observed in aged Tg2576 mice (Wilcock et al. 2006). The age-dependent β-amyloidosis is associated with oxidative stress and inflammatory response including astrogliosis, microglial activation, and cytokine production (Irizarry et al. 1997; Frautschy et al. 1998; Pappolla et al. 1998; Smith et al. 1998; Tan et al. 1999). Hyperphosphorylated tau, but not NFT, is present as it is in PDAPP mice (Kawarabayashi et al. 2004). A variety of behavioral tasks have shown age-dependent behavioral deficits. Spatial memory impairment is robust in more aged Tg2576 mice but can be observed as early as 6 months, suggesting contribution of soluble Aβ species to cognitive dysfunction before Aβ deposits appear (Hsiao et al. 1996; Westerman et al. 2002; Lesne et al. 2006). Soluble Aβ dodecamers with a molecular weight of 56 kDa are significantly correlated with memory impairments in the Tg2576 mice (Lesne et al. 2006).

The TgCRND8 line also overexpresses APP695 transgene under the control of the PrP promoter, but the transgene has both Swedish and Indiana mutations (Chishti et al. 2001). β-amyloidosis in this model quite rapidly occurs, as early as 3 months, since the two mutations result in enhancement of both BACE and γ-secretase activities. The line exhibits high mortality with less than 50 % survival 5 months after weaning, suggesting that rapid β-amyloidosis is involved in increased susceptibility to spontaneous

Table 35.1 Representative Tg mouse models of AD

Line	Gene (mutation)	Promoter	Aβ plaques	NFTs	Cognitive deficits	Phenotype	References
PDAPP	hAPP695, 751, 770 (Ind: V717F)	PDGF-β	6–9 months	Hyperphosphorylated tau only (p-tau)	6 months	Synaptic damage, astrogliosis/ microgliosis	Games et al. (1995)
Tg2576	hAPP695 (Swe: K670N/M671L)	hamPrP	9–12 months	p-tau	6–9 months	Vascular deposits, astrogliosis/ microgliosis	Hsiao et al. (1996)
TgCRND8	hAPP695 (Swe + Ind)	hamPrP	3–5 months	p-tau	11 weeks	Rapid accumulation of Aβ42, gliosis, neuritic dystrophy	Chishti et al. (2001)
J20	hAPP695, 751, 770 (Swe + Ind)	PDGF-β	5–6 months	p-tau	7 months	Synaptic damage, astrogliosis/ microgliosis, neuritic dystrophy	Mucke et al. (2000)
APP23	hAPP751 (Swe)	mThy-1	3 months	p-tau	6 months	Astrogliosis/microgliosis, neuritic dystrophy, neuronal loss (14 months)	Sturchler-Pierrat et al. (1997)
mThy-1 hAPP751	hAPP751 (Swe + Lon: V717I)	mThy-1	3–4 months	p-tau	6–9 months	Synaptic loss, neuritic loss	Rockenstein et al. (2001, 2003)
ARC6, ARC48	hAPP695, 751, 770 (Swe + Ind + Arc: E693G)	PDGF-β	2–3 months	p-tau	3–4 months (ARC48)	Increased oligomers, highly fibrillogenic	Cheng et al. (2004, 2007)
APPDutch	hAPP751 (Dutch: E693Q)	mThy-1	22 months (CAA)	–	–	Few parenchymal plaques	Herzig et al. (2004)
PSAPP	hAPP695 (Swe) × PS-1 (M146L)	hamPrP PDGF-β	5–6 months	p-tau	5–7 months	Accelerated Aβ, astrogliosis/ microgliosis, impaired neurogenesis	Holcomb et al. (1998), Morgan et al. (2000)
APPswe/ PS-1 ΔE9	m/hAPP695 (Swe) PS-1 ΔE9	mPrP mPrP	6–7 months	p-tau	7 months	Accelerated Aβ, neuritic dystrophy impaired neurogenesis	Jankowsky et al. (2004)
5XFAD	hAPP695 (swe) (Swe, Lon, Flo) PS-1 (M146L, L286V)	mThy-1 mThy-1	2 months	p-tau	4 months	Intraneuronal Aβ, neuronal loss (9 months), synaptic loss, increased p25	Oakley et al. (2006)
JNPL3	4R0N MAPT (P301L)	mPrP	–	9 months	–	Motor dysfunction, neuronal loss (6–7 months)	Lewis et al. (2000)
Htau	hMAPT on mMapt KO	hTau	–	15 months	12 months	Neuronal loss (15 months)	Andorfer et al. (2003, 2005)
TAPP	hAPP695 (Swe) × 4R0N MAPT (P301L)	hamPrP mPrP	6–9 months	9 months	6–7 months	Enhanced NFTs in the brain	Lewis et al. (2001)
3xTg-AD	hAPP695 (Swe) 4R0N MAPT (P301L) on PS-1 (M146V) KI	mThy-1.2 mThy-1.2	6 months	10–12 months	6 months	Enhanced NFTs in the brain	Oddo et al. (2003a, b), Billings et al. (2005)

seizures leading to death. Aβ deposition starts in the subiculum and the frontal cortex, continuing in rest areas of the hippocampus and cortex and eventually spreading to the thalamus, striatum, cerebellum, and brainstem by 8 months. The model also shows neuritic dystrophy and gliosis at 5 months and CAA at 6 months. Impaired cognitive function is observed at 3 months of age, determined by the Morris water maze task.

The J20 line is a PDAPP-based Tg mouse model overexpressing APP minigene with both Swedish and Indiana mutations under the control of the PDGF promoter (Mucke et al. 2000). This mouse model produces higher levels of Aβ in the brain than PDAPP and Tg2576 mice, resulting in early-onset deficits in behavior and synaptic transmission (Hsia et al. 1999; Poirier et al. 2006; Saganich et al. 2006; Venkitaramani et al. 2007). J20 mice exhibit Aβ deposition starting around 6 months, accompanied with neuritic dystrophy and astrogliosis/microgliosis. Synaptic loss and deficit in synaptic plasticity are observed as early as 2–4 months. Notably, behavior and synaptic abnormalities are associated with molecular alterations in calcium- and synaptic activity-related proteins such as calbindin (calcium-binding protein), Arc and Fos (immediate-early gene products), and α-actinin-2 (dendritic spine actin-binding protein) (Palop et al. 2005). Chronic electroencephalographic recordings revealed aberrant excitatory neuronal activity that seems to be induced by high amounts of Aβ and increases compensatory inhibitory mechanisms in the hippocampus, which correlates with learning and memory deficits (Palop et al. 2007; Palop and Mucke 2009).

Sturchler-Pierrat et al. reported the APP23 mouse line, which has APP751 isoform with Swedish mutation driven by the murine Thy-1 promoter, with sevenfold Aβ overexpression (Sturchler-Pierrat et al. 1997). This mouse shows Aβ deposition at 6 months in the hippocampus and cortex and is associated with neuritic dystrophy, gliosis, and age-dependent increase of hyperphosphorylated tau, but not NFTs. Degeneration processes including neuritic dystrophy and distortion of cholinergic fibers are observed surrounding the plaques (Calhoun et al. 1998). The authors also reported another mouse line, APP22, having Swedish and London (V717I) mutations with only twofold overexpression of APP. Thus, Aβ deposition is not observed until 8 months of age.

The mThy-1 hAPP 751 line overexpresses APP751 isoform with Swedish and London mutations driven by the murine Thy-1 promoter (Rockenstein et al. 2001). The line shows high levels of Aβ42 and develops mature plaques in the frontal cortex as early as 3–4 months. The plaques spread to the hippocampus, thalamus, and olfactory regions at 5–7 months, accompanied with dystrophic neurites, gliosis, and a decrease in the number of presynaptic terminals.

35.4 Mouse Models Increasing Aβ Oligomers and Cerebral Amyloid Angiopathy

Although Tg mouse models overexpressing hAPP with mutations present just outside the amino acid sequence of Aβ are characterized by enhanced Aβ production with Aβ plaque formation, mutations inside the sequence of Aβ show increased oligomer formation and CAA. Which Aβ species is responsible for the neurodegenerative process in AD is still controversial, but recent studies support a critical role for small Aβ oligomers (2–12 mers) (Walsh and Selkoe 2004a; Glabe 2005; Glabe and Kayed 2006). The ARC6 and ARC48 mouse lines predominantly produce such oligomers in the brain (Cheng et al. 2004). The lines were generated based on the J20 line. The Arctic mutation (E693G, E22G in Aβ) was introduced in APP transgene for J20, resulting in triple mutations (Swedish, Indiana, and Arctic mutations). Production of Aβ with the Arctic mutation, which is highly fibrillogenic in vivo, causes more oligomers and protofibrils compared to wild-type Aβ. The ARC48 mice show high levels of both oligomeric Aβ and plaque deposition, as well as cognitive deficits (Cheng et al. 2007). Other mutations such as Dutch (E693Q, E22Q), Italian (E693K, E22K), and Iowa (D694N, D23N) located within the amino acid sequence of Aβ also increase oligomers and protofibril formation (Demeester et al. 2001; Lashuel et al. 2003; Betts et al. 2008). A possible explanation of the increase in the number of oligomers has been proposed as a proteolytic resistance of Aβ (Tsubuki et al. 2003).

Most AD patients also develop amyloid deposits around blood vessels, resulting in a condition known as CAA that can cause vascular fragility and hemorrhages (Pezzini et al. 2009). Approximately 20 % of AD patients have severe CAA at autopsy, and about 33 % of cognitively normal elderly populations have some degree of CAA (Jellinger 2002; Zhang-Nunes et al. 2006). Some mouse models such as Tg2576 and TgCRND8 develop parenchymal Aβ accumulation and CAA in the brain. However, Herzig et al. created the APP Dutch mutant mouse, which develops significant CAA with only parenchymal diffuse Aβ (Herzig et al. 2004). The mouse is a model of hereditary CAA and overexpresses hAPP751 with Dutch mutation (E693Q, E22Q) under the control of the Thy1 promoter. This mutation causes vascular Aβ deposition observed at 22 months, increased hemorrhage incidence, smooth muscle cell degeneration, perivascular microgliosis, and astrocyte activation. These findings are similar to the case of hereditary cerebral hemorrhage with amyloidosis-Dutch type (HCHWA-D) (Herzig et al. 2006). The ratio of the AβDutch (E22Q-mutant Aβ)40 to AβDutch42 is significantly higher than the ratio of Aβwt (wild-type Aβ)40 to Aβwt42 in the other APP Tg mice; thus, CAA rather than parenchymal Aβ deposits is developed.

The same team developed another model of CAA by crossing the APP Dutch mice with APP23 mice (see above) (Herzig et al. 2009). The double-Tg mice produce both Aβwt and AβDutch, but vascular deposits and hemorrhages are twice as high, and parenchymal Aβ deposition is reduced, as compared to APP23 mice, suggesting that AβDutch40 predominantly increases vascular deposits and reduces parenchymal amyloidosis.

35.5 Mouse Models with Presenilin Mutations and APP

Presenilin (PS-1, PS-2) forms the catalytic part of the γ-secretase complex responsible for the intramembranous cleavage of transmembrane proteins, including APP. Mutations in the encoding gene *PSEN1* were discovered to be pathogenic for early-onset FAD (Sherrington et al. 1995). While mutations in a related gene *PSEN2* were also linked to FAD (Ertekin-Taner 2007), those in *PSEN1* are more common, with greater than 180 mutations found to date. FAD mutations in *PSEN1* and *2* consistently show elevated levels of amyloidogenic Aβ42 production, supporting the amyloid hypothesis (Lemere et al. 1996; Scheuner et al. 1996; Citron et al. 1997). Several Tg lines with PS FAD mutants have been created in parallel with creation of APP mice (Borchelt et al. 1996; Duff et al. 1996; Citron et al. 1997).

The first Tg mouse with PS-1 mutation was reported in 1996. The mouse overexpresses a human PS-1 transgene containing the Italian (M146L) or Finnish (M146V) mutation under the control of a PDGF promoter, resulting in elevation of endogenous mouse Aβ42 production in the brain (Duff et al. 1996). However, single FAD transgenic PS mice do not develop Aβ plaques. A possible explanation why the mice fail to form the plaques is most likely structural differences between mouse and human Aβ. In the comparison of both mouse and human Aβ, three amino acid substitutions distinguish mouse and human Aβ that might contribute to their differing properties in vivo (Jankowsky et al. 2007). Therefore, the PS mice are usually crossed with hAPP lines, demonstrating that the PS mutations cause accelerated Aβ plaque formation.

Several varieties of PSAPP (PS×APP bigenic) mice have been reported. A common feature in the bigenic mice is not only increased Aβ42 production as observed in models of single PS mutations both in vivo and in vitro but also accelerated formation of Aβ deposits in the brain. A cross of Tg2576 mice with M146L PS-1 mutant mice showed increased Aβ42/Aβ40 ratio, Aβ deposition around 5–6 months accompanied by astrogliosis/microgliosis, and cognitive deficits at 5–7 months with more robust deficits at 15–17 months (Holcomb et al. 1998; Morgan et al. 2000; Arendash et al. 2001). A cross of TgCRND8 mice with M146L/L286V PS-1 mutant mice results in quick formation of Aβ deposits at 1–1.5 months (Citron et al. 1997; Chishti et al. 2001).

Line 85 is another Tg mouse expressing both hAPP Swedish and PS-1 mutants (APPswe/PS-1ΔE9: the exon-9-deleted variant of PS-1) (Jankowsky et al. 2004). The line was created by co-injection of vectors with APPswe and PS-1ΔE9 controlled by independent mouse PrP promoter elements. An advantage of the line is that the two transgenes co-integrated and co-segregate as a single locus.

5XFAD mice quickly show the most severe neuropathological aspects of AD including amyloidosis, gliosis, and neurodegeneration (Oakley et al. 2006). The mice overexpress APP695 with Swedish, London, and Florida (I716V) mutations and PS-1 with M146L and L286V mutations under the independent control of murine Thy-1 promoter elements. The mice show exclusive Aβ42 generation, intraneuronal Aβ accumulation, and Aβ deposition around 1.5–2 months, accompanied by reactive gliosis, hyperphosphorylated tau (no NFTs), and neuronal loss. Memory impairment is also observed at 4–6 months.

Although PSAPP mice are well studied, the parental PS mutant lines are not because they lack a robust AD-like neuropathology. However, the lines show excessive neuronal loss in the entorhinal cortex (Lazarov et al. 2006), increased protein oxidation and lipid peroxidation (Mohmmad Abdul et al. 2004; Schuessel et al. 2006), impaired hippocampal adult neurogenesis (Wen et al. 2002, 2004; Wang et al. 2004; Chevallier et al. 2005; Choi et al. 2008), impaired spine morphology, and synaptic plasticity in hippocampal pyramidal cornu ammonis 1 (Auffret et al. 2009). Age-related NFT-like inclusions are also observed in a PS-1 knock-in line (Tanemura et al. 2006).

35.6 Mouse Models of Tau Pathology

NFTs are intraneuronal aggregates composed of hyperphosphorylated tau, another neuropathological feature of AD. NFTs are significantly involved in neurodegeneration and the progressive memory deficit of AD (Ballatore et al. 2007), since distribution of the progressive NFTs is closely linked to the brain regions involved in AD including the hippocampus and the cortex (Ball 1977; Arnold et al. 1991; Gomez-Isla et al. 1996). Six isoforms of tau are produced by alternative splicing from a single *MAPT* (microtubule-associated protein tau) gene (Ballatore et al. 2007). The isoforms are combination of two regions: three or four repeats of a microtubule-binding domain (3R or 4R) and two, one, or no amino-terminal inserts (2 N, 1 N, or 0 N).

A number of tau mutations have been reported to be associated with frontal temporal dementia with parkinsonism linked to chromosome 17 (FTDP-17) (Lee et al. 2001; Roberson 2006). Consistent with these findings, Tg mice models expressing human tau with FTDP-17 mutations mostly show NFTs, a disruption of axonal transport, neuronal death, and behavioral deficits (Gotz et al. 2000; Murakami et al. 2006; Leroy et al. 2007; Yoshiyama et al. 2007). For example, the JNPL3 is the first Tg mouse showing that a mutant tau causes neuronal damage and cell loss (Lewis et al. 2000). The mouse expresses the PrP promoter-driven 4R0N MAPT with the P301L mutation, one of the most common mutations associated with FTDP-17, and develops progressive behavioral abnormalities, in particular motor dysfunction, with NFTs and neuronal loss in the spinal cord, brainstem, and cerebellum. Pretangles are also observed in the brain.

However, none of the mutations directly parallel to the entire AD process. As described above, most of the APP Tg mice show increase of endogenous hyperphosphorylated tau in the brain, but not NFTs. In PDAPP mice, the tau epitopes accumulate within dystrophic neuritis with 12- to 15-nm filaments, but no paired helical filaments and lesions are observed in the aged mice (Masliah et al. 2001). Whether failure of NFT development is due to differences between human and mouse tau or the shorter life span of the mouse, however, is still unclear.

For a more ideal model of AD with NFTs, a human tau (Htau) mouse was generated as a new model expressing human genomic *MAPT*, from which all six human tau isoforms are produced, on a murine *Mapt* knockout background (Andorfer et al. 2003). Both hyperphosphorylated tau and NFTs are observed in the cortex and the hippocampus as is seen in human AD but little or none in the striatum, cerebellum, or spinal cord. The mouse shows extensive cell death and develops age-related cognitive deficits (Polydoro et al. 2009).

35.7 Mouse Models Expressing Mutant APP and Tau

A cross of Tg JPNL3 mice with Tg2576 mice is the first model to exhibit both NFTs and Aβ plaques (tau and APP: TAAP mice) (Lewis et al. 2001). The TAAP mice start showing Aβ plaque formation around 6 months with numerous plaques by 9 months. NFTs are also observed in the spinal cord and brainstem as shown in the Tg JNPL3 line; however, NFTs are more enhanced in forebrain areas compared to JNPL3, suggesting synergistic effects of Aβ on NFT generation.

The 3xTg-AD (triple Tg) line was created by co-injection of two mutant transgenes (APPswe and tau P301L) into fertilized eggs obtained from mutant PS-1 (M146V) knock-in mice. This line shows increased Aβ40 and Aβ42 levels, Aβ plaques at 6 months, NFT-like lesions around 10–12 months, intraneuronal Aβ accumulation, age-dependent deficits in long-term synaptic plasticity, and spatial memory (Oddo et al. 2003a, b).

Both TAPP and 3xTg-AD mice seem to be complete models of AD, since they exhibit a wide spectrum of AD pathogenesis with β-amyloidosis and tau pathology. However, they are still not perfect. Accelerated Aβ production followed by Aβ plaque formation is caused by the FAD-linked APP and/or PS-1 mutations, whereas the tau pathology is driven by the tau mutation, which is not directly linked to AD. Thus, other models not yet reported, such as a cross of APP mice with Htau mice, are desired to more accurately parallel and encapsulate all of the featured aspects of AD pathogenesis.

35.8 Conclusion

Animal models mimicking human diseases allow us to test therapeutic potentials and investigate disease progression, contributing to improvement of quality of life for humans. Tg models are now the most popular for AD therapeutics. Since the first murine Tg model of AD was reported in 1995, many other modified models that reproduce many critical aspects of the disease have been created, though no one model can completely emulate all aspects of AD. These models have improved our understanding of AD pathogenesis and have contributed to numerous success stories for new therapeutic innovations, although none have been successfully translated into effective therapies for humans yet (Zahs and Ashe 2010). All the Tg mouse models are based on the expression of genes with FAD-related mutations found in early-onset and a relatively small subset of sporadic AD patients. However, clinical diagnosis and neuropathology in FAD patients are similar to those in late-onset sporadic cases; therefore, the models are still useful. The FAD mutations introduced into these mouse models may have unwanted effects that are not observed in the sporadic cases. Since no model perfectly matching human AD is currently available, it is important to choose an appropriate model to fit a specific research target. Especially as advancements arise, animal models of AD will continue to play crucial roles in preclinical testing for AD for many years to come.

References

Andorfer C, Acker CM, Kress Y, Hof PR, Duff K, Davies P (2005) Cell-cycle reentry and cell death in transgenic mice expressing nonmutant human tau isoforms. J Neurosci 25:5446–5454

Andorfer C, Kress Y, Espinoza M, de Silva R, Tucker KL, Barde YA, Duff K, Davies P (2003) Hyperphosphorylation and aggregation of tau in mice expressing normal human tau isoforms. J Neurochem 86:582–590

Arendash GW, King DL, Gordon MN, Morgan D, Hatcher JM, Hope CE, Diamond DM (2001) Progressive, age-related behavioral impairments in transgenic mice carrying both mutant amyloid precursor protein and presenilin-1 transgenes. Brain Res 891:42–53

Arnold SE, Hyman BT, Flory J, Damasio AR, Van Hoesen GW (1991) The topographical and neuroanatomical distribution of neurofibrillary tangles and neuritic plaques in the cerebral cortex of patients with Alzheimer's disease. Cereb Cortex 1:103–116

Auffret A, Gautheron V, Repici M, Kraftsik R, Mount HT, Mariani J, Rovira C (2009) Age-dependent impairment of spine morphology and synaptic plasticity in hippocampal CA1 neurons of a presenilin 1 transgenic mouse model of Alzheimer's disease. J Neurosci 29:10144–10152

Ball MJ (1977) Neuronal loss, neurofibrillary tangles and granulovacuolar degeneration in the hippocampus with ageing and dementia. A quantitative study. Acta Neuropathol 37:111–118

Ballatore C, Lee VM, Trojanowski JQ (2007) Tau-mediated neurodegeneration in Alzheimer's disease and related disorders. Nat Rev Neurosci 8:663–672

Bertoli-Avella AM, Oostra BA, Heutink P (2004) Chasing genes in Alzheimer's and Parkinson's disease. Hum Genet 114:413–438

Betts V, Leissring MA, Dolios G, Wang R, Selkoe DJ, Walsh DM (2008) Aggregation and catabolism of disease-associated intra-Abeta mutations: reduced proteolysis of AbetaA21G by neprilysin. Neurobiol Dis 31:442–450

Billings LM, Oddo S, Green KN, McGaugh JL, LaFerla FM (2005) Intraneuronal abeta causes the onset of early Alzheimer's disease-related cognitive deficits in transgenic mice. Neuron 45:675–688

Borchelt DR, Thinakaran G, Eckman CB, Lee MK, Davenport F, Ratovitsky T, Prada CM, Kim G, Seekins S, Yager D, Slunt HH, Wang R, Seeger M, Levey AI, Gandy SE, Copeland NG, Jenkins NA, Price DL, Younkin SG, Sisodia SS (1996) Familial Alzheimer's disease-linked presenilin 1 variants elevate Abeta1-42/1-40 ratio in vitro and in vivo. Neuron 17:1005–1013

Calhoun ME, Wiederhold KH, Abramowski D, Phinney AL, Probst A, Sturchler-Pierrat C, Staufenbiel M, Sommer B, Jucker M (1998) Neuron loss in APP transgenic mice. Nature 395:755–756

Chen G, Chen KS, Knox J, Inglis J, Bernard A, Martin SJ, Justice A, McConlogue L, Games D, Freedman SB, Morris RG (2000) A learning deficit related to age and beta-amyloid plaques in a mouse model of Alzheimer's disease. Nature 408:975–979

Cheng IH, Palop JJ, Esposito LA, Bien-Ly N, Yan F, Mucke L (2004) Aggressive amyloidosis in mice expressing human amyloid peptides with the Arctic mutation. Nat Med 10:1190–1192

Cheng IH, Scearce-Levie K, Legleiter J, Palop JJ, Gerstein H, Bien-Ly N, Puolivali J, Lesne S, Ashe KH, Muchowski PJ, Mucke L (2007) Accelerating amyloid-beta fibrillization reduces oligomer levels and functional deficits in Alzheimer disease mouse models. J Biol Chem 282:23818–23828

Chevallier NL, Soriano S, Kang DE, Masliah E, Hu G, Koo EH (2005) Perturbed neurogenesis in the adult hippocampus associated with presenilin-1 A246E mutation. Am J Pathol 167:151–159

Chishti MA et al (2001) Early-onset amyloid deposition and cognitive deficits in transgenic mice expressing a double mutant form of amyloid precursor protein 695. J Biol Chem 276:21562–21570

Choi SH, Veeraraghavalu K, Lazarov O, Marler S, Ransohoff RM, Ramirez JM, Sisodia SS (2008) Non-cell-autonomous effects of presenilin 1 variants on enrichment-mediated hippocampal progenitor cell proliferation and differentiation. Neuron 59: 568–580

Citron M, Oltersdorf T, Haass C, McConlogue L, Hung AY, Seubert P, Vigo-Pelfrey C, Lieberburg I, Selkoe DJ (1992) Mutation of the beta-amyloid precursor protein in familial Alzheimer's disease increases beta-protein production. Nature 360:672–674

Citron M et al (1997) Mutant presenilins of Alzheimer's disease increase production of 42-residue amyloid beta-protein in both transfected cells and transgenic mice. Nat Med 3:67–72

Cruts M, Van Broeckhoven C (1998) Molecular genetics of Alzheimer's disease. Ann Med 30:560–565

De Felice FG, Wu D, Lambert MP, Fernandez SJ, Velasco PT, Lacor PN, Bigio EH, Jerecic J, Acton PJ, Shughrue PJ, Chen-Dodson E, Kinney GG, Klein WL (2008) Alzheimer's disease-type neuronal tau hyperphosphorylation induced by A beta oligomers. Neurobiol Aging 29:1334–1347

DeKosky ST, Scheff SW (1990) Synapse loss in frontal cortex biopsies in Alzheimer's disease: correlation with cognitive severity. Ann Neurol 27:457–464

Demeester N, Mertens C, Caster H, Goethals M, Vandekerckhove J, Rosseneu M, Labeur C (2001) Comparison of the aggregation properties, secondary structure and apoptotic effects of wild-type, Flemish and Dutch N-terminally truncated amyloid beta peptides. Eur J Neurosci 13:2015–2024

Dodart JC, Mathis C, Saura J, Bales KR, Paul SM, Ungerer A (2000) Neuroanatomical abnormalities in behaviorally characterized APP(V717F) transgenic mice. Neurobiol Dis 7:71–85

Drechsel DN, Hyman AA, Cobb MH, Kirschner MW (1992) Modulation of the dynamic instability of tubulin assembly by the microtubule-associated protein tau. Mol Biol Cell 3:1141–1154

Drewes G (2004) MARKing tau for tangles and toxicity. Trends Biochem Sci 29:548–555

Drubin DG, Kirschner MW (1986) Tau protein function in living cells. J Cell Biol 103:2739–2746

Duff K, Eckman C, Zehr C, Yu X, Prada CM, Perez-tur J, Hutton M, Buee L, Harigaya Y, Yager D, Morgan D, Gordon MN, Holcomb L, Refolo L, Zenk B, Hardy J, Younkin S (1996) Increased amyloid-beta42(43) in brains of mice expressing mutant presenilin 1. Nature 383:710–713

Ertekin-Taner N (2007) Genetics of Alzheimer's disease: a centennial review. Neurol Clin 25:611–667, v

Frautschy SA, Baird A, Cole GM (1991) Effects of injected Alzheimer beta-amyloid cores in rat brain. Proc Natl Acad Sci U S A 88:8362–8366

Frautschy SA, Yang F, Irrizarry M, Hyman B, Saido TC, Hsiao K, Cole GM (1998) Microglial response to amyloid plaques in APPsw transgenic mice. Am J Pathol 152:307–317

Games D, Adams D, Alessandrini R, Barbour R, Berthelette P, Blackwell C, Carr T, Clemens J, Donaldson T, Gillespie F et al (1995) Alzheimer-type neuropathology in transgenic mice overexpressing V717F beta-amyloid precursor protein. Nature 373:523–527

Glabe CC (2005) Amyloid accumulation and pathogenesis of Alzheimer's disease: significance of monomeric,

oligomeric and fibrillar Abeta. Subcell Biochem 38:167–177
Glabe CG, Kayed R (2006) Common structure and toxic function of amyloid oligomers implies a common mechanism of pathogenesis. Neurology 66:S74–S78
Gomez-Isla T, Price JL, McKeel DW Jr, Morris JC, Growdon JH, Hyman BT (1996) Profound loss of layer II entorhinal cortex neurons occurs in very mild Alzheimer's disease. J Neurosci 16:4491–4500
Gotz J, Barmettler R, Ferrari A, Goedert M, Probst A, Nitsch RM (2000) In vivo analysis of wild-type and FTDP-17 tau transgenic mice. Ann N Y Acad Sci 920:126–133
Gotz J, Chen F, van Dorpe J, Nitsch RM (2001) Formation of neurofibrillary tangles in P301l tau transgenic mice induced by Abeta 42 fibrils. Science 293:1491–1495
Haass C (2004) Take five–BACE and the gamma-secretase quartet conduct Alzheimer's amyloid beta-peptide generation. EMBO J 23:483–488
Hardy J, Selkoe DJ (2002) The amyloid hypothesis of Alzheimer's disease: progress and problems on the road to therapeutics. Science 297:353–356
Haughey NJ, Liu D, Nath A, Borchard AC, Mattson MP (2002a) Disruption of neurogenesis in the subventricular zone of adult mice, and in human cortical neuronal precursor cells in culture, by amyloid beta-peptide: implications for the pathogenesis of Alzheimer's disease. Neuromolecular Med 1:125–135
Haughey NJ, Nath A, Chan SL, Borchard AC, Rao MS, Mattson MP (2002b) Disruption of neurogenesis by amyloid beta-peptide, and perturbed neural progenitor cell homeostasis, in models of Alzheimer's disease. J Neurochem 83:1509–1524
Herzig MC, Winkler DT, Burgermeister P, Pfeifer M, Kohler E, Schmidt SD, Danner S, Abramowski D, Sturchler-Pierrat C, Burki K, van Duinen SG, Maat-Schieman ML, Staufenbiel M, Mathews PM, Jucker M (2004) Abeta is targeted to the vasculature in a mouse model of hereditary cerebral hemorrhage with amyloidosis. Nat Neurosci 7:954–960
Herzig MC, Van Nostrand WE, Jucker M (2006) Mechanism of cerebral beta-amyloid angiopathy: murine and cellular models. Brain Pathol 16:40–54
Herzig MC, Eisele YS, Staufenbiel M, Jucker M (2009) E22Q-mutant Abeta peptide (AbetaDutch) increases vascular but reduces parenchymal Abeta deposition. Am J Pathol 174:722–726
Holcomb L, Gordon MN, McGowan E, Yu X, Benkovic S, Jantzen P, Wright K, Saad I, Mueller R, Morgan D, Sanders S, Zehr C, O'Campo K, Hardy J, Prada CM, Eckman C, Younkin S, Hsiao K, Duff K (1998) Accelerated Alzheimer-type phenotype in transgenic mice carrying both mutant amyloid precursor protein and presenilin 1 transgenes. Nat Med 4:97–100
Hou L, Liu Y, Wang X, Ma H, He J, Zhang Y, Yu C, Guan W, Ma Y (2011) The effects of amyloid-beta(42) oligomer on the proliferation and activation of astrocytes in vitro. In Vitro Cell Dev Biol Anim 47:573–580
Hsia AY, Masliah E, McConlogue L, Yu GQ, Tatsuno G, Hu K, Kholodenko D, Malenka RC, Nicoll RA, Mucke L (1999) Plaque-independent disruption of neural circuits in Alzheimer's disease mouse models. Proc Natl Acad Sci U S A 96:3228–3233
Hsiao K, Chapman P, Nilsen S, Eckman C, Harigaya Y, Younkin S, Yang F, Cole G (1996) Correlative memory deficits, Abeta elevation, and amyloid plaques in transgenic mice. Science 274:99–102
Hutton M, Hardy J (1997) The presenilins and Alzheimer's disease. Hum Mol Genet 6:1639–1646
Irizarry MC, McNamara M, Fedorchak K, Hsiao K, Hyman BT (1997) APPSw transgenic mice develop age-related A beta deposits and neuropil abnormalities, but no neuronal loss in CA1. J Neuropathol Exp Neurol 56:965–973
Jan A, Gokce O, Luthi-Carter R, Lashuel HA (2008) The ratio of monomeric to aggregated forms of Abeta40 and Abeta42 is an important determinant of amyloid-beta aggregation, fibrillogenesis, and toxicity. J Biol Chem 283:28176–28189
Jankowsky JL, Fadale DJ, Anderson J, Xu GM, Gonzales V, Jenkins NA, Copeland NG, Lee MK, Younkin LH, Wagner SL, Younkin SG, Borchelt DR (2004) Mutant presenilins specifically elevate the levels of the 42 residue beta-amyloid peptide in vivo: evidence for augmentation of a 42-specific gamma secretase. Hum Mol Genet 13:159–170
Jankowsky JL, Younkin LH, Gonzales V, Fadale DJ, Slunt HH, Lester HA, Younkin SG, Borchelt DR (2007) Rodent A beta modulates the solubility and distribution of amyloid deposits in transgenic mice. J Biol Chem 282:22707–22720
Jellinger KA (2002) Alzheimer disease and cerebrovascular pathology: an update. J Neural Transm 109:813–836
Jin M, Shepardson N, Yang T, Chen G, Walsh D, Selkoe DJ (2011) Soluble amyloid beta-protein dimers isolated from Alzheimer cortex directly induce Tau hyperphosphorylation and neuritic degeneration. Proc Natl Acad Sci U S A 108:5819–5824
Kawarabayashi T, Shoji M, Younkin LH, Wen-Lang L, Dickson DW, Murakami T, Matsubara E, Abe K, Ashe KH, Younkin SG (2004) Dimeric amyloid beta protein rapidly accumulates in lipid rafts followed by apolipoprotein E and phosphorylated tau accumulation in the Tg2576 mouse model of Alzheimer's disease. J Neurosci 24:3801–3809
Klein WL, Krafft GA, Finch CE (2001) Targeting small Abeta oligomers: the solution to an Alzheimer's disease conundrum? Trends Neurosci 24:219–224
Klyubin I, Betts V, Welzel AT, Blennow K, Zetterberg H, Wallin A, Lemere CA, Cullen WK, Peng Y, Wisniewski T, Selkoe DJ, Anwyl R, Walsh DM, Rowan MJ (2008) Amyloid beta protein dimer-containing human CSF disrupts synaptic plasticity: prevention by systemic passive immunization. J Neurosci 28:4231–4237
Lacor PN, Buniel MC, Furlow PW, Clemente AS, Velasco PT, Wood M, Viola KL, Klein WL (2007) Abeta oligomer-induced aberrations in synapse composition, shape, and density provide a molecular basis for loss of connectivity in Alzheimer's disease. J Neurosci 27:796–807

Larson J, Lynch G, Games D, Seubert P (1999) Alterations in synaptic transmission and long-term potentiation in hippocampal slices from young and aged PDAPP mice. Brain Res 840:23–35

Lashuel HA, Hartley DM, Petre BM, Wall JS, Simon MN, Walz T, Lansbury PT Jr (2003) Mixtures of wild-type and a pathogenic (E22G) form of Abeta40 in vitro accumulate protofibrils, including amyloid pores. J Mol Biol 332:795–808

Lazarov O, Marr RA (2010) Neurogenesis and Alzheimer's disease: at the crossroads. Exp Neurol 223:267–281

Lazarov O, Peterson LD, Peterson DA, Sisodia SS (2006) Expression of a familial Alzheimer's disease-linked presenilin-1 variant enhances perforant pathway lesion-induced neuronal loss in the entorhinal cortex. J Neurosci 26:429–434

Lee VM, Goedert M, Trojanowski JQ (2001) Neurodegenerative tauopathies. Annu Rev Neurosci 24:1121–1159

Lemere CA, Lopera F, Kosik KS, Lendon CL, Ossa J, Saido TC, Yamaguchi H, Ruiz A, Martinez A, Madrigal L, Hincapie L, Arango JC, Anthony DC, Koo EH, Goate AM, Selkoe DJ (1996) The E280A presenilin 1 Alzheimer mutation produces increased A beta 42 deposition and severe cerebellar pathology. Nat Med 2:1146–1150

Leroy K, Bretteville A, Schindowski K, Gilissen E, Authelet M, De Decker R, Yilmaz Z, Buee L, Brion JP (2007) Early axonopathy preceding neurofibrillary tangles in mutant tau transgenic mice. Am J Pathol 171:976–992

Lesne S, Koh MT, Kotilinek L, Kayed R, Glabe CG, Yang A, Gallagher M, Ashe KH (2006) A specific amyloid-beta protein assembly in the brain impairs memory. Nature 440:352–357

Lewis J, McGowan E, Rockwood J, Melrose H, Nacharaju P, Van Slegtenhorst M, Gwinn-Hardy K, Paul Murphy M, Baker M, Yu X, Duff K, Hardy J, Corral A, Lin WL, Yen SH, Dickson DW, Davies P, Hutton M (2000) Neurofibrillary tangles, amyotrophy and progressive motor disturbance in mice expressing mutant (P301L) tau protein. Nat Genet 25:402–405

Lewis J, Dickson DW, Lin WL, Chisholm L, Corral A, Jones G, Yen SH, Sahara N, Skipper L, Yager D, Eckman C, Hardy J, Hutton M, McGowan E (2001) Enhanced neurofibrillary degeneration in transgenic mice expressing mutant tau and APP. Science 293:1487–1491

Maezawa I, Zimin PI, Wulff H, Jin LW (2011) Amyloid-beta protein oligomer at low nanomolar concentrations activates microglia and induces microglial neurotoxicity. J Biol Chem 286:3693–3706

Magdesian MH, Carvalho MM, Mendes FA, Saraiva LM, Juliano MA, Juliano L, Garcia-Abreu J, Ferreira ST (2008) Amyloid-beta binds to the extracellular cysteine-rich domain of Frizzled and inhibits Wnt/beta-catenin signaling. J Biol Chem 283:9359–9368

Mandelkow EM, Biernat J, Drewes G, Steiner B, Lichtenberg-Kraag B, Wille H, Gustke N, Mandelkow E (1993) Microtubule-associated protein tau, paired helical filaments, and phosphorylation. Ann N Y Acad Sci 695:209–216

Masliah E (2001) Recent advances in the understanding of the role of synaptic proteins in Alzheimer's disease and other neurodegenerative disorders. J Alzheimers Dis 3:121–129

Masliah E, Sisk A, Mallory M, Games D (2001) Neurofibrillary pathology in transgenic mice overexpressing V717F beta-amyloid precursor protein. J Neuropathol Exp Neurol 60:357–368

Mattson MP (2004) Pathways towards and away from Alzheimer's disease. Nature 430:631–639

McKee AC, Carreras I, Hossain L, Ryu H, Klein WL, Oddo S, LaFerla FM, Jenkins BG, Kowall NW, Dedeoglu A (2008) Ibuprofen reduces Abeta, hyperphosphorylated tau and memory deficits in Alzheimer mice. Brain Res 1207:225–236

Mohmmad Abdul H, Wenk GL, Gramling M, Hauss-Wegrzyniak B, Butterfield DA (2004) APP and PS-1 mutations induce brain oxidative stress independent of dietary cholesterol: implications for Alzheimer's disease. Neurosci Lett 368:148–150

Morgan D (2007) Amyloid, memory and neurogenesis. Exp Neurol 205:330–335

Morgan D (2011) Immunotherapy for Alzheimer's disease. J Intern Med 269:54–63

Morgan D, Diamond DM, Gottschall PE, Ugen KE, Dickey C, Hardy J, Duff K, Jantzen P, DiCarlo G, Wilcock D, Connor K, Hatcher J, Hope C, Gordon M, Arendash GW (2000) A beta peptide vaccination prevents memory loss in an animal model of Alzheimer's disease. Nature 408:982–985

Mucke L, Masliah E, Yu GQ, Mallory M, Rockenstein EM, Tatsuno G, Hu K, Kholodenko D, Johnson-Wood K, McConlogue L (2000) High-level neuronal expression of Abeta 1–42 in wild-type human amyloid protein precursor transgenic mice: synaptotoxicity without plaque formation. J Neurosci 20:4050–4058

Murakami T et al (2006) Cortical neuronal and glial pathology in TgTauP301L transgenic mice: neuronal degeneration, memory disturbance, and phenotypic variation. Am J Pathol 169:1365–1375

Oakley H, Cole SL, Logan S, Maus E, Shao P, Craft J, Guillozet-Bongaarts A, Ohno M, Disterhoft J, Van Eldik L, Berry R, Vassar R (2006) Intraneuronal beta-amyloid aggregates, neurodegeneration, and neuron loss in transgenic mice with five familial Alzheimer's disease mutations: potential factors in amyloid plaque formation. J Neurosci 26:10129–10140

Oddo S, Caccamo A, Kitazawa M, Tseng BP, LaFerla FM (2003a) Amyloid deposition precedes tangle formation in a triple transgenic model of Alzheimer's disease. Neurobiol Aging 24:1063–1070

Oddo S, Caccamo A, Shepherd JD, Murphy MP, Golde TE, Kayed R, Metherate R, Mattson MP, Akbari Y, LaFerla FM (2003b) Triple-transgenic model of Alzheimer's disease with plaques and tangles: intracellular Abeta and synaptic dysfunction. Neuron 39:409–421

Oddo S, Caccamo A, Tran L, Lambert MP, Glabe CG, Klein WL, LaFerla FM (2006) Temporal profile of amyloid-beta (Abeta) oligomerization in an in vivo

model of Alzheimer disease. A link between Abeta and tau pathology. J Biol Chem 281:1599–1604
Palop JJ, Mucke L (2009) Epilepsy and cognitive impairments in Alzheimer disease. Arch Neurol 66:435–440
Palop JJ, Chin J, Bien-Ly N, Massaro C, Yeung BZ, Yu GQ, Mucke L (2005) Vulnerability of dentate granule cells to disruption of arc expression in human amyloid precursor protein transgenic mice. J Neurosci 25:9686–9693
Palop JJ, Chin J, Roberson ED, Wang J, Thwin MT, Bien-Ly N, Yoo J, Ho KO, Yu GQ, Kreitzer A, Finkbeiner S, Noebels JL, Mucke L (2007) Aberrant excitatory neuronal activity and compensatory remodeling of inhibitory hippocampal circuits in mouse models of Alzheimer's disease. Neuron 55:697–711
Pappolla MA, Chyan YJ, Omar RA, Hsiao K, Perry G, Smith MA, Bozner P (1998) Evidence of oxidative stress and in vivo neurotoxicity of beta-amyloid in a transgenic mouse model of Alzheimer's disease: a chronic oxidative paradigm for testing antioxidant therapies in vivo. Am J Pathol 152:871–877
Pastor P, Goate AM (2004) Molecular genetics of Alzheimer's disease. Curr Psychiatry Rep 6:125–133
Pezzini A, Del Zotto E, Volonghi I, Giossi A, Costa P, Padovani A (2009) Cerebral amyloid angiopathy: a common cause of cerebral hemorrhage. Curr Med Chem 16:2498–2513
Poirier R, Wolfer DP, Welzl H, Tracy J, Galsworthy MJ, Nitsch RM, Mohajeri MH (2006) Neuronal neprilysin overexpression is associated with attenuation of Abeta-related spatial memory deficit. Neurobiol Dis 24:475–483
Polydoro M, Acker CM, Duff K, Castillo PE, Davies P (2009) Age-dependent impairment of cognitive and synaptic function in the htau mouse model of tau pathology. J Neurosci 29:10741–10749
Price DL, Wong PC, Borchelt DR, Pardo CA, Thinakaran G, Doan AP, Lee MK, Martin LJ, Sisodia SS (1997) Amyotrophic lateral sclerosis and Alzheimer disease. Lessons from model systems. Rev Neurol (Paris) 153:484–495
Raff MC, Whitmore AV, Finn JT (2002) Axonal self-destruction and neurodegeneration. Science 296:868–871
Roberson ED (2006) Frontotemporal dementia. Curr Neurol Neurosci Rep 6:481–489
Rocchi A, Pellegrini S, Siciliano G, Murri L (2003) Causative and susceptibility genes for Alzheimer's disease: a review. Brain Res Bull 61:1–24
Rockenstein E, Mallory M, Mante M, Sisk A, Masliaha E (2001) Early formation of mature amyloid-beta protein deposits in a mutant APP transgenic model depends on levels of Abeta(1–42). J Neurosci Res 66:573–582
Rockenstein E, Adame A, Mante M, Moessler H, Windisch M, Masliah E (2003) The neuroprotective effects of cerebrolysin in a transgenic model of Alzheimer's disease are associated with improved behavioral performance. J Neural Transm 110:1313–1327
Saganich MJ, Schroeder BE, Galvan V, Bredesen DE, Koo EH, Heinemann SF (2006) Deficits in synaptic transmission and learning in amyloid precursor protein (APP) transgenic mice require C-terminal cleavage of APP. J Neurosci 26:13428–13436
Scheff SW, Price DA (2001) Alzheimer's disease-related synapse loss in the cingulate cortex. J Alzheimers Dis 3:495–505
Scheuner D et al (1996) Secreted amyloid beta-protein similar to that in the senile plaques of Alzheimer's disease is increased in vivo by the presenilin 1 and 2 and APP mutations linked to familial Alzheimer's disease. Nat Med 2:864–870
Schuessel K, Frey C, Jourdan C, Keil U, Weber CC, Muller-Spahn F, Muller WE, Eckert A (2006) Aging sensitizes toward ROS formation and lipid peroxidation in PS1M146L transgenic mice. Free Radic Biol Med 40:850–862
Selkoe DJ (1991) Alzheimer's disease. In the beginning. Nature 354:432–433
Selkoe DJ (2008) Soluble oligomers of the amyloid beta-protein impair synaptic plasticity and behavior. Behav Brain Res 192:106–113
Shankar GM, Li S, Mehta TH, Garcia-Munoz A, Shepardson NE, Smith I, Brett FM, Farrell MA, Rowan MJ, Lemere CA, Regan CM, Walsh DM, Sabatini BL, Selkoe DJ (2008) Amyloid-beta protein dimers isolated directly from Alzheimer's brains impair synaptic plasticity and memory. Nat Med 14:837–842
Sherrington R et al (1995) Cloning of a gene bearing missense mutations in early-onset familial Alzheimer's disease. Nature 375:754–760
Smith MA, Hirai K, Hsiao K, Pappolla MA, Harris PL, Siedlak SL, Tabaton M, Perry G (1998) Amyloid-beta deposition in Alzheimer transgenic mice is associated with oxidative stress. J Neurochem 70:2212–2215
Sturchler-Pierrat C, Abramowski D, Duke M, Wiederhold KH, Mistl C, Rothacher S, Ledermann B, Burki K, Frey P, Paganetti PA, Waridel C, Calhoun ME, Jucker M, Probst A, Staufenbiel M, Sommer B (1997) Two amyloid precursor protein transgenic mouse models with Alzheimer disease-like pathology. Proc Natl Acad Sci U S A 94:13287–13292
Tan J, Town T, Paris D, Mori T, Suo Z, Crawford F, Mattson MP, Flavell RA, Mullan M (1999) Microglial activation resulting from CD40-CD40L interaction after beta-amyloid stimulation. Science 286:2352–2355
Tanaka S, Shiojiri S, Takahashi Y, Kitaguchi N, Ito H, Kameyama M, Kimura J, Nakamura S, Ueda K (1989) Tissue-specific expression of three types of beta-protein precursor mRNA: enhancement of protease inhibitor-harboring types in Alzheimer's disease brain. Biochem Biophys Res Commun 165:1406–1414
Tanemura K, Chui DH, Fukuda T, Murayama M, Park JM, Akagi T, Tatebayashi Y, Miyasaka T, Kimura T, Hashikawa T, Nakano Y, Kudo T, Takeda M, Takashima A (2006) Formation of tau inclusions in knock-in mice with familial Alzheimer disease (FAD) mutation of presenilin 1 (PS1). J Biol Chem 281:5037–5041
Tanzi RE, Bertram L (2005) Twenty years of the Alzheimer's disease amyloid hypothesis: a genetic perspective. Cell 120:545–555

Tatebayashi Y, Lee MH, Li L, Iqbal K, Grundke-Iqbal I (2003) The dentate gyrus neurogenesis: a therapeutic target for Alzheimer's disease. Acta Neuropathol 105:225–232

Town T (2009) Alternative Abeta immunotherapy approaches for Alzheimer's disease. CNS Neurol Disord Drug Targets 8:114–127

Townsend M, Shankar GM, Mehta T, Walsh DM, Selkoe DJ (2006) Effects of secreted oligomers of amyloid beta-protein on hippocampal synaptic plasticity: a potent role for trimers. J Physiol 572:477–492

Townsend M, Mehta T, Selkoe DJ (2007) Soluble Abeta inhibits specific signal transduction cascades common to the insulin receptor pathway. J Biol Chem 282:33305–33312

Tsubuki S, Takaki Y, Saido TC (2003) Dutch, Flemish, Italian, and Arctic mutations of APP and resistance of Abeta to physiologically relevant proteolytic degradation. Lancet 361:1957–1958

Venkitaramani DV, Chin J, Netzer WJ, Gouras GK, Lesne S, Malinow R, Lombroso PJ (2007) Beta-amyloid modulation of synaptic transmission and plasticity. J Neurosci 27:11832–11837

Verret L, Jankowsky JL, Xu GM, Borchelt DR, Rampon C (2007) Alzheimer's-type amyloidosis in transgenic mice impairs survival of newborn neurons derived from adult hippocampal neurogenesis. J Neurosci 27:6771–6780

Volles MJ, Lansbury PT Jr (2002) Vesicle permeabilization by protofibrillar alpha-synuclein is sensitive to Parkinson's disease-linked mutations and occurs by a pore-like mechanism. Biochemistry 41: 4595–4602

Volles MJ, Lee SJ, Rochet JC, Shtilerman MD, Ding TT, Kessler JC, Lansbury PT Jr (2001) Vesicle permeabilization by protofibrillar alpha-synuclein: implications for the pathogenesis and treatment of Parkinson's disease. Biochemistry 40:7812–7819

Walsh DM, Selkoe DJ (2004a) Oligomers on the brain: the emerging role of soluble protein aggregates in neurodegeneration. Protein Pept Lett 11:213–228

Walsh DM, Selkoe DJ (2004b) Deciphering the molecular basis of memory failure in Alzheimer's disease. Neuron 44:181–193

Walsh DM, Klyubin I, Fadeeva JV, Cullen WK, Anwyl R, Wolfe MS, Rowan MJ, Selkoe DJ (2002) Naturally secreted oligomers of amyloid beta protein potently inhibit hippocampal long-term potentiation in vivo. Nature 416:535–539

Wang R, Dineley KT, Sweatt JD, Zheng H (2004) Presenilin 1 familial Alzheimer's disease mutation leads to defective associative learning and impaired adult neurogenesis. Neuroscience 126:305–312

Wen PH, Shao X, Shao Z, Hof PR, Wisniewski T, Kelley K, Friedrich VL Jr, Ho L, Pasinetti GM, Shioi J, Robakis NK, Elder GA (2002) Overexpression of wild type but not an FAD mutant presenilin-1 promotes neurogenesis in the hippocampus of adult mice. Neurobiol Dis 10:8–19

Wen PH, Hof PR, Chen X, Gluck K, Austin G, Younkin SG, Younkin LH, DeGasperi R, Gama Sosa MA, Robakis NK, Haroutunian V, Elder GA (2004) The presenilin-1 familial Alzheimer disease mutant P117L impairs neurogenesis in the hippocampus of adult mice. Exp Neurol 188:224–237

Westerman MA, Cooper-Blacketer D, Mariash A, Kotilinek L, Kawarabayashi T, Younkin LH, Carlson GA, Younkin SG, Ashe KH (2002) The relationship between Abeta and memory in the Tg2576 mouse model of Alzheimer's disease. J Neurosci 22:1858–1867

Wilcock DM, Colton CA (2008) Anti-amyloid-beta immunotherapy in Alzheimer's disease: relevance of transgenic mouse studies to clinical trials. J Alzheimers Dis 15:555–569

Wilcock DM, Alamed J, Gottschall PE, Grimm J, Rosenthal A, Pons J, Ronan V, Symmonds K, Gordon MN, Morgan D (2006) Deglycosylated anti-amyloid-beta antibodies eliminate cognitive deficits and reduce parenchymal amyloid with minimal vascular consequences in aged amyloid precursor protein transgenic mice. J Neurosci 26:5340–5346

Yoshiike Y, Chui DH, Akagi T, Tanaka N, Takashima A (2003) Specific compositions of amyloid-beta peptides as the determinant of toxic beta-aggregation. J Biol Chem 278:23648–23655

Yoshiyama Y, Higuchi M, Zhang B, Huang SM, Iwata N, Saido TC, Maeda J, Suhara T, Trojanowski JQ, Lee VM (2007) Synapse loss and microglial activation precede tangles in a P301S tauopathy mouse model. Neuron 53:337–351

Zahs KR, Ashe KH (2010) 'Too much good news'—are Alzheimer mouse models trying to tell us how to prevent, not cure, Alzheimer's disease? Trends Neurosci 33:381–389

Zhang-Nunes SX, Maat-Schieman ML, van Duinen SG, Roos RA, Frosch MP, Greenberg SM (2006) The cerebral beta-amyloid angiopathies: hereditary and sporadic. Brain Pathol 16:30–39

Animal Models: Behavior and Pathology: Preclinical Assessment of the Putative Cognitive Deficits in HAND

36

Landhing M. Moran, Rosemarie M. Booze, and Charles F. Mactutus

Abstract

Despite the reduced prevalence of HIV-1-associated dementia in the era of combination antiretroviral therapy (CART), nearly half of all HIV-1-positive individuals on CART are afflicted with mild to severe HIV-1-associated neurocognitive disorders (HAND). A greater understanding of the progression of HAND and the development of potential therapeutics require preclinical studies that utilize an integrative profile of cognitive function, from which valid inferences can be drawn about underlying processes. We propose a set of preclinical behavioral tasks that tap various components of executive function, the cognitive domain which shows the greatest decline in the progression of HAND. Fronto-striatal circuitry and dopaminergic systems are integral to executive function and are also particularly vulnerable to injury by HIV-1 infection. The tasks described in this chapter provide measures that are both sensitive to alterations in the function of these systems and relevant to typical cognitive deficits observed in HAND. Further, each of the tasks presently described has a very clear human analogue. Prepulse inhibition of the auditory startle response is a measure of preattentive processing and sensory gating. The multi-choice serial reaction time task assesses different types of attentional processes, including sustained attention, selective attention, and set-shifting. The Morris water maze provides a profile of reference and working memory. Within each behavioral measure, there are a number of variables, which can be manipulated to provide information on several components of executive function, affording the opportunity to model the specific cognitive deficits of HAND.

L.M. Moran • R.M. Booze • C.F. Mactutus (✉)
Behavioral Neuroscience Program, Department of Psychology, University of South Carolina,
1512 Pendleton Street, Columbia, SC 29208, USA
e-mail: mactutus@mailbox.sc.edu

H. Xiong and H.E. Gendelman (eds.), *Current Laboratory Methods in Neuroscience Research*, Springer Protocols Handbooks, DOI 10.1007/978-1-4614-8794-4_36,

Keywords
Sensorimotor gating • Selective attention • Sustained attention • Shifting attention • Reference memory • Working memory

36.1 Introduction/Background

36.1.1 Profile of HAND

By the end of 2009, the worldwide prevalence of HIV-1 was estimated to be 33 million (UNAIDS/WHO 2010). HIV-1-related neuropsychological impairment occurred in approximately 30 % of individuals with asymptomatic HIV-1 infection and in 50 % of those with acquired immunodeficiency virus (AIDS) before the introduction of combination antiretroviral therapy (CART) (Heaton et al. 1995). During the CART era, which brought about controlled HIV-1 viremia and improved immune function, the incidence of the most severe form of neuropsychological impairment, dementia, was reduced (Sacktor et al. 2002). Pre-CART prevalence estimates of dementia were approximately 16 % of AIDS cases (McArthur et al. 1993), whereas more recent estimates are less than 5 % (Heaton et al. 2010). However, mild to severe cognitive and motor impairments continue to afflict up to 50 % of patients on CART and disrupt their functioning in daily activities, and pose a greater risk for early mortality (Mayeux et al. 1993; Ellis et al. 1997; Heaton et al. 2011).

HIV-1-associated neurocognitive disorders (HAND) have been characterized in HIV-1 patients by determining performance in a number of cognitive domains. Several studies report typical cognitive impairments in patients with HAND, as well as the neurological structures and pathways that may be altered. However, preclinical studies that specifically focus on how behavioral and cognitive performance in animals relates to HAND are scarce and in need of clarification in how certain tasks will selectively measure performance for each cognitive domain implicated in HAND. Studies on cognitive deficits characteristic of HAND that use rodents, including HIV-1 transgenic (Tg) rats, rats and mice injected with HIV-1 proteins gp120 and Tat, and immunodeficient mice, are mostly limited to assessment of motor function (Hill et al. 1993; June et al. 2009), spatial memory (Zink et al. 2002; Griffin et al. 2004; Vigorito et al. 2007; Lashomb et al. 2009; Tang et al. 2009), and memory related to fear conditioning (Pugh et al. 2000). We propose a set of behavioral tasks that provide an integrative profile of cognitive function, relating more generally to executive function, which is mediated by the prefrontal cortex (PFC) and its connections to subcortical regions (Cummings 1993; Bonelli and Cummings 2007). Executive function, along with motor function and information processing speed, shows the greatest decline as a function of HIV-1 disease progression compared to other cognitive domains (Reger et al. 2002). In the CART era of lessened severity of neurocognitive impairment, executive function, which requires intact attentional, memory, and learning abilities, remains an area in which HIV-1-positive individuals show the greatest deficits (Cysique et al. 2004; Garvey et al. 2009; Heaton et al. 2011).

36.1.2 Dopaminergic Hypothesis of HAND

Several lines of evidence have implicated the dopamine (DA) system as a target in chronic HIV-1 infection (Fig. 36.1). Parkinsonian symptoms and sensitivity to DA receptor antagonists, as well as radiological studies, metabolic uptake studies, and postmortem brain autopsies that show abnormalities in basal ganglia structure and function, were among the first pieces of evidence that HIV-1 infection disrupts the DA system (Berger and Nath 1997; Koutsilieri et al. 2002). Alterations in the DA system and related brain regions found in human imaging and postmortem examinations

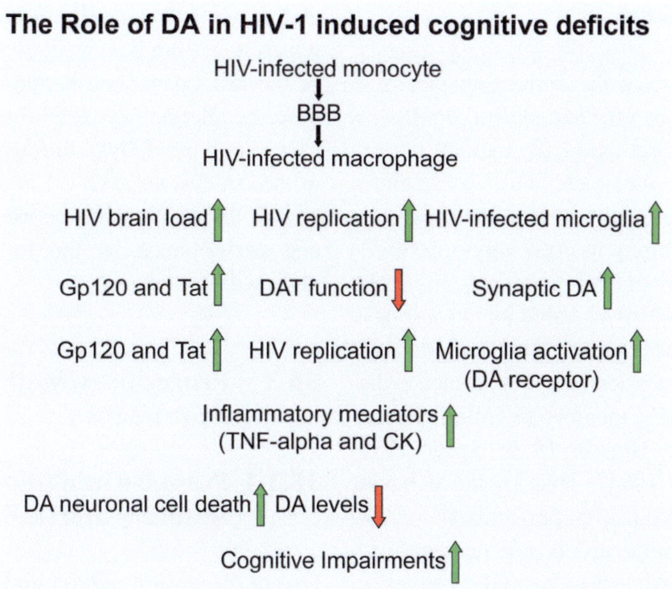

Fig. 36.1 Role of dopamine (DA) in the development of HIV-1 induced cognitive impairments (adapted from Purohit et al. 2011). HIV-infected monocytes cross the blood–brain barrier (BBB) with widespread distribution in the brain and become macrophages, which further infect microglia. As a result, viral load in the brain increases, and HIV-1 proteins gp120 and Tat are released, binding to and impairing DAT function. This impairment leads to greater synaptic concentrations of DA, which binds to DA receptors on adjacent microglia, promoting further HIV-1 replication and release of gp120 and Tat, and production of inflammatory mediators such as cytokines (*TNF-alpha*) and chemokines. Both inflammatory mediators and HIV-1 proteins induce apoptosis of DA neurons, causing a significant dysfunction of the DA system and ensuing cognitive impairment. These proteins can bind to and impair functions of DAT, leading to elevated levels of DA in the dopaminergic synapses. Synaptic dopamine may diffuse and bind to the adjacent microglia through DA receptors and activate these cells to promote further HIV-1 replication and stimulate production of inflammatory mediators such as cytokines (TNF-alpha) and chemokines. Increased HIV-1 replication further elevates brain viral load and further increases shedding of gp120 and Tat. Both of these proteins can cause DA neuronal cell death. Inflammatory mediators can also induce apoptosis of DA neurons. Increased cell death of these neurons causes a significant dysfunction of the DA system and an impairment of cognitive functions

have been correlated with neurocognitive deficits (Chang et al. 2004, 2008; Kumar et al. 2011; Meade et al. 2011). In particular, the dopamine transporter (DAT) is reduced in HIV-1 patients, especially in those with cognitive and motor deficits (Wang et al. 2004). It is well-established that the DAT is critical for cognitive function, with roles in DA homeostasis, maintenance of stable synaptic DA concentrations, and regulation of disease and brain processes (Nieoullon 2002; Chudasama and Robbins 2006). As documented in our own work, the DAT is targeted by HIV-1 proteins, Tat and gp120, resulting in transporter impairment (Aksenov et al. 2008; Zhu et al. 2009, 2011; Ferris et al. 2010), and these effects are due to direct protein–protein interactions (Zhu et al. 2009) involving an allosteric modulation of DAT by the Tat protein (Zhu et al. 2011). In addition, DA-dependent signaling has been identified as a mechanism of HIV-1 protein neurotoxicity (Aksenova et al. 2006; Wallace et al. 2006; Silvers et al. 2007). This work explicitly supports Tat modulation of the DAT. The DA system is a clinically relevant target in chronic HIV-1 infection; dysfunction of the DA system is the proposed neurochemical mechanism underlying the trajectory to neurocognitive impairment. Targeting a clear neurobiological mechanism is fundamental for devising a translational data-based therapy. Clearly, although other neural systems may be involved, the mapping onto DA system dysfunction with HIV-1 infection is quite striking.

36.2 Objectives

One main objective of the current chapter is to illustrate how the careful and skillful configuration of the behavioral assay, the explicit control of environmental variables, and a thoughtful experimental design with relevant control groups provide the opportunity to draw valid inferences about underlying cognitive processes. The interpretation of alterations in behavior on any task purported to assess cognitive processes must be mindful of Tolman's learning-performance distinction and Spear's memory-retention distinction (Tolman and Honzik 1930; Spear 1978; Spear and Riccio 1994). That is, the behavior observed during acquisition performance or witnessed on a retention test is not necessarily a demonstration of what was learned or stored in memory. In other words, it is important to recognize that the dependent variable is an indirect measure of a psychological construct. In each of the behavioral paradigms discussed herein, the utilization of a psychological dose–response is inherent in the experimental design. Thus, for example, the interstimulus interval (ISI) in prepulse inhibition (PPI), stimulus duration in selective attention, and retention interval in working memory all provide a varying degree of load on a cognitive process that permits a functional assessment of the relevant construct. In each of the paradigms described herein, multiple sensory systems may be assayed. This is particularly advantageous with the use of transgenic animals that often have some sensory system impairment. Notably, each of the tasks presently described has a very clear human analogue.

The second main objective of this chapter is to describe an integrative set of tasks that are well-established in assessing cognitive function that can be applied to preclinical studies on HAND. Rodent studies are the primary focus of the present review; studies that employed large animal species, such as those with nonhuman primates, were not included due to the increasingly restricted accessibility of these latter species for scientific study. There is compelling evidence for the face (that a test seems to measure what it needs to), construct (that a test measures what it claims to), and predictive validity (the extent to which a test predicts a future outcome), of each of the tasks described herein. Each of the tasks described herein taps cognitive domains relevant to the study of HAND. Further, based on the role of the DA system in HIV-1 infection as described above, the relationship between the DA system and performance on the following behavioral tasks will be stressed.

36.3 Procedures/Methodological Approaches

36.3.1 Prepulse Inhibition of the Auditory Startle Response

One of the earliest neurological alterations found in the progression of HIV-1 infection is the disruption of brainstem-evoked potentials with reduced amplitude and increased latency in response to auditory and visual stimuli (Ollo et al. 1991; Gil et al. 1992; Pagano et al. 1992; Fein et al. 1995; Goodwin et al. 1996; Schroeder et al. 1996; Castello et al. 1998; Vigliano et al. 2000). Such alterations can be found before any higher-order cognitive deficits are detected. The apparent damage to the midbrain and pontine auditory pathways suggested in clinical patients provides a basis for the study of the auditory startle response (ASR) and PPI of the ASR in assessing the cognitive domains of HAND. PPI is a widely accepted operational measure of the pre-attentive process of sensorimotor gating, or the ability to filter extraneous sensory stimuli. Ison and colleagues have conducted some of the first studies on reflex modification in the rat, which includes the phenomena of PPI (Ison et al. 1973; Hoffman and Ison 1980; Ison and Hoffman 1983). PPI is exhibited when a stimulus, which elicits a startle reflex, is preceded by a weaker prepulse stimulus. The presence of the weak prepulse stimulus inhibits the startle response to varying degrees, based on parameters such as the ISI between the prepulse and the startle stimulus, the sensory modality of the stimuli, and the intensity of the stimuli.

There are a number of advantages of using the ASR and PPI of the ASR as indices of sensorimotor integration and gating, respectively. First, the ASR is relatively easy and unambiguous to measure, the result of a sudden, intense stimulus; PPI is also readily observable with quantifiable reduction of the response to a startling stimulus when it is preceded approximately 30–200 ms by a weak stimulus (Hoffman and Ison 1980). Second, the ASR and PPI may be observed in all mammals, allowing for comparative studies across different species and facilitating the difficult task of cross-species extrapolation. Third, the ASR and PPI may be studied throughout the lifespan, albeit their development is constrained by the maturation of the auditory system. Fourth, the general consensus on both mediating and regulatory neural circuitry facilitates the determination of how different drugs or toxins may affect the ASR and PPI (Koch 1999; Fendt et al. 2001). Fifth, the animal does not have to learn the response, as it is a reflex, thus eliminating the need for a training period. Finally, PPI is an operational measure of sensorimotor gating: a preattentive mechanism that serves to protect the integrity of sensory and cognitive information processing.

36.3.1.1 Materials/Apparatus

Although rat (and mouse) strain differences in PPI may be revealed by various parametric manipulations, including prepulse modality and intensity, and ISI, the cross-species ubiquity of PPI suggests they are rather superfluous. Age constraints are primarily attributed to the developmental time course of sensory capacity.

A number of commercially available startle systems can be used to measure the startle reflex (e.g., Columbus Instruments, Hamilton-Kinder; San Diego Instruments). The startle platform should be enclosed in a sound-attenuating environment; we employ a 10-cm thick double-walled, 81×81×116 cm isolation cabinet (external dimensions), which minimizes reflected sound and provides well over 30 dB(A) (A-weighted decibels) of sound attenuation relative to the external environment with an ambient sound level in the chamber, in the absence of any stimuli presentation, of ~22 dB(A). The high-frequency loudspeaker of the startle system, employed to deliver the auditory stimuli (frequency range of 5–16 kHz), is mounted inside the chamber 30 cm above the perspex animal test cylinder. The speaker has a maximum intensity of 120 dB(A) measured at the source (100 dB(A) inside the test cylinder). For adult rats, use a test cylinder in which they fit comfortably (e.g., 8.75 cm in internal diameter). Rest the cylinder on its accompanying 2.5×20 cm perspex stand, locating it centrally within the test chamber. The animal's response to the auditory stimulus will produce a deflection of the test cylinder, which is converted into analogue signals by a piezoelectric accelerometer affixed to the bottom of the cylinder. The response signals are digitized (12 bit A to D) and saved to a hard disk. Sound levels should be measured and calibrated using a sound level meter with a microphone, preferably placed inside the test cylinder.

36.3.1.2 Auditory Startle Response Habituation

The animals are exposed to a 36-trial session to habituate them to the startle stimulus and test procedures, which reduces response variability in subsequent PPI testing. All white noise stimuli are passed as broad-band through the range possible by the horn tweeter (5–16 kHz) to which rats are sensitive. For example, Sprague–Dawley rats have a range of hearing from 250 Hz to 80 kHz, with the greatest sensitivity to tones of 8 kHz (Kelly and Masterton 1977). Each session begins with a 5-min acclimation period of 70 dB(A) background white noise, followed by 36 trials of a 120 dB(A) white noise stimulus with a 20 ms duration and a 10 s intertrial interval (ITI). If more gradual rates of habituation are desired, extend the ITI accordingly. All test sessions are conducted in the dark. Care must be taken to avoid handling the animal in any way that would increase its sensitivity to the startle stimuli, such as accidentally pinching its tail in the door of the test cylinder. If the animal is in a greater state of arousal due to such a mishap, one should be circumspect about the validity of that animal's data.

36.3.1.3 PPI Testing

After one or two daily sessions of ASR habituation, animals are tested for approximately 20 min with the PPI procedure, in which they are given a 5-min acclimation period in the dark with 70 dB(A) background white noise, followed by six ASR trials with a 10 s ITI. Next, there are 36 PPI trials with a pulse only stimulus, or 8, 40, 80, 120, and 4,000 ms ISIs between prepulse and startle stimuli, presented in a Latin square design. A range of ISIs is implemented, as PPI is a temporal gating phenomenon that relies on the particular timing of the prepulse and startle stimulus presentations. Evaluation of response measures across ISIs provides a way to determine changes in temporal sensitivity that may occur in the case of HIV-1 infection or other treatments. The pulse only and 4,000 ms ISI trials are control trials used to calculate percent inhibition on PPI trials. The 4,000 ms ISI is used in order to provide a control that includes both the prepulse and the startle stimulus. Long ISIs in the range of 800–4,000 ms, which were used in the seminal PPI studies that included ISI as a factor, do not produce inhibition of the startle response as shorter ISIs do (e.g., Ison and Hammond 1971; Hoffman and Ison 1980). The ISI is conventionally defined as the time from prepulse stimulus onset to startle stimulus onset (note: software limitations may preclude strict adherence to that convention if ISIs shorter than the prepulse duration are used). The startle stimulus intensity is routinely ~120 dB(A) measured at the level of the speaker. We have used auditory prepulse intensities of 75 and 85 dB(A); prepulse intensities ranging from 3 to 15 dB(A) above the background noise produce inhibition of the startle response. The San Diego Instruments equipment provides a prepulse stimulus rise/fall time of 2 ms each, with the 20 ms broad-band white noise stimulus employed. Although a measureable startle response will occur with an acoustic signal that reaches an intensity of about 90 dB(A) within 12 ms of its onset (Fleshier 1965), significantly shorter rise/fall times in the order of 0.1 ms are available. Visual prepulses (a brief flash of light) or tactile prepulses (an air puff to the animal's back) also inhibit the ASR. A variable 20 s ITI is used for PPI trials (range of 15–25 s). Peak response amplitude across the 100 ms window after startle stimulus onset, and latency, measured as the time from the onset of the startle stimulus to the time of the peak response, are recorded and used in analyses. Further measures can be derived from the amplitude data: percent PPI and a temporal sensitivity index. Percent PPI is calculated for each animal as follows: 100 × {[(amplitude during control trials) − (amplitude during prepulse + pulse trials)]/amplitude during control trials}. Control trial amplitude is calculated as the average of the pulse only and 4,000 ms ISI trials. The temporal sensitivity index is calculated by taking the average amplitude for all animals across all prepulse ISIs (8–120 ms) within a particular treatment (e.g., drug dose), subtracting this average from each individual amplitude score, and calculating the average of the absolute value of each of the difference scores (Moran et al. 2009). This measure varies with prepulse intensity and modality, and with various drug treatments (Fig. 36.2).

36.3.1.4 Typical Results

Given the impressive evidence that alterations in brainstem-evoked potentials are among the earliest readily quantifiable alterations observed in the progression of HIV-1 (Ollo et al. 1991; Gil et al. 1992; Pagano et al. 1992; Fein et al. 1995; Goodwin et al. 1996; Schroeder et al. 1996; Castello et al. 1998; Vigliano et al. 2000), our focus on preattentive processes has been particularly fruitful. A striking parallel is suggested with our recent preclinical documentation of alterations in the preattentive process of sensorimotor gating in HIV-1 transgenic (Tg) rats (Moran et al. 2013) and in rats exposed to viral proteins such as gp120 and Tat. Specifically, intrahippocampal administration of gp120 alters peak ASR latency on the PPI test in 24-day-old rats (Fitting et al. 2007), and intrahippocampal administration of Tat was associated with a shift in ISI for maximal inhibition of the ASR in adult rats (Fitting et al. 2006a). The dopaminergic component of HIV-1 infection has also been evident in our finding that administration of the D1/D2 receptor agonist apomorphine (APO) restores normal inhibition

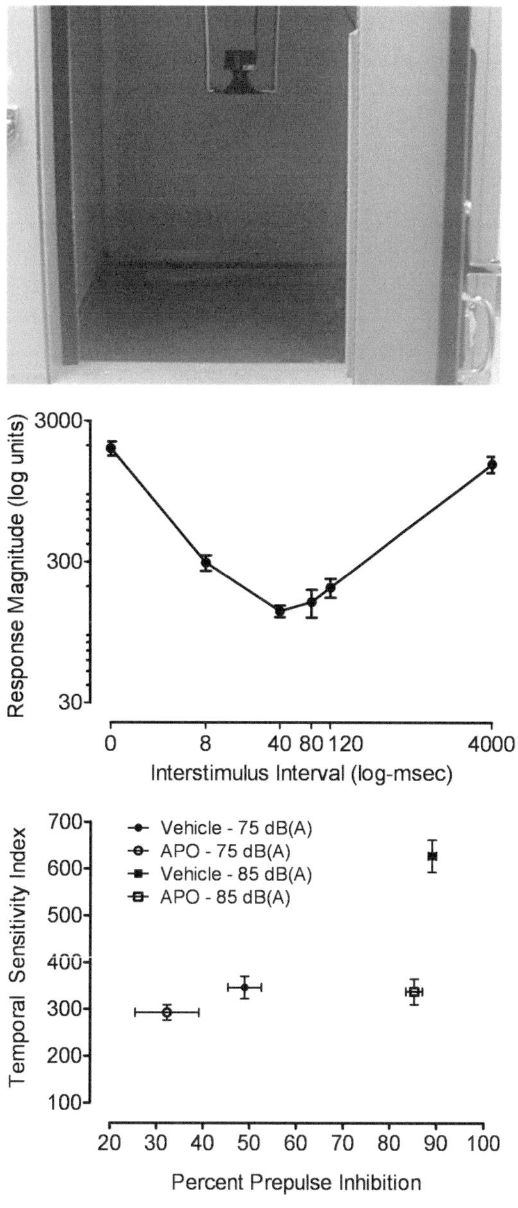

Fig. 36.2 (*Top panel*) Prepulse inhibition of the auditory startle response is tested in a 10-cm thick double-walled, 81 × 81 × 116 cm isolation cabinet (external dimensions), which minimizes reflected sound and provides well over 30 dB(A) of sound attenuation relative to the external environment with an ambient sound level in the chamber, in the absence of any stimuli presentation, of ~22 dB(A). The high-frequency loudspeaker of the startle system employed to deliver the auditory stimuli (frequency range of 5–16 kHz) is mounted inside the chamber 30 cm above the perspex animal test cylinder one would locate centrally within the test chamber. (*Middle panel*) In normal Sprague–Dawley rats, the mean response magnitude ($\log_{10}$ units) follows a quadratic trend across interstimulus interval (ISI), with peak inhibition of the startle response at the 40 ms ISI. (*Bottom panel*) Percent PPI and the temporal sensitivity index are altered differentially by the D1/D2 receptor agonist apomorphine (APO) depending on prepulse intensity. The temporal sensitivity measure decreases with APO treatment when using a high intensity prepulse (85 dB(A)), reflected as a flattening of the ISI function, but does not change with the low intensity prepulse (75 dB(A)). Percent PPI decreases with APO treatment when using the low intensity but not the high intensity prepulse

after neonatal gp120 exposure, which alone reduces PPI in adult rats (Fitting et al. 2006c, 2007, 2008b). Most recently, a linear dose-dependent attenuation of PPI as a function of increasing methamphetamine dose was noted with HIV-1 Tg rats, which was significantly different than controls (Moran et al. 2012). Thus, chronic viral protein exposure appears to produce DA-mediated alterations in the process of sensorimotor gating, as we had detected with neonatal viral protein injection (Fitting et al. 2006a, b, c, 2007, 2008b). The clinical utility of this paradigm in a number of neuropsychiatric disorders (e.g., schizophrenia and Huntington's disease) (Castellanos et al. 1996; Braff et al. 2001) further suggests its potential utility with HIV-1 alterations in cognitive processes.

36.3.1.5 Neurochemical Substrates

Pharmacological manipulations have been fundamental in assessing alterations in PPI and the underlying mechanisms. Dopaminergic systems in particular are one well-established regulatory substrate of sensorimotor gating (Davis 1980). Administration of APO has been used to disrupt PPI in preclinical studies of schizophrenia (Braff and Geyer 1990). It is generally accepted that APO, in doses of 0.03–1.0 mg/kg, yields dose-dependent decreases in PPI across several prepulse intensities, with or without changes in startle baseline (Peng et al. 1990; Schwarzkopf et al. 1993; Varty and Higgins 1994; Lipska et al. 1995; Jones and Shannon 2000). The D2 receptor in particular is associated with impaired PPI, as shown by the reversal of its disruption by D2 antagonists (Mansbach et al. 1988; Rigdon and Viik 1991; Hoffman and Donovan 1994).

PPI disruption by APO has generally been explained by the theory of reduced detectability of the prepulse, because it has been found at lower signal-to-noise ratios rather than at higher signal-to-noise ratios. However, this may be a result of using a single ISI of 100 ms. In our studies on PPI, we implemented a range of ISIs and found that APO does disrupt PPI using a high intensity prepulse (85 dB(A)), which was reflected in a flattening of the ISI function with APO (Fig. 36.2).

36.3.1.6 Problems/Pitfalls and Solutions

As mentioned earlier, the startle response measured in PPI studies is sensitive to factors outside of the procedure that may induce stress in the animal. Therefore, adequate and careful handling of animals prior to testing is necessary, as well as separating the animals from the startle stimuli when they are not being tested. This is accomplished with the use of the sound-attenuating double-walled chamber, which prevents animals outside of the chamber from hearing either the stimuli or the ultrasonic vocalizations of animals being tested. In addition, although the animal's movement is necessarily restrained by the confines of the test cylinder, various sizes are available to minimize discomfort. Plastic test cylinders allow for this, but keep animals in the same general position relative to the speaker, which makes them preferable to wire mesh restraints, which are also more difficult to clean prior to each animal's test session. The use of the accelerometer as a response transducer allows measurement of the startle response based on force or acceleration, without the weight of the animal being a confounding variable, as it is with systems that use load cells. Stimulus delivery and response sensitivity should be routinely calibrated in order to prevent unintentional alterations in response measures.

36.3.2 Attention: Multi-choice Serial Reaction Time Task

The multi-choice serial reaction time task (MCSRTT) taps fronto-striatal circuitry and recruits the executive processes of both attention and inhibitory control (Robbins 2002; Pezze et al. 2007; Bushnell and Strupp 2009; Besson et al. 2010). With respect to neuropsychological effects, it is widely accepted that HIV-1 infection is associated with "executive dysfunction," comprising a component of nearly all of the observed cognitive impairment profiles (Ances and Ellis 2007; Dawes et al. 2008). Intact executive functions are most notably dependent on the frontal cortex as well as the basal ganglia and posterior parietal cortex (Stuss and Levine 2002). Within the MCSRTT, multiple measures can be examined (premature errors, perseverative errors, omission errors, and errors of commission) to assess different aspects of attention (e.g., sustained attention, selective attention, shifting attention). As transgenic animals often express some sensory system impairments, one may choose among several sensory modalities for stimulus presentation: vision, audition, and/or olfaction.

Advantages of the MCSRTT include the range of distinct performance measures that can be obtained and the variety of stimulus parameters that can be manipulated to alter these measures. The functions of the PFC, striatum, and dopaminergic systems have been implicated in this task, making it a suitable candidate for studies on HAND (Robbins 2002). Although the five-choice version of this task is commonly used, a three-choice design represents an alternative, which minimizes the variability in response time that may occur when there are several ports of varying distance from the rat (Bayer et al. 2000; Morgan et al. 2002; Gendle et al. 2004b). A second distinction of the three-choice design is not requiring the rat to turn around to obtain the reinforcers at the back wall; rather they may be delivered under the center response port.

36.3.2.1 Materials/Apparatus

Testing is conducted in a perspex chamber enclosed in sound-attenuating wooden box (Fig. 36.3). The chamber consists of a square waiting area (26.5 × 25 × 30 cm), separated by a thin metal door from an adjacent testing alcove containing three funnel-shaped ports. A light-emitting diode (LED) is mounted above each port. The left and right ports are each located 45° from the center port, and are 8 cm from each other. Trial onset is indicated when the door between the two compartments is raised. Infrared photocells monitor the entrance to the alcove and each port. A 1-s nose poke into a port indicates a response. To provide olfactory distractors, bottles containing liquid odorants (commercially available food flavorings or fragrances such as the following: cinnamon, black walnut, pineapple, sweet pea, butter, coconut, oeillet, lime, and peppermint) are connected via a tube to each port, with solenoid valves controlling the presentation of compressed air through a specific odorant and port (airflow of 1.0 L/min). Air in the chamber is cleared through small centrifugal fans mounted on the outside of the chambers at a rate of four complete exchanges per minute.

36.3.2.2 Training

Animals are first trained to make a 1-s nose poke response into a port to receive a food pellet reinforcer, and then trained to make the nose poke response into the port which is indicated by a particular stimulus, such as illumination of an LED located over the port, or the release of an odor from one of the ports. The door separating the waiting area and the alcove is raised, indicating the beginning of the trial. When the animal breaks the infrared beam at the alcove entrance, the stimulus indicating the correct port is presented until the rat makes a 1-s nose poke in any port or 60 s elapses. A correct response is a nose poke of at least 1 s into the port indicated by the stimulus. Correct responses are reinforced with a 45 mg food pellet, delivered into a small opening underneath the center port from a pellet dispenser. The stimulus location is presented in a semi-random sequence so that any given location was not correct more than four times in a row. Training continues until the subjects reach a criterion of 80 % correct out of a maximum 250 trials in a session.

36.3.2.3 Sustained Attention Task

After training, a sustained attention task can be implemented in which the interval between trial onset and cue presentation (pre-stimulus delay; 0–9 s) and the duration of the stimulus are varied. One approach to this task is to have three sets of sessions; the first two each have a fixed stimulus duration (e.g., 700 ms for the first set and 400 ms for the second set), and the last set has a variable stimulus duration ranging from 200 to 700 ms. In this way, specific deficits in sustained attention can be observed regarding whether it is the brevity or the unpredictability of the stimulus duration that affects performance. Animals are tested for 200 trials per session, for several sessions until they attain a preestablished acquisition criterion. The stimulus location and pre-stimulus delay, as well as the stimulus duration for the last set of sessions, are presented in a semi-random schedule in order to maintain balance across all variable levels.

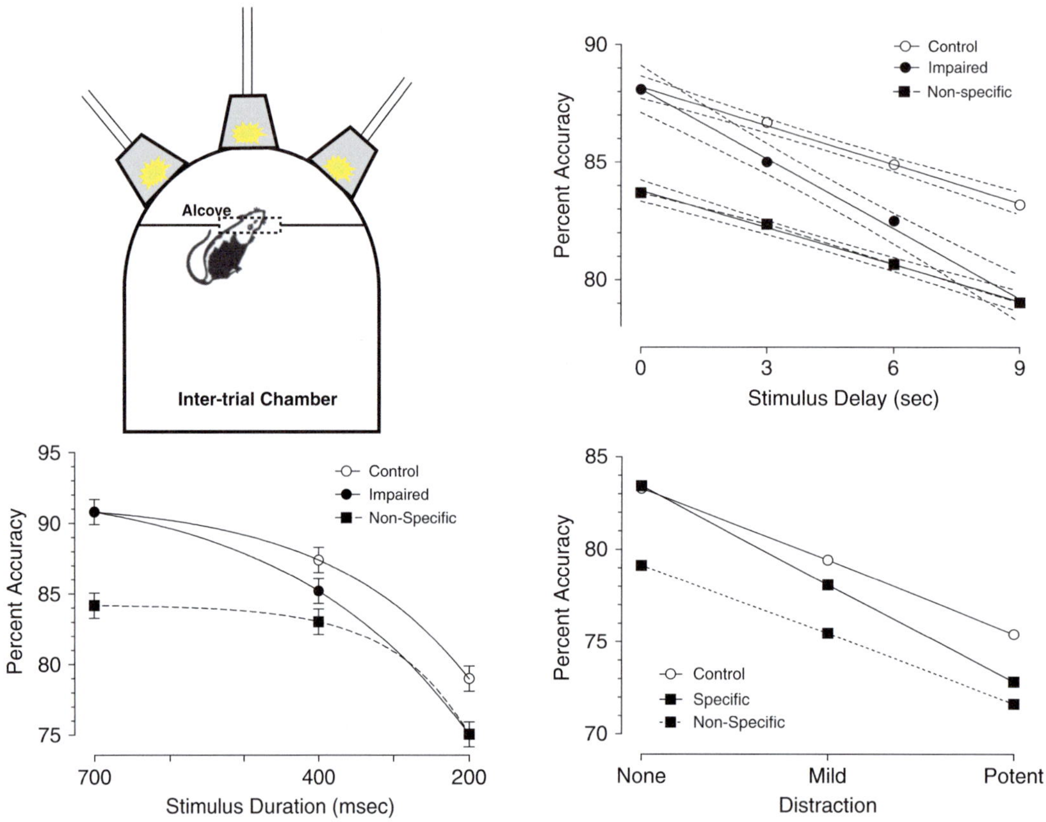

Fig. 36.3 (*Top left*) The multi-choice serial reaction time task is conducted in a perspex chamber enclosed in sound-attenuating chamber. The chamber consists of a square intertrial waiting area (26.5 × 25 × 30 cm), separated by a thin metal door from an adjacent testing alcove containing three funnel-shaped ports. An LED is mounted above each port. To provide olfactory distractors, bottles containing liquid odorants can be connected via tube to each port, with solenoid valves controlling the presentation of compressed air through a specific odorant and port (airflow of 1.01/min). Air in the chamber is continuously cleared (4×/min). (*Top right*) Percent accuracy is shown as a function of stimulus delay (mean ±95 % CI) for adult Long-Evans untreated rats. A specific attentional impairment is also shown as diverging from comparable accuracy at the 0 s delay (idealized data). In contrast, also shown is a 5 % reduction in overall performance attributable to nonspecific effects (idealized data). (*Bottom left*) Percent accuracy is shown as a function of stimulus duration for adult Long-Evans untreated rats (the single phase decay fit with an $r^2>0.99$). A specific attentional impairment is also shown as diverging from comparable accuracy at the 700 msec stimulus duration (idealized data). In contrast, a 5 % reduction in overall performance attributable to nonspecific effects is also portrayed (idealized data). (*Bottom right*) Percent accuracy is shown as a function of the presence of distracting stimuli for adult Long-Evans untreated rats. A specific attentional impairment is shown as diverging from comparable responding in the no-distracter condition (idealized data). In contrast, a 5 % reduction in overall performance attributable to nonspecific effects is also portrayed (idealized data). The use of parametric functions is integral to making valid inferences about the integrity of cognitive processes

36.3.2.4 Selective Attention Task

Selective attention can be assessed by measuring the animal's ability to respond to the stimulus cues during presentation of salient, nonpredictive cues. The presentation and timing of distractors, such as an olfactory cue, are varied, as well as the stimulus location, pre-stimulus delay, and stimulus duration for the predictive cue. On one-third of trials in each session, the distractor cue is presented at one of the three ports, either 1 or 2 s prior to the cue onset, for a duration of 1 s. The remainder of the trials does not include distractors. Delay, duration, cue location, and distraction condition for each trial should be selected pseudorandomly, balanced for each session.

36.3.2.5 Shifting Attention

Shifting attention, or attentional set-shifting, can be assessed with intradimensional shift (IDS) and extra-dimensional shift (EDS) tasks. An IDS requires learning to respond to a different cue of the same modality, and an EDS requires learning to respond on the basis of a completely different modality. After animals have learned to respond on the basis of a single cue, they can begin the IDS task. This task can initially involve the simultaneous presentation of two cues of the same modality, such as two distinct odors, each emitted from a different port in a pseudorandom fashion such that there are no more than four successive presentations of the same odor from the same port. Left–right presentation of each cue should also be balanced. One of the two cues is randomly determined to be correct for a particular session, so that if the animal makes a 1 s nose poke into the port associated with that cue, it will receive a food pellet. After the rat leaves the alcove, the door is closed and the trial ends. The door opens again 10 s later to begin a new trial. Each session continues until 250 trials are completed or 150 min elapse, whichever comes first; daily sessions are conducted until the established acquisition criterion is reached. At this point, the correct and incorrect cues are reversed and sessions are conducted again until the acquisition criterion is reached. Another two reversals may follow. Once IDSs between two cues have been learned, three new cues of the same modality may be used in a new set of sessions, utilizing all three of the ports in the alcove. The other procedural details are identical to those for the two-cue IDS task.

For the EDS task, the modality of the stimulus cue is switched from one set of sessions to the next. If the first modality used is odor, then a specific odor is presented at each randomly selected port to indicate the correct response. Once the acquisition criterion is reached, the modality switches to a randomly selected spatial location, so that only a nose poke at a specific port is rewarded, while each of the three odors is randomly presented at the correct port. Two additional EDSs can follow, so that there are two shifts from odor to spatial location (the first of which is a switch from the IDS task to the EDS task), and two from spatial location to odor.

36.3.2.6 Response Types

For all task types, a nose poke of greater than 1 s into any port prior to the stimulus cue onset (premature response), into an incorrect port following the cue (inaccurate response), or a failure to respond within 15 s of cue offset (omission error) ends the trial with no reinforcement. If the animal does not enter the alcove within 30 s after the door opens, the door closes and a nontrial is recorded. After the door closes to mark the end of the trial, an ITI of 5 s follows before the door opens again. Percentage of correct responses (number of correct responses divided by the number of response trials on which the rat waited to respond until the light cue was presented and responded within 15 s of cue onset) and the percentage of errors as described above (premature responses, inaccurate responses, omission errors, and nontrials) are recorded and used for analyses. Premature responses are interpreted as a measure of impulsivity or inhibitory control. Inaccurate responses and omission errors are considered to result from a lapse in attention. Nontrials may reflect motivational levels. Response latency during correct trials can be recorded to measure information processing speed. Performance following an incorrect choice is also of interest, as a way to assess emotional regulation and reactions to stress induced by the error and difficult stimulus parameters.

For the IDS and EDS tasks, errors made before reaching the acquisition criterion are recorded and analyzed. Comparisons between blocks of trials may also be made after determining the trial on which a certain number of consecutive correct responses had been made, to split the trials into two blocks. This can give information on differences in the rate of acquiring the new response rule. Perseverative responding should also be examined in the IDS task. A certain degree of persistence in a response choice is expected if the animal had prior experience with reinforcement following that choice, but gradual learning of the new response rule is normally observed. A standard way to measure this change in responding is based on the rate at which the animal makes pairs of incorrect responses. The average number of instances of an incorrect response following an incorrect response within a set of 30 trials can be compared with those of consecutive sets, i.e., trials 1–30 are compared to trials 2–31, and so on. When the rate of an incorrect response following an incorrect response drops below 50 %, the end of the perseverative phase is marked as the 15th trial within that particular set. The degree of perseverative responses tends to drop in relation to experience with IDSs, or reversals. After the perseverative phase, changes in the rate of correct responses can be divided into a chance category and two or three post-chance categories, determined by a chosen cut-off (e.g., 23–43 % correct for chance, 43–66 % for the first post-chance category, 66–88 % for the second post-chance category, and 88 % as the acquisition criterion). During the three-choice task, an error may be made by responding to the previously correct cue or to the third cue which was neither the correct one in the previous trial nor in the current trial. The former is a reflection of a deficit in inhibitory function (prepotent cue error), and the latter shows a deficit in associative learning (neutral error). Prepotent cue errors are more common than neutral errors in normal rats.

36.3.2.7 Typical Results

Performance during attention tasks tends to improve greatly over 20 testing sessions (200 trials per session), with close to 100 % correct responses. The rate of error types (premature responses, inaccurate responses, and omission errors) increases with the increase in stimulus delay and decrease in stimulus duration. Omission errors and inaccurate responses provide measures of sustained attention, and premature responses measure the degree of inhibitory control. Percentage correct tends to be lower on trials that follow an error than on those that follow a correct response. Premature responses, omission errors, nontrials, and latency to enter the alcove or to respond also tend to increase following an incorrect response. Mean correct response latency is longer when the cue is presented immediately after trial onset and when the longest stimulus duration is used (Morgan et al. 2002). Inaccurate responses are greater during trials in which a distracter cue is presented (Bayer et al. 2000), and premature responses are greater following an incorrect response when a distractors had been presented (Gendle et al. 2004b). Figure 36.3 illustrates idealized data for typical parametric functions observed in the MCSRTT.

36.3.2.8 Neurochemical Substrates

Among other neurotransmitter systems, the dopaminergic system has been found to be integral to performance on the multi-choice reaction time task. Depletion of dopamine with the neurotoxin 6-hydroxydopamine (6-OHDA) in the ventral striatum reduces the general strength of response, as measured by errors of omission and response latency. When the target region for DA depletion is the medial PFC, response accuracy is impaired when short and variable ITIs are used (Robbins 2002). In contrast, medial PFC infusion of D1 agonists significantly improves accuracy. Intracellular recordings of PFC neurons have shown that neuronal activity follows an inverted "U"-shaped function with increasing dose of D1 receptor agonists, reflecting performance on an attention task, in which a moderate amount of D1 receptor stimulation improved attention, but high levels reduced attention to both relevant and irrelevant stimuli (Vijayraghavan et al. 2007). Systemic administration of the D1 receptor agonist SKF81297 is associated with a decrease in premature responses but also an increase in latency to respond and response bias (responding to the same port as on the previous trial) (Bayer et al. 2000).

Additional pharmacological studies involving prenatal cocaine have also provided information on the effects of dopaminergic system alterations on attention. Prenatal cocaine exposure produces detectable differences in performance in the absence of alterations in maternal weight gain, gestation length, pup birth weight, or litter size. Percentage correct is significantly reduced in rats exposed prenatally to cocaine compared to controls when the 200 ms (briefest) cue is presented and when the cue is presented immediately after trial onset (Morgan et al. 2002). Rats that are exposed to cocaine prenatally perform as well as controls when cue duration is constant but cue onset and location are unpredictable. However, they do tend to have a greater percentage of premature responses during selective attention trials. In some cases, they also show a smaller effect of a previous error on omission error rate, compared to control animals (Morgan et al. 2002), but another study found that male rats that had been prenatally exposed to cocaine had a higher rate of omission errors following an incorrect response, both with and without distractors (Gendle et al. 2004b). The ability to selectively attend to the predictive cue and ignore the distractor cue was compromised in these animals. The differences in performance between treatment groups in these studies may be a result of the length of training the animals received prior to testing. Interactions between prenatal cocaine exposure and later treatment with the D1 agonist SKF81297 have also been observed; rats exposed to prenatal cocaine have a higher rate of omission errors after SKF81297 administration than rats that were not exposed to prenatal cocaine, suggesting that prenatal cocaine produced a lasting alteration on sustained attention as modulated by mesocortical dopaminergic systems (Bayer et al. 2000).

36.3.2.9 Problems/Pitfalls and Solutions

The MCSRTT is a relatively time-intensive procedure, requiring significant training periods before animals attain the acquisition criterion. Also, careful analysis of response measures must be made to rule out sensory or motivational deficits. As stated earlier, transgenic rats may have some sensory impairments, which should guide the selection and use of visual, auditory, and/or olfactory cues.

36.3.3 Hippocampal- and Cortical-Dependent Cognitive Processes: Morris Water Maze

The Morris water maze task, which affords well-established procedures for assessing spatial navigation, was introduced in 1979, and appeared in print shortly thereafter (Morris 1981). The fact that rats can learn to navigate rather directly to a fixed, but hidden, goal from a novel starting location is putatively held to reflect the animal's flexible use of its spatial location relative to visible distal (i.e., stationary, extramaze) cues (Tolman 1948). Normal rats will swim to and persistently search across the target location. Variations of the task can provide information on the use of both working and reference memory in successfully navigating to a hidden platform using spatial cues. Spatial navigation in the Morris water maze is extremely sensitive to hippocampal lesions (Morris et al. 1982), suggesting that spatial learning is a type of "representational memory" putatively held to be mediated by the medial temporal lobe (O'Keefe and Nadel 1978). However, it is also clear that rats with damage to the medial frontal cortex also show profound impairment of spatial mapping ability, suggesting that both of these structures form a functionally integrated system for learning and using spatial representations of aspects of environments (Sutherland et al. 1982).

Advantages of the MWM include the relatively rapid acquisition of the task by normal rats, in no more than a couple of days, a lack of required pretraining, and no need for deprivation to provide motivation to perform the task; motivation is provided by escape from water. The extensive evidence of its face, construct, and predictive validity for the study of hippocampal- and cortical-sensitive memory processes involved in reference and working memory (e.g., Brandeis et al. 1989; McNamara and Skelton 1993; D'Hooge and De Deyn 2001) along with the virtual human analogues task (e.g., Hamilton et al. 2002; Skelton et al. 2006; Antonova et al. 2011) provide a compelling rationale for its cross-species translational relevance.

36.3.3.1 Materials/Apparatus

Acquire rats as required by federally regulated procedures. Either pigmented or albino strains may be used; strain differences may be observed with spatial navigation deficits associated with albinism and inbreeding, but not domestication (Tonkiss et al. 1992; Harker and Whishaw 2002). Animals may be of a broad range of ages for study in this task, from preweanling to senescent animals, albeit at either end of the age continuum, due respect must be given to the animals' response repertoire and the integrity of its visual system (O'Steen et al. 1995; Spencer et al. 1995; Carman and Mactutus 2001, 2002; Carman et al. 2002, 2003).

A large circular water-tight pool should be purchased or constructed. Stainless steel, aluminum, or fiberglass materials are all appropriate; however, the wall of the tank should be seamless or else rendered so with fiberglass. This is necessary to avoid providing a significant proximal cue to guide the animals' swimming behavior in the maze. The inside of the tank should be finished in a non-reflected paint, contrasting to the rat strain chosen; both of these parameters are critical to accurate contrast object tracking. Although a broad range of tank sizes have been employed and reported in the literature, for the most common study of adult animals the typical tank size ranges from 1.2 to 2.1 m in diameter. The height of the tank is typically 0.40–0.60 m, with a water level maintained at approximately a 2/3 of wall height, 0.25–0.40 m above the base of the tank. The height of the water level is another critical parameter; water levels too low will likely obscure a number of environmental cues from a significant area of the tank, whereas water levels too high are more likely to engender escape attempts. The variable of water temperature also deserves serious consideration; 26 ± 1 °C was originally used and is quite common. Colder water temperatures are more likely to represent a confounding factor, especially with very young or very old animals, animals under the influence of a drug, or otherwise physiologically compromised with respect to thermoregulatory function (Panakhova et al. 1984; Rauch et al. 1989). The task motivation is provided by escape from the water; the addition of a thermal stress is unnecessary and only likely to confound the interpretation of any observed performance deficit. A wide range of escape platform sizes has also been reported, ranging from 2.5 cm square to a 27.3×25.7 cm rectangle (Lindner and Schallert 1988; Carman and Mactutus 2001), with the more typical sizes affording a surface area of 10–12 cm diameter. Perhaps the decision about platform size may be best made by scaling relative to the chosen tank size, a decision which influences the extent to which a spatial strategy is utilized by the rat (Mactutus and Booze 1994). The ratio of surface area of platform size:tank size has varied across a broad range, from 1:42 to 1:565 (Sutherland et al. 1982; Morris 1984); ratios in the range of 1:100–1:300 support robust spatial learning. The escape platform is rendered invisible by a twofold approach. First, the platform is positioned hidden below the surface of the water, typically 1.0–2.5 cm. Second, the water is made opaque with the addition of milk, nondairy powder, or nontoxic tempera paint. As one will want to change the tank water daily, adequate gravity drain plumbing or availability of a self-priming pump is a necessity. If milk is used and someone forgoes the daily draining, the furor of other laboratory personnel the next morning will assure it will not happen again!

The tank should be placed in a room such that adequate visual cues will be available in the surrounding environment. These cues may be inherent to the structure of the laboratory (e.g., doorways, windows, and ceiling), or those commonplace in laboratories (e.g., file cabinets, sinks, bookcases), or added specifically for the present purpose (e.g., curtains, posters, and/or hanging objects suspended from the ceiling). The use of cues that are removable is advantageous for conducting specific control procedures and one will most likely find preferable. If the animal's view of the extramaze cues is blocked by a set of curtains surrounding the maze, the animal's performance will fall to chance levels (Morris 1984); unrestricted access to all areas of the pool appears necessary for the animals to form a cognitive map (Sutherland et al. 1987). Finally, light levels appropriate for the rat are used in the task; we have found light levels of 10–25 lx, as measured at the water surface, work well.

For preweanling pups, the tank may be modified in size, with a 0.4 m diameter and height of 0.25 m. The water level is 0.20 m, and the platform is 2.5-cm-squared, submerged 1.0 cm below the surface of the water. Note the relative ratio of platform size:tank size is maintained at ~1:200, providing a searching task of comparable area and difficulty to that typically employed with adult rats.

Video tracking software was introduced over 20 years ago for use with the Morris water maze. There are many systems commercially available (e.g., Columbus Instruments, HVS Image, Noldus, San Diego Instruments). To the best of our knowledge, these systems rely on edge contrast, and, as such, dictate specific considerations be made to maximize reliable object (animal) tracking. The manufacturers provide sufficient detail to appropriately configure the tracking system. The use of indirect lighting is of paramount concern.

36.3.3.2 Reference Memory

The most common procedure or protocol employed in the water maze provides an assessment of reference memory. The basic strategy here involves training the animal to a single hidden platform location across several days, such that it forms a long-term or reference memory of the platform location relative to visible distal extramaze cues. The protocol is relatively straightforward, assuming that the basic technical set up of the video tracking system has been performed and demonstrated to be functional. The use of a battery-operated toy boat with a light on its bow has been useful in identifying the circumference of the tank for the computer software. A perspex base in the form of a cross, with slotted sections on each of the four arms, may be useful in positioning and securing the escape platform in a consistent and effortless manner. This base is laid on the bottom of the tank, centered, and weighted with lead. The balancing of platform location across groups is greatly facilitated with such a base.

36.3.3.3 Acquisition Training

1. Fill the tank with water of the appropriate temperature, add opaque agent, verify computer tracking function, adjusting light levels as necessary, set up the relevant computer files for the experimental animals, and perform all other typical pre-experiment procedures. Animal handling procedures may be utilized as necessary. Although rats are great swimmers, familiarity with their characteristic swimming posture is recommended (Schapiro et al. 1970).
2. In the prototypical case, sets of training trials spaced across days are typically employed. As in Morris (1981), we typically employ 20 acquisition trials administered across 3 days (8, 8, and 4) followed by a probe test. The usual caveats about massed trial vs. spaced trial should be considered when selecting the training trial protocol.
3. One may find it advantageous to train animals in a squad of four (assuming four different treatment groups in the experiment), thus providing an ITI of approximately 5 min between successive trials for any one animal. Good experimental design dictates that platform location be balanced across subjects; that is, one-fourth of each treatment group should be trained to a platform in each of the four compass locations, N, E, S, and W (Fig. 36.4). The same location is used for each squad across all days/trials. It is very critical to use more than one starting location to avoid the animals learning a simple turn preference or directional path to the platform. Two, three, four, or up to eight different starting positions have been reported in the literature (but see also probe test caveats). As in Morris (1981), utilizing two start locations in a counterbalanced order is sufficient to disrupt such a nonspatial strategy; e.g., counterbalancing of two diagonally opposite starting locations is NE, SW, SW, NE, NE, SW, SW, NE. The same sequence of starting locations may be used for each animal in the squad; however, starting positions are also assigned via a Latin square across squads. The rat is placed in the pool, facing the wall, at the assigned starting location. Importantly, lower, do not drop, the rat into the water facing the side wall while starting the trial.
4. One potentially important control to consider, as integral to the experimental design, is a group of animals trained to a hidden platform which is not in a fixed location, but rather varies from trial to trial. The location of the platform is changed according to a Latin square design. The purpose of including this random place group of animals, who will still learn that escape from the water is possible by climbing on the hidden platform, is to assure

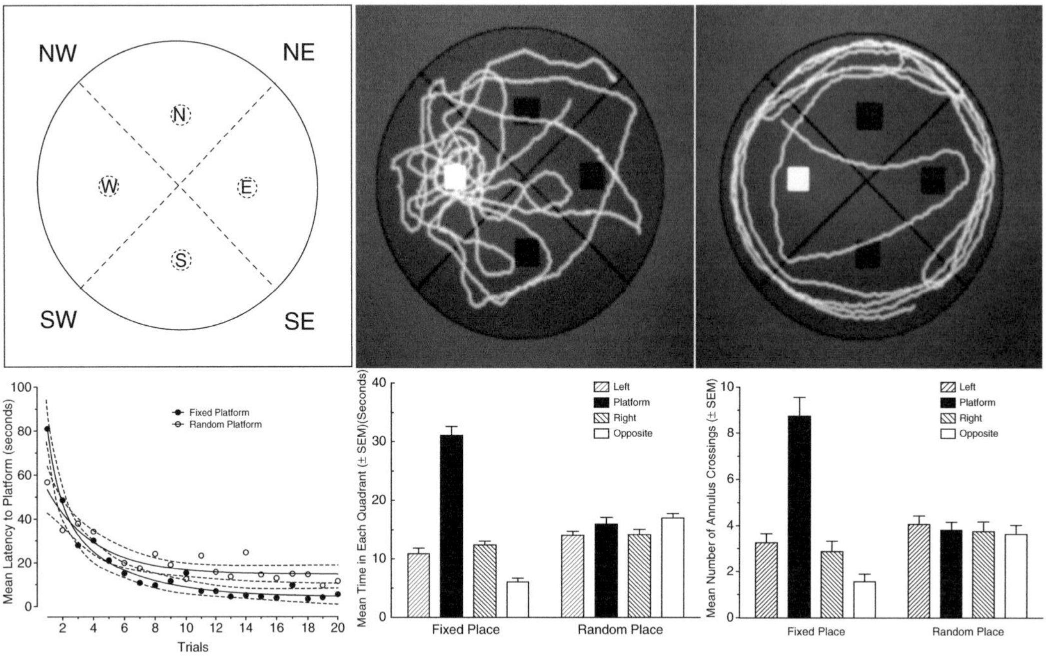

Fig. 36.4 (*Top left*) An overhead view of a spatial water maze. The hidden platform is located in one of the four quadrants (N, S, W, or E). The animal is placed in the pool facing the wall at one of the four locations indicated by the *dotted lines*: NE, SW, NW, or SE. (*Top middle*) Probe trial (60 s) search path of an adult Long-Evans rat, starting from the NE position, demonstrating an accurate and persistent searching of the area where the platform was located during training trials. (*Top right*) Probe trial (60 s) search path of an adult Long-Evans rat that had received an injection of scopolamine (0.7 mg/kg) prior to training; this performance impairment is striking, but cannot be interpreted without additional controls (see text for discussion of control procedures). (*Bottom left*) A comparison of performance during fixed platform and random platform conditions can highlight the animals' use of spatial cues in navigation. Mean response latencies (±95 % CI) of adult Long-Evans rats decrease in a biphasic manner (r^2>0.99) during fixed platform trials, compared to the single rate decay (r^2>0.99) under conditions in which the platform location does not have a constant relation to the spatial cues. These results are consistent with the former animals learning the "rule" and then the platform location whereas the latter learned only the "rule." A terminal acquisition probe test (60 s) demonstrates that time spent in each quadrant (*bottom middle*) and platform (annulus) crossings (*bottom right*) also varied based on whether the platform location was fixed or random. In contrast to the preferential and accurate targeted searching for the platform in the original training area for the fixed platform condition, those animals in the random condition demonstrated preferential and accurate searching of all four platform locations (search path not illustrated). The total number of platform crossings was 16 vs. 15 for the fixed and random platform groups. The animals in both conditions learned well what they were instructed to do, search preferentially one location, or search in four possible locations, suggesting that the random control condition actually encourages a highly successful nonspatial strategy to solve the task

that no local cues are available from the hidden platform or the environment to guide, and hence confound, the search behavior of the animals.

5. Remove yourself as a cue while the animal is swimming. Wearing cologne or perfume, playing loud music through earbuds, and providing any other point sources of sensory stimulation for the animals to use as cues should be explicitly avoided.
6. At a minimum, for training trials, the latency to find and climb onto the submerged platform and path length (swim distance) are recorded. Although not commonly reported, measures of accuracy, independent of the animal's swim speed, should also be considered. For example, consider the initial heading error after swimming approximately one body length (12–14 cm or some other criterion of one's choice, Sutherland et al. 1982), and search

error, i.e., the cumulative distance of the animal to the platform, computed at the a rate of 10 times/s (corrected for differences in direct path length from the various starting positions to the goal) (Gallagher et al. 1993).

7. If the animal does not find the platform within 60–120 s (depending on tank size) manually guide the rat to the platform. Critically, do not pick the animal up out of the water and place it on the platform; this will likely retard and interfere with its acquisition of the task as you have just reinforced the animal for not finding the platform.
8. Allow the rat to remain undisturbed on the escape platform for a period of 30 s. One will note the animal's behavior on the platform will include much rearing and turning, particularly on the initial trials. Such behavior will gradually habituate across trials. While there is ample evidence to suggest that this platform interval is particularly important period during which the animals acquire distal cue information (Sutherland and Linggard 1982; Sutherland and Dyck 1984), there is also compelling evidence that rats obtain distal cue information while en route to the platform (Sutherland et al. 1987; Devan et al. 1992).
9. Remove the animal from the platform, towel dry, and cycle through the next animal until all animals in the squad have received their eight (or four) trials for the day. Utilize the same procedures for all the other squads to be run in the experiment.

36.3.3.4 Probe Test of Terminal Acquisition

After the fourth trial on the third day, the level of terminal acquisition should be assessed. It is critically important to establish terminal acquisition performance prior to consideration of the construct of spatial memory. The most typical procedure is a probe test, one in which the platform is removed from the pool and the searching behavior of the animal recorded for a period of 60 s. Again, remove yourself as a cue during the test. There are three important caveats to conducting this immediate probe test. One, the animal should be started from a location not previously used during training, i.e., a novel starting location. This is necessary if one is to draw any valid inference about the animal's flexible use of the information it has learned, e.g., its "cognitive map." Two, at the end of the 60 s period, reinsert the platform in the pool in its original location and permit the animal to escape accordingly. Recognize that the probe test is actually an extinction trial; the reinsertion of the escape platform minimizes the consequences of the "escape failure." More likely than not the animal will again display rearing and turning behavior on the platform, as during early learning trials. Remove the animal from the platform, towel dry, and cycle through the remaining animals in the squad. Three, to establish that the distal visual cues are responsible for the animal's target searching behavior, run the probe test again, but with a uniform-colored curtain drawn around the maze to obscure the environmental cues. The animals should again be started from a location not previously used, the last of the four available starting locations. Again, reinsert the platform in the pool in its original location and permit the animal to escape.

36.3.3.5 "Reversal" Acquisition Training

An alternative approach to the probe test is the use of a transfer test, often referred to as a "reversal" procedure, where the animal is asked to reacquire the task with the sole change in procedure being that the escape platform is now located in the quadrant diagonally opposite to that of original training. As the animals have already mastered the escape "rule," i.e., swim outward and find the platform, they now have only to encode the new location to the distal cues. With normal animals this reversal procedure is learned rapidly, typically within about four trials. To the extent that the animals learned well the original platform location, their retention of that original learning will retard the learning of the reversal task. A lack of interference of original learning on the reversal procedure would suggest an impairment of retention processes. The same dependent measures employed during original acquisition training are appropriate for reversal learning. In addition, on the first trial that the platform is repositioned, the proportion of swim distance within the previously correct quadrant has also been reported (Sutherland et al. 1982). Sometimes multiple reversals may be desired; the function relating spatial bias to retention interval is relatively insensitive to extinction effects (Morris 1984).

36.3.3.6 Probe Test of Spatial Memory

The assessment of spatial memory can subsequently be conducted, having established the magnitude of original spatial learning. Different groups of animals tested after different intervals will provide a retention function; to the extent that alterations in the slope of the retention function vary with the experimental treatments, valid inferences may be drawn about the integrity of spatial memory and/or memory retrieval processes. The choice of retention intervals may benefit from the observations that accurate spatial navigation is well retained through 14 days, and may not significantly decline over a 90-day period (Sutherland and Dyck 1984). The quadrant preference (relative distribution of swimming time in each quadrant) and search accuracy indexed by platform crossings (number of crossings through the conceptual location of the platform in each of the quadrants) are typically measured. Other accuracy measures worth analyzing include search error (Gallagher et al. 1993) and initial heading error (Sutherland et al. 1982), as discussed above.

36.3.3.7 Working Memory

The assessment of spatial working memory may be performed with a "match to sample" procedure (Morris 1983). Access to the extramaze cues must be available for this procedure. The essence of the procedure is to run two trials per day, providing a novel sample trial followed by the opportunity of a matching trial. Specifically, the platform is hidden in one of the four quadrants (N, E, S, or W, balanced across groups) for trial 1, and remains in that same location for trial 2. The rat is given 30 s on the platform once it successfully escapes. One may infer the integrity of working memory from the extent to which the rats benefit on trial 2 from the new location information learned on trial 1 (Morris 1983). It is also reasonable to interpolate a delay between the sample and match trial to permit one to infer the retention function for spatial working memory. The rat must use a "win-stay" strategy for successful task solution. Another possible extension of this procedure is to incorporate a measure of choice accuracy, i.e., the use of a two-platform procedure (see below).

36.3.3.8 Visible Platform, Nonspatial, Control Task

Impairments in performance of the hidden platform task may arise because of alterations in motoric ability, visual system integrity, and motivation rather than due to any deficit(s) in the cognitive processes underlying spatial learning per se. The visible platform task is not a hippocampally dependent task. In the visible platform task, the escape platform is elevated above the surface of water, thus providing an experimental control for the influence of such noncognitive deficits. That is, if the animals are impaired in both the hidden and visible platform tasks, the most parsimonious inference cannot be one of a specific spatial learning deficit. The prototypical procedure involves training to the visible platform, which is moved to a new position (N, E, S, or W) on every trial. The platform may be made more visible by fixing a high contrast sponge block to its top surface. The use of the sponge provides some attempt to equate that escape is nevertheless to a wet surface, as in the hidden platform task. Again, more than a single starting position is required, for the same rationale as specified in the hidden platform procedure. It is critical to hide the extramaze cues with a curtain surrounding the maze so that they are not available for use by the animal. The goal for the animal is simply to use the single proximal cue provided by the elevated platform to guide its escape behavior. A caveat, although this control task is commonly used, is that the visual system requirements of using a proximal intramaze cue are not truly equated with those required for using multiple distal, extramaze cues.

36.3.3.9 Additional Tasks

The platform discrimination procedure is another test of spatial memory that involves two platforms in the water tank, only one of which is stable enough for the rat to climb on to it. Again, the rat must use spatial cues on the walls to navigate to the correct platform. Nonspatial learning can also be assessed if the platforms are visible, and the distal cues are blocked by a curtain surrounding the tank (Stewart and Morris 1993). The animal has to remember the location of the stable platform based on some discriminating feature of the platform, such as being painted black or with stripes. In order to determine that the animal has learned to discriminate between the stable and the floating platforms based on their appearance, the location of the platforms is varied across trials. Escape latency is typically measured, but the particular choice that the animal makes to directly swim to either the stable platform or the floating one is the main measure of interest. Differences in spatial and nonspatial memory have been further delineated by lesion studies, which have shown that caudate lesions impair visual discrimination but not spatial navigation, while fornix lesions have the reverse effect (Packard and McGaugh 1992).

36.3.3.10 Typical Results

Latency to reach the hidden platform decreases over training trials in a curvilinear fashion, indicating the ability to learn the task of finding and climbing onto the platform (Fig. 36.4). The decrease in latency occurs whether the platform is kept in a fixed location or moved pseudorandomly among multiple locations. More precise measurement of memory and the ability to use spatial cues to find the platform is assessed with quadrant preference and platform crossings during probe trials. For probe tests, that time spent in the target quadrant, where the platform had been located, is typically far greater than that spent in any other quadrant. Platform crossings in the quadrant where the platform had been located during training should also be greater than that in other quadrants. Note that those rats trained to a randomly positioned platform do not fail to search for the platform; rather they adopt a highly successful circular path that permits them to search all four locations. If the animal in the random platform condition shows any particular preference, the most parsimonious explanation is the likely presence of a local cue, such as a tank seam. In a series of studies on the ontogeny of spatial learning, we found it useful to compare the target and opposite quadrants for a discrimination measure; the target and adjacent quadrants for a preference measure; and the opposite and adjacent quadrants for an evasion measure (Carman and Mactutus 2001, 2002; Carman et al. 2002, 2003).

Neurochemical Substrates

Pharmacological studies have provided information on the effects of various drugs on spatial navigation (McNamara and Skelton 1993; D'Hooge and De Deyn 2001). The function of forebrain cholinergic systems is also integral to successful performance in the water maze (Morris et al. 1982; Cassel et al. 1998; Sloan et al. 2006), and is illustrated in Fig. 36.4 by the obvious alterations in swim path evident following 0.7 mg/kg of scopolamine. The role of dopaminergic systems in spatial learning and memory has been demonstrated with D1 receptor mutant mice, which perform poorly on the task compared to both D3 receptor mutant mice and wild type controls, with longer latency to reach the platform and lack of a bias for the target quadrant during probe trials (Xing et al. 2010). D3 receptor mutant mice perform normally, however. Dopamine (DA) neurotransmission in the PFC is known to play an essential role in mediating executive functions such as the working memory, behavioral flexibility, and decision-making (Seamans and Yang 2004; Floresco and Magyar 2006).

Problems/Pitfalls and Solutions

The basic premise of assessing reference memory in the hidden platform task is that the animal learns the location of a submerged platform relative to distal extramaze cues. Auditory, olfactory, gustatory, and proximal visible cues are putatively held as irrelevant to the solution. The case for validity of inferences about hippocampal- and/or cortical-mediated processes is most compelling if the latter statement is demonstrated to be the case. Specifically, it is important to determine whether the animal is using the spatial cues to solve the task. The inclusion of the probe tests in the presence and absence of extramaze cues provides such a demonstration.

Lighting is another source of variance that is all too often not even mentioned in a procedure section. This concern has two tremendous ramifications. The advantages provided by video tracking of a robust set of dependent measures cannot be taken for granted. The standard video tracking software typically uses edge contrast to define the animal in a black/white environment. Thus, use of an albino strain dictates a black maze, whereas the pigmented strains dictate a white maze. But it is not that simple; the use of overhead lighting poorly located and without appropriate diffusing will generate "hot spots" as reflected by the water surface, and thus trick the camera to follow or bounce from animal to "wave." Bad lighting will easily wreck your experiment.

Visual system deficits can be confounding; gross deficits may be ruled out with visible platform tests. Multiple auditory cues, instead of visual cues, can support spatial navigation (Sutherland and Dyck 1984).

Although the commonly accepted theoretical view is that the use of spatial cues located distally in the room is the main strategy employed by the rat, i.e., via a "cognitive map," it has become apparent that animals can use other strategies as well. Other possibilities include learning the relationship between distance and direction in the pool and learning associations based on configurations of cues seen from the platform or other distinct areas of the pool (Stewart and Morris 1993). In many cases, it appears that strategies are not mutually exclusive and that several can be used by the animal as complementary or alternating. Sutherland and Hamilton (2004) suggest that navigational processes that are studied with spatial memory tasks such as the water maze need to be examined in the context of three dimensions: reference frame, movement control, and information. Each of these consists of several features that the animal can use to navigate successfully to the platform. Given this growing appreciation of rodent spatial navigation strategies, the inclusion of appropriate controls begs one's close attention.

36.4 Discussion/Summary

With respect to neuropsychological effects, it is widely accepted that HAND is associated with "executive dysfunction," comprising a component of nearly all of the observed cognitive impairment profiles (Ances and Ellis 2007; Dawes et al. 2008). Intact executive functions are most notably dependent on the frontal cortex as well as the basal ganglia and posterior parietal cortex (Stuss and Levine 2002). Empirical tasks that permit assessment of executive function and attention are thus crucial to any experimental model of HIV-1/drug abuse cognitive impairment. We have outlined several tasks in this chapter, which load these cognitive domains.

Given the impressive evidence that alterations in brainstem-evoked potentials are among the earliest readily quantifiable alterations observed in the progression of HIV-1 (Ollo et al. 1991; Gil et al. 1992; Pagano et al. 1992; Fein et al. 1995; Goodwin et al. 1996; Schroeder et al. 1996; Castello et al. 1998; Vigliano et al. 2000), a focus on preattentive processes might also be particularly fruitful. A striking parallel is suggested with our recent preclinical documentation of alterations in the preattentive process of sensorimotor gating, involving these very same brain regions, based on exposure to viral proteins such as gp120 and Tat (Fitting et al. 2006a, b, c, 2007, 2008b). It is well-established that dopamine circuitry is one regulatory substrate of sensorimotor gating (Davis 1980; Braff et al. 2001) and the clinical utility of this paradigm in a number of neuropsychiatric disorders (e.g., schizophrenia, Huntington's disease; Braff and Geyer 1990; Castellanos et al. 1996) suggests its potential utility with HIV-1 HAND.

With sophisticated behavioral paradigms, lasting cognitive dysfunction can be detected in specific functional domains. The MCSRTT taps fronto-striatal circuitry and recruits the executive processes of both attention and inhibitory control (Robbins 2002; Pezze et al. 2007; Bushnell and Strupp 2009; Besson et al. 2010). We have used these techniques in assessing cognitive profiles and attention deficits (e.g., with cocaine treatment, Mactutus 1999; Bayer et al. 2000, 2002;

Garavan et al. 2000; Morgan et al. 2002; Gendle et al. 2003, 2004a, b; Foltz et al. 2004), and thus this task may be particularly useful for assessment of impaired cognitive domains in HAND.

The Morris water maze is commonly used to study spatial memory and the formation of a cognitive map in rodents. Damage to the hippocampus or medial frontal cortex produces severe impairments of spatial mapping ability. Variations in the maze task can be used to study both working and reference memory in rodents. In humans with HAND, executive function, which requires intact learning and memory abilities, is impaired, which may make this task particularly suited for exploring the neural basis for these deficits (Fitting et al. 2008a, b).

Interestingly, the DA system as a clinically relevant HAND target is increasingly supported by human imaging (Chang et al. 2008; Meade et al. 2011), neurocognitive (Kumar et al. 2011; Meade et al. 2011), and postmortem examinations (Kumar et al. 2009; Silvers et al. 2007) of the DA system. In particular, the DAT is critical for DA homeostasis, is key in maintaining stable synaptic DA concentrations, and regulates disease (executive/frontal; Nieoullon 2002; Chudasama and Robbins 2006) and brain (reward/striatal; Sulzer 2011) processes. The behavioral tasks described here are all sensitive to disruption by dopaminergic agents or loss of DA system integrity within the brain. Indeed, executive system function in humans is sensitive to DA fluctuations, and these tasks may be used to probe DA function.

Despite the fact that there is compelling evidence for the face, construct, and predictive validity of each of the tasks described herein, we must acknowledge the reality that no one behavioral measure will ever be synonymous with a cognitive process or function. The validity of our inferences will be only as compelling as our willingness to be relentless in the critical analysis of the measures we employ.

Acknowledgments This work was supported, in part, by the National Institute on Drug Abuse [Grants DA013137, DA031604, and DA035714] and by the National Institute of Child Health and Human Development [Grant HD043680].

References

Aksenov MY, Aksenova MV, Silvers JM, Mactutus CF, Booze RM (2008) Different effects of selective dopamine uptake inhibitors, GBR 12909 and WIN 35428, on HIV-1 Tat toxicity in rat fetal midbrain neurons. Neurotoxicology 29:971–977

Aksenova MV, Silvers JM, Aksenov MY, Nath A, Ray PD, Mactutus CF, Booze RM (2006) HIV-1 Tat neurotoxicity in primary cultures of rat midbrain fetal neurons: changes in dopamine transporter binding and immunoreactivity. Neurosci Lett 395:235–239

Ances BM, Ellis RJ (2007) Dementia and neurocognitive disorders due to HIV-1 infection. Semin Neurol 27:86–92

Antonova E, Parslow D, Brammer M, Simmons A, Williams S, Dawson GR, Morris RG (2011) Scopolamine disrupts hippocampal activity during allocentric spatial memory in humans: an fMRI study using a virtual reality analogue of the Morris water maze. J Psychopharmacol 25:1256–1265

Bayer LE, Brown A, Mactutus CF, Booze RM, Strupp BJ (2000) Prenatal cocaine exposure increases sensitivity to the attentional effects of the dopamine D1 agonist SKF81297. J Neurosci 20:8902–8908

Bayer LE, Kakumanu S, Mactutus CF, Booze RM, Strupp BJ (2002) Prenatal cocaine exposure alters sensitivity to the effects of idazoxan in a distraction task. Behav Brain Res 133:185–196

Berger JR, Nath A (1997) HIV dementia and the basal ganglia. Intervirology 40:122–131

Besson M, Belin D, McNamara R, Theobald DE, Castel A, Beckett VL, Crittenden BM, Newman AH, Everitt BJ, Robbins TW, Dalley JW (2010) Dissociable control of impulsivity in rats by dopamine d2/3 receptors in the core and shell subregions of the nucleus accumbens. Neuropsychopharmacology 35:560–569

Bonelli RM, Cummings JL (2007) Frontal-subcortical circuitry and behavior. Dialogues Clin Neurosci 9:141–151

Braff DL, Geyer MA (1990) Sensorimotor gating and schizophrenia. Human and animal model studies. Arch Gen Psychiatry 47:181–188

Braff DL, Geyer MA, Swerdlow NR (2001) Human studies of prepulse inhibition of startle: normal subjects, patient groups, and pharmacological studies. Psychopharmacology (Berl) 156:234–258

Brandeis R, Brandys Y, Yehuda S (1989) The use of the Morris water maze in the study of memory and learning. Int J Neurosci 48:29–69

Bushnell PJ, Strupp BJ (2009) Assessing attention in rodents. In: Buccafusco J (ed) Methods of behavior analysis in neuroscience, 2nd edn. CRC, Boca Raton, FL, pp 119–143

Carman HM, Mactutus CF (2001) Ontogeny of spatial navigation in rats: a role for response requirements? Behav Neurosci 115:870–879

Carman HM, Mactutus CF (2002) Proximal versus distal cue utilization in spatial navigation: the role of visual acuity? Neurobiol Learn Mem 78:332–346

Carman HM, Booze RM, Mactutus CF (2002) Long-term retention of spatial navigation by preweanling rats. Dev Psychobiol 40:68–77
Carman HM, Booze RM, Snow DM, Mactutus CF (2003) Proximal versus distal cue utilization in preweanling spatial localization: the influence of cue number and location. Physiol Behav 79:157–165
Cassel JC, Cassel S, Galani R, Kelche C, Will B, Jarrard L (1998) Fimbria-fornix vs selective hippocampal lesions in rats: effects on locomotor activity and spatial learning and memory. Neurobiol Learn Mem 69:22–45
Castellanos FX, Fine EJ, Kaysen D, Marsh WL, Rapoport JL, Hallett M (1996) Sensorimotor gating in boys with Tourette's syndrome and ADHD: preliminary results. Biol Psychiatry 39:33–41
Castello E, Baroni N, Pallestrini E (1998) Neurotological auditory brain stem response findings in human immunodeficiency virus-positive patients without neurologic manifestations. Ann Otol Rhinol Laryngol 107:1054–1060
Chang L, Lee PL, Yiannoutsos CT, Ernst T, Marra CM, Richards T, Kolson D, Schifitto G, Jarvik JG, Miller EN, Lenkinski R, Gonzalez G, Navia BA (2004) HIV MRS consortium. A multicenter in vivo proton-MRS study of HIV-associated dementia and its relationship to age. Neuroimage 23:1336–1347
Chang L, Wang GJ, Volkow ND, Ernst T, Telang F, Logan J, Fowler JS (2008) Decreased brain dopamine transporters are related to cognitive deficits in HIV patients with or without cocaine abuse. Neuroimage 42:869–878
Chudasama Y, Robbins TW (2006) Functions of frontostriatal systems in cognition: comparative neuropsychopharmacological studies in rats, monkeys and humans. Biol Psychol 73:19–38
Cummings JL (1993) Frontal-subcortical circuits and human behavior. Arch Neurol 50:873–880
Cysique LA, Maruff P, Brew BJ (2004) Prevalence and pattern of neuropsychological impairment in human immunodeficiency virus-infected/acquired immunodeficiency syndrome (HIV/AIDS) patients across pre- and post-highly active antiretroviral therapy eras: a combined study of two cohorts. J Neurovirol 10:350–357
D'Hooge R, De Deyn PP (2001) Applications of the Morris water maze in the study of learning and memory. Brain Res Brain Res Rev 36:60–90
Davis M (1980) Neurochemical modulation of sensory-motor reactivity: acoustic and tactile startle reflexes. Neurosci Biobehav Rev 4:241–263
Dawes S, Suarez P, Casey CY, Cherner M, Marcotte TD, Letendre S, Grant I, Heaton RK (2008) HNRC group. Variable patterns of neuropsychological performance in HIV-1 infection. J Clin Exp Neuropsychol 30:613–626
Devan BD, Blank GS, Petri HL (1992) Place navigation in the Morris water task: effects of reduced platform interval lighting and pseudorandom platform positioning. Psychobiology 20:120–126
Ellis RJ, Deutsch R, Heaton RK, Marcotte TD, McCutchan JA, Nelson JA, Abramson I, Thal LJ, Atkinson JH, Wallace MR, Grant I, Kelly M, Chandler JL, Spector SA, Jernigan T, Masliah E, Dupont R (1997) Neurocognitive impairment is an independent risk factor for death in HIV infection. Arch Neurol 54: 416–424
Fein G, Biggins CA, MacKay S (1995) Delayed latency of the event-related brain potential P3A component in HIV disease. Progressive effects with increasing cognitive impairment. Arch Neurol 52:1109–1118
Fendt M, Li L, Yeomans JS (2001) Brain stem circuits mediating prepulse inhibition of the startle reflex. Psychopharmacology (Berl) 156:216–224
Ferris MJ, Frederick-Duus DF, Fadel J, Mactutus CF, Booze RM (2010) Hyperdopaminergic tone in HIV-1 protein treated rats and cocaine sensitization. J Neurochem 115:885–896
Fitting S, Booze RM, Hasselrot U, Mactutus CF (2006a) Intrahippocampal injections of Tat: effects on prepulse inhibition of the auditory startle response in adult male rats. Pharmacol Biochem Behav 84:189–196
Fitting S, Booze RM, Mactutus CF (2006b) Neonatal hippocampal Tat injections: developmental effects on prepulse inhibition (PPI) of the auditory startle response. Int J Dev Neurosci 24:275–283
Fitting S, Booze RM, Mactutus CF (2006c) Neonatal intrahippocampal glycoprotein 120 injection: the role of dopaminergic alterations in prepulse inhibition in adult rats. J Pharmacol Exp Ther 318:1352–1358
Fitting S, Booze RM, Mactutus CF (2007) Neonatal intrahippocampal gp120 injection: an examination early in development. Neurotoxicology 28:101–107
Fitting S, Booze RM, Gilbert CA, Mactutus CF (2008a) Effects of chronic adult dietary restriction on spatial learning in the aged F344 x BN hybrid F1 rat. Physiol Behav 93:560–569
Fitting S, Booze RM, Mactutus CF (2008b) Neonatal intrahippocampal injection of the HIV-1 proteins gp120 and Tat: differential effects on behavior and the relationship to stereological hippocampal measures. Brain Res 1232:139–154
Fleshier M (1965) Adequate acoustic stimulus for startle reaction in the rat. J Comp Physiol Psychol 60:200–207
Floresco SB, Magyar O (2006) Mesocortical dopamine modulation of executive functions: beyond working memory. Psychopharmacology (Berl) 188:567–585
Foltz TL, Snow DM, Strupp BJ, Booze RM, Mactutus CF (2004) Prenatal intravenous cocaine and the heart rate-orienting response: a dose-response study. Int J Dev Neurosci 22:285–296
Gallagher M, Burwell R, Burchinal M (1993) Severity of spatial learning impairment in aging: development of a learning index for performance in the Morris water maze. Behav Neurosci 107:618–626
Garavan H, Morgan RE, Mactutus CF, Levitsky DA, Booze RM, Strupp BJ (2000) Prenatal cocaine exposure impairs selective attention: evidence from serial reversal and extradimensional shift tasks. Behav Neurosci 114:725–738

Garvey LJ, Yerrakalva D, Winston A (2009) Correlations between computerized battery testing and a memory questionnaire for identification of neurocognitive impairment in HIV type 1-infected subjects on stable antiretroviral therapy. AIDS Res Hum Retroviruses 25:765–769

Gendle MH, Strawderman MS, Mactutus CF, Booze RM, Levitsky DA, Strupp BJ (2003) Impaired sustained attention and altered reactivity to errors in an animal model of prenatal cocaine exposure. Brain Res Dev Brain Res 147:85–96

Gendle MH, Strawderman MS, Mactutus CF, Booze RM, Levitsky DA, Strupp BJ (2004a) Prenatal cocaine exposure does not alter working memory in adult rats. Neurotoxicol Teratol 26:319–329

Gendle MH, White TL, Strawderman M, Mactutus CF, Booze RM, Levitsky DA, Strupp BJ (2004b) Enduring effects of prenatal cocaine exposure on selective attention and reactivity to errors: evidence from an animal model. Behav Neurosci 118:290–297

Gil R, Breux JP, Neau JP, Becq-Giraudon B (1992) Cognitive evoked potentials and HIV infection. Neurophysiol Clin 22:385–391 (French)

Goodwin GM, Pretsell DO, Chiswick A, Egan V, Brettle RP (1996) The Edinburgh cohort of HIV-positive injecting drug users at 10 years after infection: a case-control study of the evolution of dementia. AIDS 10:431–440

Griffin WC, Middaugh LD, Cook JE, Tyor WR (2004) The severe combined immunodeficient (SCID) mouse model of human immunodeficiency virus encephalitis: deficits in cognitive function. J Neurovirol 10:109–115

Hamilton DA, Driscoll I, Sutherland RJ (2002) Human place learning in a virtual Morris water task: some important constraints on the flexibility of place navigation. Behav Brain Res 129:159–170

Harker KT, Whishaw IQ (2002) Place and matching-to-place spatial learning affected by rat inbreeding (Dark-Agouti, Fischer 344) and albinism (Wistar, Sprague-Dawley) but not domestication (wild rat vs. Long-Evans, Fischer-Norway). Behav Brain Res 134:467–477

Heaton RK et al (1995) The HNRC 500—neuropsychology of HIV infection at different disease stages. HIV Neurobehavioral Research Center. J Int Neuropsychol Soc 1:231–251

Heaton RK et al (2010) HIV-associated neurocognitive disorders persist in the era of potent antiretroviral therapy: CHARTER Study. Neurology 75:2087–2096

Heaton RK et al (2011) HIV-associated neurocognitive disorders before and during the era of combination antiretroviral therapy: differences in rates, nature, and predictors. J Neurovirol 17:3–16

Hill JM, Mervis RF, Avidor R, Moody TW, Brenneman DE (1993) HIV envelope protein-induced neuronal damage and retardation of behavioral development in rat neonates. Brain Res 603:222–233

Hoffman DC, Donovan H (1994) D-1 and D-2 dopamine-receptor antagonists reverse prepulse inhibition deficits in an animal model of schizophrenia. Psychopharmacology (Berl) 115:447–453

Hoffman HS, Ison JR (1980) Reflex modification in the domain of startle: I. Some empirical findings and their implications for how the nervous system processes sensory input. Psychol Rev 87:175–189

Ison JR, Hammond GR (1971) Modification of the startle reflex in the rat by changes in the auditory and visual environments. J Comp Physiol Psychol 75:435–452

Ison JR, Hoffman HS (1983) Reflex modification in the domain of startle: II. The anomalous history of a robust and ubiquitous phenomenon. Psychol Bull 94:3–17

Ison JR, McAdam DW, Hammond GR (1973) Latency and amplitude changes in the acoustic startle reflex of the rat produced by variation in auditory prestimulation. Physiol Behav 10:1035–1039

Jones CK, Shannon HE (2000) Effects of scopolamine in comparison with apomorphine and phencyclidine on prepulse inhibition in rats. Eur J Pharmacol 391:105–112

June HL, Tzeng Yang AR, Bryant JL, Jones O, Royal W 3rd (2009) Vitamin A deficiency and behavioral and motor deficits in the human immunodeficiency virus type 1 transgenic rat. J Neurovirol 15:380–389

Kelly JB, Masterton B (1977) Auditory sensitivity of the albino rat. J Comp Physiol Psychol 91:930–936

Koch M (1999) The neurobiology of startle. Prog Neurobiol 59:107–128

Koutsilieri E, Sopper S, Scheller C, ter Meulen V, Riederer P (2002) Parkinsonism in HIV dementia. J Neural Transm 109:767–775

Kumar AM, Fernandez JB, Singer EJ, Commins D, Waldrop-Valverde D, Ownby RL, Kumar M (2009) Human immunodeficiency virus type 1 in the central nervous system leads to decreased dopamine in different regions of postmortem human brains. J Neurovirol 15:257–274

Kumar AM, Ownby RL, Waldrop-Valverde D, Fernandez B, Kumar M (2011) Human immunodeficiency virus infection in the CNS and decreased dopamine availability: relationship with neuropsychological performance. J Neurovirol 17:26–40

LaShomb AL, Vigorito M, Chang SL (2009) Further characterization of the spatial learning deficit in the human immunodeficiency virus-1 transgenic rat. J Neurovirol 15:14–24

Lindner MD, Schallert T (1988) Aging and atropine effects on spatial navigation in the Morris water task. Behav Neurosci 102:621–634

Lipska BK, Swerdlow NR, Geyer MA, Jaskiw GE, Braff DL, Weinberger DR (1995) Neonatal excitotoxic hippocampal damage in rats causes post pubertal changes in prepulse inhibition of startle and its disruption by apomorphine. Psychopharmacology (Berl) 122:35–43

Mactutus CF (1999) Prenatal intravenous cocaine adversely affects attentional processing in preweanling rats. Neurotoxicol Teratol 21:539–550

Mactutus CF, Booze RM (1994) Accuracy of spatial navigation: the role of platform and tank size. Soc Neurosci Abstr 20:1014

Mansbach RS, Geyer MA, Braff DL (1988) Dopaminergic stimulation disrupts sensorimotor gating in the rat. Psychopharmacology (Berl) 94:507–514

Mayeux R, Stern Y, Tang MX, Todak G, Marder K, Sano M, Richards M, Stein Z, Ehrhardt AA, Gorman JM (1993) Mortality risks in gay men with human-immunodeficiency-virus infection and cognitive impairment. Neurology 43:176–182

McArthur JC, Hoover DR, Bacellar H, Miller EN, Cohen BA, Becker JT, Graham NMH, McArthur JH, Selnes OA, Jacobson LP, Visscher BR, Concha M, Saah A (1993) Dementia in AIDS patients: incidence and risk-factors. Neurology 43:2245–2252

McNamara RK, Skelton RW (1993) The neuropharmacological and neurochemical basis of place learning in the Morris water maze. Brain Res Brain Res Rev 18:33–49

Meade CS, Lowen SB, Maclean RR, Key MD, Lukas SE (2011) FMRI brain activation during a delay discounting task in HIV-positive adults with and without cocaine dependence. Psychiatry Res 192:167–175

Moran LM, Mactutus CF, Booze RM (2009) Generality of disruption of prepulse inhibition by the dopamine agonist apomorphine [abstract]. In: College on Problems of Drug Dependence 71st annual meeting abstract book (Abstract #422), Reno/Sparks, NV, 20–24 June 2009, p 106

Moran LM, Aksenov MY, Booze RM, Webb KM, Mactutus CF (2012) Adolescent HIV-1 transgenic rats: evidence for dopaminergic alterations in behavior and neurochemistry revealed by methamphetamine challenge. Curr HIV Res 10:415–424

Moran LM, Booze RM, Webb KM, Mactutus CF (2013) Neurobehavioral alterations in HIV-1 transgenic rats: evidence for dopaminergic dysfunction. Exp Neurol 239:139–147

Morgan RE, Garavan HP, Mactutus CF, Levitsky DA, Booze RM, Strupp BJ (2002) Enduring effects of prenatal cocaine exposure on attention and reaction to errors. Behav Neurosci 116:624–633

Morris RGM (1981) Spatial localisation does not depend on the presence of local cues. Learn Motiv 12:239–260

Morris RGM (1983) An attempt to dissociate 'spatial mapping' and 'working-memory' theories of hippocampal function. In: Seifert W (ed) Molecular, cellular and behavioural neurobiology of the hippocampus. Academic, London

Morris R (1984) Developments of a water-maze procedure for studying spatial learning in the rat. J Neurosci Methods 11:47–60

Morris RG, Garrud P, Rawlins JN, O'Keefe J (1982) Place navigation impaired in rats with hippocampal lesions. Nature 297:681–683

Nieoullon A (2002) Dopamine and the regulation of cognition and attention. Prog Neurobiol 67:53–83

O'Keefe J, Nadel L (1978) The Hippocampus as a cognitive map. Oxford University Press, New York, NY

O'Steen WK, Spencer RL, Bare DJ, McEwen BS (1995) Analysis of severe photoreceptor loss and Morris water-maze performance in aged rats. Behav Brain Res 68:151–158

Ollo C, Johnson R Jr, Grafman J (1991) Signs of cognitive change in HIV disease: an event-related brain potential study. Neurology 41:209–215

Packard MG, McGaugh JL (1992) Double dissociation of fornix and caudate nucleus lesions on acquisition of two water maze tasks: further evidence for multiple memory systems. Behav Neurosci 106:439–446

Pagano MA, Cahn PE, Garau ML, Mangone CA, Figini HA, Yorio AA, Dellepiane MC, Amores MG, Perez HM, Casiró AD (1992) Brain-stem auditory evoked potentials in human immunodeficiency virus-seropositive patients with and without acquired immunodeficiency syndrome. Arch Neurol 49:166–169

Panakhova E, Buresova O, Bures J (1984) The effect of hypothermia on the rat's spatial memory in the water tank task. Behav Neural Biol 42:191–196

Peng RY, Mansbach RS, Braff DL, Geyer MA (1990) A D2 dopamine receptor agonist disrupts sensorimotor gating in rats: implications for dopaminergic abnormalities in schizophrenia. Neuropsychopharmacology 3:211–218

Pezze MA, Dalley JW, Robbins TW (2007) Differential roles of dopamine D1 and D2 receptors in the nucleus accumbens in attentional performance on the five-choice serial reaction time task. Neuropsychopharmacology 32:273–283

Pugh CR, Johnson JD, Martin D, Rudy JW, Maier SF, Watkins LR (2000) Human immunodeficiency virus-1 coat protein gp120 impairs contextual fear conditioning: a potential role in AIDS related learning and memory impairments. Brain Res 861:8–15

Purohit V, Rapaka RS, Schnur P, Shurtleff D (2011) Potential impact of drugs of abuse on mother-to-child transmission (MTCT) of HIV in the era of highly active antiretroviral therapy (HAART). Life Sci 88:909–916

Rauch TM, Welch DI, Gallego L (1989) Hypothermia impairs performance in the Morris water maze. Physiol Behav 46:315–320

Reger M, Welsh R, Razani J, Martin DJ, Boone KB (2002) A meta-analysis of the neuropsychological sequelae of HIV infection. J Int Neuropsychol Soc 8:410–424

Rigdon GC, Viik K (1991) Prepulse inhibition as a screening test for potential antipsychotics. Drug Dev Res 23:91–99

Robbins TW (2002) The 5-choice serial reaction time task: behavioural pharmacology and functional neurochemistry. Psychopharmacology (Berl) 163:362–380

Sacktor N, McDermott MP, Marder K, Schifitto G, Selnes OA, McArthur JC, Stern Y, Albert S, Palumbo D, Kieburtz K, De Marcaida JA, Cohen B, Epstein L (2002) HIV-associated cognitive impairment before and after the advent of combination therapy. J Neurovirol 8:136–142

Schapiro S, Salas M, Vukovich K (1970) Hormonal effects on ontogeny of swimming ability in the rat: assessment of central nervous system development. Science 168:147–150

Schroeder MM, Handelsman L, Torres L, Jacobson J, Ritter W (1996) Consistency of repeated event-related potentials in clinically stable HIV-1-infected drug users. J Neuropsychiatry Clin Neurosci 8:305–310

Schwarzkopf SB, Bruno JP, Mitra T (1993) Effects of haloperidol and SCH 23390 on acoustic startle and prepulse inhibition under basal and stimulated conditions.

Prog Neuropsychopharmacol Biol Psychiatry 17: 1023–1036

Seamans JK, Yang CR (2004) The principal features and mechanisms of dopamine modulation in the prefrontal cortex. Prog Neurobiol 74:1–58

Silvers JM, Aksenova MV, Aksenov MY, Mactutus CF, Booze RM (2007) Neurotoxicity of HIV-1 Tat protein: involvement of D1 dopamine receptor. Neurotoxicology 28:1184–1190

Skelton RW, Ross SP, Nerad L, Livingstone SA (2006) Human spatial navigation deficits after traumatic brain injury shown in the arena maze, a virtual Morris water maze. Brain Inj 20:189–203

Sloan HL, Good M, Dunnett SB (2006) Double dissociation between hippocampal and prefrontal lesions on an operant delayed matching task and a water maze reference memory task. Behav Brain Res 171:116–126

Spear NE (1978) The processing of memories: forgetting and retention. Lawrence Erlbaum Associates, Hillsdale, NJ

Spear NE, Riccio DC (1994) Memory: phenomena and principles. Allyn & Bacon, Needham Heights, MA

Spencer RL, O'Steen WK, McEwen BS (1995) Water maze performance of aged Sprague-Dawley rats in relation to retinal morphologic measures. Behav Brain Res 68:139–150

Stewart CA, Morris RGM (1993) The watermaze. In: Sahgal A (ed) Behavioural neuroscience, vol I, a practical approach. Oxford University Press, Oxford, pp 107–122

Stuss DT, Levine B (2002) Adult clinical neuropsychology: lessons from studies of the frontal lobes. Annu Rev Psychol 53:401–433

Sulzer D (2011) How addictive drugs disrupt presynaptic dopamine neurotransmission. Neuron 69:628–649

Sutherland RJ, Dyck RH (1984) Place navigation by rats in a swimming pool. Can J Psychol 38:322–347

Sutherland RJ, Hamilton DA (2004) Rodent spatial navigation: at the crossroads of cognition and movement. Neurosci Biobehav Rev 28:687–697

Sutherland RJ, Linggard R (1982) Being there: a novel demonstration of latent spatial learning in the rat. Behav Neural Biol 36:103–107

Sutherland RJ, Kolb B, Whishaw IQ (1982) Spatial mapping: definitive disruption by hippocampal or medial frontal cortical damage in the rat. Neurosci Lett 31:271–276

Sutherland RJ, Chew GL, Baker JC, Linggard RC (1987) Some limitations on the use of distal cues in place navigation by rats. Psychobiology 15:48–57

Tang H, Lu D, Pan R, Qin X, Xiong H, Dong J (2009) Curcumin improves spatial memory impairment induced by human immunodeficiency virus type 1 glycoprotein 120 V3 loop peptide in rats. Life Sci 85:1–10

Tolman EC (1948) Cognitive maps in rats and men. Psychol Rev 55:189–208

Tolman EC, Honzik CH (1930) Introduction and removal of reward, and maze performance in rats. University of California publications in psychology, vol 4. University of California Press, Berkeley, CA, pp 257–275

Tonkiss J, Shultz P, Galler JR (1992) Long-Evans and Sprague-Dawley rats differ in their spatial navigation performance during ontogeny and at maturity. Dev Psychobiol 25:567–579

UNAIDS/WHO (2010) Report on the global AIDS epidemic 2010. UNAIDS/WHO, Geneva, ISBN 9789291738717

Varty GB, Higgins GA (1994) Differences between 3 rat strains in sensitivity to prepulse inhibition of an acoustic startle response: influence of apomorphine and phencyclidine pretreatment. J Psychopharmacol 8:148–156

Vigliano P, Boffi P, Bonassi E, Gandione M, Marotta C, Rainò E, Russo R, Rigardetto R (2000) Neurophysiologic exploration: a reliable tool in HIV-1 encephalopathy diagnosis in children. Panminerva Med 42:267–272

Vigorito M, LaShomb AL, Chang SL (2007) Spatial learning and memory in HIV-1 transgenic rats. J Neuroimmune Pharmacol 2:319–328

Vijayraghavan S, Wang M, Birnbaum SG, Williams GV, Arnsten AF (2007) Inverted-U dopamine D1 receptor actions on prefrontal neurons engaged in working memory. Nat Neurosci 10:376–384

Wallace DR, Dodson S, Nath A, Booze RM (2006) Estrogen attenuates gp120- and tat1-72-induced oxidative stress and prevents loss of dopamine transporter function. Synapse 59:51–60

Wang GJ, Chang L, Volkow ND, Telang F, Logan J, Ernst T, Fowler JS (2004) Decreased brain dopaminergic transporters in HIV-associated dementia patients. Brain 127:2452–2458

Xing B, Kong H, Meng X, Wei SG, Xu M, Li SB (2010) Dopamine D1 but not D3 receptor is critical for spatial learning and related signaling in the hippocampus. Neuroscience 169:1511–1519

Zhu J, Mactutus CF, Wallace DR, Booze RM (2009) HIV-1 Tat protein-induced rapid and reversible decrease in [3H]dopamine uptake: dissociation of [3H]dopamine uptake and [3H]2beta-carbomethoxy-3-beta-(4-fluorophenyl)tropane (WIN 35,428) binding in rat striatal synaptosomes. J Pharmacol Exp Ther 329:1071–1083

Zhu J, Ananthan S, Mactutus CF, Booze RM (2011) Recombinant human immunodeficiency virus-1 transactivator of transcription(1-86) allosterically modulates dopamine transporter activity. Synapse 65:1251–1254

Zink WE, Anderson E, Boyle J, Hock L, Rodriguez-Sierra J, Xiong HG, Gendelman HE, Persidsky Y (2002) Impaired spatial cognition and synaptic potentiation in a murine model of human immunodeficiency virus type 1 encephalitis. J Neurosci 22:2096–2105

Correction to: Stereological Analysis

Kristi M. Anderson, Adam M. Szlachetka,
and R. Lee Mosley

Correction to:
Chapter 4 in: H. Xiong and H.E. Gendelman (eds.), *Current Laboratory Methods in Neuroscience Research*, Springer Protocols Handbooks, https://doi.org/10.1007/978-1-4614-8794-4_4

An equation in Chapter 4, page 78, of the original version of this book was inadvertently published incorrectly, which has now been updated:

The incorrect equation "CT number (HU) = [(μm − μwater) / μwater] × 1,00" has now been updated as "Number = count × [(section thickness) / (section thickness + cell height)]."

The updated version of this chapter can be found at
https://doi.org/10.1007/978-1-4614-8794-4_4

H. Xiong and H.E. Gendelman (eds.), *Current Laboratory Methods in Neuroscience Research*,
Springer Protocols Handbooks, https://doi.org/10.1007/978-1-4614-8794-4_37

Index

H. Xiong and H.E. Gendelman (eds.), *Current Laboratory Methods in Neuroscience Research*,
DOI 10.1007/978-1-4614-8794-4, © Springer Science+Business Media New York 2014

D

W

X

Z